Structure–Property Relations in Polymers

ADVANCES IN CHEMISTRY SERIES **236**

Structure–Property Relations in Polymers

Spectroscopy and Performance

Marek W. Urban, EDITOR
North Dakota State University

Clara D. Craver, EDITOR
Craver Chemical Consultants

Developed from a symposium sponsored
by the Division of Polymeric Materials:
Science and Engineering
at the 201st National Meeting
of the American Chemical Society,
Atlanta, Georgia,
April 14–19, 1991

American Chemical Society, Washington, DC 1993

Library of Congress Cataloging-in-Publication Data

Structure–property relations in polymers : spectroscopy and
  performance / Marek W. Urban, editor, Clara D. Craver, editor.
  p.   cm.—(Advances in chemistry series, ISSN 0065–2393; 236)
  "Developed from a symposium sponsored by the Division of Polymeric
Materials: Science and Engineering at the 201st National Meeting of
the American Chemical Society, Atlanta, Georgia, April 14–19, 1991."
  Includes bibliographical references and index.

  ISBN 0–8412–2525–7
  1. Polymers—Analysis—Congresses.   2. Spectrum analysis—Congresses.

  I. Urban, Marek W., 1953-   .   II. Craver, Clara D.   III. American Chemical Society.
Division of Polymeric Materials: Science and Engineering.   IV. Series.

QD1.A355 no. 236
[QD139.P6]
540 s—dc20
[547.7′046]   93-15964
                                                                                      CIP

The paper used in this publication meets the minimum requirements of American National
Standard for Information Sciences—Permanence of Paper for Printed Library Materials, ANSI
Z39.48–1984.   ∞

FOREWORD

The ADVANCES IN CHEMISTRY SERIES was founded in 1949 by the American Chemical Society as an outlet for symposia and collections of data in special areas of topical interest that could not be accommodated in the Society's journals. It provides a medium for symposia that would otherwise be fragmented because their papers would be distributed among several journals or not published at all.

Papers are reviewed critically according to ACS editorial standards and receive the careful attention and processing characteristic of ACS publications. Volumes in the ADVANCES IN CHEMISTRY SERIES maintain the integrity of the symposia on which they are based; however, verbatim reproductions of previously published papers are not accepted. Papers may include reports of research as well as reviews, because symposia may embrace both types of presentation.

ABOUT THE EDITORS

PROFESSOR MAREK W. URBAN received his M.S. degree in chemistry from Marquette University in 1979, followed by the Ph.D. degree from Michigan Tech in 1984. Before joining North Dakota State University in 1986, he spent two years as a research associate in the Department of Macromolecular Science at Case Western Reserve University in Cleveland.

He is the author of over 130 research papers, numerous review articles, and several book chapters, mostly published in the ACS Advances in Chemistry Series. He is the author of the book entitled *Vibrational Spectroscopy of Molecules and Macromolecules of Surfaces*, currently in production by Wiley-Interscience. His research interests range from characterization of polymer networks and surfaces using FTIR spectroscopy to structure–property relationships at surfaces and interfaces, from inorganic polymers and coatings to mobility of small molecules in polymer networks, and from spectroscopic measurement of diffusion in polymer networks to nonequilibrium thermodynamics. For his pioneering work on rheophotoacoustic spectroscopy, he was awarded the 1990 Megger's Award presented for the most outstanding paper published in the *Applied Spectroscopy* Journal. The award is given by the Federation for Analytical Chemistry and Spectroscopy Societies. He also is credited with a patent in this area. Five years in a row, from 1986 to 1991, he was honored by the 3M Company with the young faculty award. From 1987 to 1988, he served as a chairperson for the Society for Applied Spectroscopy, Minnesota Chapter. He is an invited speaker at many international conferences, Gordon Research conferences, and industrial laboratories. He serves as a consultant to several chemical companies.

His involvement in the ACS symposia initiated in 1991, when he cochaired the International Symposium on Spectroscopy of Polymers in Atlanta, from which this book was derived. He is a co-chair of the 1993 Symposium on Hyphenated Techniques in Polymer Characterization, held in Chicago 1993 during the ACS National Meeting, and will serve as a chair of the International Symposium on Polymer Spectroscopy, to be held in Washington in 1994. He is also a lecturer in the spectroscopy workshops offered by the ACS Division of Polymeric Materials: Science and Engineering. At North Dakota State University, he is a director of the summer coatings science short courses and serves as a Director of the Graduate Program in the Polymers and Coatings Department.

CLARA D. CRAVER is the president of Craver Chemical Consultants. Until 1990 she was president of Chemir Laboratories, which she founded in 1958, and the vice president of Polytech Laboratories. She received her B.Sc. degree in chemistry (cum laude) from Ohio State University in 1945. She is a member of Phi Beta Kappa and was awarded the honorary degree of Doctor of Science by Fisk University in 1974. Her early work at Esso Research Laboratories, 1945–1949, resulted in patents on spectroscopic characterization of complex hydrocarbon mixtures. She established the molecular spectroscopy laboratory at Battelle Memorial Institute in 1949, and as group leader through 1958, she conducted spectroscopic studies for Battelle's research projects and original research in the areas of drying oils, asphalts, paper, and rosins. Her research on organic coatings won her the 1955 Carbide and Carbon Award of the ACS Division of Organics Coatings and Plastics Chemistry. In 1975 she became chairman of that division and has been active in ACS governance continually since then, including 15 years service as counselor.

In the early 1980s, Craver initiated workshops on Fourier transform infrared spectroscopy at ACS national meetings to offer educational opportunities and chaired three prior major ACS symposia on new developments in polymer characterization by instrumental methods. She began the Coblentz Society's spectral publication program and is editor of five books of Special Collections of Infrared Spectra and of an 11-volume spectral data collection. She served as consultant to ASTM for their evaluated infrared spectral publication program, supported by the Office of Standard Reference Data of the National Institute of Standards and Technology. She was chairman of the Joint Committee on Atomic and Molecular Physical Data, on which she now serves as Executive Committee member-at-large. She is past chairman of the ASTM Committee on Molecular Spectroscopy, and was named a Fellow of ASTM in 1982 when she received that organization's highest award, the Award of merit. She is a Fellow of the American Institute of Chemists and Certified Professional Chemist. She was honored by being named the 1989 National Honorary Member of the Women's Professional Honorary Chemistry Sorority, Iota Sigma Pi.

She is director of the Fisk Infrared Institute held annually at Vanderbilt University of IR, Raman, and FTIR spectroscopy. She also lectures at other short courses and is an ACS tour speaker. She directs short courses including FTIR laboratories at national ACS meetings for the Division of Polymeric Materials: Science and Engineering. This book is her fourth volume on spectroscopic and other instrumental methods of polymer characterization.

She co-chaired the ACS International Symposium on Spectroscopy of Polymers held in Atlanta in 1991 from which this book is derived, and chaired the two-day FTIR workshop that preceded the symposium.

CONTENTS

PREFACE

A LARGE VARIETY OF ANALYTICAL TECHNIQUES for polymer characterization has been developed and reviewed under various conditions. However, no formal attempt has been made to correlate macromolecular structures with polymer properties. This book is intended to address these issues in perspective. It outlines the capabilities of Fourier transform infrared and Raman, fluorescence, and mass spectroscopies of polymers, and builds a bridge between the molecular-level information obtained in spectroscopic measurements and specific polymer properties. Molecular-level understanding of the macroscopic processes responsible for the behavior of polymers and other materials is needed to enhance our knowledge and to lead to further advancements in the chemistry of polymeric materials.

This volume continues the coverage of the prior books edited by Clara Craver on many aspects of polymer physical chemistry, chemical physics, and spectroscopy. The book is designed to assist newcomers to the field of polymer spectroscopy, as well as to serve as a reference source for more experienced scientists. This goal is accomplished by starting with a few introductory tutorial chapters on fundamental aspects of each technique, followed by sections addressing specific polymer problems such as Crystalline Polymers and Copolymers, Surfaces and Interfaces of Polymers, Spectroscopic Approaches to Polymers in Solutions and Polymer Networks, Spectroscopy and Thermally Induced Processes in Polymers, and Polymer Analysis and Surface Modifications. The chapters present discussions ranging from basic to applied topics in polymers, coatings, and interfacial concepts.

We would like to express our sincere appreciation to all who participated in the 1991 Atlanta Symposium on Polymer Characterization and to the reviewers whose hard work hopefully enhanced the overall book quality. A special note of thanks is addressed to Professor Thomas Brenna of Cornell University, Professor Gregory Gillispie of North Dakota State University, and Dr. Bill Simonsick of DuPont, who agreed to contribute to the book without being involved in the Symposium. The Division of Polymeric Materials; Science and Engineering and the Petroleum Research Fund of the American Chemical Society are acknowledged for the financial support of the symposium.

MAREK W. URBAN
Department of Polymers and Coatings
North Dakota State University
Fargo, ND 58105

CLARA D. CRAVER
Craver Chemical Consultants
Box 265
French Village, MO 63036

June 16, 1991

FUNDAMENTAL CONCEPTS IN SPECTROSCOPY OF POLYMERS

Although the analysis of polymers by infrared, Raman, and fluorescence spectroscopy combined with mass spectrometry has been known for many years, the techniques continue to develop, providing better resolution and sensitivity. This section is intended to familiarize the reader with the physical principles of each technique. Selected examples of the fundamental concepts used in the analysis of polymers illustrate the specific techniques and their sensitivities.

Fourier Transform Infrared and Fourier Transform Raman Spectroscopy of Polymers

Principles and Applications

Marek W. Urban

Department of Polymers and Coatings, North Dakota State University, Fargo, ND 58105

This chapter covers the fundamental principles and current applications of Fourier transform (FT) infrared and Fourier transform Raman spectroscopies as utilized in the analysis of polymeric materials. The primary emphasis of the first part is on the principles and advantages of these interferometric methods, whereas the remaining sections illustrate numerous applications focusing on the structure–property considerations in polymers. Particular attention is given to the most recent developments in FT analysis and includes examples of structural and conformational analysis of polymers, biological studies, and the applications of FT infrared and Raman microscopy to remote measurements. The differences and the complementary nature of infrared and Raman spectroscopies are also presented.

EXPERIMENTAL SCIENCES HAVE BEEN PROFOUNDLY INFLUENCED by the development of novel instrumentation. Virtually all scientific instrumentation is now under computer control, and sophisticated, faster data collection allows scientists to channel their resources more effectively toward particular goals. The sophistication of many current physical approaches mandates the use of highly sensitive, fast instruments and reasonably powerful data acquisition computers to attain insights about fundamental aspects of processes under investigation. Examples of such sophisticated interplay are infrared and

0065–2393/93/0236–0003$10.50/0

Raman spectrometers, which, used in conjunction with the fast Fourier transform (FT) algorithms, are key instruments in modern vibrational spectroscopy. This chapter will first review the fundamental principles that govern infrared and Raman activity and then discuss the principles and applications of FT detection in vibrational spectroscopy.

Infrared and Raman Processes

The normal modes of vibration of any molecule can be divided into three classes. Some modes may be observed in the Raman spectrum, some in the infrared, and some may not be seen in either spectrum. For a molecule that possesses a high degree of symmetry, the rule of mutual exclusion states that no vibrational mode may be observed in both the infrared and Raman spectra. This high symmetry is defined by a center of inversion operation. As the symmetry is reduced, and the molecule no longer contains a center of inversion, some vibrational modes may be seen in both the infrared and in the Raman spectra. However, the mode will often have quite different intensity in the two spectra. The quantum mechanical selection rules state that observation of a vibrational mode in the infrared spectrum requires a change in dipole moment during the vibration. In other words, the vibration is infrared active if the following condition is fulfilled:

$$[\mu]_{v',v''} \neq \int \Phi_{v'}(Q_a)\mu\Phi_{v''}(Q_a)\,dQ_a \tag{1}$$

where $[\mu]_{v',v''}$ is the dipole moment in the electronic ground state; Φ is the vibrational eigenfunction; v' and v'' are the vibrational quantum numbers before and after transition, respectively; and Q_a is the normal coordinate of the vibration (1).

Infrared spectroscopy is based on an absorption process and involves measuring the amount of energy that passes through or is reflected off a sample and comparing this amount to that transmitted or reflected from a perfect transmitter or reflector, respectively. The plot of the relative transmitted or reflected energy as a function of energy is an infrared spectrum. In recent years, this old spectroscopic method has diverged into two apparently different approaches to measuring the infrared spectrum; one uses dispersive optics, and the second uses an interferometric technique. The interferometric approach, combined with fast Fourier transform algorithms, provides several distinct advantages including higher resolution, higher energy throughput, and better precision, and hence the technique called Fourier transform infrared spectroscopy had dominated the field.

The observation of a vibrational mode in the Raman spectrum requires a change in the electron polarizability resulting from the movement of atoms. Thus, a given vibrational mode will be Raman active if the following condition is fulfilled:

$$[\alpha]_{v',v''} \neq \int \Phi_{v'}(Q_a)\alpha\Phi_{v''}(Q_a)\,dQ_a \tag{2}$$

Here, $[\alpha]$ is the polarizability tensor of the vibration and the remaining parameters are the same as for the infrared activity (eq 1) (*1–6*). These apparent differences in the principles governing both effects have led to the development of two physically distinct experimental approaches to obtain infrared and Raman spectra. The molecular information obtained in one experiment complements the other.

As indicated in equation 2, the detection of Raman scattering involves a completely different set of problems and is based on entirely different experimental principles. When monochromatic radiation of frequency ν_o strikes a transparent sample, the light is scattered. Most of this scattered light consists of radiation at the frequency of the incident light but differs in the direction of propagation and polarization state. This portion of the light is called Raleigh scattering. However, approximately 1 out of 10^6 photons that impinge upon the sample can be inelastically scattered, and this portion of the scattered light is called Raman scattering. This inelastically scattered fraction of the light is composed of new modified frequencies $(\nu_o \pm \nu_k)$, where $(\nu_o - \nu_k)$ is referred to as Stokes scattering, and $(\nu_o + \nu_k)$ is anti-Stokes scattering. Figure 1 illustrates a schematic representation of the absorption and scattering processes leading to infrared and Raman spectra. The energy level diagram shows that the anti-Stokes scattering requires that the molecule start in an excited vibrational state. The only means of populating these excited vibrational states is thermally; therefore, the anti-Stokes intensities will be very temperature dependent and normally quite weak at room temperature. If the anti-Stokes scattering can be observed, the sample temperature can be determined by the ratio of the Stokes to anti-Stokes intensities.

The selection rules for Raman scattering allow only those transitions to be detected for which one of the elements of the polarizibility tensor $[\alpha]_{v',v''}$, or a combination thereof, belongs to a species of the point group to which the normal coordinate Q_a also belongs. However, the induced dipole moment **M** of the ground state of a molecule is not only proportional to the strength of electric field **E**, but also depends on nonlinear terms, such as hyperpolarizibility β:

$$\mathbf{M} = \alpha\mathbf{E} + \tfrac{1}{2}\beta\mathbf{EE} + \cdots \tag{3}$$

Although simple calculations show that the hyperpolarizibility β is typically one million times smaller than α, in some situations β and Q_a can transform identically, whereas α does not. Under these circumstances, some vibrations

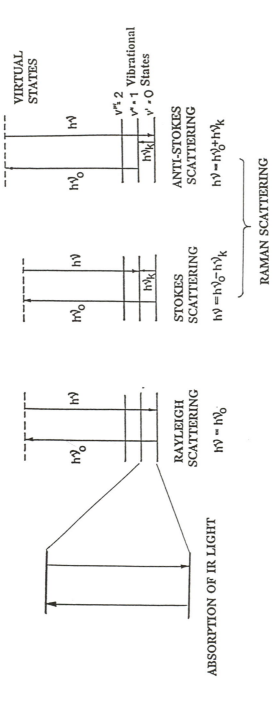

Figure 1. Schematic representation of Rayleigh, Stokes, and anti-Stokes scattering.

can be detected in the hyper-Raman effect, but not in an ordinary Raman scattering experiment.

Various factors may alter the effectiveness of Raman scattering. The overall intensity of the Raman scattering (I_R) is proportional to the fourth power of ν_k and is given by the following equation:

$$I_R = \frac{2^3 \pi (\nu_o \pm \nu_k)^4}{3c^4} I_o \sum_{\nu', \nu''} |\alpha_{\nu', \nu''}|^2 \qquad (4)$$

where I_o is the laser field intensity and c is the speed of light. This process, just like the infrared absorption, is considered to be a linear process because the scattering intensity I_R is proportional to the incident laser intensity I_o. Under certain experimental conditions, however, nonlinear processes are also possible. If, instead of one quantum of energy ν_o, two quanta of incident laser beam $2\nu_o$ are employed in a pulse sequence, they may give rise to a two-photon absorption process. As a result of perturbation of the induced dipole moment of a molecule, a nonlinear optical process called hyper-Raman scattering (HRS) (7) can be induced. It produces a $(2\nu_o + \nu_k)$ shift with the intensity I_{HR} proportional to the square of molecular hyperpolarizibility β derivative with respect to the vibrational normal coordinates:

$$I_{HR} = \frac{2^6 \pi^2 (2\nu_o \pm \nu_k)^4}{3c^5} I_o^2 \sum_{\nu', \nu'', \nu'''} |\beta_{\nu', \nu'', \nu'''}|^2 \qquad (5)$$

Although the first observation of HRS was made well over 20 years ago (6), HRS had been used only occasionally in chemical application (8–10), and only recently its use was demonstrated in the surface-enhanced hyper-Raman effect (11). In spite of the fact that the HRS process is 10^6 weaker than already weak Raman scattering, its attractiveness comes from the differences in selection rules. Special importance of the HRS lies in the fact that many fundamental frequencies inactive in both infrared and Raman may happen to be active in HRS. One of the most illustrative examples is benzene, where no less than six frequencies are active in hyper-Raman only, which can be distributed as $2B_{1u} + 2B_{2u} + 2E_{2u}$ normal vibrational mode representation (12). Another example is the simple SF_6-type octahedral (O_h) molecule, which possesses a triply degenerate frequency in F_{2u} which is hyper-Raman active, and inactive in both infrared and Raman. In spite of many attractive features, the use of hyper-Raman for polymer analysis is still yet to be explored and should open new avenues for the future.

Dispersive and Interferometric Detection

The limitations on sensitivity and detection limits imposed by dispersive infrared instruments in the past led to the development of Fourier transform interferometric infrared spectrometers. Interferometric detection is based upon splitting the source radiation into two equal beams of approximately equal intensity. This beam splitting is shown in Figure 2, which also illustrates that the radiation from each path is reflected back by mirrors onto the divider called the beam splitter. The recombined two beams may either interact constructively or destructively depending upon the phase difference of the two optical paths. This constructive or destructive interference will vary as the path length in one of the arms is varied. The resulting pattern forms the interferogram that represents the relationship between the energy and the path difference in the two arms of the interferometer. The interference pattern clearly will be a function of the wavelength of light because the relative path differences will be expressed as different integral values of wavelength. Interferograms represent the interference of the incident wavelengths, and hence they are converted into a spectrum by using a Fourier transform algorithm. The Michelson interferometer, such as illustrated in Figure 2, was the first instrument to split a source radiation into two separate beams, change the path length of one of the paths, and recombine the radiation on the beam splitter to cause interference. References 13 and 14 give further details regarding FT spectroscopy.

Although the advantages of Fourier transform over other techniques are well documented, the fact that almost all incident radiation is used simultaneously in the transmission–absorption experiments is the major feature. Thus,

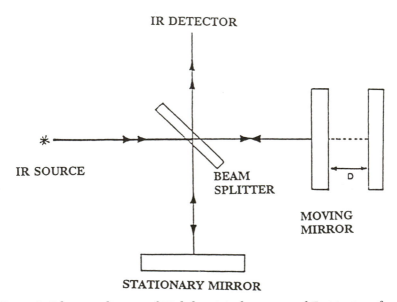

Figure 2. Schematic diagram of Michelson interferometer and Fourier transform spectrometer.

in comparison to a scanning, single-channel dispersive experiment, all of the light is being used rather than a very small fraction (the light passed by the slits of a monochromator), and this feature resulted in the selectivity and sensitivity enhancement of the surface techniques.

The high sensitivity of interferometric instruments has in turn brought about the development of new sensitive spectroscopic probes and enhanced the existing ones. The sensitivity enhancements of attenuated total reflectance (ATR), reflection–absorption (R–A), and further developments of diffuse reflectance (DRIFT), emission, photoacoustic, photothermal beam deflection, or surface electromagnetic wave (SEW) spectroscopies are primarily attributed to the enhanced sensitivity of the FT instruments. A schematic diagram of each technique along with a brief description is given in Figure 3. Because many of the presented techniques are capable of surface-depth profiling, Table I provides the approximate depth-penetration ranges, and other characteristics for the surface techniques are illustrated in Figure 3.

In spite of the fact that the step-scan interferometry has been well documented in the past (*15*, *16*), only recently has the coupling between dynamic FTIR and step-scan interferometer been documented (*17*). In an ordinary, continuous FT operation, a movement of the Michelson interferometer mirror modulates each IR wavelength at the Fourier domain frequency, $f = 2v/\lambda$, where v is the mirror velocity of the Michelson interferometer and λ is the wavelength of IR light. The step-scan mode of detection uses the same moving mirror, but the mirror moves in steps, and upon completion of each step the mirror stops and data are collected. Although the mirror may move slightly during the data collection, the most recent designs provide a fixed mirror position. The utility of such step-scan interferometry was extended to two-dimensional infrared spectroscopy, earlier introduced by Noda (*18*, *19*), and photoacoustic depth profiling of polymer laminates (*20*). In the recent studies (*21*) the phase photoacoustic signal analysis, which gives a constant thermal diffusion length across the photoacoustic spectrum, was examined. As opposed to the continuous FT scanning interferometer, such an approach makes the surface depth of penetration at all wave numbers the same and thus makes surface-depth profiling analysis of polymers simpler. This condition is not true for the spectra obtained with the modulation frequency variations.

In Raman spectroscopy, the situation is quite different. This difference lies in the nature of a nonresonant versus a resonant process. In the Raman scattering experiment, only a small fraction of the incident photons are inelastically scattered with a change in energy. Thus they contain information about the normal modes of vibration of the molecule, roughly one in a million. Because these incident photons can also participate in other photophysical processes such as absorption and fluorescence, often unavoidable interference in a form of fluorescence may result. It usually comes from impurities or the sample itself and can completely dominate the weak Raman

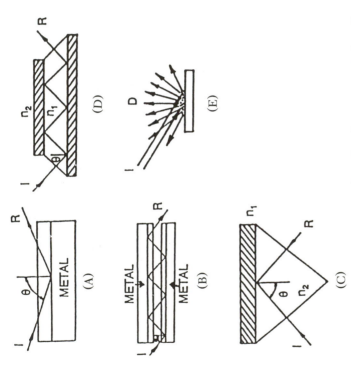

Figure 3. Surface-sensitive Fourier transform infrared (FTIR) techniques: A, single reflection–absorption (R–A) setup; incident light (I) penetrates the sample and is reflected (R) by the metal mirror (Θ should be between 75° and 89.5°); B, multiple-reflection setup; incident light (I) penetrates the sample and is reflected (R) by a metal mirror; C, single internal reflection; incident light (I) passes through the internal reflection element and is totally reflected (R) at Θ > Θ_c (n_1 and n_2 are the refractive indices of the sample and the internal element, respectively); D, multiple-reflection setup in attenuated total reflection (ATR) mode; E, diffuse reflectance (DRIFT) setup; the incident light (I) is diffusively scattered in all directions (D), collected by hemispherical mirrors, and redirected to the detector.

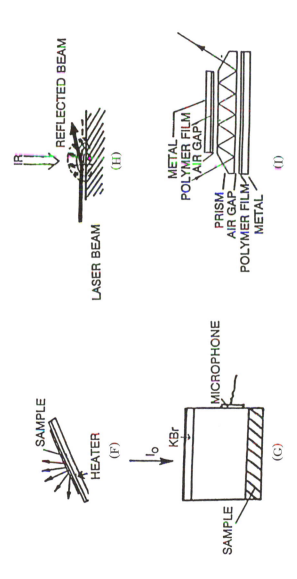

Figure 3. Continued. F, emission setup; the source of the IR light is replaced by a heated sample and emitted light is analyzed by the infrared detector; G, photoacoustic setup; the incident modulated light with the intensity I_o impinges upon the sample surface; the light is absorbed and as a result of reabsorption, heat is released to the surface which, in turn, generates periodic acoustic waves that are detected by a sensitive microphone; H, photothermal beam deflection setup; the incident IR light generates heat waves at the surface (just like in photoacoustics) causing reflections of the parallel to the surface laser beam; the laser beam is deflected as a result of refractive index changes caused by temperature changes of the surrounding gas; and I, surface electromagnetic wave (SEW) spectroscopy setup; when a surface-coated metallic substrate is examined by ATR at the near critical angle, the intensity enhancement of the spectrum of the surface species is produced because of excitement of SEW at the metal surface.

Table I. Approximate Depth Penetration Ranges for Various Surface FTIR Techniques

Technique	Depth Penetration Range	Preferred Surface Characteristics
ATR	40 Å–7 μm	smooth, in good contact with an ATR crystal
R–A	monolyer–2 μm	coated, shiny metallic
DRIFT	monolayer to a few micrometers	powder, rough
Photoacoustic	100 Å–150 μm	no restrictions
Photothermal	similar to PA	
SEW	monolayer	metallic
Emission	40 Å–few micrometers	no restriction

scattering. When a laser source with an excitation frequency in the ultraviolet or visible range is used to produce Raman scattering, absorption processes that may cause fluorescence can be often several orders of magnitude stronger than the actual Raman component of the scattering process. Many approaches to avoid this problem have been tried, ranging from the addition of quenching agents to time-based discrimination between the fluorescence and Raman events. Notable successes have been achieved, but no universal solution. The only way to completely avoid fluorescence is to avoid the absorption process giving rise to the fluorescence. If the excitation source is shifted into the near-infrared region of the spectrum, the incident photons may not have enough energy to exceed the threshold for absorption. This condition can be ensured by designing a Raman experiment using a long-wavelength near-infrared laser. As a result, a Raman spectrum that is relatively free from fluorescence interference can be produced, although difficulty arises from the poor sensitivity of conventional near-infrared spectrometers. As a matter of fact, the first such experiments, which were conducted by Chantry et al. (22) in the 1960s, showed that Raman scattering due to iodine in carbon tetrachloride can be achieved with infrared excitation and an interferometer. Although the quality of the spectra and resolution left much to be desired, this was the first reported study in the infrared region. However, good shot-noise-limited detectors available in the visible region are no longer operative beyond 1 μm. To compensate for the poor detector performance, an obvious instrument to employ is an interferometer, which has a large multiplex advantage when operated in a detector-noise-limited experiment.

As already stated, to induce fluorescence-free Raman scattering, monochromatic light in the near-IR region must be used. This light source can be achieved by replacing an ordinary infrared source of radiation with a Nd:YAG

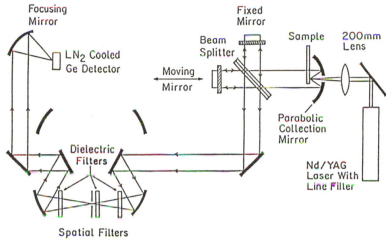

Figure 4. Schematic diagram of Fourier transform Raman spectrometer. (Reproduced from reference 24. Copyright 1986 American Chemical Society.)

laser and the use of near-IR detector, such as that illustrated in Figure 4. Similarly with FTIR spectroscopy, the fact that almost all light is used in a FT Raman experiment is advantageous, although it has several associated problems. All frequencies of light fall on the detector at once, and hence in an FT Raman experiment, the dominant Rayleigh line (the photons scattered at the incident frequency) would cause detector saturation. This one strong line must be optically filtered out before striking the detector. The filter should be selective enough that associated Raman lines are not filtered out. Current technology using either dielectric filters or holographic filters allows Raman spectra down to 100 cm^{-1}. Although the FT Raman approach minimizes fluorescence, the sensitivity is diminished by a factor of 16 from that of a visible-based Raman experiment done with an argon laser operating at 514.5 nm. The Raman scattering has a v^4 dependence (*see* eq 4 $(v_0 \pm v_k)^4$), so use of a long-wavelength laser involves loss in scattered intensity. In spite of these difficulties, recent studies (*21*) showed that obtaining Raman spectra using a Nd:YAG laser as an excitation source along with proper filtering and detection systems, such as presented in Figure 4, are indeed feasible. The importance of effective filtering of the Rayleigh line is demonstrated in Figure 5, which shows FT Raman spectrum of bis(phenylimino)-terephthaldehyde (BPT) with single, double, and triple filtering.

The experiments in this field by Hirschfeld and Chase (*23, 24*) opened new vistas in the field of Raman spectroscopy by creating a Fourier counterpart of infrared, which appears to overcome certain limitations of the conventional Raman instrumentation. Subsequent improvements have led to

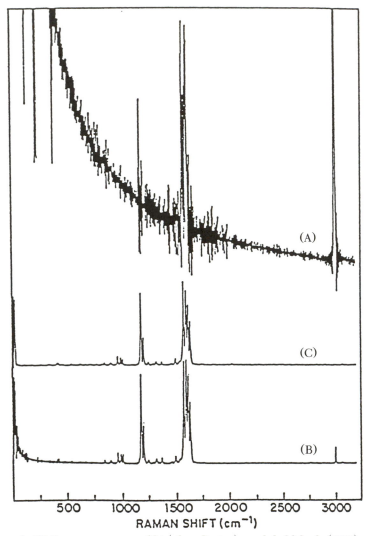

Figure 5. FT Raman spectrum of bis(phenylimino)terephthaldehyde (BPT) with single-stage filtering (A), double-stage filtering (B), and triple-stage filtering (C). (Reproduced from reference 24. Copyright 1986 American Chemical Society.)

the use of FT Raman spectroscopy for qualitative and quantitative analysis of polymers and other materials, complementing FTIR spectroscopy established in the early 1970s. The frequency diagram illustrating the location of visible, near-infrared, mid-infrared, and far-infrared energy ranges is shown in Figure 6.

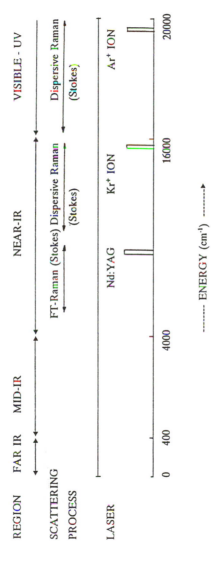

Figure 6. A schematic diagram illustrating the excitation energy ranges for IR and Raman experiments.

FTIR Spectroscopy in Polymer Analysis

An immense amount of literature deals with the analysis of polymer microstructures and characterization of polymers using both dispersive and interferometric infrared spectroscopy. As a matter of fact, characterization of various polymer functional groups through analysis of their vibrational group frequencies of individual polymers is a routine analytical task. The analysis of more subtle microstructural differences and changes within the same polymer using infrared spectroscopy may provide such information as determination of distribution of copolymerized units in polymer chains, determination of sequence length, configurational and conformational isomerisms, chain branching, end-group analysis, hydrogen bonding, chain order and crystallinity, chain folding, and molecular orientations. The enhanced signal-to-noise ratio available through the use of interferometry has resulted in an extensive amount of work in the infrared and Raman spectroscopy of polymers. Both the theory of vibrational spectroscopy and its applications to polymeric systems have been treated by many useful monographs (25–27). We focus on polymer analysis, which would be difficult without interferometric detection.

One of the rules of thumb in vibrational spectroscopy is that the fewer the bands, the more ordered the structures are expected. An example is the FTIR spectrum of polyethylene shown in Figure 7, trace A. In essence, it consists of three regions of fundamental vibrations, due to C–H stretching (3000 cm^{-1}), bending (1400–1500 cm^{-1}), and wagging (700–800 cm^{-1}) normal modes of the CH_2 and CH_3 units. By following the $3n - 6$ formula (or $3n - 5$ for linear molecules, where n is the number of atoms in one molecule) and calculating a number of vibrational degrees of freedom for a polyethylene molecule with a molecular weight of 100,000 that consists of more than 7000 CH_2 units containing approximately $n = 21,000$ atoms, more than 63,000 vibrational modes would be predicted. This number is of course, not observed because a high concentration of the long-range ordering of the CH_2 groups and the fewer terminal CH_3 groups is present. However, as the molecular weight decreases, the concentration of the CH_3 groups increases, and as a result of local environment changes, new symmetries impose new selection rules leading to more complex spectra. Furthermore, when defects such as kinks, folding, or bending of the polymer chains occur, a local symmetry of the chain units contributing to the defects will be disturbed. This disturbance will be reflected not only in the changes of the selection rules leading to new bands in the 1300–1400-cm^{-1} region, but also in the band intensity changes. These changes are evidenced by several increasing bands in traces B through E of Figure 7, which illustrates DRIFT transmission spectra of the model compounds, $C_{72}H_{144}$, $C_{16}H_{32}$, $C_{15}H_{30}$, and $C_{14}H_{28}$, respectively.

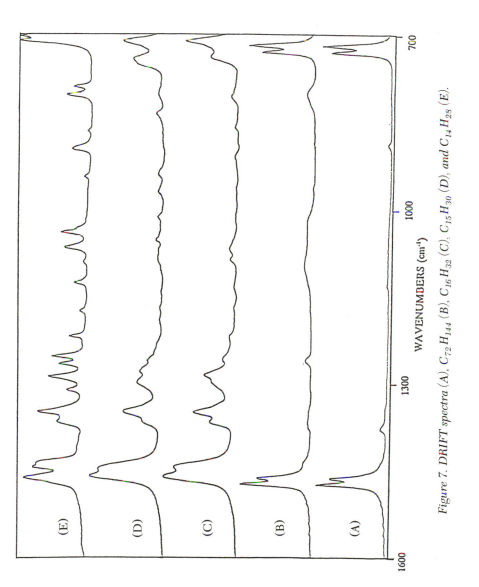

Figure 7. DRIFT spectra (A), $C_{72}H_{144}$ (B), $C_{16}H_{32}$ (C), $C_{15}H_{30}$ (D), and $C_{14}H_{28}$ (E).

The sensitivity of FT detection can be advantageous when it comes to the detection of a small number of the structural defects compared to a total number of the main species in a polymer. One of the illustrative examples is chain folding in polyethylene, in which concentration of the CH_2 units involved in folding is significantly smaller than the remaining CH_2 units. Whereas theoretical foundations of infrared band intensities in polymers with and without chain folding were addressed and carefully executed by Abbate et al. (28), experimental studies were possible when model compounds became available. As was predicted by Abbate et al. and confirmed experimentally, the C–H bending modes due to CH_2 units at the fold point will absorb in the 1300–1400-cm^{-1} region. By taking integrated intensity of the defect band or bands in the 1300–1400-cm^{-1} region, and ratioing it to the defect-free C–H bending modes at 1460 and 1470 cm^{-1}, a calibration plot can be made between this intensity ratio and a total number of carbons in the model compounds. With such a correlation, the unknown polyethylene sample can be analyzed. Figure 8 depicts a calibration plot obtained using model $C_{14}H_{28}$, $C_{15}H_{30}$, $C_{16}H_{32}$, $C_{72}H_{144}$, $C_{96}H_{196}$, and $C_{168}H_{334}$ compounds (29). The low-density polyethylene film sample is also marked on Figure 8 and appears to contain approximately 90 CH_2 units between each chain folding.

Because of preferential orientation of polymer chains, usually induced by processing, transmission infrared measurements of polymer films often exhibit a certain degree of anisotropy. This anisotropy is reflected in a higher concentration of normal vibrations in the polymer film plane perpendicular to the direction of the incident beam. If the incident light is polarized parallel to the transition moment responsible for IR activity, the intensity of such transition will be at maximum, provided that the transition is allowed by IR selection rules. When the same electric vector is perpendicular to the transition, the transition will not be seen in the IR spectrum because absorption of IR radiation is not only dependent upon the selection rules and concentration, but also on the transition moment orientation. This issue is particularly important in polymer films because on drawing, polymer chains may become oriented in a draw direction. For various practical reasons such orientations are often desirable; in particular, the orientation of polymer chains perpendicular to a draw direction is important because it may affect not only surface morphological properties, but also relaxation behavior, dimensional stability, and others.

Although in the past IR spectroscopy was extensively used to determine orientation of the polymer chains through dichroic ratio calculations[1] in oriented polymer systems (30–35), only recently the issue of surface crys-

---

[1] Dichroic ratio is defined as $D = A_{\parallel}/A_{\perp}$ where D is the dichroic ratio and $A_{\parallel}$ and $A_{\perp}$ are absorbances obtained parallel and perpendicular, respectively, to the polymer draw directions. For two-dimensional studies, Fresner's orientation function defined as $f = (D - 1)/(D + 2)(D_0 + 2)/D_0 + 1)$ is usually used; $D_0 = 2\cot^2\alpha$; α is the transition-moment angle.

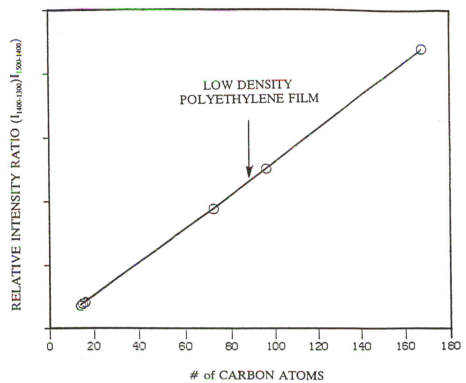

Figure 8. *Intensity ratio plotted as a function of the number of carbon atoms in model compounds.*

tallinity has been addressed (*36*). According to these studies, using direct intensity measurements of strong bands due to CH_2 bending and rocking normal vibrations obtained in ATR, the crystallinity content at the surface is higher than that of the bulk of polymer. However, the analytical approach used in these studies did not account for the surface optical effects, which can often cause distortions of intense ATR bands. When this possibility is taken into account using Kramers–Kronig transformation (KKT) and Frensel's relationships for parallel and perpendicular orientations (*37–39*), it appears that the crystallinity content near the surface may be different (*40*). Figure 9 illustrates the CH_2 bending modes at 1474 (a) and 1464 (b) cm^{-1}, along with the rocking modes at 730 (a) and 720 (b) cm^{-1} of polyethylene. Spectrum A was recorded in a transmission mode of detection, with the IR incident light polarized perpendicular (*s*-polarized) to a draw direction of the sample, whereas spectrum B was in the parallel direction (*p*-polarized). Traces C and D represent the spectra recorded in the ATR mode at a 45° angle of incidence. A comparison of the transmission traces A and B indicates that the

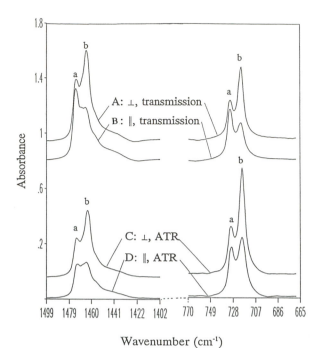

Wavenumber (cm⁻¹)

Figure 9. FTIR spectra of polyethylene films: A, transmission perpendicular polarization; B, transmission parallel polarization; C, ATR perpendicular polarization; and D, ATR parallel polarization. Peaks a and b at 1474 and 1464 cm⁻¹ are CH_2 bending modes; peaks a and b at 730 and 720 cm⁻¹ are polyethylene rocking modes. (Reproduced with permission from reference 40. Copyright 1992 Butterworth.)

intensity ratios of the 730- and 720-cm⁻¹ bands are inverted when going from perpendicular (A) to parallel (B) orientations; namely $(I_A/I_B) \perp <$ $(I_A/I_B) \parallel$. Although a similar trend is observed in the ATR measurements (traces C and D), the changes are not as pronounced. In either case, the intensity ratio is sensitive to the orientation effect, but this sensitivity is diminished in the surface-sensitive ATR experiment because a higher content of the randomly oriented amorphous phase or less crystalline phase near the surface is expected. Further considerations on the use of KKT in ATR can be found in references 41 and 42.

To establish the surface crystallinity content, scrupulous characterization of the surface molecular orientations is necessary. Several approaches, including ATR measurements using *p*-polarized light (*43*), orientation function measurements with the tilted angles (*44*), or polarization modulation (*45*), were used. As an example, the dichroic ratios obtained from the intensity measurements of the 730- and 720-cm⁻¹ bands obtained in transmission are plotted as a function of the penetration depth at 720 cm^{-1} in Figure 10, **A** and **B**, and are compared to the corresponding ATR measurements shown in Figure 10, A′ and B′. Plot A of Figure 10 shows that within a given range of the penetration depths, the dichroic ratio of the 730-cm^{-1} band changes slightly but oscillates around unity. Because the transition-moment vector of this band coincides with the orthorhombic unit cell *a* axis, these results indicate that the *a* axis is either randomly oriented along the film plane or preferentially oriented perpendicular to the film plane direction. In contrast, plot A′ indicates that the dichroic ratios of this band measured from the transition spectra are much greater, reaching almost 1.4 values. These observations suggest that, in the core of the film, the *a* axis is predominantly oriented along the film draw direction. Similar results for ATR measurements are illustrated by curves B and B′ in Figure 10.

The use of dichroic and trichroic ratios to determine polymer orientation provides an additional dimension to the polarization experiments and was demonstrated in the past. For example, the three-dimensional technique is based on a sample tilting with respect to the incident polarized radiation. In recent studies, Fina and Koenig (*44, 46, 47*) extended previous trichroic measurements by establishing the effect of refractive index dispersion and determination of the limiting factors. The trichroic ratio technique was applied to poly(ethylene terephthalate) (PET) films to calculate the trichroic

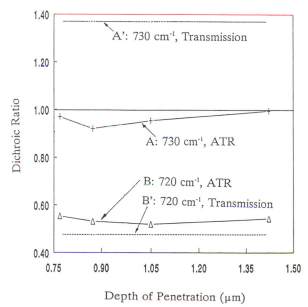

Figure 10. Dichroic ratios plotted as a function of penetration depth. (Reproduced with permission from reference 40. Copyright 1992 Butterworth.)

absorption on one-way drawn PET films and to establish the distribution of vibrational modes in the plane transverse to the draw direction. For example, Figure 11 illustrates the calculated thickness direction spectrum along with parallel and perpendicular experimental spectra of PET. On the basis of these measurements, the morphological model of one-way drawn PET was proposed.

Hydrogen bonding in polymers has been of interest for quite a while. Coleman and co-workers (48, 49) initiated spectroscopic studies on the compatibility of polymer blends in which hydrogen bonding was a key player because its strength and concentration significantly affected compatibility of two polymers. Carbonyl stretching vibrations have proven to be a straightforward quantitative probe of molecular mixing for such blend systems as poly(4-vinylphenol)–poly(vinyl acetate) (PVPh–PVAc), polyethylene–poly(vinyl acetate) (PE–PVAc), and polyacetones. For the ethylene–vinyl acetate (EVA) blends, carbonyl normal stretching vibrations were employed as a probe of molecular mixing and showed high sensitivity of this band to intermolecular mixing. Because the strength of intermolecular interactions through hydrogen bonding is also affected by a kinetic energy of the system, simultaneously heating and cooling the sample during infrared measurements

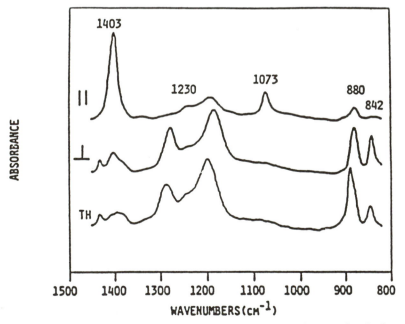

Figure 11. Parallel and perpendicular polarized spectra of PET. The thickness direction spectrum (TH) was generated from the parallel and perpendicular measurements. (Reproduced with permission from reference 47 Copyright 1986 Wiley Intersciences.)

provides a means of determining the strength of the interactions and miscibility of polymers. For example, Figure 12 illustrates a series of FTIR transmission spectra recorded during a heating cycle of 80–20% (w/w) resol-EVA blend and shows that with increasing temperature the amount of hydrogen-bonded carbonyl groups decreases. This effect is demonstrated by vanishing intensities of the band at 1710 cm^{-1} at the expense of the increasing free carbonyl band at 1735 cm^{-1}.

Elastomeric polymer networks are often prepared from oligomeric molecules by chemically joining prepolymer chains in a fairly random fashion. FTIR spectroscopy has been used in various polymeric studies, but the detection of network structures that develop during cross-linking has often been limited because a relatively small number of bonds forms, compared to the total number of bonds present in the system. Although under such circumstances photoacoustic FTIR detection described in this book might be a useful alternative and, in many cases, a more sensitive approach, recent transmission FTIR studies of polyester–styrene (PEs–S) cross-linking have shown that a fraction of styrene (S) monomer homopolymerizes to form the network consisting of S–PEs and PS–PEs branches (50). Furthermore, the homopolymerization process results in atactic polystyrene (PS), which forms physical cross-linking along with the chemical reactions between C=C bonds of PEs and S. This process is demonstrated in Figure 13, which compares FTIR transmission spectra of isotactic, syndiotactic, and atactic stereoisomers with the S–PEs cured film spectrum. Clearly, the 760-cm^{-1} band due to the C–H bending out-of-plane normal vibrations of the phenyl groups in spectra C and D is the same in both spectra. Other spectral regions also confirm the presence of atactic polystyrene. These studies are in contrast to the previous electron microscopy studies, which assumed polyester gel formation upon cross-linking and neglected the possibility of styrene homopolymerization.

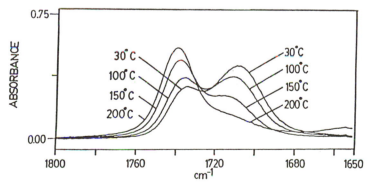

Figure 12. Intensity changes of the carbonyl groups resulting from temperature increase. (Reproduced from reference 48. Copyright 1987 American Chemical Society.)

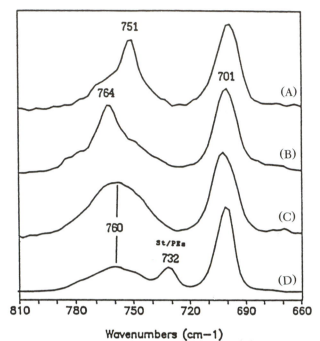

Figure 13. Transmission FTIR spectra of isotactic PS (A), syndiotactic PS (B), atactic PS (C), and styrene–polyester (D). (Reproduced with permission from reference 50. Copyright 1991 Butterworth.)

FT Raman Spectroscopy in Polymer Analysis

As indicated in the introduction, the primary motivation for the development of FT Raman spectroscopy is the minimization of background fluorescence. Once this step is accomplished, FT Raman spectroscopy can be utilized fully not only for the analysis of polymer structures, but also in monitoring their formation. This use is particularly important when the band-intensity changes due to totally symmetric vibrations (for example, C=C stretching) can be used as a measure of certain processes. To illustrate this use, Figure 14 shows the spectra of methyl methacrylate polymerization done in the presence of a highly fluorescent diazo initiator (B. Chase, private communication). The loss of double-bond intensity is clearly seen after 4 h of reaction. Figure 15 shows the intensity of the C=C band normalized to the intensity of the $C-CH_3$ deformation. This normalization is required because the scattering power of the reacting solution changes with time and such change must be normalized out in order to observe the true loss in intensity due to reaction. Rapid data acquistion as well as ease of sample handling make FT Raman spectroscopy an ideal tool for the study of such reacting systems. The advantage of no

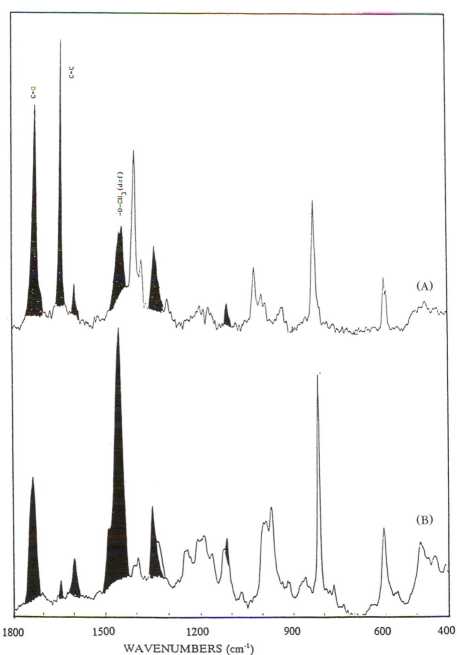

Figure 14. FT Raman spectra of methyl methacrylate before polymerization (A) and 4 h later (B).

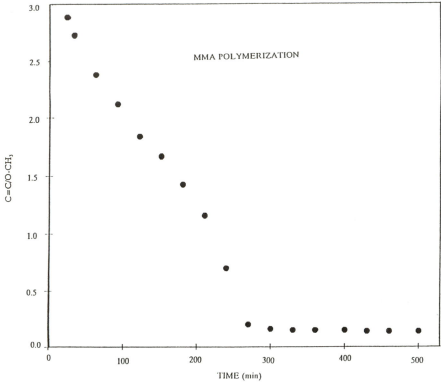

Figure 15. A percentage conversion plot for the polymerization of methyl methacrylate at 65 °C in which C=C band intensity at 1640 cm^{-1} is plotted as a function of time.

sample preparation and the nondestructive nature of measurements place FT Raman and photoacoustic FTIR measurements in a similar sampling category; namely, in both cases minimal sample preparation is required. Because of that feature, the analysis of polyurethane- and polyurea-based forms as well as advanced composites (51) was possible using FT Raman spectroscopy.

Polyamides, commercially known as nylons, are synthesized by a self-condensation polymerization process to form $-[-NH-CO-(CH)_n]-$, where n usually ranges from 3 to 12 (52, 53). The behavior of secondary amide groups is well known (54), but the length of the CH sequence introduces an additional dimension to the properties of nylons, because the CH_2 sequence may have a different degree of crystallinity in the different stereoisomer structures (55). FT Raman spectroscopy was used (56) to confirm previous spectroscopic findings on nylon fibers and illustrated that indeed FT Raman spectroscopy can produce fluorescence-free Raman spectra. Typical spectra of nylon-6,6 and nylon-6,12 are shown in Figure 16. Even without detailed

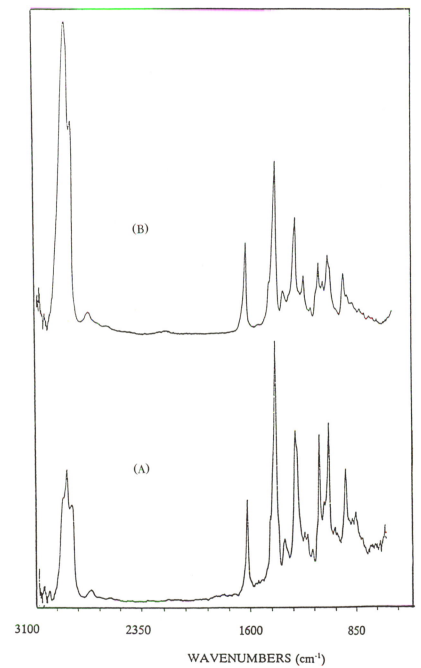

WAVENUMBERS (cm⁻¹)

Figure 16. FT Raman spectra of nylon-6,6 (A) and nylon-6,12 (B).

identification of all spectroscopic features, the presence of higher content of CH_2 groups is reflected in th increased intensity of the C–H stretching bands around 3000 cm^{-1} and relatively weak bands due to the C–O stretching bands at 1500 cm^{-1}. This content will certainly reflect the properties of both polymers.

Many commercial polymers contain additives, such as chain terminators, antioxidants, and UV absorbers, and therefore the fluorescence problem is often quite severe in such "less than pure" systems. In the past this problem has inhibited the use of Raman techniques for the study of such systems. The near-infrared (NIR) approach to Raman spectroscopy has permitted the study of such systems. Another polymeric system that is of current interest in many fields is the polyimides. These materials are finding applications in composites and electronic materials. The high level of fluorescence encountered in these materials has made Raman spectroscopy impossible.

Because of its nondestructive nature, FT Raman spectroscopy can be used for many sampling situations that do not permit sample modifications or in situ studies. In the first category, FT Raman spectroscopy was used in the analysis of forest products (57). The representative FT Raman spectra of redwood (soft wood) and oak (hard wood) are shown in Figure 17. The band at 1269 cm^{-1} in trace B is attributed to rosin (58), a characteristic component of soft woods. These studies showed that it is possible not only to differentiate between hard and soft woods, but the degree of hardness can be assessed by monitoring the intensity changes of the 1269-cm^{-1} band. Previous methods required the use of destructive extraction techniques. In the second category, when sampling situations do not permit in situ studies, an illustrative example is the analysis of reaction-injection-molded (RIM) polyurethanes (59) shown in Figure 18. The absence of fluorescence enabled a comparison of FT Raman polyurethane spectra that contain different hard-to-soft-segment ratios. The bands at 2974 cm^{-1} (CH_3 stretch), 2936 and 2876 cm^{-1} (CH_2 stretch), and 1456 cm^{-1} (CH_2 bend) are attributed to soft segments consisting of polyethylene oxide and polypropylene. Other bands at 640, 900, 1184, and 1617 cm^{-1} are due to aromatic rings of hard segments, those at 1712 cm^{-1} are due to carbonyl groups, and those at 1523 cm^{-1} are due to amide I.

When the Raman effect was recognized as a useful tool for studying biological systems, it was quickly realized that the band intensities were often too weak to be used for analysis. As a matter of fact, under many circumstances the bands due to solvent molecules would be the only spectral feature detected (unless weakly scattering water was a solvent). The situation changed very quickly when resonance Raman (60) was brought up as one of the alternatives (61–63). In this case, when the electronic band would match with the excitation laser line and a given transition was vibronically allowed, a strong band enhancement would be observed. Because many biologically active molecules or their models exhibit electronic transitions in ultraviolet–

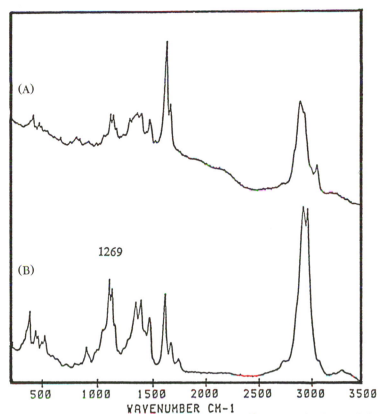

Figure 17. FT Raman spectra in the 3500–80-cm^{-1} region of redwood (A) and oak (B). (Reproduced with permission from reference 57. Copyright 1990 Society for Applied Spectroscopy.)

visible (UV–vis) region, Raman bands recorded by using the UV–vis excitation laser line can be enhanced. However, when the sample or impurities in the sample induce strong fluorescence, even resonance Raman conditions cannot help. Because the near-IR excitation line is used to produce FT Raman spectra, leaving fluorescence behind in the UV–vis region, FT Raman measurements may be an attractive alternative for biological studies. A useful application of this technique was recently reported on photoactive proteins, in which the photoability limits the laser power used to generate resonance excitation in the UV–vis region (64). An interesting feature found in these studies was that the spectrum of bactriorhodopsin is dominated by a nonresonant Raman scattering component due to the protein-bound pigment retinal, whereas the dominating features in the reaction centers are bacteriopheophylin, bacteriochlorophyll, and carotenoids. Furthermore, the relative band

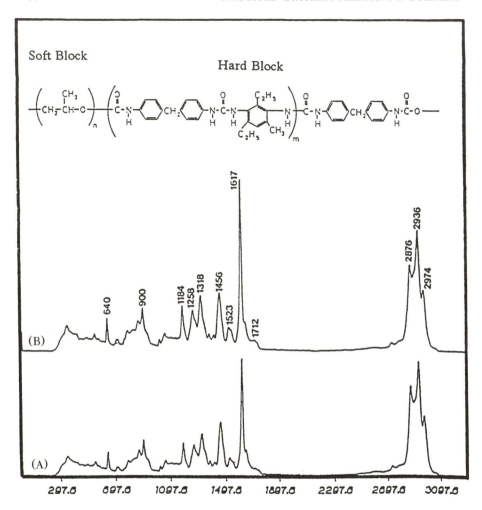

RAMAN SHIFT (cm⁻¹)

Figure 18. FT Raman spectra of RIM polyurethanes containing 35% (A) and 12.5% (B) (w/w) hard segments. (Reproduced with permission from reference 59. Copyright 1990 Society for Applied Spectroscopy.)

intensities in the spectrum of retinyline recorded under nonresonant conditions were identical to those observed in the resonance Raman spectrum of bacteriorhodopsin.

Many biological macromolecules contain chromophores that are intrinsic parts of the species of interest. Upon exposure to visible light, such as in a typical non-FT Raman experiment (65), chromophoric entities often fluo-

DPPC/AMPHOTERICIN B (24:1)

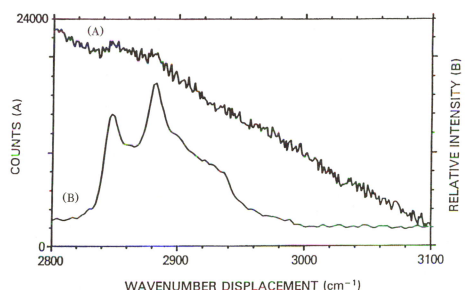

Figure 19. Raman spectra of DPPC with amphotericin (mole ratio = 24:1): A, dispersive spectrum; B, FT spectrum. (Reproduced with permission from reference 70. Copyright 1977 Elsevier.)

resce and thus give rise to undesirable background that usually prohibits measurement of a detectable Raman signal. For these reasons, FT Raman spectroscopy was chosen to study membrane structures, specifically as a sensitive probe of molecular fluidity manifested in a form of rotational isomerism of the hydrocarbons in phospholipid membrane components (66). In another study (67), the same authors further enhanced the FT advantage to chromophoric species and utilized 90° scattering, which appears to produce results similar to those obtained with 180° geometry.

The use of ordinary Raman spectroscopy of biological systems has been extensive (68). In an effort to elucidate molecular-level mechanisms responsible for interactions between drugs and membranes, Lewis et al. (69) used fluorescence-free FT Raman spectroscopy and examined model polyene antibiotics in the presence of model membrane lipid bilayers. Although specific findings are beyond the scope of this chapter, Figure 19 illustrates a useful example of Raman spectra in the C–H stretching region of polyene ring system in amphotericin B (70), a potential antifungal antibiotic. Interactions of this drug with dipalmitoylphosphatidylcholine (DPPC) liposomal bilayers may provide valuable insights into the drug mode of action as a putative membrane channel former. The dispersive Raman spectrum shown by trace A of Figure 19 gives only fluorescence background, whereas FT Raman measurements shown in trace B provide a good signal-to-noise ratio Raman spectrum. The effect of poly(L-alanine) sequence lengths on structural properties was examined by using FT Raman spectroscopy (71). In contrast to the previous studies (72), shorter oligomers do not exhibit β-sheet secondary structures, but contain a significant amount of α-helix. Apparently, the presence of strong bands at 440 and 240 cm^{-1} is responsible for β structures that are found in oligopolypeptides containing 10 poly(L-alanine) units, and the species with 40 units have characteristic bands at 375, 335, and 120 cm^{-1} attributed to the α-helical conformers (73). Because the Co–C bond in photolabile methylcoenzyme B plays a key role in many biological processes, the determination of its strength was of primary importance (74). Studies using FT Raman spectroscopy showed that the Co–C vibrational frequency is at 500 cm^{-1} for methylcobalamin. This assignment was confirmed by detection of the band at 470 cm^{-1} due to Co–C stretch in deuterated methylcobalamin (75). In other studies, Sommer and Katon (76) addressed the issue of Rayleigh rejection and spatial resolution with several examples of polyamide (Kevlar) and wood fibers.

The problem of thin-film analysis by Raman spectroscopy is basically one of sensitivity. The scattering volume is quite low in a thin film, and when the scattering cross section is reduced by going to the near-IR, detection of the small number of scattered photons becomes extremely difficult. Either the scattering volume must be increased or the cross section increased. Two alternatives exist for increasing the cross section: resonance Raman and surface-enhanced Raman spectroscopy. Resonance excitation requires an electronic absorption near the laser frequency. Although the move to the near-IR region was prompted by moving away from electronic absorption bands with a premise that it is not likely to find systems with near-IR excitations, in some cases the presence of absorption bands may be advantageous. For example, macrocyclic phthalocyanines and their polymeric counterparts have electronic transition bands in the near-IR region, and the only possibility of obtaining IR and Raman spectra is to use photoacoustic FTIR and FT Raman spectroscopy, respectively (77–79). Surface-enhanced Raman spectroscopy (SERS) has been shown to exist with near-IR excitation. In fact, the enhancement factors for gold and copper are larger in the near-IR than in the visible region. SERS has also been used extensively to understand molecular mechanisms relevant to adhesion of polymers and polymer–metal interactions. The major contributions in this area are obviously credited to Boerio and co-workers (80–84), who successfully established the effect of substrates on various metal–polymer chemical reactions. With the use of FT Raman, SERS may become an even more powerful approach to the analysis of polymers. Although local electromagnetic effects on rough Ag, Au, or Cu surfaces can enhance the Raman scattering signal from monolayers up to 10^6 stronger (85), the enhanced near-IR excited SERS spectra were reported to be 10^5–10^6 stronger than a normal Raman spectrum for pyridine (86).

FTIR and FT Raman Microscopy: Remote Measurements

Continually increasing demand for in situ and remote measurements of various chemical processes presents new opportunities for both FT vibrational spectroscopies. In many sampling situations the amount of sample is limited and thus does not permit the use of common modes of detection. Although a spectral collection using a microscope involves the same measurements, the difference is the necessity to use focused light to a desired diameter. Such applications require the use of a focusing apparatus, which narrows the incident beam to a diameter of the transmitting fiber optics and applies such modified light to a sample of interest. Although focusing can be also accomplished with a focusing microscope objective, the spatial resolutions for IR and Raman measurements vary substantially. For example, it is impossible to obtain IR spectra of objects smaller than 6–10 μm because of interference limits, as the diameter of the aperture begins to match the wavelength of IR light. The situation is different in Raman microscopy if visible light is the excitation source and a commonly accepted 1–3 μm spatial resolution is expected. The use of FTIR microscopy to view and record IR spectra of Kevlar fibers (87) and polymer laminates (88) has been reported. As an example, Figure 20 illustrates a series of FTIR spectra recorded from the cross section of the multilayer polymer sample. FT Raman measurements have also been demonstrated (89), although further modifications are warranted. As another example, Figure 21 illustrates optical arrangements required to obtain FT Raman spectra. With this approach, good quality FT Raman spectra of 12-μm thick Kevlar fibers, as well as other polymers, have been shown (D. B. Chase, private communication).

Once the feasibility of FTIR and FT Raman microscopy was demonstrated, a fiber-optic probe was developed (90, 91). Figure 22 illustrates the principle of fiber-optic arrangements for FT Raman experiments. In this system, a single fiber at the center is used to deliver the laser light to the sample. This single fiber is surrounded by 18 collecting fibers, which, upon collection of scattered light, send the scattered light to the interferometer. Although a primary limitation in the mid-FTIR range is still limited throughput of the fiber optics, FTIR fiber-optic measurements have received quite a bit of attention in the near-IR region. Archibald et al. (92) demonstrated remote near-IR reflectance measurements with the use of a pair of optical fibers and a FT spectrometer. Remote FT Raman measurements using near-IR have been also shown (93).

Summary and Conclusions

The primary objective of this chapter was to identify and outline the fundamental principles and selected applications of FT vibrational spectroscopy for polymer analysis that are not covered in the remaining chapters of this book. Although the advantages of the Fourier transform techniques are very

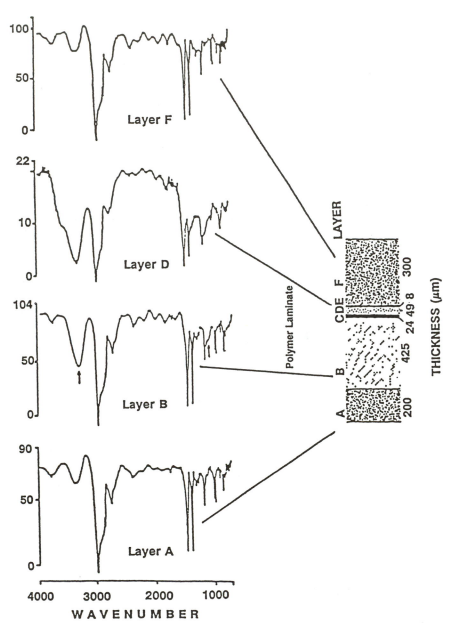

Figure 20. FTIR spectra of a cross section of a multilayer polymer laminate. (Reproduced with permission from reference 88. Copyright 1986 Society for Applied Spectroscopy.)

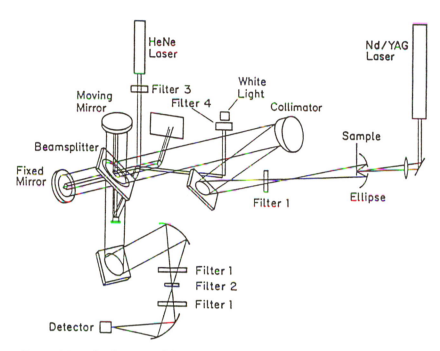

Figure 21A. The diagram of the macroscopic Raman collection system used for microscopic experiments shown in Figure 21B. (Reproduced with permission from reference 89. Copyright 1989 Society for Applied Spectroscopy.)

powerful, new developments in Hadamard transform spectroscopy or rediscovery of step-scan mode interferometry and coupling it to a dynamic, along with a rapid-scan FTIR spectrometer, make the field of polymer vibrational spectroscopy appealing not only to those who are engaged in designing new instruments, but also to those involved in structural and quantitative studies of polymers.

   Although the advantages of using the near-IR excitation and interferometric approach in the FT Raman experiment have been well documented, and new developments leading to a better signal-to-noise ratio are expected to occur, continuous developments are reported in dispersive Raman spectroscopy. The main disadvantage of the FT method comes from an incomplete rejection of the Rayleigh line in the near-IR region, resulting in a reduced overall signal-to-noise ratio and in the low-frequency region loss. Although the use of Chevron[2] filters may improve FT detection (*94*), the use

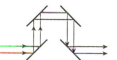

[2] Chevron filter consists of four mirrors in the following configuration. They can be purchased from the Omega Optical Co. Brattleboro, VT.

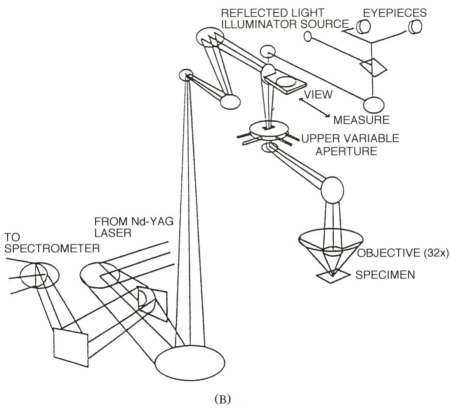

(B)

Figure 21B. For microscopic experiments B, the sample and ellipse shown in Figure 21A are replaced by this apparatus, which consists of an FTIR microscope accessory operating in the reflectance mode. (Reproduced with permission from reference 89. Copyright 1989 Society for Applied Spectroscopy.)

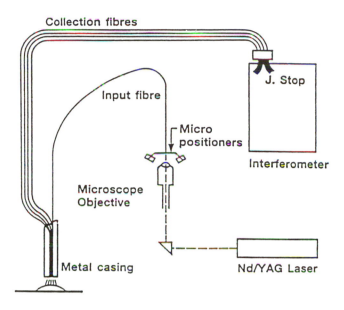

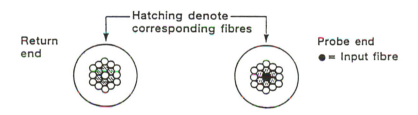

Figure 22. A schematic diagram illustrating the fiber-optic probe used with the FT Raman spectrometer. (Reproduced with permission from reference 91. Copyright 1990 Elsevier.)

of near-IR excitation (95) and multichannel detectors for wavelengths shorter than 1 μm is possible (96, 97).

With the continually increasing computer capabilities, data processing, and information theories, two- and three-dimensional experiments both in IR and Raman techniques will become fairly common. The instruments will become even faster and allow monitoring processes not observable.

Acknowledgments

The author is thankful to the National Science Foundation (EPSCoR Program) and numerous industrial sponsors including 3M Company (St. Paul,

MN), Eastman Kodak (Rochester, NY), Shell Chemical Company (Houston, TX), NRECA (Washington, DC), Medtronic (Minneapolis, MN), BYK Chemie (Germany), and Hitachi Chemical Company (Japan) for partial support. D. B. Chase is acknowledged for providing some of his data and perhaps useful comments on the manuscript.

References

1. Szymanski, H. A. *Theory and Practice on Infrared Spectroscopy*; Plenum Press: New York, 1964.
2. Barrow, G. M. *Introduction to Molecular Spectroscopy*; McGraw-Hill: New York, 1962.
3. Wilson, E. B.; Decius, J. C.; Cross, P. C. *Molecular Vibrations*; McGraw-Hill: New York, 1955.
4. Colthup, N. B.; Daly, L. H.; Wiberley, S. E. *Introduction to Infrared and Raman Spectroscopy*; Academic Press: Orlando, FL, 1964.
5. Griffiths, P. R. In *Advances in Infrared and Raman Spectroscopy*; Vol. 9.; Clark, R. J. H.; Hester, R. E., Eds.; Heyden: London, England, 1981.
6. Nakamoto, K. *Infrared and Raman Spectra of Inorganic and Coordination Compounds*; 4th ed.; John Wiley: New York, 1986.
7. Tehhune, R. W.; Maker, P. D.; Savage, C. M. *Phys. Rev. Lett.* **1965**, *14*, 681.
8. Syvin, S. J.; Rauch, J. E.; Decius, J. C. *J. Chem. Phys.* **1965**, *43*, 4083.
9. Christie, J. H.; Lockwood, D. J. *J. Chem. Phys.* **1971**, *54*, 1141.
10. Fanconi, B.; Peticolas, W. L. *J. Chem. Phys.* **1969**, *50*, 2244.
11. Golab, J. T.; Sprague, J. R.; Carron, K. T.; Schatz, G. S.; Van Duyne, R. P. *J. Chem. Phys.* **1988**, *88*, 7942.
12. The group theory for predicting IR and Raman activities based on symmetry considerations is discussed in Cotton, F. A. *Chemical Applications of Group Theory*; 2nd ed.; Wiley Interscience: New York, 1971.
13. Griffiths, P. R. In *Transform Techniques in Chemistry*; Griffiths, P. R., Ed.; Plenum Press: New York, 1978.
14. Urban, M. W.; Koenig, J. L. In *Vibrational Structure and Spectra*; Vol. 18; Durig, J. R., Ed.; Elsevier: New York, 1990.
15. Chamberlain, J. E. *The Principles of Interferometric Spectroscopy*; John Wiley: Chichester, England, 1979.
16. Debarre, D.; Boccara, A. C.; Fournier, D. *Appl. Opt.* **1981**, *20*, 4281.
17. Smith, M. J.; Manning, C. J.; Palmer, R. A.; Chao, J. L. *Appl. Spectrosc.* **1988**, *42*, 546.
18. Noda, I. *J. Am. Chem. Soc.* **1989**, *111*, 8116.
19. Noda, I. *Appl. Spectrosc.* **1990**, *44*, 550.
20. Palmer, R. A.; Manning, C. J.; Chao, J. L.; Noda, I.; Dowrey, A. E.; Marcott, C. *Appl. Spectrosc.* **1991**, *45(1)*, 12.
21. Dittmar, R. M.; Chao, J. L.; Palmer, R. A. *Appl. Spectosc.* **1991**, *45(7)*, 1104.
22. Chantry, G. W.; Gebbie, H. A.; Hilsum, C. *Nature (London)* **1964**, *203*, 1052.
23. Hirschfeld, T.; Chase, B. *Appl. Spectrosc.* **1986**, *40(2)*, 133.
24. Chase, B. *J. Am. Chem. Soc.* **1986**, *108*, 7485.
25. Painter, P. C.; Coleman, M. M.; Koenig, J. L. *The Theory of Vibrational Spectroscopy and Its Applications to Polymeric Systems*; Wiley Interscience: New York, 1982.
26. Bower, D. I.; Maddams, W. F. *The Vibrational Spectroscopy of Polymers*; Cambridge University Press: New York, 1989.

27. Ward, I. M. *Adv. Polym. Sci.* **1985**, 66, 81.
28. Abbate, S.; Gussoni, M.; Zerbi, G. *J. Chem. Phys.* **1979**, 70(8), 3577.
29. Urban, M. W.; Koenig, J. L., unpublished results.
30. Tobin, M. C.; Carraro, M. J. *J. Polym. Sci.* **1957**, 24, 93.
31. Read, B. E.; Stein, R. S. *Macromolecules* **1968**, 1, 116.
32. Glenz, W.; Peterlin, A. *J. Macromol. Sci. Phys.* **1970**, B4(3), 473.
33. Masetti, G.; Zerbi, G. *Macromolecules* **1973**, 6(5), 700.
34. Jasse, B.; Koenig, J. L. *J. Macromol. Sci. Rev. Macromol. Chem.* **1979**, C17, 61.
35. Burchell, D. J.; Lasch, J. E.; Farris, R. J.; Hsu, S. L. *Polym. Bull.* **1982**, 23, 965.
36. Zerbi, G.; Gallino, G.; Fanti, N. D.; Baini, L. *Polymer* **1989**, 30, 2324.
37. Bardwell, J. A.; Dignam, M. *Spectrochim. Acta* **1988**, 44A, 1435.
38. Bertie, J. E.; Eysel, H. H. *Appl. Spectrosc.* **1985**, 39, 392.
39. Graf, R. T.; Koenig, J. L.; Ishida, H. In *Fourier Transform Infrared Characterization of Polymers*; Ishida, H., Ed.; Plenum Press: New York, 1987; p 385.
40. Huang, J. B.; Hong, J. W.; Urban, M. W. *Polymer* **1992**, 33(24), 5173.
41. Huang, J. B.; Urban, M. W. *Appl. Spectrosc.* **1992**, 46(6), 1014.
42. Urban, M. W.; Huang, J. B. *Appl. Spectrosc.* **1992**, 46(11).
43. Hobbs, J. P.; Sung, S. C. P.; Krishnan, K.; Hill, S. *Macromolecules* **1983**, 16, 193.
44. Fina, L. J.; Koenig, J. L. *J. Polym. Sci. B Polym. Phys.* **1986**, 24, 2509.
45. Stein, R. S. J. *J. Appl. Phys. Appl. Polym. Phys.* **1961**, 5, 95.
46. Fina, L. J.; Koenig, J. L. *J. Polym. Sci. B Polym. Phys.* **1986**, 24, 2525.
47. Fina, L. J.; Koenig, J. L. *J. Polym. Sci. B Polym. Phys.* **1986**, 24, 2541.
48. Coleman, M. M.; Painter, P. C. *Macromolecules* **1987**, 20, 226.
49. Lee, J. Y.; Moskala, E. J.; Painter, P. C.; Coleman, M. M. *Appl. Spectrosc.* **1986**, 40(7), 991.
50. Urban, M. W.; Gaboury, S. R.; Provder, T. *Polym. Commun.* **1991**, 32(6), 171.
51. Hay, J. N.; Boyle, J. D.; Parker, S. F.; Wilson, D. *Polymer* **1989**, 30, 1032.
52. Williams, K. P. J.; Mason, S. M. *Trends Anal. Chem.* **1990**, 9(4), 119.
53. Clark, E. S.; Wilson, F. C. *Nylon Plastics*; Wiley Interscience: New York, 1973.
54. Jakes, J.; Krimm, S. *Spectrochim. Acta* **1971**, 27A, 19.
55. Hendra, P. J.; Maddams, W. F.; Royaud, I. A. M.; Willis, H. A.; Zichy, V. *Spectrochim. Acta* **1990**, 46A, 747.
56. Angelo, R. J.; Miura, H.; Gardner, K. H.; Chase, D. B.; English, A. D. *Macromolecules*, **1989**, 22, 117.
57. Kenton, R. C.; Rubinovitz, R. L. *Appl. Spectrosc.* **1990**, 44(8), 1377.
58. Atalla, R. H.; Agarwal, U. P. *Science (Washington, DC)* **1985**, 227, 636.
59. Miller, C. E.; Archibald, D. D.; Myrick, M. L.; Angel, S. M. *Appl. Spectrosc.* **1990**, 44(8), 1297.
60. Koningstein, J. A. *Introduction to the Theory of Raman Spectroscopy*; D. Reidel: Dordrecht, Netherlands, 1972; and reference therein.
61. Felton, R. H.; Yu, N.-T. *The Porphyrins*; Dolphin, D., Ed.; Physical Chemistry, Part A; Academic Press: Orlando, FL, 1978.
62. Spiro, T. G. *Iron Porphyrins*; Lever, A. B. P.; Gray, H. B., Eds.; Addison-Wesley: Reading, MA, 1983.
63. Aher, S. A. *Methods in Enzymology*; Colowick, S. P.; Kaplan, N. O., Eds.; Vol. 76, Academic Press: Orlando, FL, 1981.
64. Johnson, C. K.; Rubinovitz, R. *Appl. Spectrosc.* **1990**, 44(7), 1103.
65. Tu, A. T. *Raman Spectroscopy in Biology: Principles and Applications*; Wiley-Interscience: New York, 1982.
66. Hallmark, V. M.; Zimba, C. G.; Swalen, J. D.; Rabolt, J. F. *Microchim. Acta* **1988**, II, 121.

67. Zimba, C. G.; Hallmark, V. M.; Swalen, J. D.; Rabolt, J. F. *Appl. Spectrosc.* **1989**, *43(7)*, 231.
68. Levin, I. W. *Advances in Infrared and Raman Spectroscopy*; Clark, R. J. H.; Hester, R. E., Eds.; Wiley Heyden: New York, 1984; Vol. 11, Chapter 1.
69. Lewis, E. N.; Kalasinsky, V. F.; Levin, I. W. *Appl. Spectrosc.* **1988**, *42(7)*, 1188.
70. Bunow, M. R.; Levin, I. W. *Biochim. Biophys. Acta* **1977**, *464*, 202.
71. Hallmark, V.; Rabolt, J. F. *Macromolecules* **1989**, *22*, 500.
72. Sutton, P.; Koenig, J. L. *Biopolymers* **1970**, *9*, 615.
73. Shotts, J.; Sievers, A. *Biopolymers* **1974**, *13*, 2593.
74. Hallmark, V. M.; Zimba, C. G.; Swalen, J. D.; Rabolt, J. F. *Appl. Spectrosc.* **1990**, *44(2)*, 321.
75. Nie, S.; Marzilli, P. A.; Marzilli, L. G.; Yu, N-Teng *J. Chem. Soc. Chem. Commun.* **1990**, *770*, 25.
76. Sommer, A. J.; Katon, J. E. *Appl. Spectrosc.* **1991**, *45(4)*, 527.
77. Exsted, B. J.; Urban, M. W. *J. Organomet. Polym. Chem.*, in press.
78. Exsted, B. J.; Urban, M. W. *J. Appl. Polym. Sci.* **1993**, *47*, 2019.
79. Chapter 35, this book.
80. Young, J. T.; Tsai, W. H.; Boerio, F. J. *Macromolecules* **1992**, *25*, 887.
81. Boerio, F. J.; Hong, P. P.; Tsai, H. W.; Young, J. T. *Surf. Interface Anal.* **1991**, *17*, 448.
82. Boerio, F. J.; Hong, P. P.; Clark, P. J.; Okamoto, Y. *Langmuir* **1990**, *6*, 721.
83. Ondrus, D. J.; Boerio, F. J. *J. Coll. Interf. Sci.* **1988**, *124*, 349.
84. Ondrus, D. J.; Boerio, F. J.; Grannen, K. J. *J. Adhesion* **1989**, *29*, 27; and references therein.
85. Angel, S. M.; Katz, L. F.; Archibal, D. D.; Lin, L. T.; Honigs, D. E. *Appl. Spectrosc.* **1988**, *42*, 1327.
86. Angel, S. M.; Myrick, M. L. *Anal. Chem.* **1989**, *61*, 1648.
87. Urban, M. W.; Koenig, J. L. *Digilab Letters* **1986**, *2(1)*.
88. Harthcock, M. A.; Lentz, L. A.; Davie, B. L.; Krishnan, K. *Appl. Spectrosc.* **1986**, *40*, 210.
89. Messerschmidt, R. G.; Chase, D. B. *Appl. Spectrosc.* **1989**, *43(1)*, 11.
90. Williams, K. P. J. *J. Raman Spectrosc.* **1991**, *21*, 328.
91. Williams, K. P. J.; Mason, S. M. *Trends Anal. Chem.* **1990**, *9(4)*, 119.
92. Archibald, D. D.; Miller, C. E.; Lin, L. T.; Honigs, D. E. *Appl. Spectrosc.* **1988**, *42(8)*, 1549.
93. Archibald, D. D.; Lin, L. T.; Honigs, D. E. *Appl. Spectrosc.* **1988**, *42(8)*, 1558.
94. Radziszewski, J. G.; Michl, J. *Appl. Spectrosc.* **1990**, *44(3)*, 414.
95. Porterfield, D. R.; Campton, A. *J. Am. Chem. Soc.* **1988**, *110*, 408.
96. McCreary, R.; Wang, Y. *Anal. Chem.* **1989**, *62*, 2647.
97. Pemberton, J.; Sobocinski, R. *J. Am. Chem. Soc.* **1989**, *111*, 432.

RECEIVED for review March 18, 1992. ACCEPTED revised manuscript June 9, 1992.

Vibrational Spectroscopy of Polymers

Analysis, Physics, and Process Control

H. W. Siesler

Department of Physical Chemistry, University of Essen, D 4300 Essen, Germany

Improvements in rapid-scanning Fourier-transform infrared (FTIR) spectroscopy, the recent introduction of Fourier-transform Raman spectroscopy, and the more efficient exploitation of the near-infrared region launched vibrational spectroscopy into a new era of polymer chemical and physical applications. On the one hand, increased sensitivity led to breakthrough sampling techniques such as photoacoustic spectroscopy and diffuse reflectance measurements; on the other hand, improved time resolution largely enhanced the potential of FTIR spectroscopy for on-line combination with other techniques such as gas and liquid chromatography or thermal analysis. Significant progress also has been made in the characterization of time-dependent phenomena by the simultaneous acquisition of spectral and other relevant physical data (e.g., during mechanical measurements). Furthermore, multivariate data evaluation, which was restricted to near-IR multicomponent analysis of agricultural products for more than a decade, is increasingly applied in the field of polymer analysis. Last, but not least, the development of fiber optics for the near-IR wavelength range has opened up completely new areas for process control and remote sensing.

V IBRATIONAL SPECTROSCOPY IS AN IMPORTANT TOOL FOR CHARACTERIZATION of the chemical and physical nature of polymers. In principle, the complementary techniques of infrared and Raman spectroscopy provide qualitative

0065–2393/93/0236–0041$12.75/0

and quantitative information about the following structural details of a polymer:

- **Chemical nature and composition:** structural units, type and degree of branching, end groups, additives, impurities
- **Steric order:** cis/trans (Z/E) isomerism, stereoregularity
- **Conformational order:** physical arrangement of the polymer chains (planar or nonplanar), regular conformations
- **Three-dimensional state of order:** crystalline and amorphous phases, number of chains per unit cell, intermolecular forces, lamellar thickness
- **Orientation:** type and degree of preferential polymer chain and side group alignment in anisotropic materials

Despite the uncontested importance of vibrational spectroscopy for the characterization of macromolecular structure, it should be emphasized that only a limited number of problems may be solved by its exclusive application. In the majority of cases maximum information on the structural details in question can be obtained only by an appropriate choice and combination of chemical and physical methods. In this respect the introduction of rapid-scanning Fourier-transform IR (FTIR) and FT-Raman spectroscopy and the advantages of these methods over conventional dispersive instrumentation have revitalized the utilization of vibrational spectroscopy in polymer research. In combination with other techniques and the application of destruction-free sampling procedures, FTIR and FT-Raman spectroscopy allow a better correlation of the spectroscopic data with the results obtained by other methods from the original sample.

The intention of the present chapter is to review some aspects of mid- and near-infrared and Raman spectroscopy in polymer research with special emphasis on selected examples of diffuse reflectance measurements, comparison of photoacoustic FTIR and FT-Raman spectroscopy, dynamic FTIR characterization of time- and temperature-dependent phenomena, and the application of light-fiber optics to process control and remote sensing.

Experimental Details

The principles, theory, and instrumentation of FTIR spectroscopy have been covered in detail in several books and reviews (1–4). The FTIR spectra presented here were measured on spectrometers (Nicolet 7199 and Bruker IFS88).

Since the introduction of laser sources during the 1960s, a vast amount of work has been undertaken using Raman spectroscopy to probe a wide range of polymeric structures (4, 5–7). Conventional Raman spectrometers that were used to carry out this work generally consist of a visible laser source, an efficient double or triple monochromator, and a photomultiplier detection system. However, a number of problems are inherent in the application of conventional laser Raman spectroscopy to polymer systems. For a large proportion of samples, irradiation with visible light caused strong fluorescence of additives and impurities that was superimposed on the weak Raman signal. A number of (sometimes very time-consuming) approaches were attempted to circumvent this problem, including prior purification, burning out the fluorescence, shifting the excitation line, or using pulsed lasers. These approaches, at best, have been only partially successful. A further problem with conventional Raman spectroscopy is that highly colored samples or polymers that contain fillers may absorb the Raman photons which prevents the photons from reaching the detector and leads to thermal degradation of the investigated polymer. The foregoing difficulties are some of the reasons why the Raman technique is not as familiar as infrared techniques, despite the fact that, in certain circumstances, Raman spectroscopy has a number of advantages over infrared spectroscopy. The recent development of FT-Raman spectroscopy (*8–12*), however, increases the likelihood that this technique will become standard instrumentation in spectroscopic laboratories.

The main features of an FT-Raman instrument are as follows:

- a neodymium–yttrium aluminum garnet (Nd–YAG) laser, operated at 1064 nm with a power output range of 0–2 W
- an efficient dielectric filter system to remove the Rayleigh component of the scattered radiation
- a FTIR spectrometer equipped with a quartz or calcium fluoride beam splitter for operation in the near-infrared region
- a photoelectric detector; usually a cooled germanium unit (at 77 K) or an indium–gallium–arsenide (InGaAs) detector operating at ambient temperature
- a Raman sampling compartment with 180° or 90° lens-based optics

The use of near-infrared excitation confers a number of advantages on a FT-Raman system. Both fluorescence and self-absorption of the Raman signal are very much reduced and, due to the lower energy of the exciting light, thermal degradation is also less of a problem. A further breakthrough in this field is the development of a FT-Raman microscope system for small samples in the micrometer range that offers the application of mapping procedures (*13*). However, compared to conventional Raman, FT-Raman does have some disadvantages and limitations. It is less sensitive because, due to the proportionality of the Raman scattering cross section to the fourth power of the exciting frequency, the shift from the visible to the near-infrared region reduces the intensity of Raman scattering. Furthermore, as a consequence of detector and filter limitations, the Raman shift can presently only be measured between ~ 3500 and 100 cm^{-1}. This long-wavelength cutoff excludes the investigation of longitudinal acoustic modes in lamellar polymer structures (4, 7).

The Raman spectra presented here were carried out using a FT-Raman accessory (Bruker FRA106) that was interfaced to the FTIR unit (Bruker IFS88).

Diffuse Reflectance FTIR Studies of the Glass Powder–Coupling Agent Interface of a Composite

The interface between the reinforcement and the polymer material is of critical importance to the mechanical properties and performance of composites, and coupling agents are used to improve the interfacial bonding (14, 15). Because of its nondestructive sample preparation procedure, and high sensitivity to surfaces, diffuse reflectance FTIR spectroscopy has proved to be one of the most successful techniques to study the reaction of a coupling agent with either the resin matrix or the reinforcement material (16, 17). Frequently, organofunctional silanes are used to promote the adhesion between the inorganic reinforcement and the organic polymer. In feasibility studies concerning the substitution of amalgam-based dental filling materials by glass powder–polyacrylate composites, vibrational spectroscopy proved to be an extremely important method for investigation of the glass powder–coupling agent interface and the progress of the photoinitiated curing of the monomeric diacrylate (18). For the composite under investigation, an applied γ-methacryloxypropyltrimethyloxysilane coupling agent reacted with the SiOH functionalities of the glass surface by condensation and formed bonds with the diacrylate by polymerization of the C=C double bonds (Structure 1).

With respect to a more detailed understanding of the microstructural changes at the glass powder–coupling agent interface, the diffuse reflectance FTIR spectra of Aerosil OX 50 (average particle diameter 50 nm; 50 m^2/g) were studied as a function of increasing coupling agent coating. Attention was focused on spectroscopic changes of non-hydrogen-bonded and associated SiOH groups of the glass surface, the increase of OCH_3 and CH_2 functionalities, and the variation of the proportion of hydrogen-bonded and non-hydrogen-bonded C=O groups. The amount of coupling agent was characterized by the organic carbon content determined independently by elemental analysis (1% C corresponded to approximately 100 mmol of coupling agent on 100-g glass powder).

The spectra were measured in a diffuse reflectance cell (Harrick) with KCl as the inert reference matrix [sample:matrix about 1:2 (w/w)] and 500

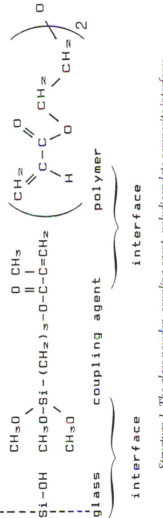

Structure 1. The glass powder–coupling agent–polydiacrylate composite interfaces.

scans were accumulated at a resolution of 2 cm^{-1} (0.964-cm^{-1} data point interval). Spectra of three selected samples with organic carbon contents of 0, 0.95, and 1.8% are shown in Figure 1 and reflect the following intensity and frequency changes with increasing coating (Figure 2; ν denotes stretching vibration) (18):

- The intensity of the sharp $\nu(OH)$ absorption (3740 cm^{-1}) of non-hydrogen-bonded SiOH groups decreases drastically (Figure 2a).

- The intensity of the $\nu(OCH_3)$ absorption of the SiOCH$_3$ functionalities at 2848 cm^{-1} increases (Figure 2b).

- The intensity of the $\nu(CH_2)$ and $\nu(C=O)_{total}$ (Figure 2c) absorptions in the 2900- and 1700–1725-cm^{-1} regions, due to the coupling agent, naturally increase. An absorption that can be assigned to free C=C resonance stabilized carbonyl groups gradually increases at 1718 cm^{-1} at the expense of the initial $\nu(C=O)_{associated}$ band at 1705 cm^{-1} (Figure 2d).

- The $\nu(OH)$ absorption of hydrogen-bonded SiOH functionalities at 3666 cm^{-1} slightly decreases.

Thus, the mechanism for the chemisorption of the coupling agent at the glass powder interface can be outlined. In a first step, the free OH groups of the glass react completely with the available SiOCH$_3$ functionalities of the siliconemethacrylate. As consumption of free SiOH groups increases, the hydrogen-bonded SiOH groups take over this reaction role (although to a smaller degree) and the percentage of unreacted SiOCH$_3$ groups increases. During this process the percentage of non-hydrogen-bonded C=O groups of the coupling agent increases at the cost of the hydrogen-bonded moieties.

The application of near-IR spectroscopy to determine the residual C=C double-bond content and enable estimation of the progress of the diacrylate curing process will be discussed in the section on near-IR spectroscopy.

FTIR Photoacoustic Spectroscopy Compared to FT-Raman Spectroscopy

The request for rapid identification of unknown polymeric materials with varying morphologies is a very common problem in analytical laboratories. Due to the broader availability of FTIR instrumentation, the task is often accomplished using infrared spectroscopy even though there are frequently difficulties associated with finding a suitable sampling technique. Photoacoustic FTIR as well as FT-Raman spectroscopy offer a simple approach to this problem. Independently of the morphology, almost any polymer sample can

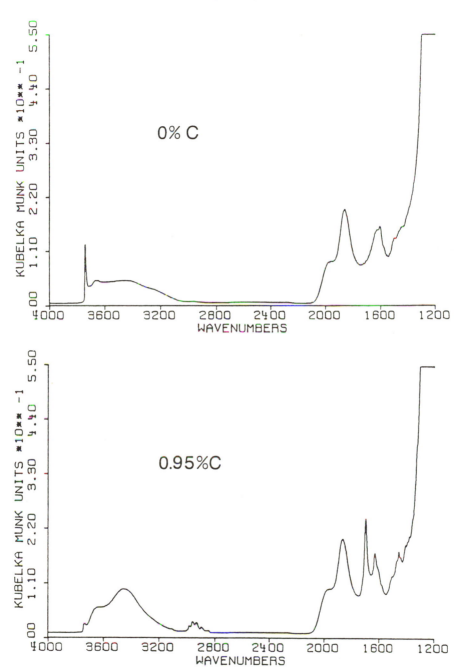

Figure 1. *Diffuse reflectance FTIR spectra of Aerosil OX 50 coated with increasing amounts (0, 0.95, and 1.8% C) of coupling agent.* Continued on next page.

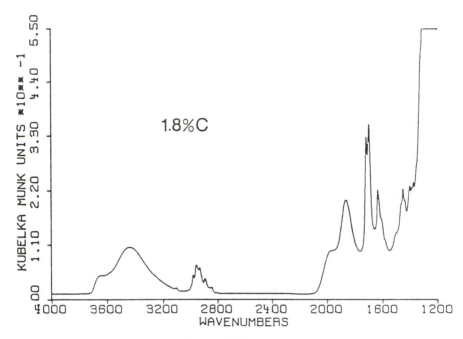

Figure 1. Continued.

be placed in the photoacoustic (PA) cell or in the laser beam to collect a spectrum in a matter of minutes. To demonstrate the preferential applicability of FTIR-PA and FT-Raman spectroscopy for the destruction-free identification of polymer samples, some selected examples will be discussed after a short introduction to the basic principles of photoacoustic FTIR spectroscopy.

The photoacoustic effect was first reported by Bell in 1881 (*19*), but the method remained in the background until a transducer more sensitive and selective than the ear could be found. It was not until 1973 that a renaissance of photoacoustic spectroscopy occurred (*20*), and since then numerous interesting applications of this technique have been reported in the ultraviolet, visible, and near-infrared regions (*21–25*). The introduction of FTIR spectroscopy expanded the availability of PA spectroscopy to the mid-infrared region (*25–28*) where it is applied quite frequently to the investigation of polymers (*29–30*).

In FTIR-PA spectroscopy the solid sample is placed in an enclosed cell with an IR-transparent entrance window and a built-in sensitive microphone (Figure 3). The cell contains air or, for increased sensitivity, another coupling gas, such as helium, under atmospheric pressure (*25, 31, 32*). This cell is mounted in the sample compartment of the FTIR spectrometer and the microphone takes over the function of a conventional FTIR detector. The

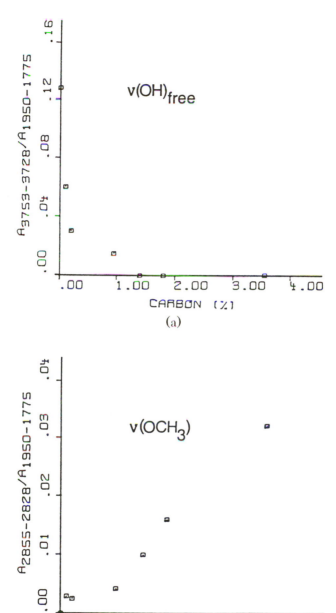

Figure 2. Intensity changes of selected absorption bands as a function of the amount of coating represented by the content of organic carbon. Continued on next page.

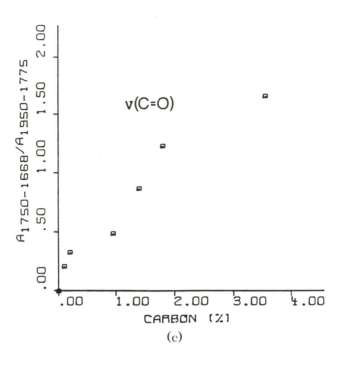

(c)

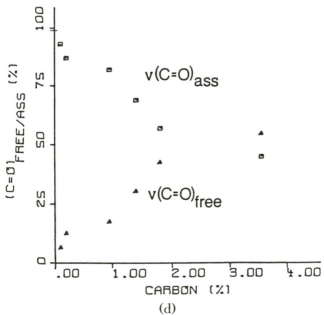

(d)

Figure 2. Continued

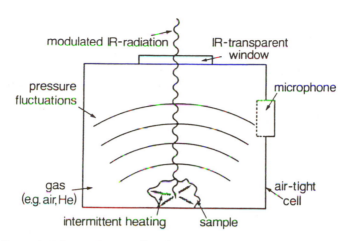

Figure 3. Scheme of a PA cell and principle of the photoacoustic effect.

sample is irradiated with polychromatic light that is modulated with frequency f:

$$f = 2\nu\upsilon \tag{1}$$

where ν is the wavenumber (per centimeter) and υ is the mirror velocity (centimeters per second) of the Michelson interferometer. With a typical mirror velocity of 0.1 cm/s, the modulation frequencies vary between 800 and 80 s^{-1} in the 4000–400-cm^{-1} region, respectively.

When radiation of a certain wavelength is absorbed, the sample is excited to a higher energy level whence it can return to the ground state by emission of light or (more commonly) by nonradiative decay processes or heat release. The photoacoustic effect is based on the heat release phenomenon. The released heat is transferred to the sample surface at a rate that depends on the thermal characteristics of the sample. Due to the modulation of the absorbed light, the nonradiative deactivation processes lead to a periodic sample heating that is coupled into the surrounding fill gas at the sample–gas interface and produces a pressure fluctuation within the PA cell. The acoustic wave is then propagated through a small connecting passage to the microphone chamber where the resulting PA signal—recorded as a function of wavenumber—is correlated with the optical spectrum.

PA spectroscopy has some limitations insofar as the frequency positions of absorption bands are reproduced but the band intensities cannot be interpreted according to the conventional rules of transmission spectroscopy. The signal saturation is a phenomenon that is important for understanding the PA effect and the interpretation of PA spectra (*21, 23, 26, 32–34*). The amplitude of the previously mentioned periodic heating is decreased by thermal damping processes during propagation in the investigated medium. It can be shown (*21, 23*) that only the part of the sample within a defined distance from the sample surface contributes to the heat transfer to the surrounding gas (*23*) and the intensity of the PA signal depends on the coefficients of optical absorption and thermal diffusion.

In the practical application of the PA spectroscopy method the spectrum of carbon-black is recorded as background and then ratioed against the PA spectrum of the sample under investigation. Because of the lower signal-to-noise ratio for PA spectroscopy, longer scanning times (in the range of minutes) are required compared to transmission spectroscopy. The PA spectra presented in Figures 4, 5a, and 6 were taken in a Gilford—Nicolet photoacoustic cell with a mirror velocity of 0.11 cm/s and about 500 scans were accumulated with a resolution of 4 cm^{-1} (1.928-cm^{-1} data point interval). Recent reproduction of the data in a PA cell (MTEC Model 200) on a FTIR spectrometer (Perkin-Elmer 1760X) with a mirror velocity of 0.10 cm/s and a resolution of 8 cm^{-1} demonstrated that, despite a comparable signal-to-noise ratio, the number of scans could be pushed down to 32, thereby reducing the analysis time considerably (*35*).

To demonstrate the potential of photoacoustic FTIR and FT-Raman spectroscopy, the spectra of selected polymers in different morphologies

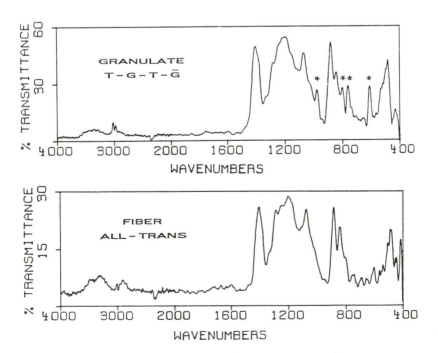

Figure 4. PA spectroscopy spectra of poly(vinylidene fluoride) granulate in the II (α) modification and fibers in the I(β) modification. Asterisks denote characteristic absorption bands of the II(α) form.

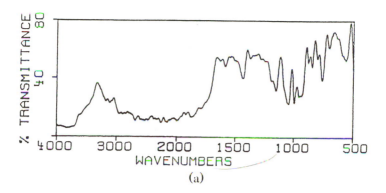

(a)

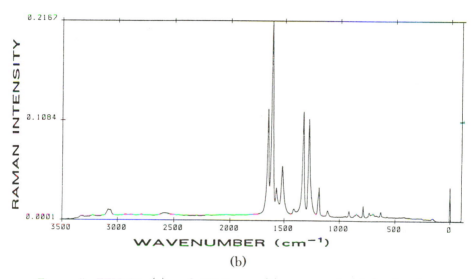

(b)

Figure 5. *FTIR-PA (a) and FT-Raman (b) spectra of poly(p-phenylene terephthalamide) fibers.*

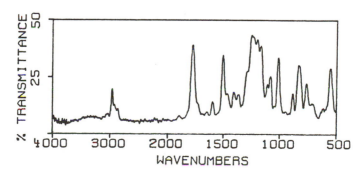

Figure 6. *FTIR-PA spectrum of a carbon-black-filled polycarbonate sheet.*

were examined "as received." Notwithstanding the simplicity of sample preparation for both techniques, the FTIR-PA and FT-Raman spectra of poly(vinylidene fluoride) granulate and fibers shown in Figures 4 and 7, respectively, clearly reflect absorption bands characteristic of specific conformational states of order due to the differences in the mechanical pretreatment of the polymer during the production process. Whereas the granulate occurs primarily in the $TGT\overline{G}$ conformation of the II(α) form, the fibers have been transformed largely to the all-trans conformation of the I(β) modification.

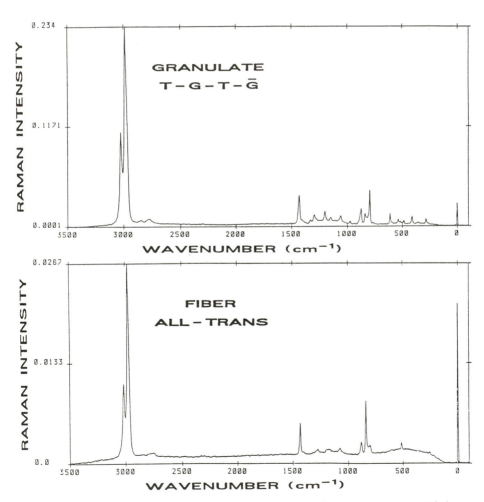

Figure 7. FT-Raman spectra of poly(vinylidene fluoride) granulate in the II(α) form and fibers in the I(β) modification.

The phenomenon of saturation is demonstrated by the strongly reduced intensities (compared to a transmission spectrum) of the $\nu(C{=}O)$ and $\nu(C{-}N) + \delta(N{-}H)$ absorptions of the PA spectrum of poly(p-phenylene terephthalamide) fibers (Figure 5a). The excellent quality of the corresponding FT-Raman spectrum shown in Figure 5b recommends FT-Raman spectroscopy as the method of choice for such samples where saturation effects dominate the PA spectrum. Depending on the nature of the filler, either FTIR-PA or FT-Raman spectroscopy has advantages. In the case of carbon-black-filled polymers, FT-Raman spectroscopy usually fails to yield useful spectra, whereas the excellent FTIR-PA spectrum obtained from a sheet of carbon-black-filled polycarbonate (Figure 6) emphasizes its value as a valuable alternative. On the other hand, highly infrared active filler materials such as glass fibers will strongly overlap the actual polymer spectrum from 1400 cm^{-1} to longer wavelengths, whereas they do not significantly contribute to the FT-Raman spectrum (35).

Depth-Profiling in Polymers by Attenuated-Total-Reflection FTIR Spectroscopy

Although attenuated-total-reflection (ATR) spectroscopy is a well-established method for the characterization of polymer surfaces (3, 4, 36), comparatively few available publications elucidate the detailed theoretical and practical background for depth profiling with this technique (37–39).

The principle of ATR spectroscopy is based on the phenomenon of total internal reflection. For the experiment the sample (refractive index n_2) is brought into direct contact with the surface of a reflection element of high refractive index n_1 ($n_1 > n_2$). When the angle of incidence of the light beam at the reflection element–sample interface exceeds the critical angle, total internal reflection takes place and the radiation slightly penetrates the optically rarer medium (4, 36). The penetration of the radiation into the sample is usually on the order of a few micrometers, which is sufficient to constitute a short absorbing path. Therefore, total internal reflection will be attenuated in the wavelength regions of sample absorption and the recorded spectrum will be very similar to the transmission spectrum.

The penetration depth (d_P) for which the electric field component decreased to $1/e$ of its value in the reflection element–sample interface was derived theoretically by Harrick (36):

$$d_P = \frac{\lambda/n_1}{2\pi\sqrt{\sin^2\Theta - (n_2/n_1)^2}} \qquad (2)$$

Here, Θ is the angle of incidence, λ is the wavelength of radiation in vacuum, and n_1 and n_2 are the refractive indexes. This equation demon-

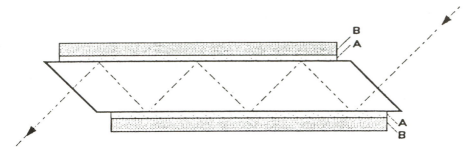

Figure 8. Setup of the two-layer ATR measurements: surface layer of variable thickness, A; and detection layer, B.

strates the dependence of penetration depth on the wavelength of the radiation, the angle of incidence and the refractive indexes of the sample, and the reflection element. For the practical application of the ATR method in polymer–polymer interdiffusion experiments, this theoretically derived penetration depth is of limited value. What is actually required is a measure of the depth to which the composition of the investigated polymer can be reliably determined by ATR spectroscopy under specified experimental conditions. To this end, polymer films with varying, definite thicknesses (surface layer) and a coating detection layer of "infinite" thickness were measured by ATR-FTIR spectroscopy to determine the thickness of the surface layer (later referred to as information depth d_I) at which characteristic absorption bands of the detection layer are no longer observable (*40*). The evaluation of these systematic measurements should then reveal the dependence of the information depth on the wavelength of the incoming radiation.

The arrangement of the surface and detection layers on the ATR crystal is shown in Figure 8, and detailed results will be presented here with reference to the system with a polystyrene (PS) surface layer and a polyamide 12 detection layer. The reflection element was a KRS5 parallelepiped with a refractive index of 2.39, an angle of incidence of 45°, and a total number of six reflections. The intensities of the ν(NH), amide I, amide II, and δ(NH) absorption bands of the polyamide 12 detection layer were determined as a function of increasing thickness of the polystyrene surface layer (Figure 9) and their normalized intensities (Table I) plotted versus PS film thickness. Examples of such plots are shown for the ν(NH) and amide I bands in Figure 10. A linear wavelength dependence of the information depth d_I can be derived from these data (Figure 11), and comparison of the theoretically derived penetration depth d_P for polystyrene (1.59 refractive index) with the experimentally determined information depth d_I yields the relation (*40*)

$$d_I = 1.29 \, d_P \tag{3a}$$

This relation is very close to the relation determined for poly(2,6-dimethyl-1,4-phenylene ether) (PPE) with polyamide 12 as the detection layer (*40, 41*),

$$d_I = 1.33\ d_P \qquad\qquad (3b)$$

and shows that a proportionality constant of about 1.31 can be used for any polymer system under the specified experimental conditions. In more detailed investigations this information depth was used as the experimental basis for interdiffusion experiments of low-molecular weight polystyrene into poly(2,6-dimethyl-1,4-phenylene ether) (*41*).

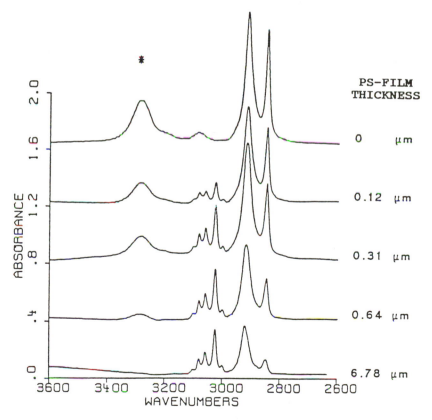

Figure 9. ATR spectra of polystyrene (PS) surface layers with different thickness and a polyamide 12 detection layer. Asterisks denote evaluated absorption bands. Continued on next page.

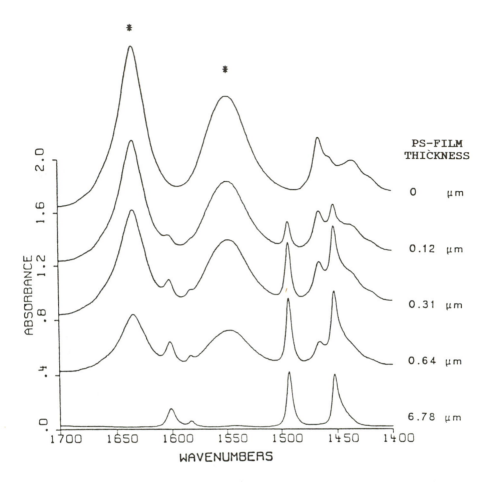

Figure 9. Continued

Variable-Temperature FTIR Measurements and Hydrogen Bonding

The role played by hydrogen bonds in the structure and properties of polymers has been the subject of numerous investigations (3, 4, 42–44). Hydrogen bonds are of particular importance for the chemical, physical, and mechanical properties of polymers that contain amide (polypeptides, proteins, polyamides), urethane (polyurethanes), and hydroxyl [cellulose, poly(vinyl alcohol), poly(acrylic acid)] functional groups.

Although the energies of hydrogen bonds are weak (20–50 kJ/mol) in comparison to covalent bonds (of the order of 400 kJ/mol), this type of molecular interaction is large enough to produce appreciable frequency and

Table I. Normalized Integral (*i*) or Peak-Maximum (*p*) Absorbances of Selected Polyamide 12 Absorption Bands as a Function of Polystyrene Film Thickness (*x* and *y* are integration limits)

Sample Number	PS Film Thickness	Ai/Ai_{max}			Ap/Ap_{max} $\delta(N-H)$
		$\nu(N-H)$	$\nu(C=O)$	$\nu(C-N)$ $+\delta(N-H)$	
0	0	1	1	1	1
1	0.05	0.49	0.75	0.79	0.77
2	0.07	0.51	0.82	0.72	0.84
3	0.12	0.47	0.65	0.64	0.84
4	0.31	0.10	0.34	0.33	0.53
5	0.63	0.04	0.26	0.24	0.49
6	0.64	0.09	0.27	0.25	0.53
7	1.09	—	0.08	0.05	0.37
8	1.33	—	0.09	0.08	0.42
9	1.35	—	0.09	0.08	0.42
10	1.49	—	0.03	0.02	0.37
11	1.53	—	0.02	0.01	0.32
12	1.79	—	0.04	0.01	0.28
13	2.13	—	—	—	0.30
14	2.31	—	—	—	0.23
15	2.43	—	—	—	0.28
16	2.45	—	—	—	0.23
17	6.78	—	—	—	—

[a] $x = 3412$ cm^{-1} $y = 3160$ cm^{-1} $Ai_{max} = 24.28$
[b] $x = 1687$ cm^{-1} $y = 1610$ cm^{-1} $Ai_{max} = 26.05$
[c] $x = 1590$ cm^{-1} $y = 1503$ cm^{-1} $Ai_{max} = 27.29$
[d] $\bar{\nu}_{max} = 720.6$ cm^{-1} $Ap_{max} = 0.43$

intensity changes in the vibrational spectra. In fact, the disturbances are so significant, that mid-infrared and Raman spectroscopy have become one of the most informative sources of criteria for the presence and strength of hydrogen bonds (3, 4, 44–46). Any variation of spectroscopic parameters, such as intensity, wavenumber position, and band shape, directly reflects the temperature dependence of the vibrational behavior of the investigated polymer as a consequence of changes in the inter- and intramolecular interactions. Unfortunately, only few vibrational spectroscopists are aware that near-infrared spectroscopy offers a powerful alternative to gain insight into these phenomena. To demonstrate the potential of variable temperature measurements for the elucidation of hydrogen-bonding effects, the results obtained from mid- and near-infrared spectroscopic studies of thermoplastic polyamide 12 will be outlined.

In polyamides and polyurethanes hydrogen bonding involves the proton-donating group NH and the proton acceptor C=O. Many of the investigations so far reported deal with the observed frequency shift and intensity increase of the ν(NH) stretching vibration band when hydrogen bonding

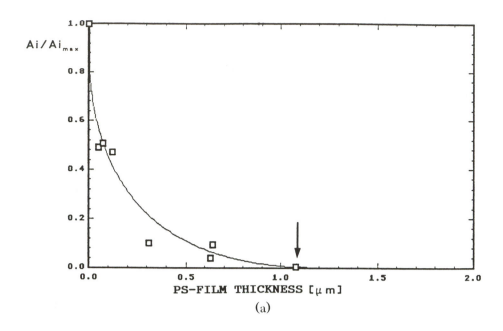

(a)

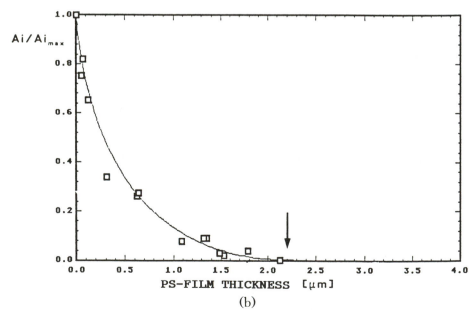

(b)

Figure 10. Normalized integrated absorbances (A/Ai_{max}) *of the* $\nu(NH)$ *(a) and* $\nu(C{=}O)$ *(b) absorption bands of polyamide 12 versus polystyrene (PS) film thickness. The arrow marks the information depth* d_I.

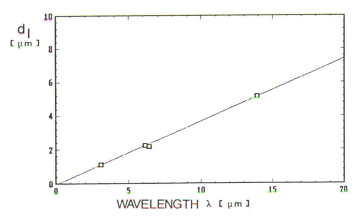

Figure 11. Information depth d_I *of polystyrene as a function of wavelength (the detection layer is polyamide 12).*

occurs (*44–46*). The frequency shift $\Delta\nu(NH)$, in particular, has been correlated with various chemical and physical properties of the hydrogen bond (e.g., enthalpy of formation, hydrogen-bond distance) (*3, 4, 45, 46*). Increasing hydrogen bond strength decreases the N $\cdots$ O distance, and this decrease is reflected by an increase in the shift between the $\nu(NH)$ absorption frequency of the associated and nonassociated NH functionalities.

The $\nu(NH)$ region of the mid-infrared spectrum of the investigated polyamide is shown in Figure 12. According to differential scanning calorimetric (DSC) measurements polyamide 12 melts at about 450 K and the following spectral changes of the $\nu(NH)$ band are observed with increasing temperature:

- The intensity of the $\nu(NH)_{associated}$ band decreases.
- The peak maximum of the $\nu(NH)_{associated}$ band shifts toward higher wavenumbers.
- The half-bandwidth of the $\nu(NH)_{associated}$ band increases considerably.
- A $\nu(NH)_{free}$ band appears at higher wavenumbers.

The wavenumber shift and increase in bandwidth of the $\nu(NH)_{associated}$ band at higher temperatures are the result of a general weakening of the hydrogen bonds and a concomitant broader distribution of the hydrogen bond energies. Unfortunately, the intensity of the $\nu(NH)_{free}$ band is very low in the mid-infrared range and, additionally, at elevated temperatures the $\nu(NH)$ absorptions of the free and associated functionalities overlap. A total $\nu(NH)$ absorbance procedure for the quantitative assessment of hydrogen

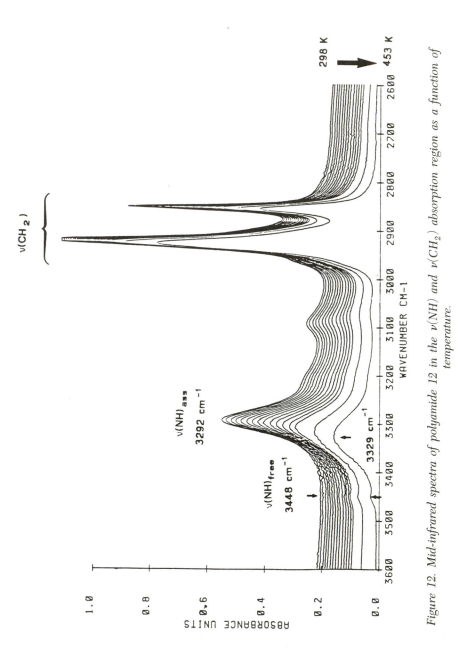

Figure 12. Mid-infrared spectra of polyamide 12 in the $\nu(NH)$ and $\nu(CH_2)$ absorption region as a function of temperature.

bonding in polyamides and polyurethanes has been proposed by Sricha-trapimuk and Cooper (47). However, for aliphatic polyamides, Coleman et al. (48) have shown that when the strong dependence of the absorption coefficient upon band frequency (and, in turn, the strength of the hydrogen bond) is neglected, the proportion of nonbonded NH groups at elevated temperatures derived with the foregoing procedure from mid-infrared spectra in the $\nu(NH)$ region is too high. Inspection of Figure 13 shows that near-infrared spectroscopy certainly offers an alternative by evaluating the high-frequency wing of the $2 \times \nu(NH)_{free}$ overtone absorption. For a quantitative evaluation of nonbonded NH groups, however, the phenomenon pointed out by Coleman et al. (48) as well as the reversal of the intensity ratio between the fundamentals of the free and associated NH groups and their first overtones must be taken into account (44). Additionally, contrary to mid-infrared spectroscopy, where the $\nu(OH)$ absorption of water is superimposed by the dominating $\nu(NH)$ band of the polyamide, the loss of water during thermal treatment of the polymer can readily be quantified by the intensity decrease of the $\nu(OH) + \delta(OH)$ combination band at about 5150 cm^{-1}.

Recent investigations (49; H. W. Siesler, unpublished results) showed that variable-temperature FTIR spectroscopy with polarized radiation can also be advantageously applied to study the temperature-induced reorientational motion of the mesogenic groups in liquid-crystalline side-chain polymers with polyacrylate main chains (50). The structure of the investigated polymer, poly(1-(6-(4'-cyanophenyl-4-benzyloxy)hexyloxycarbonyl)-ethylene), is shown in Figure 14 along with the polarization spectra measured at room temperature of a sample whose mesogenic groups were oriented in the friction direction of polyimide-coated and mechanically pretreated KBr-windows. The molecular weight of the polymer was 20,000 and the transition temperature from the nematic to the isotropic phase was 133 °C (49). To study the relative orientational motion of the mesogenic groups, the spacer and the main chain as a function of heating, the $\nu(C\equiv N)$ and $\delta(CH_2)$ absorption bands (Figure 14) were evaluated in terms of their orientation functions (*see* eq 5). The sample was subjected to a programmed heating of 1 °C/min and 64 scans were taken in 1-min intervals with the polarization direction alternately parallel and perpendicular to the mechanical reference (friction) direction. Depending on the thermal pretreatment, in repeated measurements a strong reorientational behavior was observed between the glass transition (33 °C) and clearing point (133 °C) beyond which the sample became isotropic. Although this behavior is not clearly understood, the orientation function–temperature plot proves that, compared to the mesogenic groups, the main chains exhibit only relatively small orientational effects obviously as a consequence of small coupling. The spacers demonstrate an intermediate behavior compared to the mesogenic groups and the main chain.

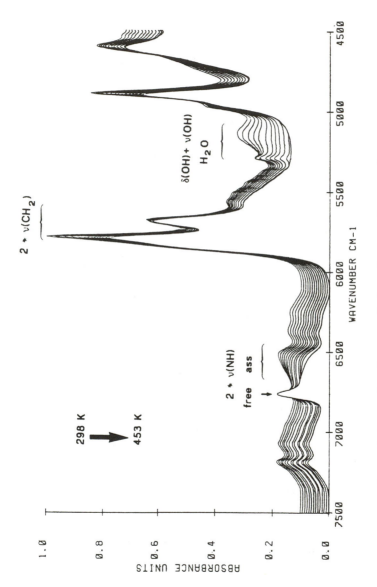

Figure 13. Near-infrared spectra of polyamide 12 as a function of temperature.

Rheooptical FTIR Spectroscopy

The mechanical properties of polymeric materials are of considerable importance to their engineering applications. Apart from the chemical structure and the thermal history, molecular orientation has a major influence on the mechanical properties of a polymer. The increased need for more detailed data and a better understanding of the mechanisms involved in polymer deformation has led to the search for new experimental techniques to characterize transient structural changes during mechanical processes. With the advent of rapid-scanning FTIR spectroscopy, simultaneous vibrational spectroscopic and mechanical (so-called rheooptical) measurements during the deformation of polymers emerged. These rheooptical measurements were used as informative probes for the study of deformation and relaxation phenomena in polymer films in the late 1970s; since then they have been applied to obtain data on strain-induced crystallization and orientational and conformational changes during mechanical treatment of a wide variety of polymers (*4, 46, 51*).

Contrary to the technique of time-resolved FTIR spectroscopy, which has improved time resolution down to the microsecond range due to ordered interferometric sampling techniques (*52*), the application of rheooptical FTIR spectroscopy is not restricted to the characterization of reversible structural changes caused by small-amplitude oscillatory strains; therefore, irreversible phenomena over large elongation and recovery scales may be studied (*46, 51*).

The experimental principle of rheooptical FTIR spectroscopy is illustrated in Figure 15. The technique is restricted to a film geometry of the sample and the specimen to be tested is uniaxially drawn and recovered in the sample compartment of the FTIR spectrometer. During the mechanical treatment interferograms can be acquired in small time intervals (down to 50 ms). Upon completion of the experiment the interferograms are transformed to the corresponding spectra for further processing of the conventional data. The electromechanical apparatus constructed for the deformation and relaxation measurements is shown in Figure 16. For orientation measurements the polarization direction of the incident radiation is alternately adjusted parallel and perpendicular to the stretching direction by a pneumatically rotatable wire-grid polarizer that is also controlled by the computer. The construction of a heating cell as a closed, nitrogen-purged system allows the deformation and stress relaxation to be studied under controlled temperature conditions (± 0.5 K) up to 523 K.

Presently we are testing the application of FT-Raman spectroscopy for rheooptical measurements. For this purpose, the stretching machine shown in Figure 16 is utilized in back-scattering geometry by rotating the clamp mechanism for 90 ° with the Raman beam impinging onto the sample through the front window of the stretching machine. To improve the back-scattering

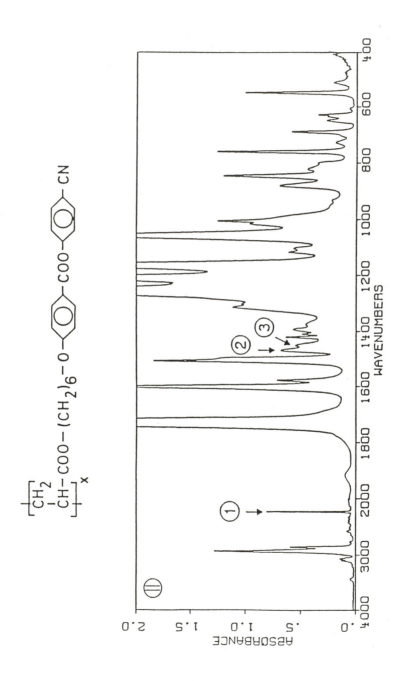

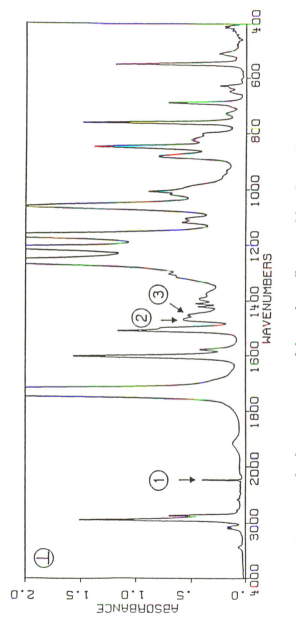

Figure 14. Structure and polarization spectra of the mechanically oriented liquid crystalline side-chain polymer: $\nu(C\equiv N)$ mesogenic unit, 1; $\delta(CH_2)$ main chain, 2; $\delta(CH_2)$ spacer, 3.

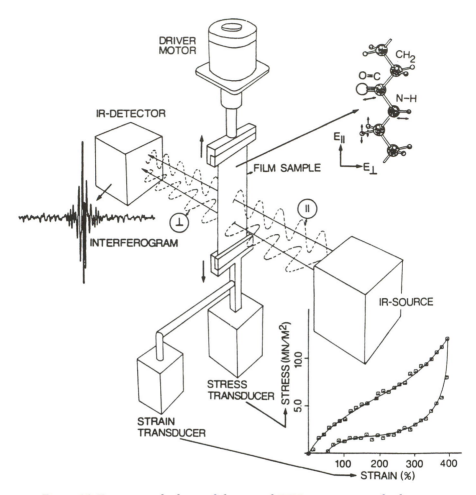

Figure 15. Experimental scheme of rheooptical FTIR spectroscopy of polymers.

efficiency, a mirror was mounted behind the sample. Here, of course, the sample morphology is no longer restricted to a thin film, but conventional polymer tensile test samples with thicknesses up to several millimeters or polymer cables and fibers can be investigated. With the increase of sample thickness, a stress transducer for higher loads must be utilized. To obtain an acceptable signal-to-noise ratio of the spectra acquired during the mechanical tests, longer acquisition times up to 30–40 s must be applied. Thus, for a Raman spectral data density comparable to rheooptical FTIR investigations, the elongation rate must be reduced accordingly. Nevertheless, preliminary results indicate that this technique is also very valuable for the characterization of phase transitions in thermoplastic systems and the phenomenon of strain-induced crystallization in polymer networks (H. W. Siesler, unpublished results).

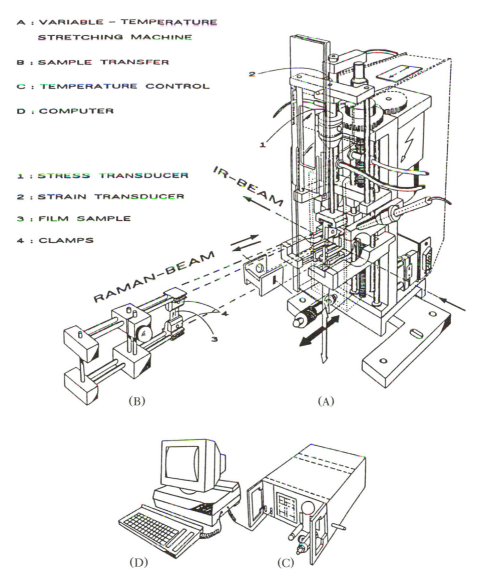

A : VARIABLE – TEMPERATURE
    STRETCHING MACHINE

B : SAMPLE TRANSFER

C : TEMPERATURE CONTROL

D : COMPUTER

1 : STRESS TRANSDUCER

2 : STRAIN TRANSDUCER

3 : FILM SAMPLE

4 : CLAMPS

IR-BEAM

RAMAN-BEAM

(B)   (A)

(D)   (C)

*Figure 16. Variable-temperature and computer-controlled stretching machine
for rheooptical FTIR and FT-Raman spectroscopy.*

Based on a uniaxial orientation model, the present FTIR data have been evaluated from the original spectra with programs that automatically calculate (1) the structural absorbance A_0 (*4, 46*)

$$A_0 = (A_{\parallel} + 2 A_{\perp})/3 \tag{4}$$

and (2) the orientation functions f (*4, 46*) of absorption bands. The transition moment directions of these absorption bands are assumed to be parallel or perpendicular, respectively, to the polymer chain axis

$$f_\parallel = (R - 1)/(R + 2) \tag{5a}$$

and

$$f_\perp = -2(R - 1)/(R + 2) \tag{5b}$$

The dichroic ratio R is defined as $R = A_\parallel/A_\perp$. For perfect parallel chain alignment $f = 1$, for perpendicular alignment $f = -1/2$, and for random orientation $f = 0$. The structural absorbance has been chosen as the intensity parameter because it eliminates the influence of changing orientation on the actual intensity of an absorption band (4, 46). Changes in sample thickness during elongation were compensated by comparison against a suitable reference band.

The potential of rheooptical FTIR spectroscopy can very instructively be demonstrated with reference to the elucidation of orientation phenomena and strain-induced crystallization of segmented polyurethanes and polyamides. In fact, these classes of polymers are particularly suited to such investigations because they contain functional groups with characteristic IR absorptions that can be assigned to specific domain locations of the polymer under examination. The value of infrared spectroscopy, in general, and orientational measurements with polarized radiation, in particular, has been widely recognized for this class of polymers, and a sound basis of functional group–frequency correlations has been established in the literature (47, 53). Additional information becomes available from measurements at elevated temperature and investigations of NH-deuterated samples because the isotope exchange offers a means to differentiate the hard segments into phase-separated species and moieties dispersed in the soft segments (46, 51).

The characterization of segmental orientation during cyclic elongation–recovery procedures of poly(ether-*block*-amides) (PEBA) based on chain-extended polyamide 12 (hard segments) and oligotetrahydrofuran (soft segments) will be discussed. The structure of these polymers is given in Structure 2 (54). A broad range of mechanical properties can be produced by the variation of the ratio of hard to soft segments. The sample detailed here

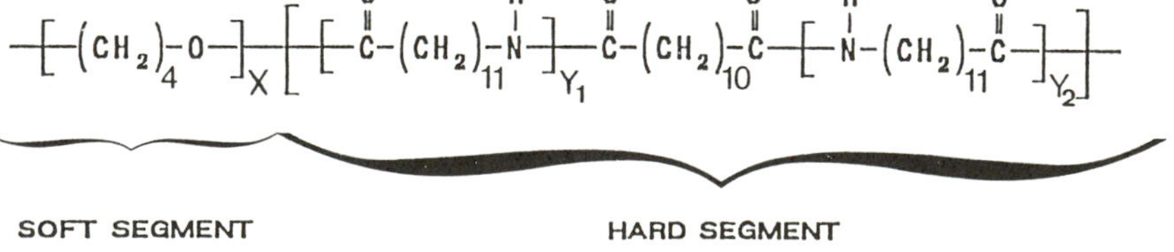

SOFT SEGMENT   HARD SEGMENT

Structure 2. General structure of poly(ether-block-amides).

has a hard to soft segment composition of 39.6:49.7 % (w/w) with an oligotetrahydrofuran molecular weight of 1000. The spectral assignment of absorption bands that are characteristic of different domains is outlined in Figure 17.

The stress–strain diagram of a cyclic loading–unloading procedure run at room temperature with an elongation and recovery rate of 100% strain per minute and the corresponding polarization spectra recorded during this mechanical treatment in 5-s intervals (10 scans with a resolution of 4 cm^{-1}) are shown in Figures 18 and 19, respectively. The orientation functions of the $\nu(NH)$, $\nu(CH_2)$, $\nu(C=O)_{ester}$, and $\nu(C=O)_{amide}$ absorption bands were evaluated for transition moment directions perpendicular to the chain axis, whereas the transition moment of the $\nu(C-O-C)$ absorption of the oligotetrahydrofuran segments was assumed parallel to the polymer chain. In Figure 20 the orientation functions of the evaluated absorption bands are plotted versus strain; they reflect the different behavior of the individual chain segments. Thus, the initial transverse orientation ($f < 0$) is only observed for functionalities located in the hard segments (N–H, $C=O_{amide}$). In accordance with Bonart's model (55), this behavior is characteristic for a lamellar hard segment morphology that is eventually broken up into smaller fibrillar subunits and leads to a positive orientation ($f > 0$) at higher elongations. The orientation function of the $\nu(C-O-C)$ absorption characterizes the soft segments that orient into the direction of stretch upon elongation and disorient upon recovery. The contribution of hard segments to the $\nu(CH_2)$ absorption and the interfacial character of the $C=O_{ester}$ functionality results in deviations from the orientation function behavior of the pure soft segments. The irreversible structural changes—elongations and orientations—are reflected by the spectroscopic as well as the mechanical data.

The ability to monitor strain-induced crystallization in the soft-segment phase of the investigated polymer is demonstrated in Figure 21. For this purpose the integral structural absorbance ratio of the 997-cm^{-1} conformational-regularity band of crystalline oligotetrahydrofuran segments (56) and the 2800-cm^{-1} thickness-reference band has been plotted along with the intensity of the reference band versus strain. Apart from thickness changes, the graph clearly demonstrates the appearance and disappearance of strain-induced crystallization as a function of loading and unloading, respectively.

Near-Infrared Spectroscopy

Although the wavelength region of the near-infrared (10,000–4000 cm^{-1} or 1000–2500 nm) has been used over many decades for the quantitative analysis of compounds containing OH, NH, and CH functionalities (e.g., determination of OH number, water, protein, or residual double-bond content) (4, 57–59), it has never been established as a widespread analytical and physical tool comparable to mid-infrared spectroscopy.

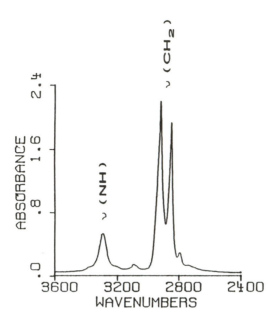

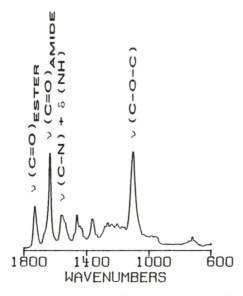

Figure 17. Spectral assignment of the IR spectrum of the investigated
poly(ether-block-amide).

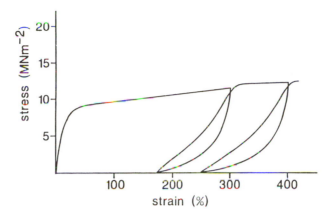

Figure 18. Stress–strain diagram of the elongation–recovery cycle.

In the course of the 1970s and 1980s however, two new developments initiated a renaissance of near IR-spectroscopy in analytical chemistry. On the one hand, chemometric data evaluation techniques in combination with diffuse–reflection measurements have opened up the possibility of nondestructive, rational multicomponent analysis of solid samples. On the other hand, the introduction of optical light fibers has contributed to an enormous instrumental expansion of conventional near-IR spectroscopy in terms of remote control. Fiber-optic probes allow a separation of the spectrometer and the sample measurement location over several hundred meters and greatly alleviate the analysis of toxic or otherwise critical samples including process and reaction control.

The basic difference between near- and mid-infrared (4000–400 cm^{-1}) spectroscopy is attributed to the fact that the absorption bands observed in the mid-IR spectrum can (with few exceptions) be assigned to fundamental vibrations of the investigated molecule, whereas the near-IR absorptions belong to overtone and combination vibrations, primarily of OH, NH, and CH functionalities. This statement, however, may lead to an underestimation of the information content of a near-IR spectrum because (in analogy to a mid-IR spectrum) the near-IR spectrum also contains information on the (1) temperature, (2) inter- and intramolecular interactions, (3) thermal and mechanical pretreatment, (4) ionic concentration (aqueous solutions), (5) viscosity/molar mass (polymers), (6) density, and (7) particle size–fiber diameter (in the case of diffuse–reflection spectra) of the sample under investigation.

The intensity of an absorption band decreases by a factor of about 10–100 in going from the fundamental to the first overtone, which necessitates larger optical pathlengths for the near-IR spectral region (about 1–2 mm for undiluted samples; up to 100 mm for solutions). However, this leads to a considerable advantage in sample handling compared to conventional mid-IR spectroscopy (4, 57). Thus, Figures 22 and 23 demonstrate the monitoring of the UV-induced polymerization of a diacrylate by mid- and near-IR spectroscopy, respectively. In both cases, information on the amount of residual double bonds in the cured resin can be readily derived from the decrease of the 1600-cm^{-1} ν(C=C) absorption in the mid-IR region and the

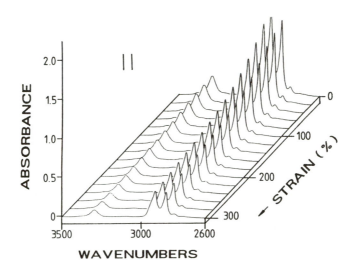

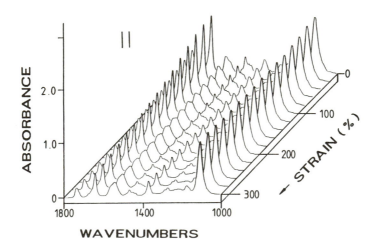

Figure 19. Polarization spectra recorded during elongation of the investigated poly(ether-block-amide) up to 300% strain.

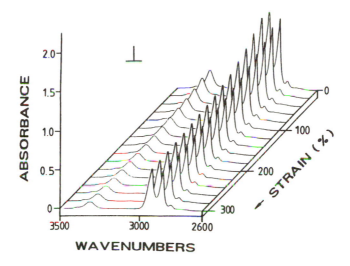

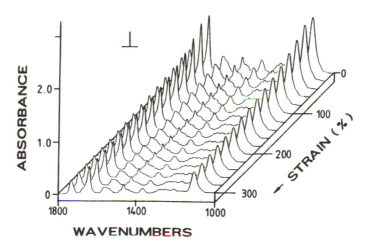

Figure 19. Continued

$2\nu(\text{=C--H})$ absorption at 1625 nm in the near-IR region with the advantage of a 200-fold increase in thickness for the near-IR sample and a concomitant ease of sample handling. Furthermore, glass or quartz, which are insensitive to water, may be used as window materials in the near-IR region. Instrumental advantages of the light source and the detector allow near-IR spectra to be recorded with a signal-to-noise ratio $\gg 10,000$ compared to the mid-IR signal-to-noise ratio $< 10,000$. This observation supports the application of

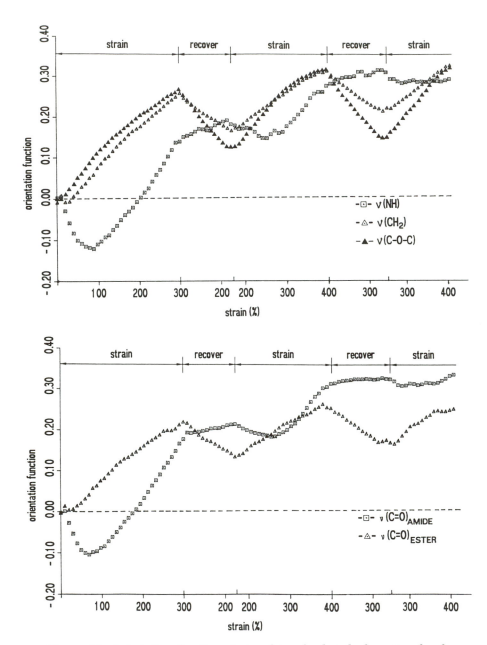

Figure 20. Orientation function–strain plots of selected absorption bands corresponding to the mechanical treatment outlined in Figure 19.

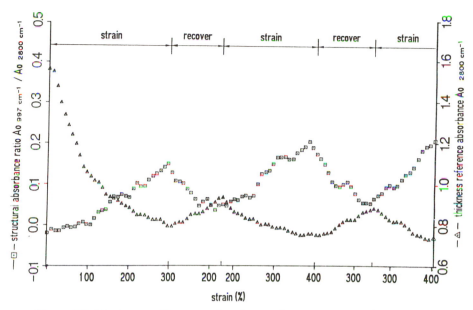

Figure 21. Monitoring of strain-induced crystallization in a poly(ether-block-amide) by rheooptical FTIR spectroscopy.

chemometric data evaluation techniques (60). Last, but not least, the materials conventionally used as light fibers (glass, quartz) have an attenuation minimum in the near-infrared region (61, 62).

Near-Infrared Diffuse-Reflection Spectroscopy. Near-infrared spectroscopy has been used for almost two decades in diffuse-reflection measurements for the analysis of agricultural and food products by filter instruments (58, 59). Technological progress has developed to a point where this technique can be applied with scanning instruments in the area of chemical and pharmaceutical multicomponent analysis of liquid and solid formulations and in polymer analysis. In principle, the method belongs to the discipline of chemometrics, which has been recognized since the mid-1970s (63, 64). The purpose of this discipline is to generate correlations between experimental data (e.g., absorption intensities in the present case) and the chemical composition or physical properties of the investigated samples by mathematical and statistical procedures [e.g., multilinear wavelength regression (MLWR), principal component analysis (PCA), partial least squares (PLS)].

The principle of the measurement procedure for quantitative analysis is based on recording the near-IR transmittance or diffuse-reflection spectra of reference samples (the number depending on the number of components or parameters to be determined) of known composition. These spectra are

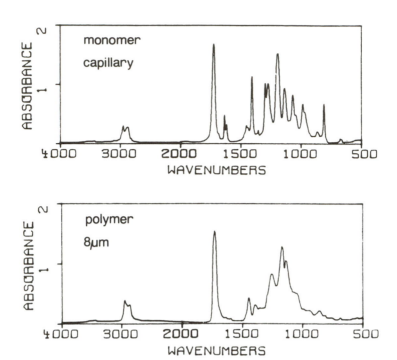

Figure 22. Mid-infrared spectra of the UV-induced polymerization of a diacrylate.

stored in the computer of the spectrometer, and the levels of the constituents or the physical parameters are determined by independent, conventional analytical or physical methods. Then the set of reference spectra and the independently determined values of the parameters under investigation are used by the selected statistical method to build a calibration (57–60). Use of this existing standardization and the near-IR spectra enables the unknown samples to be evaluated with regard to the individual parameters of interest. Once the calibration has been performed, the analysis time required for an unknown sample is drastically reduced to a few minutes in comparison to the several hours or even days previously required.

Promising applications have been reported (57) for the destruction-free analysis of polymers of widely varying morphology including the determination of characteristic chemical and physical parameters of synthetic fibers. The experimental results on polyacrylonitrile fibers demonstrated here were obtained on a near-IR spectrometer (Bran and Luebbe 500) with the fiber bundles mounted in a sample cup that is positioned under the integrating sphere. The measurement of opaque solids is performed in diffuse reflection, whereas highly transparent solids, pastes, and liquids can be measured with the transflection method, which is based on a diffusely reflecting cell bottom

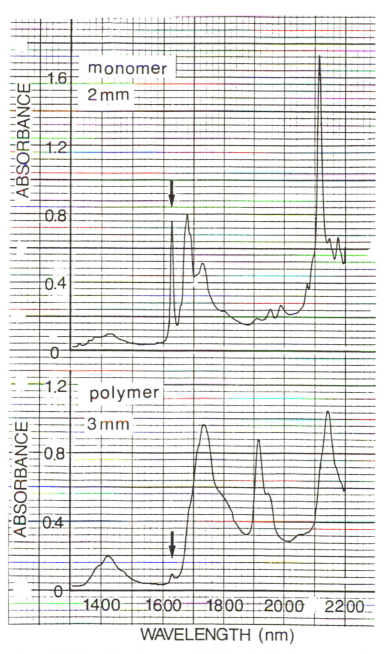

Figure 23. Near-infrared spectra of the UV-induced polymerization of a diacrylate.

that passes the radiation through the sample twice. The radiation directly reflected from the gold-coated inner surface of the integrating sphere serves as a reference intensity (Figure 24). Around 100 s are required to take a spectrum in the 1100–2500-nm wavelength range. The concentration proportional value of $\log 1/R$ is used as the ordinate scale:

$$\log 1/R = 1/s \sum a_i c_i \qquad (6)$$

where $R = I/I_0$ represents the intensity ratio of light reflected by the sample and the reference, respectively, s is the scattering coefficient, a_i are the absorptivities, and c_i are the concentrations of the individual components. Frequently the first derivative of $\log 1/R$ is used to reduce the influence of sample inhomogeneity (57, 59, 60).

The calibration was based on the $\log 1/R$ spectra of about 70 reference samples. The spectrum of such a reference sample is shown in Figure 25. In Table II, the actual (A) and near-IR-predicted (P) values and their residuals (R) for the parameters under investigation are shown for three test samples. These values demonstrate the potential of this technique for rational chemical and physical multicomponent analysis.

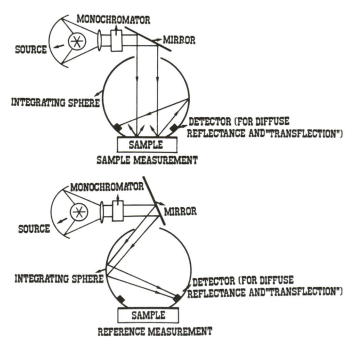

Figure 24. Simplified optical scheme of a scanning near-infrared diffuse-reflectance spectrometer.

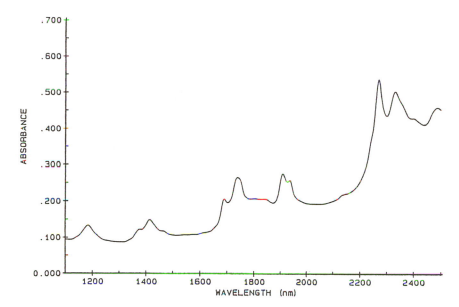

*Figure   25.   Near-infrared   diffuse-reflectance   spectra   of   as-received
polyacrylonitrile fiber bundles.*

**Table II. Actual (A) and Near-IR-Predicted (P) Values and Their Residuals
R (P − A) of the Dimethylformamide (DMF) Solvent Residues and the
Preparation, Shrinkage, and Strain of Polyacrylonitrile Fibers
Measured as Fiber Bundles in Diffuse Reflection**

Sample	DMF (%)	Preparation (%)	Shrinkage (%)	Strain (%)
1 P	0.9342	0.2715	4.7389	22.9778
A	0.9200	0.2800	5.0000	22.8000
R	0.0142	−0.0085	−0.2611	0.1778
2 P	0.6661	0.2655	4.0705	20.4069
A	0.7000	0.2600	4.0000	21.0000
R	−0.0339	0.0055	0.0705	−0.5931
3 P	0.7107	0.2797	3.9485	22.8066
A	0.7300	0.2700	3.9000	23.0000
R	0.0193	0.0097	0.0485	−0.1934

Near-Infrared   Light-Fiber   Spectroscopy.   In   conventional
transmission spectroscopy, the sample of interest is measured in the sample
compartment of the spectrometer, whereas the principle of fiber-optic spectroscopy is based on the transfer of light from the spectrometer via a suitable
device—the light fiber—to the sample and back to the spectrometer after
transmission of, or reflection from, the sample. Such light fibers usually
consist of quartz with a length ranging up to several hundred meters. Details

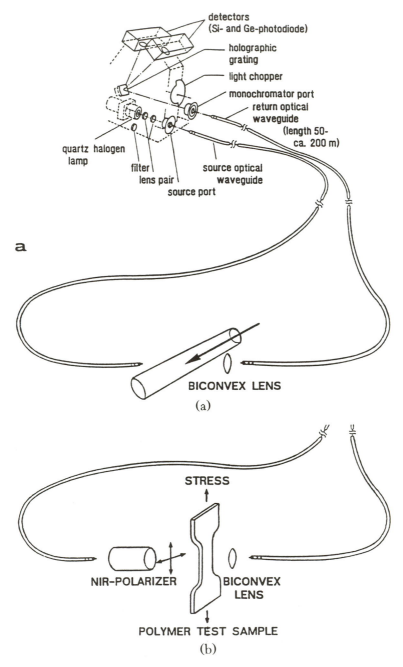

Figure 26. Optical scheme of a near-infrared light-fiber spectrometer with two applications of at-line monitoring: (a) At-line monitoring of process streams and (b) at-line monitoring of molecular orientation.

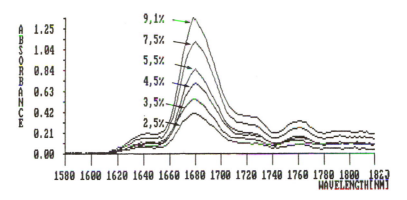

Figure 27. Near-infrared spectra of polystyrene–CCl₄ solutions with different polystyrene concentrations measured on the flow.

of light-fiber geometries, substrates, and optical properties are available in the literature (*61, 62*). The obvious advantage of such a geometry is the ability to separate and vary the location of measurement from the spectrometer within the limits given by the length of the light fiber. Various probes that can be integrated into different reaction vessels or bypasses offer a multiplicity of remote sensing on- and at-line process-control applications (*65*).

The optical scheme of one of the instruments used for our studies (the Guided Wave M 200 spectrometer) is shown in Figure 26. Figure 26 also includes two applications of transmission monitoring liquid streams in quartz tubes (a) and molecular orientation in solid polymer test samples (b). In both cases the transmitted light is refocused into the return waveguide via a biconvex lens.

The spectra obtained by monitoring polystyrene–CCl₄ solutions of various concentrations on the flow in a 19-mm quartz tube are shown in Figure 27 (*57, 65*). In combination with a near-IR polarizer (Glan–Thompson prism) the same optical configuration offers a very elegant approach for the characterization of anisotropy in polymeric solids. This anisotropy characterization is achieved by evaluation of the dichroic effects of selected functionalities measured with light polarized parallel and perpendicular to the drawing direction of the investigated polymer. Figure 28 shows the near-IR polarization spectra of a 260% drawn polyamide 12 specimen (thickness 1.3 mm). This spectra reflect significant dichroic effects on the individual absorption bands (*57, 65*). Comparison of the results derived from samples with different draw ratios shows that the dichroic ratios obtained from one overtone [e.g., $2\nu(\text{CH}_2)$] can also be transferred to the next overtone $(3\nu(\text{CH}_2))$. Near-IR spectroscopy is unique for this purpose in that it offers a destruction-free investigation even of thick specimens in their original morphology.

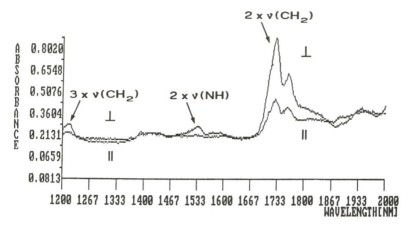

Figure 28. Near-infrared polarization spectra of a 260% drawn polyamide 12 test specimen measured with light polarized parallel (∥) and perpendicular (⊥) to the stretching direction.

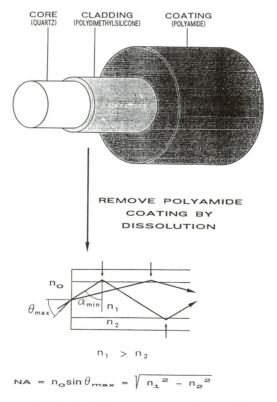

Figure 29. Near-infrared light-fiber ATR sensor (see text). NA denotes the numerical aperture; n_0, n_1, and n_2 are the refractive indexes of air, the core, and the cladding, respectively.

An application that shows promise is the modification of a commercial light fiber based on a quartz core, a poly(dimethylsiloxane) cladding, and a polyamide coating (Figure 29) to act as an ATR sensor in the near-infrared region (66). Removal of the polyamide coating by dissolution (over distances of several meters) allows the remaining light fiber to be wound around a Teflon holder and immersed into the solution for analysis. This method is based principally on the effect that the compound to be determined can be enriched in the poly(dimethylsiloxane) cladding (with no interference from water, for example) and detected via total internal reflection of the near-IR beam at the core–cladding interface. To compensate for the absorptions of the poly(dimethylsiloxane) cladding, a background has to be measured against air. A similar application has been described for the mid-infrared region with a polymer-coated internal reflection element (67) and a Teflon-cladding fluoride glass optical fiber (68).

Acknowledgments

The financial and instrumental support of Deutsche Forschungsgemeinschaft, Bonn, Germany, Ministerium für Wissenschaft und Forschung des Landes Nordrhein-Westfalen, Düsseldorf, Germany, Fonds der Chemischen Industrie, Frankfurt, Germany, Hüls AG, Marl, Germany, Bayer AG, Dormagen, Germany, and Bran and Luebbe GmbH, Norderstedt, Germany, are gratefully acknowledged. The author also thanks S. Dekiert, B. Feldhäuser, U. Becker, P. Wu, and I. Grose for experimental assistance.

References

1. Geick, R. In *Topics in Current Chemistry*; Springer: Berlin, Germany, 1975; Vol. 58, p 73.
2. Griffiths, P. R.; deHaseth, J. A. *Fourier-Transform Infrared Spectroscopy*; Chemical Analysis Series 83; Wiley: New York, 1986.
3. Koenig, J. L. *Adv. Polym. Sci.* **1983**, *54*, 87.
4. Siesler, H. W.; Holland-Moritz, K. *Infrared and Raman Spectroscopy of Polymers*; Dekker: New York, 1980.
5. Gilson, T. R.; Hendra, P. J. *Laser-Raman Spectroscopy*; Wiley: New York, 1970.
6. Koenig, J. L. *Appl. Spectrosc. Rev.* **1971**, *4(2)*, 233.
7. Cutler, D. J.; Hendra, P. J.; Fraser, G. In *Developments in Polymer Characterization*; Dawkins, J. V., Ed.; Applied Science Publishers: London, England, 1980; Vol. 2, p 71.
8. Hirschfeld, T.; Chase, B. *Appl. Spectrosc.* **1986**, *40*, 133.
9. Zimba, C. G.; Hallmark, V. M.; Swalen, J. D.; Rabolt, J. F. *Appl. Spectrosc.* **1987**, *41*, 721.
10. Hendra, P. J., Ed. *Spectrochim. Acta* **1990**, *46A(2)*, 121.
11. Schrader, B.; Hoffmann, A.; Simon, A.; Sawatzki, J. *Vibrational Spectrosc.* **1991**, *1*, 239.
12. Hendra, P. J.; Jones, C.; Warnes, G. *Fourier Transform Raman Spectroscopy*; Ellis Horwood: Chichester, England, 1991.
13. Sawatzki, J.; Simon, A. In *XXIIth International Conference on Raman Spec-

troscopy, August 13–17, 1990, Columbia, South Carolina; Durig, J. R.; Sullivan, J. F., Eds.; Wiley: Chichester, England, 1990.

14. Plueddemann, E. P. *Silane Coupling Agents*; Plenum: New York, 1983.
15. Ishida, H.; Koenig, J. L. *Polym. Eng. Sci.* **1978**, *18*, 128.
16. Graf, T. R.; Koenig, J. L.; Ishida, H. *Anal. Chem.* **1984**, *56*, 773.
17. Davis, J. A.; Sood, A. *Makromol. Chem.* **1985**, *186*, 1631.
18. Siesler, H. W. *Mikrochim. Acta* (Wien) **1988**, *I*, 319.
19. Bell, A. G. *Philos. Mag.* **1881**, *11*, 510.
20. Rosencwaig, A. *Opt. Commun.* **1973**, *7*, 305.
21. Rosencwaig, A. *Photoacoustics and Photoacoustic Spectroscopy*; Wiley: New York, 1980.
22. Pao, Y.-H., Ed. *Optoacoustic Spectroscopy and Detection*; Academic: Orlando, FL, 1977.
23. Somoano, R. B. *Angew. Chem.* **1978**, *90*, 250.
24. Hunter, T. F.; Turtle, P. C. In *Advances in IR and Raman Spectroscopy*; Clark, R. J. H.; Hester, R. E., Eds.; Heyden: London, England, 1980.
25. Coufal, H.; McClelland, J. F. *J. Mol. Struct.* **1988**, *173*, 129.
26. Graham, J. A.; Grim, W. M., III; Fateley, W. G. In *FTIR Spectroscopy—Applications to Chemical Systems*; Ferraro, J. R.; Basile, L. J., Eds.; Academic: Orlando, FL, 1985; Vol. 4, p 345.
27. Vidrine, D. W. *Appl. Spectrosc.* **1980**, *34*, 314.
28. Rockley, M. G. *Appl. Spectrosc.* **1980**, *34*, 405.
29. Yang, C. Q.; Fateley, W. G. *Anal. Chim. Acta* **1987**, *194*, 303.
30. Urban, M. W. *J. Coat. Technol.* **1987**, *59*, 29.
31. Adams, M. J.; King, A. A.; Kirkbright, G. F. *Analyst* **1976**, *101*, 73.
32. McClelland, J. F.; Kniseley, R. N. *Appl. Opt.* **1976**, *13*, 2658.
33. Rosencwaig, A.; Gersho, A. *Science* **1975**, *19*, 556.
34. Rosencwaig, A.; Gersho, A. *J. Appl. Phys.* **1976**, *47*, 64.
35. Grose, R. I.; Hvilsted, S.; Siesler, H. W. *Makromol. Chem., Makromol. Symp.* **1991**, *52*, 175.
36. Harrick, N. J. *Internal Reflection Spectroscopy*; Interscience: New York, 1980.
37. Gidaly, G.; Kellner, R. *Fresenius Z. Anal. Chem.* **1980**, *302*, 257.
38. Ohta, K.; Iwamoto, R. *Appl. Spectrosc.* **1985**, *39*, 418.
39. Blackwell, C. S.; Degen, P. J.; Osterholtz, F. D. *Appl. Spectrosc.* **1978**, *32*, 480.
40. Becker, U. M.Sc. Thesis, University of Essen, Essen, Germany, 1988.
41. Machate, Ch., Ph.D. Thesis, University of Münster, Münster, Germany, 1989.
42. Painter, P. C.; Coleman, M. M.; Koenig, J. L. *The Theory of Vibrational Spectroscopy and Its Applications to Polymeric Materials*; Wiley-Interscience: New York, 1982.
43. *The Hydrogen Bond*; Schuster, P.; Zundel, G.; Sandorfy, C., Eds.; North-Holland: New York, 1976.
44. Vinogradov, S. N.; Linnell, R. H. *Hydrogen Bonding*; Van-Nostrand Reinhold: New York, 1972.
45. Murthy, A. S. N.; Rao, C. N. R. *Appl. Spectrosc. Rev.* **1968**, *2*, 69.
46. Siesler, H. W. *Adv. Polym. Sci.* **1984**, *65*, 1.
47. Srichatrapimuk, V. W.; Cooper, S. L. *J. Macromol. Sci. Phys.* **1978**, *B15*, 267.
48. Skrovanek, D. J.; Painter, P. C.; Coleman, M. M. *Macromolecules* **1985**, *18*, 299 and 1676.
49. Bürkle, K.-R. Ph.D. Thesis, University of Ulm, Ulm, Germany, 1987.
50. Zentel, R.; Benalia, M. *Makromol. Chem.* **1987**, *188*, 665.
51. Siesler, H. W. *Makromol. Chem., Makromol. Symp.* **1992**, *53*, 89.
52. Noda, I. *Appl. Spectrosc.* **1990**, *44(4)*, 550.

53. Ishihara, H.; Kimura, I.; Saito, K.; Ono, H. *J. Macromol. Sci. Phys.* **1974**, *B10*, 591.
54. Lohmar, J.; Meyer, K.; Goldbach, G. *Makromol. Chem.* **1988**, *189*, 2053.
55. Bonart, R.; Hoffmann, K. *Colloid Polym. Sci.* **1982**, *260*, 268.
56. Tadokoro, H.; Kobayashi, M. In *Polymer Spectroscopy*; Hummel, D. O., Ed.; Verlag Chemie: Weinheim, Germany, 1974; p. 1.
57. Siesler, H. W. *Makromol. Chem., Makromol. Symp.* **1991**, *52*, 113.
58. Weyer, L. G. *Appl. Spectrosc. Rev.* **1985**, *21*, 1.
59. Osborne, B. G.; Fearn, T. *Near Infrared Spectroscopy in Food Analysis*; Wiley: New York, 1986.
60. Hirschfeld, T.; Stark, E. In *Analysis of Food and Beverages*; Charamboulos, G., Ed.; Academic: Orlando, FL, 1984; p 505.
61. Herbrechtsmeier, P. *Chem. Ing. Tech.* **1987**, *59*, 637.
62. Frank, W. *Fernmelde-Ing.* **1990**, *44(3)*, 1.
63. Sharaf, M. A.; Illman, D. L.; Kowalski, B. R. *Chemometrics*; Wiley-Interscience: New York, 1986.
64. Martens, H.; Naes, T. *Multivariate Calibration*; Wiley: Chichester, England, 1989.
65. Feldhäuser, B. M.Sc. Thesis, University of Essen, Essen, Germany, 1988.
66. Bürck, J.; Conzen, J.-P.; Ache, H.-J. *Fresenius J. Anal. Chem.* **1992**, *342*, 394.
67. Heinrich, P.; Wyzgol, R.; Schrader, B.; Hatzilazaru, A.; Lübbers, D. W. *Appl. Spectrosc.* **1990**, *44(10)*, 1641.
68. Ruddy, V.; McCabe, S. *Appl. Spectrosc.* **1990**, *44(9)*, 1461.

RECEIVED for review May 14, 1991. ACCEPTED revised manuscript September 22, 1992.

Spectroscopic Characterization of Polymers

Fluorescence Principles

Gregory D. Gillispie

Department of Chemistry, North Dakota State University, Fargo, ND 58105

The principles of fluorescence that apply to the determination of polymer physical properties are covered in tutorial fashion. Fluorescence is a sensitive technique for probing the microenvironment in large molecules and for following conformational changes and fast (microsecond or less) internal motion. This survey focuses on the spectroscopy of aromatic molecules commonly used as fluorescent tags on synthetic polymers. Individual sections cover the electronic structure of aromatic molecules, absorbance and fluorescence spectroscopy, and the dynamical processes that control fluorescence intensity.

The conformations of macromolecules in dilute solution are typically characterized by such terms as flexible coils, rigid rods, and globular particles. When a polymer folds upon itself, chemically distinct microdomains and microenvironments are created. The number and distribution of such domains are a function of temperature, pressure, solvent dielectric constant, pH, ionic strength, and the concentration of the macromolecule itself. There is, therefore, a premium on characterization techniques that (1) can be directly applied to the sample without any manipulation that might disrupt the polymer; (2) are highly specific in their ability to sense different microenvironments and the changes in the microenvironment distribution as bulk conditions are modified; and (3) are extremely sensitive.

Spectroscopic methods, including infrared, Raman, NMR, and fluorescence, are often superior to macroscopic techniques, such as viscosity and molecular weight determination, in this regard. Fluorescence is the topic of

0065–2393/93/0236–0089$10.75/0

this chapter; the other optical spectroscopies are covered elsewhere in this volume.

The vibrational spectroscopies, infrared and Raman, are applicable to polymer structure determination (structure in the sense of chemical bond properties) because the stretching and bending vibrations in a polymer closely resemble those in monomers. Of course, if the vibrational spectra were *completely* transferable, these spectroscopies would be worthless as microenvironmental probes. Because the spectral changes with environment are rather subtle, accurate measurement of small changes in band position is required. The greatest source of model uncertainty is often the validity–reliability of the proposed relationship between spectral position and physical property. Details can be found in the companion overviews of vibrational spectroscopy in this volume. We also note the extensive use of Raman spectroscopy as a probe of biopolymer conformation (1).

Fluorescence, in contrast, is an electronic spectroscopy that, as such, offers little information about individual bond properties, but otherwise has some extremely favorable features:

- Fluorescence is extremely sensitive.
- Fluorescence is inherently a multidimensional, selective technique.
- The fluorescence excitation spectrum can readily be determined for strongly scattering or even opaque samples (amorphous solids, turbid and frozen solutions, cracked glasses, etc.) for which the conventional absorbance approach is difficult or even impossible.
- Fluorescence provides dynamical information on a time scale relevant to polymer internal motions.
- Well-developed and well-understood models are available for data interpretation.

Fluorescence is the most sensitive optical spectroscopy. Useful signals routinely can be detected at nanomolar concentrations, and at the subpicomolar level in favorable cases. This sensitivity is crucial for experiments in which a fluorescence probe is chemically attached to a nonfluorescent polymer backbone because the probe concentration can be kept low enough to avoid influencing the properties of the polymer itself. The multidimensional feature, that is, the intensity depends on two wavelengths (excitation and emission) instead of just one, confers increased detection specificity and provides a wider choice of experimental configurations.

The dynamical information is directly realized in pulsed excitation experiments, but rates of some processes can also be inferred from steady-state polarization measurements. Almost every spectroscopic method can be applied to reaction kinetic studies. However, the information fluorescence provides on internal motions and rotational dynamics in the microsecond to picosecond regime is unsurpassed. Finally, fluorescence experiments can be interpreted with standard equations derived from essentially first principles models, such as those that describe Stern–Volmer quenching, orientational depolarization, and Forster and Dexter energy transfer processes (2–4).

On the debit side, whereas infrared absorption, Raman scattering, and nuclear magnetic resonance phenomena are exhibited by virtually every molecule, most synthetic polymers are not intrinsically fluorescent and must be tagged with covalently attached fluorescence probes. Fluorescence poses an expanded menu of experimental configurations and a sometimes daunting array of new concepts and terminology. Note that fluorescence intensities are inextricably linked to molecular dynamics considerations (e.g., excited state radiationless processes, quenching, excimer formation, etc.) as opposed to the simple and direct intensity versus wavelength picture of infrared or UV–visible absorbance spectroscopy.

This chapter was written as a tutorial on fluorescence principles, not as a review article on physical characterization of polymers by fluorescence; that aspect is covered by the research papers in this volume. The goal here is to provide a compact introduction that serves as a reliable starting point for consultation of more advanced treatments of polymer photophysics (4). Whether or not polymer scientists perform fluorescence measurements themselves, they need a basic understanding of the technique to follow the literature. Most research papers are written at a more specialized level than what is found in the standard chapters on fluorescence and phosphorescence in instrumental analysis textbooks. The question "why" more than "what" or "how" is often the main stumbling block to the nonspecialist trying to understand a fluorescence paper. Thus, the writing here has been guided by an attempt to anticipate and directly address those concepts that experience has shown to be potentially the most confusing.

Owing to space limitations, selected from the entire range of luminescence techniques are those initiated by photon absorption; other light emission methods (chemiluminescence, pulse radiolysis, bioluminescence) follow similar principles. For convenience, fluorescence is used as a generic term for photoluminescence and only a few comments are made about phosphorescence. Our major emphasis is on polymers that have been "tagged" or labeled with polycyclic aromatic or heteroaromatic fluorescence probes such as naphthalene, anthracene, phenanthrene, pyrene, and carbazole. The probes can be placed directly into the polymer backbone, attached as pendant groups, or located as end caps on chain polymers.

After a brief introduction to absorbance spectroscopy and electronic structure notation, the sequence of steps that generate a Boltzmann vibronic population distribution in the first excited singlet state, from which the fluorescence process originates, are outlined. These steps include photoabsorption, vibrational relaxation within an electronic state, and internal conversion between excited singlet electronic states. This is followed by a discussion of the fluorescence spectral distribution, the degree of structure in the spectrum, what this structure does (or does not) mean, the possible mirror-image relationship to the absorbance spectrum, and wavelength shifts as a function of solvent.

The next section covers the decay processes (fluorescence emission, internal conversion to S_0, intersystem crossing to T_1, quenching, excimer formation, etc.) that depopulate the emitting excited state. The relative magnitude of the radiative and nonradiative rate constants controls the fluorescence efficiency (quantum yield) whereas their absolute magnitudes determine the lifetime. The quantitative Strickler–Berg equation for radiative decay rates and the empirical energy gap law for radiationless transitions are mentioned. Environmental effects on radiationless transition rates are addressed briefly.

Absorbance Spectroscopy of Aromatic Molecules

Jablonski Diagrams and Platt Notation. Photoabsorption is the usual mode for creating the excited electronic states from which fluorescence occurs. Absorption and fluorescence are complementary photoprocesses that connect upper and lower electronic states:

Absorption: Low-energy state + photon → high-energy state

Fluorescence: High-energy state → low-energy state + photon

Each electronic state has its own set of internal vibrational energy levels. Interstate vibrational transitions (as distinguished from the intrastate vibrational transitions of infrared spectroscopy) occur simultaneously with the electronic excitation or deexcitation and are responsible for structure (bands, peaks, shoulders) in the electronic spectra. Such vibronic (*vib*rational–elec*tronic*) structure is never fully resolved in solution spectra and is sometimes completely lost.

The creation of population in the state that ultimately emits fluorescence begins with absorption of a photon by the ground electronic state, which is conventionally labeled S_0 (S is the singlet and 0 is the numerical index of relative energies). Although the fluorescence almost always occurs only from S_1, the first excited singlet state, there are other, higher electronic states into which the molecule can be excited. A key concept of *monomer* solution fluorescence is the following:

> The fluorescence spectral distribution (i.e., relative intensity as a function of wavelength) does not depend on which electronic state is initially excited.

The number of excited molecules created, and, hence, the number of fluorescence photons subsequently generated, is nearly always a function of the excitation wavelength, but the fluorescence spectral shape is not.

The standard molecular orbital model considers that an excited state is derived from the ground state via promotion of an electron from an occupied orbital to a vacant, higher energy orbital. If all the electrons are spin-paired in S_0, which is usually the case for organic molecules, a one-electron promotion creates two singly occupied orbitals. If the electrons in these two orbitals are spin antiparallel, a singlet state results. In triplet states the two electron spins are parallel. The excited singlet states in order of increasing energy are labeled $S_1, S_2, S_3, \ldots$; the excited triplet states are similarly labeled $T_1, T_2, \ldots$. The lowest triplet state is conventionally designated T_1, not T_0, as might be expected, because the first excited singlet state and the lowest triplet state are nominally derived from the same electron configuration:

$$\ldots (\mathrm{HOMO})^1 (\mathrm{LUMO})^1$$

where HOMO and LUMO refer to the highest energy occupied molecular orbital and lowest energy unoccupied molecular orbital, respectively (in the ground electronic state). In reality, S_1 and T_1 often arise from different electron configurations, especially if S_1 is an n-π^* state. This is commonly the case for aromatic carbonyls.

The electronic absorbance spectra of organic molecules are safely interpreted solely in terms of transitions to excited singlet states owing to the extreme weakness of the spin-forbidden singlet–triplet transitions. Incorporation of "heavy atoms" into the molecular framework increases spin–orbit coupling and thereby partially lifts the spin-forbiddenness, but not nearly enough to invalidate the previous sentence. Note that although the phenomenon is referred to as the heavy atom effect, it is a nuclear charge dependence, not a mass effect, per se. An external heavy atom effect can be realized with solvents that contain one or more high atomic number atoms (e.g., bromoform, ethyl iodide); however, the internal heavy atom effect is usually more effective. Extensive discussions and data tabulations on heavy atom effects and spin–orbit coupling are available (5).

The wide variation in absorption strengths for the various electronic transitions connected to S_0 is evident from the spectra of phenanthrene and anthracene shown in Figure 1. For example, the peak molar absorptivity is about 200 times lower in the phenanthrene $S_1 \leftarrow S_0$ transition between 350

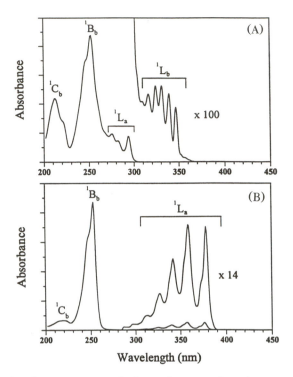

Figure 1. Absorbance spectra of phenanthrene and anthracene in heptane solution at room temperature. The upper states of the various electronic transitions are labeled according to the Platt notation. Phenanthrene, A; anthracene, B.

and 300 nm than in the transition that peaks at 251 nm. Two other electronic transitions of intermediate strength can be identified. One begins with a well-defined band at 293 nm and the other is centered around 215 nm.

Attempts to systematically account for the UV–visible absorbance spectra of aromatic molecules have a long history. Clar (6) and others (see literature cited in reference 7) achieved some successful empirical classifications, but the real breakthrough in understanding came when Platt and his co-workers refined the perimeter-free electron model and provided a simple quantum mechanical underpinning to the observations (7, 8). The perimeter-free electron model is essentially just a particle-on-a-ring elaboration of the particle-in-the-box analysis taught in undergraduate physical chemistry. It is remarkable that after the passage of more than four decades, during which time theory has ever increased in sophistication and predictive capability, the Platt analysis has maintained its vitality. The reader is strongly urged to consult the collected original papers (9) of the Platt group for details;

reference 5 is also recommended for its clear discussion and data presentation.

In the Platt scheme, the first two excited singlet states in an unsubstituted aromatic hydrocarbon are labeled ^1L_a and ^1L_b. Their relative order is not fixed. As the chain length of the linear polyacenes is increased, the energy of ^1L_a drops relative to ^1L_b, which is the lower energy state in the series for only benzene and naphthalene. Absorption to the ^1L_a state is predicted to be moderately strong, whereas the transition to ^1L_b is weak. A very strong transition to the ^1B_b state lies at shorter wavelength than the ^1L_a and ^1L_b transitions and there are no intervening states expected between them and ^1B_b. Additional transitions to other states (^1C_b, ^1B_a, ^1C_a) at shorter wavelengths from ^1B_b are unimportant for our purposes. As a guideline, transitions to ^1L_b, ^1L_a, and ^1B_b from S_0 are, respectively, weak [$100 < \epsilon < 1000$ L/(mol cm)], medium ($\epsilon \approx 10,000$), and strong ($\epsilon \approx 10^5$).

The counterparts to three of the electronic transitions observed for phenanthrene are easily identified in anthracene (Figure 1B). The missing absorption to the ^1L_b state in anthracene is unquestionably buried in the short wavelength tail of the much stronger ^1L_a transition. Identification of the ^1L_b transition is much easier when it is S_1 than when it is S_2. Several of the other aromatic chromophores commonly used to label polymers also have ^1L_b as the lowest excited singlet state; they include naphthalene, pyrene, carbazole, and fluorene. The ^1L_b state is also the lowest excited state in benzene and its methyl- and halo-substituted derivatives, but the S_1-S_2 gap is so large in those molecules that there is no potential for confusion.

As the spectra are viewed from longer to shorter wavelength, a reasonably well-defined onset to most of the electronic transitions is observed. The longest wavelength band in a structured absorbance region is commonly referred to as the 0-0 band. The notation implies that it represents a transition between the zero-point (vibrationless) levels of the upper and lower states. With the further reasonable approximation that the zero-point energy is the same in the two states, the photon energy corresponding to the 0-0 band equals the electronic energy of the excited state. The energy level diagrams shown in Figure 2 were constructed in this fashion from the spectra of Figure 1 and are commonly referred to as Jablonski diagrams. The energy of the lowest triplet state, relevant to S_1 deactivation via intersystem crossing, is determined from the phosphorescence spectrum. Once T_1 is positioned, higher triplet states can be placed with the aid of triplet–triplet transient absorbance data.

Note the structure from the simultaneous vibrational transitions that accompany electronic transitions. For example, the $S_1(^1L_a) \leftarrow S_0$ absorbance spectrum of anthracene (Figure 1B) contains a series of bands separated by 1460 ± 40 cm^{-1}, which is in the appropriate range to be assigned to an aromatic ring stretching vibration. At least five members of the progression

Jablonski Diagrams

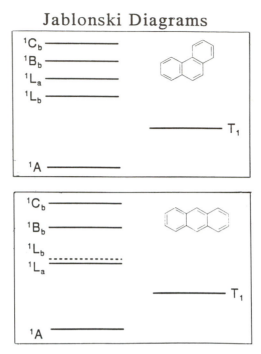

Figure 2. State energy level diagrams for phenanthrene (top) and anthracene (bottom). States positioned from the spectra of Figure 1 and the phosphorescence spectra. The horizontal lines represent the electronic energies of the states.

are readily observed and more can be found with second-derivative spectral processing. With the standard assumption that only the zero-point level in S_0 is appreciably populated at thermal equilibrium, it is common practice to calculate S_1 vibrational energy levels from the band spacings. An example is shown in Figure 3 for anthracene. The corresponding S_0 vibrational levels could be deduced from the fluorescence spectrum. There are six vibronic bands in the phenanthrene $S_1(^1L_b) \leftarrow S_0$ transition (Figure 1A). The spacing is smaller than in anthracene and the structure is less regular.

The extent of vibronic structure varies widely from one electronic transition to another. Whether to assign a given spectral region to two or more excited state transitions or to vibronic structure within a single electronic transition is not always an easy decision. Polarized absorbance and fluorescence excitation spectra, low-temperature measurements to enhance the resolution, solvent shifts, the results of molecular orbital calculations, and the infamous "chemical intuition" all play a role. An early paper by Becker and co-workers (10), which exemplifies the application of most of these techniques, is still representative of the approaches taken today.

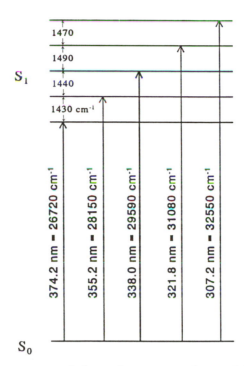

Figure 3. Representation of the anthracene S_1 vibrational levels from the solution absorbance spectrum.

Any quasidiatomic (i.e., single vibrational mode) depiction similar to Figure 3 grossly underestimates the total number of available vibrational levels. The true vibronic structure is far more complicated than the solution spectra suggest (*vide infra*) owing to incomplete resolution. Moreover, only a fraction of the total set of S_1 levels can be optically accessed from the S_0 zero-point level. The inaccessible "dark" levels play a key role in the various vibrational relaxation and radiationless transition processes.

Solvent Effects. The solvent effects on an absorbance spectrum can be separated into three categories: wavelength shift of the entire spectrum, modification of the vibronic structure (spectral shape), and a change in the overall absorption strength. The last factor requires an extremely strong interaction between solvent and solute, one that significantly distorts the solute's electron distribution. There is also the possibility that solvent interactions lift the symmetry forbiddenness of a given molecular transition (*see* the subsequent discussion of the Ham effect), but such cases are usually better viewed as changes in vibronic structure because the integrated absorption strength of the entire electronic transition is little affected. The magnitude of the wavelength shifts depends on the polarity of the solvent and the dipole moment difference between the S_0 and S_1 states. The literature on charge transfer transitions (both intra- and intermolecular) and donor−acceptor complexes is extensive (11), but comments are limited here to the milder interactions evident for aromatic chromophores.

For characterization of the physical properties of polymers, interest is often in the difference between hydrocarbon and aqueous solution environments. Figure 4 is inspired by the classic work of Lawson et al., who compared the absorbance spectra of benzene in the gas phase, and in perfluorohexane and hexane solution (12). We have remeasured the solution spectra and added one for water as solvent. The similarity between the absorbance in the gas phase and in perfluorohexane solvent (12) is remarkable and indicative of almost no interaction of benzene with the fluorocarbon. Even though hexane is usually considered a "noninteracting" solvent (and it certainly is when compared, for example, to methanol), it does wash out the vibronic structure and shift the entire spectrum about 2 nm to longer

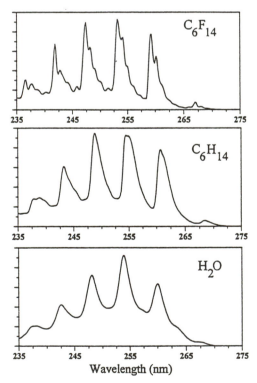

Figure 4. Absorbance spectra of benzene in perfluorocyclohexane and hexane solvent and in aqueous solution.

wavelength. The higher polarity of water as a solvent further reduces the vibronic structure (the valleys between the maxima are not as deep), but the band positions agree to within 1 nm (relative to alkane solvent) and the overall intensity pattern is not markedly affected.

As a component of an environmental analysis project in our laboratory, we have extensively studied the electronic spectroscopy of aromatic hydrocarbons in aqueous solutions. The results for the other polycyclic aromatic hydrocarbons are similar to those for benzene with regard to solvent dependence of the absorbance spectra.

Absorbance Spectra in Labeled Polymers. The foregoing principles for monomers apply nearly without modification to polymer systems tagged with aromatic substituents. Shifts of the absorbance spectra, relative to those of the model monomeric chromophores in the same solvent, are small. For example, consider the almost completely superimposable absorbance spectra of polystyrene and ethylbenzene shown in Figure 5. The compendium of absorbance and fluorescence spectra by Berlman (*13*) includes additional monomer spectra for toluene, *n*-propylbenzene, isopropyl benzene, *sec*-butylbenzene, diphenylmethane, bibenzyl, and phenylcyclohexane; all monoalkyl derivatives of benzene absorb UV light in similar fashion. The presence of a single methylene spacer between a phenyl ring and another aromatic group (as, for example, in diphenylmethane) isolates them electronically to a large degree. The electronic isolation is even greater if there are additional methylene spacers.

Nonetheless, the existence of at least a modicum of electronic communication across a single $-CH_2$ group is revealed in the diphenylmethane and ethylbenzene spectra shown in Figure 6; the 9-nm relative shift to longer wavelength in diphenylmethane is indicative of some conjugation extension

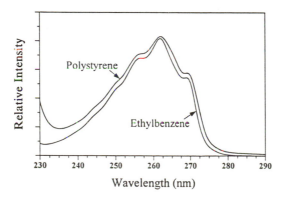

Figure 5. Comparison of the absorbance spectra of polystyrene (MW = 600) and ethylbenzene in heptane solution.

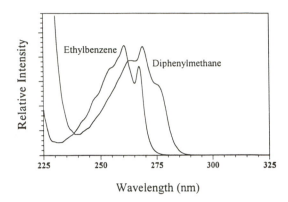

Figure 6. *Comparison of the absorbance spectra of ethylbenzene and diphenylmethane in heptane solution. Note that the two compounds exhibit very similar spectra but that the diphenylmethane spectrum lies about 9 nm to longer wavelength.*

from one phenyl group to the other across the methylene spacer. Comparison of either the benzyl acetate or benzyl alcohol spectra in Berlman (13) to the ethylbenzene absorbance shows similar shifts. Admittedly, the effect is not very dramatic in any of these cases.

Many bichromophoric molecules have been synthesized for spectroscopic study. For example, pyrene is commonly used to end cap polymers to explore end-to-end cyclization dynamics via the pyrene excimer fluorescence (14). The absorbance spectrum of a nonconjugated polymer labeled with pyrene closely resembles that of the corresponding alkyl-substituted pyrene.

Energy transfer studies of a bichromophoric system by the Guillet group (15, 16) provide the spectra for Figure 7. The absorbance spectrum of polymer I contains separate naphthyl and anthracenyl contributions (the absorption of the phenyl substituent lies to shorter wavelength of the range shown in the figure), but only the naphthyl absorption applies in polymer II. The composite (superimposed) spectrum generated from separate spectra for polymer II and 9-methylanthracene fits the spectrum of polymer I extremely well. The authors did find it necessary to reduce the molar absorptivity of 9-methylanthracene by 12% to achieve the best fit in the composite spectrum, but this could easily reflect the influence of the nearby carbonyl group on the molar absorptivity of the anthracene moiety (per the preceding comments about electronic communication across a methylene spacer).

Relaxation Steps Leading to the Emitting State

Vibrational Relaxation after S_1 Excitation. An individual photon absorption event populates an excited state vibronic level. Consider first the case of direct excitation into the S_1 state. According to the Franck–Con-

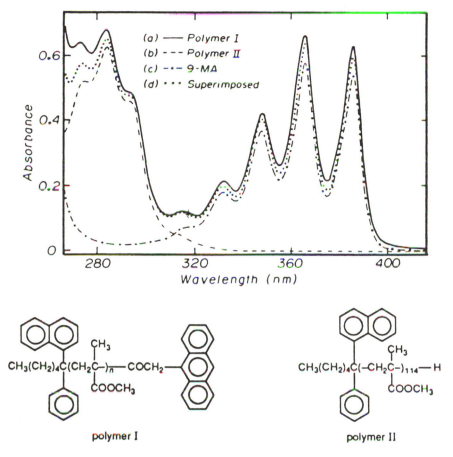

Figure 7. The absorbance spectrum of a bichromophoric polymer and its decomposition into monomer components. (Reproduced with permission from reference 15. Copyright 1990.)

don principle, the photoexcitation step occurs over such a short time interval (less than a femtosecond) that the sluggish nuclei lag behind the redistribution of electronic charge associated with the $S_1 \leftarrow S_0$ transition.

All molecular properties sensitive to electronic charge distribution (e.g., dipole moment, chemical reactivity) are changed by the photoexcitation. Even though the connectivity of the atoms in S_1 is the same as in S_0 (in the absence of photochemistry or photoisomerization), the force constants for bond stretches, bends, and torsions and the bond lengths can change significantly. Thus, the so-called Franck–Condon or vertical excitation creates an imbalance because the initially created excited state electron distribution is prepared on the framework of the ground electronic state nuclear geometry.

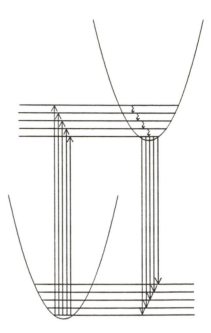

Figure 8. Caricature illustrating the origin of Stokes shift between the fluorescence and excitation transitions. Straight lines signify radiative transitions and wiggly lines in excited state illustrate relaxation cascade. Note that O–O band is the only vibronic transition common to the excitation and fluorescence.

The molecule must rearrange its relative nuclear positions to accommodate the new electronic distribution. Note that the nuclei are driven to their new positions by the changed electron distribution, not vice versa.

The notion of a specific, single ground-state geometry is an oversimplification, albeit usually not a crucial one. Classically the molecule spans a *range* of geometries as it executes its vibrational motions. [The quantum mechanical description invokes vibrational wavefunctions, the squares of which yield the probability that the molecule is in a certain geometry (conformation).] A familiar example is that of a ground state with a (vibrationally averaged) planar arrangement, say about a C=C double bond, but an intrinsically twisted excited state; ethylene is the obvious paradigm. Despite the foregoing quibble, there is little harm in saying that the ethylene S_1 excited state is created planar by photoexcitation and must then relax to its nonplanar equilibrium geometry. Biphenyl represents the reverse case; the torsional angle about the central C–C bond is substantial in S_0, but the molecule tends toward planarity in the excited state.

A photoexcited molecule achieves greater energy stability as it relaxes into its optimal excited state geometry. The process is commonly depicted as a cascade down the manifold of vibrational levels (Figure 8) and is referred to as a vibrational relaxation. In condensed media, vibrational relaxation causes dissipation of a certain amount of energy to the solvent on approximately the picosecond time scale. The vibrational relaxation is complete when a Boltzmann population distribution over the excited state vibronic levels is reached.

Again, recall that the density of S_1 vibronic levels available to participate in the relaxation is enormously greater than that which is revealed by the absorbance spectrum.

Internal Conversion from S_n ($n > 1$) to S_1. The possibility of excitation into higher energy states than S_1 also exists. The empirical observation that fluorescence from singlet states above S_1 is generally not observed is referred to as Kasha's rule. Although several "violations" of Kasha's rule (which never was promulgated as an absolute edict) have been observed (*see* reference 17, for example), truly marked departures are rare. The phenomenological interpretation for the weak fluorescence from S_n ($n > 1$) is that radiationless processes rapidly take the molecule down the ladder of excited singlet states to S_1 before the radiative decay can occur. The time scale of these internal conversions is generally in the subpicosecond range, so the $S_n \rightarrow S_1$ internal conversion occurs simultaneously with intrastate vibrational relaxations. Radiationless processes will be discussed in more detail in a later section. We note here that the fast internal conversions down to S_1 are consistent with the small electronic energy gaps between the higher excited singlet states. Internal conversion from S_1 to S_0 is much slower because of the large S_1-S_0 energy gap; refer to Figure 2.

As a photoexcited molecule in fluid solution rapidly relaxes to an S_1 Boltzmann distribution, any molecular memory of how the emitting state was created is lost. Accordingly, nearly all of the following text applies equally well to fluorescence emission in chemiluminescence, bioluminescence, and pulse radiolysis experiments, in addition to photoluminescence.

Fluorescence Spectroscopy of Aromatic Molecules

Stokes Shift and Mirror Image Symmetry. The degree of structure in a room temperature solution fluorescence spectrum often, but not always, resembles that of the corresponding $S_1 \leftarrow S_0$ absorbance spectrum. The collective excited state population undergoes statistical decay on a time scale that is typically nanoseconds, but the interval over which an *individual* photon emission event occurs is still governed by the Franck–Condon principle; that is, the emission takes place in vertical fashion. Immediately following the photon emission, the ground (electronic) state reassumes its normal electronic charge distribution, albeit mapped onto the excited state nuclear geometry. Subsequent rearrangement of the nuclear positions to the ground-state equilibrium geometry releases energy to the solvent via another vibrational relaxation.

Figure 8 is a typical illustration of the vertical photoprocesses and vibrational relaxation for a highly stylized polyatomic with a single geometric coordinate. Molecular cartoons of this type are often drawn to rationalize the vibronic structure of spectra similar to those in Figure 1. However, poly-atomic molecules formally require multidimensional geometric coordinate descriptions because there are 3N-6 vibrational modes, where N is the number of atoms in the molecule. At best, Figure 8 has some sort of qualitative validity if the abscissa is identified as some total (and usually totally undefined) geometrical coordinate. The reader is urged to remember that polyatomic molecules possess a dense manifold of vibrational levels in any given electronic state.

Owing to the Franck−Condon principle (vertical photon transitions) the excitation takes place to higher energy from the S_1-S_0 electronic energy gap (the energy difference between the two zero-point levels), whereas the fluorescence occurs at lower energy from the electronic energy gap. Conse-quently, the emission is red-shifted relative to the excitation (absorbance). The larger the geometry difference between S_0 and S_1, the greater the displacement of the $S_1 \leftarrow S_0$ absorbance pattern to higher energy and the $S_1 \rightarrow S_0$ emission to lower energy, relative to the electronic energy gap. The separation between the absorbance and emission maxima is commonly re-ferred to as the Stokes shift. Qualitatively, the greater the Stokes shift, the greater the S_1-S_0 geometry difference. [See Berlman (13) for more precise definitions of the Stokes shift.] Unfortunately, it is a practical impossibility to turn the qualitative inference into quantitative interpretation for any but the smallest molecules owing to the multidimensional nature of the vibrational motions.

Figure 8 further indicates that the 0-0 transition is the only one common to both the absorbance and the emission spectra. Another concept implicit in the figure is that of mirror-image symmetry between the absorbance (excita-tion) and fluorescence spectra. The better the mirror-image symmetry, the more nearly identical the S_0 and S_1 geometries and bonding (and hence their vibrational frequencies) are assumed to be. Note that the only meaningful assessment of the mirror-image symmetry is comparison of the fluorescence with the $S_1 \leftarrow S_0$ absorbance (or excitation) spectrum, not with absorbance to any higher state. A depiction of normalized absorbance and fluorescence spectra on the same graph is useful for assessment of the Stokes shift and mirror-image symmetry. Berlman's book (13) includes over 100 examples and the reader is urged to consult this resource, whose value is enhanced because all the spectra presented therein were taken under the same conditions.

Excitation versus Absorbance Spectra. In the literature, it is often the *excitation* spectrum, not the absorbance spectrum, that is pre-sented. We shall, therefore, first identify the similarities and differences between the two presentations of essentially the same information before turning to actual examples.

An excitation spectrum is operationally obtained by fixing the emission wavelength of the detection system and varying the wavelengths of light to which the sample is subjected. Because the fluorescence spectral shape (i.e., the relative intensity at various wavelengths) is independent of excitation wavelength, the emission intensity for a given solute and solvent is directly proportional to the number of molecules photoexcited via absorbance within the collection volume of the detection system. This number, in turn, depends on the intensity of the light source as a function of wavelength and the sample absorbance, which is also wavelength dependent.

The fluorescence intensity, I_F, at emission wavelength λ_{em} and excitation wavelength λ_{ex} can be written as

$$I_F(\lambda_{em}, \lambda_{ex}) = I_{ex}(\lambda_{ex}) f_{abs}(\lambda_{ex})(QY)(GC) g(\lambda_{em}) d(\lambda_{em}) \qquad (1)$$

"Reading" the right-hand side of the equation from left to right (which is also the order of the photon paths), we have the number of photons incident on the sample (I_{ex}), the fraction actually absorbed (f_{abs}) to create potentially fluorescent molecules, the fraction of excited molecules that actually emit photons (QY is the quantum yield), the fraction of emitted photons that enter the emission monochromator (GC is the geometrical collection factor), the fraction g of the total emission spectrum that falls within the bandpass of the emission monochromator, and finally the wavelength-dependent sensitivity of the detection system (d).

The fraction of incident photons absorbed is $1 - T$ where T, the transmittance, is related to the absorbance A by

$$T = 10^{-A} = \exp(-2.303 A) \qquad (2)$$

For dilute solutions (say absorbances less than 0.05), the series expansion of the exponential function is adequately truncated at the linear term

$$\exp(-2.303 A) \approx 1 - 2.303 A \qquad (3)$$

and I_F is proportional to $A(\lambda_{ex})$:

$$I_F(\lambda_{em}, \lambda_{ex}) = 2.303 A(\lambda_{ex}) I_{ex}(\lambda_{ex})(QY)(GC) g(\lambda_{em}) d(\lambda_{em}) \qquad (4)$$

If the fluorescence intensity is then divided by the excitation intensity, the resulting ratio I_F/I_{ex} is directly proportional to the absorbance. Note that the choice of emission wavelength does not affect the λ_{ex} dependence of the excitation spectrum apart from an overall scale factor. The usual practice is to choose the monitoring wavelength to correspond to the maximum fluorescence intensity, but other choices are sometimes convenient.

Why would one go to all this extra trouble to simply recover the absorbance spectrum? Two important reasons are sensitivity and sample versatility. The excitation spectrum of a moderate to strongly emitting molecule can be measured at concentrations several orders of magnitude lower than that for which an absorbance measurement is feasible. The advantages of measuring a small signal against an even smaller background (as in the fluorescence-detected excitation spectrum) versus searching for a small change in a large light signal, as in absorbance, are considerable. Moreover, as noted in the introduction, the excitation spectrum can readily be determined for strongly scattering or even opaque samples (amorphous solids, turbid and frozen solutions, cracked glasses, etc.).

Discrepancies between the excitation and absorbance spectra can usually be traced to either instrumental factors or the presence of more than one emitting species. The most common instrumental factor is noncorrection of the excitation spectrum for wavelength variation in source intensity. Commercial spectrofluorimeters with quantum counter accessories are now routinely available for the acquisition of corrected spectra. Many older instruments lack this capability, but an inability to correct the excitation spectrum is generally not a great handicap.

The second instrumental source that causes deviation of the excitation spectrum from the absorbance spectrum is associated with absorbance values being sufficiently large to invalidate truncation of the exponential function expansion at the linear term. As the excitation spectrum is scanned, the exciting light penetrates to different depths in the cell according to the varying absorbance values. [In very dilute solutions, the exciting light attenuation is negligible and the excitation is uniform along the beam path.] The geometry of the photon collection system is fixed and the collection efficiency is not uniform over the entire excitation volume, so a spectral distortion is introduced. Here, too, this poses little problem for data interpretation as long as one realizes the mechanism is operative. For very high concentrations, such as can be obtained for doped polymers, front face (as opposed to right angle) detection is a good way to eliminate most of the variation in geometric collection efficiency.

A molecular source for differences between excitation and absorption spectra is dependence of the fluorescence quantum yield on excitation wavelength, but true examples for monomers in solution are extremely rare. Whenever this mechanism is suspected, one is advised to scrupulously eliminate other possibilities first. The presence of two or more absorbing species with different fluorescence efficiencies at the monitoring wavelength is far more likely. If the fluorophor of interest is not a strong emitter or present at low concentration, then the possibility of impurity emission must always be considered. In fact, recommended procedure is to compare the absorbance and excitation spectra whenever possible to test for impurities. Similarly, it is a good idea to verify that the emission spectral shape is independent of excitation wavelength and that the excitation spectrum is independent of the wavelength at which the emission is monitored.

Multicomponent absorbance and fluorescence can still arise in a high purity sample owing to ground-state dimerization. For example, the dimer of benzoic acid formed by hydrogen bonding absorbs at slightly different wavelength than the monomer and there is a similar fluorescence shift. Of course, in bichromophoric systems like those shown in Figure 6, differences between the absorbance and excitation spectra are highly likely.

Excitation and Emission Spectra of Aromatic Chromophores. A few examples will serve to illustrate the concepts introduced to this point. The anthracene spectra in Figure 9 represent a high degree of mirror-image symmetry. The 0-0 bands in emission and excitation are easily identified and coincide to within 2 nm. Perylene, which also has an $^1L_a S_1$ state, is another example of excellent mirror-image symmetry (*18*).

The phenanthrene spectra in Figure 10 illustrate how problems can easily arise in the interpretation of fluorescence data. If the ^1L_b transition between 350 and 300 nm is carelessly overlooked (or if it gets lost in the noise), the erroneous conclusion that there is a very large Stokes shift and hence, large S_1-S_0 geometry difference, might easily be reached. A complete gap between the excitation (absorbance) and emission spectrum does occur, for example, in molecules that undergo excited state proton transfer (*19*), but that is not the case here. In reality, anthracene and phenanthrene have nearly identical Stokes shifts. When the ^1L_b features of phenanthrene are shown on an expanded basis in the inset to Figure 10, we find that the 0-0 bands are actually even more coincident than in anthracene; the separation is just over 1 nm. The mirror-image symmetry, however, breaks down considerably. Finally, we note the excellent agreement between the phenanthrene excitation

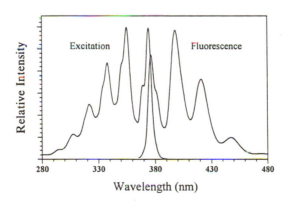

Figure 9. Fluorescence and fluorescence excitation spectra of anthracene in heptane solution at room temperature, illustrating a high degree of mirror-image symmetry.

spectra in Figure 10 with the absorbance spectrum shown in Figure 1 even though the solvents are not the same. As noted in an earlier section, many aromatics (naphthalene, pyrene, carbazole, and fluorene) commonly used to label polymers also have ^1L_b as the lowest excited singlet state.

Our final fluorescence spectrum example is biphenyl (Figure 11), which has been termed unusual (20) because there is more structure in the emission than in the excitation spectrum. It is tempting to associate this behavior with the nonplanar ground state and the fact that the excited state is

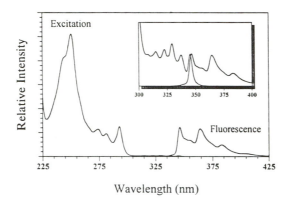

Figure 10. Fluorescence and fluorescence excitation spectra of phenanthrene in aqueous solution at room temperature. On the scale of this figure the 1_{L_b} transition between 350 and 300 nm in excitation is barely discerned. Inset: Expansion of the initial portion of the excitation spectrum with renormalization of the fluorescence reveals a small Stokes shift but a significant departure from mirror-image symmetry.

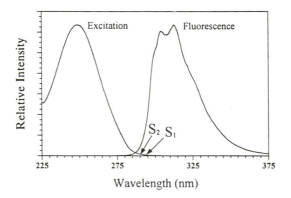

Figure 11. Fluorescence and fluorescence excitation spectra of biphenyl in aqueous solution at room temperature. Arrows indicate origin positions of the S_1 and S_2 states, as revealed in the low-temperature fluorescence excitation spectrum.

expected to tend toward planarity. Actually, the state with maximum near 248 nm is *at least* S_3 (and perhaps even a higher singlet state!); the very weak transitions to S_1 and S_2 do not show up in the low-resolution fluid solution spectra, but are clearly revealed in low-temperature work (*21*). One should be extremely cautious about ascribing too much significance to vibrational structure (or lack thereof) in fluid solution spectra.

A Further Caution on the Interpretation of Vibronic Structure.   The vibronic structure provides the most conveniently obtained, and sometimes virtually the only, experimental information on excited state geometries. Application of the structure-yielding diffraction techniques (neutron, electron, X-ray) to species that exist for only nanoseconds poses obvious problems. Some excited electronic state infrared and Raman spectra have been measured, particularly for the relatively long-lived triplet states, but those experiments require highly specialized instrumentation. Thus, the temptation to attempt analysis of the vibronic structure from solution absorbance and fluorescence spectra is great, but nonetheless one that should be resisted. The higher resolution afforded by low-temperature matrix or supersonic jet studies reveals that the actual vibronic structure is far more complicated than what the room temperature spectra indicate.

This point is strikingly made in Figure 12, which compares the fluid solution anthracene fluorescence spectrum with the corresponding emission of anthracene in a frozen *n*-heptane solution (Shpol'skii matrix) at 10K. The 1450-cm^{-1} progression-forming "mode" shown in Figure 3 is actually the

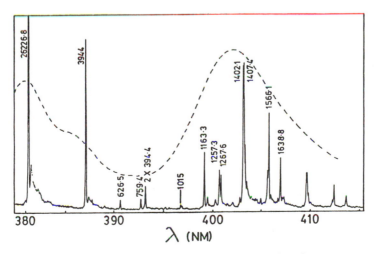

Figure 12. Laser site-selected fluorescence spectrum of anthracene in an n-heptane Shpol'skii matrix at 10 K. The dashed line indicates the room temperature fluorescence spectrum.

overlap of many active fundamental vibrations with frequencies in the range $1100-1650$ cm^{-1}.

One commonly finds statements in physical chemistry textbooks along the lines of "the population of excited vibrational levels at 298 K is usually very low." When made in the context of diatomic molecules, the statement is true. However, when the statement is extended to polyatomics with more than just a few atoms, it is quite erroneous, and there is probably insufficient appreciation for the breadth of the thermal distribution in a large molecule. For example, even though the anthracene zero-point level for a 300-K Boltzmann distribution is more populated than any other vibrational level, its fraction of the total population is *far* less than 1%. The room-temperature spectrum of a large molecule represents the overlap of individual transitions from hundreds (and even thousands) of levels, each of which can participate in absorption or fluorescence transitions to several vibrational levels in the other electronic state.

Along the same vein, one must be similarly cautious about assessments of mirror-image symmetry. Several years ago we reported fluorescence and excitation spectra for 1-aminoanthraquinone in a low-temperature *n*-heptane matrix (*22*). The spectra under these conditions are highly structured with bands that are only a few-hundredths of a nanometer wide and the mirror-image symmetry appeared excellent. When we recently examined the corresponding gas phase spectra, it was discovered that despite the mirror-image symmetry, there must be significant differences between the bonding and vibrational properties in S_0 and S_1 (*23*).

Solvent Dependence. Fluorescence properties can be markedly sensitive to the choice of solvent. The literature on solvent effects is extensive. For example, biochemists use solvent effects to probe the local environment in biological macromolecules that have been labeled with fluorescence probes (*20*); a similar approach is clearly applicable to synthetic polymers. The variation in the fluorescence can take the form of a wavelength shift of the entire fluorescence spectrum, a modification of intensity (fluorescence quantum yield), and in a few cases, such as pyrene, a *predictable* variation in the vibronic structure.

With regard to the fluorescence wavelength shift, the solvent dependence is associated with differences between the ground- and excited-state electron distributions. The usual starting point for discussion is the Lippert equation (*24*), which draws attention to the dipole moment difference between S_0 and S_1. More often than not there is a substantial dipole moment *increase* associated with photoexcitation. By once again applying the Franck–Condon vertical excitation principle, this time to a "supermolecule" of the solute and its solvation sheath, one deduces that the effectively instantaneous creation of the excited electronic state electron distribution with the solvent sheath at its ground-state arrangement is destabilizing. As the solvent shell rearranges to accommodate the increased solute dipole moment, the supermolecule relaxes to lower energy. Consequently, the fluorescence exhibits a shift to longer wavelength. Figure 13 shows a particularly dramatic example of a popular fluorescence solvent polarity probe. An extensive discussion of solvent dependence, the Lippert equation, and related concepts can be found in Lakowicz (*20*).

For polycyclic aromatic hydrocarbons with alkyl or other relatively non-polar substituents, the dipole moment difference between S_0 and S_1 is small and may even be restricted to zero by symmetry. For these molecules, solvent effects on the fluorescence wavelength distribution are small, as may be verified by comparing the aqueous solution spectra for phenanthrene and biphenyl in Figures 10 and 11 with the corresponding spectra for cyclohexane solvent (*13*).

The solvent effect on fluorescence *intensities* is not as easily predicted or explained. For example, in unpublished work from our laboratory, we showed that the fluorescence quantum yield of benzene in aqueous solution is much lower than in aliphatic solvents, yet other aromatics are little affected or their fluorescence even intensified by the change from a nonpolar organic solvent to water. The solvent effects on fluorescence quantum yields are primarily related to changes in the radiationless transition rates, for which quantitative theories are not nearly as well developed as the Lippert equation.

We note, however, a well-understood solvent effect on intensities for heterocyclic aromatics, which often have close-lying n-π^* and π-π^* states as S_1 and S_2. Hydrogen bonding solvents interact strongly with the nonbonding lone-pair electrons of the heteroatom, but to a different degree for each

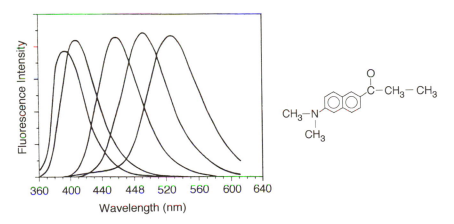

Figure 13. Solvent dependence of the fluorescence spectrum of Prodan, a popular solvent polarity probe. From left to right the solvents are cyclohexane, chlorobenzene, dimethylformamide, ethanol, and water. (Sprectra reproduced from reference 24. Copyright 1990 American Chemical Society.)

electronic state. For example, in acridine there are two in-plane lone pair electrons on the nitrogen atom in the ground state and π-π* excited state, but only one nonbonding electron in the n-π* excited state. Thus, if the solvent is changed from a nonhydrogen-bonding solvent such as cyclohexane to a strongly hydrogen bonding solvent such as methanol, the π-π* excited state is lowered preferentially relative to the n-π* state. This differential stabilization may be enough to invert the n-π* and π-π* states (Figure 14). The π-π* state is much more intrinsically fluorescent owing to both its higher radiative decay rate and lower radiationless $S_1 \rightarrow T_1$ decay rate (*see* following discussion of El-Sayed's rule below), and the increase in fluorescence efficiency can be dramatic. Enhancements by several orders of magnitude are possible.

Pyrene is often used as a photophysical probe for polymers owing to its proclivity for excimer formation, a topic discussed in the next section. A different aspect is noted here. Pyrene exhibits a reasonable degree of structure in its fluorescence spectrum (Figure 15). Five major bands, labeled I through V, are readily identified. Band I, the shortest wavelength major feature, is conventionally called the 0-0 band and nominally represents the transition between the S_0 and S_1 zero-point levels. As was discussed earlier, this "band" is actually the overlap of many vibronic transitions. Nevertheless,

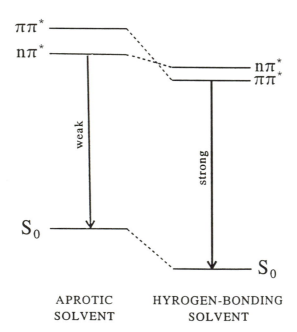

Figure 14. Illustration of the inversion of the n-π* and π-π* states of a heteroaromatic molecule as the solvent is changed from an aprotic to a hydrogen-bonding solvent.

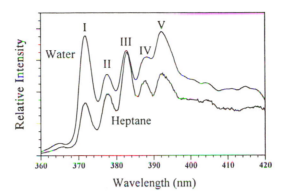

Figure 15. Fluorescence spectra of pyrene in water and in heptane. The spectra have been normalized to the same intensity at band III.

it retains considerable true 0-0 character. Now according to molecular symmetry, the 0-0 band in an isolated molecule (i.e., gas phase) of pyrene is forbidden. Solvent perturbation can lift this forbiddeness, and in strongly polar solvents band I is considerably intensified relative to the other bands. The phenomenon is sometimes referred to as the Ham effect in honor of its discoverer (25). The standard practice is to use the ratio of band I to band III intensity; this ratio forms the basis of a solvent polarity scale (26). In this fashion pyrene can reveal the polarity of microenvironments, as in a labeled polymer system, or when the pyrene is solvated in a micelle.

Deactivation of the Emitting State

The Jablonski Diagram Revisited. The discussion now turns from the energetic aspects (state diagrams, wavelength distribution) of fluorescence to a consideration of the rate processes that deactivate the emitting state. Fluorescence temporal behavior yields the most detailed picture available on the motion of polymer systems on the fast (10^{-6}–10^{-12}-s) time scale. This section provides an overview of the important radiationless processes that compete with fluorescence emission and how the rate constants are determined from experiment.

The kinetic processes that deactivate the S_1 state can be separated into radiative (fluorescence), unimolecular nonradiative (internal conversion to S_0 and intersystem crossing to T_1), and bimolecular nonradiative (quenching, excimer–exciplex formation, energy transfer) contributions. The fluorescence (k_F), intersystem crossing (k_{ISC}), and internal conversion (k_{IC}) processes follow first-order kinetics and each is therefore associated with a time-independent rate constant, as is illustrated in Figure 16A. The bimolecular steps are vital aids in revealing the details of macromolecule internal motion, but

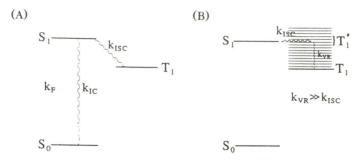

Figure 16. A: Conventional "vertical" representation of radiationless and radiative intramolecular decay paths from S. B: Schematic energy level diagram indicating that radiationless transitions are horizontal (isoenergetic) processes, but they are followed in solution by much faster vibrational relaxation processes with average rate constant K_{VR}.

we will first consider the case of a single chromophore in dilute, degassed solution, in which the intermolecular processes can be neglected.

The directly measurable fluorescence parameters are quantum yield and lifetime. The fluorescence quantum yield, Φ_F, is the ratio of number of photons emitted to the number of photons absorbed, or equivalently, the ratio of the number of emitted photons to the number of excited molecules initially created by photoexcitation. Fluorescence quantum yield is related to the first-order radiative and nonradiative rate constants by

$$\Phi_F = k_F/(k_F + k_{IC} + k_{ISC}) = k_F/(k_F + k_{NR}) \tag{5}$$

where k_{NR} is the total nonradiative decay rate constant. Equation 5 can be readily derived from a photostationary state treatment, but it is applicable to pulsed experiments as well. The sum in the denominator of eq 5 represents the total rate constant for the exponential decay of the S_1 population. The reciprocal of that sum is the fluorescence lifetime (τ_F):

$$\tau_F = (k_F + k_{IC} + k_{ISC})^{-1} = (k_F + k_{NR})^{-1} \tag{6}$$

Note that the term "fluorescence lifetime" is slightly misleading. More properly, τ_F is the S_1 lifetime. Reference is made to τ_F as the fluorescence lifetime simply because fluorescence is the usual (and by far the most convenient) way to follow the S_1 population decay. If transient absorption from S_1 were followed, the same S_1 lifetime should be found.

Equation 5 can be written more compactly:

$$\Phi_F = k_F \tau_F \tag{7}$$

At this stage there are two measurable quantities (Φ_F, τ_F), but three rate constants (k_F, k_{IC}, k_{ISC}). The radiative decay rate, k_F, can be determined from eq 7; then the total nonradiative decay rate, k_{NR}, can be determined from eq 6. A third measurement, usually that of the quantum yield for triplet formation via intersystem crossing, is necessary for a determination of the individual nonradiative rate constants. Many times the approximation is made that internal conversion is negligibly slow compared to intersystem crossing in which case

$$k_{ISC} \cong k_{NR} = (1 - \Phi_F)/\tau_F \qquad (8)$$

Results for selected polycyclic aromatic hydrocarbons derived on this basis are given in Table I. However, note that there are cases for which nonradiative decay is dominated by $S_1 \to S_0$ internal conversion (e.g., stilbenes, polyenes, β-carotene, etc.).

In a strict sense, internal conversion and intersystem crossing ought to be represented as isoenergetic processes, as opposed to the "vertical" depiction of Figure 16A. For example, in the $S_1 \to T_1$ intersystem crossing, electronic energy equal to the S_1-T_1 electronic energy gap is converted into T_1 excess vibrational energy (Figure 16B). However, the radiationless transitions are followed by much faster vibrational relaxation. The internal conversion and intersystem crossing might be viewed as rate-limiting steps in a conventional kinetic picture of the sequential "reactions":

$$S_1 \to S_0^\# \to S_0$$

$$S_1 \to T_1^\# \to T_1$$

where $S_0^\#$ and $T_1^\#$ represent vibrationally excited states with the same total vibronic energy as that of the S_1 state undergoing the radiationless transition(s).

Table I. Photophysical Data for Aromatic Molecules[a]

Molecule	Φ_F	τ_F (ns)	k_F (s^{-1})	k_{NR} (s^{-1})	$\epsilon(S_1 \leftarrow S_0)$
Benzene	0.07	29	2.4×10^6	3.2×10^7	210
Naphthalene	0.23	96	2.4×10^6	8.0×10^6	300
Anthracene	0.36	4.9	7.3×10^7	1.3×10^8	10,000
Phenanthrene	0.13	57.5	2.3×10^6	1.5×10^7	220
Pyrene	0.32	290	1.1×10^6	2.3×10^6	
Biphenyl	0.18	16.0	1.1×10^7	5.5×10^7	
Fluorene	0.80	10	8×10^7	$2. \times 10^7$	10,000
Perylene	0.94	6.4	1.5×10^8	6×10^6	39,000
Benzo[*ghi*]perylene	—	107	—	—	—
Carbazole	0.38	16.1	2.3×10^7	3.9×10^7	4,300
Chrysene	0.14	44.0	3.1×10^6	1.9×10^7	700

[a] Data taken from reference 13.

Two points to note:

1. The experimentally determined nonradiative rate constants are truly those of the intersystem crossing and internal conversions themselves because the vibrational relaxation is so much faster.

2. The radiationless transitions are fundamentally intramolecular processes (i.e., solvent-assisted vibrational relaxation is not necessary) as is demonstrated by supersonic jet studies of gas phase isolated molecules (27).

Fluorescence Quantum Yield and Lifetime Measurements. The fluorescence quantum yield is a branching ratio between radiative and nonradiative decay, and its value must therefore lie between 0 and 1. Although this chapter is not intended as a methods document, a few words on quantum yield measurements are appropriate here. Most often fluorescence quantum yields are determined in a relative sense by comparing the integrated fluorescence intensity for the molecule of interest to that of a standard. One approach is to balance (i.e., match) the absorbances of sample and reference compound at a wavelength where both absorb well, to record their separate fluorescence spectra excited at that wavelength (thereby ensuring that sample and reference each absorb the same number of photons), and to measure the areas under the respective fluorescence curves. Be aware, however, that subleties abound in the process. Unless the sample and reference fluorescences are closely overlapped in wavelength, the spectra must be corrected for wavelength response of the detection system as well as converted to an intensity versus wavenumber format (which, in turn, introduces a slit correction). Other sources of error include too great a variation of the absorbance of either sample or standard over the excitation bandpass or differential sensitivity of the sample and standard to oxygen or self-quenching. References 28 and 29 should be consulted for details.

Moreover, the issue still remains of how to determine the *absolute* fluorescence quantum yield of the standard, which is usually accomplished via integrating sphere techniques. Two examples emphasize the intricacies of absolute quantum yield determination. The first is the case of 9,10-diphenylanthracene, used as the standard by Berlman (13). Over the years the recommended value has jumped back and forth between 0.83 and 1.00 and there still appears to be less than unanimity on the subject (30). The second, and in many ways more striking, example is the recent meticulous work by Johnston and Lipsky (31) that suggests the long-accepted fluorescence quantum yield of benzene vapor may be too high by a factor of almost 4. Nevertheless, with reasonable care it should be possible to generally obtain fluid solution relative fluorescence quantum yields accurate to within 25%.

The most straightforward way to determine lifetimes is to populate S_1 much faster than the subsequent excited state depopulation occurs via fluorescence and radiationless decay. If all the decay paths are first order, the S_1 population decreases exponentially in time. A plot of the natural logarithm of fluorescence counting rate versus time is then linear and the lifetime is simply extracted as the negative reciprocal of the slope.

Whether the excitation step is sufficiently separated in time from the subsequent decay depends on the lifetime value and the nature of the excitation source. There are many different possible excitation sources and the reader is directed to Demas (32) for details. We will mention four types here: (1) gas-filled discharge lamps that operate at a few kilohertz with a pulse duration of a few nanoseconds; (2) low-repetition rate pulsed excimer, nitrogen, and Nd–YAG lasers with typical pulse durations also in the few nanosecond range; (3) high repetition rate picosecond duration laser systems (e.g., mode-locked and cavity dumped ion and Nd–YAG systems; (4) continuous excitation sources, commonly xenon arc, with their output sinusoidally modulated at tens of megahertz with acoustooptic or electrooptic devices. Sources of this type form the basis of phase-resolved fluorescence spectroscopy (33). The most attractive feature of phase resolved methods is that with a commercial spectrofluorimeter one can measure both conventional luminescence and excitation spectra as well as lifetimes.

If the fluorescence lifetime is not long compared to the excitation pulse duration, it is necessary to extract the lifetime by deconvolution, a mathematical processing of the data to give the best agreement between the experimental decay curve and those calculated for various assumed lifetimes. Under favorable conditions, lifetimes as short as $1/10$ the excitation pulse width have been extracted. The precision with which lifetimes can be measured is much higher than for quantum yields. Uncertainties of less than 1% for lifetimes above 10 nanoseconds are routinely achieved.

Figure 17 presents illustrative data for anthracene in solution. In this experiment the laser pulsewidth has a duration comparable to the excited state lifetime so deconvolution would be required to extract the lifetime. Note that the decay is extended in time when the solution is degassed to eliminate oxygen quenching (see later discussion of the Stern–Volmer equation).

The Radiative Decay (Fluorescence) Process.   The rate constant for fluorescence radiative decay from S_1 to S_0 can often be accurately estimated because the spontaneous fluorescence and absorbance processes are related through the Einstein A and B coefficients (34). Strickler and Berg (35) generalized the Einstein treatment to molecular systems and derived an equation for k_F that contains the so-called integrated absorbance spectrum (necessarily limited to the $S_1 \leftarrow S_0$ transition). Agreement between the Strickler–Berg k_F value and that derived directly from experiment via eq 6 is

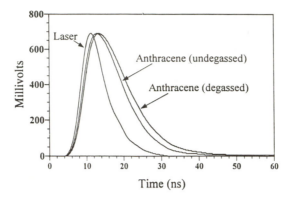

Figure 17. Fluorescence lifetime traces for anthracene in degassed and air-saturated cyclohexane solution, along with the laser excitation time profile. Excitation source was a dye laser and temporal data acquired with a digital oscilloscope.

excellent. If the experimental value is more than a factor of 2 lower than the calculated value, a weakly absorbing S_1 state buried in the long wavelength tail of a stronger transition to S_2 should be suspected. A number of misassignments have been cleared up in this fashion.

A simplified version of the Strickler–Berg equation suitable for rough estimates is

$$k_F(s^{-1}) = 3 \times 10^{-9} \nu^2 \epsilon_{max} \, \Delta\nu_{1/2} \qquad (9)$$

where ν is the average wavenumber of the fluorescence, ϵ_{max} is the maximum molar absorptivity in units of liters per mole per centimeters, and $\Delta\nu_{1/2}$ is the half-width of the electronic transition. The average half-width reported by Berlman (13) is 3600 cm^{-1}. We may take $\epsilon_{max} = 100$ as the minimum molar absorptivity to be expected in aromatics. For a typical emission wavelength of 350 nm, the radiative decay rate is therefore approximately 10^6 s^{-1}; equivalently, the lifetime would be 1 µs if fluorescence were the sole S_1 decay path. Any higher molar absorptivity (if S_1 is ^1L_a, for example) or any contribution of nonradiative decay will shorten the lifetime. Thus, the *long time* limit of processes that can be monitored by fluorescence is about 1 µs. Slower processes can be monitored via the longer-lived phosphorescence. In part, the popularity of pyrene as a photophysical probe for polymer physical properties is its long intrinsic fluorescence lifetime [ca. 300 ns for the monomer in dilute solution (36)].

The reciprocal of k_F is sometimes referred to as the radiative lifetime or pure radiative lifetime. In other words, it is the excited state lifetime that would follow in the hypothetical case where the nonradiative decay is suppressed.

Intramolecular Decay: Radiationless Transitions. There is no reliable radiationless transition rate constant formula analogous to the Strickler–Berg equation; a priori predictions of internal conversion rates and intersystem crossing rates are beyond reach now and the situation is not expected to change soon. Fortunately, there do exist very useful empirical correlations and we will touch upon these briefly.

The theoretical description of a radiationless transition from an initial state i to a final state f begins with Fermi's Golden Rule:

$$k_{NR} = 2\pi/h|V_{if}|^2\rho_f(E_i) \tag{10}$$

in which the coupling matrix element V_{if} is a measure of the strength of the interaction between the initial and final states and ρ_f is the density of final vibronic levels at the energy of the initial state (E_i). The modern era of radiationless transition theory dates to the seminal Robinson and Frosch paper (37). Many formal theory papers and review articles (38) were written on the topic in the 1970s and early 1980s. Unfortunately, first principle calculations have proven impossible. There is no reliable theoretical alternative to experimental measurements or estimates based on experimental values in chemically similar systems.

In this regard, the correlation of radiationless transition rate with electronic energy gap, first presented by Siebrand (39) has been extremely useful. As discussed earlier, internal conversion and intersystem crossing are energy-conserving processes, wherein the electronic energy difference between the initial and final states must be converted into vibrational energy in the final (accepting) state. The conversion efficiency is enhanced by the presence of a large number of final states to accept the energy (the density of states term in the Golden Rule) and by a strong coupling between the initial and final states (the matrix element term).

The coupling matrix elements can be cast into a form that contains Franck–Condon factor type terms, again between the initial state (with small vibrational quantum numbers) and the final states, which are characterized by large vibrational quantum numbers. The larger the electronic energy difference between the two states, the larger the change in vibrational quantum number required by the radiationless transition, and, accordingly, the smaller the Franck–Condon factor. The density of states and coupling matrix elements, as a function of electronic energy gap, follows opposite trends. For electronic energy gaps more than a few thousand cm^{-1}, the electronic energy difference factor dominates, and in this range the radiationless transition rates decrease nearly exponentially with increasing electronic energy gap.

The electronic energy gap in the Siebrand correlation is scaled to account for the relative proportion of C–H to C–C oscillators in the molecule. The importance of C–H stretches is a consequence of their efficiency in accepting energy. Owing to their high frequency, fewer accepting quanta are required.

Simplified energy gap plots without scaling are shown in Figure 18. Even though the conformance to a straight line is not quite as impressive as for the scaled plots, the energy gap law is still very clearly revealed.

Despite the fact that it is a spin-forbidden transition, $S_1 \to T$ intersystem crossing in aromatic molecules competes very effectively with the $S_1 \to S_0$ internal conversion and, in many cases, dominates. There are both obvious and subtle reasons why this is so. The obvious reason is that the T_1 state has a much smaller energy gap with S_1 than does S_0. A more subtle reason is that the intersystem crossing from S_1 can proceed through higher triplet states than T_1.

We had noted earlier that the rapid internal conversion cascade from higher singlet states slows down at S_1 instead of proceeding all the way to S_0. The S_1-S_0 electronic energy gap is almost always much greater than the gaps between excited singlet states. Note, too, the possibility of spectroscopically unobserved singlet states contributing to the $S_n \to S_1$ internal conversion cascade.

The invariably cited and most striking exception to Kasha's rule is azulene in which the $S_2 \to S_0$ fluorescence quantum yield is itself reasonably strong, 0.03, and, moreover, orders of magnitude stronger than the $S_1 \to S_0$ quantum yield (41). However, azulene is really less anomalous for its S_2 fluorescence than it is for its exceptionally large S_2-S_1 energy gap. In fact, the *most* anomalous feature of azulene is that the $S_1 \to S_0$ internal conversion is about 2 orders of magnitude faster than the energy gap correlation predicts (42).

The energy gap correlation shown in Figure 18 applies to planar aromatic hydrocarbons. When n-π^* states are involved, intersystem crossing from S_1 is often greatly enhanced. El-Sayed's rule states that intersystem crossing

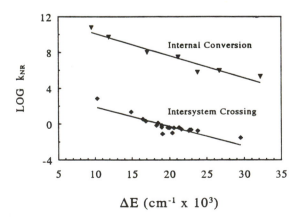

Figure 18. Illustrations of energy gap law for $T_1 \to S_0$ *intersystem crossing and* $S_1 \to S_0$ *internal conversion. Data taken from references 39 and 40.*

between an n-π^* singlet and π-π^* triplet state (or vice versa) is much faster than intersystem crossing between two n-π^* states or two π-π^* states (*43*).

Environmental Effects on Radiationless Transitions. The capacity of an aromatic solute to absorb radiation and become electronically excited is not a strong function of temperature or solvent. Although individual vibronic bands become narrower and better resolved at reduced temperature, the integrated absorption strength is nearly constant. According to the Strickler–Berg equation, therefore, any substantial environmental effects on luminescence intensities must be ascribed to radiationless transitions or bimolecular factors. In this section we comment briefly on the role of environment on radiationless transitions.

A notion that seems to have become rather firmly entrenched is that radiationless transitions are strongly affected by the rigidity of the medium or the structural rigidity of the emitter itself. Several authors have commented on the "loose bolt" effect (*13*), which expresses the notion that floppy motion enhances radiationless transition rates, whereas the imposition of rigidity suppresses nonradiative decay. In our opinion, too much has been made of this point. For every comparison of two structurally similar molecules in which the more rigid one has a higher fluorescence quantum yield, a counterexample can be offered. Witness the very high quantum yield of 9,10-diphenylanthracene (0.8–1.0) versus the value of ca. 0.3 for anthracene.

The preceding paragraph is not a denial of the experimental fact that luminescence yields are often higher in rigid media, but in most cases "floppiness," per se, is not the reason. First consider the observation of phosphorescence, for which the usual experimental condition is "immobilization" of the molecule in a rigid matrix such as an organic glass at liquid nitrogen temperature; 3-methylpentane, methylcyclohexane, and the polar mixture of solvents that is commonly referred to as EPA are popular choices for glass-forming solvents. By way of contrast, phosphorescence in fluid solution is a comparatively rare phenomenon. In general, however, freezing the sample to liquid nitrogen temperature does *not* markedly slow the $T_1 \rightarrow S_0$ intersystem crossing. The impact of the rigidity imposed by freezing the sample is to prevent diffusion of oxygen, which otherwise quenches the long-lived phosphorescence; triplet-triplet annihilation, similarly a diffusional process, is also suppressed (*44*). Samples that are very carefully degassed will show phosphorescence in fluid media. In fact, *any* change in conditions that inhibits oxygen diffusion enhances the phosphorescence: The possibilities include immobilization of the molecule in a polymer host (which is, of course, just another form of organic glass), adsorption on filter paper or other substrates, or incorporation into cyclodextrins. By and large it is the oxygen quenching that is inhibited, as opposed to a significant reduction of the intramolecular radiationless decay rate constants.

Temperature effects on fluorescence quantum yields can be quite large and the usual (but not exclusive) direction of change is that increasing temperature tends to reduce the fluorescence. There is simply not enough space here for a full discussion, but we will mention two classic examples: *meso*-substituted anthracenes and *trans*-stilbene (and higher polyenes). In both cases there is a strong temperature variation of fluorescence quantum yield, but the cause is the presence of an excited state that promotes efficient radiationless decay lying just above the emitting state.

Most *meso*-anthracenes have fluorescence quantum yields of unity to within experimental error at liquid nitrogen temperature, but values of only 0.01–0.1 at room temperature. For the parent anthracene the $S_1 \rightarrow T_1$ intersystem crossing, if forced to occur over the full energy gap of about 11,000 cm^{-1}, would be quite slow. However, a higher triplet state lies just below S_1 and acts as an intermediate state in the intersystem crossing. Changes in solvent do not change the S_1-T_2 energy gap substantially and therefore have little effect on the intersystem crossing rate. The room temperature fluorescence quantum yield of anthracene is about the same in hexane as in ethanol and not much different in either solvent at much lower temperatures. In *meso*-anthracenes the S_1 state is shifted to lower energy owing to extension of conjugation along the short in-plane axis (45). The shift is just enough to bring the S_1 state below T_2. At room temperature the broad Boltzmann distribution still allows T_2 to be accessed and the intersystem crossing can efficiently proceed. At low temperature this channel is closed off and only the direct (and slow) $S_1 \rightarrow T_1$ intersystem crossing remains.

In the case of *trans*-stilbene, the B_u excited state that carries high oscillator strength is just slightly lower in energy than an A_g state, which is radiatively one-photon forbidden but two-photon allowed with the ground state. As the molecule is twisted about the central C–C bond, the B_u state rises in energy and the A_g state falls (of course the g and u symmetry labels only apply for the planar molecule). Thermal activation from B_u to A_g leads to very efficient internal conversion. The literature on the topic is very extensive (46) and we will not discuss it further here.

Intermolecular Decay Processes. The electronically excited S_1 state can easily "live" long enough to undergo bimolecular interactions, as was discussed in the context of phosphorescence in the last section. Of course, all solvent interactions with an excited molecule are essentially of a bimolecular nature, but we will make some very brief comments on the interactions with other solutes (including self-interaction of a bichromophore). There are three main cases: quenching of the excited state (especially by dissolved molecular oxygen), excimer or exciplex formation, and energy transfer. Full chapters on these topics can be found in Guillet's monograph (4).

Quenching. The data in Table I show that typical fluorescence lifetimes for polycyclic aromatic hydrocarbons are in the range 1–100 ns, corresponding to total decay rates of 10^7–10^9 s^{-1}. The diffusion controlled second-order rate constant in fluid media is on the order of 10^{10} M^{-1}/s. Thus, if quenching occurred on every collision, a quencher concentration of 10^{-3} M would be sufficient to cut the fluorescence intensity in half. The concentration of dissolved oxygen in most air-saturated solvents is of that magnitude and oxygen quenching of fluorescence is not inconsequential. The extent of fluorescence quenching strongly correlates with fluorescence lifetime (*13*). In alkane solvent at room temperature, the fluorescence reduction for air-saturated solutions compared to degassed solutions is about 20% for anthracene (lifetime near 5 ns), a more than 10-fold reduction for naphthalene (lifetime near 100 ns), and still much larger yet for the very long-lived pyrene.

Quenching processes are easily incorporated into the kinetic analysis. The S_1 (fluorescence) lifetime is shortened by the quenching for which a pseudo-first-order rate constant can be written in terms of a second-order rate constant, K_Q, and the quencher concentration $[Q]$. Equation 6 is rewritten to incorporate the additional deactivation mechanism:

$$\tau_F = \{k_F + k_{NR} + k_Q[Q]\}^{-1} \tag{11}$$

and the fluorescence quantum yield is similarly reduced because eq 6 is still valid.

The Stern–Volmer expression

$$F_0/F = 1 + k_Q\tau_F[Q] \tag{12}$$

is often applied to steady-state data; F_0 is the fluorescence intensity in the absence of quencher, F is the intensity when the quencher is present at a concentration $[Q]$, and τ_F is the lifetime for zero quencher concentration.

Excimer Formation. The interaction of two identical polycyclic aromatic hydrocarbon molecules, each in its ground electronic state, is predominantly repulsive. When one of the molecules is electronically excited, an attractive force for its ground-state counterpart may develop. The resulting excited-state stabilized dimer that forms is known as an excimer (*excited state dimer*). The corresponding term for chemically distinct species that are attractive in the excited state but repulsive in the ground state is exciplex.

Excimer formation is a common phenomenon for aromatic molecules, especially if the singlet state lifetime is long. The popularity of pyrene for excimer fluorescence studies is thus explained; it combines a long monomer S_1 lifetime and favorable excimer emission properties. The excimer emission occurs at longer wavelength than the monomer emission by an energy roughly corresponding to the stabilization energy in the excited state and the destabilization energy in the ground state (which is expected to be larger). Because excimer emission occurs to a repulsive ground-state surface, excimer emission is characteristically unstructured.

The kinetic behavior of excimer formation and decay is described at great length in the literature and will not be discussed here. We do point out that the excimer has its own radiative and nonradiative decay rates, and when these rates are added to the previously presented decay paths, the kinetic scheme can become quite complicated. The usual starting point is the well-known Birks model (47).

As a final comment on excimers, we consider the expected excitation spectrum of a molecule that undergoes excimer formation. First note that there is only one absorbing species (the monomer). The initial chronology remains as before: excitation into some chromophore or monomer singlet state, followed by internal conversion–vibrational relaxation into a (monomer) S_1 Boltzmann distribution. From that point options for the excited molecule are fluorescence, internal conversion to S_1, intersystem crossing to T_1, quenching by dissolved oxygen, *and* excimer formation (via interaction with a ground-state species). The efficiencies of the excimer formation and decay paths are independent of excitation wavelength. Thus, the excitation spectrum will be the same whether monomer or excimer emission is monitored. If there *is* a deviation between the monomer and excimer excitation spectra, then the ground-state interactions are not negligible.

Energy Transfer. We previously considered the absorbance properties of a polymer labeled with two distinct chromophores, as shown in Figure 7. Now consider the fluorescence properties of bichromophoric systems such as an alkane chain tagged with anthracene at one end and naphthalene at the other. For excitation wavelengths of greater than 380 nm, no absorbance (and therefore no fluorescence) occurs. For excitation between 320 and 380 nm, only the anthracene absorbs and its corresponding fluorescence will be observed. However, at excitation wavelengths shorter than about 320 nm, both the naphthalene and anthracene will absorb and their relative degree of excitation is easily predicted. From their known fluorescence efficiencies, a total fluorescence spectrum can be predicted.

Usually the chromophore with the higher S_1 state energy exhibits reduced fluorescence relative to the prediction, whereas the fluorescence of the chromophore with the lower S_1 state energy is intensified. The obvious conclusion is that the electronic excitation of the naphthalene has somehow been transferred to the anthracene via a donor–acceptor (D–A) mechanism, usually represented

$$D^* + A \rightarrow D + A^* \tag{13}$$

where the asterisk denotes electronic excitation.

There is, of course, a trivial mechanism in which the naphthalene emission is simply reabsorbed by the anthracene. However, experiments have shown unequivocally that the same effect can be achieved without an intermediate photon actually being emitted. For this nonradiative energy transfer, there are two main mechanisms: resonance transfer (Forster mechanism) and electron exchange mechanism (Dexter mechanism). In both cases, theoretical expressions for the rate of transfer involve the overlap of the donor emission and acceptor absorbance spectra and the separation in space of the donor and acceptor. Electron exchange is a short-range interaction whereas the Forster mechanism occurs over longer distances, following an R^{-6} distance dependence. Measurement of the efficiency of energy transfer allows determination of the separation of donor and acceptor and this is the basis of the "spectroscopic ruler" approach (*15*). A full discussion of energy transfer fluorescence characterization of polymers can be found in Guillet (*4*).

Concluding Comments

The goal of this tutorial has been to present the principles of aromatic molecule solution fluorescence spectroscopy in a compact fashion. I have especially emphasized vibronic energy level structure and its role in determining the shape of the spectra and the rates of the radiationless transitions. Someone who has assimilated this material ought to have a reasonable background for reading the literature and further study. For anyone who wishes a more intensive program, the following approach is suggested for background reading to accompany exploration of Guillet's monograph (*4*).

First, I recommend a thorough perusal of the first 100 pages of Berlman (*13*). The fact that the instrumentation details are now outdated detracts little from the book's value. An especially profitable exercise is to leaf through the spectra and draw your own conclusions and correlations. Next, I suggest consultation of the collected papers from the Platt group (*9*) and the monograph by McGlynn et al. (*5*), which, although it is nominally directed at the triplet state, contains much material relevant to energy levels and radiationless transitions in aromatic molecules.

For those readers who are vitally interested in technical details of solution luminescence spectroscopy, the monograph by Parker (*3*) is enthusiastically recommended. Although it is more than 20 years old (and unfortunately now out of print), this book has withstood the test of time very well. Careful study of Parker will significantly aid proper execution and interpretation of fluorescence experiments.

The bible of photophysical principles is the monograph by Birks (*2*). It's a bit curious that Berlman, McGlynn et al., Parker, and Birks were all published within a span of a few years and have not been superseded in the intervening two decades.

Acknowledgments

I thank Marek Urban for the invitation to write this chapter. The actual writing represented a much larger time commitment (by about an order of magnitude!) than I had originally anticipated, but the opportunity to discuss fluorescence in a way not possible in standard research papers was irresistible. I also wish to thank my research students for their interest and contributions, my mentors (A. U. Khan and E. C. Lim) for their inspiration over the years, and the National Science Foundation, the United States Geological Survey, and the Air Force for financial support of my own luminescence research. The outstanding aid of Randy St. Germain in preparation of the figures is gratefully acknowledged.

References

1. Babcock, G. T. *Biological Applications of Raman Spectroscopy*, 1988, pp 294–346.
2. Birks J. B. *Photophysics of Aromatic Molecules*; Wiley: New York, 1970.
3. Parker, C. A. *Photoluminescence of Solutions*; Elsevier: New York, 1968.
4. Guillet, J. *Polymer Photophysics and Photochemistry*; Cambridge University Press: Cambridge, England, 1985.
5. McGlynn, S. P.; Azumi, T.; Kinoshita, M. *Molecular Spectroscopy of the Triplet State*; Prentice-Hall: Englewood Cliffs, NJ, 1969.
6. Clar, E. *Aromatische Kohlenwasserstoffe*; Springer: Berlin, 1941.
7. Klevens, H. B.; Platt, J. R. *J. Chem. Phys.* **1949**, *17* 470.
8. Klevens, H. B.; Platt, J. R. *J. Chem. Phys.* **1949**, *17*, 484.
9. Platt, J. R. et al. *Systematics of the Electronic Spectra of Conjugated Molecules*; Wiley: New York, 1964.
10. Becker, R. S.; Singh, I. S.; Jackson, E. A. *J. Chem. Phys.* **1963**, *38*, 2144.
11. Mulliken, R. S.; Person, W. B. *Molecular Complexes*; Wiley-Interscience: New York, 1969.
12. Lawson, C. W.; Hirayama, F.; Lipsky, S. *J. Chem. Phys.* **1969**, *51*, 1590.
13. Berlman, I. B. *Handbook of Fluorescence Spectra of Aromatic Molecules*; Academic: Orlando, FL, 1971.
14. Winnik, M. A. *Acc. Chem. Res.* **1985**, *18*, 73.
15. Liu, G.; Guillet, J. E. *Macromolecules* **1990**, *23*, 1388.
16. Liu, G.; Guillet, J. E.; Al-Takrity, E. T. B.; Jenkins, A. D.; Walton, D. R. M. *Macromolecules* **1990**, *23*, 1393.
17. Mahaney, M.; Huber, J. R. *J. Mol. Spectrosc.* **1981**, *87*, 438.
18. Ito, S.; Kanno, K.; Ohmari, S.; Onogi, Y.; Yamamoto, M. *Macromolecules* **1991**, *24*, 659.
19. Barbara, P. B.; Walsh, P. K.; Brus, L. E. *J. Phys. Chem.*, **1989**, *93*, 29 and references therein.
20. Lakowicz, J. R. *Principles of Fluorescence Spectroscopy*; Plenum: New York, 1983.
21. Gillispie, G. D.; Van Benthem, M. H.; Connolly, M. A. *Chem. Phys.* **1986**, *106*, 459.
22. Carter, T. P.; Van Benthem, M. H.; Gillispie, G. D. *J. Phys. Chem.* **1983**, *87*, 189.
23. Balakrishnan, N.; Gillispie, G. D. *J. Phys. Chem.* **1989**, *93*, 2337.
24. Weber, G.; Farris, F. J. *Biochemistry* **1979**, *18*, 3075.

25. Ham, J. S. *J. Chem. Phys.* **1953**, 21, 756.
26. Kalyanasundaram, K.; Thomas, J. K. *J. Am. Chem. Soc.* **1977**, 99, 2039.
27. Amirav, A.; Jortner, J. *Chem. Phys. Lett.* **1983**, 94, 545.
28. Demas, J. N.; Crosby, G. A. *J. Phys. Chem.* **1971**, 75, 991.
29. Wrighton, M. S.; Ginley, D. S.; Morse, D L. *J. Phys. Chem.* **1974**, 78, 2229.
30. Hamai, S.; Hiriyama, F. *J. Phys. Chem.* **1983**, 87, 83.
31. Johnston, D. B.; Lipsky, S. *J. Phys. Chem.* **1991**, 95, 3486.
32. Demas, J. N. *Excited State Lifetime Measurements*; Academic: Orlando, FL, 1983.
33. Bright, F. V.; Betts, T. A.; Litwiler, K. S. *Crit. Rev. Anal. Chem.* **1990**, 21, 389.
34. Barrow, G. M. *Physical Chemistry, Fifth Edition*; McGraw-Hill: New York, 1988.
35. Strickler, S. J.; Berg, R. A. *J. Chem. Phys.* **1962**, 37, 814.
36. Agarwal, U. P.; Jagannath, H.; Rao, D. R.; Rao, C. N. R. *J. Chem. Soc., Faraday Trans. 2* **1977**, 73, 1020.
37. Robinson, G. W.; Frosch, R. P. *J. Chem. Phys.* **1962**, 37, 1962.
38. Freed, K. F. *Acc. Chem. Res.* **1978**, 11, 74.
39. Siebrand, W. *J. Chem. Phys.* **1967**, 47, 2411.
40. Gillispie, G. D.; Lim, E. C. *Chem. Phys. Lett.* **1979**, 63, 193.
41. Gillispie, G. D.; Lim, E. C. *J. Chem. Phys.* **1978**, 68, 4578.
42. Gillispie, G. D.; Lim, E. C. *J. Chem. Phys.* **1976**, 65, 2022.
43. El-Sayed, M. *J. Chem. Phys.* **1963**, 38, 2834.
44. Tsai, T. C.; Robinson, G. W. *J. Chem. Phys.* **1968**, 49, 3184.
45. Gillispie, G. D.; Lim, E. C. *J. Chem. Phys.* **1976**, 65, 2022.
46. Sension, R. J.; Repinec, S. T.; Hochstrasser, R. M. *J. Chem. Phys.* **1990**, 93, 9185.
47. Birch, D. J. S.; Birks, J. B. *Chem. Phys. Lett.* **1976**, 38, 432.

RECEIVED for review April 15, 1991. ACCEPTED revised manuscript October 5, 1991.

Laser Fourier Transform Mass Spectrometry for Polymer Characterization

J. Thomas Brenna[1], William R. Creasy[2], and Jeffrey Zimmerman[2]

[1]Division of Nutritional Sciences, Cornell University, Savage Hall, Ithaca, NY 14853
[2]IBM Corporation, P.O. Box 8003, D/T67, Endicott, NY 13760

This review focuses on laser-based Fourier transform ion cyclotron resonance mass spectrometry (FTMS) for polymer structure determination and identification of industrially important polymers on surfaces. In structural studies, laser desorption has been used as a gentle ionization technique to desorb and ionize intact polymer molecular ions of intractable polymers. Spatially resolved studies involving higher laser fluences cause more fragmentation and recombination but retain sufficient information to permit identification at spatial resolution of about 10 μm. Molecular fragments appear most often in negative ion spectra. Recombination products and carbon clusters are prominent in positive ion spectra. Three distributions of carbon clusters are observed in separate mass ranges, the highest of which are identified as fullerenes first observed in laser ablation of graphite. Ongoing advances such as postionization techniques are expected to make laser FTMS an increasingly attractive and convenient tool for polymer analysis.

COMBINING LASER-INDUCED VAPORIZATION with some form of mass spectrometry to analyze solids has been practiced for over 20 years. In fact, an exhaustive bibliography of the field, now 6 years old, contains 1461 references (*1*), 60 of which are to polymer-related papers. In a number of fields, laser mass spectrometry has become the analytical method of choice including, for

0065–2393/93/0236–0129$07.50/0

instance, particle isotopic analysis. The technique continues to develop and provide answers to a widening array of questions even though it is not yet a routine tool in polymer analysis. Ongoing investigations of novel instrumentation with strategic improvements over older designs show promise for difficult analyses. In particular, the introduction of high performance analytical Fourier transform mass spectrometry (FTMS) permits many types of experimental strategies to be executed for polymer analysis that were impossible with previous systems. Although this work is in its early stages of development for polymer applications, experiments to date have yielded useful results. We present here a representative (not exhaustive) review of the field, including discussions of instrumentation for spatially resolved analysis developed in the recent past and several applications from our laboratory and others.

Laser Mass Spectrometry. Two broad classes of laser mass spectrometry experiments have developed independently. They are most conveniently classified as (1) laser-microprobe experiments, which require spatial resolution and are aimed at identification and localization of polymeric materials on the surface, and (2) laser-desorption experiments, the goal of which is to maximize the amount of structural information, including molecular weight distributions, in the mass spectrum. The two experiments are not mutually exclusive, but competing analytical issues most often force a choice between the two analysis modes for optimization. FTMS has been applied to both sorts of experiments. The traditional commercial instrumentation for laser microprobe, which is based on time-of-flight analysis that has existed for a decade, will be described briefly.

Laser-microprobe mass spectrometry (LAMMS) has been used for the past 20 years for elemental and molecular identification of solids. The field is referred to by a number of names, which include the earliest, laser-microprobe mass analysis (LAMMA) and laser ionization mass analyzer, both of which are now associated with commercial instruments. Fundamentally, LAMMS is a spatially resolved surface or near-surface (few nanometers to micrometers) sensitive technique with the chief advantage of applicability to all solids regardless of electrical conductivity. The technique was developed initially to serve the needs of the biomedical research community for localization of easily ionized elements, such as the alkali metals, alkaline earths, or halides. As commercial instrumentation appeared in the late 1970s and early 1980s, the technique was applied to a wide variety of solids. Polymers were among the first class of engineering materials to be analyzed by LAMMS (2–5). These early applications focused on the ability of the technique to provide fingerprint mass spectra for identification of small quantities of polymeric material (1 pg) on surfaces as well as to provide polymer structural information.

From the first instruments reported in the early 1970s until 1988, LAMMS systems were based almost exclusively on time-of-flight (TOF) mass spectrometry. In the LAMMS experiment, a beam of laser light (usually but not exclusively in the UV) is focused to a small spot [typically 10 μm or less (6)] and directed onto a solid surface. As a result of this single laser pulse, ions are ejected from the surface and extracted into a mass spectrometer. The laser pulse is typically about 10-ns duration, and ion emission is usually on this order. Therefore, scanning quadrupoles or magnetic sector (without array detectors) mass spectrometers cannot capture a complete mass spectrum from a single pulse. The TOF instrument does acquire an entire mass spectrum, with a theoretically unlimited mass range, from a single pulse.

To minimize fragmentation, the laser power is often adjusted to the minimum level at which ion emission occurs. However, because of the requirement for spatial resolution, ionization efficiency must be high, and the minimum power level usually causes extensive damage and molecular rearrangement during the vaporization process. This mode of laser vaporization is often referred to as "plasma ionization" (7) or "ablation" (8) to differentiate it from gentler laser experiments. Gentle ionization, used in "desorption" experiments, tends to be less efficient and, therefore, must sample a wider surface area to produce sufficient signal.

Positive ion TOF−LAMMS spectra are generally characterized by extensive fragmentation; only major structural moieties, such as aromatic rings, appear in the spectra. Often, low-mass hydrocarbon ions appear up to mass ~ 200 μ, with scattered structure-specific peaks. Because the energy spread of ions emerging from the surface is high (up to hundreds of electronvolts), the mass resolution of TOF−LAMMS spectra is rarely better than 1 μ. In many cases the poor mass resolution causes hydrocarbon peaks to obscure minor peaks of greater structural significance. It is primarily for this reason that TOF−LAMMS polymer spectra rarely are used for structural studies and only can be used to distinguish polymers with major structural differences by gross fingerprint. In fact, recent reports suggest pattern-recognition techniques are most effective for this task (9, 10). These limitations suggest that a laser microprobe interfaced to a more powerful form of mass spectrometry will yield more detailed information.

Fourier Transform Ion Cyclotron Resonance Mass Spectrometry. FTMS in its present form was first demonstrated in 1974 by Comisarow and Marshall (11), and commercial instrumentation first appeared in 1981. Outstanding reviews of the fundamentals of the technique are available (12−17). Briefly, FTMS is an ion-trapping technique in which ions formed by any means are trapped in an intense magnetic field [typically 3 tesla (T)] crossed with a weak electric field (1−10 V). It is a fact of elementary physics that charged particles in a uniform magnetic field orbit with a characteristic motion. This orbital frequency is termed the "cyclotron" frequency, and in a static magnetic field it depends inversely on the mass-to-charge ratio of the particle as

$$w = \frac{qB_e}{m}$$

where w is the cyclotron frequency, q is the electrical charge carried by the ion, m is the ionic mass, and B_e is the static magnetic field. In the FTMS experiment, the frequency is independent of the ion kinetic energy and dependent only on m/e, to the first order. (Accurate mass calibration requires that a second-order term involving the electric field be included.) This fact permits high- and ultrahigh-resolution measurements of the cyclotron frequency and hence the ion mass. Commercial instrumentation achieves mass resolution in excess of 1 part in 1 million and accurate masses to < 1 ppm. These capabilities allow resolution of ions at the same nominal mass and assignment of elemental composition on the basis of mass alone. Furthermore, because the FTMS is an ion trap, a complete mass spectrum can be obtained from a single pulsed event such as a laser pulse. Complex manipulations of ions also are possible, for example, selective ion ejection and collision-induced dissociation, or photodissociation with a second laser. All these capabilities are used to derive structural information about ions.

Prior to the availability of commercial instrumentation, FTMS with home-built systems became one of the methods of choice in the field of gas phase ion/molecule chemistry. Since the advent of commercial instruments in 1981, gas (18), liquid, and supercritical fluid chromatography (19) interfaces have been developed for chemical applications. Numerous sources for high molecular weight analysis have been designed and implemented for FTMS including Cs ion secondary ion mass spectrometry (SIMS) (20, 21) and fast atom bombardment; ^{252}Cf fission fragment plasma desorption (22–24); and electrospray ionization (25–27). For the most part, these sources have been applied to biological samples, particularly peptides, nucleotides, and oligosaccharides, although a polystyrene spectrum has been reported for the SIMS source (20).

Laser sources have been used in conjunction with FTMS since the 1970s, and a CO_2 laser (λ = 10.6 μm) source was available soon after introduction of the first commercial instrument. This source directed a broad beam of ~ 10-μm diameter onto the surface, but had no provision for in situ specimen viewing.

FTMS–LAMMS. In 1988, three laboratories, working independently, reported the design of a laser-microprobe system for Fourier transform mass spectrometry (FTMS) (28–30). The instruments at IBM-Endicott and IBM-San Jose have similar optical paths. The University of Metz system (30) is a very different design that uses an excimer laser (KrF; 247 nm) with a helium–neon pilot laser for visualization of the laser ablation spot, Cassegrain optics for focusing the laser onto the specimen, and different provisions for specimen handling. We will discuss in detail the IBM-Endicott instrument (28), the optics of which are similar to a commercial system offered by Extrel-Millipore FTMS (Madison, WI).

Instrumentation. A schematic of the instrument is shown in Figure 1. An Nd:YAG laser system (Quantel International, which is now Continuum, Santa Clara, CA) operated at $\lambda = 266, 532$, or 1064 μm is directed through an optical attentuator, which permits continuous adjustment of power level without moving the beam. The output passes through a telescope for fine focusing, and then enters the vacuum system. The beam passes through a 75-mm focal length objective lens, reflects off a mirror, passes through the space between the cell plates, and is focused onto the sample surface. The sample is positioned about 3 mm from one of the trap plates, and, during ablation, ions enter the FTMS cell and are trapped for analysis. A separate high-vacuum compatible fiber optic directs illumination light through the cell and onto the sample at the corner opposite to the entry of the laser light.

For viewing, light reflected off the sample follows the laser optical path in the opposite direction. After leaving the vacuum chamber, light is directed to an ocular piece for viewing by a sliding mirror. Alternatively, a dichroic mirror can be used instead of the fully reflecting mirror to allow the sample to be viewed during the 266-nm ablation. Typically, an uninteresting area of the sample is laser-ablated several times to make a crater to allow determination of the precise position of the beam. The crater then can be located

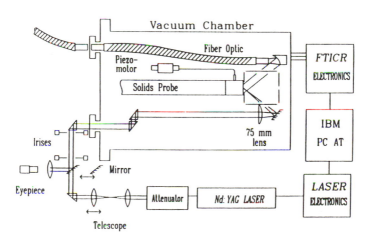

Figure 1. Schematic diagram of the IBM-Endicott FTMS–LAMMS instruments. Details given in the text. (Reproduced with permission from reference 46. Copyright 1989 San Francisco Press, Inc.)

relative to crosshairs in the ocular piece and interesting areas can be positioned for analysis.

Briefly, we mention two systems in addition to the optics that are necessary for operation of the microprobe. The electronics of the laser and the FTMS must be timed under computer control for optimal performance of each device. Communication between the electronics is mediated by a personal computer (IBM PC AT) equipped with a laboratory interface board. The board contains programmable counter/timers to trigger the laser flash-lamps and the Q-switch. The PC is interfaced to the FTMS computer (Nicolet 1280) via an RS232 line, and interfaced to the laser electronics by three lines that trigger pulse sequences. This arrangement permits the FTMS electronics to demand a laser pulse with 1-μs accuracy, which is necessary for carefully timed pulse sequences.

Specimen motion is another issue of importance for the FTMS micro-probe. From a handle outside the vacuum system, the sample can be moved by rotation of the solids probe on which it is mounted. This motion does not affect the focus of the laser beam on the sample surface. A second degree of freedom can be conveniently obtained by adjusting the sample distance from the trap plate, which has the effect of sweeping the probe beam across the surface because it impinges on the surface at an angle. This convenience comes at a price, however, because the beam focus is compromised when this stage position is adjusted. We have pursued a superior but mechanically more complex approach in which a miniature spring-loaded sample stage is moved in the lateral direction by a plunger system. This approach works well in practice but is somewhat delicate.

Structural Studies: Laser-Desorption FTMS

Rapid and sensitive analysis of polymer molecular structure is critical to determination of functional properties. Specific structural data provide information for optimizing synthetic schemes and understanding the relationships of macroscopic properties to molecular characteristics. This information is the primary objective of laser desorption (LD) experiments, although it is obtained occasionally with microprobe instruments. Instrumentation for LD–FTMS is similar to that previously described for FTMS–LAMMS except that no provision is made for specimen viewing or precise laser focusing.

Structure/Cure Conditions: Polycyclic Polymers.   In a series of several papers, C. Brown and co-workers (31–38) reported substantial structural and mechanistic information for a variety of polymers analyzed in LD–FTMS. These studies were conducted with a commercial system interfaced to a CO_2 laser ($\lambda = 10.6$ μm) and were specifically aimed at obtaining structural information without regard to spatial resolution. This work repre-

sents the most comprehensive series of polymer FTMS experiments published to date.

The most carefully studied polymer among those considered by C. Brown et al. is poly(*para*-phenylene) (PPP), an intractable highly conjugated polymer that is resistant to environmental attack and that attains electrical conduction upon doping. The structure and molecular weight of PPPs prepared by various routes with various monomers were investigated by LD–FTMS. Sample preparation consisted of producing a pellet of PPP by compression with a 10-lb sledgehammer.

PPP is desorbed predominantly as molecular ions (without fragmentation), which means that oligomer distributions are directly determined with each laser pulse and number and weight average molecular weights can be calculated from the mass spectra. Figure 2 is a section of the mass spectrum from a PPP 12-mer, wherein the unbranched dodecamer (12-mer) H$-(\phi_{12})-$H appears at mass 914. The peaks appearing at masses 912 and 913 are apparently analogues of this molecule, which are deficient in two hydrogen atoms (at mass 912), plus its ^{13}C isotope peak. The peaks provide evidence for a chain termination structure generated by intramolecular ring closure as shown in Structure 1. The proposed PPP end-group configuration shown in Structure 1 is based on the mass spectrum in Figure 2. The data also show variable degrees of halogenation dependent on synthetic details. These studies led to the conclusion that electrical conductivities of various

910   920   930   940   950

Figure 2. Section of the positive ion LD–FTMS mass spectrum of poly(para-phenylene) (PPP) in the region of the dodecamer (12-mer).

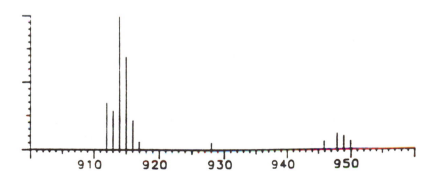

Structure 1. PPP end-group configuration.

polymers cannot be explained by models of electron interaction that rely exclusively on homogeneous linear PPP structures. Similar conclusions were reached in studies of heterocyclic aromatic polymers (with N, S, and Se substitution). In related LD–FTMS spectra of aromatic polymers containing heteroatoms [e.g., poly(phenylene-sulfide)], high-mass carbon clusters (n = 150–400) that resemble fullerenes (discussed later) were observed. No anomalously stable ions were observed in these spectra, although ions did appear at exclusively even masses as is observed in the laser vaporization of graphite.

Extent of Cure: Polyimide. The extent of cure is often monitored spectroscopically; for example, by selected monitoring of IR bands specific to the uncured or cured material. These experiments often are limited in sensitivity and reveal differences in cure state greater than 1%; however, unwanted selectivity based on three-dimensional bond orientation in space may be exhibited. Although laser microprobe FTMS polymer spectra are complex (to be discussed), they are also highly sensitive to subtle details of polymer structure and may be used as a sensitive monitor of cure state.

An example of the use of 266-nm FTMS–LAMMS to distinguish poly-imide cure state was reported by Creasy and Brenna (39, 40). Polymerized [molecular weight (MW) ≈ 25,000] but uncured polyamic acid of pyromellitic dianhydride–oxydianiline (PMDA–ODA) was compared to the cured poly-imide. The structures of polyamic acid and polyimide are shown in Structures 2 and 3, respectively. The negative ion mass spectra are shown in Figures 3a and 3b, respectively. The degree of polymerization is identical for cured and uncured polymer because the curing process involves a dehydration reaction with accompanying changes in intermolecular packing. However, the mass spectra are quite different because many higher mass peaks are present from ablation of polyamic acid than from the polyimide. The extra peaks in Figure 3a can be assigned to fragments of the polymer chain smaller than a monomer (40). These negative ions may be stabilized by a large number of carbonyl or carboxyl groups. The polyimide spectrum in Figure 3b, on the other hand, contains only the characteristic CN^-, OCN^-, and C_nN^- peaks,

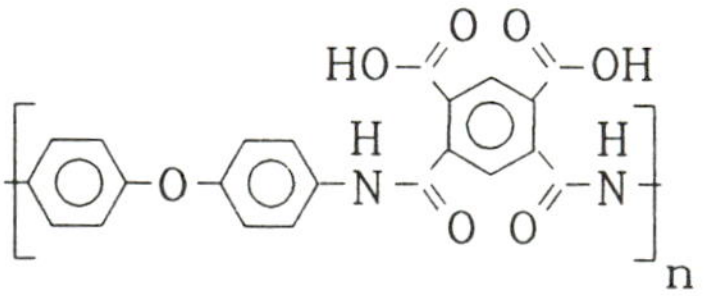

Structure 2. Polyamic acid.

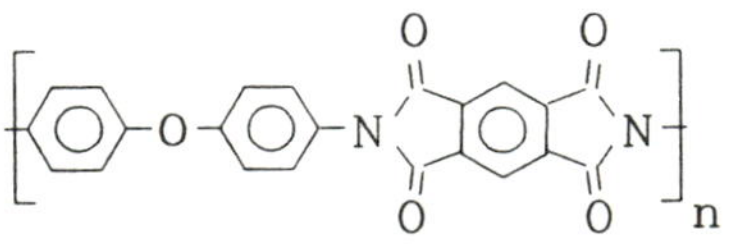

Structure 3. Polyimide.

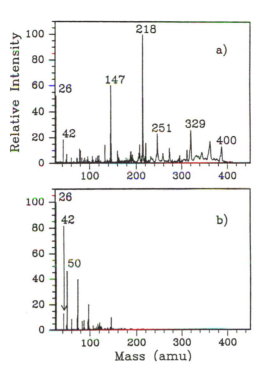

Figure 3. Negative ion LAMMS–FTMS spectra of pyromellitic dianhydride–oxydianiline (PMDA–ODA) polyamic acid (a) and polyimide (b). High-mass ions can be assigned to fragments of the polymer. (Reproduced with permission from reference 39. Copyright 1988 Elsevier.)

along with some carbon cluster peaks. The reason for the smaller number of fragment peaks for the imide is not known, although it may be related to the greater structural rigidity of the imide compared to the acid, which may give rise to greater fragmentation. These studies show that molecular information about the polymer morphology is obtained from the FTMS–LAMMS mass spectrum.

Molecular Weight Distributions. In addition to the oligomer distributions described for PPP, several other groups have published mass spectra from which molecular weights can be calculated. R. Brown and co-workers (36) have published spectra of polyethylene glycols (PEG) of several molecular weights (600, 1000, 1450, 3350, 6000), PEG methyl ester (PEGME; MW = 5000), polystyrene (PS; MW = 2000), poly(caprolactone-diol) (MW = 2000), poly(propylene glycol) (PPG; MW = 4000), and poly(ethylenimine) (PEI; MW = 600, 1200). In most cases, number and weight average molecular weights are within 10% of the values obtained by conventional methods, and in no case is the variance more than 20%. In related work, molecular weight distributions of alcoxylated pyrazole and hydrazine polymers (*41*) (with varying degrees of polymerization) yield molecular weights that compete favorably with weights obtained via other methods. Copolymer distributions were also reported by this method (*42*).

It should be emphasized that the molecular weight *distributions* obtained in these analyses reveal far more than average molecular weight. Some oligomers appear more abundant than their nearest neighbors, which may reveal subtle property differences induced by details of the synthesis. We conclude from the literature that molecular weight distributions can be obtained for a wide range of polymers using CO_2 laser desorption–FTMS for polymers of modest molecular weight.

The preceding studies were conducted with a 3-T magnet, which is the most common magnet field employed in FTMS at this time. In a study specifically aimed at testing the mass range of a 7-T magnet, molecular weight 8000 PEG was desorbed by a CO_2 laser (*43*). Ions up to mass 10,000 were observed, and the spectrum peaks around mass 8600 (molecular weight calculations were not reported). The mass limit of FTMS is more correctly reported as an m/e limit, which is on the order of about 38,000 for thermal ions in a 3-T magnet (*44*). At least one technique, electrospray ionization (to be discussed), routinely results in multiply charged ions (up to 150^+ have been reported) that effectively extend the mass range by that factor. For instance, ions of mass 100,000 with a charge state of 100^+ would appear at $m/e = 1000$. These techniques may permit much higher molecular weight distributions to be determined. However, care must be taken for quantitative determinations of distribution at high accuracy using known standards of the same polymer. Several effects, such as a dependence of cation attachment or volatility on molecular weight, may shift the measured distributions systematically. Instrument-dependent artifacts, such excitation or detection conditions, and even the positioning of the sample near the FTMS cell may affect the measurement. Investigations of these issues are beginning to appear (*45*).

Structural Studies: Laser-Microprobe FTMS

In a series of papers published over the past few years, we have described mass spectra generated from laser-microprobe analysis of polymers with the IBM-Endicott FTMS–LAMMS instrument (*8, 39, 40, 46–49*). The aim of these studies is to characterize the distribution of ions formed by focused laser irradiation of well-defined polymers, with the goal of *identification* of unknown polymeric particulates. For this reason, the laser spot size is never larger than 10-μm diameter, and spectra are recorded as a result of the lowest irradiance level that generates sufficient signal. This data is also of interest in studies of the economically important but poorly understood process of laser ablation/drilling, which finds application in fields as diverse as retinal surgery and printed circuit board fabrication. We discuss here the major features of mass spectra observed in these analyses, organized by phenomenon.

Negative Ions: Structure. Most early studies of LAMMS of polymers have focused on positive ion spectra, though negative ion spectra usually contain more structural information. An excellent example of this is the case of poly(methyl methacrylate) (PMMA). The ablation characteristics of PMMA are somewhat unusual in that no signal is obtained during the first hundred or so laser pulses. This is consistent with observations of UV–laser etching of PMMA (50). The negative ion mass spectrum of PMMA is shown in Figure 4. Almost all the ions in this spectrum can be assigned to fragments formed by direct scission of the polymer chain. Assignments are given in Table I. No higher mass ions are observed in this spectrum. This spectrum is in sharp contrast to that observed for positive ions, which is discussed in more detail in the following text.

Odd Mass Ion Series and Stable Subunits. A positive ion spectrum of polyethylene glycol (PEG) is shown in Figure 5. This spectrum is very different from many PEG spectra presented previously in the literature and highlights the stark contrast between experiments that require spatially resolved information and those optimized for structural information. Previous

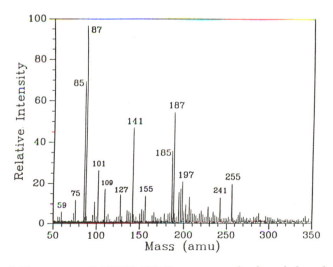

Figure 4. Negative ion LAMMS–FTMS spectrum of polymethyl methacrylate (PMMA) showing primarily structural ions (Reproduced with permission from reference 49. Copyright 1991 Society for Applied Spectroscopy.)

Table I. Fragment Ions Observed in Negative-Ion Spectrum of PMMA[a].

M/E	Structure
59	$CHCOO_3$
85	$COOC(CH_3)CH_2$
87	CH_3COOCH_2CH
101	$CH_2C(CH_3)(COOCH_3)-H[\{R\}-H]$
127	$R-CH_2C$
141	$R-CH_2CCH_3$
155	$R-CH_2C(CH_3)CH_2$
185	$R-CH_2C(CH_3)COO$
187	$C(CH_3)COOCH_3-RH$
197	$HC(CH_3)(CO)-R-CH_2CCH_3$
241	$H-CCO-(R_2H)$
255	$C(CH_3)(CO)-R_2$

[a] $CH_2C(CH_3)(COOCH_3) \equiv R$ (repeat unit).

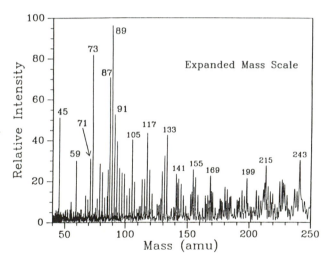

Figure 5. Positive ion LAMMS–FTMS spectrum of polyethylene glycol (PEG). The spectrum is dominated by an odd-mass ion series consisting of rearrangement products. (Reproduced with permission from reference 49. Copyright 1991 Society for Applied Spectroscopy.)

spectra concentrated on desorption of intact *n*-mers in an attempt to observe distributions present in the solid (*36, 43*). By defocusing the laser material is gently desorbed. We have accomplished desorption with lower molecular weight PEG (Creasy, W. R.; Brenna, J. T., unpublished observations). Due to differences in absorption, laser wavelength (UV versus IR) probably plays a major role in the way polymers are desorbed from the surface. However, the requirement for fine focus imposed by geometric constraints or the requirement for high ionization efficiency because of limited sample size impose the need for high power densities that cause a high degree of fragmentation.

The mass spectrum is dominated by odd-mass ions in the mass range $m/z = 40–170$. Each peak corresponds to the formula $H(CH_2)_xO_y^+$. There are no other combinations of C, H, and O, the constituent elements of PEG that fit this ion series. This formula suggests that the laser-induced plasma consists of major proportions of CH_2 and O, perhaps in radical form, that condense and protonate upon cooling and give rise to the observed spectrum. Similar effects, which give rise to different products, are observed in positive ion spectra generated from poly(vinyl acetate) (PVAc), polystyrene (PS), poly(methyl methacrylate) (PMMA), and under special conditions of irradiation of PPS. In each of these cases the spectra are distinct and easily differentiated from one another.

The positive ion spectrum of PMMA is presented in Figure 6. This spectrum is dominated by a dense series of rearrangement products of the type observed for PEG, along with intermediate-mass carbon clusters. The observed series fits the elemental formula $HCO_2(CH_2)_m–C_n$, which is written to suggest that an acid series may be coupled to a bare carbon cluster series. Other structures are likely, including a highly unsaturated series.

Reports of other positive ion studies of PMMA using a CO_2 laser (no spatial resolution) have appeared. C. Brown et al. mixed PMMA with KBr to

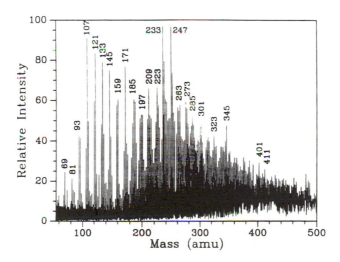

Figure 6. Positive ion LAMMS–FTMS spectrum of PMMA. The spectrum is dominated by odd-mass ions and a series of intermediate mass carbon clusters. (Reproduced with permission from reference 49. Copyright 1991 Society for Applied Spectroscopy.)

promote cation attachment, and oligomers up to the 22-mer, with some evidence of hydrogen rearrangement (37), were reported. Hsu and Marshall studied polymer dyes in PMMA and obtained ions from the host polymer (51). These authors report protonated and cationized dimer, trimer, and tetramer ions as the dominant species in the spectrum. Nuwaysir et al. (42) failed to observe odd-mass ion series in their studies of methacrylate copolymers. The absence of odd-mass ion series in these spectra may reflect the higher energy density required for efficient ionization conditions in the laser-microprobe experiment.

Carbon Clustering.   Carbon cluster formation is a common observation in LAMMS–FTMS of polymers. Although no theories have been advanced to relate carbon cluster formation to polymer structure, it is of utility for fingerprint purposes.

The most representative case of carbon cluster formation is that of polyimide (PI). The negative ion spectrum of PI was discussed previously; the positive ion mass spectrum of PI is shown in Figure 7. Although there are over 200 peaks in this spectrum, no ions characteristic of the structure of PI are observed. The spectrum consists exclusively of carbon cluster ions, some of which appear associated with H atoms. Separate distributions of carbon clusters appear in three distinct mass ranges. We have labeled each region as low, middle- (or intermediate-), and high-mass distributions (39, 46). Each of these distributions has been observed in spectra of polymers of widely varying covalent structure, although PI is the only polymer for which all three

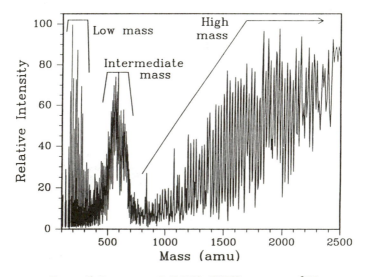

Figure 7. Positive ion LAMMS–FTMS spectrum of PI.

distributions are observed in the same spectrum. Further, we note that comparison of these carbon clusters with published spectra of carbon clusters formed upon laser ablation of graphite, with or without cooling using a pulsed valve, reveals a high degree of correspondence for the low- and high-mass distributions. Observation of spectra from a solid of high heteroatom content such as polyimide, which is indistinguishable from that for pure carbon, is evidence for similar mechanisms and a strong driving force for carbon cluster formation (*48*).

Low-Mass Carbon Clusters. An expanded section of the positive ion spectrum showing the low-mass distribution is shown in Figure 8. Carbon cluster ions appear at every carbon number from C_{10} through C_{25}, and are often associated with a peak at 2 μ in excess of the pure carbon peak. Peak intensities show distinct maxima with a period of four carbons and peaks at 11, 15, 19, and 23. This is precisely the same pattern observed with carbon clusters generated by graphite ablation. Ablation of perdeuterated PI reveals that the peaks shifted out 2 μ are due to selective addition of two hydrogen atoms, rather than substitution of a nitrogen atom, which is also present in abundance in the laser-induced plasma (*47*).

Intermediate-Mass Carbon Clusters. This distribution of clusters is shown in Figure 9. There is a peak at every mass over the range $m/\delta m = 150\text{–}400$. The intensities oscillate with a period equal to 12 μ. Valleys occur at mass numbers corresponding to bare carbon clusters, and peaks corre-

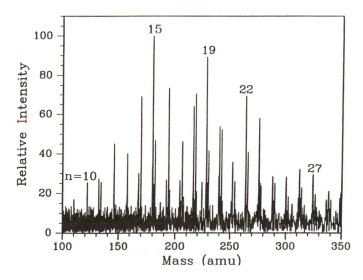

Figure 8. Low-mass carbon clusters in the position ion spectrum of PI. (Reproduced with permission from reference 39. Copyright 1988 Elsevier.)

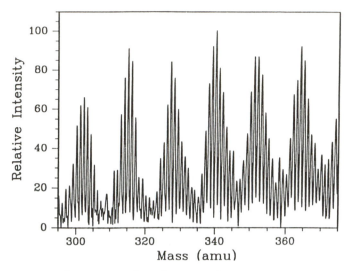

Figure 9. Intermediate-mass carbon clusters in the positive ion spectrum of PI.
(Reproduced with permission from reference 39. Copyright 1988 Elsevier.)

spond to addition of six to seven hydrogen atoms. The composition $C_n H_{1-12}$ can then be assigned to these ions. Identical experiments with perdeuterated PI confirm these observations (47). This distribution is not observed for graphite ablation because no hydrogen atoms are present. It is interesting to note that the valleys in the present spectrum occur at the bare carbon clusters, which is consistent with the formation of some type of polyaromatic hydrocarbon. This distribution is the most common distribution observed in FTMS–LAMMS polymer spectra. In the PEG spectrum discussed previously, the odd mass ion series merges at higher masses with intermediate-mass carbon cluster ions similar to those observed for PI. Intermediate-mass carbon clusters also occur for PVAc, PMMA, and poly(vinyl chloride) (PVC).

High-Mass Carbon Clusters. Figure 10 gives the distribution of high-mass carbon clusters from PI. Ions appear at exclusively even carbon numbers and range out past mass 5000, where the sensitivity of the instrument falls off. This set of ions appears to be identical to the ion set observed for laser ablation of graphite, which are now commonly referred to as fullerenes (52, 53). C. Brown and co-workers (38) also reported spectra that included even-numbered clusters for some polymers.

Magic Numbers. Magic numbers are defined as those carbon cluster numbers that exhibit anomalous stability relative to neighboring clusters. No magic number were observed for PI or any of the polymers studied by C. Brown and co-workers (38).

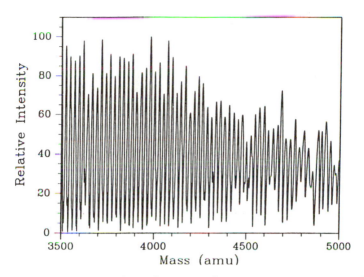

Figure 10. High-mass carbon clusters in the positive ion spectrum of PI. (Reproduced with permission from reference 39. Copyright 1991 Elsevier.)

A positive ion mass spectrum of poly(phenylene sulfide) (PPS) is shown in Figure 11. The sole signal observed in this spectrum arises from the high-mass carbon cluster distribution. This spectrum differs from the PI spectrum because C_{60} at 720 is observed at anomalously high intensity. This molecule, whose structure has been confirmed to be that of a truncated icosahedron (or soccer ball), has been named buckminsterfullerene and is currently the subject of tremendous interest. Buckminsterfullerene was first observed during laser ablation of graphite (53, 54), and is known to be very stable due to a closed shell structure. Its formation in the presence of heteroatoms (e.g., sulfur) is evidence for a strong driving force for formation, as well as stability once generated.

In over 20 polymers studied thus far, the variety of phenomena observed has permitted differentiation of FTMS–LAMMS spectra by inspection of positive ion spectra. In TOF–LAMMS differentiation was performed most successfully using computer-based pattern-recognition techniques, primarily because TOF–LAMMS positive ion spectra are dominated by low-mass hydrocarbons for most carbon-based polymers. Differences in extraction conditions and plasma interactions in the two experiments may cause this domination. It is clear that FTMS–LAMMS has unique capabilities for spatially resolved analysis of polymeric particles.

The generation potential for oligomer distributions by FTMS–LAMMS has not been fully investigated; it may be possible, with only a modest sacrifice of spatial resolution, by the use of other wavelengths (e.g., 532 or 1064 nm) available from a Nd:YAG laser. The full capabilities of FTMS, such

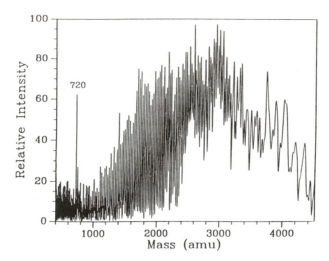

Figure 11. Positive ion LAMMS–FTMS spectrum of poly(phenylene sulfide) (PPS). The spectrum consists entirely of fullerenes. Buckminsterfullerene appears as a magic number. (Reproduced with permission from reference 49. Copyright 1991 Society for Applied Spectroscopy.)

as tandem MS and ultrahigh-mass resolution, have not been exploited to definitively assign ion structure in laser-microprobe studies. Broad ion kinetic energy distributions resulting from laser ablation may give rise to unexpected effects and may contribute to difficulties in obtaining high-performance mass spectra. Such problems are known to exist for electrospray ionization FTMS of proteins and were addressed experimentally before high-resolution spectra were obtained by this technique (26). Means to reduce the ion kinetic energies in laser ablation studies are under investigation.

Contaminant Analysis

Analysis of polymer products often involves the detection of impurities or contaminants that could affect performance. LAMMS methods allow contamination analysis of small spatial locations and can also provide depth profiling of polymer films. Either technique can be helpful for understanding problems associated with manufacturing processes.

An example of a depth profile that was made by FTMS–LAMMS is shown in Figure 12 (40). The figure shows depth profiles of K^+ in a polyimide film. The samples were prepared by immersing a free-standing film of PMDA–ODA polyimide in a 1.0 M KCl solution. The water and K^+ ion diffused into the film from the solution. The depth profiles of K^+ were measured after 0.5 and 20 h in the solution. The signal level of the K^+ ion is

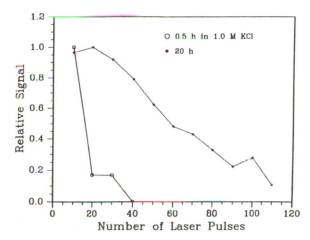

Figure 12. Depth profiles of K^+ ions in a polyimide film. The K^+ signal was detected using LM–FTMS by signal-averaging every 10 laser pulses. Ten pulses corresponds to a depth of 0.5–1.0 μm. Eight runs were averaged for each curve.

detected with the FTMS by ablating the polyimide by sequential laser pulses on one spot. Ten laser pulses were averaged for each data point. Eight runs from different spots were averaged to obtain the data in Figure 12. By measuring the crater depth, it was found that 10 laser pulses correspond to a depth of 0.5–1.0 μm. Figure 12 shows that after 0.5 h, the K^+ is near the surface of the film, and after 20 h, the K^+ has penetrated more than 5–10 μm into the film. Because the crater that is made by the laser has a rounded rather than a flat bottom, the depth profile cannot be used to obtain quantitative diffusion information, although it does demonstrate that diffusion occurs on such a time scale.

LAMMS is quite sensitive to K^+ and other alkali metal ions because of their low ionization potential. Similarly, it is also sensitive to ionic organic species. For example, Figure 13 shows a spectrum from a film of cadmium arachidate ($Cd^+C_{19}H_{39}COO^-$). The film is a five-monolayer-thick Langmuir–Blodgett film (55). Judging from the signal-to-noise ratio of the spectrum, it should be possible to detect a single monolayer. Other ionic surfactants, such as alkyl sulfonates, can be detected with good sensitivity. For example, we have detected a perfluorinated alkyl sulfonate contaminant in a 2-nm polyimide layer.

On the other hand, nonionic surfactants or impurities are more difficult to detect in low concentrations or thin films, probably because they are laser desorbed as neutrals rather than as ions. Asamoto and co-workers (56) analyzed additives in polyethylene by extracting the additives with solvent prior to LD–FTMS. Blease and co-workers (57) studied polymer additives

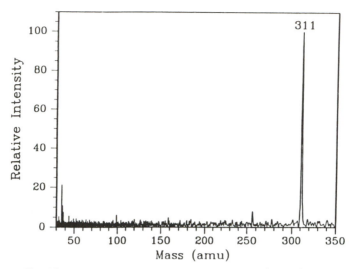

Figure 13. Negative ion LM–FTMS spectrum of a five monolayer Langmuir–Blodgett film of cadmium arachidate using 266-nm laser radiation. The peak at mass 311 corresponds to the arachidate anion.

with LD–FTMS and reported collisionally activated dissociation (CAD) results. As previously mentioned, Hsu and Marshall reported the detection of dyes in PMMA to concentrations of 0.1% (51). Liang and co-workers determined the molecular weight distributions of some pure samples of nonionic ethoxylate-based surfactants (58). Xiang and co-workers studied phosphite additives as antioxidants using laser desorption–electron-impact ionization, both as neat samples and in polymers at concentrations of 0.1% (59). Finally, Johlman and colleagues (60) analyzed a number of different additives, principally by mixing pure or solvent-extracted samples with a potassium salt for cation attachment. However, detection of nonionic additives or curing agents at low concentration in a polymer matrix is generally a difficult problem that must be attempted on a case-by-case basis.

In general, analysis of unknown polymers or quantitative analysis using LAMMS involve some fundamental difficulties. One problem is due to the different optical absorption characteristics of different materials. The different responses of materials to laser radiation causes different amounts of material to be ablated and can even lead to entirely different ionization mechanisms (7, 61, 62). The measurements of polymer etch depths per laser pulse at various laser energies have been measured only for a small number of polymers (63). For some polymers, it is possible to find laser conditions for which the laser ablation forms cleanly etched features and flat-bottomed craters, which are important for quantitative depth profiling. However, the variations that exist between polymers imply that an unknown or arbitrary polymer cannot be quantitatively analyzed without an extensive amount of characterization of its ablation properties.

Another factor associated with the LAMMS technique is varying sensitivity to elements with different ionization potentials. Sensitivity to atoms or molecules with low ionization potentials can be orders of magnitude greater than for higher ionization potential species. This effect causes the greater sensitivity to K^+ or ionic organic molecules that was mentioned previously. Matrix effects on ionization probabilities analogous to those observed in SIMS may be relevant (64). LAMMS studies have the additional disadvantage that a dense gas-phase plasma is formed, which causes additional complications from gas-phase ionization and reaction processes, some of which have been discussed.

Due to these complicating effects, quantitative analysis using LAMMS is possible only in specific cases in which standards are available that closely resemble the unknown. In many cases, this type of standard is difficult to accomplish.

Various experimental approaches can be used to circumvent some of these problems. One example is the use of a secondary ionization method to ionize neutral species that are ablated from the surface by the laser. By separating the ablation and the ionization steps, greater selectivity or control can be achieved. For FTMS, a simple method is the use of electron-impact ionization synchronized to the laser pulse (65). For example, Figure 14 shows a spectrum of polystyrene that was ablated using 266-nm laser radiation and ionized by electron impact (EI) (66). The mass spectrum corresponds closely to the EI spectrum of styrene monomer. McIver and co-workers used a similar technique to detect fractions of a monolayer of molecular species adsorbed on a metal surface (67, 68). Thus, instrumental improvements can be used to allow greater selectivity, better reproducibility, or simpler identification of unknowns.

Future Work

From the foregoing discussions it should be clear that this novel approach to polymer analysis has demonstrated a number of unique capabilities. Ongoing instrumental developments aimed at high-mass analysis, generally focused at biological polymers (usually proteins), have not been applied yet to synthetic polymers. We mention here a few instrumental approaches currently being pursued vigorously in biological areas.

The major difficulty in fitting FTMS instruments with novel ion sources is the position of the trapping cell, typically surrounded by a 1-m-long dewar for a superconducting magnet. A solution to this problem is to generate ions from the sample outside the magnetic field and inject them into the FTMS cell. Two instrument manufacturers, IonSpec (Irvine, CA) and Bruker

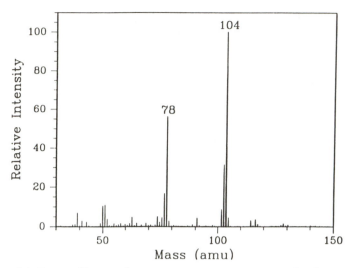

Figure 14. Laser ablation–electron-impact mass spectrum of polystyrene. A 266-nm laser pulse was used to ablate neutral material and a simultaneous electron beam was used for ionization. The mass spectrum corresponds closely to an EI spectrum of the styrene monomer.

(Billerica, MA), offer an "external" ion source (69) for their FTMS systems. As far as we are aware at this writing, no laser microprobe has been implemented for an external ion source system, although such a device should be relatively straightforward because external ion sources are made specifically to permit novel ion source applications for FTMS. The spectra made in this fashion may be dependent on the extraction field employed, and so spectra presented here may differ in some details from such a source.

Determination of molecular weight distributions by LD–FTMS is inherently limited in ability to produce sufficient populations of high-mass ions and by the upper mass limitation of the mass spectrometer. Recent developments in the field of matrix-assisted laser desorption show promise toward forming high-mass ions by direct laser interaction and have been demonstrated in the FTMS (70). This method couples well with the low operating pressures necessary in FTMS, and thus no external ion source or differential pumping is required. The successful implementation of electroscopy ionization sources to study polymeric species may also lead to a decrease in the degree of polymer fragmentation as well as an effective decrease in the mass-to-charge ratio of ions due to the addition of multiple charges. Spectra of PEG with average molecular weights up to 17,500 (71) have been obtained and electrospray ionization recently has been executed on FTMS (25). Unfortunately, electrospray requires an external source with differential pumping and both techniques—matrix assisted desorption and electrospray—require extensive sample preparation that can alter the original nature of sample, whereas direct laser vaporization can examine samples in situ. The ability to trap and mass-isolate ions within the FTMS enables the identification of unknown ions formed by laser vaporization with a variety of dissociative and reactive methods. Collisionally induced dissociation (CID) (17, 72, 73), photodissociation (74–76) and ion–molecule reactions (77) all can be used in conjunction with accurate mass analysis to provide conclusive identification of unknown ions. CID has traditionally provided structural information at laboratory frame kinetic energies up to several kiloelectron volts (72) and has been used successfully with laser desorbed ions (17, 73). Quantitative determination of bonding and binding energies for a number of ionic species have also been obtained (78–80). The maximum kinetic energy available is dependent on the magnitude of the magnetic field, the size of the ion cyclotron resonance cell (the maximum radius of the excited ions), and the mass-to-charge ratio of the ion (44). This limits the technique for large-mass ions because the center-of-mass kinetic energy available is greatly reduced and the ability to introduce sufficient energy to induce fragmentation is decreased.

Laser photodissociation offers ion identification using fragmentation information and wavelength-dependent spectra. These approaches can overcome some of the shortcomings of CAD (81), although they require some a priori knowledge of the ion structure in order to select the proper laser wavelengths (74–76). These methods have been used successfully to differentiate isomeric species. Infrared photodissociation, which often yields the lowest energy dissociation pathway, has been applied to laser desorbed ions as large as 1500 u in the FTMS (82, 83). However, as the mass-to-charge ratio increases, radiative emission competes favorably with photon absorption, thereby limiting the upper mass limit (74). The fragmentation of laser-desorbed ions via UV radiation has also been demonstrated (82). Ion–molecule reactions can also be employed to aid in the identification of ions through determination of electron affinities, ionization potentials, and the selective reactivity of one isomer relative to another (77). All the foregoing techniques aid in the identification of ion structure, but are specific on a case-by-case basis (84).

Laser-desorption and laser-microprobe methods combined with FTMS have been used for a variety of polymer studies. The literature suggests that the technique possesses a unique capability to discern polymer structure in a rapid and precise manner. Further improvements in instrumentation and, particularly, the application of existing strategies to polymer problems will greatly expand the flexibility of laser FTMS for polymer studies.

Acknowledgments

William Creasy thanks John Rabolt, IBM Almedan Research Center, for providing the Langmuir–Blodgett film of cadmium arachidate. The authors thank an anonymous reviewer for many helpful comments.

References

1. Conzemius, R. J.; Junk, G. A. *Bibliography of Laser Mass Spectrometry for Characterization of Synthetic and Natural Solids*; U.S. Department of Energy: Washington, DC, 1986; Report DOE/NBM–7001005.
2. Gardella, J. A., Jr.; Hercules, D. M.; Heinen, H. J. *Spectrosc. Lett.* **1980**, *13*, 347.
3. Gardella, J. A., Jr.; Hercules, D. M. *Fresenius Z. Anal. Chem.* **1981**, *308*, 297.
4. Graham, S. W.; Hercules, D. M. *Spectrosc. Lett.* **1982**, *15*, 1.
5. Mattern, D. E.; Lin, F. T.; Hercules, D. M. *Anal. Chem.* **1984**, *56*, 2762.
6. Simons, D. S. *Appl. Surf. Sci.* **1988**, *31*, 103.
7. Vertes, A.; Juhasz, P.; Wolf, M. D.; Gijbels, R. *Int. J. Mass Spectrom. Ion Process.* **1989**, *94*, 63–85.
8. Brenna, J. T. In *Analytical Fourier Transform Ion Cyclotron Resonance Spectroscopy*; Asamoto, B., Ed.; VCH Publishers: New York, 1992; pp 187–213.
9. Radicati, di Brozolo, F.; Odom, R. W.; Harrington, P. D. B.; Vorhees, K. J. In *36th ASMS Conference on Mass Spectrometry and Allied Topics*; San Francisco, CA 1988; p 287.
10. Radicati di Brozolo, F.; Odom, R. W.; Harrington, P. D. B.; Vorhees, K. J. *J. Appl. Polym. Sci.* **1990**, *41*, 1737–1752.
11. Comisarow, M. B.; Marshall, A. G. *Chem. Phys. Lett.* **1974**, *25*, 282.
12. Asamoto, B.; Dunbar, R. C. *Analytical Applications of Fourier Transform Ion Cyclotron Resonance Mass Spectrometry*; VCH Publishers: New York, 1991.
13. Marshall, A. G.; Grosshans, P. B. *Anal. Chem.* **1991**, *63*, 215A–229A.
14. Nibbering, N. M. M. *Acc. Chem. Res.* **1990**, *23*, 279–285.
15. Wancek, K. P. *Int. J. Mass Spectrom. Ion Phys.* **1989**, *95*, 1–38.
16. *Lasers in Mass Spectrometry*; Lubman, D. M., Ed.; Oxford University Press: New York, 1990; Chaps. 7 and 11–15.
17. Marshall, A. G. *Adv. Mass Spectrom.* **1989**, *11A*, 651–669.
18. Ledford, E. B.; White, R. L.; Ghaderi, S.; Wilkins, C. L.; Gross, M. L. *Anal. Chem.* **1980**, *52*, 2450.
19. Lee, E. D.; Henion, J. D.; Cody, R. B.; Kinsinger, J. A. *Anal. Chem.* **1987**, *59*, 1309–1312.
20. Amster, I. J.; Loo, J. A.; Furlong, J. J. P.; McLafferty, F. W. *Anal. Chem.* **1987**, *59*, 313–317.
21. Castro, M. E.; Mallis, L. M.; Russell, D. H. *J. Am. Chem. Soc.* **1985**, *107*, 5652–5657.
22. Loo, J. A.; Williams, E. R.; Amster, I. J.; Furlong, J. J. P.; Wang, B. H.; McLafferty, F. W.; Chait, B. T.; Field, F. H. *Anal. Chem.* **1987**, *59*, 1880.
23. Williams, E. R.; McLafferty, F. W. *J. Am. Soc. Mass Spectom.* **1990**, *427*.
24. Williams, E. R.; Henry, K. D.; McLafferty, F. W. *J. Am. Chem. Soc.* **1990**, *112*, 6157.
25. Henry, K. D.; Williams, E. R.; Wang, B. H.; McLafferty, F. W.; Shabanowitz, J.; Hunt, D. F. *Proc. Nat. Acad. Sci.* **1989**, *86*, 9075–9078.
26. Henry, K. D.; McLafferty, F. W. *Org. Environ. Mass Spectrom.* **1990**, *25*, 490–492.
27. Henry, K. D. Ph.D. Thesis, Cornell University, 1991.
28. Brenna, J. T.; Creasy, W. R.; McBain, W.; Soria, C. *Rev. Sci. Instrum.* **1988**, *59*, 873.
29. Ghaderi, S. In *36th ASMS Conference on Mass Spectrometry and Allied Topics*, San Francisco, CA, 1988; p 1126.
30. Pelletier, M.; Krier, F.; Muller, J. F.; Weil, D.; Johnston, M. *Rapid Commun. Mass Spectrom.* **1988**, *2*, 146.

31. Brown, C. E.; Kovacic, P.; Wilkie, C. A.; Cody, R. B., Jr.; Kinsinger, J. A. *J. Polym. Sci., Polym. Lett. Ed.* **1985**, *23*, 453.
32. Brown, C. E.; Kovacic, P.; Wilkie, C. A.; Kinsinger, J. A.; Hein, R. E.; Yaniger, S. I.; Cody, R. B., Jr. *J. Poly. Sci, Poly. Chem. Ed.* **1986**, *24*, 255.
33. Brown, C. E.; Kovacic, P.; Wilkie, C. A.; Cody, R. B., Jr.; Hein, R. E.; Kinsinger, J. A. *Synth. Metals* **1986**, *15*, 265.
34. Brown, C. E.; Kovacic, P.; Cody, R. B., Jr.; Hein, R. E.; Kinsinger, J. A. *J. Polym. Sci., Polym. Lett. Ed.* **1986**, *24*, 519.
35. Brown, C. E.; Kovacic, P.; Wilkie, C. A.; Cody, R. B., Jr.; Hein, R. E.; Kinsinger, J. A. *ACS Polym. Mater. Sci. Eng.* **1986**, *54*, 306.
36. Brown, R. S.; Weil, D. A.; Wilkins, C. L. *Macromolecules* **1986**, *19*, 1255.
37. Brown, C. E.; Wilkie, C. A.; Smukalla, J.; Cody, R. B., Jr.; Kinsinger, J. A. *J. Polym. Sci., Polym. Chem. Ed.* **1986**, *24*, 1297.
38. Brown, C. E.; Kovacic, P.; Welch, K. J.; Cody, R. B., Jr.; Kinsinger, J. A. *Arabian J. Sci. Technol.* **1988**, *13*, 163.
39. Creasy, W. R.; Brenna, J. T. *Chem. Phys.* **1988**, *126*, 453.
40. Creasy, W. R.; Brenna, J. T. In *Polyimides: Materials, Chemistry and Characterization*; Feger, C.; Khojasteh, M. M.; McGrath, J. E., Eds.; Elsevier: Amsterdam, Netherlands, 1989.
41. Nuwaysir, L. M.; Wilkins, C. L. *Anal. Chem.* **1988**, *60*, 279.
42. Nuwaysir, L. M.; Wilkins, C. L.; Simonsick, W. J., Jr. *J. Am. Soc. Mass Spectrom.* **1990**, *1*, 66–71.
43. Ijames, C. F.; Wilkins, C. L. *J. Am. Chem. Soc.* **1988**, *110*, 2687.
44. Marshall, A. G.; Verdun, F. R. *Fourier Transforms in NMR, Optical and Mass Spectrometry*; Elsevier: New York, 1990.
45. Laude, D. A.; Beu, S. C.; Hogan, J. D. In *39th ASMS Conference on Mass Spectrometry and Allied Topics*; Nashville, TN, 1991; p 459.
46. Brenna, J. T. *Microbeam Analysis* **1989**, *306*.
47. Brenna, J. T.; Creasy, W. R.; Volksen, W. *Chem. Phys. Lett.* **1989**, *163*, 499.
48. Creasy, W. R.; Brenna, J. T. *J. Chem. Phys.* **1990**, *92*, 2269.
49. Brenna, J. T.; Creasy, W. R. *Appl. Spectrosc.* **1991**, *45*, 80–91.
50. Srinivasan, R. *Laser Processing and Laser Diagnostics*; Springer: New York, 1984; pp 343–354.
51. Hsu, A. T.; Marshall, A. G. *Anal. Chem.* **1988**, *60*, 932.
52. Kratchmer, W.; Lamb, W. D.; Fostiropoulos, K.; Huffman, D. R. *Nature* **1990**, *347*, 354–357.
53. Curl, R. F.; Smalley, R. E. *Science* **1988**, *242*, 1017.
54. Kroto, H. W. *Science* **1988**, *242*, 1139.
55. Naselli, C.; Rabolt, J. F.; Swalen, J. D. *J. Chem. Phys.* **1985**, *82*, 2136.
56. Asamoto, B.; Young, J. R.; Swalen, J. D. *Anal. Chem.* **1990**, *62*, 61.
57. Blease, T. G.; Scrivens, J. H.; Monaghan, J. J.; Weil, D. A. In *36th ASMS Conference on Mass Spectrometry and Allied Topics*; San Francisco, CA 1988; p 357.
58. Liang, Z.; Marshall, A. G.; Westmoreland, D. G. *Anal. Chem.* **1991**, *63*, 815.
59. Xiang, X.; Dahlgren, J.; Enlow, W. P.; Marshall, A. G. In *39th ASMS Conference on Mass Spectrometry and Allied Topics*; Nashville, TN, 1991; p 412.
60. Johlman, C. L.; Wilkins, C. L.; Hogan, J. D.; Donovan, T. L.; Laude, D. A., Jr.; Youssefi, M. *J. Anal. Chem.* **1990**, *62*, 1167.
61. Karas, M.; Bachmann, D.; Bahr, U.; Hillencamp, F. *Int. J. Mass Spectrom. Ion Process.* **1987**, *78*, 53.
62. Vertes, F.; Wolf, M. D.; Juhasz, P.; Gijbels, R. *Anal. Chem.* **1989**, *61*, 1029–1035.
63. Srinivasan, R.; Braren, B. *Chem. Rev.* **1989**, *89*, 1303.

64. Wilson, R. G.; Novak, S. W. *J. Appl. Phys.* **1991**, *69*, 466.
65. Shomo, R. E.; Marshall, A. G.; Weisenberger, R. C. *Anal. Chem.* **1985**, *57*, 2940.
66. Creasy, W. R. In *38th ASMS Conference on Mass Spectrometry and Allied Topics*; Tuscon, AZ, 1990; p 850.
67. Land, D. P.; Tai, T. L.; Lindquist, T. M.; Hemminger, J. C.; McIver, R. T., Jr. *Anal. Chem.* **1987**, *59*, 2927.
68. Land D. P.; Pettiette-Hall, C. L.; McIver, R. T., Jr.; Hemminger, J. C. *J. Am. Chem. Soc.* **1989**, *11*, 5970.
69. Kofel, P.; Allemann, M.; Kellerhals, H.; Wanczek, K. P. *Int. J. Mass Spectrom. Ion Process.* **1989**, *87*, 237–247.
70. Hettich, R. L.; Buchanan, M. V. *J. Am. Soc. Mass Spectrom.* **1991**, *2*, 22.
71. Wong, S. F.; Meng, C. K.; Feng, J. B. *J. Phys. Chem.* **1988**, *92*, 546.
72. Cody, R. B., Jr.; Burnier, R. C.; Freiser, B. S. *Anal. Chem.* **1982**, *54*, 96.
73. McCrery, J.; Peake, D. A.; Gross, M. L. *Anal. Chem.* **1985**, *54*, 1181.
74. Dunbar, R. C. *Anal. Instrum.* **1988**, *17*, 113.
75. van der Hart, W. J. *Mass Spectrom. Rev.* **1989**, *8*, 237.
76. Watson, C. H.; Baykut, G.; Eyler, J. R. In *Fourier Transform Mass Specrtrometry: Evolution, Innovation, and Applications*; Buchanan, M. V., Ed.; ACS Symposium Series 359; American Chemical Society: Washington, DC, 1987; pp 140–154.
77. Kemper, P. R.; Bowers, M. T. In *Techniques for the Study of Ion–Molecule Reactions*; Farrar, J. M.; Sanders, J. W. H., Eds.; Wiley: New York, 1988.
78. Bensimon, M.; Houriet, R. *Int. J. Mass Spectrom. Ion Process.* **1986**, *72*, 93.
79. Katritzky, A. R.; Watson, C. H.; Dega-Szafran, Z.; Eyler, J. R. *J. Am. Chem. Soc.* **1990**, *112*, 2471.
80. Hop, C. E. C. A.; McMahon, T. B.; Willet, G. D. *Int. J. Mass Spectrom. Ion Process.* **1990**, *101*, 191.
81. Bowers, W. D.; Delbert, S. S.; Hunter, R. L.; McIver, R. T., Jr. *J. Am. Chem. Soc.* **1984**, *106*, 7288.
82. Watson, C. H.; Baykut, G.; Eyler, J. R. *Anal. Chem.* **1987**, *59*, 1133.
83. Zimmerman, J. A.; Watson, C. H.; Eyler, J. R. *Anal. Chem.* **1990**, *63*, 361.
84. Bohme, D. K.; Wlodek, S.; Zimmerman, J. A.; Eyler, J. R. *Int. J. Mass Spectrom. Ion Process.*, in press.

RECEIVED for review July 15, 1991. ACCEPTED revised manuscript May 19, 1992.

CRYSTALLINE POLYMERS
AND COPOLYMERS

The high sensitivity and specificity of FTIR and FT–Raman spectroscopy make them ideal characterization tools for studying complex crystalline polymers and copolymers. Whereas internal and longitudinal modes of a Raman spectrum can be used for structural analysis of crystalline polymers or topotacticity changes during solid-state polymerization, phase behavior and molecular-level interactions of polymer blends can be successfully studied by FTIR. The applicability of near-IR spectroscopy for qualitative analysis of thermoplastic polymers is also discussed.

Use of Raman Spectroscopy in Characterizing the Structure and Properties of Crystalline Polymers

Leo Mandelkern and Rufina G. Alamo

Department of Chemistry and Institute of Molecular Biophysics, Florida State University, Tallahassee, FL 32306

An overview of some of the applications of Raman spectroscopy to the understanding of the structure of crystalline, flexible chain polymers is presented. Selected physical properties are discussed in terms of these structural parameters. The major emphasis is on the polyethylenes, but the relationships between structure and properties are easily generalized to other polymer systems. The three main spectral regions of interest in the study of the polyethylenes are the $5-50$-cm^{-1} longitudinal acoustic mode (LAM) region, the DLAM region in the vicinity of about 200 cm^{-1}, and the $900-1500$-cm^{-1} internal mode region.

$\mathbf{A}$N OVERVIEW OF THE APPLICATIONS OF RAMAN SPECTROSCOPY to the understanding of the structure of crystalline, flexible chain polymers is given in this chapter. These structural parameters are then related to selected physical properties. Polyethylenes are used as the vehicle for our discussion for several reasons: These polymers have a richness of spectroscopic information and have admirably served as model systems for crystalline polymers. We find that the key structural factors that are found in the polyethylenes are also present in other crystalline systems.

A basic understanding of the thermodynamics of fusion of crystalline polymers as well as their crystallization kinetics is crucial to understanding the crystallization behavior of polymers (1). However, to understand the properties that actually develop in a system, it is necessary to identify and quantify

0065-2393/93/0236-0157$09.50/0

the key structural quantities that describe the crystalline state because polymers can only crystallize in a reasonable time scale at temperatures well below their melting points. For kinetic reasons, the crystallization process is very rarely or never close to completion. For homopolymers, the level of crystallinity that can be attained ranges from 40 to greater than 90% depending on the described nature of the repeating unit, the molecular weight, and the crystallization conditions. Hence, a nonequilibrium state invariably develops. The resulting system is thus polycrystalline and morphologically complex.

Extensive experimental studies have identified some of the key independent structural variables that describe the crystalline state (*2–4*). These variables are the degree of crystallinity, the structure of the residual noncrystalline region, the crystallite thickness distribution, the structure and relative amount of the interfacial region, details of the crystallite structure beyond that of the lamellar habit, and the supermolecular structure (the arrangement of the crystallites). These quantities can be varied over wide limits by control of molecular weight and crystallization conditions (*2, 4–6*). Although Raman spectroscopy cannot identify and quantify all of these structural variables, this technique can make a significant contribution. The spectral regions that can be used to describe specific aspects of the crystallization behavior have been summarized (*3*), but for present purposes, it will be helpful to briefly describe them once again.

There are three main spectral regions that give structural information about the crystalline state of the polyethylenes. The internal modes, which give quantitative information with respect to the elements of phase structures, are in the region of approximately $900-1500$ cm^{-1} (*7–9*). The longitudinal acoustic mode (LAM), which gives the ordered sequence length distribution, lies in the range of about $5-50$ cm^{-1} (*10–17*). The disordered LAM (DLAM) is in the region of 200 cm^{-1} and gives a measure of the long-range conformational disorder (*18–21*).

Structure

Phase Structure. A quantitative description of the phase structure of semicrystalline polymers is of primary importance in understanding properties (*2, 4*). Three distinct structural regions are involved: the ordered crystalline region that, for polyethylene, represents the orthorhombic unit cell, the isotropic, liquidlike or disordered region, and the interphase that is comprised of chain units that connect these two conformationally very different structural regions. In polymer crystallization, a chain can traverse all three regions or, in some cases, can be restricted to the crystalline and interfacial regions. The participation of a given molecule in all three regions is a unique and important feature of polymer crystallization. In 1949, Flory pointed out that a demarcation between the ordered crystalline region and the disordered liquidlike region is not sharply defined because for most polymers the flux of chains that emerges from the crystallite cannot be immediately dissipated in the liquidlike region (22). Consequently, an interphase or interfacial region develops that involves a partially ordered set of chain units. Theoretical analyses of the interfacial structure are in general agreement with one another (23–29). It is apparent from these studies that the detailed structure and extent of the interphase are specific to a given polymer.

Many different kinds of measurement can be used to quantitatively describe the phase structure. Utilizing the principle of the additivity of a particular property of the pure liquid and crystalline states, measurements of the density and enthalpy of fusion can be used to determine the degree of crystallinity. We shall designate these levels of crystallinity as $(1 - \lambda)_d$ and $(1 - \lambda)_{\Delta H}$, respectively, where λ is the noncrystalline fraction, d is the density, and ΔH is the enthalpy of fusion. Wide- and small-angle X-ray diffraction, as well as several different types of NMR measurements, can also be profitably used for this purpose.

Analysis of the Raman internal modes for the purpose of quantitative elucidation of the phase structure was originally given by Strobl and Hagedorn (7). The core degree of crystallinity, α_c, is obtained by this method because the only contribution to this quantity comes from the structure of the orthorhombic unit cell. α_c is calculated from the CH_2 bending band at 1416 cm^{-1} and is the component of this vibration that is split by the crystal field. The degree of liquidlike material, α_a, is obtained from the twisting region at 1303 cm^{-1}. The total integrated intensity of the twisting region, 1295–1303 cm^{-1}, is independent of the phase structure (7, 8). Methods of analyzing the spectra were given in detail by Strobl and Hagedorn (7), as well as in reports from this laboratory (8, 9). The analysis of a large amount of experimental data yielded a significant finding: $\alpha_a + \alpha_c \neq 1$ (2, 8). The spectra of the completely liquidlike and the completely crystalline polymers cannot be superposed and proportioned to represent an observed spectra; a partially ordered, primarily trans, anisotropic region must also be included. This contribution has been defined as the interfacial region α_b.

Figure 1 is a plot of the core crystallinity α_c against $(1 - \lambda)_{\Delta H}$ for an extensive set of data of linear polyethylenes, branched polyethylenes, and random ethylene copolymers. The data points for the linear polymers fall on the 45° straight line over the very large range of 0.4–0.9 in the level of crystallinity. We can therefore conclude that the core crystallinity and $(1 - \lambda)_{\Delta H}$ are identical to one another. For the branched polymers and copolymers, α_c is about 5% less than $(1 - \lambda)_{\Delta H}$. This small disparity can be attributed to the broad melting range of the structurally irregular chains. $(1 - \lambda)_{\Delta H}$ includes the contribution of a small amount of crystallinity that has already disappeared at room temperature. Because α_c is measured at ambi-

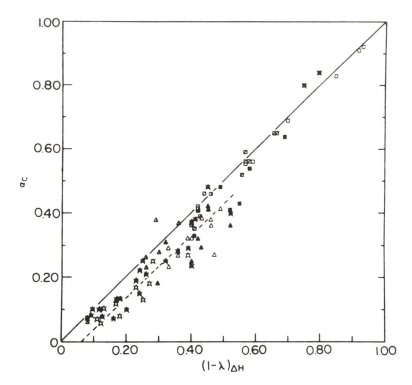

Figure 1. Plot of degree of crystallinity, α_c, as determined from the Raman internal modes, against $(1 - \lambda)_{\Delta H}$ for linear (—) and branched (- -) polyethylene and ethylene copolymers. (Reproduced with permission from reference 2. Copyright 1985 Society of Polymer Science, Japan.)

ent temperature, this contribution is not included in the internal mode analysis.

In contrast to the concordance between these two methods, there is a significant quantitative difference between $(1 - \lambda)_d$ and $(1 - \lambda)_{\Delta H}$. As is illustrated in Figure 2, $(1 - \lambda)_d$ for linear polyethylene is always greater than $(1 - \lambda)_{\Delta H}$. A similar result is also observed with structurally irregular copolymers (2).

The basis for the discrepancy between $(1 - \lambda)_d$ and $(1 - \lambda)_{\Delta H}$ can be found in the plot of Figure 3, where $(1 - \lambda)_d$ is plotted against the sum $(\alpha_c + \alpha_b)$. All the data fall quite well on the 45° straight line, which indicates a one-to-one correspondence between the two quantities. Because α_c measures the core crystallinity and α_b measures the interfacial content, a self-consistent and physically satisfying interpretation of these results is found in the conclusion that the density measures both the core crystallinity and the partially ordered anisotropic interfacial region. On the other hand, the enthalpy of fusion measures only the core crystallinity.

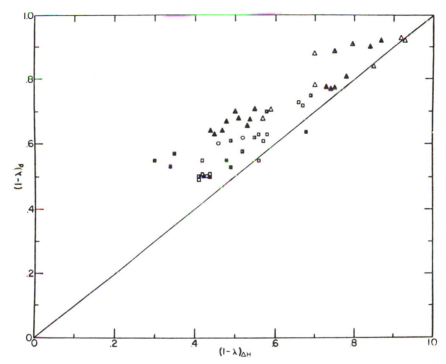

Figure 2. Plot of degree of crystallinity obtained from density, $(1 - \lambda)_d$, against the value obtained from the enthalpy of fusion, $(1 - \lambda)_{\Delta H}$, for linear polyethylene. (Reproduced with permission from reference 2. Copyright 1985 Society of Polymer Science, Japan.)

Although the degree of crystallinity is a well-established quantitative concept, different methods of measurement yield different results. These methods have the same functional dependence with respect to molecular constitution and crystallization conditions. However, there are small but significant differences that are due to the contributions from the different elements of phase structure. For example, Magill and co-workers have found that for poly(tetramethyl-*p*-silphenylene siloxane), $(1 - \lambda)_d$ is slightly higher than $(1 - \lambda)_{\Delta H}$ (30).

Studies with other polymers, as well as other methods of measurement, give similar results and give strong support to the quantitative analysis of the Raman internal modes that has been given for the polyethylenes. For example, using the methods of high resolution solid-state carbon-13 NMR, it is possible to decompose the T_1 relaxation decay curves into three components that correspond to the crystalline, liquidlike, and interfacial regions (31, 32). A plot of α_c, determined from the Raman internal modes, against the degree of crystallinity determined by NMR is given in Figure 4 for a variety

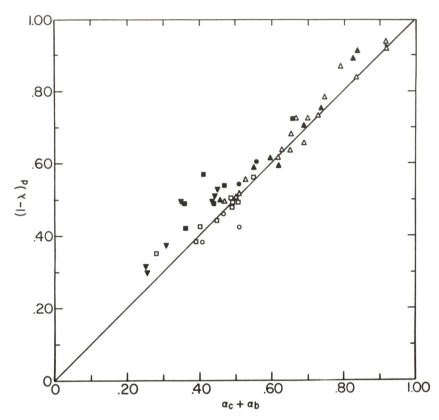

Figure 3. Plot of $(1 - \lambda)_d$ against the sum of $(\alpha_c + \alpha_b)$ for linear and branched polyethylenes and ethylene copolymers. (Reproduced with permission from reference 2. Copyright 1985 Society of Polymer Science, Japan.)

of polyethylene samples. We find a one-to-one correspondence between the two methods similar to that found between α_c and $(1 - \lambda)_{\Delta H}$. From the decomposition of the T_1 magnetization decay curves of a set of polyethylenes the interfacial components were found to be quantitatively identical to α_b obtained from the Raman internal modes (*31*).

The line shapes of the carbon-13 resonances in the solid state of polyethylene (*33*), polypropylene (*34*), and poly(tetramethylene oxide) (*35*) have been decomposed into three components that correspond to the components under discussion. The results are summarized in Table I. Several important features are summarized in this table. The fractions of the interfacial regions are quite similar for the three polymers; all are in the range 0.16–0.30. We can conclude, therefore, that chain molecules, in general, are characterized by significant interfacial contents. For each of the two poly(propylene) specimens (the same sample crystallized in two different

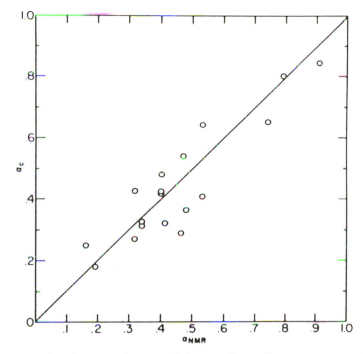

Figure 4. Plot of Raman-determined degree of crystallinity α_c against NMR degree of crystallinity α_{NMR}. (Reproduced with permission from reference 31. Copyright 1983 John Wiley & Sons, Inc.)

Table I. Phase Structures from Carbon-13 Line Shapes

Polymer	$(1 - \lambda)_d$	$(1 - \lambda)_{\Delta H}$	$(1 - \lambda)_{NMR}$	Δ^a	$\alpha_{b(NMR)}{}^b$	Ref.
Poly(ethylene)	0.761		0.70		0.16	34
Poly(propylene)						
Sample A	0.718	0.454	0.480	0.26	0.30	34
Sample B	0.828	0.540	0.570	0.29	0.27	34
Poly(tetramethylene oxide)			0.60		0.22	35

^{a} $\Delta = (1 - \lambda)_d - (1 - \lambda)_{\Delta H}$.
^{b} Directly observed.

ways) there is very good agreement between $(1 - \lambda)_{\Delta H}$ and $(1 - \lambda)_{NMR}$. Moreover, the difference between $(1 - \lambda)_d$ and $(1 - \lambda)_{\Delta H}$ corresponds, in each case, to the interfacial fraction deduced from NMR. Thus, the NMR results for polypropylene parallel the Raman internal mode analysis of the polyethylenes. Studies with other polymers would further clarify the generality of these results.

The analysis of the broad line proton NMR resonance also requires a three-component phase structure for its decomposition (*36*). The magnitude of the interfacial content (and its molecular weight dependence) is similar to that deduced from the Raman analysis (37).

The interfacial content of linear polyethylene is very dependent on chain length for rapidly crystallized molecular-weight fractions (38). At low molecular weights, the interfacial content is relatively small—about 5%. However, an appreciable interfacial content, on the order of 15–20%, is observed at much higher molecular weights. The change in interfacial content with molecular weight parallels the variation in the interfacial free energy that is associated with the basal plane of the lamellae. This significant proportion of interphase that is deduced from the analysis of the experimental data has important ramifications in understanding certain macroscopic properties.

The phase structure of random copolymers depends not only on the molecular weight, but also on the co-unit content and, in certain specific cases, on the chemical nature of the co-unit itself. The introduction of noncrystallizing co-units into the chain leads to a rapid and continuing decrease in the crystallinity level with increasing side-group content. The level of the core crystallinity varies from about 48% for 0.5 mol% of branches to about 7% for 6 mol% of branch points. The chemical nature of a specific branch type has virtually no influence on the crystallinity level as long as it does not enter the lattice. In cases where the branches enter the lattice, such as CH_3 and Cl, higher levels of crystallinity will be observed. The changes in α_c with molecular weight, at a fixed co-unit content, follow a pattern that is similar to homopolymers, although the level of crystallinity is much reduced (39).

The interfacial content of random copolymers at a fixed co-unit content is independent of molecular weight. This conclusion is evidenced by studies of a set of ethylene–hexene copolymers that have a most probable molecular-weight distribution with branch contents that range from 1.2 to 1.7 mol% (39) and a set of hydrogenated polybutadienes (randomly ethyl-branched ethylene copolymers) of slightly higher branch content (39). In contrast, the interfacial content of homopolymers increases substantially with molecular weight (8). The interfacial content of random copolymers is a strong monotonically increasing function of the co-unit content. The value of α_b approaches 15–20%, while at the same time α_c is reduced to as low as 5–8% (40). Therefore, the interfacial region can represent an appreciable portion of the total crystallite structure. More extensive experiments are necessary to determine whether details of the interfacial structure change as the chemical nature of the co-unit is varied.

Analysis of the Raman internal modes has pioneered the quantitative determination of the proportion of interphase in crystalline polymers: The existence and importance of an interfacial region is now well established. Analysis of this spectral region also allows for the quantitative establishment of the complete phase structure. Other experimental methods, such as NMR previously cited, have quantitatively confirmed the results obtained by Raman spectroscopy for polyethylene and other crystalline polymers.

Ordered Sequence Length.   The ordered sequence length can be obtained from the Raman low-frequency longitudinal acoustic vibrational mode (LAM) by using the relation given by Shimanouchi and co-workers (*10*, *11*). The appropriate relation can be expressed as

$$\Delta\bar{\nu} = \frac{m}{2cL}\left(\frac{E}{\rho}\right)^{1/2} \tag{1}$$

Here $\Delta\bar{\nu}$ is the mode frequency, m is the mode order ($m = 1, 3, 5, \ldots$), c is the velocity of light, ρ is the density of the vibrating sequence, and E is the Young's modulus in the chain direction. The ordered sequence length L is thus inversely related to the mode frequency. This length can be related, in a straightforward manner, to the lamellar thickness. The LAM band has been observed in many crystalline polymers and has been widely used in studying their structure (*12–17*, *41*) .

We focus our attention on the use of this band in studying the structure of the polyethylenes and the *n*-alkanes. An abundance of experimental data is available for these systems. However, before analyzing experimental data, the question of the direct applicability and validity of eq 1 must be addressed. Equation 1 was derived on the basis of a uniform elastic rod with the *n*-alkane as the proper molecular analogue (*10*, *11*). The applicability of eq 1 to real polymer systems can thus be properly questioned (*41*). For crystalline polymers, which characteristically possess a lamellar morphology, there has been the concern that the noncrystalline units attached to the crystalline sequences or stems will require serious modification of eq 1. The *n*-alkanes themselves, in either extended or folded forms (*42*, *43*), also have a lamellar habit. In this case, the question arises as to what influence the molecular end groups have on eq 1 and if, in fact, the ordered sequence lengths can be obtained from the observed LAM (*31*, *44*).

One matter of immediate concern for any system (the polyethylenes and *n*-alkanes, in particular) is the correct value of E. For polyethylene the values used range from $E = 2.9 \times 10^{12}$ dynes/cm^2 (*14*) to 3.6×10^{12} dynes/cm^2 (*10*, *11*). Because the modulus enters eq 1 as the square root, the uncertainty in L will be 5% for this range of E values. In some applications, this uncertainty will not be of serious concern, whereas in others it will be important (*44*).

There are at least two methods by which to test the validity of eq 1 and to assess the conditions of molecular and crystallite structure under which corrections, if any, are necessary. One method is to measure the mode orders and see if the proper frequency ratio is observed. The other method is to measure the crystallite thickness by an independent method and relate it to the ordered sequence length obtained by the Raman LAM. The analysis of the small-angle X-ray scattering maxima has been a traditional way to calculate the crystallite thickness. To obtain the thickness, the directly observed periodicity has to be corrected for the level of crystallinity. To compare this dimension with the ordered sequence length determined by LAM, a correction must be made for chain tilt (45). Thus, a series of corrections, each subject to experimental error, must be made to enable a comparison of the small-angle X-ray scattering result with those from Raman. Therefore, in many cases when the need for the required corrections is not recognized, it is not surprising that agreement is not obtained and modification is made to eq 1. However, when care is taken in making the corrections to properly compare crystallite thicknesses, eq 1 is shown to be valid (46). (If the data in reference 45, where the chain tilt was taken into account, were also corrected for the crystallinity level, then good agreement is found between the LAM and small-angle X-ray diffraction maximum for the lamellar thickness.)

The analysis of mode orders has been shown to be effective in assessing eq 1 as is illustrated in Table II (16). Here, several molecular weight fractions of linear polyethylene were crystallized in various ways to obtain different

Table II. Longitudinal Acoustical Mode Frequencies and Properties of Crystalline Polyethylene Samples (16)

M_w	L_R	Crystallization Conditions	Degree of Crystallinity[a]	$\Delta\bar{\nu}_1$ (cm^{-1})	$\Delta\bar{\nu}_3$ (cm^{-1})	$\beta_3 = \Delta\bar{\nu}_3 / \Delta\bar{\nu}_1$
93,000	120	Solution crystal $T_c' = 85.8\ °C$	0.72	23.6 ± 0.2	68.0 ± 0.5	2.88
93,000	124	Solution crystal $T_c' = 87.2\ °C$	0.78	22.8 ± 0.2	64.5 ± 0.5	2.83
93,000	135	Solution crystal $T_c' = 89.5\ °C$	0.81	21.0 ± 0.2	59.5 ± 0.5	2.83
11,500	189	Quenched dry ice–isopropanol	0.62	15.0 ± 0.2	44.0 ± 0.5	2.93
8,400	334	Isothermal crystallization $T_c' = 125\ °C$ 30 days	0.94	8.5 ± 0.1	24.0 ± 0.5	2.82
27,800	405	Isothermal crystallization $T_c' = 126\ °C$ 30 days	0.89	7.0 ± 0.2	20.5 ± 0.5	2.93

[a] Determined from enthalpy of fusion.

levels of crystallinity and the first and third mode orders, $\Delta\bar{\nu}_1$ and $\Delta\bar{\nu}_3$, were determined. As indicated in the table, the ratio $\Delta\bar{\nu}_1/\Delta\bar{\nu}_3$ is close to the theoretically expected value for this range in ordered sequence lengths, crystallinity levels, and molecular weight. Hence, we can conclude from these data that for the different crystallization modes, and $L = 100$ Å or more, eq 1 gives a very good representation of the experimental data. Therefore, eq 1 can be used in its simple form to determine the ordered sequence length. Other kinds of experiments have also shown that there is no appreciable perturbation of the LAM frequencies by the noncrystalline portions of the molecule (*47*, *48*). There is, therefore, no need to develop complex theories to account for corrections to eq 1 that do not, in fact, exist or are negligible (*49–55*).

There is, however, one situation for polymers where the application of eq 1 in its simple form fails. A study of the thermodynamic and structural properties of a series of molecular weight fractions of hydrogenated polybutadienes with a fixed value of 2.3-mol% branch points pointed out this failure (*56*). Here, the observed small-angle diffraction maximum, found to be 155 Å, is independent of molecular weight. When this maximum is corrected by the level of core crystallinity, the crystallite or lamellar thickness is obtained. The thickness is about 60 Å for $M_w \leq 1.6 \times 10^4$ (where M_w is the weight-average molecular weight) and monotonically decreases to 30 ± 5 Å for $M_w = 4.2 \times 10^5$, the highest molecular weight studied. Thin-section electron microscopy studies confirm these lamellar thicknesses (*56*). The most probable ordered sequence length, L_R, obtained from the LAM, for the same set of fractions crystallized in the same manner, is about 70 Å and independent of chain length. When corrected with a chain tilt angle of 30°, the crystallite or lamellar thickness is found to be 61 Å for all of the fractions. Comparison of the experimental results shows that the LAM, small-angle diffraction maxima, and electron microscopy give excellent agreement in the crystallite thickness on the order of 60 Å or higher. However, when L_c decreases from the 60 to 30 Å, there is a serious error in the LAM-determined values. In this range of extremely small crystallite thickness, the LAM values are about a factor of 2 too large. Clearly, eq 1 is not applicable in this case. It must be recognized, however, that these results present a very unique situation. Not only are we dealing with small crystallite sizes, but concomitantly the thickness of the interphase is about one-quarter to one-third of the crystallite thickness (*56*). We thus have an example where small ordered sequence lengths, about 20–50 carbon atoms, are joined with a partially ordered interphase. In addition, the interfacial region will contain a higher than nominal concentration of branch points. In this special case, the pertinent vibrations between the crystalline sequences are not uncoupled. Usually for polymer systems, the vibrations between the ordered sequences are uncoupled and eq 1 can be properly applied. It is clear that theory will have to be modified to account for this very special situation.

Another example where there are major complications in relating the LAM frequencies to ordered sequence lengths is in interpreting the structure of the n-alkanes in the solid state. Recently, n-alkanes containing up to 400 carbon atoms have been synthesized (42, 43). There is interest in quantitatively describing their structures and using LAM in this endeavor. Elucidating the details of the folded structures that can develop (42, 44, 57, 58) is of particular importance. The problem to be resolved is whether sharp folds that lead to integral-valued fold lengths are formed or whether, in contrast, a disordered overlayer is present. This is the same problem that has already been resolved for polymers crystallized from dilute solution (46, 59). In this case, it has been established that a significant interfacial and disordered overlayer is present.

The problem that exists with the n-alkane can be briefly outlined as follows. In the extended form (i.e., for molecular crystals) the mode orders are not in the ratio of 5:3:1 (11, 14, 44). The deviations decrease with increasing chain length and become very small at carbon numbers greater than about 200. There is, therefore, a definite influence of the end group of the molecule on the observed frequency. However, a calibration can be established between $\Delta\bar{\nu}$ and n that allows for the determination of the ordered sequence lengths of extended n-alkanes (44). The problem with folded structures arises because the frequencies of the ordered sequence lengths are in the region where there is a serious discrepancy in the ratio of mode orders. Hence, an unknown correction must be made. Even if the assumption is made for the purpose of this problem that the ordered sequences are uncoupled, their determination from $\Delta\bar{\nu}_1$ is ambiguous at best (44). Therefore, until the basic theoretical problems are solved, ordered sequence lengths of the folded n-alkanes cannot be obtained with the certainty necessary to resolve the structural problem of concern.

The LAM frequency not only allows for the determination of the most probable ordered sequence length in crystalline polymers; the distribution of these lengths can also be obtained. (The true position of the LAM band is not only displaced by the distribution of ordered sequence lengths, but also by temperature and frequency factors.) Following Snyder et al. (60, 61) the distribution function $f(L)$ can be directly related to the observed intensity, I_ν^{obs}, at frequency $\Delta\bar{\nu}$ by the relation

$$f(L) \propto \left[1 - \exp\left(\frac{-hc\,\Delta\bar{\nu}}{kT} \right) \right] \Delta\bar{\nu}^2\, I_\nu^{obs} \qquad (2)$$

Here, $f(L)$ is proportional to the number of crystalline sequences of length L. The ordered sequences of length L that give rise to scattering at a frequency $\Delta\bar{\nu}$ can be derived from eq 1. Equation 2 has been successfully applied to a variety of systems (17, 62, 63). (The displacement of the observed $\Delta\bar{\nu}$ depends on the width of the LAM band. This displacement is negligible for LAMs that appear at frequencies of 15 cm^{-1} or higher and have half-widths less than about 5 cm^{-1} (*61*). This condition is usually fulfilled by the *n*-alkanes. However, for some polymers, particularly branched or ethylene copolymers, where the observed half-width is on the order of 10–15 cm^{-1}, displacements to higher frequencies may exceed 5 cm^{-1} or more.) An example given in Figure 5 typifies the situation where the sample has a narrow ordered sequence distribution (*62*). In such cases, the agreement between the Raman LAM analysis and thin section electron microscopy is excellent (*62, 63*). As was pointed out earlier, to compare the lamellar thickness obtained from either electron microscopy or small-angle X-ray scattering with the LAM-determined ordered sequence length, correction must be made for the tilt angle. The Raman LAM can also be very effectively employed in cases where the lamellae are not organized into regular stacks (*63*). Small-angle X-ray scatterings do not give any meaningful results in this situation. Although electron microscopy indicates the presence of lamellae, a quantitative measure is difficult.

For systems that have broad distributions (such as asymmetric or bimodal), precise results are difficult to obtain from the LAM (*63*). The main difficulty in analysis of more complex distributions is the resolution of the small ordered sequence lengths, which tend to merge with the background and thus make the correct baseline difficult to draw. In general, as the length

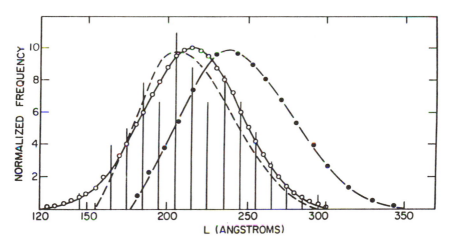

Figure 5. Normalized frequency distributions of crystallite thickness for sample $M_n = 76,700$, $M_w = 80,800$ *crystallized at 118 °C for 2 min. Key: electron micrograph histogram,* ——; *smooth distribution from electron micrograph,* - - -; *uncorrected Raman LAM distribution,* —●—; *Raman distribution corrected for 30° tilt,* —○—. (*Reproduced with permission from reference 62. Copyright 1989 John Wiley & Sons, Inc.*)

distribution becomes broader, the observed intensity is spread out over a larger frequency range; thus resolution of the extremes is more difficult. These problems for broad distributions are not problems of principle, but rather of practice, which will eventually be overcome.

DLAM.   The DLAM is a low-frequency Raman band that measures long-range conformational disorder. DLAM has been observed in all of the noncrystalline polymers studied, including molten polyethylene (64, 65), and it can be related to the average of the conformational disorder in a long-chain molecule (64). A basic and important question is whether the DLAM band is found in semicrystalline polymers. A DLAM band identical to that of the molten polymer has been observed in crystalline linear polyethylene (66). Studies at ambient temperature with samples that have varying degrees of crystallinity show that the intensity of the DLAM bands is proportional to $(1 - \lambda)_{\Delta H}$ and that the intensities extrapolate to the correct value for the completely disordered system at $(1 - \lambda)_{\Delta H} = 0$ (66). These results give very strong and independent support of previous conclusions that the disordered interlamellar regions have the same structural features as the pure completely molten polymer (3, 4). This conclusion is obviously of major importance in interpreting the properties of crystalline polymers. Furthermore, studies of the temperature dependence of the DLAM band do not show any discontinuities in the position or shape of the DLAM in going from the crystalline state to the pure melt (3, 66).

In the discussion, heretofore, we have demonstrated the validity and utility of specific spectral regions in describing the main structural regions of semicrystalline polymers. These results now can be applied to develop a better understanding of the crystalline behavior and properties of such systems. Some examples have been selected from among the many examples that illustrate how the information obtained from Raman spectroscopy yields important and detailed structural features. When these examples are coupled with the procedures outlined, a better molecular understanding of certain properties emerges.

Applications

Secondary Crystallization.   After nonisothermal crystallization, the density of linear polyethylene slowly increases with long-time storage at room temperature. The tacit assumption is that this density increase reflects an increase in the level of crystallinity. Consequently, this process has been termed secondary crystallization. In a recent study, the density changes were accompanied by an analysis of the Raman internal modes (67). For rapidly crystallized high molecular weight linear polyethylenes, branched polyethylene, and ethylene copolymers, a density increase of about 5% is observed after long-term crystallization (i.e., about several hundred days) at room temperature. Contrary to expectations, the Raman data indicate that the core crystallinity remains constant over this time period. Instead, the density increase reflects an increase in the partially ordered interfacial region at the expense of a decrease in the disordered liquidlike region. This behavior is illustrated in Figure 6 for a fraction $M_\eta = 8 \times 10^6$ (67), where M_η is the viscosity average molecular weight. After 392 days, the density increases from 0.9260 to 0.9310; α_c remains constant between 0.40 and 0.41, while α_a decreases from 0.48 to 0.41 and α_b increases from 0.11 to 0.20. Similar effects are observed with branched polymers and copolymers. Lower molecular weight linear polymers, for example, $M_w = 1.5 \times 10^5$ only show very small changes in any of these quantities. It will be of interest to analyze the Raman internal modes during the secondary crystallization process that accompanies isothermal crystallization to determine what changes occur in the phase structure. Under these crystallization conditions, significant crystallite thickening takes place (68, 69).

Crystallite Thickening.   A long-standing controversy exists with regard to whether crystallites thicken in the chain direction during either the primary or secondary portions of isothermal crystallization. Contrary experimental evidence had been presented (*see* references 68 and 69 for a summary of literature on the subject. In the present context, we do not consider the rapid thickening that occurs immediately after nucleation). Raman LAM measurements on molecular weight fractions, which were crystallized under controlled isothermal crystallization condition, have resolved this problem. Because the Raman experiment only requires very small amounts of sample, extensive studies of the influence of molecular weight and crystallization temperatures, using fractions, could be carried out (69). A typical example is given in Figure 7 for a linear polyethylene fraction $M_w = 188,000$, $M_n = 179,000$ crystallized at 128 °C for the indicated times. The most probable ordered sequence length increases from 340 to 500 Å for the time scale and temperature in this example. It is quite clear from the figure that, at the same time, there is a significant broadening of the size distribution, even in the secondary crystallization region.

The crystallite thickening process is dependent on both molecular weight and the temperature. The thickening rate decreases with molecular weight and increases with temperature. The rather drastic temperature effect is illustrated in Figure 8 for a fraction $M_w = 70,000$, $M_n = 65,000$ (68). The strong influence of the annealing temperature is clearly indicated in the figure. Data of the type illustrated can serve as a base to develop a better molecular understanding of the thickening process (69).

Mixtures.   When two different species, which are homogeneous in the melt, are crystallized, it is important to establish if co-crystallization or some

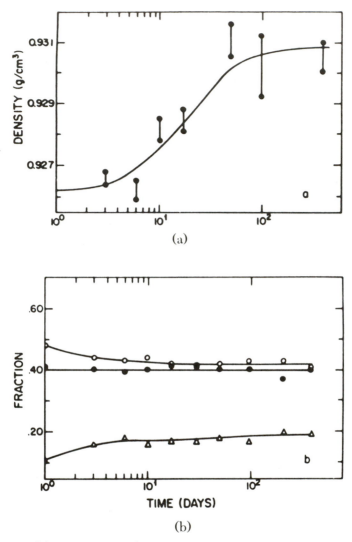

(a)

(b)

Figure 6. (a) Plot of variation of density with time after storage of a linear polyethylene, $M_\eta = 8 \times 10^6$, at room temperature. (b) Plot of variation in the liquidlike (○), crystalline (●), and interfacial content (△) with time for the same sample. (Reproduced with permission from reference 67. Copyright 1989 Springer-Verlag.)

type of segregation occurs. Any changes in the crystallite and phase structures are of interest. Co-crystallization can be detected by analysis of the endothermic peaks (obtained by differential scanning calorimetry) and selective extraction (70, 71). Both methods must be applied concurrently to obtain unambiguous results (71). The changes, if any, in the crystallite and phase

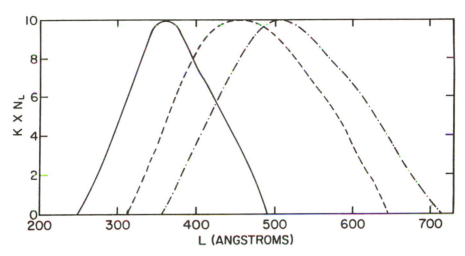

Figure 7. Normalized crystallite size distribution for fraction (M_w = 188,000, M_n = 179,000) crystallized at 128 °C for different times. Key: 135 min, ——; 1140 min, - - -; 10,000 min, ·–·–·. (Reproduced with permission from reference 68. Copyright 1982 Springer-Verlag.)

structure can be determined by Raman LAM and internal mode analyses. Mixtures of linear polyethylene fractions with compositional and molecular weight fractions of ethylene copolymers have been studied in this manner, and it has been established that some type of species segregation occurs during isothermal crystallization. However, upon rapid crystallization there is a range of copolymer compositions that co-crystallize with the homopolymer of corresponding chain length. The ordered sequence length distributions are quite different for the two cases as is illustrated in Figure 9 for mixtures of an equal amount of a linear polyethylene fraction $M_w = 1.05 \times 10^5$ and a hydrogenated polybutadiene $M_w = 1.08 \times 10^5$ that has 2.2-mol% branch points. The ordered sequence length distribution for the isothermally crystallized 50:50 mixture (Figure 9A) displays two distinct maxima, each of which corresponds to that of the pure species. This result is expected for a mixture that does not co-crystallize. In contrast, for the 50:50 quenched mixture where co-crystallization takes place, the size distribution is continuous, which is consistent with the formation of only one type of crystallite (Figure 9B). Surprisingly, in this case the size distribution resembles that of the pure linear species. This characteristic of the distribution is confirmed by scanning electron microscopy (71).

Internal mode analysis indicates that some interesting changes take place in the relative amounts of the phases upon the co-crystallization of linear polyethylene with random ethylene copolymers (71, 72). Although only a very limited amount of data are available for fractions (71), when these data are taken together with those for unfractionated systems (72), a general pattern

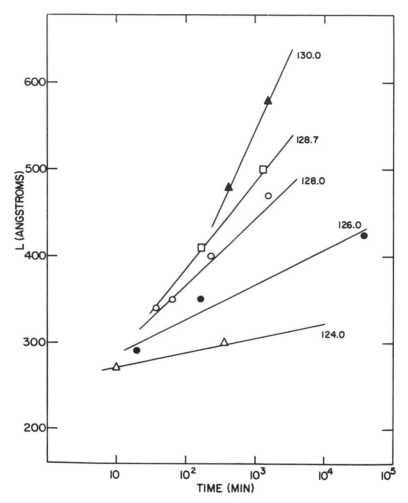

Figure 8. Plot of the most probable thickness (maximum in distribution function) as a function of log time for fraction (M_w = 70,000, M_n = 65,000) crystallized at the indicated temperatures. (Reproduced with permission from reference 68. Copyright 1982 Springer-Verlag.)

emerges. The fraction of core crystallinity, and the fraction of the liquidlike region change monotonically, but not linearly, with the composition of the mixture. On the other hand, as illustrated in Figure 10, the relative amount of the interfacial region has a maximum with mixture composition. The intermingling of the two kinds of ordered sequences enhances the fraction of units in the interfacial region and presumably increases the thickness of the interphase. A wide variation in the phase structure can then be envisioned by the co-crystallization of a linear polymer with a copolymer.

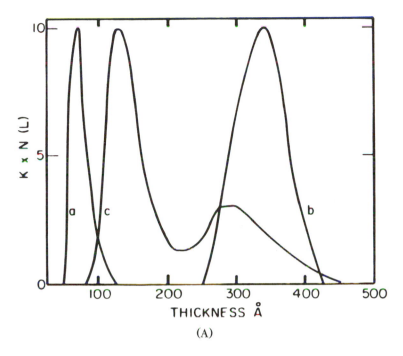

(A)

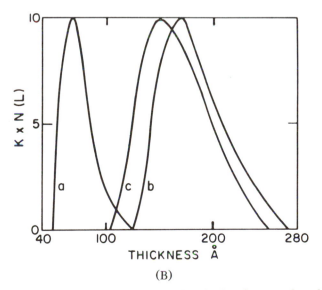

(B)

Figure 9. Normalized ordered sequence length distribution of isothermally crystallized (A) and rapidly crystallized (B) mixtures of linear polyethylenes and hydrogenated polybutadiene determined from Raman LAM. Key: pure hydrogenated polybutadiene, a; pure linear polyethylene, b; 50:50 mixture, c. (Reproduced with permission from reference 71. Copyright 1988 John Wiley & Son.)

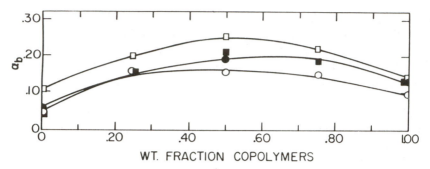

Figure 10. Plot of fraction interphase, α_b, for homopolymer–copolymer mixtures as function of the weight fraction of copolymer. Key: fraction of linear polyethylene and hydrogenated polybutadiene ●, (71); mixtures of unfractioned linear polyethylene and linear low-density polyethylenes, ○, □, ■ (72).

Thickness of Structural Regions. The thicknesses of the three major structural regions are of obvious importance in interpreting the properties of crystalline polymers. LAM and internal mode measurements can give the thicknesses of the core crystallite, the interlamellar region, and the interfacial regions in a straightforward manner. For narrow size distributions the maximum value, L_R, in the LAM-determined ordered sequence distribution can be identified with the crystallite thickness L_c by the relation $L_c = L_R \cos \theta$, where θ is the angle of inclination between the chain axis and the normal to the basal phase of the lamellar crystallite. If we define L_a as the thickness of the liquidlike region and L_b as the thickness of the interface that is associated with one of the crystallite basal planes, it follows that the crystalline fraction is given by

$$\alpha_c = \left[\frac{L_c}{(L_a + 2L_b) + L_c} \right] \times \frac{\rho_c}{\rho_T} \qquad . \qquad (3)$$

the interfacial fraction is given by

$$\alpha_b = \left[\frac{2L_b}{(L_a + 2L_b) + L_c} \right] \times \frac{\rho_b}{\rho_T} \qquad (4)$$

and the liquidlike fraction is given by

$$\alpha_a = \left[\frac{L_a}{(L_a + 2L_b) + L_c} \right] \times \frac{\rho_a}{\rho_T} \qquad (5)$$

where ρ_a, ρ_b, and ρ_c are the density of the liquidlike, interfacial, and crystalline regions, respectively. ρ_T is the density of the sample. Because α_a, α_c, α_b, and L_c are determined experimentally from Raman spectroscopy, the values of L_a and L_b can be calculated. The tilt angle θ is taken to be 30°; if a slightly higher tilt angle is used, the basic conclusions remain essentially unaltered. The denominator in these equations represents the long period, which can also be obtained by either small-angle X-ray scattering or electron microscopy, and includes the statement that there are two interfaces per crystallite.

The results for linear polyethylene fractions crystallized by quenching to -78 °C are illustrated in Figure 11 (73). The core crystallite thickness remains constant at 140 Å as the molecular weight increases from 10^4 to 10^6. Concomitantly, over the same molecular weight range, the thickness of the disordered interlamellar layer increases from ~ 75 to 175 Å. These results are in accord with electron microscopy studies where it was found that for molecular weight fractions crystallized under similar conditions, the average crystallite thickness remained constant whereas the long period increased with chain length (74). Similar results also have been obtained by small-angle X-ray scattering (75). Of particular interest in this analysis is the determination of the interfacial thickness. L_b is molecular weight dependent with values ranging from 14 Å at $M = 10^4$ to ~ 25 Å for $M = 10^6$. The

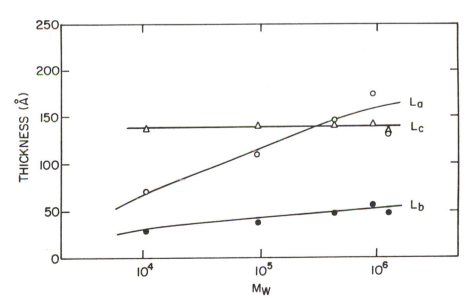

Figure 11. Plot of thickness values as a function of molecular weight for linear polyethylene fractions quenched to -78 °C. Key: crystallite core thickness L_c, $\triangle$; interlamellar thickness L_a, $\bigcirc$; interfacial thickness L_b, ●. (Reproduced from reference 73. Copyright 1990 American Chemical Society.)

interfacial thickness thus makes a significant contribution to the total structure. Similar results have been obtained qualitatively with samples that have a most probable molecular weight distribution (73). Isothermal crystallization (73) and the analysis of carbon-13 solid state NMR data, coupled with small-angle X-ray diffraction, also give comparable values for L_b (33, 35).

A major conclusion with respect to melt-crystallized samples is that there is a significant molecular weight dependent interfacial region in linear polyethylene. The values of the interfacial thickness are in agreement with theoretical expectations of 10–30 Å (28). A discussion of the reasons for these values and their molecular weight dependence is beyond the scope of this chapter; they are, however, detailed elsewhere (4, 28, 29). It suffices to state that these results support the conclusion that the polyethylenes possess an irregularly structured interphase. The deduced interfacial structure is incompatible with a set of regularly folded chains or minor variants thereof.

A similar analysis can be performed on crystals formed in dilute solutions. The results are summarized in Table III. The data in the last two columns of Table III contrast strongly with the results obtained for the bulk-crystallized fractions. At a fixed crystallization temperature, T_c, the quantities L_c, L_a, and L_b are all independent of molecular weight. The value of L_b of ~ 5 Å is very similar to that deduced from a comparison of Raman LAM and small-angle X-ray scattering (46). It is very important to note that a significant disordered overlayer of ~ 20–25 Å is associated with these crystallites. Thus, polyethylene crystals formed in dilute solutions, as well as other polymers, are not completely crystalline (59). The invariance of L_b with molecular weight and the reduced value of L_b relative to bulk-crystallized systems are characteristic features of crystals formed in dilute solution. These features can be explained by the reduced influence of topological restraints to the crystallization process in dilute systems. At the same time, however, about 20–30% of the system is not contained within the crystallite core.

Table III. Structural Parameters of Linear Polyethylene Fractions Crystallized from Dilute Solution (73)

M_w ($\times 10^{-4}$)	T_c (°C)	α_c	α_a	α_b	L_R (Å)	$L_c{}^a$ (Å)	L_b (Å)	L_a (Å)
5.26	87	0.80	0.16	0.04	127	110	3	22
11.5	87	0.77	0.14	0.09	127	110	6	20
16.1	87	0.78	0.16	0.06	136	118	5	24
22.5	87	0.78	0.19	0.03	127	110	2	27
80.0	87	0.79	0.15	0.06	134	116	4	22
150.0	87	0.73	0.18	0.09	126	109	7	27

[a] Tilt angle taken as 30°.

Crystals from Dilute Solution. The data in Table III indicates that the LAM values for crystallites formed from dilute solution are independent of molecular weight for a given isothermal crystallization temperature. These results are illustrated graphically in Figure 12 (*46*). It is apparent from the plot that the crystallites have a relatively narrow size distribution and the ordered sequence lengths are independent of molecular weight. Small-angle X-ray studies have shown that the total lamellar periodicity is constant over a wide range of molecular weights for different isothermal crystallization temperatures (*76, 77*). Because α_c is essentially constant with chain length, the small-angle X-ray and LAM data are in agreement with respect to the crystallite thickness dependence on molecular weight. These experimental findings are contrary to a theoretical treatment that predicts a substantial change in crystallite thickness with molecular weight (*78*). These results are also important because they demonstrate that properties are independent of molecular weight, but are dependent on the crystallization temperature and, thus, the crystallite thickness (*79, 80*).

n-Alkanes. The recent synthesis of high molecular weight *n*-alkanes has shown that both extended and folded chain structures can be formed in the crystalline state (*42, 43*). The lengths of the ordered sequences are close to an integral fraction of the extended chain lengths. An integral value implies a sharp fold that involves the minimum number of units in specific orientations. On the other hand, if a disordered folded structure existed, the ordered sequence length would be slightly less than the integral value. The difference in sequence length between the two structures is between 5–10 Å. The theoretical difficulty of relating the observed LAM frequency to the ordered sequence length when there is an end-group effect, even when the vibrating entities are uncoupled, was discussed in the introductory section. In the problem under discussion here, a further complication is that the sequences

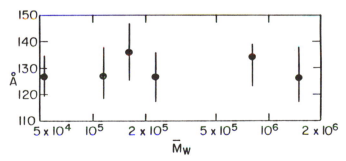

Figure 12. Plot of ordered sequence length from LAM of polyethylene crystals formed in dilute solution against $\overline{M}_w$. Bars show the half-height range. (Reproduced with permission from reference 46. Copyright 1986 Springer-Verlag.)

may be coupled to one another. Thus, within our present theoretical understanding, there are major impediments to using the LAM to analyze the
folded structure of the n-alkanes. Contrary reports notwithstanding, the use
of this method will not enable the determination of the ordered sequence
length or the establishment of the nature of the fold structure.

However, analysis of the internal mode regions does show quantitative
differences between the extended and folded structures. These differences
are illustrated in the spectra given in Figure 13 for $C_{192}H_{386}$ in both the
extended and singly folded form (81). For comparative purposes, the spectrum of the folded structure has been shifted along the ordinate so that the
intensity of its peak at $\Delta\bar{\nu} = 1416$ cm^{-1} coincides exactly with the intensity
of the extended structure at the same frequency. It is clear from this
superposition that the folded structure is less crystalline than the extended
structure. The amorphous bands ($\Delta\bar{\nu}$: 1080, 1303, and 1440 cm^{-1}) are
indicated by arrows in the figure. A more quantitative analysis, carried out in
the conventional manner (81), indicates that α_c for the folded structure is
only 0.80–0.85. This result is only compatible with a disordered overlayer for
the folded structure. In this respect, the structure is very similar to the
corresponding structure of polymer solution crystals (see Table III).

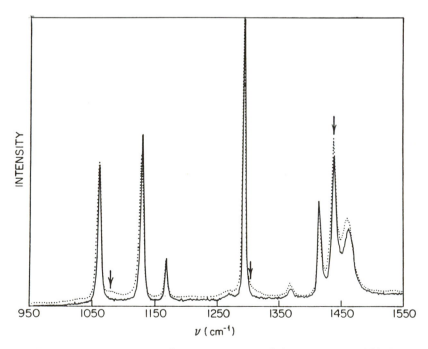

*Figure 13. Raman spectra for $C_{192}H_{386}$: extended structure, —; folded
structure, ...*

Specific Applications

We shall discuss three specific examples where Raman spectroscopy plays a unique role in the understanding of properties.

Carbon-13 Spin Relaxation Time.   Carbon-13 spin lattice relaxation times, T_1, of crystalline polyethylenes can vary over wide limits (*31*, *82–86*). T_1 values ranging from 2 to 400 s have been reported, but no rationale has been given for this wide variance. Utilizing LAM and internal mode data, the structural basis for these observations can be explained. The results are summarized in Figure 14, where the ordered sequence lengths obtained by LAM are plotted against T_1 (*31*). The T_1 values reported here range from 40 to 4500 s. For the main body of data, there is a smooth monotonic increase of the crystalline T_1 with crystallite thickness; the curve appears to increase without limits. Even much larger T_1s are expected on this basis if larger crystallite thickness is attained. Four data points—two for selectively oxidized samples and two for low molecular weight linear polyethylenes—do not lie on the curve delineated by the main body of results. These four samples have the common feature that the crystallite thickness is comparable to the extended chain length; thus these samples will have different interfacial structures relative to those of the main body of the samples.

The experimental finding that T_1 increases with L_R explains the divergent results that were initially reported. Closer examination of the structural features reveals that the lower values of T_1 are characterized by larger ratios of α_b/α_c whereas specimens with very high values of T_1 have very low values of α_b/α_c. Even a modest interfacial content is sufficient to profoundly influence T_1. Thus, there is a strong coupling of the interfacial structure with the motions within the crystalline region that yield T_1. When these connections, or junctions, are severed, the relaxation within the crystalline region becomes more retarded. This point is clarified in Figure 14 by the data for the four extended chain specimens where T_1 is found to be much larger than expected based solely on crystallite thickness considerations. The importance of the interfacial region in this problem is further emphasized by comparison of T_1 for the original and the selectively oxidized samples. With only very modest increases in crystallite thickness, T_1 increases twofold for an aircooled sample ($\alpha_c = 0.64$) and sevenfold for a quenched sample ($\alpha_c = 0.48$) when the specimens are selectively oxidized. These results elucidate the role of the interfacial region in modulating the motions within the crystalline region. The crucial role played by Raman spectroscopy in sorting out the different variables involved and focusing attention on the key structural features is apparent.

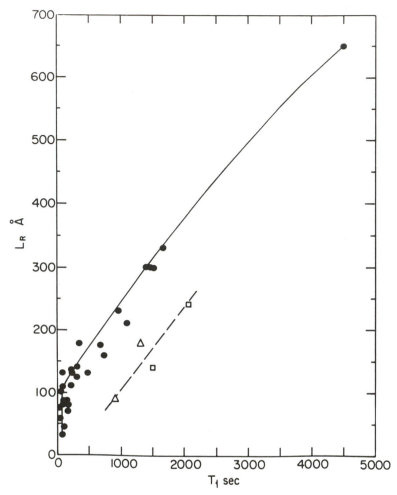

Figure 14. Plot of crystallite thickness L_R *against crystalline* T_1 *for linear polyethylenes. The dashed curve represents samples that are either selectively oxidized or low molecular weight fractions. (Reproduced with permission from reference 31. Copyright 1983 John Wiley & Sons, Inc.)*

Dynamic Mechanical Relaxation. The linear and branched polyethylenes, as well as copolymers of ethylene, display a series of well-known relaxation transitions that can be detected by a variety of physicochemical techniques (87, 88). In order of decreasing temperature below the melting temperature, these transitions are designated conventionally as the α, β, and γ transitions (87). The γ transition is usually observed in the range of −150 to −120 °C and can be identified with the glass-transition temperature. The β transition is found in the range of −30 to +10 °C, depending on the molecular constitution of the chain. The molecular basis for this relaxation remained elusive until recently (87, 89). The α transition has been observed over a range of at least 100 °C, depending on the structural details of the polyethylenes. All the methods used to characterize the α relaxation show that it results from motions of chain units that lie within the crystalline portion of the polymer. An increase of the temperature of the α transition, T_α, with crystallite thickness is suspected. The application of Raman spectroscopy to the problem has greatly clarified the molecular and structural basis for the α and β transitions (88, 89).

Studies of polyethylene samples that have ordered sequence lengths that vary from ~ 60 to > 300 Å, quantitatively established the dependence of T_α on crystallite thickness. This point is illustrated in Figure 15 (88). The location of the α transition is found to depend primarily on the crystallite thickness. More detailed study indicates the distinct possibility that the interfacial structure exerts an important influence on this relaxation. The level of crystallinity and the supermolecular structures do not play any significant role in determining the location of T_α. A strong correlation is found between the α-transition and the carbon-13 NMR crystalline T_1 (31). We can also note from Figure 15 that, in contrast to T_α, the temperature of the β transition, T_β, is independent of crystallite thickness.

An understanding of the structural basis for the β transition has been elusive and controversial (88). This transition is widely observed in branched

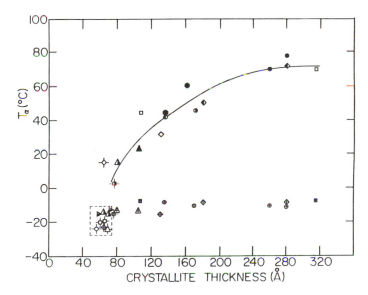

Figure 15. Plot of α- and β-transition temperatures, from tan δ, at 3.5 Hz against the crystallite thickness. Details of super molecular structure are described in reference 88.

polyethylenes and ethylene copolymers, but has only been observed in very high molecular weight linear polyethylene (these observations are summarized in reference 88). This transition is invariant with crystallite thickness and does not depend directly on either the crystallinity level or the supermolecular structure. Correlation of the dynamic mechanical experiments with the Raman internal modes has established that the β transition results from the relaxation of chain units that are located within the interfacial region. This transition is very intense for random ethylene copolymers and can also be discerned in the linear polymer as is illustrated in Figure 16, where the loss modulus E'' is plotted against the temperature in the β-transition region (88). The plot in Figure 16 shows that the intensity of the β transition can change significantly for samples that have the same degree of crystallinity. The intensity of this transition varies with the magnitude of the interfacial content, α_b. For example, the existence of a transition in sample C is questionable. However, for the two higher molecular weight samples (3×10^6 and 8×10^6) the β transition is centered about -10 °C and is well defined. For these samples, α_b varies from 7–13%. An interfacial content greater than about 10% appears to be sufficient to permit clear detection of the β transition (88). Assignment of the β transition to the relaxation of chain units in the interfacial region explains its intensity in random copolymers, which typically have large interfacial contents. It also explains the very unique copolymer composition relations that are observed and separates this transition from the glass-transition temperature (89). The reality of a β transition in linear polyethylene has been established by this type of structural characterization; its elusiveness in the past can be related to the low interfacial content of the lower molecular weights that were studied most intensively. A

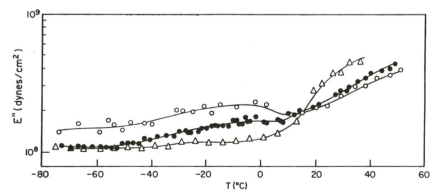

Figure 16. Loss modulus E'' vs. temperature for linear polyethylenes in the β relaxation region (88). Sample A (○): unfractionated $M_w = 8 \times 10^6$, $\alpha_a = 0.35$, $\alpha_b = 0.13$, $(1 - \lambda)_d = 0.63$; *sample B (●): fractionated* $M_w = 3.16 \times 10^5$, $\alpha_a = 0.31$, $\alpha_b = 0.10$, $(1 - \lambda)_d = 0.65$; *sample C (△): unfractionated* $M_w = 1.5 \times 10^5$, $\alpha_a = 0.37$, $\alpha_b = 0.07$, $(1 - \lambda)_d = 0.63$.

compilation of the experimental results for the β transition in linear and branched polyethylene is given in Table IV. This summary emphasizes the relation between the interfacial content α_b and the intensity necessary for the existence of the β transition. When the interfacial content is small, less than 5–7%, the β transition is not observed. This conclusion is exemplified by the results for the solution-formed crystals and for low and medium molecular weight bulk-crystallized linear polyethylene. When the interfacial content increases above 10%, as is found in high molecular weight bulk-crystallized linear polyethylene and in both solution- and bulk-crystallized branched polyethylenes, well-defined β transitions are observed. These conclusions are consistent with the results for different types of chlorinated polyethylenes (88). It is also clear that just a high noncrystalline content is not a sufficient requirement for a β relaxation to be observed.

The relation that has been deduced between the β transition in the polyethylenes and the relaxation of chain units in the interfacial region should also hold for other semicrystalline polymers. The dynamic mechanical behavior of polyoxymethylene is very similar to that of polyethylene. The introduction of small amounts of ethylene oxide co-units into the chain greatly enhances the intensity of the originally weak β transition (90, 91). These results parallel those for copolymers of ethylene and point to a common origin. Because the ethylene oxide co-units are effectively excluded from the crystal lattice, an enhanced interfacial structure is expected.

Small-angle X-ray and neutron scattering studies, as well as studies involving dielectric relaxation, have demonstrated the existence of an interfacial region in poly(ethylene oxide) (92) and poly(vinylidene fluoride) (93, 94). Thus, the concept of an interfacial region is now well-documented for several quite different polymer types and can be taken to be a universal feature of all crystalline systems comprised of flexible chains. The concept of

Table IV. Summary of β Transitions in Polyethylenes (88)

Sample	Observation	Interfacial content α_b (%)
Solution crystals of linear polyethylene	Not observed	< 5
Solution crystals of branched polyethylene	Transition observed[a]	11–17
Bulk-crystallized linear polyethylene	Not observed for low molecular weights ($< 2 \times 10^5$)	< 7
	Observed for high molecular weights ($> 2 \times 10^5$)	> 10
Bulk-crystallized branched polyethylene	Strong relaxation always observed	11–21

[a] No report of dynamic mechanical studies on such systems. Transition is observed by indirect measurement of thermal expansion coefficient.

an interfacial region that is unique to polymeric crystals was initially proposed by Flory (22). The analysis of the phase structure of the polyethylenes by Raman internal modes provides an experimental base for the phenomenon and has stimulated research on other polymers.

Tensile Properties.   The use of Raman spectroscopy for characterization purposes has been extremely useful for interpreting tensile properties of the polyethylenes (95–97). A good example is in the analysis of the yield stress. It is well known that the yield stress for the polyethylenes, as well as other polymers, depends on the crystallinity level (95–97). However, analysis by means of the Raman internal mode gives a deeper insight into the nature of the phenomenon as is illustrated in Figures 17 and 18 (96). These figures give extensive results for unfractionated random ethylene copolymers. In Figure 17, the yield stress is plotted against $(1 - \lambda)_d$ for a compilation of available data from the literature (96). Although the yield stress varies with the crystallinity level, it falls to zero when 15–20% crystallinity remains. In contrast, Figure 18 shows that the yield stress is directly proportional to α_c, the core crystallinity level. Very similar results are obtained with linear polyethylene (97). These results clearly imply that yielding is governed by some process, or processes, that involve the deformation or transformation of part or all of the crystalline regions. Raman spectroscopy has thus focused attention on the pertinent element of phase structure that is involved in the yielding process.

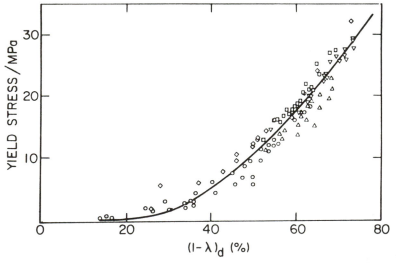

Figure 17. Compilation of literature data for yield stress as a function of degree of crystallinity, $(1 - \lambda)_d$, calculated on a density basis. (Reproduced with permission from reference 96. Copyright 1990 John Wiley & Sons, Inc.)

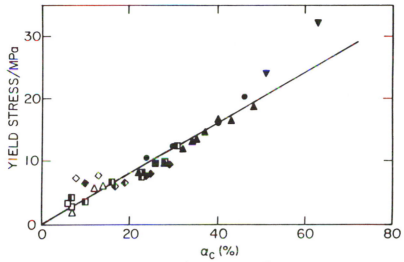

Figure 18. Plot of yield stress against core crystallinity for random ethylene copolymers. (Reproduced with permission from reference 96. Copyright 1990 John Wiley & Sons, Inc.)

Summary

We have summarized the application of Raman spectroscopy to the characterization of certain structural elements of crystalline polymers of flexible chains. The polyethylenes were chosen as models for such systems because they served admirably in the study of other properties. Certain key structural features are delineated and quantitatively described by this method, both in principle and in practice. Different spectral regions are involved in this process. The main structural elements that are quantitatively described are the phase structure, the distribution of ordered sequence lengths, and the general conformational structures of the interlamellar regions. The values of these quantities can be varied over wide limits by varying the molecular weight and crystallization conditions.

The structural parameters that can be determined by Raman spectroscopy are very important to the determination of a variety of microscopic and macroscopic properties. A few selected examples have been discussed for illustrative purposes.

Acknowledgment

The support of this work by the National Science Foundation Polymer Program grant DMR−89−14167 is gratefully acknowledged. We also thank Dr. A. J. Peacock for obtaining the spectra that are illustrated in Figure 13.

References

1. Mandelkern, L. *Crystallization of Polymers*; McGraw-Hill: New York, 1964.
2. Mandelkern, L. *Polym. J.* **1985**, *17*, 337.
3. Mandelkern, L. In *Polymer Characterization: Physical Properties, Spectroscopic, and Chromatographic Methods*; Craver, C. D.; Provder, T., Eds.; Advances in Chemistry 227; American Chemical Society: Washington, DC, 1990; p 377.
4. Mandelkern, L. *Acc. Chem. Res.* **1990**, *23*, 380.
5. Mandelkern, L. In *Characterization of Materials in Research, Ceramics and Polymers*; Syracuse University Press: Syracuse, NY, 1975; p 369.
6. Maxfield, J.; Mandelkern, L. *Macromolecules* **1977**, *10*, 1141.
7. Strobl, G. R.; Hagedorn, W. *J. Polym. Sci. Polym. Phys. Ed.* **1978**, *16*, 1181.
8. Glotin, M.; Mandelkern, L. *Colloid Polym. Sci.* **1982**, *260*, 182.
9. Mandelkern, L.; Peacock, A. J. *Polym. Bull.* **1986**, *16*, 529.
10. Mizushima, S. Shimanouchi, T. *J. Am. Chem. Soc.* **1949**, *71*, 1320.
11. Schaufele, R. F.; Shimanouchi, T. *J. Chem. Phys.* **1967**, *47*, 3605.
12. Hartley, A.; Leung, Y. K.; Booth, C.; Shepherd, I. W. *Polymer* **1976**, *17*, 354.
13. Hendra, P. J.; Majid, H. A. *Polymer* **1977**, *18*, 573.
14. Strobl, G. R.; Eckel, R. *J. Polym. Sci., Polym. Phys. Ed.* **1976**, *14*, 913.
15. Fraser, G. V. *Polymer* **1978**, *19*, 857.
16. Glotin, M.; Mandelkern, L. *J. Polym. Sci., Polym. Phys. Ed.* **1983**, *21*, 29.
17. Glotin, M.; Mandelkern, L. *J Polym. Sci., Polym. Lett. Ed.* **1983**, *21*, 807.
18. Snyder, R. G. *J. Chem. Phys.* **1982**, *76*, 3921.
19. Snyder, R. G.; Wunder, S. L. *Macromolecules* **1986**, *19*, 2404.
20. Snyder, R. G.; Schlotter, N. E.; Alamo, R.; Mandelkern, L. *Macromolecules* **1986**, *19*, 621.
21. Mattice, W. L.; Snyder, R. G.; Alamo, R.; Mandelkern, L. *Macromolecules* **1986**, *19*, 2404.
22. Flory, P. J. *J. Chem. Phys.* **1949**, *17*, 223.
23. Mansfield, M. L. *Macromolecules* **1983**, *16*, 914.
24. Flory, P. J.; Yoon, D. Y.; Dill, K. A. *Macromolecules* **1984**, *17*, 862.
25. Yoon, D. Y.; Flory, P. J. *Macromolecules* **1984**, *17*, 868.
26. Marqusee, J. A.; Dill, K. A. *Macromolecules* **1986**, *19*, 2420.
27. Marqusee, J. A. *Macromolecules* **1989**, *22*, 472.
28. Kumar, S. K.; Yoon, D. Y. *Macromolecules* **1989**, *22*, 3458.
29. Zuniga, I.; Rodrigues, K.; Mattice, W. L. *Macromolecules* **1990**, *23*, 4108.
30. Okui, N.; Li, H. M.; Magill, J. H. *Polymer* **1978**, *19*, 411.
31. Axelson, D. E.; Mandelkern, L.; Popli, R.; Mathieu, P. *J. Polym. Sci., Polym. Phys. Ed.* **1983**, *21*, 2319.
32. Axelson, D. E.; Russell, K. E. *Prog. Polym. Sci.* **1985**, *11*, 221.
33. Kitamaru, R.; Horii, F.; Murayama, K. *Macromolecules* **1985**, *19*, 636.
34. Saito, S.; Moteki, Y.; Nakagawa, M.; Horii, F.; Kitamaru, R. *Macromolecules* **1990**, *23*, 3256.
35. Hirai, A.; Horii, F.; Kitamaru, R.; Fatou, J. G.; Bello, A. *Macromolecules* **1990**, *23*, 2913.
36. Kitamaru, R.; Horii, F.; Hyon, S. H. *J. Polym. Sci., Polym. Phys. Ed.* **1977**, *15*, 821.
37. Mandelkern, L. *Physical Properties of Polymers*; Mark, J. E., Ed.; American Chemical Society: Washington, DC, 1984; p. 155.
38. Mandelkern, L.; Peacock, A. J. *Studies in Physical and Theoretical Chemistry*; Lacher, R. C., Ed.; Elsevier Science Publishers, B. V.: Amsterdam, The Netherlands, 1988; Vol. 54, pp. 201–227.

39. Alamo, R. G.; Mandelkern, L. *Macromolecules* **1991**, *24*, 6480.
40. Alamo, R. G.; Mandelkern, L. *Macromolecules* **1989**, *22*, 1273.
41. Rabolt, J. F. *CRC Crit. Rev. Solid State Mater. Sci.* **1985**, *12*, 165.
42. Ungar, G.; Stejny, J.; Keller, A.; Bidd, I.; Whiting, M. C. *Science* **1985**, *229*, 386.
43. Lee, K. S.; Wegner, G. *Makromol. Chem. Rapid Commun.* **1985**, *6*, 203.
44. Snyder, R. G.; Alamo, R. G.; Mandelkern, L., submitted for publication in *J. Phys. Chem.*
45. Folkes, M. G.; Keller, A.; Stejny, J.; Goggin, P. L.; Fraser, G. V.; Hendra, P. J. *Colloid Polym. Sci.* **1975**, *253*, 354.
46. Mandelkern, L.; Peacock, A. *J. Polym. Bull.* **1986**, *16*, 529.
47. Hendra, P. J.; Marsden, E. P.; Cudby, M. E. A.; Willis, H. A. *Makromol. Chem.* **1975**, *176*, 2443.
48. Fanconi, B.; Rabolt, J. F. *J. Polym. Sci., Polym. Phys. Ed.* **1985**, *23*, 1201.
49. Olf, H. G.; Peterlin, A.; Peticolas, W. L. *J. Polym. Sci., Polym. Phys. Ed.* **1974**, *12*, 359.
50. Hsu, S. L.; Krimm, S. *J. Appl. Phys.* **1976**, *47*, 4265.
51. Hsu, S. L.; Ford, G. W.; Krimm, S. *J. Polym. Sci., Polym. Phys. Ed.* **1977**, *15*, 1769.
52. Hsu, S. L.; Krimm, S. *J. Appl. Phys.* **1977**, *48*, 4013.
53. Krimm, S. *J. Polym. Sci., Polym. Phys. Ed.* **1978**, *16*, 2105.
54. Chang, C.; Krimm, S. *J. Polym. Sci., Polym. Phys. Ed.* **1979**, *17*, 2163
55. Peterlin, A. *J. Appl. Phys.* **1979**, *50*, 838; *J. Mater. Sci.* **1979**, *14*, 2994.
56. Alamo, R. G.; Mandelkern, L.; Chan, E. K.; Voigt-Martin, I. G. *Macromolecules*, in press.
57. Mandelkern, L.; Stack, G. M. *Macromolecules* **1988**, *21*, 510.
58. Stack, G. M.; Mandelkern, L.; Kröhnke C.; Wegner, G. *Macromolecules* **1989**, *22*, 4351.
59. Mandelkern, L.; Organization of Macromolecules in the Condensed Phase; Young, D. A., Ed. *Faraday Discuss. Chem. Soc.* **1979**, *68*, 310.
60. Snyder, R. G.; Krause, S. J.; Scherer, J. R. *J. Polym. Sci.; Polym. Phys. Ed.* **1978**, *16*, 1593.
61. Snyder, R. G.; Scherer, J. R. *J. Polym. Sci., Polym. Phys. Ed.* **1980**, *18*, 421.
62. Voigt-Martin, I. G.; Stack, G. M.; Peacock A. J.; Mandelkern, L. *J. Polym. Sci., Polym. Phys. Ed.* **1989**, *27* 957.
63. Voigt-Martin, I. G.; Mandelkern, L. *J. Polym. Sci., Polym. Phys. Ed.* **1989**, *27*, 967.
64. Snyder, R. G. *J. Chem. Phys.* **1982**, *76*, 3921
65. Snyder, R. G.; Wunder, S. L. *Macromolecules* **1986**, *19*, 204.
66. Snyder, R. G.; Schlotter, N. E.; Alamo, R.; Mandelkern, L. *Macromolecules* **1986**, *19*, 621.
67. Alamo, R. G.; McLauglin, K. W.; Mandelkern, L. *Polym. Bull.* **1989**, *22*, 299.
68. Stack, G. M.; Mandelkern, L.; Voigt-Martin, I. G. *Polym. Bull.* **1982**, *8*, 421.
69. Stack, G. M. Ph.D. Dissertation, Florida State University, Tallahassee, FL, 1983.
70. Glaser, R. H.; Mandelkern, L. *J. Polym. Sci., Polym. Phys. Ed.* **1988**, *26*, 221.
71. Alamo, R. G.; Glaser, R. H.; Mandelkern, L. *J. Polym. Sci., Polym. Phys. Ed.* **1988**, *26*, 2169.
72. Wang, J.; Pang, D.; Huang, B. *Polym. Bull.* **1990**, *24*, 241.
73. Mandelkern, L.; Alamo, R. G.; Kennedy, M. A. *Macromolecules* **1990**, *23*, 4721.
74. Voigt-Martin, I. G.; Mandelkern, L. *J. Polym. Sci., Polym. Phys. Ed.* **1984**, *22*, 1901.
75. Robelin-Souffaché, E.; Rault, J. *Macromolecules* **1989**, *22*, 3581.
76. Jackson, J. F.; Mandelkern, L. *Macromolecules* **1968**, *1*, 546.

77. Ergoz, E.; Mandelkern, L. *J. Polym. Sci.* **1972**, *10B*, 631.
78. Sanchez, I. C.; DiMarzio, E. A. *Macromolecules* **1971**, *4*, 677.
79. Mandelkern, L. *Progress in Polymer Science*; Jenkins, A. D., Ed.; Pergamon Press: New York, 1970; p. 165.
80. Mandelkern, L. *Annual Review of Material Science*; Huggins, R. A., Ed.; Annual Reviews: Palo Alto, CA, 1976; Vol. 6, p 119.
81. Peacock, A. J.; Mandelkern, L. unpublished results.
82. Schröter, B.; Posern, A. *Makromol. Chem.* **1981**, *182*, 675.
83. Fleming, W. W.; Lyerla, J. R.; Yannoni, C. S. *Polym. Prepr.* (*Am. Chem. Soc., Div. Polym. Sci.*) **1981**, *22*, 275.
84. Vander Hart, D. L. *J. Am. Chem. Soc.* **1979**, *12*, 1232.
85. Axelson, D. E. *J. Polym. Sci., Polym. Phys. Ed.* **1982**, *20*, 1427.
86. Schröter, B.; Posern, A. *Makromol. Chem., Rapid Commun.* **1982**, *3*, 623.
87. McCrum, N. G.; Read, B. E.; Williams, G. *Anelastic and Dielectric Effects in Polymer Solids*; Wiley: New York, 1967.
88. Popli, R.; Glotin, M.; Mandelkern, L.; Benson, R. S. *J. Polym. Sci., Polym. Phys. Ed.* **1984**, *22*, 407.
89. Popli, R.; Mandelkern, L. *Polym. Bull.* **1983**, *9*, 260.
90. Bohn, L. *Kolloid Z.* **1965**, *201*, 20.
91. Papir, Y. S.; Baer, E. *Mater. Sci. Eng.* **1971**, *8*, 310.
92. Russell, T. P.; Ito, H.; Wignall, G. D. *Macromolecules* **1988**, *21*, 1703.
93. Hahn, B.; Wendorff, J.; Yoon, D. Y. *Macromolecules* **1985**, *18*, 718.
94. Hahn, B.; Herrmann-Schönherr, O.; Wendorff, J. H. *Polymer* **1987**, *28*, 201.
95. Popli, R.; Mandelkern, L. *J. Polym. Sci., Polym. Phys. Ed.* **1987**, *25*, 441.
96. Peacock, A. J.; Mandelkern, L. *J. Polym. Sci., Polym. Phys. Ed.* **1990**, *28*, 1917.
97. Mandelkern, L.; Kennedy, M. A.; Peacock, A. J. in preparation.

RECEIVED for review May 14, 1991. ACCEPTED revised manuscript July 22, 1992.

Fourier Transform IR Spectroscopic Analysis of the Molecular Structure of Compatible Polymer Blends

M. Sargent[1] and J. L. Koenig*

Department of Macromolecular Science, Case Western Reserve University, 10900 Euclid Avenue, Cleveland, OH 44106–7202

Compatible polymer blends are formed by the combination of two or more polymers to produce a homogeneous, single phase mixture. The ability of polymers to form compatible blends requires that the interaction between the unlike polymer chains be at least as favorable as the self-association of each of the component polymers. Compatible polymer blends can therefore be characterized by the formation of intermolecular interactions between specific chemical groups of the component polymers. Infrared spectroscopy has been used to study polymer blend compatibility on the molecular level. The presence of molecular interactions is determined by examining the differences between the blend spectrum and the spectra of the component polymers. These spectral differences include shifts in the absorption frequency, increases in the band width, and changes in the absorptivity of the bands. This chapter reviews the application of spectral data processing techniques, such as factor analysis, difference spectroscopy, and least squares curve fitting, that characterize these interactions.

THE USE OF POLYMER BLENDS in a variety of scientific and industrial applications has been clearly established over the last several decades. New polymeric materials with superior chemical and physical properties may be

---

[1] Current address: Miami Valley Laboratories, Proctor and Gamble Company, P.O. Box 397707, Cincinnati, OH 45239–8707.
* Corresponding author

0065–2393/93/0236–0191$08.25/0

developed by selectively combining two or more homopolymers to produce a compatible blend. The ability of polymers to form compatible blends requires that the interaction between the unlike polymer chains be at least as favorable as the self-association of each of the component polymers. Reviews of the initial studies of polymer compatibility were published by Flory (1) and Tompa (2). More current reviews of compatible polymer blends include those by Bohn (3), Rosen (4), Krause (5), Barlow and Paul (6), Olabisi et al. (7), and Paul and Newman (8).

Incompatible polymer blends can be characterized as having components that exist in isolated phases or domains. In contrast, the components of a compatible blend are intimately mixed to form a single homogeneous phase. Compatibility of a polymer blend is dependent on many variables, including blend composition, temperature, and method of mixing. Phase separation of compatible blends may also occur over a period of time.

Relatively few techniques have been developed that are capable of examining intermolecular interactions between polymer components on the molecular level. Infrared spectroscopy has been successfully applied to detect such molecular interactions through analysis of changes in specific bands of the blend spectrum. Reviews of the application of infrared spectroscopy to the study of polymer blend compatibility have been published by Coleman and Painter (9, 10). A variety of spectral data processing techniques have been developed to assist in the characterization of the interactions (11). The presence of intermolecular interactions can be confirmed from factor analysis results. The spectrum of the interaction can be isolated using difference spectroscopy and the degree of interaction in the blend can be quantified by least squares curve fitting. The specific type of interaction may also be identified from differences between the blend spectrum and those of the pure components. Such changes include shifts in the absorption frequency, increases in the band width, and changes in the absorptivity of the bands.

Thermodynamics of Polymer Blends

The degree of miscibility of a mixture is determined by the Gibbs free energy of mixing, ΔG_{mix}, according to the equation

$$\Delta G_{mix} = \Delta H_{mix} - T \Delta S_{mix}$$

where ΔH_{mix} is the enthalpy of mixing, ΔS_{mix} is the entropy of mixing, and T is the temperature of the mixture. ΔG_{mix} can vary with the composition of the overall mixture in several ways, as shown in Figure 1. To achieve complete miscibility over all compositions, two conditions must be satisfied: ΔG_{mix} must be less than zero and the second derivative of the free energy with respect to the two components, $\delta^2 \Delta G_{mix}/\delta\phi^2$, must be greater than zero (ϕ is the volume fraction of each component in the mixture). Blends that can be

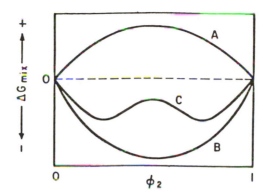

Figure 1. Possible free energy of mixing diagrams for binary mixtures. (Reproduced with permission from reference 6. Copyright 1991 Society of Plastics Engineers, Inc.)

described by curve A are completely immiscible because they have a free energy of mixing that is always positive. Blends that follow curve B satisfy both conditions and are therefore miscible over all compositions. Curve C describes blends that are partially miscible. When the blend is composed of compositions in the central range, the free energy of the system may be reduced by separating into two phases whose compositions are given by the two minima to the left and right of center.

The thermodynamic model most frequently used to describe the mixing of polymers is the Flory–Huggins theory (12), which assumes a lattice on which the polymer molecules can be arranged. Scott (13) applied the Flory–Huggins theory to mixtures of dissimilar polymers and obtained the following equation for the Gibbs free energy of mixing:

$$\Delta G_{mix} = \frac{RTV}{V_r}\left[\left(\frac{\phi_A}{X_A}\right)\ln\phi_A + \left(\frac{\phi_B}{X_B}\right)\ln\phi_B + \chi_{AB}\phi_A\phi_B\right]$$

Here V is the total volume, V_r is a reference volume taken as close as possible to the molar volume of the smallest polymer repeat unit, ϕ_A and ϕ_B are the volume fractions of polymers A and B, respectively, X_A and X_B are the degrees of polymerization of polymers A and B in terms of the reference volume. χ_{AB} is the interaction parameter that is related to the enthalpy of interaction of the polymer repeat units, each of molar volume V_r, R is the universal gas constant, and T is the temperature of the mixture.

Because a mixing process increases the randomness or disorder of the system, the change in entropy is always positive. However, the entropy of mixing is a function of the molecular sizes of the component polymers and approaches zero as the degree of polymerization increases. Therefore, because the entropy of mixing is very small for polymer mixtures, the free energy of mixing is essentially determined by the sign and magnitude of the enthalpy of mixing. This enthalpy of mixing depends on the energy change associated with nearest neighbor contacts during mixing. The free energy of mixing will be negative if the enthalpy of mixing is either negative or zero or if it is positive but less than the entropy term.

A negative enthalpy of mixing indicates that heat is evolved during mixing and occurs if the component polymers attract each other more than they attract their own kind. Such a situation is encountered very infrequently in polymer blends because the intermolecular energy is determined mostly by dispersive interaction in which the energy of a contact pair of dissimilar polymers is approximated by the geometric mean of the self-association energies of mixing. Therefore, negative enthalpy can only occur if strong, nondispersive interactions are formed between the different polymer components.

Determination of the Molecular Structure Using IR

From the previous discussion of the thermodynamics of polymer blends, it is clear that polymer compatibility can only occur when a strong molecular interaction occurs between the two components. This intermolecular interaction must be greater than the homopolymer intramolecular interactions of the components.

One of the aspects of infrared spectroscopy that is widely known is the ability to detect differences in molecular structure and interactions. It is on this basis that infrared is used to study polymer compatibility in blends (14). For such studies, it is necessary to generate and interpret the "interaction" spectrum. This interaction spectrum is the difference between the spectrum of the blend and the spectra of the component polymers, and it reflects the difference in the molecular interactions constituting the blend. Factor analysis methods can be used to verify the interaction spectrum. Interpretation of the interaction spectrum in terms of the molecular structure and interactions depends on the system under examination.

Factor analysis is a mathematical procedure that determines the number of spectroscopically identifiable, linearly independent components in a series of mixtures. One of the first applications of factor analysis to the infrared spectra of mixtures was conducted by Antoon et al. (15). An excellent summary of the mathematical principles involving this procedure can be found in a review by Gillette et al. (11). Essentially, the number of pure components is found by determining the rank of a covariance matrix $[C]$, which is the product of the data matrix of the spectra of mixtures $[M]$ multiplied by its transpose $[M]^t$:

$$[C] = [M][M]^t$$

The rank of this covariance matrix is determined by solving the eigenvalue problem

$$[C][E] = [E][\lambda]$$

where $[E]$ is the eigenvector matrix and $[\lambda]$ is the diagonal eigenvalue matrix.

Ideally, the number of pure components corresponds to the number of nonzero eigenvalues. However, experimental random noise in the data will also produce nonzero eigenvalues. A theory of error in factor analysis was therefore developed by Malinowski (16) to determine the correct number of nonzero eigenvalues resulting from the pure components. The difference between the built error-free data and the actual experimental results is expressed as the real error (RE)

$$\mathrm{RE} = \left[\frac{\sum_{j=p+1}^{j=m}\lambda_j}{n(mp)} \right]^{1/2}$$

where m is the number of spectra, p is the number of pure components, n is the number of points per spectrum, and λ is the eigenvalue. An indicator function (IND) can then be calculated from the equation

$$\mathrm{IND} = \frac{\mathrm{RE}}{(m-p)^2}$$

This indicator function attains its minimum value when the correct number of nonzero eigenvalues has been selected.

Factor analysis can be applied to a polymer blend system to determine whether a compatible or an incompatible mixture has been formed. For a binary mixture that is incompatible, the results of factor analysis will indicate the presence of only two components in the blend. However, for a compatible mixture an interaction will occur between the two component polymers and its presence will be indicated in the factor analysis results as a third component in the blend.

Once factor analysis has been used to positively determine the compatibility of a polymer blend, difference spectroscopy (or spectral subtraction) can be used to isolate the infrared spectral changes resulting from the interaction between the component polymers. A detailed description of the application of digital subtraction to infrared spectra has been published by Koenig (17). The infrared spectrum of a compatible polymer blend is actually composed of contributions from the component polymers plus an additional contribution resulting from the intermolecular interactions formed between the components. The spectral contributions from these interactions can be identified using digital subtraction. This technique involves subtracting the sealed spectra of each of the component polymers from the spectrum of the blend. This interaction spectrum results from frequency shifts, band broadening, and changes in peak intensity.

After the interaction spectrum of a compatible blend has been isolated by difference spectroscopy, least squares curve fitting can be applied to determine the concentration of the components present in the mixture. Blackburn (18) has developed a least squares method that uses the pure component spectra to determine the relative amounts of each component present in the mixture spectrum. The fitting equation presented by Blackburn is as follows:

$$\sum_{i=1}^{N} R_{ij} = \sum_{j=1}^{M} \left[\sum_{i=1}^{N} W_i R_{ij} R_{ik} \right] x_j$$

$$S_i = \sum_{j=1}^{M} x_j R_{ij}$$

where N is the number of data points in each spectrum, M is the number of component spectra used in the fitting procedure, S_i is the data for the spectral range of the mixture, W_i is a statistical weighting factor equal to the inverse of S_i, R_{ik} is the absorbance data for the ith spectral element of the kth component spectrum, and x_j is the multiplier used by the least squares procedure that gives the best fit of the standard spectra to the mixture spectrum. It is from the x_j values that the volume fractions of the components in the mixture are determined.

Accuracy of the least squares curve fitting procedure can be measured by the multiple correlation coefficient, R, with 1.0 corresponding to perfect correlation. This coefficient can be calculated from the following equation:

$$R^2 = \frac{\Sigma (M_c - M_m)^2}{\Sigma (M_o - M_m)^2}$$

where M_o is the observed spectrum, M_c is the calculated spectrum, and M_m is the mean spectrum.

The methods of factor analysis, difference spectroscopy, and least squares curve fitting have been applied by Koenig and Rodriquez (19) to the study of compatible poly(phenylene oxide) (PPO) and polystyrene (PS) blends. The indicator function from factor analysis reached a minimum value when the number of components equaled three. These three independent components were proposed to be PS and two different PPO conformations, which is in agreement with results of a study by Wellinghoff et al. (20). Wellinghoff et al. determined from spectroscopic analysis that a strong interaction between the phenyl ring of PS and the phenylene ring of PPO was responsible for PPO conformational changes that occurred upon blending with PS. The interaction spectra of PPO–PS blends of various compositions were obtained by difference spectroscopy and least squares curve fitting was then applied to determine the concentration of the interaction spectrum in the blend spectrum. The interaction contribution reached a maximum value in blends having a composition of 30:70 PPO–PS. Factor analysis and least squares curve fitting were also applied to determine that partial demixing of the blends begins when the PPO–PS blend are heated to 200 °C. Partial demixing was concluded because, although the indicator function still attained a minimum value corresponding to three independent components, there was an overall decrease in the contribution of the interaction spectrum to that of the blend spectrum.

A shift of the carbonyl stretching peak to lower frequencies occurs for compatible polymer blends. Leonard et al. (*21*) demonstrated that this shift in the carbonyl peak in a compatible blend results from localized interactions between the component polymers. One percent solutions of the model compound methyl acetate (MA) (a simple model for the ester group) were prepared in the mixed solvents hexane–benzene and hexane–orthodichlorobenzene (ODB). The carbonyl stretching region of the spectra of the solutions indicates only one band, as seen in Figure 2. However, the second and fourth derivatives, as well as the Fourier self-deconvoluted spectrum, clearly show that the carbonyl peak actually is composed of two bands. Similar results were found for MA dissolved in various compositions of hexane–ODB. In contrast, when MA is dissolved in a single solvent, a doublet in the carbonyl peak cannot be detected. This observation suggests, therefore, that the shift of the carbonyl stretching peak is caused by localized solvent–solute interactions rather than by a bulk property of the medium, such as the refractive index or dielectric effects. The strength of the intermolecular interactions between MA and each of the solvents varies with each system and is reflected by the different magnitudes in the frequency shifts of the carbonyl peak.

Blend compatibility may also be studied through examination of changes in the width at half-height of the carbonyl stretching peaks. A study of various polyesters (PE) blended with poly(vinyl halide) (PVX) was conducted by Cousin and Prud'homme (*22*). The width at half-height of the PE carbonyl band increased in the compatible PE–PVX blends, whereas no changes in the width were detected for the incompatible blends. This increase in width for the compatible blends was attributed to the rigidity and random coil conformation of the PE molecule. Not all the PE carbonyl groups are favorably disposed to interact with the PVX and will therefore experience no change in their vibrational frequency. Band broadening, therefore, results from a distribution in the strength of the interactions, ranging from strong hydrogen bonds that produce the greatest shift in frequency to an absence of any interactions, which results in no frequency shifts.

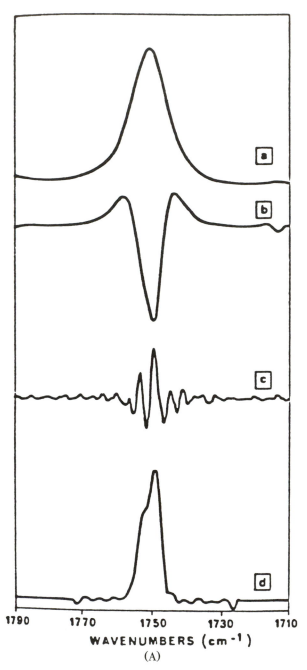

Figure 2. Carbonyl stretching absorption of MA in a mixture of hexane–benzene (50:50 vol%) (A) and in a mixture of orthodichlorobenzene–hexane (8:92 vol%) (B). Continued on next page.

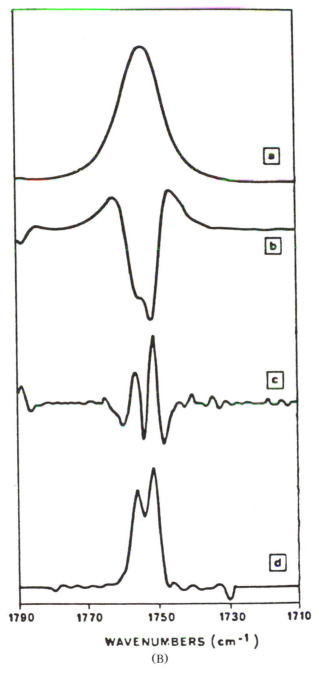

1790   1770   1750   1730   1710

WAVENUMBERS (cm⁻¹)

(B)

Figure 2. Continued. *Key: Curves represent the FTIR spectrum (a); second derivative of the spectrum (b); fourth derivative of the spectrum (c); and self-deconvoluted spectrum (d). (Reproduced with permission from reference 21. Copyright 1985 Butterworth.)*

A Fourier transform (FT) IR study of blend compatibility was conducted by Coleman and Zarian (23) on blends of poly(ε-caprolactone) (PCL) and poly(vinyl chloride) (PVC). Figure 3 shows the carbonyl stretching region of the spectra of the blends recorded at 75 °C. Two changes in this peak occur as the concentration of PVC is increased: a shift to lower frequency and an increase in the width at half-height. The change in width at half-height as a function of PVC concentration is plotted in Figure 4. The resulting S-shaped curve indicates that the magnitude of the interactions of the carbonyl with PVC approaches saturation at a concentration of approximately 4:1 PVC–PCL molar ratio, which corresponds to approximately 60-wt% PVC. This saturation effect was explained by considering the relative lengths of the structural repeat units of PCL and PVC. Assuming a planar zigzag conformation, the $-(CH_2)_5COO-$ unit of PCL is approximately 3.4 times as large as the $-CH_2CHCl-$ unit of PVC. From this approximation it was determined that a molar excess of about 4:1 PVC–PCL is necessary for saturation to occur.

Further evidence that a change occurs in the carbonyl structure of PCL can be seen in Figure 5. The 1161-cm^{-1} peak in the PCL spectrum has been assigned by Kirkpatrick (24) as the result of contributions from C–O stretching and O–C–H bending vibrations. However, after blending PVC and PCL in a 5:1 molar ratio, this peak shifts to 1165 cm^{-1}.

Room temperature studies were also conducted on the PCL–PVC blends, with emphasis placed on the presence of a crystalline PCL component in conjunction with the amorphous components of PCL and PVC. In the carbonyl stretching region of the blends, two peaks occur for the semicrystalline PCL. A peak at 1724 cm^{-1} is assigned to the crystalline PCL component and a peak at 1737 cm^{-1} results from the PCL amorphous

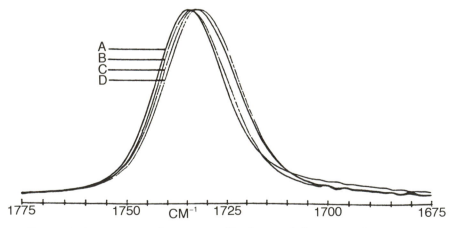

Figure 3. FTIR spectra of PVC–PCL blends recorded at 75 °C for pure PCL (A), 1:1 (B), 3:1 (C), and 5:1 (D) molar PVC:PCL, respectively. (Reproduced with permission from reference 23. Copyright 1979 Wiley.)

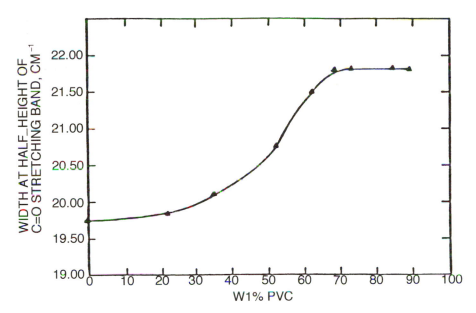

Figure 4. Plot of the width at half-height of the carbonyl stretching frequency as a function of PVC concentration for PVC–PCL blends recorded at 75 °C. (Reproduced with permission from reference 23. Copyright 1979 Wiley.)

component. For blends having PVC concentrations of 3:1 molar ratio PVC–PCL or greater, the peak at 1724 cm^{-1} cannot be detected, which indicates that blends of these ratios exist in an essentially amorphous state. A slight shift of the amorphous peak to a lower frequency occurs as the concentration of PVC is increased. These shifts are comparable to the shifts detected at 75 °C. Additionally, for the semicrystalline blends, the crystalline peak at 1724 cm^{-1} shifts to higher frequencies as the PVC concentration is increased. Both frequency shifts for the crystalline and amorphous carbonyl peaks indicate the existence of a specific interaction between the two polymers that involves the carbonyl group of PCL.

The other significant occurrence in the carbonyl region of the infrared spectra of the blends recorded at room temperature is that for blends that contain a PVC–PCL molar ratio of 3:1 or greater, the width at half-height of the band is identical within experimental error to that of the blends studied at 75 °C. This result indicates that band width is not a function of temperature.

Although Coleman and Zarian (*23*) established that the interaction in the PCL–PVC blends involves the carbonyl group of the PCL, the specific type of interaction between these two polymers was unclear. Two types of interactions may be occurring: an interaction between the PCL carbonyl group with either the α-hydrogen or the carbon–chlorine bond of PVC.

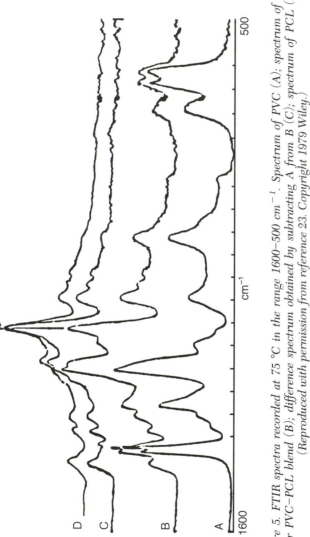

Figure 5. FTIR spectra recorded at 75 °C in the range 1600–500 cm^{-1}. Spectrum of PVC (A); spectrum of 5:1 molar PVC–PCL blend (B); difference spectrum obtained by subtracting A from B (C); spectrum of PCL (D). (Reproduced with permission from reference 23. Copyright 1979 Wiley.)

Varnell et al. (25) undertook further investigations of this system to determine which of these two interactions is occurring.

No shifts of the C—Cl peaks of PVC were observed in the blend spectra. Based on studies of low molecular weight analogues conducted by Varnell et al. (25), initial conclusions are that the interaction does not involve the C—Cl bond of PVC. However, such a conclusion cannot be made because not all the C—Cl groups can interact with the PCL carbonyl groups and the unreacted groups may, therefore, hide any slight shifting.

To find conclusive evidence of a specific interaction involving the PVC α-hydrogen, α-deuterated PVC was blended with PCL. Such an interaction could be detected in this system because the C—D stretching peak is well separated from the C—H peak. Based on results obtained from studies of the model compound system methyl acetate and deuterated chloroform, the C—D stretching vibration is expected to shift to a higher frequency if there is an interaction between the PCL carbonyl group and the deuterium atom of the α-deuterated PVC. The C—D stretching peak does, in fact, shift to higher frequencies as the PCL concentration is increased. Thus it can be concluded that the specific interaction in the PCL—PVC blends is that of a hydrogen bond between the PCL carbonyl group and the α-hydrogen of PVC.

In the foregoing studies of polymer blends, the occurrence of a carbonyl band shift to lower frequencies or an increase in the width at half-height of the band was considered to be evidence of a specific interaction between the two component polymers. To support this conclusion, it should be proved that no such changes in the carbonyl peak occur for incompatible systems. This, in fact, was demonstrated in the study of the poly(β-proprolactone) (PPL) and PVC blend system conducted by Coleman and Varnell (26). The spectra of PPL—PVC blends were recorded at 80 °C, which is above the melting point of PPL and, therefore, ensures that both systems are in the amorphous state. Unlike the previously studied compatible blends, the PPL carbonyl band appears within experimental error to remain unchanged in frequency or shape, despite changes in the PVC concentration. Further evidence can be seen in the plot of the width at half-height of the carbonyl band as a function of PVC concentration. Unlike the S-shaped curve found for the compatible blends, PPL—PVC blends show virtually a straight horizontal line. Similar behavior was found for the PPL—PVC blends studied in the solid state at room temperature. Thus the absence of either a shifting of the carbonyl peak to lower frequencies or an increase in the width at half-height may be considered evidence that a polymer blend system is incompatible and that the chains of one component do not recognize the existence of the second component.

A study was conducted by Garton (27) to determine if the specific nature of the intermolecular interaction can be determined by the degree of frequency shift of the carbonyl stretching peak. Polyester—chlorinated polymer blends can experience three different types of interactions: a hydrogen bonding interaction between the carbonyl group of the polyester and either the α-hydrogen or the β-hydrogen of the chlorinated polymer or a dipole–dipole interaction between the polyester carbonyl group and the C–Cl group of the chlorinated polymer. As shown by Varnell et al. (25), PCL–PVC blends produce a shift in the carbonyl peak as a result of a hydrogen-bonding interaction between the PCL carbonyl group and the α-hydrogen of PVC. In Garton's study, polyester was blended with a random copolymer of 80:20 PVC-poly(acrylonitrile) (PAN). Because this copolymer does not possess any α-hydrogen, it is evident that the type of intermolecular interaction formed would be between the polyester carbonyl and either the β-hydrogen or the C–Cl group of the copolymer. The shift of the carbonyl peak to lower frequencies in the polyester PVC–PAN blend is nearly identical to the shift that occurs in the PCL–PVC blends. Thus the degree of shifting experienced by the carbonyl stretching peak of a compatible blend cannot be used to determine the specific type of interaction that has been formed.

Coleman et al. (28) next examined the carbonyl stretching region of the infrared spectrum of a blend composed of two crystallizable components to determine the presence or absence of intermolecular interactions; specifically, blends of poly(bisphenol A carbonate) (PC) and PCL were studied. This system is unique because both polymers are crystallizable, although large differences exist in their crystalline melting points (approximately 230 °C and 70 °C, respectively) as well as in their glass transition temperatures (approximately 149 and −71 °C, respectively). Chemical interactions similar to those found previously in the PCL–PVC system (25) were identified in this blend. Specifically, in the spectra recorded at 75 °C (above the crystalline melting point of PCL), the PCL amorphous carbonyl band shifts to lower frequencies upon addition of PC. This band shift indicates that a specific chemical interaction is occurring between the two polymers and it involves the carbonyl group of PCL.

Utilization of the carbonyl region in an infrared spectrum to determine blend compatibility for a group of poly(vinyl phenol) (PVPh) blends was demonstrated by Moskala et al. (29). The carbonyl region for various blend compositions of PCL–PVPh, PPL–PVPh, and poly(vinyl pyrrolidone) (PVPr) and PVPh were obtained. The PCL–PVPh blends were cast from tetrahydrofuran (THF) and recorded at 75 °C, which is above the PCL melting point. As the composition of PVPh increases, the intensity of the 1708-cm^{-1} band increases, while that of the 1734-cm^{-1} band decreases. The 1708-cm^{-1} band results from the hydrogen-bonding of PCL carbonyl groups to PVPh phenolic hydroxyl groups, whereas the 1734-cm^{-1} band is assigned to the self-associated carbonyl groups in the amorphous PCL. Thus, as the PVPh composition in the band increases, there is a corresponding increase in the degree of interaction between the PVPh and PCL. Blends of PPL–PVPh were cast from THF and recorded at 89 °C, which is above the melting point of PPL. The spectrum of amorphous PPL in the carbonyl region is composed of a broad band located at 1741 cm^{-1}. Upon blending with PVPh, the PPL carbonyl groups form hydrogen bonds with the hydroxyl groups of PVPh. This bonding is demonstrated by the formation of a new band located at 1722 cm^{-1}, which increases in intensity with the amount of PVPh in the blend. Blends of PVPr–PVPh were cast from THF and recorded at room temperature. Pure PVPh contains a broad band located at 1682 cm^{-1} in the carbonyl region of the spectrum. However, this band actually results from a combination of carbonyl stretching and N–C stretching vibrations. A second band located at 1658 cm^{-1} appears in the blend spectrum and is attributed to PVPr carbonyl groups that have formed hydrogen bonds with the hydroxyl groups of PVPh.

Benedetti et al. (*30*) conducted an experiment to determine the relative strengths of various types of interactions by examining the amount of shift in the carbonyl peak. Functionalized ethylene–propylene copolymers (FEP) were prepared by reacting the molten polymer with diethylamylate (DEM) in the presence of dicumylperoxide (DCP). Solutions of FEP were dissolved in *n*-heptane (n=C_7H_{16}), tetrahydrofuran (THF), carbon tetrachloride (CCl_4), 1,1,1-trichloroethane (CCl_3–CH_3), and chloroform ($CHCl_3$), and the resulting carbonyl stretching regions of their spectra were recorded. The strength of the interaction between the carbonyl group in the DEM unit of FEP and the various solvents was determined from the amount of shift of the carbonyl peak to lower frequencies. The solvents were then ranked in the following order according to decreasing strength: n-C_7H_{16} > THF > CCl_3–CH_3 > CCl_4 > $CHCl_3$. These results indicate that the strongest interaction occurs between the DEM carbonyl group and a methine hydrogen.

Further studies on the shift of the carbonyl peak experienced by compatible blends were conducted by Garton (*31*). PCL or its low molecular weight model compound methyl acetate was dissolved in a mixture of two solvents that duplicate the possible interacting centers in chlorinated polymers. These two solvents, A and B, will both interact with the PCL carbonyl group. It should then be possible to resolve the carbonyl stretching band into the two components that result from the two types of interactions and then calculate the area of each component. An equilibrium constant, K, may then be calculated according to the following equation:

$$K = \frac{[C=O \cdots A][B]}{[C=O \cdots B][A]}$$

Thus, a comparison of the strength of several possible interacting centers may be established from the equilibrium constants.

A comparison of the interaction strength between the model compound MA and several α-hydrogenated chlorocarbons and heptane is shown in Figure 6, where the slope of the line is equal to the equilibrium constant

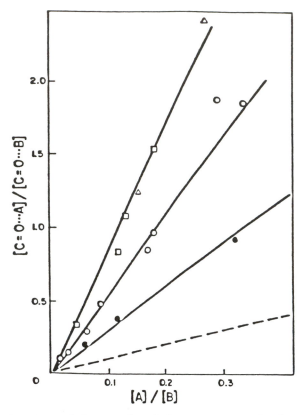

Figure 6. Association behavior of methyl acetate in mixed solvents. Solvent A = 1122TCE (□), dichloromethane (△), chloroform (○), carbon tetrabromide (●). Solvent B = heptane. The dashed line corresponds to no preferential association of methyl acetate (i.e., K = 1). (Reproduced with permission from reference 31. Copyright 1983 Society of Plastics Engineers, Inc.)

defined previously. A preference for association of the MA to the α-hydrogenated chlorocarbons is clearly established. Figure 7 compares the relative interaction strengths of an α-hydrogenated chlorocarbon (1,1,2,2-tetrachloroethane) (1122TCE), a β-hydrogenated chlorocarbon (1,1,1-trichloroethane) (111TEC), and carbon tetrachloride (CCl_4). These three solvents were chosen because they represent the possible interacting sites in poly(vinyl chloride) and poly(vinylidene chloride), both of which form compatible polymer blends with polyester. The strength of interacting abilities in decreasing order was established to be α-hydrogenated chlorocarbon > β-hydrogenated chlorocarbon > carbon tetrachloride. Furthermore, because the equilibrium constants for the 1122TCE heptane system and the 1122TCE–CCl_4 system were found to be nearly identical, the interaction

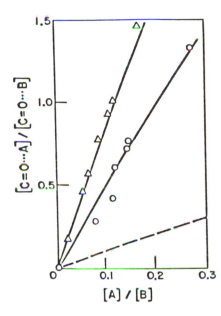

Figure 7. Association behavior of methyl acetate in mixed solvents. Solvent A = 1122TCE. Solvent B = carbon tetrabromide (△), 1,1,1-trichloroethane (○). The dashed line corresponds to no preferential association of methyl acetate (i.e., K = 1). (Reproduced with permission from reference 31. Copyright 1983 Society of Plastics Engineers, Inc.)

strength of both heptane and carbon tetrachloride to the ester are essentially equal.

Similar studies by Garton (*31*) revealed that the strength of the interaction of ester with carbon tetrabromide is much greater than with heptane. This observation is attributed to the highly polarizable C–Br bond. It was also reported that in a system of α-hydrogenated chlorocarbon (1122TCE) and a cyclic ether (THF), the equilibrium only slightly favors the 1122TCE; thus, this establishes that α-hydrogenated chlorocarbons form only slightly stronger interactions than cyclic ethers.

From the model compound and solution studies, Garton (*31*) established that these solvents interact with a model ester according to the following strengths: α-hydrogenated chlorocarbons > THF > β-hydrogenated chlorocarbons ≃ carbon tetrabromide > carbon tetrachloride ≃ heptane.

The preference to interact with α-hydrogenated chlorocarbons was then duplicated by Garton (*31*) in studies using PCL, rather than its model compound. However, when the PCL was dissolved in the 1122TCE–CCl$_4$ system, the equilibrium constant for MA was greater than that for PCL. This result was attributed to stiffness, steric limitations, or the conformation of the PCL polymer chain, all of which inhibit the ester groups from hydrogen-bonding with the chlorocarbon.

Coleman and Moskala (*32*) performed studies on the polymer blend system of poly(hydroxy ether of bisphenol A) (phenoxy) and PCL, placing emphasis on the dependence of intermolecular interactions on blend composition. Results were obtained first at 75 °C, which is above the melting point of PCL. In the carbonyl stretching region, the intensity of a shoulder at approximately 1720 cm^{-1} increased with increasing concentration of phenoxy and was attributed to an intermolecular interaction between the carbonyl group of PCL and the hydroxyl group of the phenoxy. Curve resolving studies were performed on the carbonyl region of the blends to reveal two components: a relatively narrow band centered at 1734 cm^{-1} and a relatively broad band centered at 1720 cm^{-1}. These two components corresponded to isolated PCL carbonyl groups and hydrogen-bonded carbonyl groups, respectively. The blend composition was determined from the relative fraction of hydrogen-bonded carbonyl groups. This estimate was obtained by taking the ratio of the area under the 1720-cm^{-1} peak divided by the sum of the areas under the 1720- and 1734-cm^{-1} peaks. A plot of the relative fraction of hydrogen-bonded carbonyl groups as a function of mole percent phenoxy reveals that the fraction of hydrogen-bonded carbonyls increases linearly with increasing phenoxy concentration.

The hydroxyl stretching region of the phenoxy-PCL blends was also examined to determine the relative strength of intermolecular interactions. As seen in Figure 8, the pure phenoxy in the spectrum consists of two components: a relatively narrow peak centered at 3570 cm^{-1} and a broad peak at 3450 cm^{-1}. These two peaks correspond to the free hydroxyl groups and the hydrogen-bonded hydroxyl groups, respectively. The peak due to the hydrogen-bonded hydroxyl shifts to higher frequencies upon blending of the phenoxy with PCL. However, the peak assigned to the free hydroxyls remains unchanged upon addition of the PCL. Purcell determined that the difference between the frequencies of the peak due to the free hydroxyls and that of the peak due to the hydrogen-bonded hydroxyls is a measure of the average strength of the intermolecular hydrogen bonding. Based on this conclusion, the changes in the hydroxyl stretching region of the phenoxy–PCL blend indicate that the hydrogen bonding between the PCL carbonyl group and the phenoxy hydroxyl group is weaker than the hydrogen-bonding interaction in pure phenoxy.

A study was also conducted by Coleman and Moskala (32) on phenoxy–poly(ethylene oxide), (PEO) blends to determine the relative strength of intermolecular interactions; the same technique as employed for the phenoxy–PCL system (32) was used. In contrast to the phenoxy–PCL system, the hydrogen-bonded hydroxyl peak of phenoxy–PEO blends shifts to lower frequencies with the addition of PEO. Thus it was concluded that the hydrogen-bonding interaction between the PEO carbonyl group and the phenoxy hydroxyl group is stronger than the intermolecular hydrogen bonding in pure phenoxy.

Moskala and Coleman (33) expanded their study of phenoxy blends by examining blends of phenoxy with poly(vinyl alkyl ethers); specifically, the compatible blend phenoxy–poly(vinyl methyl ether) (PVME) and the incompatible blends phenoxy–poly(vinyl ethyl ether) (PVEE) and phenoxy–

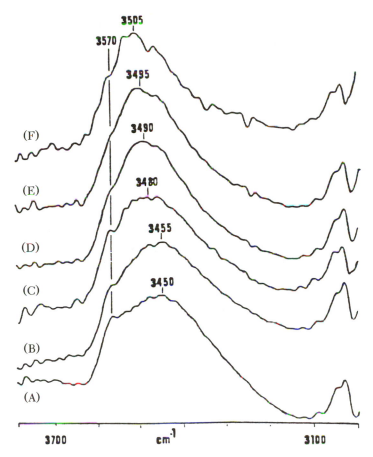

Figure 8. FTIR spectra recorded at 75 °C of PCL–phenoxy blends containing 0 (A), 10 (B), 20 (C), 30 (D), 40 (E), and 50 (F) weight percent PCL. (Reproduced with permission from reference 32. Copyright 1983 Butterworth.)

poly(vinyl isobutyl ether) (PVIE). As was seen previously for the phenoxy–PCL blends (32), the phenoxy hydrogen-bonded hydroxyl stretching peak in phenoxy–PVME blends shifts to higher frequencies upon addition of PVME. However, this shift to higher frequencies is accompanied by a corresponding decrease in relative broadness of the peak. The width at half-height for this peak decreases from 260 cm^{-1} for pure phenoxy to 150 cm^{-1} for a 20:80 wt% phenoxy–PVME blend. By contrast, the spectra of the incompatible phenoxy–PVEE and the phenoxy–PVIE blends show neither a shift in frequency nor a narrowing of the phenoxy hydrogen-bonded hydroxyl peak upon addition of the respective poly(vinyl ether). This narrowing of the self-associated hydroxyl peak in the phenoxy–PVME blend was explained in the following way. The bulky benzene ring of the phenoxy creates steric problems and prevents a consistency of hydrogen bond distances and geometries. However, the hydroxyl stretching mode narrows upon blending with PVME because the hydroxyl group can form a more homogeneous distribution of hydrogen bond lengths and geometries with the relatively flexible PVME molecule.

Moskala et al. (34) demonstrated the use of the hydroxyl stretching region to determine blend compatibility for a group of PVPh blends; specifically, PCL–PVPh, PVPr–PVPh, PEO–PVPh, and blends of PVPh with two different poly(vinyl alkyl ethers). The hydroxyl stretching region of the PCL–PVPh blends can be resolved into three peaks. The non-hydrogen-bonded hydroxyls of PVPh are responsible for a peak at 3525 cm^{-1}. A second peak at 3370 cm^{-1} results from the self-association of PVPh hydroxyl groups and a peak at 3420 cm^{-1} is assigned to PVPh hydroxyl groups hydrogen-bonded to PCL carbonyl groups. As the PCL concentration in the blend is increased, there is a corresponding decrease in intensity in the first two peaks and an increase in the 3420-cm^{-1} peak. Measuring the intermolecular interaction strength as the difference in frequency between the peak resulting from the interaction and that of the free hydroxyls leads to the conclusion that the strength of the self-associated hydroxyl groups in PVPh ($\Delta\nu = 165$ cm^{-1}) is stronger than the hydrogen bond formed between the PVPh hydroxyl groups and the PCL carbonyl group ($\Delta\nu = 105$ cm^{-1}). Similar results were obtained for PPL–PVPh blends, with the peak due to the hydroxyl groups hydrogen-bonded to carbonyl groups shifted slightly to 3440 cm^{-1}. Unlike the PCL–PVPh and PPL–PVPh blends, the self-associated hydroxyl peak occurs at a higher frequency (3360 cm^{-1}) than the hydrogen-bonded hydroxyl peak (3230 cm^{-1}) in the PVPr–PVPh blends. It is therefore concluded that the intermolecular interaction between the PVPh hydroxyl group and the PVPr carbonyl group is stronger than the hydroxyl–hydroxyl interaction. For the PEO–PVPh blends, a peak at 3200 cm^{-1} results from the PVPh hydroxyl group hydrogen-bonded to the ether oxygen of PEO. The frequency difference between this peak and that due to the free hydroxyl groups is 325 cm^{-1}. This frequency difference is greater than the difference found in the previously studied (32) phenoxy–PEO blends (270 cm^{-1}) and reflects the fact that the PVPh hydroxyl groups have a greater affinity to hydrogen bond than do the phenoxy hydroxyls. Further evidence of the greater affinity of the PVPh hydroxyl groups can be seen by comparing the blends of PVME–PVPh and phenoxy–PVME. The difference in frequency between the peak assigned to the hydroxyl groups hydrogen-bonded to ether oxygens and that due to the free hydroxyl groups is 205 cm^{-1} for the PVME–PVPh blend and 150 cm^{-1} for the phenoxy–PVME blend.

Cousin and Prud'homme (22) conducted a miscibility study of several polyesters with poly(vinyl halides) and concluded that intermolecular interactions occur on a mole-to-mole basis for compatible PE–PVX blends. The degree of frequency shift in the carbonyl stretching band is plotted as a function of weight percent poly(vinyl halide) in the blend, as shown in Figure 9. Within experimental error, no shifting of the carbonyl peak occurs for the PCL–poly(vinyl fluoride) (PVF) blends. Thus it can be concluded that the PCL carbonyl peak does not interact with the PVF methine hydrogen and the PCL–PVF blend is, therefore, incompatible. In contrast, a shifting of the carbonyl peak in the PCL–PVC and PCL–poly(vinyl bromide) (PVB) blends indicates that these two mixtures are compatible. It can also be seen that the carbonyl band frequency shift differs slightly when the PCL–PVC and PCL–PVB blends are compared. However, as seen in Figure 10, if this frequency shift is plotted as a function of molar composition of poly(vinyl halide) rather than weight percent, the degree of shifting experienced by the two blends is nearly identical. This observation leads to the conclusion that the interactions between the PCL and PVX occur on a mole-to-mole basis. Similar results were obtained when PVC and PVB were blended with other polyesters, namely, poly(hexamethylene sebacate) (PHMS) and poly(valero-lactone) (PVL).

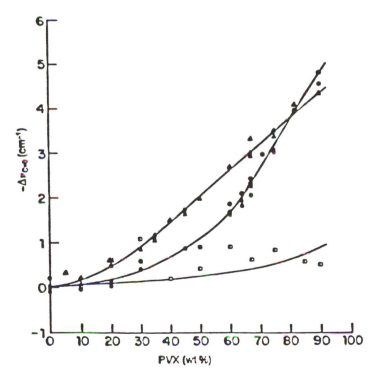

Figure 9. Frequency shift of the carbonyl group of PCL as a function of the PVX weight percent in the mixture. Measurements made at 80 °C. Key: □, PVF; ●, PVB; and ▲, PVC. (Reproduced with permission from reference 22. Copyright 1983 Butterworth.)

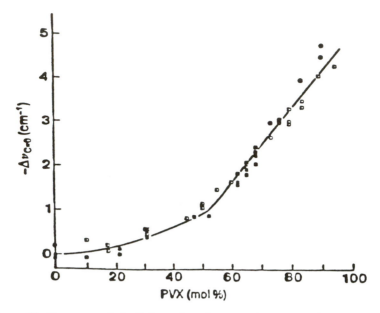

Figure 10. Frequency shift of the carbonyl group of PCL as a function of the PVX mole percent in the mixture. Measurements made at 80 °C. Key: □ , PVC; ■ , PVB. (Reproduced with permission from reference 22. Copyright 1983 Butterworth.)

For all six compatible blends studies by Cousin and Prud'homme (22), the amount of shift of the carbonyl band increased with increasing PVX concentration. This relationship occurs because at low PVX concentrations, only a small amount of PVX molecules are intermixed within the polyester matrix. Therefore, the number of interactions formed between the PE carbonyls and the PVX methine hydrogens is small. However, at high PVX concentrations a small amount of PE molecules are intermixed with the PVX matrix, which allows a larger percentage of the available carbonyl groups to form hydrogen bonds with the PVX and results in a larger shift of the carbonyl stretching peak.

Blends of poly(butylene adipate) (PBA) with PVC and PVB were studied in a similar manner. Although a decrease in the carbonyl frequency occurred with increasing PVC content for PBA–PVC blends, the amount of shift was small, especially at low PVC concentrations. For the PBA–PVB blends, no frequency shift was detected for blends with less than 60-mol% PVB. Thus it was concluded that there was only partial miscibility in the PBA–PVB blends.

To establish a relationship between the degree of miscibility and the molecular structure of the various poly(vinyl halides), a comparison of the amount of carbonyl shifting at high PVX concentrations within the respective blends was made. The carbonyl shifts of the various polyesters decreased in the following order: PHMS < PCL < PVL < PBA. This ranking corresponds to a decrease in the order of carbonyl concentration on the repeat unit of the polyester and an increase in the order of rigidity of the polyester chain. Both trends result in a decrease in the number of intermolecular interactions that may occur between the carbonyl groups and the PVX hydrogens.

Coleman et al. (35) further expanded the study of the shift in the carbonyl band by examining the dependence of the shift on temperature. Studies of PVC–ethylene vinyl acetate (EVA) copolymer blends and PVC-chlorinated polyethylene (CPE) blends were undertaken to determine whether a correlation exists between the strength of the intermolecular interaction as determined by the frequency shift of the carbonyl band and the onset of phase separation at the lower critical solution temperature (LCST). The infrared spectrum of the carbonyl band for CPE–EVA and PVC-EVA blends were recorded at room temperature. As expected, the occurrence of specific intermolecular interactions is revealed through a shift to lower frequencies and an increase in the width at half-height. Walsh et al. (36) previously reported that the LCST for CPE–EVA blends is below 130 °C. This value was confirmed by Coleman et al. by examination of the carbonyl peak of the infrared spectra recorded after heating the blends for approximately 3 h at 130 °C. Incompatibility at this temperature was demonstrated by the fact that the spectra of the pure EVA and blends containing 40- and 80-wt% CPE are nearly identical within experimental error.

Attempts were then made to determine the LCST from changes in the frequency of the carbonyl band as a function of temperature. Studies conducted to determine frequency changes in the pure EVA carbonyl band with increasing temperature showed that the frequency increased slightly in a linear relationship with increasing temperature. Spectra of an 80:20-wt% CPE–EVA blend obtained in a room temperature to 160 °C range showed that the frequency of the carbonyl peak increased with temperature. Figure 11 shows a plot of the frequency versus temperature for both the pure EVA and the 80:20-wt% CPE–EVA blend. The relative strength of the interaction at any temperature is determined by the difference between the carbonyl frequency in the blend and in the pure EVA. This difference becomes smaller as the temperature is increased, which indicates a decrease in the strength of the interaction between the two components of the blend. At temperatures ranging from 35 to 90 °C, the strength of the intermolecular interaction as determined by the difference in frequencies is great enough to result in a compatible blend. However, above approximately 110 °C, the interaction has decreased to such a degree that phase separation occurs. Thus the predicted LCST of CPE–EVA blends occurs between 90 and 110 °C, which corresponds to a critical value of the strength of the interaction.

Similar behavior was observed for 80:20-wt% PVC–EVA blends, and the LCST of this system was predicted to be between 110 and 130 °C. These results suggest that changes in the intermolecular interactions between blend

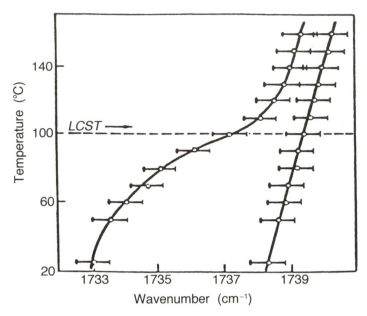

Figure 11. Plot of the temperature versus the carbonyl peak position for an 80:20-wt% CPE–EVA blend. (Reproduced with permission from reference 35. Copyright 1983 Butterworth.)

components as a function of temperature can be determined by FTIR analysis.

The effect of temperature on hydrogen bonding in compatible blends was studied by Ting et al. (37). A poly(styrene-*co*-vinylphenyl trifluoromethyl carbinol) (PFA) and PEO blend and a poly(styrene-*co*-vinylphenol hexafluorodimethyl carbinol) (PHFA) and PEO blend in a 4:1 ratio were studied at temperatures of 25, 75, 125, and 175 °C. At 75 °C, the hydrogen-bonded hydroxyl peak of the PFA–PEO blend decreased in intensity, while that of the PHFA–PEO blend did not change. This observation was attributed to the fact that the strong acidity of the PHFA hydroxyl group resulted in the formation of a stronger hydrogen bond. At 175 °C, hydrogen-bond dissociation in both blends is evidenced by the significant decrease in the intensity of the hydrogen-bonded hydroxyl group and the formation of peaks attributed to free hydroxyl groups at 3550 cm^{-1} in the PFA blend and at 3600 and 3520 cm^{-1} in the PHFA blend. These samples were then slowly cooled to room temperature, during which time the hydrogen bonds were reformed, as indicated by the reappearance of the hydrogen-bonded hydroxyl peaks. It was concluded that the phase separation of PFA–PEO and PHFA–PEO blends that occurred upon heating to 175 °C was reversible.

A study was undertaken by Skrovansk and Coleman (38) to determine the ability of a strongly self-associated polymer to interact with another polymer at the molecular level. Before intermolecular interactions between the two component polymers can be formed, the interactions present in the pure polymers must be broken. Therefore, a large negative enthalpy of mixing will only occur if the strength of the association between the two polymers exceeds the average of the strength of self-association of each of the component polymers. Therefore, a strongly self-associated polymer should react with another polymer that is weakly self-associated but contains a chemical moiety that has the potential to form a relatively strong intermolecular interaction.

For this study, Skrovansk and Coleman (38) chose polyamide as the strongly self-associated polymer because of extensive intermolecular hydrogen bonding of the amide group. The polyamide was blended with poly-(vinyl-2-vinyl pyridine) (P2VP), which is a weakly self-associated polymer that has intermolecular forces of a dispersive nature. In addition, P2VP has a nitrogen atom that contains a lone pair of electrons that can act as an excellent site for hydrogen bonding to a labile proton.

Evidence that intimate mixing has occurred in the blend of polyamide and P2VP is seen in the N–H stretching region of the infrared spectrum of the blend. The spectrum of amorphous polyamide at room temperature consists of two peaks; the most prominent peak is centered at 3310 cm^{-1}. The extreme broadness of this band is caused by the wide distribution in strengths of the hydrogen-bonded N–H groups. The second peak is located at 3444 cm^{-1} and has been assigned to the free N–H groups. The hydrogen-bonded N–H peak appears to shift to lower frequencies when the polyamide is blended with P2VP, but actually the overlapping of two major components is being observed. The first component is attributed to the self-association of polyamide (between the N–H group and the carbonyl group) and the second component results from the association of the polyamide N–H group to the P2VP nitrogen atom. The apparent relative shift in frequency after blending leads to the conclusion that the association of polyamide to P2VP is stronger than the self-association of polyamide.

Further evidence of an interaction between the polyamide and P2VP can be seen in the amide I and amide II bands of the infrared spectra obtained at room temperature after removal by subtraction of the spectral contributions from P2VP. The amide I band, centered at 1640 cm^{-1}, results from carbonyl stretching and is composed of two major contributions: one at 1640 cm^{-1} from the carbonyl groups that are hydrogen-bonded to N–H groups and the other from free carbonyl groups that appear at 1670 cm^{-1}. The peak at 1670 cm^{-1} becomes more pronounced after blending, which indicates an increase in the fraction of free carbonyl groups. As expected, the contribution of the free carbonyl band was calculated by a curve fitting technique to be approximately twice that found in the pure amorphous polyamide. This increase in the number of free carbonyl groups was interpreted as a result of a specific interaction between the two components of the blend. In pure amorphous polyamide at room temperature, nearly all the amide groups are hydrogen-bonded and there are an equal number of hydrogen-bonded N–H and C=O groups. When polyamide is blended with P2VP, the number of free carbonyl groups increases because an interaction between a polyamide N–H group and a P2VP nitrogen atom must be preceded by a corresponding break of a hydrogen bond between a polyamide N–H group and a polyamide carbonyl group.

The amide II band located at 1542 cm^{-1} results from N–H in-plane bending. This band also indicates the presence of an intermolecular interaction in the blend because it shifts to a higher frequency and broadens. This frequency shift and broadening is interpreted to be the result of an increase in the strength of the interaction involving the N–H groups.

Compatibility studies of PS–PVME were conducted by Lu et al. (39). A 50:50 molar composition of high molecular weight PS–PVME produces a compatible blend when cast from toluene. However, when the same polymer system is cast from chloroform or trichloroethylene (TCE), an incompatible blend results. Spectroscopic differences between the compatible and incompatible blends can be found in the 1100- and 700-cm^{-1} regions. The PS spectrum was subtracted from the 1100-cm^{-1} region of both types of blends. The doublet present in the resultant difference spectrum is composed of two peaks located at 1107 and 1085 cm^{-1}. The relative intensities of these two peaks is determined by whether the blend is compatible or incompatible. The 1085-cm^{-1} band has greater intensity in the compatible blends whereas the 1107-cm^{-1} band dominates in the incompatible blends. Snyder and Zerbi (40) concluded that the difference between the intensities of the two peaks in this doublet results from changes in the C–O–C asymmetric stretching in the COCH$_3$ group of PVME. The spectrum of PS shows the most significant changes with blending in the 700-cm^{-1} region. The C–H out-of-plane bending vibration of the phenyl ring is located at 697.7 cm^{-1} for pure PS, at 699.5 cm^{-1} for compatible PS–PVME blends of equal molar composition, and at a frequency between these two values for incompatible 50:50 PS–PVME blends. Changes in these two spectral regions indicate that the interaction that occurs in these blends is between the phenyl ring of PS and the COCH$_3$ group of PVME.

The dependence of compatibility on blend composition was also demonstrated in this experiment. As previously stated, 50:50 PS–PVME blends were compatible when cast from toluene but were incompatible when cast from chloroform or TCE. However, blends of 15:85 PS–PVME were compatible when cast from toluene, chloroform, or TCE and showed no differences in the 1100- or 700-cm^{-1} regions.

Sargent and Koenig (41) studied the compatibility of poly(vinylidene fluoride) (PVF$_2$) and poly(vinyl acetate) (PVAc) blends as a function of thermal treatment and blend composition. Films of the polymer blends with weight ratios ranging from 10:90 to 90:10 PVF$_2$–PVAc were cast from

solution onto KBr plates and placed in an oven under vacuum for 1 h. Two
sets of samples were prepared. The samples differed in the temperature at
which the solvent was evaporated. The first set of samples was placed in a
75 °C oven and the second set was placed in a 175 °C oven.

Factor analysis was used to determine whether blends of the various
compositions of the two homopolymers were compatible. Results indicated
that the PVF_2–PVAc blends heat treated at both 75 and 175 °C were
compatible. The spectroscopic changes that resulted from the intermolecular
interactions between the two homopolymers were isolated by subtracting the
spectra of the pure PVF_2 and pure PVAc from the spectra of the blends. The
relative amounts of each of these components present in the blends were
then determined by least squares curve fitting and the percent contributions
of each of the interaction spectra were analyzed as a function of the blend
composition and thermal treatment of the samples.

As shown in Figure 12, for both the samples heat treated at 75 °C and
175 °C a general increase occurs in the degree of interaction with a
corresponding increase in PVF_2 concentration. The degree of interaction

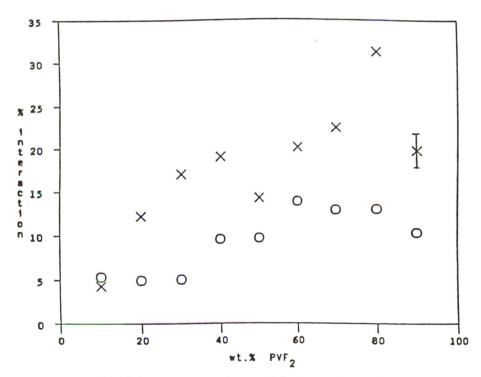

*Figure 12. Plot of the percent contribution of the interaction spectrum versus
weight percent PVF_2 for PVF_2–PVAc blends heat treated at 75 °C (○) and
175 °C (×) (41).*

within the blends was greater in the samples heat treated at 175 °C than in those heat treated at 75 °C. This difference was attributed to the fact that those samples heat treated at 175 °C were subjected to a thermal treatment that was at or above the melting point of the blends, which results in an increase in the mobility of the homopolymer molecules and allows more intimate mixing between the components. An improvement in the degree of mixing in a polymer blend system thus promotes the formation of intermolecular interaction between specific chemical groups of the component polymers, as reflected by the higher interaction contribution in the blends heat treated at 175 °C.

The percent contribution of the interaction spectrum was also followed as a function of time for the PVF_2–PVAc blends. Examination of the data revealed that within experimental error the interaction contribution did not decrease over a period of 24 days. This observation indicates that no detectable phase separation occurred within the blends during this time period of analysis.

Summary and Conclusions

Compatible polymer blends can be characterized by the formation of intermolecular interactions between specific chemical groups of the component polymers. These interactions can be studied on the molecular level using infrared spectroscopy. Spectral data processing techniques, such as factor analysis, difference spectroscopy, and least squares curve fitting, have been applied to characterize these interactions. Examination of frequency shifts, band broadening, and changes in peak intensity within the blend spectrum can also be used to identify the interactions. Additionally, the dependence of compatibility on such factors as blend composition and temperature can be determined from infrared spectroscopy studies.

References

1. Flory, P. J. *Principles of Polymer Chemistry*; Cornell University Press: Ithaca, NY, 1953.
2. Tompa, H. *Polymer Solutions*; Butterworths: London, England, 1956.
3. Bohn, L. Z. *Kolloid, Polym.* **1966**, *213*, 55.
4. Rosen, S. L. *Poly Eng. Sci.* **1967**, *7*, 115.
5. Krause, S.; *J. Macromol. Sci., Chem.* **1972**, *C7*, 251.
6. Bralow, J. W.; Paul, D. R. *Polym. Eng. Sci.* **1981**, *21*, 985.
7. Olabisi, O.; Robeson, L. M.; Shaw, M. T. *Polymer–Polymer Miscibility*; Academic: Orlando, FL, 1979.
8. *Polymer Blends*; Paul, D. R.; Newman, S., Eds.; Academic: Orlando, FL, 1978; Vols. I and II.
9. Coleman, M. M.; Painter, P. C. *J. Macromol. Sci., Chem.* **1977**, *C16*, 197.
10. Coleman, M. M.; Painter, P. C. *Appl. Spectrosc. Rev.* **1984**, *20*, 255.
11. Gillette, P. C.; Lando, J. B.; Koenig, J. L. *Fourier Transform Infrared Spectroscopy*; Academic: Orlando, FL, 1985; Vol. 4, Chapter 1.

12. Flory, P. J. *Principles of Polymer Chemistry*; Cornell University Press: Ithaca, NY, 1962; Chapter XII.
13. Scott, R. L. *J. Chem. Phys.* **1949**, *17*, 279.
14. Howe, S. E.; Coleman, M. M. *Macromolecules* **1986**, *19*, 72.
15. Antoon, M. K.; D'Esposito, L.; Koenig, J. L. *Appl. Spectrosc.* **1979**, *33*, 351.
16. Malinowski, E. R. *Anal. Chem.* **1977**, *49*, 612.
17. Koenig, J. L. *Appl. Spectrosc.* **1975**, *29*, 293.
18. Blackburn, J. A. *Anal. Chem.* **1965**, *37*, 1000.
19. Koenig, J. L.; Tovar Rodriquez, M. J. M. *Appl. Spectrosc.* **1981**, *35*, 543.
20. Wellinghoff, S. T.; Koenig, J. L.; Baer, E. *J. Polym. Sci., Polym. Phys. Ed.* **1977**, *15*, 1913.
21. Leonard, C.; Halary, J. L.; Monnerie, L. *Polymer* **1985**, *26*, 1507.
22. Cousin, P.; Prud'homme, R. E. *Multicomponent Polymer Materials*; Paul, D. R.; Sperling, L. H., Eds.; Advances in Chemistry 211; American Chemical Society: Washington, DC, 1986; pp 87–110.
23. Coleman, M. M.; Zarian, J. *J. Polym. Sci., Polym. Phys. Ed.* **1979**, *17*, 837.
24. Kirkpatrick, H. M.S. Thesis, University of Cincinnati, Cincinnati, OH, 1975.
25. Varnell, D. F.; Moskala, E. J.; Painter, P. C.; Coleman, M. M. *Polym. Eng. Sci.* **1983**, *23*, 658.
26. Coleman, M. M.; Varnell, D. F. *J. Polym. Sci., Polym. Phys. Ed.* **1980**, *18*, 1403.
27. Garton, A. *J. Polym. Sci., Polym. Lett. Ed.* **1983**, *21*, 45.
28. Coleman, M. M.; Varnell, D. F.; Runt, J. P. *Contemporary Topics in Polymer Science*; Bailey, W. J.; Tsuruta, T., Eds.; Plenum: New York, 1984; Vol. 4, p 807.
29. Moskala, E. J.; Varnell, D. F.; Coleman, M. M. *Polymer* **1985**, *26*, 228.
30. Benedetti, E.; Posar, F.; D'Alessio, A.; Vergamini, P.; Aglietto, M. M.; Ruggeri, G.; Ciardelli, F. *Br. Polym. J.* **1985**, *17*, 34.
31. Garton, A. *Polym. Eng. Sci.* **1983**, *23*, 663.
32. Coleman, M. M.; Moskala, E. J. *Polymer* **1983**, *24*, 251.
33. Moskala, E. J.; Coleman, M. M. *Polymer. Commun.* **1983**, *24*, 206.
34. Moskala, E. J.; Varnell, D. F.; Coleman, M. M. *Polymer* **1985**, *26*, 228.
35. Coleman, M. M.; Moskala, E. J.; Painter, P. C.; Walsh, D. J.; Rostami, S. *Polymer* **1983**, *24*, 1410.
36. Walsh, D. J.; Higgins, J. S.; Rostami, S. *Macromolecules* **1983**, *16*, 388.
37. Ting, S. P.; Bulkin, B. J.; Pearce, E. M.; Kwei, T. R. *J. Polym. Sci., Polym. Chem. Ed.* **1981**, *19*, 1451.
38. Skrovansk, D. J.; Coleman, M. M. *Polym. Eng. Sci.* **1987**, *27*, 857.
39. Lu, F. J.; Benedetti, A.; Hsu, S. L. *Macromolecules* **1983**, *16*, 1525.
40. Snyder, R. G.; Zerbi, G. *Spectrochim. Acta Part A* **1967**, *23A*, 391.
41. Sargent, M.; Koenig, J. L. *Vibrational Spectrosc.* **1991**, *2*, 21.

Received for review May 14, 1991. Accepted revised manuscript June 11, 1992.

Fourier Transform Infrared Spectroscopy as a Probe of Phase Behavior in Copolymer Blends

A Comparison of Theoretical Predictions to Experimental Data

Michael M. Coleman, Hongxi Zhang, Yun Xu, and Paul C. Painter

Polymer Science Program, Department of Materials Science and Engineering, The Pennsylvania State University, University Park, PA 16802

Results of theoretical and experimental studies of poly(4-vinyl phenol) (PVPh) blends with styrene-co-methyl acrylate (STMA) copolymers are reported. The calculated solubility parameters of STMA copolymers are practically independent of copolymer composition, so that the unfavorable contribution to the free energy of mixing from the "physical" intermolecular interactions remains essentially constant in all PVPh–STMA blends. PVPh is miscible with poly(methyl acrylate), but as the concentration of styrene in the STMA copolymer is increased, the contribution from the favorable hydrogen bonding interactions must decrease. Eventually a point is reached when there is an insufficient contribution to the free energy from favorable specific interactions to overwhelm the unfavorable contribution from the physical forces. Miscibility windows and maps for STMA blends with PVPh and styrene-co-vinyl phenol copolymers are readily calculated and compare favorably with experimental results performed in our laboratories.

AN EXPRESSION FOR THE FREE ENERGY OF MIXING OF POLYMERS that hydrogen-bond (*1–16*) can be obtained from a Flory-type lattice model that we have developed and tested over the past few years. As a result of this treatment we obtained a separation of the unfavorable "physical" contributions to the free energy of mixing, embodied in a Flory χ parameter, from the favorable "chemical" contributions emanating from the changing distribution of hydrogen bonds, $\Delta G_H/RT$; ΔG_H is change in free energy resulting from hydrogen bonding, R is the gas constant, and T is temperature. The free energy of mixing (ΔG_m) can be written as

$$\frac{\Delta G_m}{RT} = \frac{\Phi_A}{M_A}\ln\Phi_A + \frac{\Phi_B}{M_B}\ln\Phi_B + \Phi_A\Phi_B\chi + \frac{\Delta G_H}{RT} \qquad (1)$$

where Φ_A, Φ_B and M_A, M_B are the volume fractions and degrees of polymerization of polymers A and B, respectively.

Miscibility windows and maps are a convenient way of displaying the phase behavior of copolymer blends as a function of copolymer composition at a given temperature (8, 10–13, 16). In this chapter we consider the case of blends composed of polymers that contain vinyl phenol, methyl acrylate, and styrene segments. These studies complement recently reported studies of styrene-co-vinyl phenol (STVPh) copolymer blends with poly(n-alkyl methacrylates) (10–12), poly(ethylene-co-methyl acrylates) (13), and poly(ethylene-co-vinyl acetates) (13). Each of these prior studies emphasized experimental testing of one of the major hypotheses of our association model. This theme is continued here. The motivation for the present study comes from the recognition that the calculated solubility parameters of poly(methyl acrylate) (PMA) and polystyrene (PS) are almost identical (14), which infers that the average solubility parameters of poly(styrene-co-methyl acrylate) (STMA) copolymers are essentially independent of composition (16). This, in turn, implies that in a homopolymer–copolymer system of STMA blends the "physical" contribution to the free energy of mixing is also essentially independent of copolymer composition. Accordingly, as we dilute PMA with styrene, we are effectively only reducing the $\Delta G_H/RT$ term in eq 1. Because we know from previous studies that PVPh is miscible with PMA (7) at 25 °C, an interesting question can be posed: How much styrene can be incorporated into PMA before the system becomes immiscible? We addressed a similar question previously when we successfully predicted the miscibility windows for STVPh blends with the poly(methyl, ethyl, and butyl methacrylates) (10); the difference here is that we will be "diluting" the non-self-associating polymer PMA by copolymerization with styrene. Finally, we will turn our attention to the more complicated case of blends containing two copolymers, STVPh and STMA, where both the self-associating (VPh) and non-self-associating (MA) segments are "diluted" by copolymerization with an inert (non-hydrogen-bonding) diluent (ST).

Experimental Details

The copolymer compositions, molecular weights (M_n), and glass transition temperatures (T_g) of the polymers used in this study are summarized in Table I. The PVPh, STVPh, and PMA polymers have been described previously (13). Random STMA copolymers were synthesized by direct free-radical polymerization of styrene with methyl acrylate in toluene at 70 °C using azobisisobutyronitrile (AIBN) as the initiator. Prior to polymerization, the monomers were passed through a short column of neutral alumina and vacuum-distilled over calcium hydride. Toluene was redistilled over calcium hydride under nitrogen and the AIBN initiator was recrystallized from acetone before use. Copolymer compositions were determined by proton NMR spectroscopy in $CDCl_3$ from the relative areas of the peaks assigned to the methyl protons of the MA repeat unit and the aromatic protons of the ST repeat unit.

Solutions (1% wt/vol) of the polymers were prepared in methyl isobutyl ketone. Blends of various compositions were then made by mixing appropriate amounts of these solutions. Samples for Fourier transform infrared (FTIR) spectroscopy and differential scanning calorimetry (DSC) studies were obtained by solution casting at room temperature. The solvent was removed slowly under ambient conditions for a minimum of 24 h. The samples were then dried in a vacuum desiccator for an additional day before placement in a vacuum oven at 120 °C for 4 h to completely remove the residual solvent. To minimize water absorption, samples were stored under vacuum desiccation.

The problems associated with polymer blend sample preparation and the experimental determination of the miscibility of a particular blend have been discussed previously (11, 13). Here we only reiterate that we have been very careful to study the infrared spectra of the blend samples first, at ambient temperature after preparation, then at an elevated temperature of 150 °C

Table I. Polymers Employed in This Study

Copolymer	Symbol	M_n (GPC)	T_g (°C)
Styrene-*co*-methyl acrylate (11.2 wt%)	STMA[11]	10,700	94
Styrene-*co*-methyl acrylate (24.8 wt%)	STMA[25]	14,500	87
Styrene-*co*-methyl acrylate (38.3 wt%)	STMA[38]	8,600	74
Styrene-*co*-methyl acrylate (72.4 wt%)	STMA[72]	10,900	45
Styrene-*co*-methyl acrylate (91.2 wt%)	STMA[91]	11,100	27
Poly(methyl acrylate)	PMA	44,000	5
Styrene-*co*-4-vinyl phenol (10 wt%)	STVPh[10]	14,000	109
Styrene-*co*-4-vinyl phenol (25 wt%)	STVPh[25]	11,000	133
Styrene-*co*-4-vinyl phenol (43 wt%)	STVPh[43]	15,000	145
Styrene-*co*-4-vinyl phenol (75 wt%)	STVPh[75]	14,000	166
Poly(4-vinyl phenol)	PVPh	1500–7000	140

(above the T_g of all components) after an annealing period of approximately 15 min, and again at ambient temperature upon cooling, to ensure that every attempt has been made to approach a state of thermodynamic equilibrium.

Infrared spectra were obtained on a Fourier transform infrared (FTIR) spectrometer (Digilab FTS-60) using a minimum of 64 co-added scans at a resolution of 2 cm^{-1}. Spectra recorded at elevated temperatures were obtained using a heating cell mounted inside the sample chamber. Proton NMR spectra were recorded on FT–NMR spectrometers (Brucker WP-200 or AM-300). Molecular weights and molecular weight distributions based upon polystyrene standards were determined using a size exclusion chromatograph (Waters 150C). Thermal analysis was conducted on a differential scanning calorimeter (Perkin-Elmer DSC-7) coupled to a computerized data station. A heating rate of 20 °C/min was used in all experiments, and the glass transition temperature was taken as the midpoint of the heat capacity change.

Results and Discussion

Poly (4-vinyl phenol) Blends with Styrene-co-Methyl Acrylate Copolymers. *Theoretical Calculations*.

The polymer blend system that we describe here was deliberately chosen because the calculated solubility parameters of PS and PMA are almost identical [9.5 and 9.6 (cal/cm^3)$^{0.5}$, respectively] (*14, 16*), which infers that the solubility parameters of STMA copolymers are essentially independent of composition. This independence is illustrated schematically in Figure 1, which shows a plot of STMA solubility parameters as a function of copolymer composition. The solubility parameter of the homopolymer PVPh is also shown as a straight line parallel to the x axis.

Because we estimate the magnitude of χ by the difference in the non-hydrogen-bonded solubility parameters ($\Delta\delta$) of STMA and PVPh and the value of the reference volume, V_B (a *constant* in this case because we use the PVPh chemical repeat to define this quantity), it follows that χ is also practically independent of the copolymer composition (*3, 9, 16*). Accordingly, dilution of PMA with styrene effectively reduces only the $\Delta G_H/RT$ term in eq 1. The methodology for the calculation of the free energy of mixing, phase diagrams, miscibility windows, and maps for polymer–polymer blend systems will be mentioned only briefly here because it has been described in prior publications (*1–15*) and is presented in detail, together with appropriate computer software, in our recently completed book (*16*). The previously obtained parameters (*13*) required for this calculation are summarized in Table II. These parameters can be used with no adjustments to calculate explicitly the relative contributions from the $\Phi_A\Phi_B\chi$ and $\Delta G_H/RT$ terms of eq 1 to the total free energy $\Delta G_m/RT$. These contributions are

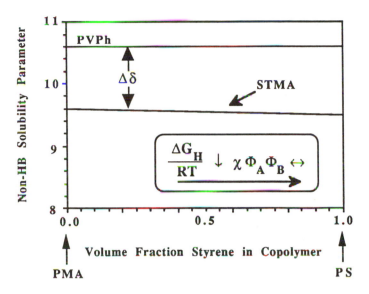

Figure 1. Schematic diagram showing the variation in solubility parameters of STMA copolymers.

Table II. Parameters Employed in This Study

Segment	Molar Volume (cm^3/mol)	Solubility Parameter $(cal/cm^3)^{0.5}$	Equilibrium Constant at 25 °C		
			K_2	K_B	K_A
MA	69.8	9.6	—	—	—
Styrene(ST)	93.9	9.5	—	—	—
VPh	100	10.6	21.0	66.8	47.5

NOTE: Enthalpy of hydrogen bond formation, $h_2 = 5.6$, $h_B = 5.2$, and $h_A = 4.0$ kcal/mol

presented graphically in Figure 2 for 50:50 (wt%) PVPh–STMA blends as a function of STMA copolymer composition at 150 °C (a temperature above the T_gs of both components of the blends, which were selected to facilitate equilibrium conditions). [For high molecular weight polymers, the contribution from combinatorial entropy (the first two terms on the right-hand side of eq 1) is negligible and is left out of Figure 2 for the sake of clarity.] Figure 2 quantitatively describes the trends suggested in the schematic diagram in Figure 1.

To calculate a miscibility window, the second derivative of the free energy of mixing of a blend of the two homopolymers, PVPh and PMA, is

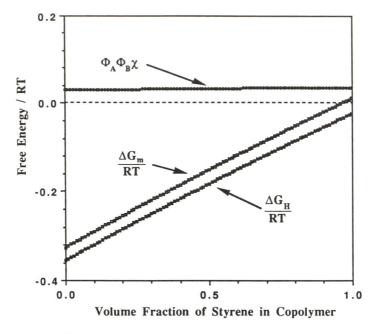

Figure 2. Contributions to the free energy of mixing as a function of copolymer composition.

initially calculated over the entire *blend* composition range at the desired temperature (150 °C, in this case). For PVPh–PMA blends the calculated second derivative curve is positive over the entire composition range for temperatures in the experimentally accessible range of −100 to +250 °C, and this blend is predicted to be miscible (single phase) (*7, 13*). Styrene is simply considered to be an inert diluent and this calculation process is now repeated for PVPh and a STMA copolymer containing 99% methyl acrylate (STMA[99]) and then repeated at 1% composition intervals down to STMA[1]. The two phase region of the phase diagram (spinodal) is defined by the area where the calculated second derivatives of the free energy assume values ≤ 0, which, in turn, sets the limits of the miscibility window. The results of such a computation are shown for the PVPh–STMA blend system at 150 °C in Figure 3. Assuming equilibrium conditions are attained, PMA and STMA copolymers that contain up to about 65 wt% styrene are predicted to be miscible with PVPh. At concentrations above 65% styrene at 150 °C, the contribution from the $\Delta G_H/RT$ term (eq 1) is not sufficient to overwhelm the unfavorable $\chi \Phi_A \Phi_B$ term.

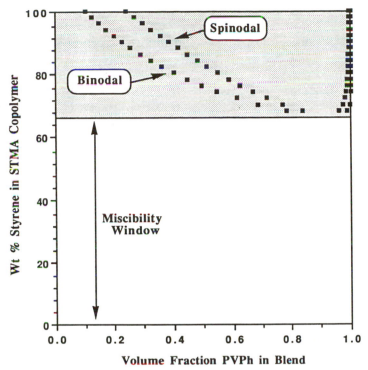

Figure 3. Theoretical miscibility window.

Experimental Results. A combination of FTIR spectroscopy and thermal analysis was employed to test the foregoing predictions. The probe size of the FTIR technique (individual chemical functional groups) is smaller than the probe size of the thermal analysis; thus the FTIR is more sensitive to mixing at the molecular level. It is important to reiterate that even the presence of a prominent infrared band attributed to an intermolecular interaction between groups of dissimilar polymers does not necessarily mean that the mixture is miscible (single phase): What is required is that the measured fraction of hydrogen-bonded groups be equal to the equilibrium distribution at that temperature. However, if we have accurate knowledge of the equilibrium constants that describe self- and interassociation, we are able to establish whether or not the system exhibits this equilibrium distribution and is therefore single phase (*11, 16*). For the PVPh–STMA polymer blend systems considered here we have such information from previous studies and we can readily calculate the fraction of hydrogen-bonded carbonyl groups that should be present in a specific blend of a particular composition at a given temperature assuming that the mixture is a single phase. We can then compare the theoretical fraction of the various intermolecular interactions with the values observed experimentally by FTIR spectroscopy. If the theoretical values equal, within error, the experimentally determined values, a miscible (single-phase) blend can be confidently inferred. Conversely, if the theoretical values deviate significantly from the experimentally determined values, an immiscible (two-phase) mixture is indicated. At the extreme, when the two phases resemble essentially pure components, the theoretical values will approach the values originally present in the pure materials.

If the polymer blend is truly miscible, the theoretical fraction of hydrogen-bonded (HB) carbonyl groups ($f_{HB}^{C=O}$) for PVPh blends with the STMA copolymers as a function of blend composition at 25 °C may be calculated from

$$f_{HB}^{C=O} = 1 - \left\{ 1 \bigg/ \left[1 + K_A \Phi_{B_1} \left[\left(1 - \frac{K_2}{K_B} \right) + \frac{K_2}{K_B} \left(\frac{1}{(1 - K_B \Phi_{B_1})} \right) \right] \right] \right\}$$

(2)

using the stoichiometric equations (11, 16)

$$\Phi_B = \Phi_{B_1} \left[\left(1 - \frac{K_2}{K_B} \right) + \frac{K_2}{K_B} \left(\frac{1}{(1 - K_B \Phi_{B_1})^2} \right) \right] \left[1 + \frac{K_A \Phi_{A_1}}{r} \right] \quad (3)$$

$$\Phi_A = \Phi_{A_1} + K_A \Phi_{A_1} \Phi_{B_1} \left[\left(1 - \frac{K_2}{K_B} \right) + \frac{K_2}{K_B} \left(\frac{1}{1 - K_B \Phi_{B_1}} \right) \right] \quad (4)$$

where Φ_{B_1} and Φ_{A_1} are the volume fractions of the totally "free monomers" of the self-associating species B and the non-self-associating species A. K_2 and K_B are equilibrium constants describing the self-association of B whereas K_A corresponds to the equilibrium constant describing the interassociation of B with A and r is the ratio of the molar volumes V_A/V_B (1–3, 10, 16).

The results of such theoretical calculations for the PVPh blends with the STMA copolymers synthesized for this study (Table I) are shown in Figure 4. These theoretical curves illustrate an important point. Consider the vertical broken line passing through the curves at a blend composition of 50% PVPh. In a miscible blend of PVPh and STMA[91], about 42% of the carbonyl groups are calculated to be hydrogen-bonded, whereas corresponding values for miscible 50:50 blends of PVPh and STMA[72], STMA[38], STMA[25], and STMA[11] are approximately 45, 52, 54, and 57%, respectively. In other words, as we dilute PMA by copolymerization with styrene, the relative proportion of phenolic hydroxyl groups to MA carbonyl groups increases for a *constant blend* composition. This, in turn, leads to an increase in the fraction

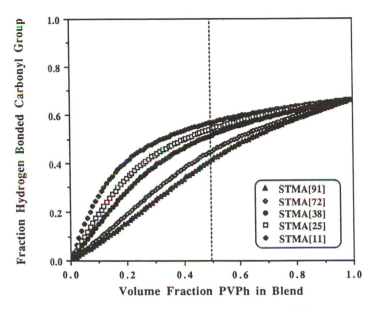

Figure 4. Theoretical fraction of hydrogen-bonded carbonyl groups in PVPh–STMA blends at 150 °C.

of hydrogen-bonded carbonyl groups in a miscible system, as dictated by the equilibrium constants.

Infrared spectra recorded at 150 °C in the carbonyl stretching region, from 1650 to 1800 cm^{-1}, of PVPh blends with four different STMA copolymers are displayed in Figures 5 and 6. For clarity, all spectra are scale expanded and plotted with respect to a relative absorbance scale, but we emphasize that all the spectra were recorded from films that were sufficiently thin to ensure that maximum absorbances did not exceed 0.6 absorbance units. The interpretation of infrared spectra of PVPh blends with carbonyl-containing polymers, such as polyesters, polyacrylates, and polymethacrylates, has been discussed in detail previously (7–13, 16) and only the essential features will be restated here. Pure amorphous STMA copolymers (denoted E in Figures 5 and 6) are characterized by carbonyl stretching vibration at approximately 1735–1739 cm^{-1}. When there is appreciable mixing at the molecular level in the PVPh–STMA blend system, an additional band is observed at approximately 1715 cm^{-1} and is attributed to hydrogen-bonded carbonyl groups (see, for example, the large contribution in the spectra denoted A in Figure 5). The large difference in the spectra recorded for PVPh blends with STMA[38] and STMA[25] is significant. Although the STMA[38] cannot definitively be pronounced a miscible system before an

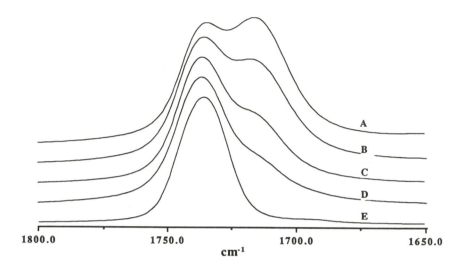

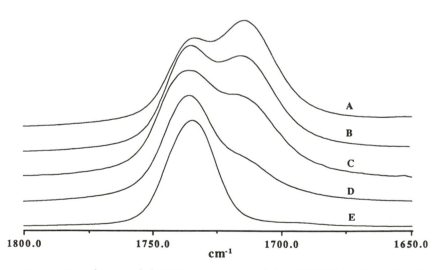

Figure 5. Scale-expanded FTIR spectra recorded at 150 °C in the carbonyl stretching region for PVPh blends with STMA[91] (top) and STMA[72] (bottom): 80:20, A; 60:40, B; 40:60, C; 20:80, D; pure STMA, E.

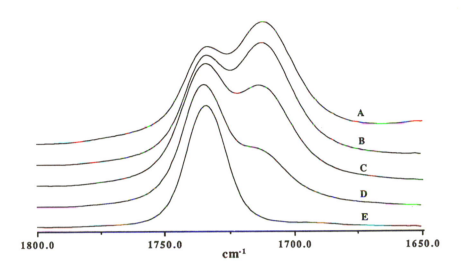

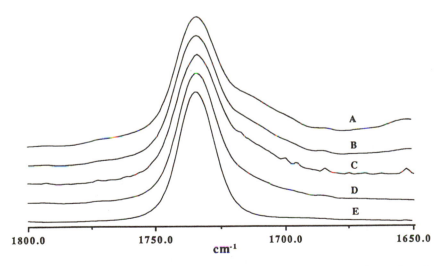

Figure 6. Scale-expanded FTIR spectra recorded at 150 °C in the carbonyl stretching region for PVPh blends with STMA[38] (top) and STMA[25] (bottom): 80:20, A; 60:40, B; 40:60, C; 20:80, D; pure STMA, E.

analysis of quantitative results, the STMA[25] is most certainly immiscible by inspection, because if it were miscible there would be an even greater fraction of hydrogen-bonded carbonyl groups present than in the PVPh–STMA[38] system.

The fraction of hydrogen-bonded carbonyl groups can be quantitatively determined by measuring the relative areas of these two bands, after due consideration is given to differences in the respective absorptivity coefficients (7, 10, 16). The results of curve fitting the carbonyl stretching region of the blends are summarized in Table III and the experimental and theoretical values of the fraction of hydrogen-bonded carbonyl groups are compared in Figure 7. Within error, the experimental values match the theoretical values for the three PVPh blends with STMA copolymers containing 91, 72, and 38% MA, which is consistent with these mixtures being miscible. In marked contrast, the PVPh–STMA[25] blend (and the corresponding STMA[11] system, which is not shown) is obviously grossly phase separated (immiscible) because the experimentally determined fraction of hydrogen-bonded carbonyl groups is significantly less than theoretically calculated for a single phase.

Corroborating evidence was obtained from thermal analysis and Table IV lists the results obtained for the PVPh–STMA blends. Two T_gs at temperatures close to those of the pure components were observed for the PVPh blends with STMA[25] and STMA[11]. Conversely, single T_gs close to those estimated from the Fox equation were observed for the corresponding blends with STMA[91], STMA[72], and STMA[38]. These results are entirely consistent with the FTIR analysis described previously.

Figure 8 shows a final comparison between the theoretically calculated miscibility window for PVPh–STMA copolymer blends at 150 °C and the experimentally determine miscibility behavior of these blends at the same temperature. The unfilled and filled circles represent miscible and immiscible blends, respectively. This encouraging comparison leads us to believe that the assumptions inherent in our association model are reasonable.

Styrene-*co*-Vinyl Phenol Copolymer Blends with Styrene-*co*-Methyl Acrylate. We now consider the case of blends composed of two copolymers. Here we calculate miscibility maps at a particular temperature and vary the composition of both copolymers. For example, we might wish to consider simultaneously the effect of diluting both vinyl phenol and methyl acrylate by copolymerization with styrene.

The calculated solubility parameters of PVPh and PS are 10.6 and 9.5 $(cal/cm^3)^{0.5}$, respectively, which sets the limits of the average solubility parameter range for STVPh copolymers. As already mentioned, the calculated solubility parameters of PMA and PS are very similar and there is practically no dependence of the average solubility parameter on copolymer composition. This behavior is illustrated graphically in Figure 9. Comparison of Figures 9 and 1 reveals that, in addition to the unfavorable (to mixing) trend

Table III. Curve Resolving Data of PVPh–STMA Blends at 150°C

Blend	"Free" C=O Band			H-Bonded C=O Band			Fraction HB C=O	
PVPh:STMA[x]	ν (cm^{-1})	$W_{1/2}$ (cm^{-1})	Area	ν (cm^{-1})	$W_{1/2}$ (cm^{-1})	Area	Expt.	Theory
PVPh:STMA[91]								
20:80	1739	22	22.4	1715	29	11.1	0.25	0.16
40:60	1739	21	14.4	1716	29	11.6	0.35	0.34
60:40	1739	21	7.73	1716	30	10.9	0.48	0.48
80:20	1739	20	15.2	1715	29	15.2	0.56	0.58
PVPh:STMA[72]								
20:80	1737	22	24.2	1713	27	11.2	0.24	0.19
40:60	1738	22	18.2	1714	28	15.3	0.36	0.38
60:40	1738	20	19.5	1714	28	27.8	0.49	0.51
80:20	1738	20	7.98	1714	27	15.4	0.56	0.60
PVPh:STMA[38]								
20:80	1736	20	14.0	1713	27	9.78	0.32	0.27
40:60	1736	21	12.4	1712	27	14.3	0.44	0.46
60:40	1736	20	4.47	1712	27	7.77	0.54	0.55
80:20	1737	20	3.64	1711	28	7.68	0.58	0.62
PVPh:STMA[25]								
20:80	1735	17	2.61	1716	25	0.66	0.14	0.33
40:60	1735	18	1.94	1715	25	0.62	0.18	0.49
60:40	1735	18	1.93	1714	24	0.61	0.17	0.57
80:20	1735	18	0.93	1713	23	0.32	0.19	0.62

due to the reduction of the contribution from the $\Delta G_H/RT$ term with increased styrene dilution, there is the counterbalancing favorable trend that arises from the decreasing difference in the solubility parameter, $\Delta\delta$, between the STVPh and STMA copolymers. The actual magnitude of χ as a function of both copolymer compositions is determined by the difference in the solubility parameters ($\Delta\delta$) of the STVPh and STMA copolymers and the value of the reference volume, V_B (a variable in the way we have defined it (for ease of computation), because it is related to the average STVPh chemical repeat) (3, 16).

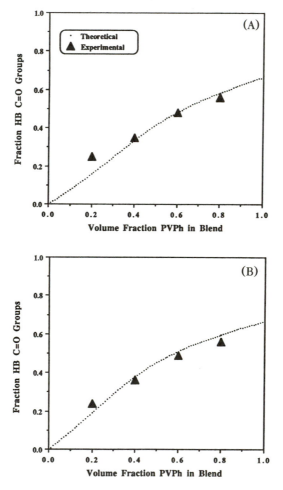

Figure 7. Comparison of the experimentally determined and theoretically calculated fraction of hydrogen-bonded carbonyl groups as a function of blend composition for blends of PVPh with STMA[91], A; STMA[72], B; STMA[38], C; and STMA[25], D, at 150 °C.

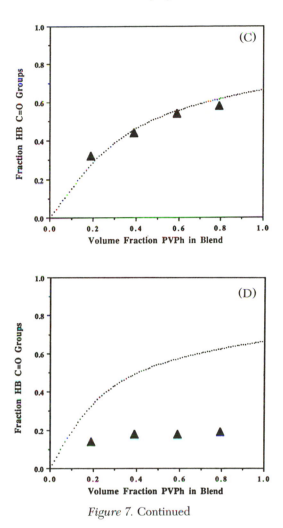

Figure 7. Continued

A miscibility map is calculated in the following manner. First, the second derivative of the free energy of mixing of a blend of the two homopolymers, PVPh (≡STVPh[100]) and PMA (≡STMA[100]), is calculated over the entire *blend* composition range at the desired temperature. If a two-phase region of the spinodal phase diagram is encountered (at *any* composition), then the blend is defined as immiscible. Conversely, if a two-phase region of the spinodal phase diagram is not encountered, then the blend is defined as miscible. This process is repeated using the appropriately computed values of the relevant molar volumes, equilibrium constants, and solubility parameters for PVPh and an STMA copolymer containing 98% methyl acrylate (STMA[98]) and then at every 2% composition to STMA[2]. Now the whole

Table IV. T_g of PVPh–STMA Blends

Blends of PVPh:STMA[x]	T_g (°C)	
PVPh:STMA[91]		
20:80	41	
40:60	45	
60:40	59	
80:20	83	
PVPh:STMA[72]		
20:80	60	
40:60	75	
60:40	102	
80:20	120	
PVPh:STMA[38]		
20:80	87	
40:60	96	
60:40	116	
80:20	129	
PVPh:STMA[25]		
20:80	87	144
40:60	87	143
60:40	86	141
80:20	86	139
PVPh:STMA[11]		
20:80	94	145
40:60	93	142
60:40	91	140
80:20	89	140

procedure is repeated for STVPh copolymers of decreasing VPh content until the matrix that corresponds to a copolymer–copolymer miscibility map is complete.

The result of such an exercise is shown for the STVPh–STMA blends at 150 °C in Figure 10. The bottom left-hand corner of the miscibility map corresponds to a blend of PVPh and PMA that is predicted to be miscible. The top right-hand corner corresponds to a blend of PS with itself. The bottom right-hand corner reflects a blend on PMA and *essentially* PS. (Actually calculations were performed down to a STVPh copolymer containing 2% VPh, and it is more accurate to state that miscibility is predicted for blends of PMA and styrene copolymers containing ≥ 2% VPh, which is not necessarily the lower limit of the concentration of VPh segments required to produce a miscible blend with PMA.) Similarly, the top left-hand corner coincides with a blend of PVPh and *essentially* PS. Elsewhere in the matrix are all the other blends of the STVPh and STMA copolymers, located according to copolymer compositions. The wedge-shaped two-phase region shown in Figure 10 predicts that as PVPh is diluted with styrene, the range of miscibility with STMA copolymers steadily increases.

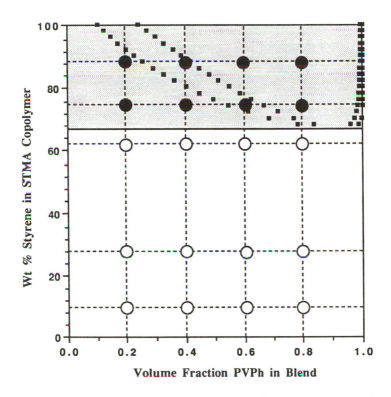

Figure 8. Comparison of theoretical miscibility window to experimental data. Filled and unfilled circles denote immiscible and miscible blends, respectively.

To test the predictions experimentally FTIR spectroscopy was again employed. Figure 11 shows typical infrared spectra of a series of 80:20 wt% STVPh blends with STMA[25] and STMA[11] recorded at 150 °C. Quantitative results from curve fitting together with theoretical calculations of the fraction of hydrogen-bonded carbonyl groups (*see* preceding text) are summarized in Table V. Examination of these results leads to the following conclusions. STMA[11] blends are miscible with STVPh copolymers containing 10 and 25% VPh, but immiscible with the corresponding copolymers containing 43 and 75% VPh. On the other hand, STMA[25] blends are miscible with STVPh copolymers containing 10, 25, and 43% VPh, but immiscible with the copolymer containing 75% VPh. Figure 12 presents a comparison between the theoretically calculated miscibility map (scale expanded for clarity) and the experimental results. It is important to reiterate that this map was calculated from transferable equilibrium constants obtained experimentally from the two homopolymers, PMA and PVPh, and there are

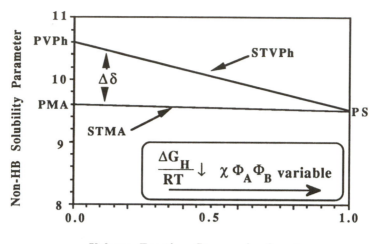

Figure 9. Schematic diagram showing the variation in solubility parameters of the two copolymers STVPh and STMA.

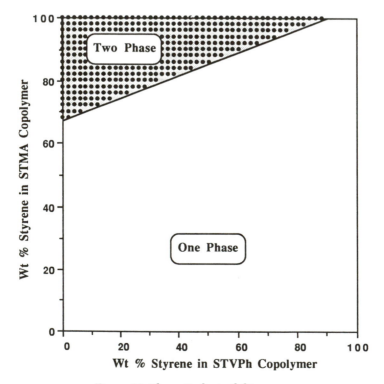

Figure 10. Theoretical miscibility map.

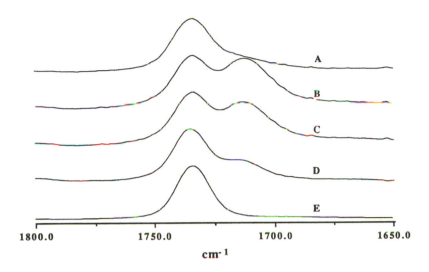

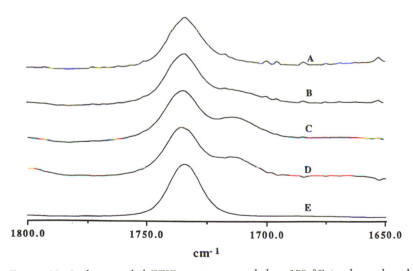

Figure 11. Scale-expanded FTIR spectra recorded at 150 °C in the carbonyl stretching region for 20:80 blends of STMA[25] (top) and STMA[11] (bottom) with STVPh[75], A; STVPh[43], B; STVPh[25], C; STVPh[10], D; and Pure STMA, E.

Table V. Curve Resolving Data of STVPh–STMA at 150°C

Blend STVPh[x]/STMA[y]	"Free" C=O Band			H-Bonded C=O Band			Fraction HB C=O	
	$v\ (cm^{-1})$	$W_{1/2}\ (cm^{-1})$	Area	$v\ (cm^{-1})$	$W_{1/2}\ (cm^{-1})$	Area	Expt.	Theory
STVPh[x]/STMA[11]								
$x = 75$	1735	18^a	0.36	1712	29^a	0.14	0.21	0.60
$x = 43$	1736	17	0.65	1713	28	0.40	0.29	0.53
$x = 25$	1736	16	0.99	1714	27	0.98	0.40	0.44
$x = 10$	1736	16	0.86	1715	23	0.52	0.29	0.29
STVPh[x]/STMA[25]								
$x = 75$	1735	19	0.93	1714	30	0.45	0.25	0.59
$x = 43$	1735	17	0.66	1712	25	0.99	0.50	0.51
$x = 25$	1736	18	0.92	1712	24	1.10	0.44	0.42
$x = 10$	1736	17	2.60	1714	25	1.78	0.31	0.28

^{a} Fixed parameters.

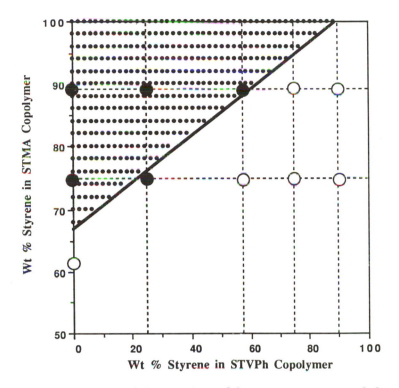

Figure 12. Comparison of theoretical miscibility map to experimental data. Filled and unfilled circles denote immiscible and miscible blends, respectively.

no adjustable parameters involved. There is excellent agreement between theoretical prediction and experimental measurement.

Acknowledgments

The authors acknowledge the financial support of the National Science Foundation, Polymers Program, the Department of Energy under grant DE–FG02–86ER13537, and the E.I. du Pont de Nemours Company.

References

1. Painter, P. C.; Park, Y.; Coleman, M. M. *Macromolecules* **1988**, *21*, 66.
2. Painter, P. C.; Park, Y.; Coleman, M. M. *Macromolecules* **1989**, 22, 570; **1989**, 22, 580.
3. Painter, P. C.; Graf, J. F.; Coleman, M. M. *J. Chem. Phys.* **1990**, 92, 6166.
4. Coleman, M. M.; Hu, J.; Park, Y.; Painter, P. C. *Polymer* **1988**, 29, 1659.

5. Coleman, M. M.; Lee, J. Y.; Serman, C. J.; Wang, Z.; Painter, P. C. *Polymer* **1989**, *30*, 1298.
6. Hu, J.; Painter, P. C.; Coleman, M. M.; Krizan, T. D. *J. Polym. Sci., Polym. Phys. Ed.* **1990**, *28*, 149.
7. Coleman, M. M.; Lichkus, A. M.; Painter, P. C. *Macromolecules* **1989**, *22*, 586.
8. Serman, C. J.; Xu, Y.; Painter, P. C.; Coleman, M. M. *Polymer* **1991**, *32*, 516.
9. Serman, C. J.; Painter, P. C.; Coleman, M. M. *Polymer* **1991**, *32*, 1049.
10. Serman, C. J.; Xu, Y.; Painter, P. C.; Coleman, M. M. *Macromolecules* **1989**, *22*, 2015.
11. Xu, Y.; Graf, J. F.; Painter, P. C.; Coleman, M. M. *Polymer* **1991**, *32*, 3103.
12. Xu, Y.; Painter, P. C.; Coleman, M. M. *Makromol. Chem. Macromol. Symp.* **1991**, *51*, 61.
13. Coleman, M. M.; Xu, Y.; Painter, P. C.; Harrell, J. R. *Makromol. Chem. Macromol. Symp.* **1991**, *52*, 75.
14. Coleman, M. M.; Serman, C. J.; Bhagwagar, D. E.; Painter, P. C. *Polymer* **1990**, *31*, 1187.
15. Bhagwagar, D. E.; Serman, C. J.; Painter, P. C.; Coleman, M. M. *Macromolecules* **1989**, *22*, 4654.
16. Coleman, M. M.; Graf, J. F.; Painter, P. C. *Specific Interactions and the Miscibility of Polymer Blends*; Technomic Publishing, Inc.: Lancaster, PA, 1991.

RECEIVED for review May 14, 1991. ACCEPTED revised manuscript May 1, 1992.

Applications of Raman Spectroscopy to the Study of Polydiacetylenes and Related Materials

S. H. W. Hankin and D. J. Sandman*

GTE Laboratories Incorporated, 40 Sylvan Road,
Waltham, MA 02254

Raman spectroscopy is a useful and sensitive probe of the structure and properties of polydiacetylene (PDA) materials. Using 1064, 632.8, 514.5, 488.0, and 457.9 nm as wavelengths of excitation, Raman spectra were observed using light polarized in the direction of the chain axis for the PDA from 1,6-di-N-carbazolyl-2,4-hexadiyne (DCH) and 1,1,6,6-tetraphenylhexadiynediamine (THD) and for chemically modified versions of these materials. From the changes in the spectra with wavelength of excitation, it was deduced that these pristine PDA single crystals have disordered surface phases. For poly-DCH that has gained six Br atoms per repeat unit, Raman spectra provide a direct indication of the extensive conversion of the PDA backbone structure to that of a mixed polyacetylene. The normal modes associated with triple- and double-bond stretching on the PDA backbone provide a useful structure–property relationship for the characterization of the products of diacetylene polymerizations where the degree of definition does not approach that of fully polymerized single crystals.

POLYDIACETYLENES ARE A CLASS OF POLYMERS with conjugated backbones available in the form of macroscopic single crystals (*1–3*) in certain cases. As single crystals, polydiacetylenes (PDA; **1**) are the best defined class of organic polymers, and it is expected that the properties of these electrically insulating

* Corresponding author. Department of Chemistry, University of Massachusetts, Lowell, MA 01854

materials might serve as models for the properties of other conjugated polymers that are less well defined. At the present time, conjugated polymers in their insulating forms are under investigation for applications such as photoconductivity, third-order nonlinear optical phenomena, and sensors based on chromic changes (4). A recent report of electroluminescence (5) may presage another application of these materials. All of these potential uses involve a knowledge of the electronic spectrum in the solid state. Therefore, the structure–property relationships associated with the solid-state spectra of conjugated polymers are important to the understanding and use of these materials.

The solid-state electronic spectra of PDA crystals are usually discussed in the same framework as neutral molecular crystals (6), that is,

$$E(k) = E_0 + D + I(k) \tag{1}$$

$E(k)$ is the observed solid-state transition energy, E_0 is the gas phase transition energy for an isolated moiety, D summarizes the energetics of the gas-to-crystal shift, and $I(k)$ summarizes the exciton transfer interaction between translationally equivalent and nonequivalent moieties. Figure 1 displays the solid-state electronic spectra of two PDA crystals with chemically related side chains: poly-DCH (1,6-di-N-carbazolyl-2,4-hexadiyne; **1a**) and poly-THD (1,1,6,6-tetraphenylhexadiynediamine; **1b**) (7). Because both crys-

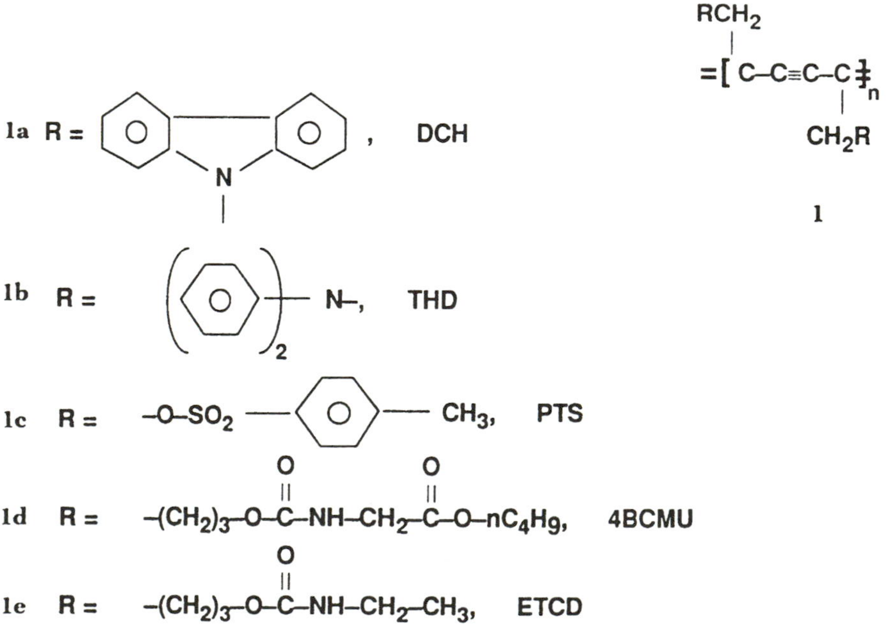

1

1

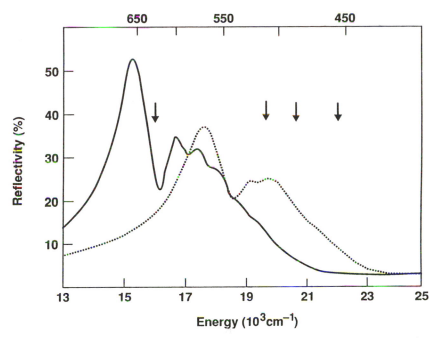

*Figure 1. Near-normal incidence b-axis reflection spectra of poly-DCH (——)
and poly-THD (- - -). Arrows indicate wavelengths of Raman excitation.
(Reproduced with permission from reference 7. Copyright 1987 Elsevier Science
Publishers B.V.)*

tals are completely polymerized, they have "infinite conjugation lengths."
Therefore, the longer wavelength absorption of poly-DCH versus poly-THD
is due to the environment surrounding the conjugated chains; that is, there
is a larger contribution to D and $I(k)$ in 1 for poly-DCH compared to
poly-THD. For the PDA from the bis-p-toluenesulfonate of 2,4-hexadiyn-
1,6-diol (PTS; **1c**), it is recognized that the environment contributes about
15% to $E(k)$ (8). In recent years, extension of the molecular crystal framework of PDA to other conjugated polymers (9, 10) has been useful.

Raman spectroscopy, especially resonance Raman (RR) spectroscopy, is
recognized as a valuable tool for the study of the structure and properties of
conjugated polymers in general (11, 12) and PDA in particular (11–13). An
important factor in this utility is the lack of interference due to strong
fluorescence from PDA crystals. The intensities of Raman bands associated
with totally symmetric vibrational modes strongly coupled to the electronically excited state may be enhanced by a factor as large as 10^6. RR is
particularly attractive for the study of conjugated polymers because the
enhanced intensity permits isolation of the Raman bands of the chromophore.
RR spectroscopy can give information about the properties of a species in an

electronic excited state. The excitation profile can be obtained by measuring the Raman cross section for a vibrational mode as a function of incident photon energy, and an estimation of the strength of the interaction between an electronic excited state and a vibrational mode may be obtained from an excitation profile. Excitation profiles have been obtained for several vibrational modes of poly-PTS (11, 12). To measure an excitation profile, a vibrational mode cannot change its Raman shift as the incident excitation energy is changed; that is, there must be no dispersion (14).

Strong dispersion of band shapes and peak positions is often found in RR spectra of conjugated polymers that are less ordered than PDA crystals, such as polyacetylene [$(CH)_x$; 2] (14, 15), polythiophenes (3) (16, 17), and disordered films of soluble PDA such as the PDA from the bis(butoxy-carbonylmethyleneurethane) of 5,7-dodecadiyn-1,12-diol (4BCMU; 1d) (18). These materials are usually studied in thin-film form.

The RR studies of poly-PTS, -DCH, and -THD have been performed with relatively thick single crystals. For such samples where the absorption coefficient exceeds 10^5 throughout the wavelengths of interest, it is important to recognize that RR spectroscopy is primarily a probe of the surface regions of the crystal (11).

Our initial motivation in applying RR spectroscopy to the study of PDA was to learn greater detail about the structural changes that occur in the course of the chemical modification of poly-DCH (19). The RR spectra of pristine poly-DCH and -THD have been studied to reveal changes that occur from chemical modification. During these studies, we found it useful to study Raman spectra of both pristine and chemically modified PDA with multiple wavelengths of excitation. We also included Fourier transform (FT) Raman excitation at 1064 nm, which is a wavelength where the materials under investigation are transparent (20). Herein we report our investigations of Raman spectra of single crystals of poly-DCH and -THD with light polarized

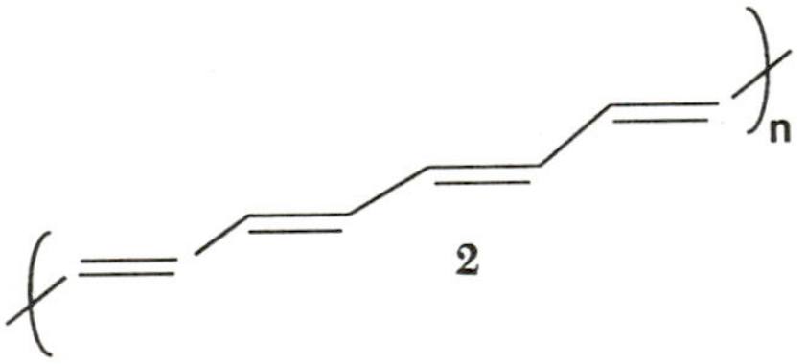

2

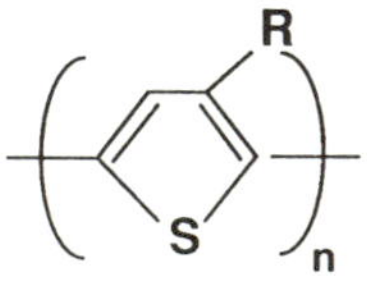

3

along the polymer chain direction as a function of wavelength of excitation, focusing primarily on the normal modes associated with triple- and double-bond stretching. From these spectra, we deduce evidence that the crystals have distinct surface phases, which is a previously unappreciated point. We discuss the changes to the Raman spectra of these materials that occur on chemical modification and how Raman spectroscopy has been used to further understanding of structure–property relationships in these materials. The crystal structures of poly-DCH and -THD are known (*21*) and, thus, spectroscopic studies of these materials add detail to substances of known crystal and molecular structure. Because all results of diacetylene polymerizations do not lead to single crystal products, it is possible, with certain assumptions (given in the following text), to use Raman spectroscopy to deduce if these less well-defined polymers have the usual en–yne backbone structure established (*21*) for the best defined PDA.

Experimental Details

General. Melting points are uncorrected. Elemental analyses were performed by Schwarzkopf Microanalytical Laboratory, Woodside, New York. The ^{1}H NMR spectrum was recorded on a 90-MHz spectrometer (Varian) in $CDCl_3$ solution with tetramethylsilane as reference. Samples of poly-DCH (*30*), poly-THD (*7*), brominated poly-DCH (*19*), and 1,6-bis(3′,6′-dibromo-N-carbazolyl)-2,4-hexadiyne (DCHBr$_4$; **4a**) (*22*) were either synthesized as previously described or were available from earlier studies. The Gammacell at Brandeis University was the source of the ^{60}Co radiation used in these studies.

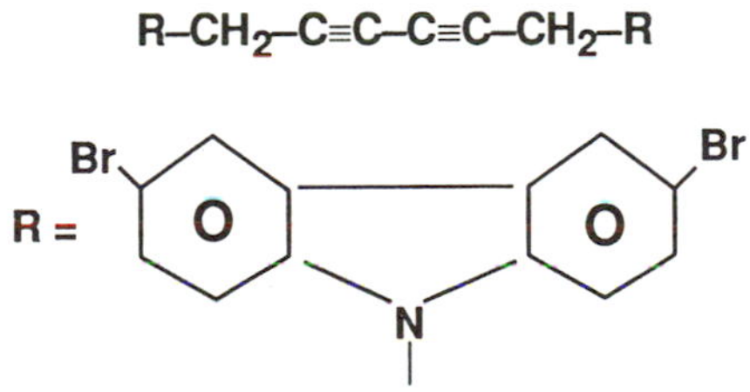

$$R–CH_2–C{\equiv}C–C{\equiv}C–CH_2–R$$

4a

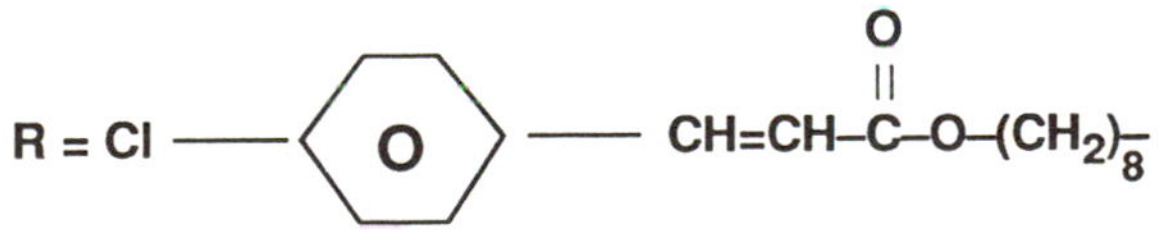

4b

In view of our deduction that PDA crystals have distinct surface phases, it was deemed appropriate to analyze our samples of PDA–DCH and –THD for trace metals by X-ray fluorescence (XRF) and for surface contamination by electron spectroscopy for chemical analysis (ESCA). A recently synthesized sample of PDA–THD contained Fe at a less than 10 ppm level and Cu and Bi at 30–40-ppm levels by XRF analysis whereas less than 10 ppm of each of these elements was found in an older sample; our samples of PDA–DCH contained less than 10 ppm of each of these elements. ESCA studies of PDA–DCH and –THD reveal atomic ratios for C and N comparable to those expected for bulk material. The atomic percentages of oxygen observed by ESCA were 7.4 and 5.0%, respectively, for PDA–THD and –DCH. No evidence for specific C–O single- or double-bond species was found.

Raman Spectroscopy.   The light sources for these experiments were an argon ion laser (Coherent Innova 70-4) with 2–5 mW of excitation at the samples in the 514.5-, 488.0-, and 457.9-nm lines and a 10-mW helium–neon laser (Spectra Physics) for 632.8-nm light with < 1.5 mW at the samples. No sample decomposition was observed with laser excitation at these powers. The dispersive device is a spectrometer (Spex 1877 Triplemate) with the entrance slit at 1 cm, the slit in the center of the bandpass monochromator at 8 mm, and the slit at the entrance to the dispersive monochromator at 50 μm. The front bandpass monochromator consists of two 600 groove/mm gratings, and spectra were obtained with the 600-groove/mm grating (bandpass ± 5 cm^{-1} at 19,436 cm^{-1}) in the third monochromator. Light was detected at 90° to the exciting laser beam by an optical multichannel analyzer (EG & G PARC OMA III). Samples were mounted on an XYZ stage and aligned parallel to the incoming light. The collection time for the multichannel analyzer was set to maximize the signal on the detector. Damage to samples was observed at excitation powers near 50 mW (normal survey scan power) during the initial experiments on these samples. This damage appeared as whitening of the sample; that is, a "bleaching out" of the sample color. Subsequent experiments using poly-THD and 488.0-nm excitation (23) demonstrated the deterioration of the Raman signal with laser powers above 17 mW at the sample that was manifested by a decrease in the ratio of the heights of the double- and triple-bond shifts with increased exposure to the laser.

Preparation of Poly-DCHBr₉.   Poly-DCH (204 mg, 0.50 mmol) and bromine (10 mL) were heated at reflux with magnetic stirring for 24 h. The mixture was cooled and diluted with carbon tetrachloride (10 mL), and the solid product was isolated by suction filtration on a sintered glass funnel. The straw-colored solid was washed with CCl₄ until the washings were colorless and then vacuum-dried to give 560-mg product. X-ray powder diffraction revealed the solid to be amorphous. The Fourier transform infrared (FTIR) spectrum was indistinguishable from that of poly-DCHBr$_8$ (*19*). *Analysis—Calculated for* $C_{30}H_{16}N_2Br_9$: *C, 32.07; H, 1.44; N, 2.49; Br, 64.00. Found: C, 32.24; H, 1.42; N, 2.23; Br, 64.28. The observed analysis corresponds to* $C_{30}H_{15.8}N_{1.8}Br_{8.99}$.

Synthesis and Polymerization of the Bis-*p*-chlorocinnamate of 10,12-Docosadiyn-1,22-diol.

To a solution of sublimed *p*-chlorocin-namoyl chloride (*24*) (14.9 g, 74 mmol) and 10,12-docosadiyn-1,22-diol (**4b**; 10.8 g, 32.3 mmol) in tetrahydrofuran (THF, 100 ml), a solution of pyridine (5 mL) in THF (20 mL) was added dropwise. This mixture was refluxed for 5 h. The cooled reaction mixture was poured into water (500 mL) and the solid product, which turned light blue in room light, was collected on a Buchner funnel. The solid was dissolved in chloroform, and this solution was extracted twice with aqueous NaHCO$_3$. The CHCl$_3$ solution was dried over MgSO$_4$, filtered, and evaporated to give a solid (20.2 g, 94% yield) that was recrystallized from acetone to give **4b**, 16.5 g (77% yield), mp 84–85 °C. The melt was stable and no evidence of reaction was noted on heating **4b** at 70–75 °C for 2 days. The IR spectrum of **4b** [mineral oil (Nujol)] exhibited the following significant absorptions (reciprocal centimeters): 1703, 1630, 1485, 1180, 1085, 1006, 980, 815, 720. The ^{1}H NMR spectrum of **4b** exhibited the following: δ 7.2–7.8 (m, 8H, aromatic), δ 6.3–6.5 (4H, –CH=CH–), δ 4.2 (t, J = 6 Hz, 4H, –OCH$_2$), δ 2.2 (t, J = 6 Hz, –CH$_2$≡C–), δ 1.1–1.8 (m, 28H). The X-ray powder pattern of **4b** exhibited the following reflections (d; angstroms): 33.1, 18.1, 11.3, 8.50, 5.60, 5.36, 5.15, 5.07, 4.89, 4.69, 4.54, 4.32, 4.26, 4.18, 3.98. *Analysis—Calculated for* $C_{40}H_{48}Cl_2O_4$: *C, 72.37; H, 7.30; Cl, 10.68. Found: C, 72.09; H, 7.37; Cl, 10.44.*

Exposure of **4b** to ^{60}Co gamma radiation (at least 50 Mrad over 30 days) converted it to a coppery brown crystalline solid. Extraction of this solid with hot hexane (in which **4b** is soluble) for 6 h did not result in a weight loss. The FTIR spectrum of this polymer was similar to that of **4b** and exhibited the following: 2920, 2851, 1714, 1638, 1592, 1492, 1466, 1407, 1321, 1274, 1253, 1233, 1224, 1203, 1184, 1171, 1108, 1091, 1013, 981, 821, 720. The X-ray powder diffraction of this polymer revealed the following reflections (d, angstroms): 6.67, 5.96, 5.56, 5.41, 4.91, 4.76, 4.56, 4.39, 4.18, 4.05, 3.84.

Raman Spectra of Single Crystals of PDA–DCH and PDA–THD

Figure 2a displays an FT-Raman spectrum of a poly-DCH single crystal obtained with 1064–nm excitation (*20*). For the study of PDA, 1064 nm is an extremely useful wavelength because PDA are transparent at this wavelength and there is no fluorescence to compete with Raman processes. Previously

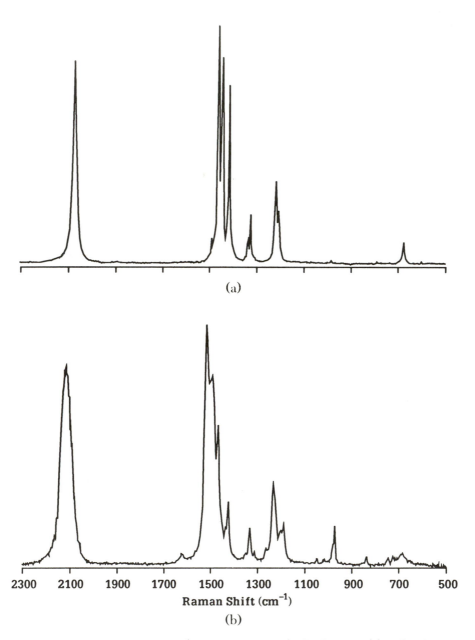

Figure 2. FT-Raman spectra (1064-nm excitation) of poly-DCH (a) and poly-DCHBr$_6$ (b). (Reproduced with permission from reference 20. Copyright 1990 Gordon and Breach.)

reported Raman spectra of poly-DCH used only 632.8-nm excitation (*12*, *25*, *26*); the spectrum in Figure 2a is in good agreement with these reports. In this spectrum the normal mode, which primarily involves triple-bond stretching (*12*), is found at 2081 cm^{-1}, and shifts at 1491, 1466, 1450, and 1420 cm^{-1} are associated with a Fermi resonance (*12*) that involves the double-bond, carbazole, and methylene groups. Bearing in mind that no two single crystals are identical, the following observations concerning spectra of poly-DCH obtained with the other wavelengths of excitation are relevant. As expected for good crystals, the spectra obtained with 632.8-, 514.5-, and 488.0-nm excitation show no dispersion of the Raman signal. With excitation at 514.5, 488.0, and 457.9 nm, some crystals show the Raman spectrum on a luminescent background. Figure 3 displays a representative spectrum of a poly-DCH crystal obtained with 457.9-nm excitation. In addition to the luminescence, the spectrum in Figure 3 reveals considerable broadening and loss of resolution of the Raman-shifted lines, especially in the double-bond region, when compared to the spectrum in Figure 2a and to spectra reported with 632.8-nm excitation. Initially, we were quite puzzled by the appearance of the spectrum in Figure 3; we will return to this issue later.

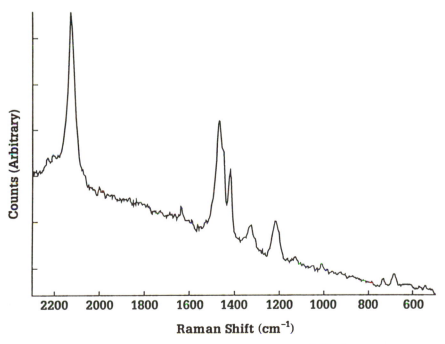

Figure 3. Raman spectrum (457.9-nm excitation) of poly-DCH.

Figure 4 shows the FT-Raman spectrum of a single crystal of poly-THD. Triple-bond shifts of 2111 and 2099 cm^{-1} and a double-bond stretch at 1485 cm^{-1} are revealed. With 632.8-nm excitation (Figure 5a), a triple-bond shift is observed at 2115 cm^{-1} and a double-bond shift is noted at 1485 cm^{-1}, in accord with an earlier observation (7).

The spectrum obtained with 514.5-nm excitation (27) is markedly different than those of poly-THD observed with 1064- and 632.8-nm light. A broad laser-induced emission peaking at a Stokes shift of about 2300 cm^{-1} completely obscures any hint of a Raman spectrum in Figure 5b. This emission is noteworthy because its maximum is found at a wavelength comparable to the maximum associated with the excitonic absorption of poly-THD (7). Additionally, fluorescence from high-quality PDA crystals is negligible (28) and observed emissions in solid PDA are due to material disordered by mechanical damage or photooxidation. If the usual "mirror image" relationship between absorption and fluorescence spectra is assumed, it is estimated that the absorption maximum leading to the emission in Figure 5b would be at 450–460 nm. An absorption maximum in this wavelength region is known to be associated with disordered PDA materials (29).

The spectrum observed with 457.9-nm excitation (Figure 5c) is clearly different from that in Figure 5a because the shifts associated with triple- and

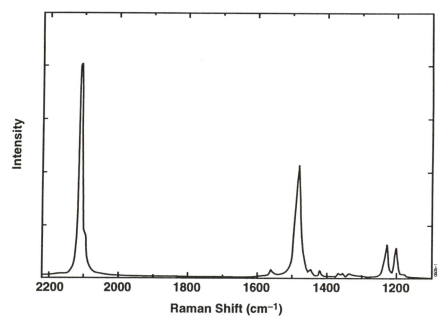

Figure 4. FT-Raman (1064-nm excitation) spectrum of poly-THD.

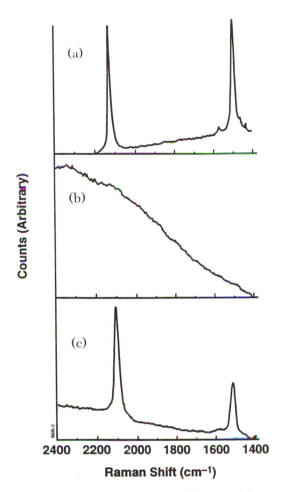

Figure 5. Raman spectra of poly-THD at 632.8 (a), 514.5 (b), and 457.9 nm (c) wavelengths of excitation. (Reproduced from reference 27. Copyright 1991 American Chemical Society.)

double-bond stretching are observed at 2105 and 1504 cm^{-1}, respectively. Spectra observed with 488.0-nm excitation reveal a Raman spectrum superimposed on an emission. The data in Figure 5 are found in crystals freshly polymerized in inert atmosphere, in crystals exposed to ambient conditions for extended time periods, and in crystals with freshly cleaved surfaces. Figure 5 displays spectra recorded at 20 °C, but spectra obtained at − 100 °C do not differ significantly. The data in Figure 5 are not the result of thermal decomposition induced by heating of the crystal by the incident laser beams. Decomposition, however, may be induced by irradiating the crystals with higher laser power (23). Note further that the data in Figure 5 do not differ significantly for poly-THD crystals polymerized thermally or with ^{60}Co gamma radiation.

Because RR spectroscopy of PDA bulk single crystals is primarily a probe of the surface region of these crystals (12), we associate the emission in Figure 5b with a disordered poly-THD surface structure. Again, well-ordered PDA crystals do not emit light (28), and the additional features in the 457.9-nm poly-THD spectrum correspond well to features found with 632.8-nm excitation (7). Therefore, the spectrum in Figure 5c is that of a disordered surface phase of poly-THD, possibly oligomeric.

A disordered surface phase of poly-THD could arise if the surface of the material fails to control the topochemical and topotactic solid-state polymerization of THD monomer to the same extent as the bulk of the crystal (21). Therefore, a surface structure for poly-THD would be distinct in origin from the photooxidized surface of poly-PTS (8). A corollary to the preceding statement on the origin of a surface phase in poly-THD is that other PDA crystals might be expected to have disordered surface structures. In view of our deductions concerning the spectra in Figure 5, we associate the broadened spectrum in Figure 3 with the presence of a disordered surface phase of poly-DCH. The spectrum of poly-PTS obtained with 406.8-nm excitation (30) reveals broadened satellites associated with both the double- and triple-bond stretching modes, and may indicate a surface phase on this crystal. In view of these observations, it is conceivable that RR spectra of PDA materials obtained with wavelengths of excitation significantly shorter than maximum absorption wavelengths may be dominated by disordered surface phases.

From the foregoing summary and earlier work on poly-PTS (12) and the PDA from the bis-ethylurethane of 5,7-dodecadiyn-1,2-diol (ETCD; le) (31, 32) a useful structure–property relationship emerges. For PDA with absorption maxima in the 620–660-nm range (blue spectra), the normal mode associated with triple-bond stretching is found at a shift of 2080–2090 cm^{-1}, whereas PDA with absorption maxima in the 540–570-nm range (red spectra) exhibit the triple-bond stretch at a shift of 2110–2120 cm^{-1}.

Raman Spectra of Chemically Modified Poly-DCH

PDA crystals have van der Waals tight-packed crystal structures and are not expected to have any particular chemical reactivity because diffusion of reagents is not facile in crystalline regions of polymers. This was the state of affairs before our observation of the reaction of poly-DCH with bromine (19). In addition to bromine, chlorine and concentrated nitric acid can also react with poly-DCH to give compositionally well-defined materials that are homogeneous when examined by electron microscopy. The reactions with bromine yield much detailed information and are of great interest. Bromine reacts with poly-DCH crystals to form new homogeneous materials that gain three to nine Br atoms per repeat unit, depending on the experimental conditions. The reactions are anisotropic, and crystallographic order is retained, especially in materials that have gained up to six Br atoms per repeat unit. Although the formation of the material with six Br atoms per repeat unit (poly-DCHBr$_6$) involves a single crystal–single crystal transformation, the product crystal is disordered. The nature of the reaction products was deduced from ^{13}C cross-polarization–magic angle spinning nuclear magnetic resonance (CP–MAS NMR) spectroscopy (22) rather than by single crystal X-ray crystallography. These NMR studies reveal that, for the case of poly-DCHBr$_6$, bromine selectively substitutes the carbazole rings in the 3 and 6 positions, and the diminution of signal in the triple-bond chemical shift region indicates the conversion of the triple bond to a brominated double bond; that is, the PDA structure of the starting polymer is converted to a mixed polyacetylene structure, as shown in Scheme I.

The conclusion that poly-DCH is converted to a mixed polyacetylene structure is an indirect conclusion in that it is deduced from a loss of a signal. Our interest in a direct indication of such a structure motivated us to study Raman spectroscopy of the brominated materials (20, 26, 33) because (CH)$_x$ had been studied extensively by Raman techniques (11, 14, 15).

Our Raman studies of brominated poly-DCH have focused on samples that have gained five to six or eight to nine Br atoms per repeat unit. These studies illustrate the advantage of using multiple wavelengths of excitation, including those in a transparent region. The solid-state electronic spectrum of the material that gained five to six Br atoms per repeat unit revealed a shoulder near 630 nm and additional absorption throughout the visible region (26, 33). We initially felt that 632.8-nm light would excite the well-crystallized regions of the sample and that 488.0-nm light would probe disordered

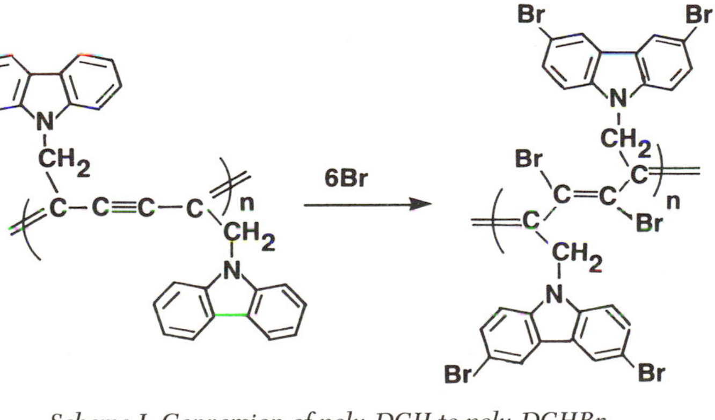

Scheme I. Conversion of poly-DCH to poly-DCHBr$_6$.

conjugated chains with 514.5 nm as a possible intermediate case. The spectra observed with 488.0- and 514.5-nm excitation are largely dominated by background emission (33). With 632.8-nm excitation (26, 33), Raman shifts of 2107 cm^{-1} with a shoulder near 2130 cm^{-1} were revealed. This hint of more than one acetylenic species is consistent with NMR work (22) that revealed more than one residual acetylenic resonance. In the region of double-bond stretching, shifts of 1427, 1454, 1469, 1486, and 1522 cm^{-1}, similar to those of the pristine polymer, were noted. For the brominated material, these vibrations are likely to be due to a composite of different species that contain brominated double bonds in the extended conjugated backbone and, possibly, Fermi resonance that involves a backbone double-bond methylene group and a 3,6-dibromocarbazolyl group. The most informative Raman spectrum of this material was obtained with 1064-nm excitation (20) and is shown in Figure 2b. Comparison of this spectrum with that of the pristine polymer (Figure 2a) reveals that the shifted lines in the brominated polymer are somewhat broader. The shift at 1518 cm^{-1} is the strongest feature in Figure 2b, and we recall that a Raman shift near 1520 cm^{-1} is characteristic of an extended polyene chain (34). We conclude, for poly-DCHBr$_6$, that the Raman spectra obtained with 632.8- and 1064-nm excitation strongly reinforce the conclusion (22) that this material has been extensively converted to a mixed polyacetylene structure.

If poly-DCH crystals are refluxed with excess bromine *without* stirring, materials that have gained approximately eight Br atoms per repeat unit are isolated (19). These materials have some crystallinity and show a shoulder near 450 nm in their solid-state electronic spectra. In contrast to the copper–bronze color of poly-DCHBr$_6$, poly-DCHBr$_8$ is straw-colored, which suggests extensive disruption of the conjugated backbone structure. Solid-state NMR studies (22) indicate that the carbazole rings are also brominated in the 3 and 6 positions and that remaining bromine atoms are involved in converting the conjugated backbone to a partially saturated backbone. Elemental analysis of a recently isolated material indicates that nine Br atoms per repeat unit are gained by refluxing poly-DCH and bromine *with* magnetic stirring. Although Raman and FTIR spectra of poly-DCHBr$_8$ and −Br$_9$ are indistinguishable, the Br$_9$ material is amorphous by X-ray diffraction. To date, an informative Raman spectrum of these materials has been obtained only with 632.8-nm excitation; the spectra obtained with shorter wavelengths are dominated by emission. Figure 6, for example, reveals the spectrum obtained at 488.0 nm; features near shifts of 2100 and 1500 cm^{-1} are barely noticeable. Figure 7 displays the spectrum obtained with 632.8-nm excitation, where broadened shifts near 2140, 1467, 1453, and 1420 cm^{-1} are noted. These shifts are sufficiently similar to the shifts of the pristine poly-DCH that we assume they are due to some low concentration of conjugated segments that remain at the end of the reaction; such segments would dominate the interaction of this solid with 632.8-nm light.

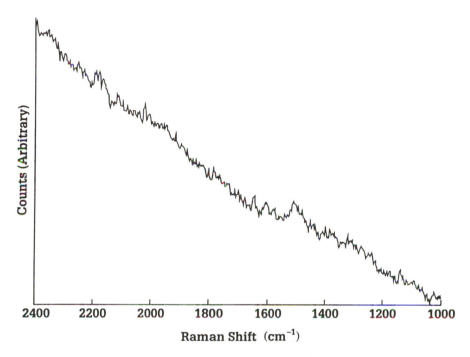

Figure 6. Raman spectrum (488.0-nm excitation) of poly-DCHBr$_8$.

Just as the monomer DCHBr$_4$ **(4a)** was used as a model compound for the solid-state NMR studies (*22*), we hoped to be able to use it as a model for Raman studies of a brominated carbazole. The spectra, however, were dominated by emission for wavelengths of excitation between 457.9 and 632.8 nm. Only the 632.8-nm spectrum gave hints of Raman lines near 1467, 1325, and 1222 cm^{-1}. The domination by emission in attempts to record Raman spectra of **4a** are not surprising in view of the reported (*35*) phosphorescence of 3,6-dibrominated carbazoles.

Raman spectra of chemically modified versions of poly-DCH are most informatively studied using multiple wavelengths of excitation. The Raman spectra of poly-DCHBr$_6$ and −Br$_8$ confirm the conclusions of the solid-state NMR studies (*22*) that although these materials are homogeneous when examined by electron microscopy, they are not homogeneous at the molecular repeat level.

Raman Spectra of PDA Materials Less Well Defined Than Single Crystal

Not all PDA materials are as well defined as the foregoing examples of fully polymerized single crystals. For example, there are many instances where

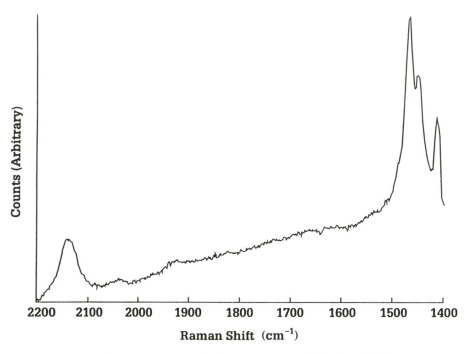

Figure 7. Raman spectrum (632.8-nm excitation) of poly-DCHBr$_8$.

PDA polymer crystals contain significant amounts of monomer that do not polymerize due to a mismatch between monomer and polymer lattices (*21*). Additionally, it does not follow that if a diacetylene monomer reacts to give a colored product, this colored product has the en–yne (*21*) backbone structure. Other reaction pathways are conceivable. Therefore, in the absence of a complete crystal structure, other means of proof of polymer backbone structure are highly desirable, especially for insoluble materials. Useful experimental techniques for this purpose include X-ray powder diffraction [observation of a 4.9-Å repeat distance (*21*)], solid-state ^{13}C NMR, and Raman spectroscopy. Application of Raman techniques for this purpose assumes that the observed spectra are representative of the bulk of a solid sample, not just the surface.

If a strongly colored solid material is the product of a diacetylene polymerization and it exhibits a Raman spectrum with shifts near 2100 and 1500 cm^{-1}, it is reasonable to conclude that the material has the en–yne backbone structure because well-defined PDA crystals have Raman spectra that exhibit the normal modes associated with triple- and double-bond stretching near 2100 and 1500 cm^{-1}. This approach is implicit in the study of PDA Langmuir–Blodgett films (*11*), where excitation profiles have been obtained in certain cases.

We recently reported the synthesis and study of the solid-state reactivity of a series of bis-p-chlorocinnamates of diacetylene diols (*24*). The ester of 10,12-docosadiyn-1,22-diol (**4b**) is thermally stable, but is polymerized by 370-nm light and ^{60}Co gamma radiation to a coppery brown solid whose X-ray powder diffraction reveals a reflection at d = 4.91 Å. The Raman spectrum (632.8-nm excitation) of this material is shown in Figure 8, and the observation of shifts at 2133 and 1448 cm^{-1} suggests the usual en–yne structure for this material.

In the course of our studies of the crystal and molecular structure and solid-state reactivity of DCHBr$_4$, (**4a**) (*36*), we also studied the properties of the dark violet largely amorphous solid formed by extended heating above 200 °C. The Raman spectrum (632.8-nm excitation) revealed shifts at 2128, 1481, and 1464 cm^{-1}, and this information is sufficient to conclude that the product of thermal solid-state polymerization of DCHBr$_4$ has the usual en–yne structure.

Conclusions

Raman spectroscopy, using excitation in both transparent and absorbing regions, will continue to be a valuable tool in the study of the structure and

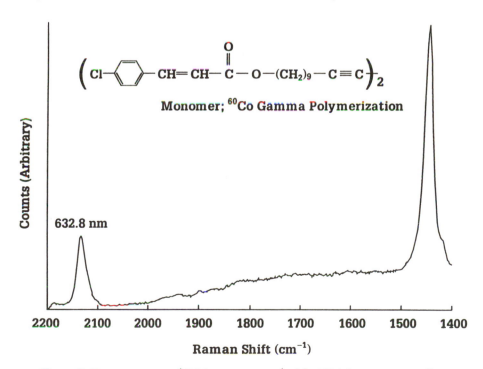

*Figure 8. Raman spectrum (632.8-nm excitation) of the PDA from monomer **4b**.*

properties of PDA materials and other conjugated polymers. The normal modes associated with triple- and double-bond stretching, found near shifts of 2100 and 1500 cm^{-1}, respectively, constitute a very useful structure–property relationship for the characterization of more complex PDA materials than those emphasized in earlier studies and herein. As an example of a more complex PDA system, a recent study of phase separation in cross-polymerized diacetylenes may be cited (37).

The initial comprehensive studies of RR spectroscopy by Batchelder and co-workers (8, 11, 12), primarily with PDA–PTS single crystals, are of major value because the measured excitation profiles allow estimation of the interaction between an electronic excited state and a vibrational mode. Additionally, these studies point the way toward the study of more complex systems. The studies with single crystals of PDA–DCH and –THD are an important extension of this work because they reveal that dispersion in RR spectra of PDA crystals may be a relatively widespread phenomenon. The deduction that disordered PDA surface phases, likely formed by failure of the crystal surface to control the topochemical and topotactic polymerization, are the structural source of the dispersion clearly raises an important issue in the study of conjugated polymers for electronic and optical applications.

Dispersion is a common feature of the RR spectra of conjugated macromolecules less structurally ordered than PDA single crystals. The three models (namely, amplitude mode, conjugation length, and effective conjugation coordinate) used to explain the features of RR spectra of less-ordered conjugated polymers were compared recently (38). With reference to eq 1, a given sample of a partially crystalline conjugated polymer may be expected to have a variety of conjugation lengths (hence differing E_0), a variety of local environments of varying crystallinity (hence variation in D), and, in the spirit of our discussions, the possibility of distinct surface structures.

There are further indications that the surface topographies of PDA crystals are distinct from the bulk and may be quite complex. An atomic force microscopy study (39) concluded that the substituent positions of a PDA crystal surface layer were different from those of the bulk. Indications that the surface structure of the PDA of the bis-isopropylurethane of 5,7-dodecadiyn-1,12-diol varies from crystal to crystal were found in our RR spectral studies using 514.5-nm excitation at ambient temperature (40). This work indicated that the number of shifted lines associated with a given normal mode, the relative intensity of the lines of a given mode, and the amount of background luminescence change from crystal to crystal. We conclude that the foregoing studies are a useful beginning to understanding the potential complexity of the surfaces of PDA crystals.

Acknowledgments

The authors thank Dr. M. J. Smith (Nicolet Instruments) for recording the FT-Raman spectra displayed in this work. Compound **4b** was synthesized by

Dr. R. A. Haaksma. The authors thank E. Yost for the synthesis of the most recent samples of THD monomer and polymer. X-ray powder patterns were furnished by Dr. G. P. Hamill and M. J. Downey. D. R. Hammond supplied FTIR spectra, F. X. Pink performed the XRF and ESCA analyses, and B. M. Foxman (Brandeis University) facilitated access to the Gammacell.

References

1. *Crystallographically Ordered Polymers*; Sandman, D. J., Ed.; ACS Symposium Series 337; American Chemical Society: Washington, DC, 1987.
2. *Polydiacetylenes*; Cantow, H. J., Ed.; Advances in Polymer Science 63; Springer: Berlin, Germany, 1984.
3. *Polydiacetylenes*; Bloor, D.; Chance, R. R. Eds.; NATO ASI Series; Martinus Nijhoff: Dordrecht, Netherlands, 1985.
4. Sandman, D. J. *1991 McGraw-Hill Yearbook of Science and Technology*; McGraw-Hill: New York, 1991; pp 71–75.
5. Burroughes, J. H.; Bradley, D. D. C.; Brown, A. R.; Marks, R. N.; Mackay, K.; Friend, R. H.; Burns, P. L.; Holmes, A. B. *Nature (London)* **1990**, *347*, 539.
6. Wright, J. D. *Molecular Crystals*; Cambridge University Press: New York, 1987; Chapter 6, p 96.
7. Morrow, M. E.; White, K. M.; Eckhardt, C. J.; Sandman, D. J. *Chem. Phys. Lett.* **1987**, *140*, 263.
8. Batchelder, D. N. In *Polydiacetylenes*; Bloor, D.; Chance, R. R., Eds.; NATO ASI Series; Martinus Nijhoff: Dordrecht, Netherlands, 1985; pp 187–212.
9. Sandman, D. J.; Chen, Y. J., *Synth. Metals* **1989**, *28*, D613.
10. Bassler, H. In *Optical Techniques to Characterize Polymer Systems*; Bassler, H., Ed.; Elsevier: New York, 1989; pp 181–225.
11. Batchelder, D. N. In *Optical Techniques To Characterize Polymer Systems*; Bassler, H., Ed.; Elsevier, New York, 1989; pp 393–427.
12. Batchelder, D. N.; Bloor, D. In *Advances in Infrared and Raman Spectroscopy*; Clark, R. J. H.; Hester, R. E., Eds.; Wiley: New York, 1984; Vol. 11, pp 133–209.
13. Chance, R. R. In *Encyclopedia of Polymer Science and Technology*; Kroschwitz, J. I., Ed.; Wiley: New York, 1986; Vol. 4, pp 767–779.
14. Lichtman, L. S.; Sarhangi, A.; Fitchen, D. B. *Solid State Commun.* **1980**, *36*, 869.
15. Schen, M. A.; Chien, J. C. W.; Perrin, E.; Lefrant, S.; Mulazzi, E. *J. Chem. Phys.* **1988**, *89*, 7615.
16. Vardeny, Z.; Ehrenfreund, E.; Brafman, O.; Heeger, A. J.; Wudl, F. *Synth. Metals* **1987**, *18*, 183.
17. Botta, C.; Luzzati, S.; Bolognesi, A.; Catellani, M.; Destri, S.; Tubino, R. In *Materials Research Society Symposium Proceedings*; Chiang, L.-Y.; Chaikin, P. M.; Cowan, D. O., Eds.; Materials Research Society: Strasbourg, France, 1990; Vol. 173, pp 397–402.
18. Zheng, L. X.; Benner, R. E., Vardeny, Z. V.; Baker, G. L. *Phys. Rev. B* **1990**, *42*, 3235.
19. Sandman, D. J.; Elman, B. S.; Hamill, G. P.; Hefter, J.; Velazquez, C. S. In *Crystallographically Ordered Polymers*; Sandman, D. J., Ed.; ACS Symposium Series 337; American Chemical Society: Washington, DC, 1987; Chapter 9, pp 118–127.
20. Hankin, S.; Sandman, D. J. *Mol. Cryst. Liq. Cryst.* **1990**, *186*, 197.
21. Enkelmann, V. In *Polydiacetylenes*; Cantow, H. J., Ed.; Advances in Polymer Science 63; Springer: Berlin, Germany, 1984; pp 91–136.

22. Eckert, H.; Yesinowski, J. P.; Sandman, D. J.; Velazquez, C. S. *J. Am. Chem. Soc.* **1987**, *109*, 761.
23. Hankin, S.; Sandman, D. J.; Yost, E. A.; Stark, T. J. *Synth. Metals* **1992**, *49*, 281.
24. Sandman, D. J.; Haaksma, R. A.; Foxman, B. M. *Chem. Mater.* **1991**, *3*, 471.
25. Elman, B. S.; Thakur, M. K.; Sandman, D. J.; Newkirk, M. A.; Kennedy, E. F. *J. Appl. Phys.* **1985**, *57*, 4996.
26. Sandman, D. J.; Chen, Y. J.; Elman, B. S.; Velazquez, C. S. *Macromolecules* **1988**, *21*, 3112.
27. Hankin, S.; Sandman, D. J. *Macromolecules* **1991**, *24*, 4983.
28. Bloor, D.; Rughooputh, S. D. D. V.; Phillips, D.; Hayes, W.; Wong, K. S. *Electronic Properties of Polymers and Related Materials*; Kuzmany, H.; Mehring, M.; Roth, S., Eds.; Springer: Berlin, Germany, 1985; pp 253–255.
29. Bloor, D. *Photon, Electron, and Ion Probes of Polymer Structure and Properties*; Dwight, D. W.; Fabish, T. J.; Thomas, H. R., Eds.; ACS Symposium Series 162; American Chemical Society: Washington, DC, 1981; pp 81–104.
30. Kuzmany, H.; Kurti, J. *Synth. Metals* **1987**, *21*, 95.
31. Sandman, D. J.; Chen, Y. J. *Synth. Metals* **1989**, *28*, D613.
32. Sandman, D. J.; Chen, Y. J. *Polymer* **1989**, *30*, 1027.
33. Hankin, S. H. W.; Sandman, D. J. *Polymer Commun.* **1990**, *31*, 22.
34. Shirakawa, H.; Ito, T.; Ikeda, S. *Polymer J.* **1973**, *4*, 460.
35. Yokoyama, M.; Funaki, M.; Mikawa, H. *J. Chem. Soc., Chem. Commun.* **1974**, 372.
36. Sandman, D. J.; Velazquez, C. S.; Hamill, G. P.; Foxman, B. M. *Polym. Prepr. Am. Chem. Soc., Div. Polym. Chem.* **1992**, *33*, 284.
37. Nitzsche, S. A.; Hsu, S. L.; Hammond, P. T.; Rubner, M. F. *Macromolecules* **1992**, *25*, 2391.
38. Kurti, J.; Kuzmany, H. *Phys. Rev. B* **1991**, *44*, 597.
39. Magonov, S. N.; Bar, G.; Cantow, H.-J.; Bauer, H.-D.; Müller, I.; Schwoerer, M. *Polym. Bull.* **1991**, *26*, 223.
40. Hankin, S. H. W.; Sandman, D. J. In *Materials Research Society Symposium Proceedings*; Chiang, L. Y.; Garito, A. F.; Sandman, D. J., Eds.; Materials Research Society: Strasbourg, France, 1992; Vol. 247, pp 661–667.

RECEIVED for review May 14, 1991. ACCEPTED revised manuscript July 15, 1992.

9

Qualitative Identification of Polymeric Materials Using Near-Infrared Spectroscopy

Huston E. Howell and James R. Davis

Research and Development, Fibers Division, BASF Corporation, Enka, NC 28728

During the past decade near-infrared (near-IR) spectroscopy has become a widely used analytical technique. Most applications have been for quantitative analysis. In the past several years, near-IR spectroscopy has been successfully used for qualitative analysis through the use of computer-assisted data processing. Two major methods of data analysis have yielded successful results: discriminant analysis and spectral searching. Discriminant analysis is useful for discrete wavelength spectra obtained with bandpass filters and continuous spectra obtained with a scanning monochromator. Spectral searching of first derivative spectra is used for continuous spectra from a scanning monochromator. In this study, we demonstrate the ability to qualitatively identify various textile fibers and thermoplastic polymers with near-IR spectroscopy using these two methods.

NEAR-INFRARED SPECTROSCOPY has been widely used for over a decade for quantitative measurements in many areas. Extensive calibration algorithms and diagnostic statistics have been developed (*1*, *2*). Recently the potential of near-IR spectroscopy as a qualitative identification technique has been recognized.

The near-IR region is comprised of overtones (1100–1800 nm) and combinations (1800–2500 nm) of the fundamental vibrations of the mid-IR region. Band assignments for many common organic compounds have been made (*1*, *3*). In general, visual interpretation of spectra is more difficult than

0065–2393/93/0236–0263$06.75/0

in the mid-IR region because of the nature of the bands (i.e., overtones and combinations). The spectral bands are often overlapped, and differentiation between similar materials appears less definitive than in the mid-IR region. In spite of these apparent disadvantages, the near-IR region provides useful qualitative information, especially when computer-assisted data analysis techniques, such as multivariate statistics and spectral searching, are used.

Mark and Tunnel (4) demonstrated the qualitative classification of various materials using discriminant analysis with Mahalanobis distances. Lo and Brown (5) demonstrated the ability to qualitatively identify materials using computerized spectral library searching. Ciurczak and co-workers (6, 7) used qualitative near-IR spectroscopy for a variety of applications in the pharmaceutical area. Ghosh and Rodgers demonstrated the ability to identify heat-set type and distinguish nylon 6 from nylon 6,6 using near-IR discriminant analysis (8).

In this study, we examine the usefulness of near-IR spectroscopy to qualitatively identify various textile fibers and thermoplastic polymers. Qualitative identification is very important when working with polymeric materials. Consistency of composition, to a large extent, determines the consistency of physical properties that impact all aspects of usage: from quality control in manufacturing to performance of the final product. Because our intent was to perform an empirical survey of the qualitative capabilities, we did not attempt to extensively interpret relationships between spectra and chemical structures. Although such a fundamental exercise would be useful, it is beyond the intended scope of this investigation. Overall, we found that near-IR spectroscopy is a very useful technique for the qualitative identification of polymer composition, especially when computer-assisted data analysis techniques are used.

Qualitative Identification Using Discriminant Analysis

Discriminant analysis using Mahalanobis distances, which is a multivariate statistical technique for classifying groups, has proved to be a powerful technique in qualitative spectroscopic identification. This technique was demonstrated and thoroughly described by Mark and Tunnel (4). Initially, a calibration set that consists of "known" materials (i.e., standards) is analyzed. The discriminant analysis program (used in this study) calculates the absorbance relationships at various wavelengths for each sample, and then mathematically selects a set of wavelengths that best reveal the differences between groups of materials that are identified as being of similar composition. In practice, three to six wavelengths usually are adequate to provide good classification. Finally, the results are expressed as the Mahalanobis distances between the various groups. An intergroup distance 15 or more generally indicates that discrimination is feasible. When the calibration is complete, "unknowns" are analyzed to determine if they are similar to any of the calibration samples.

Qualitative Identification Using Spectral Searching

Qualitative identification can also be performed by computerized spectral library searching. The spectrum of an unknown material is compared with the spectra of reference materials. This approach is widely used in the mid-IR region to identify unknown materials. Spectral searching programs may vary according to the particular vendor; however, most operate in the same general manner. The computer mathematically compares the spectrum of the unknown with a set of reference spectra. The results are expressed in a listing of reference materials, which is rank-ordered by the goodness of agreement between the spectrum of the unknown and the spectrum of each reference material. For the particular software used in this study, the goodness of the match is expressed by a numerical value called the hit quality index (HQI). A HQI = 0 indicates a perfect spectral match with no dissimilarity. The goodness of match decreases as the HQI increases. The spectral search program also displays the spectrum of the unknown material on the computer monitor together with each hit from the spectral library to allow the user to visually compare the spectra.

Experimental Details

The various textile fibers studied include most of those commonly encountered in the textile industry: poly(ethylene terephthalate) (PET), polypropylene, cotton, acrylic, wool, rayon, and nylon. Spectra of the fibers were obtained for several different forms of each fiber type, such as variable deniers, with and without titanium dioxide, yarn and fabric form, and staple and continuous filament.

Twenty-five widely used thermoplastic materials were analyzed: poly(vinyl chloride) (PVC), poly(vinyl acetate) (PVAc), poly(methyl methacrylate) (PMMA), poly(vinylidene fluoride) (PVDF), polyacetal (PA), nylon 6, nylon 12, nylon 6–aramid, polyethylene (PE), polypropylene (PP), poly(methyl pentene) (PMP), poly(ethylene terephthalate) (PET), thermoplastic polyurethane (TPU), poly(butylene terephthalate) (PBT), cellulose acetate (CA), polysulfone (PSU), poly(aryl ether sulfone) (PES), polycarbonate (PC), poly(phenylene sulfide) (PPS), poly(aryl ether ether ketone) (PEEK), poly(aryl ether ether ketone ether ketone) (PEEKEK), polystyrene (PST), acrylonitrile–butadiene–styrene (ABS), acrylonitrile–styrene–acrylate (ASA), and styrene–acrylonitrile (SAN). Thermoplastic materials received in pellet form were ground in a laboratory mill and separated into particle-size groups of 20, 30, and > 30 mesh. Powered thermoplastic materials were analyzed as received.

Near-IR spectra were collected at 4-nm intervals over the range of 1100–2500 nm using a scanning spectrophotometer (Bran + Luebbe, Infra-Alyzer model 500) connected to a personal computer (Dell model 310). Three scans were averaged for each sample. Spectral data were collected in a quartz-windowed cup containing 2–3 g of sample. All spectra were processed in the log(1/reflectance) form.

Discriminant analysis using Mahalanobis distances was performed using software provided by the instrument vendor (Bran + Luebbe). This program performs a mathematical computation that selects a set of wavelengths that best describe the differences between the various samples. Simulated filter spectra were created using the filter transform software function (IDAS-PC, Bran + Luebbe). This procedure converts the continuous scan spectra to the spectral form obtained from discrete filter instruments, which are widely used. Spectral searching was performed by importing the spectra into Spectra-Calc (Galactic Industries) software using the Joint Committee on Atomic and Molecular Physics format. Spectral libraries of regular log(1/reflectance) and first derivative spectra were prepared by averaging several spectra of the different physical forms of each material.

Results and Discussion

Near-IR Spectra. *Textile Fibers.* The near-IR spectra of various textile fibers are shown in Figures 1–3. Significant differences across the entire spectral range (1100–2400 nm) are apparent between the spectra of polypropylene, acrylic, and PET shown in Figure 1. Smaller, but distinct, differences around 1500 (O–H stretch overtone) and 2100 nm (C–O stretch overtone) are seen for the spectra of cotton and rayon in Figure 2. These materials have the same chemical structure (cellulose), but they differ, primarily, in crystalline phase morphology. Rayon exhibits the cellulose II morphology and cotton exhibits the cellulose I morphology. The spectra of nylon 6 and nylon 6,6 shown in Figure 3 exhibit very subtle, but observable, differences in the region above 2000 nm (combination band region).

From this survey of various fiber types, it is apparent that near-IR spectra contain useful qualitative information. These examples clearly show how sensitive the near-IR region is to subtle chemical differences, and even morphological differences of the same chemical composition. This result is not surprising because the bands in the near-IR region arise primarily from C–H, O–H, and N–H functional groups. Qualitative differences in the spectra are further enhanced by the first derivative spectra.

Thermoplastic Polymers. The near-IR spectra of all 25 thermoplastic materials analyzed were sufficiently different to permit qualitative identification. The near-IR spectra of two aromatic polyesters, PET and PBT, are

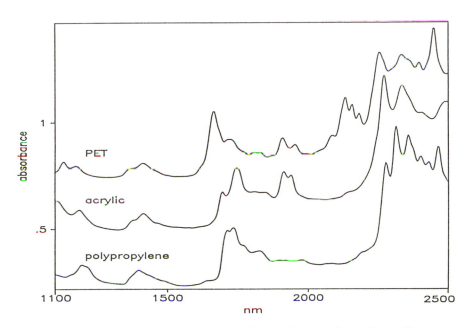

Figure 1. Near-IR spectra of uncolored polypropylene, acrylic, and PET fibers.

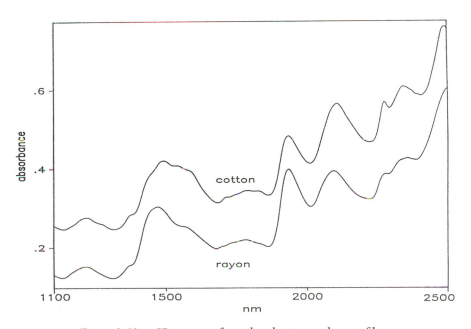

Figure 2. Near-IR spectra of uncolored cotton and rayon fibers.

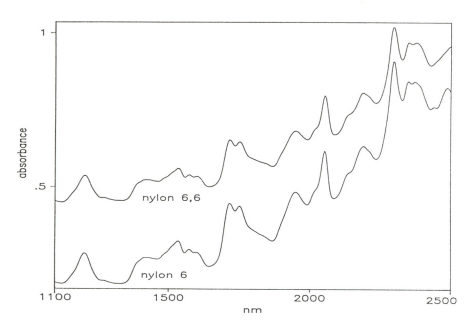

Figure 3. Near-IR spectra of uncolored nylon 6 and nylon 6,6 fibers.

shown in Figure 4. Although these materials have very similar chemical structures (PBT has two additional $-CH_2-$ groups in the chain backbone), the near-IR spectra are sufficiently different in the 1600−1800-nm region (C−H and C = H stretch overtone) to be distinguished from each other.

The near-IR spectra of three polyolefins, high density PE, PP, and PMP, are shown in Figure 5. These materials also have similar chemical structures, particularly PP and PMP. PMP has four $-CH_2-$ groups between each pendant $-CH_3$ group, whereas PP has only two $-CH_2-$ groups between each pendant $-CH_3$ group. The spectrum of PE, linear $-CH_2-$ with a minor amount of pendant $-CH_3$ groups, is significantly different from PP and PMP in the 1600−1800- (C−H stretch overtone) and 2000−2400-nm (combination bands) regions. The spectral differences between PP and PMP are very slight, but apparent in the 1600−1800-nm region (C−H stretch overtone). The subtle differences are more apparent in the first derivative spectra.

Figure 6 shows the near-IR spectra of three structurally similar high-performance thermoplastic materials: PES, PEEK, and PEEKEK. Subtle, but distinct, differences are present in the region above 1800 nm (combination bands), particularly in the first derivative spectra. Similar differences were found for other classes of similar thermoplastic materials that have similar compositions, such as polyolefins, polyesters, polystyrenes, and polyamides.

Although qualitative identification can be performed by manual inspection, computer-assisted interpretation provides enhanced capabilities. This

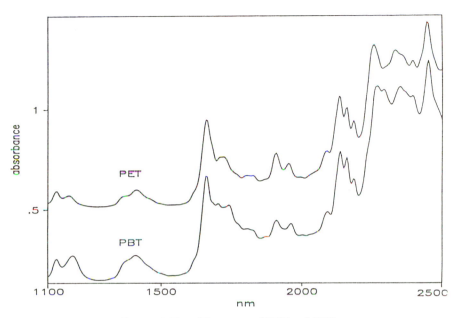

Figure 4. Near-IR spectra of PET and PBT.

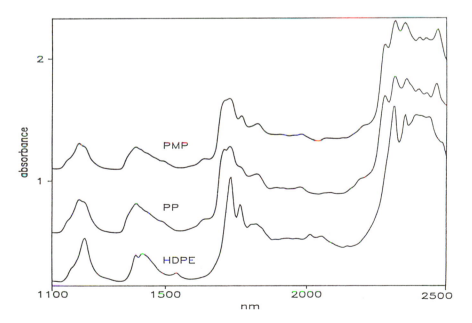

Figure 5. Near-IR spectra of PMP, PP, and HDPE.

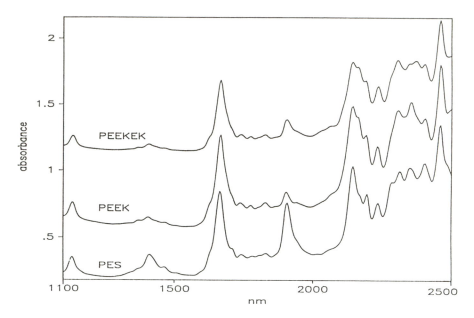

Figure 6. Near-IR spectra of PEEKEK, PEEK, and PES.

capability is particularly useful in the near-IR region because band assignments are less definitive in the near-IR region than in the mid-IR region. Bands in the near-IR region arise from overtones and combinations that tend to be broader and more overlapped than in the mid-IR region. Two types of commercially available software designed for qualitative classification–identification were evaluated: discriminant analysis and spectral searching.

Discriminant Analysis. *Fibers*. We found discriminant analysis using Mahalanobis distances very useful for qualitatively identifying uncolored or lightly colored fibers and thermoplastic polymers, but problems were encountered with dark-colored materials.

The discriminant analysis parameters and results expressed as Mahalanobis distances for the various fibers studied are shown in Table I. The magnitudes (≥ 15) of most of the intergroup distances indicate a significant discrimination between most of the groups. For continuous scan spectra, four wavelengths provide a significant discrimination between the various fiber types, except for rayon versus cotton. However, these two fibers can be easily distinguished from each other by inspection of the spectra or by using a discriminant analysis search based on standards of these two fibers only. The discriminant analysis results for the same fibers using simulated discrete filter wavelengths (i.e., Bran + Luebbe model 450) are also shown in Table I. Six wavelengths were used to provide intergroup distances comparable to those

Table I. Discriminant Analysis Results for Various Fibers Expressed as Mahalanobis Distances between Groups[a]

From	To	Continuous Scan	Discrete Filters
Acrylic	PET	38	42
	polypropylene	53	45
	rayon	31	26
	cotton	26	32
	nylon	24	20
PET	polypropylene	32	33
	rayon	34	35
	cotton	39	21
	nylon	26	41
Polypropylene	rayon	36	49
	cotton	42	41
	nylon	29	46
Rayon	cotton	9	15
	nylon	19	14
Cotton	nylon	20	23

[a] Discrimination wavelengths: continuous scan; 2480, 2260, 2440, and 2380 nm; discrete filters: 1445, 1778, 2139, 2208, 2270, and 2348 nm.

obtained from the continuous scans. These results show that discrete filter data are capable of providing qualitative information, even though visual interpretation is more difficult than with continuous scan data.

An x-y plot of two of the discriminant calibration wavelengths is shown in Figure 7. The separation of the various groups (fiber types) is observed, although not completely without overlap. Another x-y plot of two other discriminant wavelengths is shown in Figure 8. Again, resolution of the groups is observed, although the pattern is different from the pattern in Figure 7. This comparison of two-dimensional projections aids visualization of how the separation between groups is increased in multidimensional space. In this case, four wavelengths (four dimensions) were used in the discriminant calibration.

Nylon 6 and nylon 6,6 were intentionally grouped together as nylon for the discriminant calibration in the set of all fiber types. Poor resolution between these materials was obtained when they were treated separately in the set of all fibers. However, when nylon 6 and nylon 6,6 samples were treated as a separate set, the discriminant calibration showed good discrimination. The Mahalanobis distance was 21, with discrimination wavelengths of 1759, 2230, 2270, and 2310 nm. If an unknown sample were classified as nylon, then the discriminant equation developed to specifically discriminate

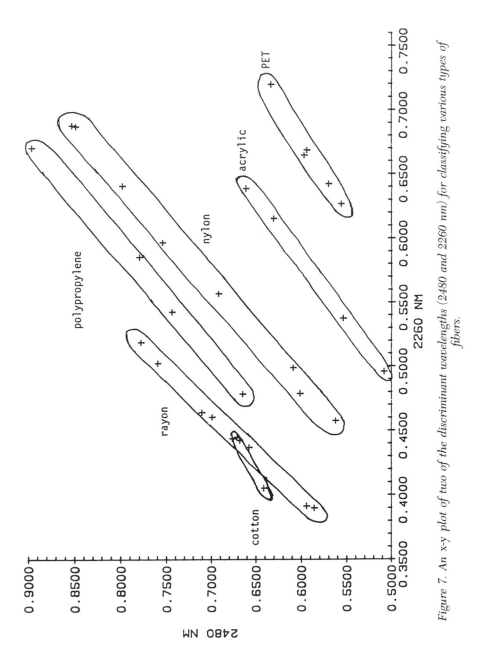

Figure 7. An x-y plot of two of the discriminant wavelengths (2480 and 2260 nm) for classifying various types of fibers.

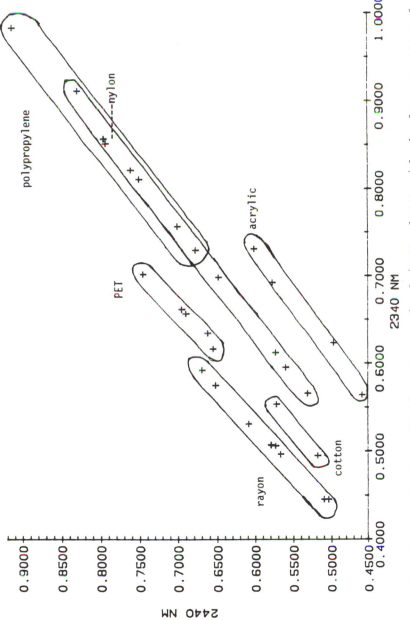

Figure 8. An x-y plot of two of the discriminant wavelengths (2440 and 2340 nm) for classifying various types of fibers.

nylon 6 versus nylon 6,6 would be used to identify the specific nylon type. This example implies that discriminant analysis is more powerful when used in a series of classifications that become progressively narrower in defining the differences between groups.

Discriminant analysis can also be used to classify groups composed of mixtures of the same materials, such as textile blends. In manufacturing, the consistency of the blend ratio is necessary to maintain constant performance attributes of the product, such as dye uptake and other physical properties. Figure 9 shows the spectra of rayon, PET, and two rayon–PET blends. The results of the discriminant classification using three wavelengths are shown in Table II. An x-y plot of two discriminant wavelengths that shows the group separation is shown in Figure 10. These results show that discriminant

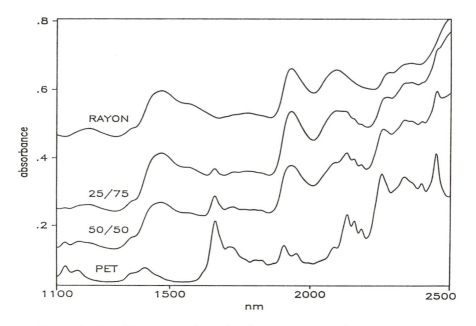

Figure 9. Near-IR spectra of uncolored rayon, PET, and two rayon/PET blends.

Table II. Discriminant Analysis Results for Rayon–PET Blends Expressed as Mahalanobis Distances between Groups[a]

	75:25	50:50	PET
Rayon	592	1539	6348
75:25 Rayon–PET		947	5756
50:50 Rayon–PET			809

[a] Discrimination wavelengths: 1722, 1734, and 2336 nm.

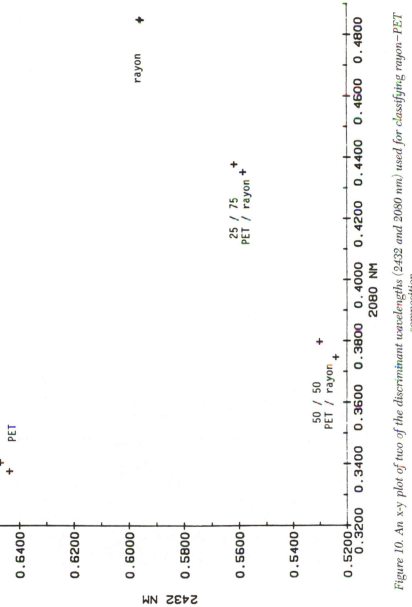

Figure 10. An x-y plot of two of the discriminant wavelengths (2432 and 2080 nm) used for classifying rayon–PET composition.

analysis may be used qualitatively, and perhaps quantitatively, to identify composition variation within a product. The technique can be used quantitatively in a process control situation by having a calibration from compositions that represent the lower limit, upper limit, and target level.

Thermoplastic Polymers. Discriminant analysis also yielded successful qualitative results for the 25 thermoplastic materials analyzed. The discriminant analysis program selected five wavelengths between 2400 and 2500 nm that provided the best differentiation between the set of 25 thermoplastic materials. The 325 intergroup Mahalanobis distances are displayed graphically, rather than in tabulated numerical form, in Figure 11. The y axis is truncated at 50 to better show the majority of intergroup values, which are between 20 and 50. Only two groups have intergroup distances of less than 10: nylon 6 versus PVDF and PMMA versus PVDF. The spectra of these materials are shown in Figure 12. The spectrum of nylon 6 is noticeably different from that of PVDF, particularly over the range of 1100–2200 nm. PVDF and PMMA have more similar spectra, although they are distinguishable in the 1500–2000-nm region (C–H and C=O stretch overtones). In both cases, the regions where the spectra have the most differences are not within the 2400–2500-nm region (combination bands) where most of the other materials were better differentiated. Twenty intergroup distances are

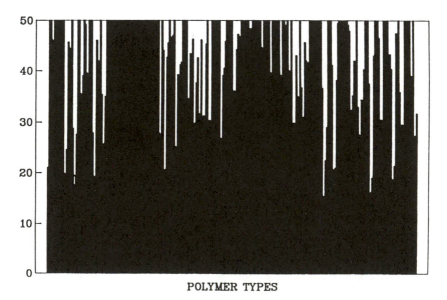

POLYMER TYPES

Figure 11. Mahalanobis distances for the 325 intergroup distances for 25 thermoplastic polymers. The upper limit is truncated at 50 to better show the lower portion of the scale. (A histogram presentation is used instead of the conventional x-y form to more clearly show all 325 intergroup values.)

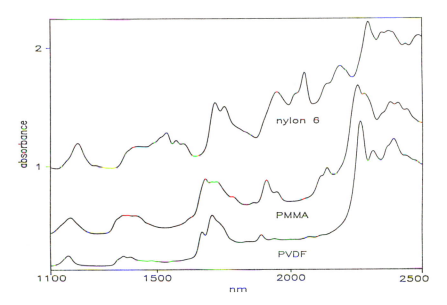

Figure 12. Near-IR spectra of nylon 6, PMMA, and PVDF. These are the three materials that were not differentiated from each other by discriminant analysis.

between 10 and 20 (poor to marginal separation), and the remaining 303 intergroup distances are greater than 20 (good separation).

A severely sloping baseline is a major pitfall for the discriminant analysis approach. Figure 13 shows the spectra of an uncolored (reference) and dark-colored (black) acrylic fiber. The dye that produces the totally absorbing (i.e., black) color in the visible region appears to be tailing into the near-IR region. This effect was found to have a severely adverse consequence on the discriminant classification.

First Derivative Near-IR Spectra. Utilization of first derivative spectra was found to greatly reduce the adverse effect of a sloping baseline due to absorbance from dyes and scattering. Figure 14 shows the first derivative spectra of the same two acrylic samples whose regular spectra are shown in Figure 13. Note that the first derivative almost eliminates the sloping baseline for the black acrylic sample. The other advantage of the first derivative is that it accentuates the subtle differences of band shoulders and band shapes that usually account for the differences between the near-IR spectra of similar materials.

Fibers. The first derivative spectra of cotton and rayon shown in Figure 15 are clearly more distinctive than the regular spectra (Figure 2). The first derivative spectra of nylon 6 and nylon 6,6 shown in Figure 16 also

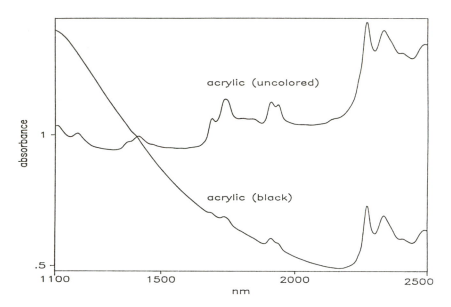

Figure 13. Near-IR spectra of an uncolored and dark-colored acrylic fiber.

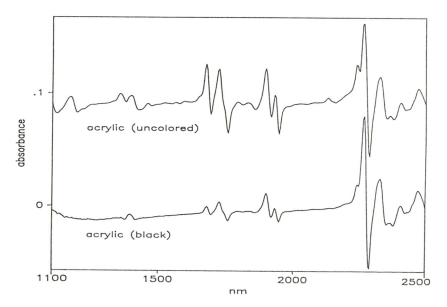

Figure 14. First derivative near-IR spectra of uncolored and dark-colored acrylic fiber (as shown in Figure 13).

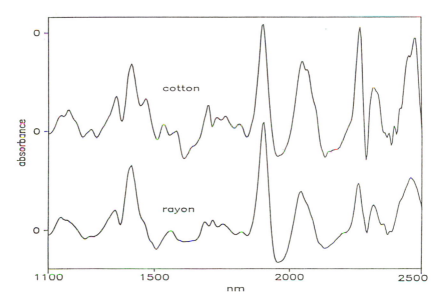

Figure 15. First derivative near-IR spectra of uncolored cotton and rayon.

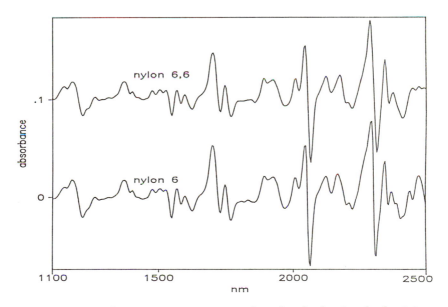

Figure 16. First derivative near-IR spectra of uncolored nylon 6 and nylon 6,6.

more clearly emphasize the spectral differences in the 2200–2500-nm region
(combination bands) than is observed in the regular spectra (Figure 3).

Thermoplastic Polymers. The first derivative near-IR spectra of PET
and PBT are shown in Figure 17. The first derivative significantly enhances
the observation of the spectral differences compared with the regular spectra
(Figure 4). The first derivative near-IR spectra of three polyolefins (HDPE,
PP, and PMP) are shown in Figure 18. The differences between PP and PMP
in the 1700–1800- (C−H stretch overtone) and 2300–2500-nm (combination
bands) regions are much more obvious in the first derivative spectra than in
the regular spectra (Figure 5). A similar accentuation in spectral differences
revealed by first derivatives is seen for the spectra of PES, PEEK, and
PEEKEK shown in Figure 19. In particular, the differences between PEEK
and PEEKEK in the 2000–2500-nm region (combination bands) are more
apparent than in the regular spectra (Figure 6). The first derivative spectra of
nylon 6, PVDF, and PMMA are shown in Figure 20. These are the three
materials that were not significantly distinguished from each other by discriminant analysis. The first derivative enhances the spectral differences between
PVDF and PMMA compared with the regular spectra (Figure 12).

Near-IR Spectral Searching. *Fibers.* We found spectral searching to be very powerful for distinguishing between materials of similar
chemical compositions, particularly when using first derivative spectra. In all
the cases that were tested, the spectral search results obtained from first

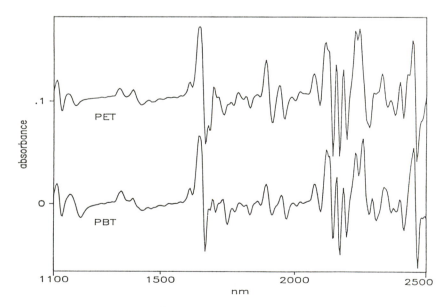

Figure 17. First derivative near-IR spectra of PET and PBT.

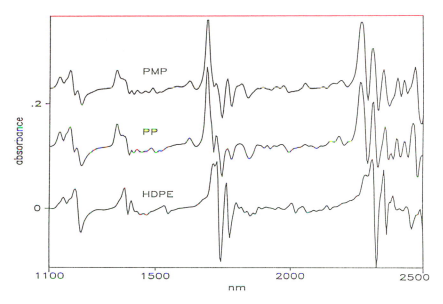

Figure 18. First derivative near-IR spectra of PMP, PP, and HDPE.

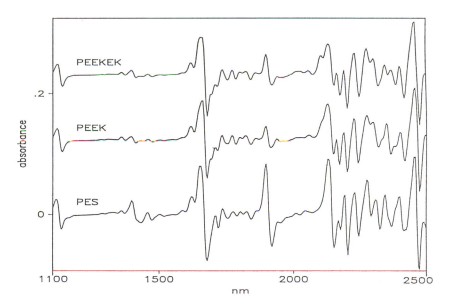

Figure 19. First derivative near-IR spectra of PEEKEK, PEEK, and PES.

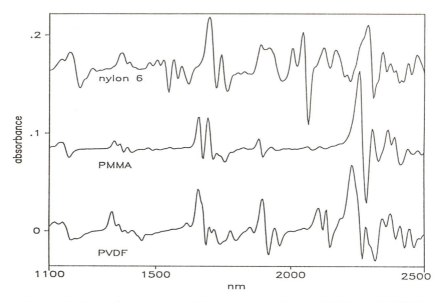

Figure 20. First derivative near-IR spectra of nylon 6, PMMA, and PVDF. These are the three materials that were not differentiated from each other by discriminant analysis.

derivative spectra were better, often significantly, than those obtained from the log(1/reflectance) form. As noted previously, the first derivative accentuates the subtle differences of band shoulders and band shapes that usually account for the differences between the near-IR spectra of similar materials, and it eliminates sloping baselines due to varying amounts of scattering and dye absorbance.

The spectral search results for the first derivative spectra of several of the fibers studied are shown in Table III. The results for the rayon "unknown" correctly identify the sample as an exact match with the rayon reference with a HQI of 0.00. Cotton is identified as the next closest match with a HQI of 0.24. Although the spectra of the two materials appear somewhat similar (Figure 15), there is a significant difference in the HQI values of rayon and cotton to indicate that rayon is by far the better match. Similar results were obtained for PET. The PET "unknown" was a perfect match with the PET reference, and no other fiber material was close according to the HQI values.

For the nylon 6 "unknown", the HQI is 0.00 for the nylon 6 reference and 0.03 for the nylon 6,6 reference. Although the numerical difference between the HQIs of the two nylon types is small, the positive identification of nylon 6 is confirmed by visually viewing the spectra. For the black acrylic with the sloping baseline (Figure 13), the HQI of 0.12 for the first derivative

Table III. First Derivative Near-IR Spectral Search Results for Fibers

Fiber Searched	Library Match	Hit Quality Index[a]
Rayon	rayon	0.00
	cotton	0.24
	wool	0.42
	acrylic	0.49
PET	PET	0.00
	acrylic	0.31
	nylon 6	0.34
	nylon 6,6	0.34
Nylon 6	nylon 6	0.00
	nylon 6,6	0.03
	polypropylene	0.30
	wool	0.32
Acrylic (black)	acrylic	0.12
	polypropylene	0.29
	PET	0.30
	nylon 6,6	0.33

[a] A smaller value indicates a better match.

spectrum (Figure 14) is significantly better for both the closeness of the match with the acrylic reference and the separation from other materials.

Thermoplastic Polymers.   Table IV shows some representative first derivative spectral search results for various thermoplastic materials. The PEEK standard in the library is consistently the best match for PEEK "unknowns", although as expected, PEEKEK and PES are also identified as close matches for PEEK (*see* Figure 6). Similarly, the first derivative spectral search results for nylon 6 are consistently correct. Even though the near-IR spectra of nylon 6,6 and nylon 12 are very similar, there are sufficient differences to permit identification of each material.

The PMP "unknown" was correctly identified with an HQI of 0.00, which indicates a significant differentiation from the PP reference with an HQI of 0.14 (Figure 18). Using first derivative spectral searching, PVDF was correctly identified and significantly distinguished from all other materials; however, using discriminant analysis, PVDF was not significantly distinguished from nylon 6 and PMMA, which have similar spectra above 2000 nm (*see* Figure 12).

**Table IV. First Derivative Near-IR Spectral Search Results for
Thermoplastic Polymers**

Fiber Searched	Library Match	Hit Quality Index[a]
PEEK	PEEK	0.02
	PEEKEK	0.06
	PES	0.10
	polycarbonate	0.14
Nylon 6	nylon 6	0.00
	nylon 6,6	0.08
	nylon 12	0.21
	nylon 6–aramid	0.21
PMP	PMP	0.00
	PP	0.14
	PE	0.29
	nylon 12	0.32
PVDF	PVDF	0.00
	PVAC	0.26
	PPS	0.27
	PBT	0.27

[a] A smaller value indicates a better match.

Summary and Conclusions

Near-IR spectroscopy can be used to qualitatively identify textile fibers and thermoplastic polymers, and probably any other organic material. Qualitative identification of uncolored or light-colored materials can be satisfactorily achieved by discriminant analysis using Mahalanobis distances with both continuous scan and discrete filter spectra. However, discriminant analysis using Mahalanobis distances seems less useful for dark-colored materials. Spectral searching using the first derivative is the more robust approach because it eliminates most of the adverse effect of sloping baselines and accentuates the subtle differences between spectra of similar materials. The use of higher order derivatives (not studied in this work) would be expected to equal, and possibly exceed, these results. Conventional near-IR spectrometers require a few grams of sample; thus, this technique may not be useful when sample size is very small. However, the technique is nondestructive. Sampling techniques and accessories capable of using much smaller sample sizes have been utilized in a few reports. This study is limited in the number of sample types. Further work in this area with a greater variety of material is needed to better understand the qualitative performance of near-IR spectroscopy.

References

1. Creaser C. S.; Davies A. M. C. *Analytical Applications of Spectroscopy*; The Royal Society of Chemistry: London, England, 1988.
2. Wetzel, D. L. *Anal. Chem.* **1983**, *55*, 1165A–1176A.
3. Weyer, L. G. *Appl. Spectrosc. Rev.* **1985**, *21*, 1–43.
4. Mark, H. L.; Tunnel, D. *Anal. Chem.* **1985**, *57*, 1449–1456.
5. Lo, S.-C.; Brown, C. W. *A Near-Infrared Spectral Library*; FTS/IR Notes No. 68; Bio-Rad: Cambridge, MA, 1989.
6. Ciurczak, E. W. *Appl. Spectrosc. Rev.* **1987**, *23*, 147–163.
7. Ciurczak, E. W.; Maldacker, T. A. *Spectroscopy* **1986**, *1*, 36–39.
8. Ghosh, S.; Rodgers, J. E. *Milliand Textilberichte* **1988**, *69*, 361–364.

RECEIVED for review May 14, 1991. ACCEPTED revised manuscript May 26, 1992.

SURFACES AND INTERFACES OF POLYMERS

Surface optical models serve to develop and establish quantitative relationships between spectroscopic data and the concentration of the surfaces species. Especially, exudation of small molecules to the film-air or film-substrate interfaces has a fairly significant effect on many macroscopic properties, including adhesion, optical properties, and chemical stability. Whereas ATR−FTIR spectroscopy is used to assess the origin of molecular processes at the film−air and film−substrate interfaces, surface-enhanced Raman spectroscopy is presented as a means of studying interactions between adsorbed surface species and surfaces. An FT−Raman experiment free of fluorescence allows acquisition of the Raman spectra during emulsion polymerization reactions, reaction kinetics, and latex film formation. As a result, polymer network growth can be predicted.

10

Optical Theory and Modeling of Gradients at Polymer Surfaces

Leslie J. Fina

Department of Materials Science and Engineering, College of Engineering, Rutgers University, P.O. Box 909, Piscataway, NJ 08855-0909

In this chapter the quantitative aspects of depth profiling with variable-angle attenuated-total-reflectance infrared spectroscopy are presented. The equivalence of the electric field intensities in transmission and attenuated total reflection are used to derive approximate equations that define the relationship between optical constant gradients and the experimental reflected intensities. The equations are compared with the exact intensities found with Fresnel coefficients as a way to define optimal experimental conditions. The treatment is developed to include optical models of concentration and structural gradients and is extended to include orientation gradients. Fresnel coefficients and electric field intensities are calculated from oriented systems and are used to model the reflection properties and to define the accuracy of the necessary approximations.

TOTAL INTERNAL REFLECTION OCCURS when the refractive index of the incident medium is greater than the refractive index of the second medium and the angle of incidence is greater than the critical angle of reflection. In the case of nonabsorbing and absorbing media, the incident and reflected light coherently interact at the interface to form a standing wave. The standing wave consists of a sinusoidal variation in the electric field amplitude in the incident medium (1, 2). The electric field amplitude has two components in the plane of the interface and one normal to it. The components of the amplitude in the plane of the interface are continuous across the interface; that is, values in the denser and rarer media are identical at the interface (3). On the other hand, the component perpendicular to the

0065–2393/93/0236–0289$06.00/0

interface displays a discontinuity because of the requirement that the displacement—not the electric field—be continuous. All three components of the electric field amplitude decay exponentially from the interface, and the decaying field has been termed an evanescent wave. When the second medium is nonabsorbing, all of the incident radiation is reflected for light both parallel and perpendicular to the plane of incidence (p- and s-polarized light, respectively). The plane of incidence is defined by the incident light direction and the normal direction to the interface. In other words, no energy from the evanescent wave is transmitted into the rarer medium. When the second medium is absorbing, the evanescent wave interacts with the medium, the reflectance is reduced, and attenuated total reflection results.

An importance difference between transmission and attenuated-total-reflection (ATR) modes involves the spatial orientation of the electric field. As pointed out by Harrick (2), Maxwell's equations for the propagation of electromagnetic radiation stipulate that electric vectors only exist perpendicular to the propagation direction. In the case of transmission spectroscopy with unpolarized light, only those dipoles that are perpendicular to the direction of propagation or have nonzero component projections on the perpendicular can be observed. This "deficiency" most often has been circumvented with tilted-film transmission spectroscopy (4–6). On the other hand, in ATR spectroscopy unpolarized incident radiation contains electric vectors with nonzero components in all three spatial directions, which is also true of the standing and evanescent waves. With s-polarized light the electric vector is confined to the direction perpendicular to the plane of incidence; with p-polarized light the electric vector is in the plane of incidence. The existence of electric vectors in three dimensions in ATR has been used successfully to probe oriented polymer systems.

The first quantitative treatment of oriented polymers was given by Flournoy and Schaffers (7), who derived equations to describe the exact reflectance from a homogeneous system of anisotropic optical constants. These expressions require no assumptions and are based on reflection coefficients equivalent to the isotropic Fresnel coefficients. With the assumptions that (1) the attenuation index (κ) is less than 0.1 (where $k = n\kappa$, n is the refractive index, and k is the absorption coefficient) and (2) the refractive index anisotropy is negligible, equations for the reflectances can be reduced to the commonly used form (7)

$$\ln R_s = -A\kappa_x$$
$$\ln R_p = -B\kappa_y - C\kappa_z$$

R_s and R_p are the experimentally observed intensities; A, B, and C are functions of the refractive index ratio and the angle of incident light. The axis system for these equations as well as the rest of this chapter is defined in Figure 1. Applications of these formulations can be found elsewhere (8–11).

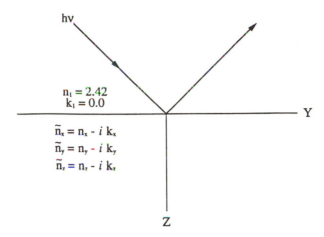

Figure 1. Axis system and optical constant definitions.

ATR spectroscopy commonly has been used to probe optical constant gradients in a qualitative or semiquantitative way. By varying the angle of incidence, the exponential decay of the evanescent wave changes. The point at which the wave decays to the negative exponent (e^{-1}) of its surface value is defined as the depth of penetration (d_p). The same depths are probed with large angles of incidence (equivalent to small d_p) as with small angles of incidence, where the angle of incidence is defined as the angle between the normal to the interface and the light propagation direction. The size of the angle of incidence does not affect the probe depths because the field amplitudes, although small at distances remote from the interface, never go to zero. However, the reflected signal is more heavily weighted from distances less than the d_p. Examples of the semiquantitative use of the d_p or the related effective thickness parameter (d_e) to a change in concentration or structure can be found in the literature (*10, 12–14*).

The first quantitative demonstration of the use of ATR spectroscopy to probe optical constant gradients was given by Hirschfeld (*15*), who expanded equations given earlier by Harrick (*1*). The absorption intensities were described by a Laplace transform of the absorption coefficients. To find concentration gradients that are represented in absorption coefficients as a function of depth from the interface, the absorption intensities were calculated by inverse Laplace transformation. A close approach to the critical angle is necessary to obtain a physically meaningful solution to the inverse transform. The methodology was applied to water on the surface of glass and silicone oil on polystyrene (*15*). An alternate derivation to that of Hirschfeld can be found elsewhere (*16*).

The remainder of this chapter is divided into two sections. The first section describes the use of optical theory to define and model the interaction of internally reflected light with materials in which the absorption coefficient varies with depth from the interface. In this form the theory can be used to model gradients in concentration or to calculate the gradients from variable-angle ATR data. The second section describes how the treatment is extended to include optical interactions in the presence of orientation gradients.

Concentration Gradients

Quantitative treatments of the interaction of infrared radiation with an absorbing material are based on the calculations of the Fresnel reflection and transmission coefficients (17), which are defined as the ratio of the complex electric field amplitudes of the reflected or transmitted waves to the incident wave. The Fresnel coefficients are applied to isotropic-plane parallel layers of arbitrary optical constant inhomogeneity between layers and can be used to calculate the reflectance or transmittance exactly. The particular advantage of the use of Fresnel coefficients is their completely general nature; that is, diverse situations of internal reflection, external reflection, refraction, and transmission can be predicted from one unifying set of equations. Additionally, the presence of an evanescent wave in internal reflection is a direct result of the treatment.

Fresnel coefficients are derived on the basis of isotropic layers and, therefore, cannot be directly used to model a gradient in optical constants perpendicular to the interface. However, it is conceivable that a matrix treatment of n plane parallel layers such as given by Heavens (18) can be used to calculate the reflection properties from an optical gradient by allowing the number of layers n to approach infinity and the layer thickness to approach zero. Due to its considerable simplicity (19), an alternate method presently is employed to model structural gradients. The starting point for the derivation of equations capable of modeling an optical gradient at an interface is the equivalence of the electric field intensities $\langle E \rangle$ in two media:

$$n_1 \langle E_1^2 \rangle = n_2 \langle E_2^2 \rangle \tag{1}$$

The absorption of energy in a transmission experiment (T) is proportional to $\langle E_T^2 \rangle d_e$, where d_e is the thickness in transmission that is equivalent to the thickness in an attenuated-total-reflection experiment. Because the absorption of energy in a reflection experiment is proportional to the integral of the electric field intensity over distance (z), we can write

$$n_1 \langle E_T^2 \rangle d_e(\theta) = \frac{n_2}{\cos \theta} \int_0^\infty \langle E_2^2 \rangle \, dz \tag{2}$$

The $1/\cos\theta$ term in eq 2 takes into account the change in beam diameter between transmission at normal incidence and reflection at angle θ. Furthermore, the electric field intensity decays exponentially from the surface value, $\langle E_{02}^2 \rangle$, in an ATR experiment. The rate of decay is defined by the depth of penetration (d_p). Equation 2 can now be modified to

$$\langle E_T^2 \rangle d_e(\theta) = \frac{n_2}{n_1 \cos\theta} \int_0^\infty \langle E_{02}^2 \rangle \exp\left(-\frac{z}{d_p(\theta)}\right) dz \qquad (3)$$

where z is the distance perpendicular to the interface and is defined as positive in the probed material. If we allow $\langle E_T^2 \rangle$ to be unity (20), then a definition of the effective thickness (d_e) results:

$$d_e(\theta) = \frac{n_2 \langle E_{02}^2 \rangle}{n_1 \cos\theta} \int_0^\infty \exp\left(-\frac{z}{d_p(\theta)}\right) dz \qquad (4)$$

In this form d_e can be used to model the reflected intensities with Beer's law. Beer's law can be written for reflection in the form

$$R = \exp\left(-\frac{4\pi k d_e}{\lambda}\right) = 1 - \frac{4\pi k d_e}{\lambda} \qquad (5)$$

for small values of the absorption coefficient k, where λ is the wavelength of incident light. Because eq 5 is an approximation, it is instructive to examine the conditions under which the approximation is accurate. Both the right- and left-hand equalities require k to be small. A comparison of exact intensities calculated with Fresnel coefficients and approximate intensities calculated with eq 5 defines the limits of accuracy. This comparison is shown in Figure 2 for three angles of incident light. The curves are based on $n_1 = 2.42$, $n_2 = 1.50$, $\lambda = 10/3$ μm, and s-polarized light, which is along the x axis in Figure 1. The figure indicates that at high angles of incidence the approximation holds in a large range of k values. Close to the critical angle of $38.3°$ the curves diverge at large k values.

The linearity of the absorption coefficient with reflection expressed in eq 5 can be combined with the definition of the effective thickness in eq 4 to yield

$$1 - R(\theta) = \frac{4\pi \langle E_{02}^2 \rangle n_{21}}{\lambda \cos\theta} \int_0^\infty k(z) \exp\left(-\frac{z}{d_p(\theta)}\right) dz \qquad (6)$$

where n_{21} is the refractive index ratio n_2/n_1. Equation 6 is based on the assumption that $k \ll 1$, which is true for most polymer absorption bands,

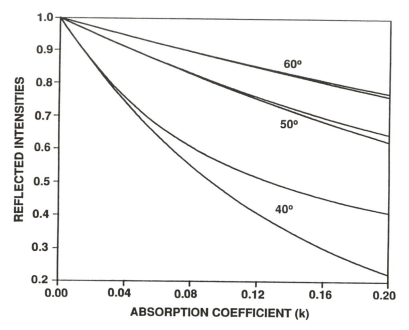

Figure 2. Comparison of the calculated s-polarized reflected intensities for two models of interaction. The top curve in each set is from the first equality in eq 5, and the bottom curve represents the exact intensities calculated from isotropic Fresnel coefficients. The angles of incident light used in the calculations are shown.

and that the refractive index gradient is negligibly small and can be repre-sented by a constant value $n_2(z) = n_2$. The electric field intensity of eq 6 in each direction of Cartesian space can be calculated from (17, 20)

$$\langle E_{2x}^2 \rangle = |t_s|^2 \exp\left(4\pi \frac{z}{\lambda} \mathrm{Im}\,\xi\right) \tag{7}$$

$$\langle E_{2y}^2 \rangle = \left|\frac{\xi}{\tilde{n}_2} t_p\right|^2 \exp\left(4\pi \frac{z}{\lambda} \mathrm{Im}\,\xi\right) \tag{8}$$

$$\langle E_{2z}^2 \rangle = \left|\frac{n_1 \sin\theta}{\tilde{n}_2} t_p\right|^2 \exp\left(4\pi \frac{z}{\lambda} \mathrm{Im}\,\xi\right) \tag{9}$$

where

$$\xi = \left(\tilde{n}_2^2 - n_1^2 \sin^2\theta\right)^{1/2}$$

In eqs 7–9, t is the Fresnel transmission coefficient found with $\bar{n} = n - ik$ and Im ξ is the imaginary part of ξ. The $\langle E^2 \rangle$ for p-polarized light contains components from both y and z directions. Both polarizations can be combined as

$$\langle E_{2p}^2 \rangle = \left(|\sin \theta_2|^2 + |\cos \theta_2|^2 \right) |t_p|^2 \exp \left(4\pi \frac{z}{\lambda} \mathrm{Im}\,\xi \right) \qquad (10)$$

where θ_2 is calculated from Snell's law $n_1 \sin \theta_1 = n_2 \sin \theta_2$. In eqs 7–10 the value of the electric field intensity at the surface and in medium 2, $\langle E_{02}^2 \rangle$, can be found by setting $z = 0$. The electric field intensities at the boundary of an internal reflection element and an absorbing material are shown in Figure 3. The solid lines are calculated for an isotropic probed material with refractive index $n_2 = 1.50$ and absorption coefficient $k_2 = 0.01$ and a nonabsorbing internal reflection element of $n_1 = 2.42$ and $k_1 = 0$. The magnitudes are relative to a value of unity for the incident beam. The large values of $\langle E_{02}^2 \rangle$ in the x and z directions near the critical angle of internal reflection

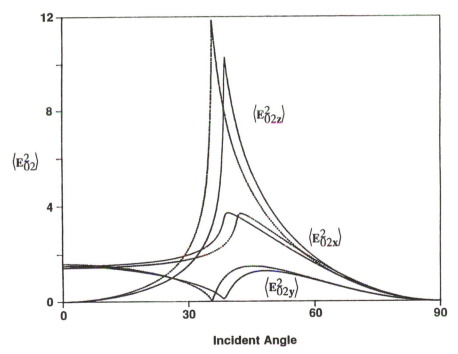

Figure 3. Electric field strengths at the interface of an internal reflection element ($n_1 = 2.42$ and $k_1 = 0$) and a probed material. Solid lines are calculated for an isotropic material ($n_2 = 1.50$, $k_2 = 0.01$) and dotted lines represent a uniaxially oriented material with chain axes in the x direction and dipoles parallel to the chain axes ($n_x = 1.60$, $n_y = n_z = 1.40$, $k_x = 0.01$, $k_y = k_z = 0.00$).

(38.3°) account for the high surface sensitivity of ATR spectroscopy. The
dotted lines are calculated for an oriented system that will be discussed later.
The effect of nonzero k values on $\langle E_{02\,x}^2 \rangle$ are shown in Figure 4 for
s-polarized light. Nonzero k values reduce the field significantly at incident
angles close to the critical angle. The critical angle occurs at the discontinuity
in curve A. Under conditions of small k values and incident angles that do
not approach the critical angle, the influence of k on the electric field and on
the depth of penetration can be neglected. Equations 7–10 can be reduced to

$$\langle E_{02\,x}^2 \rangle = \frac{4\cos^2\theta}{1 - n_{21}^2} \tag{11}$$

$$\langle E_{02\,y}^2 \rangle = \frac{4\left(\sin^2\theta - n_{21}^2\right)\cos^2\theta}{\left(1 - n_{21}^2\right)\left[\left(1 + n_{21}^2\right)\sin^2\theta - n_{21}^2\right]} \tag{12}$$

$$\langle E_{02\,z}^2 \rangle = \frac{4\sin^2\theta\cos^2\theta}{\left(1 - n_{21}^2\right)\left[\left(1 + n_{21}^2\right)\sin^2\theta - n_{21}^2\right]} \tag{13}$$

$$\langle E_{02\,p}^2 \rangle = \frac{4\cos^2\theta\left(2\sin^2\theta - n_{21}^2\right)}{\left(1 - n_{21}^2\right)\left[\left(1 + n_{21}^2\right)\sin^2\theta - n_{21}^2\right]} \tag{14}$$

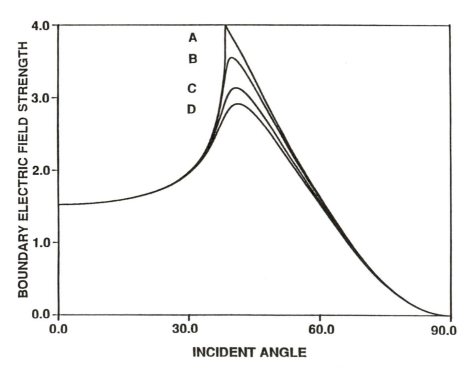

Figure 4. The dependence of the interfacial electric field strength on the
absorption coefficient: A, k = 0.00; B, k = 0.02; C, k = 0.06; D, k = 0.10.

and the depth of penetration reduces from the complex form

$$d_p(\theta) = -\frac{\lambda}{4\pi\,\text{Im}(\tilde{n}_2\cos\theta_2)}$$

to the real form

$$d_p(\theta) = \frac{\lambda}{4\pi(n_1^2\sin^2\theta - n_2^2)^{1/2}}$$

Substitution of the k-independent field strengths given in eqs 11 and 14 into 6 yields the reflectance for s- and p-polarized light:

$$1 - R_s(\theta) = \frac{16\pi n_{21}\cos\theta}{\lambda(1 - n_{21}^2)}\int_0^\infty k_s(z)\exp\left(-\frac{z}{d_p(\theta)}\right)dz \tag{15}$$

$$1 - R_p(\theta) = \frac{16\pi n_{21}\cos\theta(2\sin^2\theta - n_{21}^2)}{\lambda(1 - n_{21}^2)[(1 + n_{21}^2)\sin^2\theta - n_{21}^2]}\int_0^\infty k_p(z)\times$$

$$\exp\left(-\frac{z}{d_p(\theta)}\right)dz \tag{16}$$

When $k(z)$ is a constant, that is, when no gradient exists, eqs 15 and 16 reduce to

$$1 - R_s(\theta) = \frac{4n_{21}k_s\cos\theta}{(1 - n_{21}^2)(n_1^2\sin^2\theta - n_2^2)^{1/2}} \tag{17}$$

$$1 - R_p(\theta) = \frac{4n_{21}k_p\cos\theta(2\sin^2\theta - n_{21}^2)}{(1 - n_{21}^2)[(1 + n_{21}^2)\sin^2\theta - n_{21}^2](n_1^2\sin^2\theta - n_2^2)^{1/2}} \tag{18}$$

Equations 15–18 require k to be small so that the influence of k on $\langle E_{02}^2\rangle$, d_p, and n is negligible. When gradients do exist, the integrals in eqs 15 and 16 can be solved by assuming that the absorption coefficient gradient is composed of discrete depth increments. Under these conditions the integral is replaced by a summation and the integral becomes

$$d_p(\theta)\sum_{z=0}^{z=N}\left[\exp\left(-\frac{z}{d_p(\theta)}\right) - \exp\left(-\frac{z + \Delta z}{d_p(\theta)}\right)\right]k(z) \tag{19}$$

where Δz is the depth interval in which k is constant and N is the depth where the electric field strength is negligibly small. Equations 15 and 16 substituted with eq 19 are programmable on a computer for solutions to the inverse Laplace problem by minimization of the sum of the difference of the squares of experimental and calculated reflected intensities.

Orientation Gradients

Up to this point the optical equations necessary for understanding the attenuated-total-reflectance properties of materials that exhibit concentration gradients as a function of distance from an interface have been defined. The types of gradients that can be characterized with variable-angle ATR spectroscopy are not limited as in the past (15, 16), to the characterization of low molecular weight liquids on solid surfaces. Potentially, any gradient in which the optical constants differ between the surface and the bulk can be defined by use of the previous formulations. Two areas are worthy of further attention in the development of the methodology:

1. The influence of the refractive index gradient on the calculation of absorption coefficients. In the previous treatment the refractive index gradient was ignored, which is a reasonable approximation when the surface and bulk species do not differ significantly.

2. The untested potential of the method to define gradients in orientation.

The remainder of this section is concerned with the modeling of orientation and orientation gradient systems and the predicted accuracy of the results. Two types of model systems will be used: (1) an oriented polymer matrix containing no gradients, described as homogeneous oriented, and (2) an oriented matrix containing depth-dependent optical constant gradients, termed gradient oriented.

An early description of the use of Fresnel coefficients to define the optical properties of a homogeneous-oriented polymer in attenuated total reflectance is given by Flournoy and Schaffers (7). This work is relevant to the current study in that the exact reflectances from an oriented film can be calculated and compared with the approximate Laplace transform treatment. The exact relationships that allow the s- and p-polarized reflectances to be found in homogeneous-oriented systems are

$$\tilde{\rho}_s = \frac{\left(\tilde{n}_x^2 - n_1^2 \sin^2 \theta\right)^{1/2} - n_1 \cos \theta}{\left(\tilde{n}_x^2 - n_1^2 \sin^2 \theta\right)^{1/2} + n_1 \cos \theta} \qquad (20)$$

$$\tilde{\rho}_p = \frac{n_1\left(\tilde{n}_z^2 - n_1^2 \sin^2 \theta\right)^{1/2} - \tilde{n}_y\tilde{n}_z \cos \theta}{n_1\left(\tilde{n}_z^2 - n_1^2 \sin^2 \theta\right)^{1/2} + \tilde{n}_y\tilde{n}_z \cos \theta} \tag{21}$$

$$R = \tilde{\rho}\tilde{\rho}^* \tag{22}$$

where $\tilde{\rho}$ is a type of Fresnel coefficient. These relationships are used in Figure 5 to determine the circumstances, if any, under which the refractive index anisotropy can be ignored. The top two curves are the s-polarized reflectances with $k_x = 0.01$ and the bottom two are with $k_x = 0.10$. Within each set, the top curve is calculated with $n_x = 1.50$, which is defined as isotropic. In the bottom curve in each set, $n_x = 1.57$, which is a typical value of the refractive index along the chain axis in uniaxially oriented polymers. In the bottom set of two curves, where $k_x = 0.10$, the refractive index anisotropy is negligible only at very high angles of incident light. With small k values as in the top two curves, the refractive index anisotropy is negligible at incident angles greater than $\sim 12°$ above the critical angle (38.3°). Similar behavior is displayed for p-polarized light (not shown). The conclusion that can be drawn

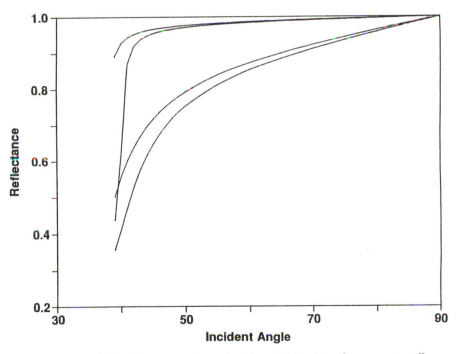

Figure 5. Calculated exact s-polarized reflected intensities from a uniaxially oriented polymer with chain axes aligned along the x direction. Equations 20–22 are used in the calculations. In the top two curves, $k_x = 0.01$; in the bottom two curves, $k_x = 0.10$. The top curve in each set has refractive index isotropy $n_x = 1.50$; the bottom curve has refractive index anisotropy $n_x = 1.57$.

from Figure 5 is that the refractive index anisotropy generally cannot be neglected in well-oriented polymers.

As in the analysis of concentration gradients, the theory has not been developed to characterize orientation gradients with Fresnel coefficients. To understand the use of Laplace transforms for orientation gradient systems, homogeneous-oriented systems are further examined. The electric field intensity at the surface is directly proportional to the reflectance in eq 6. The fields can be calculated for oriented systems with eqs 7–9 when the oriented Fresnel coefficients are found. Methods described elsewhere (18, 21–23) are used to calculate transmission coefficients from optically anisotropic polymers, and the resulting surface electric fields are shown as dotted lines in Figure 3. The oriented fields were found by using $k_x = 0.01$, $k_y = k_z = 0.0$, $n_x = 1.60$, and $n_y = n_z = 1.40$. These optical constants are representative of perfect uniaxial orientation with dipoles aligned parallel to the chain axis and the chains aligned in the x direction of the laboratory frame of reference (see Figure 1). Figure 3 indicates a substantial critical angle shift as a result of orientation. The shift is primarily a result of the refractive index anisotropy and displays very little dependence on the absorption coefficient anisotropy.

The Laplace equations 15 and 16 can be used directly to calculate reflected intensities of homogeneous uniaxially oriented systems. As in the case of concentration gradients presented earlier, the accuracy of the Laplace-calculated intensities is indicated by a comparison with the exact reflected intensities from oriented systems. Exact intensities as a function of absorption coefficient are calculated with eqs 20–22 and are shown in Figure 6 for p-polarized light along with Laplace-calculated intensities determined with eq 18. The optical parameters used in the calculations are $n_x = 1.60$ and $n_y = n_z = 1.40$. The figure is analogous to Figure 2, but a linear relationship is used to calculate the approximate intensities (eq 18) instead of an exponential relationship (eq 5). In the homogeneous-oriented systems examined here, the accuracy of the Laplace-calculated intensities follows trends similar to isotropic systems; that is, high angles of incidence and small k values provide the greatest accuracy. The use of s-polarized radiation produces calculated intensities similar to those in Figure 6.

To account completely for orientation gradients with the use of optical constants, determination of the influence of gradients in both absorption coefficient and refractive index on the reflected intensities is necessary. In this case, the general form of the Laplace transform is

$$1 - R(\theta) = \frac{4\pi \langle E_{02}^2 \rangle}{n_1 \lambda \cos \theta} \int_0^\infty n_2(z) k_2(z) \exp\left(-\frac{z}{d_p(\theta)}\right) dz \qquad (23)$$

Equation 23 assumes a negligible gradient dependence of d_p and $\langle E_{02}^2 \rangle$. The use of discrete depth intervals allows the s- and p-polarized reflectances to

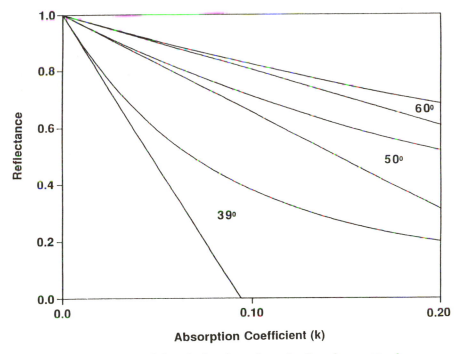

Figure 6. Comparison of the calculated p-polarized reflected intensities for two models of interaction. The curves are analogous to Figure 2 but for oriented systems. Incident light angles are shown. The top curves in each set are the exact reflected intensities calculated with anisotropic Fresnel coefficients (eqs 20–22). The bottom curves in each set are the Laplace intensities calculated with eq 18.

be modeled by assuming the form of the optical constant gradients. This modeling has been done in Figure 7 for s-polarized radiation. The gradients used in the calculations are linear and vary from a depth of 0 to 0.7 μm, which is within the depth of penetration. At distances greater than 0.7 μm the optical constants are held constant. Additionally, in all calculations the values of the optical constants at the surface are held constant to assure that observed differences in reflected intensities are due only to the absence or presence of optical constant gradients. The lowest curve represents a homo-geneous-oriented system (where no gradient exists). The highest curve is calculated for gradients in both n_2 and k_2, where n_2 decreases from a surface value of 1.50 to a bulk value of 1.40 and k_2 decreases from 0.01 to 0. The middle two curves represent hypothetically oriented systems. In the dotted curve, k_2 varies from 0.01 at the surface to 0 in the bulk while n_2 is held constant. The large-dashed curve is calculated for an n_2 variation of 1.50 at the surface to 1.40 in the bulk while k_2 is held constant. The curves in

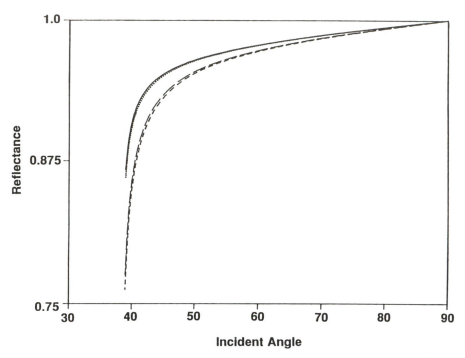

Figure 7. Calculated intensity curves for gradients in the refractive index (n_2)
and absorption coefficient (k_2). *Solid line, gradients in* n_2 *and* k_2; *small
dashes, no gradients; large dashes, gradient in* n_2 *only; dots, gradient in* k_2
only.

Figure 7 illustrate that gradients result in significant intensity changes in the
lower half of the incident angular range and, therefore, can be used to define
orientation gradients. Also, the relatively small differences between the curve
with gradients in both n_2 and k_2 (solid line) and the curve with a gradient
only in k_2 (dotted line) suggest that the refractive index gradient can be
neglected in calculations of the inverse Laplace transform with experimental
data. In contrast, the refractive index anisotropy at the surface cannot be
neglected.

Applications of the methods described in this chapter to conformational
concentration gradients at polymer surfaces can be found in reference 19.

Summary

The electric field intensity equivalence between transmission and attenuated-total-reflection modes has been used to establish an optical basis for
modeling reflectance properties of polymer interfaces that contain concentration, structure, and orientation gradients. Exact calculations of reflected

intensities in gradientless systems are found from Fresnel coefficients and are compared with calculations of approximate intensities from Laplace transforms that are capable of handling optical constant gradients. Modeling studies of homogeneous-oriented and gradient-oriented systems indicate that the Laplace treatment is accurate at low-absorption large angles of incidence and when the refractive index gradient is neglected. Conversely, the anisotropy in the refractive index at the surface of a polymer cannot be neglected.

Acknowledgments

We acknowledge the donors of The Petroleum Research Fund, administered by the American Chemical Society, for partial support of this research.

References

1. Harrick, N. J. *J. Opt. Soc. Am.* **1965**, *55*, 851.
2. Harrick, N. J. *Internal Reflection Spectroscopy*; Wiley: New York, 1967; Chapter 2.
3. Hansen, W. N. *J. Opt. Soc. Am.* **1968**, *58*, 380.
4. Fina, L. J.; Koenig, J. L. *J. Polym. Sci., Polym. Phys. Ed.* **1986**, *24*, 2509.
5. Fina, L. J.; Koenig, J. L. *J. Polym. Sci., Polym. Phys. Ed.* **1986**, *24*, 2525.
6. Fina, L. J.; Koenig, J. L. *J. Polym. Sci., Polym. Phys. Ed.* **1986**, *24*, 2541.
7. Flournoy, P. A.; Schaffers, W. J. *Spectrochim. Acta* **1966**, *22*, 5.
8. Flournoy, P. A. *Spectrochim. Acta* **1966**, *22*, 15.
9. Sung, C. S. P. *Macromolecules* **1981**, *14*, 591.
10. Hobbs, J. P.; Sung, C. S. P.; Krishnan, K.; Hill, S. *Macromolecules* **1983**, *16*, 193.
11. Stas'kov, N. I.; Golovachev, V. I.; Gusev, S. S. *Polym. Sci. USSR* **1977**, *A19–20*, 2628.
12. Carlsson, D. J.; Wiles, D. M. *Can. J. Chem.* **1970**, *48*, 2397.
13. Carlsson, D. J.; Wiles, D. M. *J. Polym. Sci., Polym. Lett. Ed.* **1970**, *8*, 419.
14. Carlsson, D. J.; Wiles, D. M., *Macromolecules* **1971**, *4(2)*, 174.
15. Hirschfeld, T. *Appl. Spectrosc.* **1977**, *31(4)*, 289.
16. Rozanov, N. N.; Zolotarev, V. M. *Opt. Spectrosc. USSR* **1980**, *49(5)*, 506.
17. Born, M.; Wolf, E. *Principles of Optics*, 6th ed.; Pergamon: New York, 1980; Chapter 1.
18. Heavens, O. S. *Optical Properties of Thin Solid Films*; Dover Publications: New York, 1965; Chapter 4.
19. Fina, L. J.; Chen, G. C. *Vib. Spectrosc.* **1991**, *1*, 353.
20. Hansen, W. N. In *Advances in Electrochemistry and Electrochemical Engineering*; Delahay, P.; Tobias, C. W., Eds.; Wiley: New York, 1973; Vol. 9, pp 1–60.
21. Schopper, H. Z. *Phys.* **1952**, *132*, 146.
22. Elsharkawi, A. R.; Kao, K. C. *J. Opt. Soc. Am.* **1975**, *65(11)*, 1269.
23. Fina, L. J.; Tung, Y. S. *Appl. Spectrosc.* **1991**, *45(6)*, 986.

RECEIVED for review July 15, 1991. ACCEPTED revised manuscript September 8, 1992.

Film–Air and Film–Substrate Interfaces of Latex Films Monitored by Fourier Transform Infrared Spectroscopy

Timothy A. Thorstenson, Kevin W. Evanson, and Marek W. Urban*

Department of Polymers and Coatings, North Dakota State University, Fargo, ND 58105

The mobility of surfactants within latex films composed of ethyl acrylate and methacrylic acid is monitored via attenuated total reflectance–Fourier transform infrared spectroscopy. Ionic surfactants are found to exhibit a degree of incompatibility with the acrylic copolymer that results in their exudation to the film interfaces. Interfacial surface tension and elongation enhance exudation of surfactant to the interfaces. Neutralization of the copolymer acid functionality, however, inhibits the surfactant exudation. This inhibition is attributed to increased surfactant–copolymer compatibility that results from the formation of surfactant–copolymer complexes that become buried in the film during coalescene. In contrast to these observations, lattices prepared with a nonionic surfactant exhibit no exudation, which is attributed to a greater degree of surfactant–copolymer compatibility.

SURFACTANTS PLAY A VITAL ROLE IN LATEX TECHNOLOGY, both in the synthesis of the latex and as postsynthesis additives. The postsynthesis additives have a major influence on the stabilization of the latex against coagulation, the modification of rheological properties, and the dispersion and stabilization of pigments. Despite their utility, these low molecular weight species can also

*Corresponding author.

0065–2393/93/0236–0305$07.75/0

give rise to a host of undesirable properties. If the surfactant is incompatible with the polymer system, it can exude to the latex film interfaces, which results in optical defects, premature degradation, or a loss of adhesion. Thus, structure–property relationships in lattices are critical.

Although the importance of surfactants and the potential problems associated with their use are well established, relatively little work has been done with regard to surfactant–polymer compatibility and surfactant dynamics within latex films. A series of styrene–butadiene (SB) lattices prepared with nonylphenol ethylene oxide surfactants were examined via electron microscopy (1). The compatibility of the surfactants with the nonpolar SB lattices was found to decrease with the increasing polarity of surfactant. On the other hand, in the more polar vinyl acetate–vinyl acrylate lattices prepared with the same nonionic surfactants, surfactant adsorption increased with the higher polarity of the latex due to greater vinyl acetate content (2). Apparently, the hydrolysis of vinyl acetate groups to form poly(vinyl alcohol) provides OH functionality with which the polyether oxygens of the surfactant interact. This mechanism is why the formation of polymer–surfactant complexes may affect copolymer–surfactant compatibility and the classification of various surfactants into penetrating and nonpenetrating types may be valid (3). This classification was proposed (4) for a series of vinyl acetate–vinyl acrylate lattices, which exhibit increasing copolymer polarity that results in induced penetration of anionic surfactants into the latex particles. These assessments were based on viscosity studies, and the effect was attributed to the formation of polyelectrolyte-type solubilized polymer–surfactant complexes. Apparently, the polymer chains uncoil, which results in acetyl groups being pushed into the aqueous phase, where surfactant molecules readily adsorb on the polymer chains; this leads to the increased solubility of these segments (3). As expected, penetration of the surfactant depended on a critical size, the charge density at the polymer–water interface, and a shape conducive to penetration. In the case of nonionic surfactants, the surfactant concentration was found to decrease with increasing polarity while, at the same time, inhibiting penetration by the anionic surfactants.

The presence of interfacial surface tension at the polymer–water interface may affect the degree of surfactant adsorption on various polymer surfaces (4, 5). The surface area per molecule of sodium dodecyl sulfate (SDS) surfactant on a latex polymer particle increases with the increasing polarity of the polymer–water interface. The driving force for the adsorption of surfactant at various polymer–water interfaces was suggested to be related to the differences in the interaction energy between the surfactant molecules and the surface in question.

On a similar note, it is reasonable to expect that the surface free energy of the substrate may influence the distribution of surfactant within the latex film. Bradford and Vanderhoff (6) prepared styrene–butadiene copolymer latex films on a variety of substrates including poly(tetrafluoroethylene)

(PTFE), poly(ethylene terephthalate) (Mylar polyester), rubber, and mercury and examined the film–air and film–substrate interfaces via electron microscopy. Although differences were observed for the film–substrate interfaces, they were assessed as being too small for further analysis and consideration. It was concluded that the surfactant exudation and film formation behavior at the film–substrate interface was closely parallel to that at the film–air interface. However, as a recent study suggests, this is not the case (7).

Zhao et al. (8, 9) employed attenuated total reflectance–Fourier transform infrared spectroscopy, (ATR–FTIR) X-ray photoelectron spectroscopy (XPS), and secondary-ion mass spectrometry (SIMS) to examine the film–air and film–substrate interfaces of butyl acrylate–methl methacrylate lattices prepared using the anionic surfactants sodium dodecyl sulfate (SDS) and sodium dodecyl diphenyl disulfonate (SDED). In this case, the latex films exhibited enrichment at both the film–air and film–substrate interfaces. The extent of enrichment was found to be dependent on the nature of the surfactant, the coalescence time, the global concentration of the surfactant, and the interface involved; the film–air interface showed a greater degree of enrichment. SDS was found to form a thick boundary layer at the film–substrate interface. This "weak boundary layer" (9), which causes major adhesion problems, was attributed to a lesser degree of compatibility with the copolymer. Surfactant enrichment was attributed to three factors: (1) initial enrichment at both interfaces to lower interfacial free energy, (2) enrichment at the film–air interface due to the transport of nonadsorbed surfactant by the water flux out of the film, and (3) longer term migration to both interfaces due to surfactant incompatibility. The molecular level interactions governing surfactant behavior were not addressed.

Surfactant–Copolymer Interactions

Although the previous studies provided some insight into the factors governing surfactant compatibility and surfactant exudation behavior, they generally were confined to the study of either a specific series of lattices or a specific series of surfactants and, as such, it is difficult to assess the ultimate chemical factors that govern surfactant–polymer interactions. Evanson and Urban (10) began a systematic examination of surfactant–copolymer interactions with the study of an ethyl acrylate–methacrylic acid (EA–MAA) latex prepared with sodium dioctyl sulfosuccinate (SDOSS) surfactant via attenuated total reflectance (ATR) and photoacoustic (PA) Fourier transform infrared (FTIR) spectroscopy. Surfactant behavior at the film–substrate interface was monitored during coalescence via circle cell ATR–FTIR spectroscopy. Figure 1 shows the 1070–960-cm^{-1} region of the spectra recorded during coalescence. Although the spectrum acquired 5 min after deposition of the latex

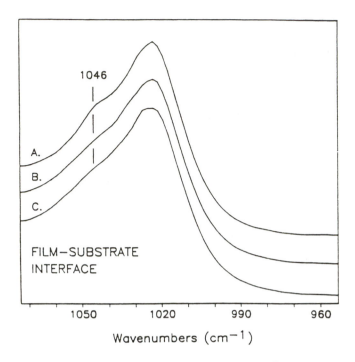

Figure 1. Circle ATR−FTIR spectra in the 1070−960-cm^{-1} region collected as a function of time for the EA−MAA latex as it coalesces: A, 5 min; B, 30 min; and C, 4 h. (Reproduced with permission from reference 10. Copyright 1991 Wiley.)

film (trace A) exhibits surfactant enrichment, as evidenced by the shoulder at 1046 cm^{-1} attributed to the symmetric S−O stretching mode of SO_3, the intensity of this band decreases with time and is no longer present after 4 h of coalescence. As shown in Figure 2, the preceding observation parallels the intensity decrease of the broad band at 3400 cm^{-1} due to water.

In contrast to these results, the spectrum of the film−air interface of the fully coalesced latex film shown in trace A of Figure 3 indicates surfactant enrichment to this interface, as demonstrated by the presence of the bands at 1046 and 1056 cm^{-1}, also attributed to the symmetric stretching mode of SO_3. Temporarily postponing the evaluation of these bands until later, it was found that the excess surfactant could be removed from the surface of the film by washing with dilute aqueous methanol. The ATR−FTIR spectrum of a surface-washed sample is shown in trace C of Figure 3, and the bands due to surfactant at 1046 and 1056 cm^{-1} are clearly absent. On the basis of these results it was concluded that although there is an initial surfactant enrichment at the film−substrate interface, this excess of water-soluble surfactant is carried to the film−air interface by the water flux that passes between the

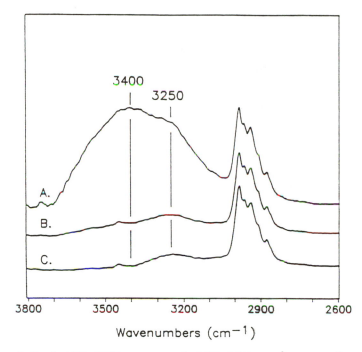

Figure 2. Circle ATR–FTIR spectra in the 3800–2600-cm^{-1} region collected as a function of time for the EA–MAA latex as it coalesces: A, 5 min; B, 30 min; and C, 4 h. (Reproduced with permission from reference 10. Copyright 1991 Wiley.)

latex particles during coalescence, and the final coalesced latex film exhibits surfactant enrichment only at the film–air interface.

It is apparent from the preceding results that the spectra of the film–substrate interface of the latex film revealed a single band due to the symmetric S–O stretch of SO_3 located at 1046 cm^{-1}, which diminishes in intensity as coalescence proceeds. The film–air interface, however, exhibited two bands located at 1046 and 1056 cm^{-1}. In contrast to these results, the transmission FTIR spectrum of neat SDOSS surfactant showed only a single band centered at 1050 cm^{-1}. These observations indicate that the S–O stretching mode is sensitive to the local environment changes around the sulfonate group that alter local symmetry in the latex–water environment.

In an effort to determine the primary causes of these local symmetry changes, spectra of solutions of SDOSS in both H_2O and EtOH were acquired, after solvent evaporation, via transmission FTIR spectroscopy. The 1050-cm^{-1} region of the spectra (with solvent contributions subtracted out) are shown in Figure 4. In the case of the EtOH solution (trace B), SDOSS exhibits the anticipated band due to the symmetric stretching mode of SO_3 at

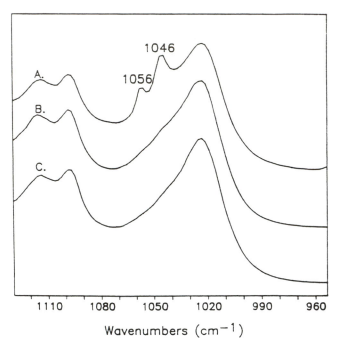

Figure 3. Circle ATR–FTIR spectra of EA–MAA latex in the 1130–950-cm^{-1} region: A, Film–air interface; B, film–substrate interface; and C, film–air interface washed with MeOH–DDI H$_2$O solution. (Reproduced with permission from reference 10. Copyright 1991 Wiley.)

1050 cm^{-1}. The aqueous solution, however, shows a band at 1046 cm^{-1} (trace A), which indicates that the band at 1046 cm^{-1} in the latex spectrum is a result of hydration of the highly hydrophilic sulfonate group.

A close analysis of transmission FTIR temperature study data (Figure 5) revealed that although there are no intensity changes of the acid dimer band at 1703 cm^{-1} on going from 25 to 165 °C, the band at 1765 cm^{-1}, which is due to "free" carboxylic acid, increases. This observation indicates that only a fraction of the hydrogen-bonded acid species is involved in hydrogen bonding interactions with other carboxylic acid species. This observation was further substantiated by the decrease of the 1735-cm^{-1} band in the same temperature range. The intensity decrease of this band in attributed to the breaking of interactions, which results in free carbonyl groups of the acid functionality (11–13). In essence, these results indicate the presence of SO$_3$ $\cdots$ HO interactions in the latex system. This assessment was further substantiated by the results of experiments in which the latex acid functionality was eliminated either through synthesis or via the addition of aqueous base. Figure 6 shows the ATR–FTIR spectrum of an "acid-free" ethyl acrylate–methyl methacry-

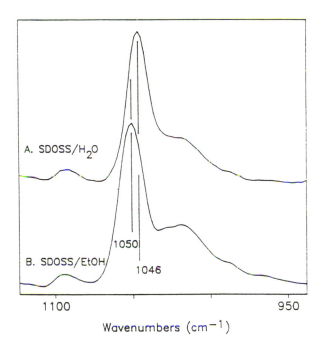

Figure 4. Transmission FTIR spectra of SDOSS after being dissolved in H₂O (A) and ethanol (B), followed by solvent evaporation. (Reproduced with permission from reference 10. Copyright 1991 Wiley.)

late (EA–MMA) latex along with the spectrum of the acid functional EA–MMA latex. Although the EA–MMA latex (trace B) exhibits the familiar splitting at 1046 and 1056 cm^{-1}, the acid-free EA–MMA latex (trace A) shows only a single band centered at 1050 cm^{-1}. Thus, it is the simultaneous presence of both the surfactant sulfonate group of SDOSS and the acid functionality of the EA–MMA copolymer that is responsible for the splitting of the 1050-cm^{-1} S–O stretching vibration to two bands at 1046 and 1056 cm^{-1}.

Further insight into the interactions between the SO$_3$ group of SDOSS and the acid functionality of the latex is gained by considering the results of an experiment in which the acid functionality of the latex suspension was neutralized via the addition of aqueous base. This neutralization process destroys the hydrogen bond-donating −OH groups present in the latex, thus allowing the assessment of the nature of interactions between surfactant SO$_3$ groups and COO$^-$ functionality of the neutralized copolymer. To investigate the effect of different cations on the observed frequency shift, two different bases (NaOH and NH$_4$OH) were employed in the neutralization procedure. Figure 7a shows the 1800–1520-cm^{-1} region of the spectra of neutralized latex films. Here, neutralization is confirmed by a broad band due to the

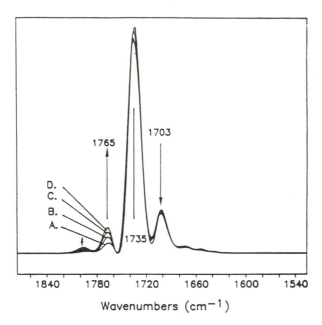

Figure 5. Maximum-entropy restored spectra at various temperatures for the EA–MAA latex in the 1860–1530-cm^{-1} region: A, 25 °C; B, 105 °C; and D, 165 °C. (Reproduced with permission from reference 10. Copyright 1991 Wiley.)

C–O stretching mode of the carboxylate salt group observed at 1590 cm^{-1} for the latex neutralized with NaOH (trace B) and at 1570 cm^{-1} for the latex neutralized with NH$_4$OH (trace C). Examination of the 1050-cm^{-1} region (Figure 7b) shows that although the spectrum of the nonneutralized film (trace A) exhibits the familiar splitting at 1046 and 1056 cm^{-1}, the neutralized films both exhibit only a single band centered at 1046 cm^{-1}. This experiment confirms that the 1056 band is due to S–O ··· H–O interactions between the copolymer acid functionality and the surfactant sulfonate groups. Furthermore, the band at 1046 cm^{-1} attributed to the S–O symmetric stretching mode of hydrated SO$_3^-$Na$^+$ ion pairs is insensitive to the ionic strength of the neutralizing agent used.

Effect of Surfactant Structure on Surfactant Mobility

The effect of surfactant structure was examined (*14*) via the preparation of EA–MAA lattices using a variety of surfactants including sodium dioctyl sulfosuccinate (SDOSS), sodium dodecylbenzene sulfonate (SDBS), sodium dodecyl sulfate (SDS), sodium nonylphenol ethylene oxide sulfonate (2 ethylene oxide units; SNP2S), and the nonionic surfactant nonylphenol ethylene oxide (40 ethylene oxide units; NP40). The structures of the

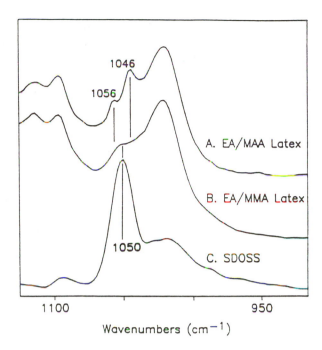

Figure 6. ATR spectra in the 1110–950-cm^{-1} region: A, EA–MAA latex; B, EA–MMA latex; and C, SDOSS surfactant. (Reproduced with permission from reference 10. Copyright 1991 Wiley.)

surfactants are shown in Chart I. To identify potential spectral features that may be used to identify the presence of surfactant at the latex film interfaces, it is first necessary to identify the relevant spectral features of the copolymer and surfactants. Figure 8 shows the 1800–500-cm^{-1} region to the spectra of the copolymer and surfactants, whereas Table I lists the tentative band assignments for these species. Examination of Figure 8 shows that, in the case of the anionic surfactants, all spectra exhibit characteristic absorbance bands in the regions of 1250–1150 (S–O asymmetric stretching), 1060–1045 (S–O symmetric stretching), and 750–550 cm^{-1} (S–O bending). As is demonstrated in Figure 8, the latter region will be particularly useful because it is essentially free of interfering absorbance bands due to the copolymer. In the case of the nonionic NP40 surfactant, identification can be readily facilitated by monitoring the band at 947 cm^{-1} due to the CH$_2$–O stretching mode of the polyether units of this surfactant.

   In an effort to investigate the effect of surfactant structure on surfactant exudation behavior, films of the lattices synthesized with the various surfactants were prepared on a poly(tetrafluoroethylene) substrate and examined via rectangular ATR–FTIR spectroscopy. Figure 9a shows the

1135–500-cm^{-1} region of the spectra of the film–air interfaces of these latex films. The SDOSS latex (trace A) exhibits slight enrichment, as demonstrated by the presence of bands at 1046 and 1056 (symmetric S–O stretch of SO$_3$), 652 (S–O bending mode of SO$_3$), and 581 cm^{-1} (SO$_3$ scissors.) Similarly, the SDBS latex (trace B) exhibits weak bands at 614 and 583 cm^{-1} due to the S–O bending and SO$_2$ scissor vibrational modes of this surfactant, respectively. The SDS latex (trace D) also shows slight surfactant enrichment at the film–air interface as shown by the weak bands at 631 and 585 cm^{-1} (S–O bending modes of SO$_4$). In contrast, the SNP2S latex (trace C) has none of the characteristic surfactant bands, which indicates that there is no surfactant enrichment at the film–air interface. The latex prepared with the nonionic surfactant NP40 also shows a trace of surfactant at the film–air interface (trace E of Figure 9a), which is evidenced by the presence of a weak band at 947 cm^{-1} that has been assigned to the CH$_2$–O (ether) stretch of this surfactant.

The spectra of the film–substrate interfaces of the same latex films are presented in Figure 9b. The spectrum of the SDOSS latex (trace A) exhibits

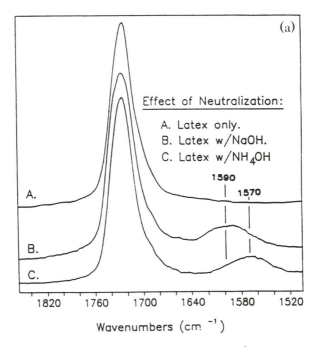

Figure 7a. ATR–FTIR spectra in the 1800–1520-cm^{-1} region of A, EA–MAA latex; B, EA–MAA latex neutralized with NaOH; and C, EA–MAA latex neutralized with NH$_4$OH. (Reproduced with permission from reference 10. Copyright 1991 Wiley.)

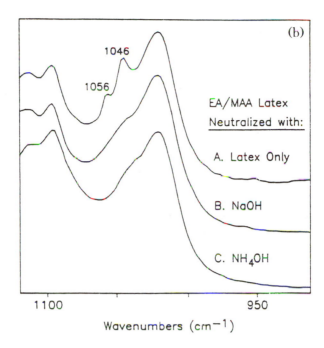

Figure 7b. ATR–FTIR spectra in the 1100–950-cm^{-1} region of A, EA–MAA latex; B, EA–MAA latex neutralized with NaOH; and C, EA–MAA latex neutralized with NH$_4$OH. (Reproduced with permission from reference 10. Copyright 1991 Wiley.)

the same characteristic absorbance bands due to surfactant that were observed in the spectrum of the film–air interface. Furthermore, the intensity of these bands is similar at both interfaces, which indicates that a similar concentration of SDOSS is present at both interfaces. A greater intensity of the bands at 616 and 583 cm^{-1} in the spectrum of the SDBS latex (trace B) indicates a somewhat higher surfactant concentration at the film–substrate interface. In addition, the band at 1046 cm^{-1} due to the symmetric S–O stretch of SO$_3$ becomes more pronounced at this interface. The SNP2S latex (trace C), which exhibited no surfactant enrichment to the film–air interface, shows considerable exudation to the film–substrate interface as illustrated by the intense bands at 1056 and 614 cm^{-1} (symmetric S–O stretching and S–O bending modes of SO$_3$, respectively). The SDS latex (trace D) also exhibits significant surfactant exudation to the film–substrate interface, again demonstrated by the intense bands due to the S–O bending mode of SO$_4$ at 631 and 585 cm^{-1}. Similarly to the spectrum at the film–air interface, the NP40 latex (trace E) shows a trace of surfactant at the film–substrate interface as evidenced by the weak band at 947 cm^{-1}. In this case it was

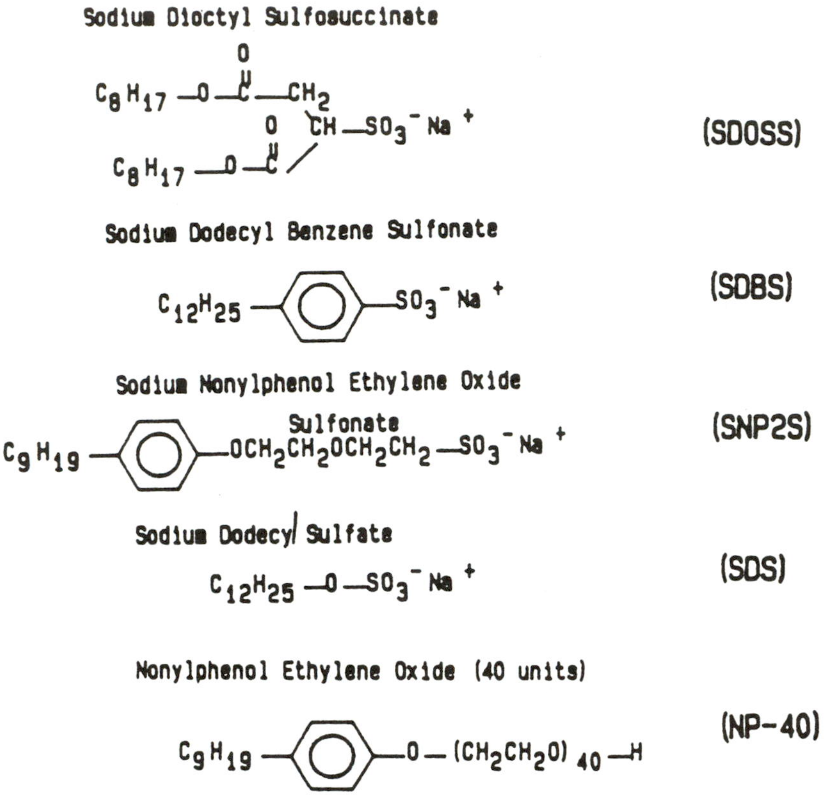

Sodium Dioctyl Sulfosuccinate

$$C_8H_{17} -O -\overset{\overset{O}{\|}}{C} -CH_2$$
$$O \quad CH -SO_3^- \ Na^+$$
$$C_8H_{17} -O -\overset{\|}{C}/$$

(SDOSS)

Sodium Dodecyl Benzene Sulfonate

$$C_{12}H_{25} -\!\!\langle O \rangle\!\!- SO_3^- \ Na^+$$

(SDBS)

Sodium Nonylphenol Ethylene Oxide
Sulfonate

$$C_9H_{19} -\!\!\langle O \rangle\!\!- OCH_2CH_2OCH_2CH_2 -SO_3^- \ Na^+$$

(SNP2S)

Sodium Dodecyl Sulfate

$$C_{12}H_{25} -O -SO_3^- \ Na^+$$

(SDS)

Nonylphenol Ethylene Oxide (40 units)

$$C_9H_{19} -\!\!\langle O \rangle\!\!- O -(CH_2CH_2O)_{40} -H$$

(NP-40)

Chart I. Chemical structures of surfactants. (Reproduced with permission from reference 14. Copyright 1991 Wiley.)

found that even vigorous washing of the surfaces of the film with dilute aqueous methanol failed to influence the intensity of the 947-cm^{-1} band, which indicates that the NP40 surfactant is uniformly distributed through the bulk of the film.

The Effect of Neutralization on Surfactant Mobility

The aforementioned results indicate that all ionic surfactants exhibit some degree of surfactant enrichment to the latex film interfaces and, with the exception of SDOSS, this enrichment appears to be greater at the film–substrate interfaces. These observations suggest that there is some inherent degree of incompatibility between the copolymer and the anionic surfactants. In contrast, the nonionic NP40 surfactant exhibits no exudation, which may indicate that the absence of a charged, highly polar hydrophilic "head" leads to increased compatibility with the relatively nonpolar latex. Additionally, the

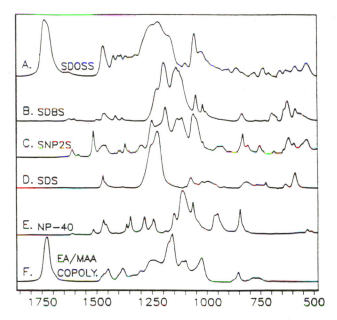

Figure 8. Transmission FTIR spectra in the 1800–500-cm⁻¹ region of the surfactants and EA–MAA copolymer: A, SDOSS; B, SDBS: C, SNP2S; D, SDS; E, NP; and F, EA–MAA copolymer. (Reproduced with permission from reference 14. Copyright 1991 Wiley.)

tendency of the nonionic surfactant to form strong hydrogen bonds with the copolymer acid functionality (*11*) leads to favorable interactions that may further enhance compatibility. As was mentioned earlier, copolymer polarity exerts significant influence upon surfactant–copolymer interactions. With this in mind, NaOH solution was added to the latex suspensions to neutralize the acid functionality and thus increase the polarity of the copolymer environment through the formation of carboxylate salt groups. Films of the neutralized lattices were prepared on a PTFE substrate and examined in the same manner as the nonneutralized lattices. Figure 10a shows ATR–FTIR spectra of the film–air interfaces of the neutralized latex films. The spectra of the lattices prepared with SDOSS, SDBS, and SNP2S surfactants (traces A, B, and C, respectively) do not exhibit any of the previously observed (Figures 9a and b) bands characteristic of the surfactants. For these lattices, neutralization of the acid functionality inhibits the exudation of surfactant to the film–air interface. In contrast, the spectrum of the SDS latex (trace D) does show weak bands at 631 and 585 cm⁻¹ due to the S–O bending modes of SO_4. This observation can be accounted for in terms of the incompatibility of SDS surfactant with acrylic copolymers, which has been identified in previous studies (*8, 9*). The NP40 latex (trace E) exhibits the same band observed in

Table I. List of the Observed Bands and Their Tentative Band Assignments[a]

SDOSS	SDBS	SNP2S	SDS	NP	Copolymer	Assignment
—	—	—	—	—	2981	asym. C–H stretch (CH_3)
2960	2958	2954	2956	2950	2960	asym. C–H stretch (CH_3)
2934	2925	2930	2919	2930	2934	asym. C–2H stretch (CH_2)
2879	2871	2877	2875	2884	2879	sym. C–H stretch (CH_3)
1735	—	—	—	—	1735	C=O stretch
—	—	—	—	—	1700	H–bonded –COOH
—	1625	—	—	—	—	Ar–S stretch
—	1603	1611	—	1609	—	p-substituent aromatic
—	—	1582	—	1580	—	C=C aromatic
—	1493	1513	—	1513	—	C=C aromatic
1464	1463	1466	1468	1466	1466	C–H deformation
1416	—	—	—	1455	1447	CH_2 scissor
1393	—	1395	—	—	—	HC–S deformation
1360	1378	1364	1380	1360	1382	C–(CH_3) sym. deformation
1314	—	1295	—	1279	1299	CH_2 wagging
—	—	—	1248	—	—	S–O stretch (SO_4)
—	—	—	1221	—	—	
1241	—	1245	—	1241	1252	C–O stretch
1216	1192	1185	—	—	—	asym. S ≥ stretch (SO_3)
1175	—	—	—	1146	1173	asym. C–O–C stretch
1094	—	—	—	1108	1098	sym. C–O–C stretch
—	—	—	1071	—	—	sym. S–O stretch (SO_4)
1050	1046	1056	—	—	—	sym. S–O stretch (SO_3)
1025	—	—	—	—	1025	C–O–C (ester)
—	1013	1013	—	—	—	=C–H in-plane deformation
—	—	942	—	947	—	CH_2–O (ether)
—	832	828	—	843	—	=C–H out-of-plane
—	—	—	830	—	—	S–O–C stretch
857	—	—	—	—	854	ester skeletal vibration
729	724	751	722	—	—	–$(CH_2)_n$– (n > 3)
—	691	683	—	—	—	CH out-of-plane
652	616	614	—	—	—	S–O bending (SO_3)
—	—	—	585	—	—	S–O bending (SO_4)
581	583	589	—	—	—	SO_2 scissor
529	—	531	—	529	—	alkyl chain
—	—	—	—	511	—	skeletal vibrations

[a] Reproduced with permission from reference 14. Copyright 1991 Wiley.

the previous spectra at 947 cm^{-1}. Because this band is almost the same intensity as that observed for the film–air interface of the nonneutralized latex film (Figure 9a, trace E), it appears that, in the case of the nonionic surfactant, the surfactant concentration at the film–air interface is not affected by neutralization.

The spectra of the film–substrate interfaces of the neutralized latex films are presented in Figure 10b. At this interface, none of the lattices prepared with anionic surfactants exhibits characteristic surfactant bands. Apparently, surfactant exudation to this interface is also inhibited. As was observed in the previous spectra, the NP40 latex again shows the presence of a surfactant band at 947 cm^{-1} with a similar intensity to that seen previously. It appears that the neutralization process again does not exert any significant effect on the distribution of this nonionic surfactant.

Although the spectra of the ionic surfactants in Figure 10b do not exhibit characteristic surfactant bands, all spectra of the film–substrate interfaces show that bands at 635 and 521 cm^{-1}, which are attributed to the wagging vibrations of the long chain α-methyl carboxylate salts formed upon neutralization of the copolymer acid functionality. The spectra presented in Figure 10a and b also indicate that these bands are detected only in the film–substrate interface spectra. This observation is likely attributed to the fact that the salt groups are quite hydrophilic and this provides a driving force for migration and orientation toward the film–substrate interface where, during coalescence, the aqueous phase remains for the longest period of time.

Based on these results it is apparent that, in the case of the ionic surfactants, neutralization results in the inhibition of surfactant exudation. In

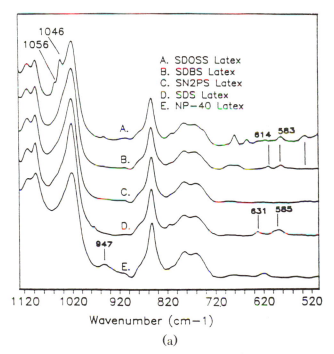

(a)

Figure 9. ATR–FTIR spectra in the 1135–500-cm^{-1} region recorded at the film–air interface (a) and at the film–substrate interface (b) of latex films prepared on PTFE: A, SDOSS latex; B, SDBS latex; C, SNP2S latex; D, SDS; and E, NP latex. (Reproduced with permission from reference 14. Copyright 1991 Wiley.) Continued on next page.

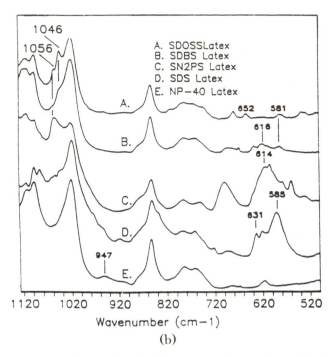

Figure 9. Continued. *(Reproduced with permission from reference 14. Copyright 1991 Wiley.)*

the nonneutralized lattices, the ionic surfactants are incompatible with the copolymer environment, and exudation to the interfaces of the latex films occurs. Here, the copolymer environment is quite nonpolar owing to the fact that it is composed primarily of ethyl acrylate. In comparison, the ionic surfactants, owing to their negatively charged sulfate or sulfonate heads, are highly polar and hydrophilic. This may lead to a certain inherent degree of incompatibility, so is is possible that during coalescence these surfactants assemble in the interstices of an inhomogeneous polymer matrix. Moreover, the ionic surfactants have relatively low molecular weight and, in comparison with the nonionic NP40, they are quite water soluble. The combination of these factors makes their diffusion to the film interfaces easy if there exists a driving force for them to do so. When the copolymer acid functionality is neutralized, however, the copolymer becomes much more polar due to the formation of carboxylate salt groups. This increased hydrophilicity causes the latex particles to swell: their hydrodynamic volume increases and chain extension occurs. As this process continues, the anionic surfactants may become displaced from the surface of the latex particles and reside in the aqueous phase. However, as chain extension proceeds further, hydrophobic ethyl groups pendant on the copolymer chains are forced out into the aqueous phase and water diffuses into the latex particles. The water-soluble anionic surfactants may then penetrate into the latex particles and associate with them through hydrophobic interactions, which leads to the formation of solubilized polymer–surfactant complexes that then become buried within the latex film as coalescence proceeds. Additionally, ionic interactions between the carboxlate salt groups and the surfactant sulfate of sulfonate groups may further enhance compatibility.

Effect of Substrate Surface Tension on Surfactant Mobility

As was mentioned earlier, another factor that may influence surfactant mobility is the surface tension of the substrate. To establish if inhibition of the exudation of the ionic surfactants due to neutralization was substrate related, films of the neutralized lattices were prepared on glass and mercury substrates. Figure 11 shows the $1135–480$-cm^{-1} region of the spectra acquired at the film–substrate interfaces of the films allowed to coalesce on glass, PTFE, and mercury substrates. In this case, the SDOSS surfactant was

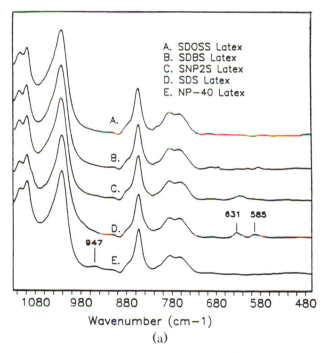

Figure 10. ATR–FTIR spectra in the $1135–480$-cm^{-1} region recorded at the film–air interface (a) and at the film–substrate interface (b) of neutralized latex films prepared on PTFE: A, SDOSS latex; B, SDBS latex; C, SNP2S latex; D, SDS latex; and E, NP latex. (Reproduced with permission from reference 14. Copyright 1991 Wiley.) Continued on next page.

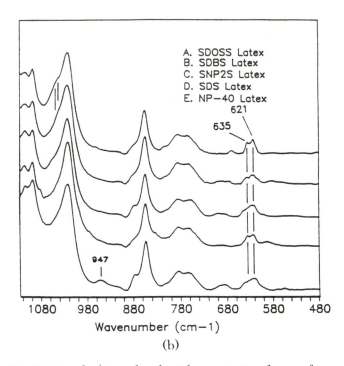

Figure 10. Continued. *(Reproduced with permission from reference 14. Copyright 1991 Wiley.)*

used. All three spectra exhibited similar features and showed no signs of absorbance bands characteristic of the SDOSS surfactant. Similar behavior was observed for the other lattices. These data indicate that the inhibition of surfactant exudation to the film–substrate interface observed upon neutralization is not substrate related. In an effort to determine if the observed inhibition of exudation following neutralization is influenced by mechanical strain, films of the neutralized lattices were subjected to elongations of up to 50% and then examined using rectangular ATR–FTIR spectroscopy. Figure 12 shows the 1135–480-cm^{-1} region of spectra acquired from the films of the neutralized SDBS latex that were subjected to mechanical elongation of 0, 30, and 50% (traces A, B, and C, respectively). As can be seen, no significant spectral changes are observed as the degree of elongation is increased. Similar results were obtained with the other lattices. As will be shown later on, elongation of the nonneutralized latex films can result in significant surfactant exudation, so it appears that neutralization of the copolymer acid functionality results in enhanced compatibility between the surfactant and the copolymer.

The effects of substrate surface tension and elongational strain are, however, much more pronounced for nonneutralized lattices (*15*). Let us

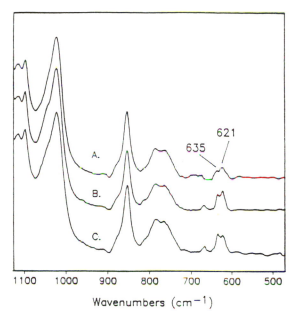

Figure 11. ATR–FTIR spectra in the 1135–480-cm⁻¹ region recorded at the film–substrate interface of neutralized SDOSS latex films prepared on different substrates: A, glass; B, PTFE; and C, mercury (Hg). (Reproduced with permission from reference 14. Copyright 1991 Wiley.)

again consider the results obtained for the film–substrate interfaces of latex films prepared on a PTFE substrate. The spectra are shown in Figure 9a and b. Here, all of the anionic surfactants except SDOSS exhibit a greater degree of exudation to the film–substrate interface. This behavior was very pronounced in the cases of SDS and SNP2S, whereas the SDBS latex showed only slight preferential enrichment to this interface. Although the migration of surfactant to this interface may be accounted for in terms of the water solubility of the surfactants, which provides a driving force for migration to the film–substrate interface where the aqueous phase is present for the longest period of time, it is necessary to also consider the effects of substrate surface tension. Because PTFE has a low surface tension (15 mN/m), deposition of a film of latex on the PTFE substrate creates a high degree of interfacial tension at the latex–PTFE interface, and this may provide a driving force for the migration of surfactant molecules to this interface so as to reduce the interfacial tension.

To establish if the interfacial surface tension does indeed provide a driving force for surfactant exudation to the film–substrate interface, latex films were prepared directly on the thallous bromide iodine (KRS-5) ATR element. KRS-5 has a surface tension similar to that of glass (approximately

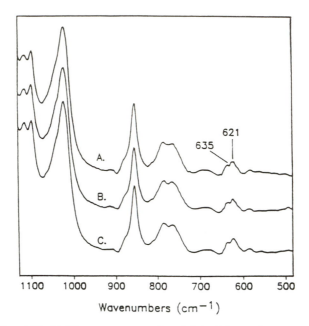

Wavenumbers (cm⁻¹)

Figure 12. ATR–FTIR spectra in the 1135–480-cm⁻¹ region for the film–substrate interface of the neutralized SDBS latex film as a function of elongation: A, 0% elongation; B, 30% elongation; and C, 50% elongation. (Reproduced with permission from reference 14. Copyright 1991 Wiley.)

70 mN/m), and this higher substrate surface tension will result in a lower degree of interfacial tension between the latex and the substrate, thus reducing the driving force for the migration of surfactant to this interface. Figure 13 shows ATR–FTIR spectra acquired at the film–substrate interfaces of these films. As can be seen, the SDOSS and SNP2S lattices (traces A and C) exhibit no previously detected characteristic surfactant bands. The SDBS and SDS films, however, do show some enrichment to this interface, but a comparison with the results observed for the PTFE substrate (Figure 9b) indicates that the degree of enrichment is significantly smaller than that for the KRS-5 substrate. In contrast to these spectral changes, the latex prepared with the nonionic NP40 surfactant shows a familiar band at 947 cm⁻¹ (CH_2–O stretching mode of the surfactant ether units) that is about the same intensity as that observed for the latex deposited on PTFE substrate.

To further investigate the observed effects of substrate surface tension, latex films were prepared on a liquid mercury substrate. ATR–FTIR spectra acquired at the film–substrate interfaces of these films are shown in Figure 14. As shown by the enhanced intensity of the bands at 1046 and 1056 cm⁻¹ (symmetric S–O stretching mode of SO_3), the SDOSS latex (trace A) reveals

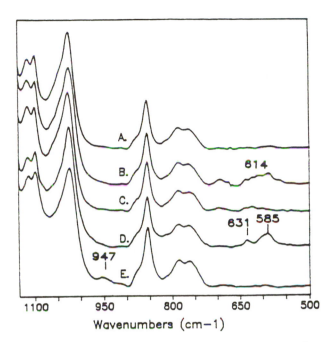

Figure 13. ATR–FTIR spectra in the region from 1150 to 500 cm^{-1} recorded at the film–substrate interface of latex films prepared on glass: A, SDOSS; B, SDBS; C, SNP2S; D, SDS, and E, NP. (Reproduced with permission from reference 15. Copyright 1991 Wiley.)

considerable exudation to the film–substrate interface. Additionally, the bands at 652 (S–O bending mode) and 581 cm^{-1} (S–O scissors) are observed. The SDBS latex (trace B) also shows significant surfactant enrichment to the mercury interface, as demonstrated by the band at 616 cm^{-1} due to the S–O bending mode of SO_3. Similar results are seen for the SDS latex (trace D), as shown by the intense bands at 631 and 585 cm^{-1} (S–O bending mode of SO_4).

In contrast to the ionic surfactants just mentioned, the SNP2S latex spectrum (trace C) exhibits a degree of enrichment, as seen by the intensity of the S–O bending mode at 614 cm^{-1}, that is less than that observed in the case of the PTFE substrate. In this case, however, a thin, cloudy film was observed on the surface of the mercury after removal of the coalesced latex film. Analysis of this material using transmission FTIR spectroscopy revealed it to be composed almost exclusively of SNP2S surfactant. Thus, the degree of enrichment to the film–substrate interface is greater than was indicated by the ATR–FTIR spectrum.

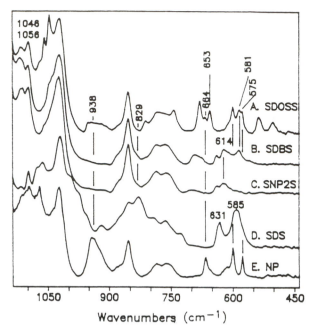

Figure 14. ATR–FTIR spectra in the region from 1150 to 500 cm^{-1} recorded at the film–substrate interface of the latex films prepared on mercury: A, SDOSS; B, SDBS; C, SNP2S; D, SDS; and E, NP. (Reproduced with permission from reference 15. Copyright 1991 Wiley.)

Again in contrast with the preceding results, the nonionic NP40 exhibits no surfactant enrichment to the mercury interface (Figure 14, trace E). This spectrum does, however, exhibit a broad band at 938 cm^{-1} attributed to the OH $\cdots$ O out-of-plane deformation vibrations of the carboxylic acid groups. Furthermore, the bands at 664, 600, and 575 cm^{-1} assigned to the O–CO in-plane vibration of α-branched aliphatic carboxylic acids are observed. These features are also observed for the SDOSS latex (trace A). Examination of the carbonyl region of this spectrum (not shown) reveals that a large portion of the acid groups present at the film–substrate interface are involved in hydrogen bonding interactions with the surfactant and other acid groups.

In the case of the incompatible anionic surfactants, the substrate employed in film preparation exerts a significant influence on the degree of surfactant enrichment that will be observed at the film–substrate interface of the latex films. This phenomenon can be understood in terms of the surface tension differences between the copolymer and various substrates. In the case of PTFE, which has a very low surface free energy of 18.5 mN/m, a considerable driving force exists for the migration of surfactant to the film–substrate interface to lower the high interfacial surface tension present there. In the case of the glass substrate, with a surface free energy approximately 70 mN/m, the polymer (surface free energy approximately 30 mN/m) may readily wet the glass substrate and thus, the driving force for surfactant migration to this interface is significantly reduced. The very high surface tension (416 mN/m) of the mercury substrate allows the polymer to initially wet the surface, but once coalescence begins, a solid latex film–liquid mercury interface exists, and this leads to a high interfacial tension that gives rise to a higher driving force for surfactant enrichment.

This hypothesis can also explain the presence of acid dimer species that are observed in the case of the NP40 latex film. As the aforementioned results show, the degree of compatibility between the NP40 surfactant and the EA–MAA copolymer is much greater than that observed for the anionic surfactants. Due to this factor, it is possible that the driving force for exudation produced by the surface free energy of mercury substrate is insufficient to induce surfactant migration to the film–substrate interface. Because the interfacial tension is not reduced by the assembly of surfactant at this interface, the copolymer may respond by orientation of its acid functionality toward the mercury. The acid groups, which have significantly greater polarity than the other species, may increase the surface free energy of the copolymer film–substrate interface, and thus reduce the interfacial excess of energy present at the solid latex film–liquid mercury interface. Similar conclusions can be tentatively drawn for the SDOSS latex. Due to the presence of two hydrophobic tails, it is possible that this surfactant is unable to properly align itself at the copolymer–mercury interface so as to effectively reduce the interfacial surface tension. Again, the orientation of acid-functional species toward this interface appears to be a primary consideration.

Effect of Elongation on Surfactant Mobility

As was indicated in the previous sections, the mobility of surfactants in neutralized latex films is insensitive to mechanical strain. In contrast, it was found (15) that considerable surfactant exudation may result upon the elongation of nonneutralized latex films. In an effort to examine surfactant mobility upon elongation, latex films were prepared on a PTFE substrate. After coalescence, the films were removed and washed with dilute aqueous methanol to remove any surfactant that may have assembled at the interfaces during coalescence. After washing, the films were elongated to the stated degree, held at this elongation for 5 min, and then allowed to relax.

The spectral results for the SDOSS latex are shown in Figure 15. At 10% elongation (trace A), the appearance of a weak band at 529 cm^{-1} (alkyl chain vibration) is detected. As the degree of elongation increases, however, additional characteristic bands, including 1046 and 1056 (symmetric S–O stretch), 653 (S–O bending mode), and 583 cm^{-1} (SO$_2$ scissors), become evident and increase considerably from 30 to 50% elongation (traces B and C, respectively). Similar behavior is observed for the SDBS latex (Figure 16) and the SNP2S latex (Figure 17).

In the case of the SDS latex (Figure 18), characteristic surfactant bands at 631 and 585 cm^{-1} (S–O bending mode of SO$_4$) are also observed and increase with elongation. However, a band at 921 cm^{-1} is observed at 50% elongation (trace C). This band is assigned to the bisulfate ion, and its presence is not surprising because SDS is known to hydrolyze during the emulsion polymerization process to form sodium bisulfate and dodecanol (16). The exudation of unhydrolyzed surfactant is confirmed by examination of the C–H stretching region of these spectra (Figure 19). The intensity of the surfactant bands at 2956, 2919, and 2852 cm^{-1} increases with the degree of elongation. This observation indicates the exudation of unhydrolyzed surfactant that results from elongation.

The exudation of anionic surfactants that accompanies film elongation may also be addressed in terms of surface tension. When the latex film is elongated, the total surface area of the film is increased and the concentration

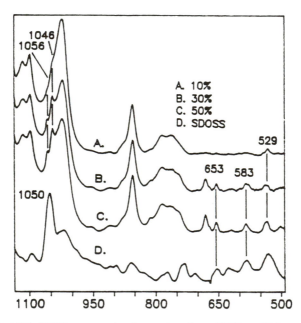

Figure 15. ATR–FTIR spectra in the region from 1135 to 500 cm^{-1} of the SDOSS latex films recorded as a function of percent elongation: A, 10%; B, 30%; and C, 50%. (Reproduced with permission from reference 15. Copyright 1991 Wiley.)

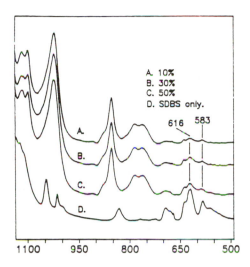

Figure 16. ATR–FTIR spectra in the region from 1135 to 500 cm^{-1} of the SDBS latex films recorded as a function of percent elongation: A, 10%; B, 30%; and C, 50%. (Reproduced with permission from reference 15. Copyright 1991 Wiley.)

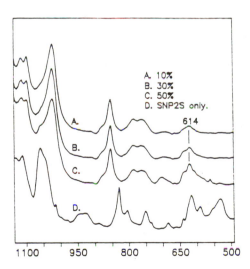

Figure 17. ATR–FTIR spectra in the region from 1135 to 500 cm^{-1} of the SNP2S latex films recorded as a function of percent elongation: A, 10%; B, 30%; and C, 50%. (Reproduced with permission from reference 15. Copyright 1991 Wiley.)

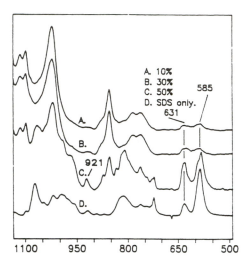

Figure 18. ATR–FTIR spectra in the region from 1135 to 500 cm^{-1} of the SDS latex films recorded as a function of percent elongation: A, 10%; B, 30%; and C, 50%. (Reproduced with permission from reference 15. Copyright 1991 Wiley.)

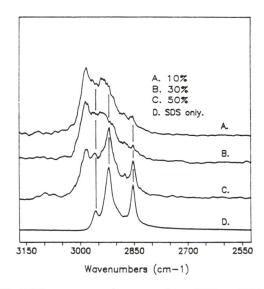

Figure 19. ATR–FTIR spectra in the region from 3150 to 2550 cm^{-1} of the SDS latex films recorded as a function of percent elongation: A, 10%; C, 30%; C, 50%; and D, SDS only. (Reproduced with permission from reference 15. Copyright 1991 Wiley.)

of surfactant at the interfaces of the film is therefore decreased. This results in a higher surface tension at the latex film interfaces, and thus provides the necessary driving force for the exudation of surfactant; the combination of copolymer–surfactant incompatibility and low molecular weight makes for ready diffusion to the interfaces of the elongated film. A similar phenomenon is observed for plasticizers in thermoplastic polymers (*17*). In contrast to the anionic surfactants, films of the NP40 latex show no exudation upon elongation (not shown). Here, the greater compatibility and the higher molecular weight of this surfactant again overwhelm the driving force for surfactant exudation.

References

1. Vanderhoff, J. W. *Br. Polym. J.* **1970**, *2*, 161–173.
2. Vijayendran, B. R.; Bone, T. L.; Gajria, C. *J. Appl. Polym. Sci.* **1981**, *26*, 1351.
3. Arai, H.; Horin, S. *J. Colloid Interface Sci.* **1969**, *30*, 372.
4. Vijayendran, B. R.; Bone, T.; Gajria, C. *J. Appl. Polym. Sci.* **1981**, *26*, 1351–1359.
5. Vijayendran, B. R. *J. Appl. Polym. Sci.* **1979**, *23*, 733–742.
6. Bradford, E. B.; Vanderhoff, J. W. *J. Macromol. Sci. Phys.* **1972**, *B6(4)*, 671–694.
7. Urban, M. W.; Evanson, K. W. *Polym. Commun.* **1990**, *31*, 279.
8. Zhao, C. L.; Holl, Y.; Pith, T.; Lambla, M. *Colloid Polym. Sci.* **1987**, 265, 823-829.
9. Zhao, C. L.; Holl, Y.; Pith, T.; Lambla, M. *Br. Polym. J.* **1989**, *21*, 155–160.
10. Evanson, K. W.; Urban, M. W. *J. Appl. Polym. Sci.* **1991**, *42*, 2287–2296.
11. Lee, Y. J.; Painter, P. C.; Coleman, M. M. *Macromolecules* **1988**, *21*, 346–354.
12. Lee, Y. J.; Painter, P. C.; Coleman, M. M. *Macromolecules* **1988**, *21*, 954–960.
13. Lichkus, A. M.; Painter, P. C.; Coleman, M. M. *Macromolecules* **1988**, *21*, 2636–2641.
14. Evanson, K. W.; Thorstenson, T. A.; Urban, M. W. *J. Appl. Polym. Sci.* **1991**, *42*, 2297–2307.
15. Evanson, K. W.; Urban, M. W. *J. Appl. Polym. Sci.* **1991**, *42*, 2309–2320.
16. Nakagaki, M.; Yokoyama, S. *Bull. Chem. Soc. Jpn.* **1986**, *59*, 935–936.
17. Ludwig, B.; Urban, M. W. *Polymer* **1992**, *33(16)*, 3343.

RECEIVED for review July 15, 1991. ACCEPTED revised manuscript September 9, 1992.

Surface Analysis of Thick Polymer Films by Infrared Spectroscopy

Mark R. Adams, KiRyong Ha, Jenifer Marchesi, Jiyue Yang, and Andrew Garton*

Polymer Program and Department of Chemistry, University of Connecticut, Storrs, CT 06269-3136

Recent advances in the application of infrared internal reflection spectroscopy (IRIRS) are described for the surface analysis of thick polymer films. Examples are chosen to illustrate the nature of the information obtained for polymer surface modifications ranging from about 5 nm thickness (plasma-modified polypropylene), through about 100 nm (sodium-etched fluoropolymers), to several hundred nanometers thickness of the modified layer (polyolefin surfaces modified by "primers"). The IRIRS data are compared to data obtained by X-ray photoelectron spectroscopy. Particular emphasis is placed on the use of derivatization reactions, and the determination of concentration profiles in the modified surfaces.

$\mathbf{V}$IBRATIONAL SPECTROSCOPY IS CAPABLE OF SURFACE ANALYSIS (1, 2) at sensitivities corresponding to monomolecular layer coverage, using techniques such as reflection–absorption infrared (RAIR) spectroscopy and surface-enhanced Raman spectroscopy (SERS). Even conventional transmission spectroscopy is a powerful surface analysis technique if the specimen is adsorbed on a high surface area substrate such as a fine powder. However, most of the nominally "surface-sensitive" techniques analyze thin films on a nonpolymeric substrate (e.g., a coating on a metallic mirror for RAIR or material sorbed on a roughened silver surface for SERS) rather than the surfaces of thick films, and many specimens of scientific and technological interest are in the form of thick films (i.e., greater that a few microns thick). The range of IR spectro-

* Corresponding author.

0065–2393/93/0236–0333$06.00/0

scopic choices for surface analysis is then much more limited. For such applications, the infrared internal reflection (IRIRS) technique [often called attenuated total reflectance (ATR)] is often the only option. The IRIRS technique is well established and widely used (3–5), but is generally perceived to be insufficiently surface sensitive to be useful for surface modifications less than a few microns in depth. This chapter describes several applications that demonstrate that IRIRS can perform surface analysis with a sensitivity comparable to X-ray photoelectron spectroscopy (XPS), and, if applied with ingenuity, can provide information unobtainable with other techniques. The applications are classified on the basis of the depth of the surface modification.

Surface Modification < 10 nm: Plasma-Treated Polypropylene Films

The adhesive properties or printability of polymer films can be improved by brief exposure to plasmas. This method is particularly useful for low surface energy polymers such as polyolefins (6–8). For brief exposure to a relatively low-energy plasma, as in this example, chemical and physical modification is limited to the first few nanometers of the surface. More prolonged exposure to high-intensity plasmas produces extensive roughening of the surface and etching of material from the specimen.

A 25-μm-thick commercial polypropylene packaging film was exposed for 5 min to a radio-frequency-generated air plasma at 2.1 torr and 20-W radio-frequency discharge. IRIRS spectra of plasma-treated film were obtained using 60° incidence germanium IRS element (50 mm long) and a Fourier transform infrared (FTIR) spectrometer (Mattson Cygnus) operated at 4 cm^{-1} resolution. Such a spectrometer is typical of the mid price range of commercial FTIRs, and other than being maintained with care, was not customized in any exceptional way. Therefore, the data presented here could be reproduced in most well-equipped laboratories.

Figure 1 shows that if the IRIRS technique is used with care, information can be obtained that is unobtainable with other techniques. Obviously, the sampling depth of the IRIRS technique (d_p, the depth at which the evanescent field has decayed to $1/e$ times its value at the surface) is greater than a few nanometers (about 300 nm at mid-IR frequencies in this case), but the evanescent field decays exponentially away from the element surface and so the first few nanometers will be sampled with greater sensitivity than the remainder of the sampling depth. The number of reflections (and hence effective thickness, d_e, i.e., the thickness of specimen that would give the same absorbance when measured in transmission) was increased by the use of a thin (1-mm) IRS element. The greatest experimental difficulty was in ensuring cleanliness, both of the IRS element and the specimen. The use of a

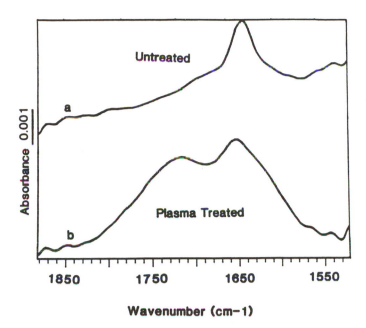

Figure 1. IRIRS spectra of polypropylene film before (a) and after (b) brief plasma treatment.

plasma cleaner was particularly effective for removing hydrocarbon contamination from the IRS element surface. The specimen was Soxhlet-extracted before exposure to the plasma, and was handled with the same degree of care as would normally be taken for X-ray photoelectron spectroscopy (XPS) analysis. Fluctuations in the water vapor content of the instrument were also problematical in the 1500–1800-cm^{-1} region, which necessitated the use of bottled nitrogen for the purge rather than dried laboratory air.

The spectra in Figure 1 were obtained by averaging 1000 scans and are unsmoothed. Optical contact between the film and the IRS element was improved by the use of a rubber backing film and moderate pressure applied by the four screws of the IRS specimen holder. The absorption at about 1650 cm^{-1} in the untreated film is probably from an antiblocking agent used in film manufacture. After brief plasma treatment, a new ketonic carbonyl absorption at 1720 cm^{-1}, with shoulders at 1630 and 1750 cm^{-1}, appears. Although Fourier deconvolution can be useful in the resolution of individual oxidized species (9), the relatively high noise level of this spectrum (note the carbonyl peak is only about 0.003 absorbance units), together with possible presence of several carbonyl species, makes deconvolution difficult. We are presently exploring the use of derivatization techniques (*see* next example) to resolve the various carbonyl species.

Quantification of IRIRS spectra can be approached in several ways. If perfect contact is assumed between the polyolefin film and the IRS element, then the "effective thickness" (d_e) can be calculated directly from the optical parameters (2, 3). However, because this assumption rarely holds true, we compared the intensity of the IRS spectrum with that of a transmission spectrum of a film of known thickness. Mandatory allowance for the wavelength dependence of d_e (3) led to a calculated d_e of 3.7 μm at ~ 1700 cm^{-1}. This and subsequent calculation were based on the well-established equations and methodology outlined by Harrick (3). The sampling depth d_p was calculated as ~ 300 nm at 1700 cm^{-1}. The ratio d_e/d_p is a measure of the enhanced sensitivity of the IRIRS technique compared with the transmission technique. Assuming a reasonable value for the extinction coefficient of carbonyl species (ca. 300 mol^{-1} cm^{-1}), the carbonyl content averaged over the sampling depth of 300 nm is therefore about 0.047 M, or about 1 carbonyl per 1400 carbons. It is theoretically possible to obtain information about the concentration profile in the surface by varying the angle of incidence, but such a sophisticated analysis could not be justified for spectra as weak as these.

One of the purposes of this chapter is to compare the IR data with those from XPS on the same specimen. The XPS data were obtained with the assistance of Professor Koberstein's group at the University of Connecticut, using a photoelectron spectrometer (Perkin Elmer PHI model 5300) equipped with a monochromatized Al K X-ray source (1486.6 eV). The use of a monochromator reduced specimen damage during long term exposure to the X-ray beam. Specimens were examined at ambient temperature and at pressures typically < 5 × 10^{-9} torr. Utility scans are presented for visual comparison of data, but atomic composition data were obtained from multiple high-resolution spectra, after comparison with appropriate standards. At present, the curve fitting is intended only as a semiquantitative aid to spectral comparison. The number of floating variables was reduced by retaining, within any series of modified specimens, a constant number of peaks, constant widths, and only slightly varying peak positions.

XPS analysis of the untreated polypropylene film showed a surface with 2–3-at% oxygen (i.e., 1 carbonyl per 30–50 carbon atoms). After brief exposure to the oxygen plasma, the oxygen content of the surface increased dramatically and a concentration gradient was established in the surface (Figure 2, carbon 1s data only). Curve fitting of the carbon 1s region indicated that for 25° takeoff (about 2.5-nm sampling depth) almost 40% of the carbon was bonded to oxygen, whereas at 62° takeoff (about 7-nm sampling depth) ~ 30% of the carbon was bonded to oxygen. The major difference between the two spectra is in the region of carbon singly bonded to oxygen (285.6 eV), as might be produced, for example, by hydroperoxide (R–O–O–H) oxidation products from polypropylene. However, it should be stressed that such curve fitting is semiquantitative, and the precise nature of

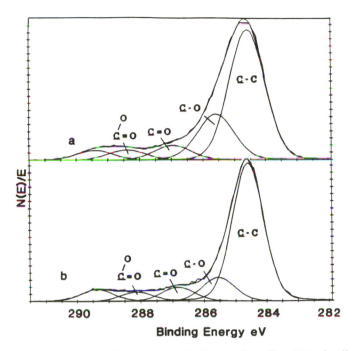

Figure 2. XPS spectra of plasma-exposed polypropylene film: 25° takeoff angle (a); 62° takeoff angle (b).

the functionality is unclear without derivatization. For example, unlike the IR case, ketones cannot be easily distinguished from aldehydes, esters, or acids. The difference in the oxygen contents averaged over 2.5 nm, and averaged over 7 nm indicates that the depth of the plasma treatment is only on the order of 5 nm, which is consistent with other analyses of plasma-treated polyolefins (6, 10).

It may be concluded from this example the IRIRS can produce spectroscopic information of a quality similar to XPS when studying surface modifications of only a few nanometers thick. The time of analysis is similar to XPS, and the limiting factor is maintaining specimen cleanliness rather than spectrometer performance. This is not, however, a routine analysis technique because considerable operator skill in specimen manipulation is required for analysis at such a high sensitivity. Therefore, IRIRS usually would be used only in this application if it had some specific advantage over XPS. In this example, it has been speculated that modification of the polyolefins by the plasma can extend further than the first few nanometers examined by XPS because of the short wavelength UV radiation associated with the plasma (11). This hypothesis will be particularly relevant in studies of adhesion to polyolefins, because a layer of cross-linked material generated just below the surface of the polyolefin may modify the failure mechanism (in either a positive or negative sense) and, hence, the fracture properties. Such an effect would be undetected by XPS because sampling depth is limited to the oxidized surface layer.

Surface Modification to a Depth of ~ 100 nm: Fluoropolymer Films Modified by Sodium Naphthalenide

It is notoriously difficult to adhesively bond fluoropolymers to other materials, which makes it difficult to exploit the outstanding dielectric properties and environmental stability of these materials in applications such as electronic circuit boards. The conventional solution to this problem is to defluorinate the polymer surface with electron donors such as sodium naphthalenide (12). The depth of defluorination of a copolymer of tetrafluoroethylene and perfluorinated alkyl vinyl ether (PFA; Du Pont) film may be determined gravimetrically because the defluorinated layer can be stripped by a chromic acid etch (13). In such a way we determined that after 1-h treatment, the depth of defluorination was about 110 nm. Because the depth of defluorination is less than the sampling depth of IRIRS, this case corresponds to the "thin film case" described by Harrick (3), and the relative concentrations of the absorbing species may be calculated directly from their relative absorbances and extinction coefficients, without correction for the frequency dependence of d_e. Figure 3 shows IRIRS spectra (germanium element, 60° incidence, 5000 scans) of the carbon–carbon double-bond region of the spectrum as a function of treatment time. Absorptions due to unsaturation (ca. 1600 cm^{-1}) and carbonyl species (ca. 1700 cm^{-1}) are readily apparent, and other regions of the spectra showed the characteristic absorptions of carbon–carbon triple bonds (ca. 2160 cm^{-1}) and hydrocarbon

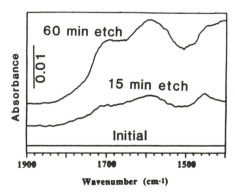

Figure 3. IRIRS spectra of PFA film as a function of sodium naphthalenide treatment time.

functionality (ca. 2900 cm^{-1}). After the treated specimens were exposed to the laboratory environment for several days, the carbonyl content of the surface increased a further 50% (*13*).

The breadth of the absorptions in Figure 3 necessitated the use of derivatization reactions to identify and quantify chemical species. As an example, derivatization of the surface with dinitrophenylhydrazine (DNPH) converts carbonyl species to hydrazones, which then can be quantified by IRIRS using aromatic absorption at 1617 cm^{-1} (Figure 4). Reaction of the surface with trifluoroacetic anhydride (TFAA) converts hydroxyl groups to trifluoroesters, which have a characteristic carbonyl absorption (Figure 5). Other derivatizations we used were reactions with bromine (for unsaturation) and sulfur tetrafluoride (for carboxylic acids). The combination of IR techniques and derivatizations allows comparative data to be obtained on the various functionalities. Therefore, absolute concentration data for all the

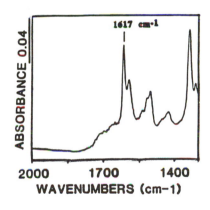

Figure 4. IRIRS spectrum of sodium-naphthalenide-treated PFA after derivatization with DNPH.

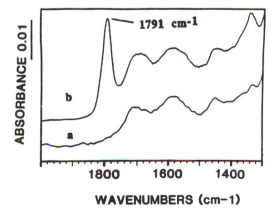

Figure 5. IRIRS spectra of sodium-naphthalenide-treated PFA before (a) and after (b) derivatization with TFAA.

preceding species can be obtained if one absolute concentration can be determined by an independent method. In the example, this absolute concentration was achieved by UV–visible spectroscopy of the hydrazone. The high extinction coefficient of hydrazone in the UV ($21,600$ mol^{-1} cm^{-1} at 351 nm) enables the hydrazone to be quantified by transmission UV–visible spectroscopy even though the hydrazone is present only in a thin surface layer. In such a way we calculated that the original carbonyl content was 3.3×10^{-9} equivalents per square centimeter of film surface [calculated assuming a perfectly flat film surface (13)]. IRIRS provides relative concentration data for ketones, hydroxyls, carbon–carbon double bonds, carbon–carbon triple bonds, carboxylic acid, and hydrocarbon functionality. These data may be visualized most easily if the functionalities are expressed as a proportion of the number of carbons in the 110-nm defluorinated layer as shown in Table I.

We, therefore, conclude that almost all the carbons in the defluorinated layer are functionalized in some fashion, with unsaturation being dominant. (Note that these data are corrected for "double counting", e.g., one triple bond per five carbon atoms implies that 40% of the carbons are triply bonded.) There were, however, several more surprising results. The hydroxyl content of the surface exceeded the carbonyl content, and we determined that this resulted largely from the washing step after etching. The hydrocarbon content greatly exceeded the content of oxygenated functionality, and we attribute this to reaction with the tetrahydrofuran (THF) solvent during the sodium naphthalenide treatment. This reaction was surprising because most previous studies were performed using XPS analysis, which is insensitive to hydrogen. The presence of hydrocarbon functionality and its exact chemical nature (e.g., CH_2 or CHF) will be crucial to the adhesion process because of its effect on the surface energy of the polymer (11, 12).

In this case, the nature of the information provided by XPS is very different from that provided by IR. Only a few nanometers at the surface are sampled by XPS for detailed surface chemical information; thus XPS alone gives an incomplete picture of the total chemical change. Figure 6 gives a survey scan of the PFA before and after sodium naphthalenide treatment.

Table I. Functional Group Content of Defluorinated Layer of PFA Averaged over the 110-nm Affected Depth

Functional Group	Functional Groups per Carbon
C=C	1/4
C≡C	1/5
C–H	1/30
–OH	1/300
C=O	1/400
COOH	< 1/2000

The first few nanometers are largely defluorinated and contain about 11-at% oxygen (ignoring hydrogen). This result is consistent with several earlier studies of similarly treated fluoropolymers (*14–16*). The use of a variable takeoff angle in XPS analysis allowed us to demonstrate that the oxygen was preferentially localized in the top few nanometers (*13*), but similar information is obtainable at a fixed takeoff angle by examination of the oxygen 2s and fluorine 2s peaks. The fluorine 2s electron has a lower binding energy than the fluorine 1s electron, and so can escape from a greater depth within the specimen. Composition data obtained from the fluorine 2s and oxygen 2s peaks (after calibration with appropriate standards) represent, therefore, an average over a greater depth than compositions obtained from the fluorine 1s and oxygen 1s peaks. For example, a poly(tetrafluoroethylene) (PTFE) surface after 1-h immersion in sodium naphthalenide contained about 15.7% oxygen and 5.7% fluorine as determined from the fluorine 1s and oxygen 1s peaks, with a precision of a few tenths of a percent. Lower oxygen (13.7%) and higher fluorine (7.7%) levels were determined when the surface was sampled at a greater depth using the fluorine 2s and oxygen 2s peaks. Note again that no information is obtainable on hydrocarbon functionality using XPS.

The XPS technique was also useful for derivatized specimens. For example, Figure 7 shows that carbon 1s region of the XPS spectrum of sodium-naphthalenide-treated PTFE before and after trifluoroacetic anhydride (TFAA) derivatization. The incorporation of CF_3 functionality (ca. 292

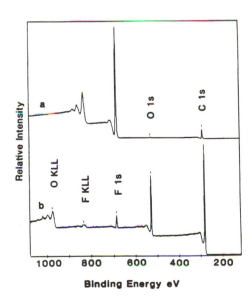

Figure 6. XPS survey scan of PTFE before (a) and after (b) 1-h sodium naphthalenide treatment.

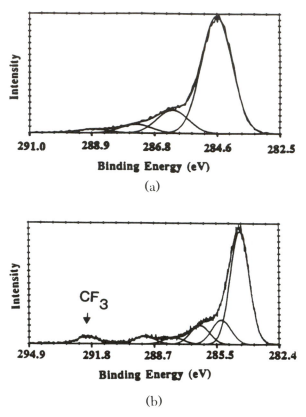

Figure 7. Carbon 1s region of XPS spectrum of PTFE after sodium naphthalenide treatment (a) and after sodium naphthalenide treatment and TFAA derivatization (b).

eV) through the trifluoroester may be used as a measure of the original hydroxyl content of the surface. Quantification of this peak indicates about 4.9% of the carbons are in this form, which implies that the original hydroxyl content was on about 1 carbon in 20. Similarly, the nitrogen $1s$ peak in the DNPH derivatized specimen could be used to determine the original carbonyl content of the surface (12). The bromine $3d_{5/2}$ peak could be used as a measure of the original unsaturation after bromination of the sodium-naphthalenide-treated surface, although there is some question whether fluorosubstituted vinyl unsaturation will react completely with bromine (13). These data provide qualitative support that the derivatization reactions were successful, and also allow comparison of the functionality of the first few nanometers of the surface with the average of the first 110 nm as determined by IRIRS.

The data in Table II summarize a series of analyses of sodium-naphthalenide-treated PFA and PTFE films. A comparison of Tables I and II allows us to conclude that the unsaturation was distributed relatively uniformly through the 110-nm affected depth, whereas the carbonyl and hydroxyl functionalities were concentrated in the top few nanometers. For example, the IRIRS data for PFA indicate an average unsaturation value of 1 functionality per 2.2 carbons, and the bromination results of XPS indicate a similar value of about 1 per 2.9 carbons. However, the XPS data indicate a hydroxyl content of 1 per 20 carbons (by TFAA derivatization) and a carbonyl content of 1 per 60 carbons (after DNPH derivatization), compared with 1 per 300 and 1 per 400 carbons, respectively, from IRIRS. The hydrocarbon content of the surface is not determinable by XPS and, therefore, cannot be compared.

One may conclude from this example, where the surface modification is about 100 nm, that IR and XPS provide very different pictures of the chemical state of the surface. Both pictures are correct, depending on how the surface is defined. In terms of the physical property that we were trying to understand (i.e., the adhesion of copper to the fluoropolymer), both pictures are informative. The XPS data reveal more about the short-range interaction forces between the copper and the modified fluoropolymer, but the IR data are essential if we are to understand the adhesion process, because failure occurs in this system within the modified layer of the fluoropolymer (*17*).

Surface Modification on a Micrometer Scale: Primers for Polyolefin–Cyanoacrylate Adhesion

For many technological applications, the formation of good adhesive bonds to polyolefins, such as polyethylene (PE) and polypropylene (PP) is desirable. However, polyolefins have a low surface energy and require additional processing steps, such as plasma or flame treatment, before even moderate adhesive bonds can be formed. Adhesion to a polycyanoacrylate would be particularly beneficial: Ethyl cyanoacrylate (Structure **1**), an example of an instant adhesive ("Krazy Glue"), has many domestic and industrial applica-

Table II. Average Composition Determined by XPS over ca. 5 nm
for Fluoropolymers

Group	PFA per Carbon	PTFE per Carbon
C=C, C≡C	1/2.9	1/3.7
−OH	1/20	1/18
−C=O	1/60	1/120

tions. We have examined the mechanism of primer action for the polyolefin surface, that is, materials that may be coated onto the polyolefin before preparation of the adhesive bond (17, 18). The primers we have examined include triphenylphosphine (TPP, 2) and cobalt acetoacetonate [Co(acac)$_2$, 3].

The method of primer application is either spraying or brushing a thin coating from dilute solution. Then the solvent is allowed to evaporate, followed by application of the adhesive. Typical data for PP/cyanoacrylate/PP adhesive bond strengths are shown in Figure 8. The term "coating thickness" is the value calculated when it is assumed that the applied primer remains on the surface of the PP as a uniform film. In fact, some mixing of polymer and primer occurs (as will be demonstrated) but the term remains a useful parameter for defining the specimen. Without primer, a PP/cyanoacrylate/PP lap shear joint (ASTM D 4501–85) has negligible strength. With increasing apparent primer thickness, the joint strength increases, but then decreases, presumably because of a change in the mode of failure when the primer layer becomes too thick (20).

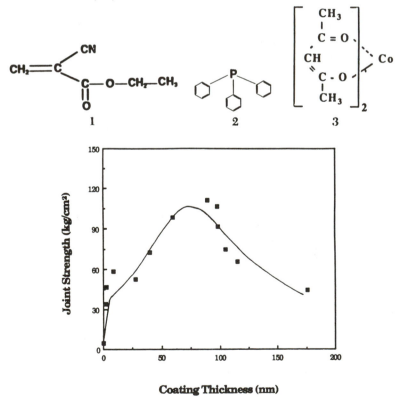

Figure 8. Adhesion strength of PP/cyanoacrylate/PP lap shear joint as a function of apparent Co(acac)$_2$ primer layer thickness.

One characteristic of these primers is that their effectiveness diminishes over a period of hours or days if the primed surfaces are allowed to age before application of the adhesive (*20*). For example, when TPP was applied to a PP surface and the primed surface was aged in the normal laboratory environment before application of the adhesive, the lap shear joint strength dropped 50% after 4 h and 80% after 24 h. Therefore, we used the IRIRS technique to monitor diffusion of the primer away from the surface on a dimension scale of fractions of a micrometer. The primer was applied to low-density polyethylene (PE) plaques in the conventional manner, and the primed surface was then pressed against a 60° incidence germanium IRS element. The change in intensity of the characteristic absorptions of the primer was then followed as a function of time.

Figure 9 shows IRIRS spectra of the PE surface before and immediately after priming with 0.03-wt% $Co(acac)_2$ solution in toluene. The integrated absorbance of the $1510\text{-}cm^{-1}$ peak was calculated using 1540 and $1489\ cm^{-1}$ as a pseudobaseline and was taken as a measure of the $Co(acac)_2$ content of the ca. 400-nm sampling depth. The IRIRS specimen remained assembled and was stored under ambient conditions for the 5 days of the test while the primer absorbance was followed with time. Diffusion of the primer away from the surface occurs over a time scale of days. The sampling depth of the IRS element (60° incidence germanium) defines the distance over which the primer must migrate to "disappear" from the IRS field of view.

Although it is theoretically possible to calculate a diffusion coefficient from such data, we felt that the boundary conditions were insufficiently well

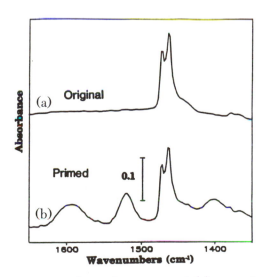

Figure 9. IRIRS spectra of PE plaques: control (a); primed with $Co(acac)_2$ applied from 0.03% toluene solution (b).

defined to justify such an analysis. Instead, we used the integrated absorbance at 1540–1489 cm^{-1} as an index of primer surface concentration as a function of aging time (Figure 10). We then showed that the primer surface concentration correlates with the lap shear strength obtained from PE/cyanoacrylate/PE joints after the same period of aging of the primed surface before fabrication of the joint. For example, after one day at 25 °C, the index of primer surface concentration had fallen about 15% whereas lap shear joints made with one-day-old Co(acac)$_2$ primed PE were about 20% weaker. Joint strength data for longer aging times and higher aging temperatures followed the trend in primer surface concentration (20). The diffusion of the Co(acac)$_2$ primer away from the PE surface and into the bulk of the polymer, therefore, can be assumed to be the factor responsible for deterioration of adhesive bonding properties of the primed surface. Separate experiments determined that the rate of evaporation of the primer under these aging conditions was insufficient to explain the surface deterioration and that the primer was chemically stable to oxidation or photolysis.

With the TPP primer, the rate of diffusion of primer away from the surface was faster than with Co(acac)$_2$, as was the rate of deterioration in adhesive bond strength. Figure 11 shows typical IRIRS spectra and integrated absorbance data for the 1430-cm^{-1} primer peak (1446–1424-cm^{-1} pseudobaseline). After one day under ambient conditions, the primer surface concentration had fallen by one-half, and joints made with one-day-old TPP-primed surfaces showed about 30% of their original strength. Lap shear joints made with TPP-primed surfaces that had been aged at 80 °C for 24 h showed negligible strength.

When diffusion gradients occur over tens of micrometers, it is possible to use IR microscopy to measure diffusion coefficients. Such a case occurs when the TPP primer is applied from the melt at 90 °C and is allowed to diffuse into plaques of polypropylene (PP). After 24 h at 90 °C the plaques were sectioned with a microtome and the spatial distribution of the TPP was determined using a microscope aperture of about 20 μm (Figure 12).

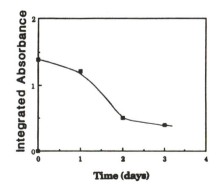

Figure 10. Change in Co(acac)$_2$ surface concentration by IRIRS (integrated absorbance at 1540–1489 cm^{-1}) as a function of time at 25 °C.

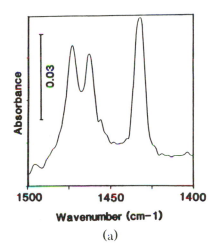

(a)

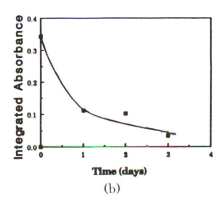

Time (days)

(b)

Figure 11. IRIRS spectrum (a) and change in primer surface concentration (b) (integrated absorbance 1446–1424 cm^{-1}) for TPP-primed PE plaques. (J. Yang, University of Connecticut, unpublished data.)

Although this was a useful way to determine the diffusion coefficient for the primer in the polymer (18), it is an unrealistic application procedure for the primer.

In response to these and other observations, we have developed a model that involves at least four necessary criteria for an effective primer for this system (18, 20): The primer must (1) mix with the polyolefin so as not to form a weak boundary layer, (2) increase the polyolefin surface energy, (3) not diffuse too rapidly into the bulk of the polyolefin, and (4) catalyze the cure of the cyanoacrylate.

Work in this area is continuing, with particular emphasis on the nature of the interaction responsible for improved adhesive bond strength, which includes the possible formation of interpenetrating polymer networks (IPN). A comparison of IR and XPS data in this example is not relevant because both TPP and $Co(acac)_2$ have significant vapor pressures, which make quantitative analysis by XPS impossible.

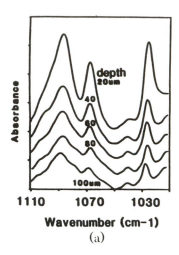

(a)

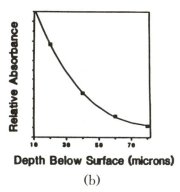

Depth Below Surface (microns)

(b)

Figure 12. IR microscopic data obtained at various depths inside a PP plaque contacted with molten TPP primer at 90 °C: mid-IR spectra (PP subtracted) (a); integrated absorbance 1100–1080 cm⁻¹ (b).

Summary

The IRIRS technique is capable of nanometer level sensitivity in surface analysis, and, in combination with derivatization techniques, can give chemical and structural information unobtainable with other techniques. Specimen cleanliness is a greater limitation than spectrometer performance. The IRIRS technique is particularly valuable for the surface analysis of thick films (where the more surface-sensitive techniques such as IRRAS and SERS are inappropriate) and when specimen limitations such as volatility preclude the use of high-vacuum techniques. The combination of IRIRS and XPS allows the determination of concentration profiles within a polymer surface.

Acknowledgments

Portions of the work described here were supported by the Donors of the Petroleum Research Fund, Connecticut Department of Higher Education, Rogers Corporation, and Loctite Corporation.

References

1. *Vibrational Spectroscopy of Molecules at Surfaces*; Yates, J. T.; Madey, T. E., Eds.; Plenum Press: New York, 1987.
2. Garton, A. *Infrared Spectroscopy of Multicomponent Polymer Materials*; Hanser Publishers: New York, 1992.
3. Harrick, N. J. *Internal Reflection Spectroscopy*; Harrick Scientific Corp.: New York, 1967.
4. Mirabella, F. M.; Harrick, N. J. *Internal Reflection Spectroscopy: Review and Supplement*; Harrick Scientific Corp.: New York, 1985.
5. Schmidt, J. J.; Gardella, J. A.; Salvati, L. *Macromolecules* **1989**, 22, 4489.
6. Boenig, H. V. *Plasma Science and Technology*; Cornell University Press: Ithaca, NY, 1982.
7. Mittal, K. L. *J. Vac. Sci. Technol.* **1976**, 13, 19.
8. Gerenser, L. J. *J. Vac. Sci. Technol.*, A **1990**, 8, 3682.
9. Maddams, W. F.; Parker, S. F. *J. Polym. Sci., Polym. Phys. Ed.* **1989**, 27, 1691.
10. Clark, D. T.; Hutton, D. R. *J. Polym. Sci., Polym. Phys. Ed.* **1987**, 25, 2643.
11. Gerenser, L. J. *J. Adhes. Sci. Technol.* **1987**, 1, 303.
12. Wu, S. *Polymer Interfaces and Adhesion*; Marcel Dekker Inc.: New York, 1982.
13. Ha, K.; McClain, S.; Suib, S. L.; Garton, A. *J. Adhes.* **1991**, 33, 169.
14. Brecht, V. H.; Mayer, F.; Binde, H. *Angew. Makromol. Chemie* **1973**, 33, 89.
15. Riggs, W. H.; Dwight, D. W. *J. Colloid Interface Sci.* **1974**, 47, 650.
16. Bening, R. C.; McCarthy, T. C. *Macromolecules* **1990**, 23, 2648.
17. Marchesi, J.; Ha, K.; Garton, A. *Proceedings of the 14th Annual Conference of the Adhesion Society*, Clearwater, FL, February 20, 1991.
18. Yang, J.; Garton, A. *Proc. ACS Div. Polym. Mater.: Sci. Eng.* **1990**, 62, 916–919.
19. Nakada, C.; Nagazawa, K., Japanese Patent Office, Public Patent Disclosure Bulletin 61–136567, June 24, 1986.
20. Yang, J.; Garton, A., *J. Appl. Polym. Sci*, in press.

RECEIVED for review July 15, 1991. ACCEPTED revised manuscript May 19, 1992.

Fundamentals and Applications of Diffuse Reflectance Infrared Fourier Transform (DRIFT) Spectroscopy

Mark B. Mitchell

Department of Chemistry, Clark Atlanta University, Atlanta, GA 30314

Diffuse reflectance spectroscopy is a powerful tool for the study of materials such as polymers and catalysts. The bulk or surface morphology of these materials is many times an important experimental parameter that can be altered by sample preparation methods used in the more common spectroscopic techniques. Diffuse reflectance infrared Fourier transform (DRIFT) spectroscopy has been shown to be more sensitive to surface species than transmission measurements and to be an excellent in situ technique. The applications of DRIFT spectroscopy to the investigation of polymer surfaces and surface structures of both fibers and films are particularly germane to this symposium and are covered in this review. The foundations of the diffuse reflectance technique in the mid-infrared range are presented, and methods that have been used to obtain spectra for polymer powders, films, and fibers are outlined, as is the use of DRIFT spectroscopy for depth-profiling studies. The use of DRIFT spectroscopy for the investigation of catalytic processes is also reviewed.

DIFFUSE REFLECTANCE INFRARED FOURIER TRANSFORM (DRIFT) spectroscopy is an infrared sampling method that involves minimal sample preparation in terms of time and sample manipulation. DRIFT spectroscopy is particularly useful for powder samples, but can also be used to investigate samples such as polymer fibers and films. The manipulations involved in preparation of a sample for transmission measurements, as an example, involve mixing the sample with an appropriate diluent and compressing the

0065-2393/93/0236-0351$07.25/0

sample into a transparent pellet. This preparation can easily result in irreversible changes in sample morphology and surface properties. Thus, minimizing sample manipulation becomes an important consideration when sample morphology and surface properties are investigated, or are important sample variables. In addition, because the sample is not compressed for DRIFT spectroscopy, good gas contact with the entire sample is possible, a distance advantage for in situ measurements. Gas flow through the sample is a capability that makes DRIFT spectroscopy useful as one segment of a hyphenated analytical system or as a technique for the study of gas–surface interactions.

Diffuse reflectance has been used extensively in the visible and ultraviolet regions of the electromagnetic spectrum for the characterization of solid materials such as inorganic powders. Many of the effects and constraints that govern diffuse reflectance spectroscopy were determined as a result of experiments with ultraviolet and visible radiation. In those investigations, the Kubelka–Munk expression was particularly useful for relating the observed reflectance spectrum to the optical constants of the material under study. References by Wendlandt and Hecht (1) and by Kortum (2) provide an excellent background and include reviews of these earlier investigations.

In 1976, Willey (3) described the optical design of an instrument developed to extend the technique of diffuse reflectance into the infrared. His Fourier transform IR (FTIR) spectrometer was specifically designed to carry out infrared diffuse reflectance measurements, although it could also be used for transmission measurements.

Since the work of Fuller and Griffiths in 1978 and 1980 (4, 5), the DRIFT spectroscopy technique has been used for the analysis of a wide variety of sample types, including polymer powders, films, and fibers; heterogeneous catalysts; and high-temperature superconductors, as well as "normal" inorganic and organic powder materials. Fuller and Griffiths laid much of the groundwork necessary for the use of the technique as an analytical tool. This review outlines the foundation laid by Fuller and Griffiths and by other investigators and presents results that demonstrate the utility of the DRIFT spectroscopy technique for a variety of sample types.

A variety of infrared spectroscopic methods are available for use to investigate polymers, and many of the applications of these techniques to polymers have been discussed (6). Transmission spectroscopy is one of the most common methods, but, as mentioned earlier, the pressure needed to form pellets of many polymers can affect surface functionality. These effects were documented by Blitz et al. (7) in their investigation of trimethoxymethylsilane on a silica surface. The authors showed that the pressures involved in making a pellet induced chemical changes in the surface functional groups. Also, many samples of interest simply are not optically thin enough to be studied by transmission. Fiber-reinforced composites are an example of such a case.

Attenuated total reflectance (ATR) is one of the best ways to obtain infrared spectra of polymer films. One of the requirements of ATR is that the sample make optical contact with an infrared crystal. Some samples are not soft enough to form good contact and are difficult to study with this method. Fibers are very difficult to measure well using ATR.

Childers and Palmer (8) and Urban et al. (9) compared photoacoustic spectroscopy (PAS) and DRIFT spectroscopy. The general conclusion is that the methods are comparable: DRIFT spectroscopy usually gives better signal to noise (S/N) ratios for most samples, but PAS is better suited for strongly absorbing or highly reflective samples. Another general conclusion is that DRIFT spectroscopy is very sensitive to sample morphology (with powder particle size being a critical parameter), whereas PAS is relatively insensitive to sample morphology and is perhaps better suited to the study of carbon-filled rubbers, for example, because it is difficult to control particle size for a DRIFT sample. PAS can be used for depth-profiling (9, 10), which is not as easily accomplished using DRIFT spectroscopy, although some concentration versus depth information can be generated using DRIFT with KBr overlayers as is discussed later.

Theory

There are two general types of reflected light: specular and diffuse. A common example of *specular* reflection (also called front-surface, regular, or Fresnel reflection) is light reflected from mirrors or other polished surfaces. Specular reflectance occurs at any interface between two materials with different refractive indices. Specularly reflected light is characterized by the rule that the angle of reflection is equal to the angle of incidence. Common examples of diffusely reflecting surfaces are the matte surfaces characteristic of certain types of paper and powders. Diffusely reflected radiation is the light reflected from a diffusely reflecting sample for which the angle of reflection does not equal the angle of incidence.

Optical Constants. The intensive property that characterizes the reflectance and transmittance of a material is the complex refractive index (n'), which is the sum of the index of refraction (n) (called the real component) and the absorption index (k') (called the imaginary component):

$$n' = n + ik' \tag{1}$$

For nonabsorbing materials, $k' = 0$. Specularly reflected light is well described by Fresnel's equations (*see* reference 1), which express the reflection of electromagnetic radiation at an interface as a function of the refractive indices of the two media that form the interface. The difference in the refractive indices leads to reflection. The differences in refractive index that lead to reflection can be due to differences in the real or the imaginary components.

In the Beer–Lambert equation, the loss of transmitted intensity, I, due to absorption processes is given by

$$I = I_0 e^{-kx} \tag{2}$$

where k is the absorption coefficient ($\neq k'$) and x is the path length through the sample. The absorption index is related to the absorption coefficient by

$$k' = k/4\pi\nu \tag{3}$$

where ν is the frequency of the radiation (per centimeter) and the absorption coefficient is related to the absorptivity, a, and the concentration, c, of an absorber by

$$k = 2.303ac \tag{4}$$

Kubelka–Munk Expression.   In any spectroscopic technique it is important to be able to characterize the response of the technique to the concentration of the analyte. A number of models have been developed that yield expressions relating observed diffuse reflectance to concentration (1, 2, 9). The number of independent experimental variables that must be evaluated for each expression is related to the complexity of the model. The most commonly used expression is the two-parameter Kubelka–Munk (KM) expression (1, 2, 11–13):

$$f(R_\infty) = (1 - R_\infty)^2/2R_\infty = k/s \tag{5}$$

where R_∞ is the absolute reflectance of an "infinitely thick" sample, $f(R_\infty)$ represents the value of the Kubelka–Munk function, k is the absorption coefficient defined previously, and s is a scattering coefficient defined for purely scattering samples by

$$I = I_0 e^{-sx} \tag{6}$$

It is generally assumed that s is a constant or at most a slowly varying function of frequency and is not a function of analyte concentration. An infinitely thick sample is one through which no light is transmitted. All light is either reflected or absorbed. The infinitely thick criterion is generally satisfied for samples that are 3–5 mm thick. This thickness does not satisfy the infinitely thick criterion for all samples or for many samples over the whole mid-infrared range; these considerations are discussed later. Nonetheless, DRIFT spectra of most compounds can be at least semiquantitative with samples of this thickness. As an aside, Fraser and Griffiths (14) argue that because the effective path length increases by a factor of 2 for diffuse reflectance, eqs 2 and 6 should be modified such that the exponents are $-2kx$ and $-2sx$, respectively. In principle, this change should have no effect on the remission function (eq 5) calculated for the infinitely thick sample, because the factor of 2 will cancel. However, this possible discrepancy should be kept in mind when comparing k data from different groups.

Typically, the absolute reflectance of eq 5 is not evaluated. Instead, the reflectance relative to some nonabsorbing standard is measured. R'_∞ is substituted for R_∞ in eq 5, where

$$R'_\infty = R_\infty(\text{sample})/R_\infty\,(\text{standard}) \tag{7}$$

R_∞ (sample) is the reflectance of the sample and R_∞ (standard) is the reflectance of a nonabsorbing standard, such as KBr or KCl. For samples that obey the Kubelka–Munk expression, plots of $f(R'_\infty)$ versus concentration (c) yield a straight line with a slope of $2.303a/s$.

In their original paper, Fuller and Griffiths (4) presented results that showed the dependence of the DRIFT spectra on particle size and concentration of the absorber. Plots of the KM intensity of a particular band showed deviations from linearity at high absorber concentrations. The KM equation is derived with the assumption that the sample is optically dilute: the concentration of the absorber is low or the absorption coefficient is small. This condition is generally met by diluting the sample of interest in a nonabsorbing matrix such as KBr or KCl, although for weak infrared absorptions, this is not necessary. For samples that satisfy the optically dilute criterion, small changes in concentration result in proportional changes in the observed KM intensities.

Fresnel Reflection.   To aid a discussion of several of the causes and possible solutions of nonlinearity associated with DRIFT spectroscopy, it is useful to consider the sample material as a composition of small grains of diluent and analyte (*see* Figure 1). When radiation interacts with the sample, three different events, which yield three classes of reflected radiation, can occur. The first class results from the simple reflection of incident radiation at a sample grain interface that is parallel to the macroscopic surface of the sample; this is called specular Fresnel reflectance, after Brimmer and Griffiths (15). This reflection from the top surface of the powder is not isotropically scattered; the angle of reflection is equal to the angle of incidence. The reflection from the top surface also retains the polarization of the incident beam.

The second class of reflected radiation is similar to the first in that it is not transmitted through any sample grains. Many of the grain interfaces that the radiation encounters will not be parallel to the surface of the macroscopic sample. Incident radiation that encounters such an interface will be reflected

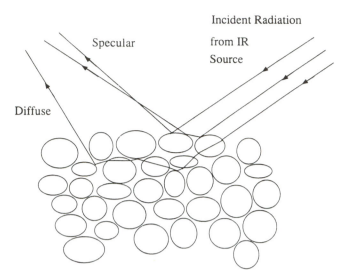

Figure 1. Schematic diagram of the DRIFT interaction with three classes of interaction shown. The top ray is used to indicate the Fresnel specular component of the reflected radiation, which is reflected from the first interface it encounters at the top of the sample. The center ray indicates the Fresnel diffuse component. After two reflective interactions and no transmission through any sample grain, the sample is reflected out of the sample. The lowest ray corresponds to the Kubelka–Munk diffuse reflectance, which is radiation that has been reflected from some grains and transmitted through others.

at an angle that is equal to the angle of incidence at the grain interface, yet not equal to the angle of incidence to the macroscopic sample. This radiation will carry no information about the analyte, but it is diffusely reflected in the sense that it is isotropically scattered and its polarization is scrambled relative to that of the incident radiation. This radiation is called diffuse Fresnel reflectance. Multiple reflection within the sample will also generate radiation emerging from the sample that has not been transmitted through any sample grain, and this radiation falls into the class of diffuse Fresnel reflectance.

The third and final class of radiation reflected from the sample is called the diffuse Kubelka–Munk reflectance. This radiation has been transmitted through at least one sample grain, contains all the information regarding the analyte that is available in the reflected radiation, is isotropically scattered, and its polarization is scrambled relative to the incident field. The intensity of this radiation obeys the Kubelka–Munk relation for the intensity of analyte absorption bands relative to concentration of the absorber, under the limitations of that relation.

The presence of Fresnel reflectance in the radiation collected by the detector leads to much of the nonlinearity associated with DRIFT spectroscopy. Specular Fresnel reflectance is relatively easy to eliminate from the collected radiation. One scheme is to collect radiation in directions that exclude the plane of incidence of the incident radiation (*16, 17*). A second scheme is to polarize the incident radiation and collect radiation through an analyzing polarizer oriented at 90° to the polarization of the incident radiation (*17, 18*). These two optical configurations are illustrated in Figure 2. Neither of the two methods will eliminate the diffuse Fresnel reflectance. One method that apparently decreases diffuse Fresnel reflectance is to grind the sample and diluent to very small particle sizes. This method decreases the fraction of incident radiation that is reflected out of the sample without being transmitted through a sample grain.

Fresnel reflectance does not obey the Kubelka–Munk relation whether specularly or diffusely reflected, and it is important to know when an observed spectrum can be characterized using the Kubelka–Munk relation. Intensity variations in the Fresnel reflectance components of the collected radiation are directly dependent on variations in both the index of refraction (n) and the absorption index (k') of the sample. In the neighborhood of an absorption band, the index of refraction fluctuates rapidly as a function of frequency and takes on the first derivative shape characteristic of a damped oscillator. This fluctuation is known as anomalous dispersion. For moderate to weak absorption bands, the Fresnel reflectance is dominated by the fluctuations of the index of refraction, and spectra that contain a large Fresnel reflectance component will show characteristic derivative-shaped bands. For absorption bands with very high absorption indices, the Fresnel reflectance is dominated by the absorption index, and the sample is actually more reflective in the neighborhood of these absorption bands than in spectral regions in which the sample is transparent. These "inverted" reflectance bands are called *reststrahlen* bands. Anomalous dispersion and *reststrahlen* bands and their effects have been discussed by several groups (*16, 18–20*).

With respect to the absorption index scale, strong infrared chromophores such as the 674-cm^{-1} C–H out-of-plane absorption of benzene have absorption indices on the order of 1.5 (*21*). The weakly absorbing C–H stretch vibrations of benzene have absorption indices of 0.01–0.05. For inorganic materials, the absorption indices associated with metal–oxygen and metal–halide vibrations can be significantly greater than 3.0. Such absorptions almost always lead to distorted peak shapes when measured using DRIFT spectroscopy.

Optical Accessories. Several companies manufacture optical accessories that can be used to obtain DRIFT spectra. The majority of these accessories correspond to one of the configurations shown in Figure 2A and B. However, the on-line configuration is used most commonly without the polarizers. Under such conditions, the on- and off-line configurations perform much differently with regard to their ability to reject Fresnel reflectance.

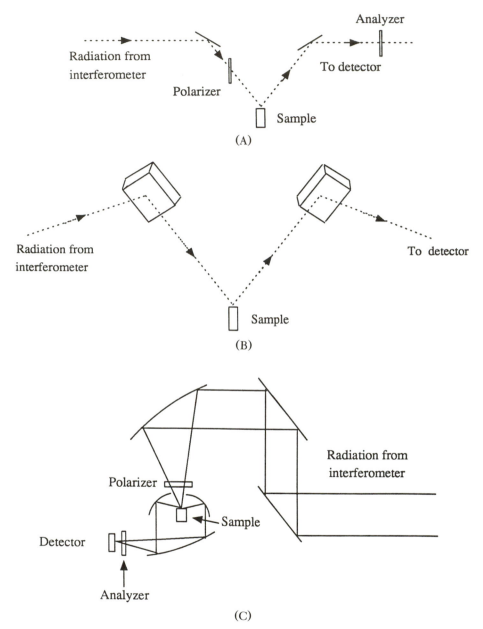

Figure 2. Schematic diagram of three commonly used optical configurations that eliminate to a great degree the contribution of Fresnel specular reflectance to the collected radiation. The top diagram (A) represents the in-line configuration in which polarizers are used to eliminate the specular component. The middle diagram (B) shows the off-line configuration in which exclusion of the plane of incidence from the collection optics is used to eliminate the specularly reflected component. The bottom diagram (C) is a configuration built by Fuller and Griffiths (3). (Reproduced with permission from reference 17. Copyright 1948.)

Comparisons of these two configurations have been carried out by at least three groups (*16–19*), and it was observed that the optical off-line configuration leads to better rejection of the Fresnel components of the reflected radiation than the on-line configuration when the polarizers are not used for the optical in-line configuration. Hembree and Smyrl (*19*) observed that even for relatively weak calcium carbonate absorptions, the on-line configuration gave nonlinear results as a function of concentration even at low concentrations. The use of a device to physically block specularly reflected radiation in the on-line configuration significantly reduced the throughput of the configuration.

Sample Preparation.   One of the limitations of diffuse reflectance spectroscopy with regard to its use as a truly quantitative technique involves the reproducibility of the technique. Because the sample materials are typically powders, the physical characteristics of the diluent and the analyte play an important role in the reproducibility of the results.

In their first paper, Fuller and Griffiths (*4*) investigated the effects of particle size on the observed spectra. The spectra were clearly shown to be dependent on particle size, and samples with smaller average particle sizes (average particle size < 10 μm) gave better spectra with reduced peak widths compared to samples with large average particle sizes (average particle size > 90 μm). An average particle size of 10 μm is a realistic goal for quantitative measurements of powders. Average particle sizes of 50–100 μm have been shown to yield useful qualitative results (*22*). Fuller and Griffiths also illustrated the requirement for optically dilute samples; they could not effectively carry out spectral subtraction over large frequency ranges unless the sample was diluted in a nonabsorbing matrix such as KCl. Later studies by Griffiths' group and by other groups showed that it is not always a requirement that the sample be diluted; polymer films and fibers are an important example of systems that can be examined without dilution, in many cases with excellent results (*see* later discussion).

With regard to the use of pressure as a way to overcome the nonreproducibility in the observed spectra associated directly with the sample preparation step, Yeboah et al. (*23*) found that the KM intensities of two of the infrared absorption bands of caffeine increased by almost an order of magnitude between a sample prepared with no pressure treatment and one subjected to a pressure of 12,000 psi for 5 min. This increase in $f(R_\infty)$ was presumed to be completely due to a corresponding decrease in the scattering coefficient, an effect that (as they point out) is commonly observed in pressing pellets for infrared analysis. This decrease occurs because the scattering at a KCl–analyte interface is less than the scattering at a combination KCl–air–analyte interface because the refractive index for KCl is similar to that for many organic analytes. For diffuse reflectance measurements though, once the sample becomes transparent, the Kubelka–Munk formulation is no longer valid because diffuse reflectance is no longer the dominant process. In general, pressure increases the spectral contrast of the observed spectra and increases the reproducibility between measurements; the standard deviation approaches $\pm 3\%$ for samples prepared with the same pressure applied for the same length of time. Hembree and Smyrl (19) also investigated the use of pressure as a way of improving reproducibility and found that by careful sample preparation without pressure, the standard deviation approached $\pm 3\%$. The use of pressure was found to decrease the magnitude of the diffusely reflected radiation and increase the specular component.

The effect of using infrared-absorbing materials as diluents was investigated by Brimmer and Griffiths (24). An important example of such a situation is a surface layer on a polymer fiber or film. Although the polymer fiber or film substrate is not technically a diluent in such a sample, many of the considerations that Brimmer and Griffiths discuss are applicable. In situ catalytic studies also provide an example of a case when the use of diluents is not desirable. As a diluent, KCl is a source of many potential uncertainties in such an investigation: Potassium is a well known promoter in many catalytic systems and chloride ions can act as a catalyst poison, so the use of these materials for in situ investigations can clearly distort results. In such cases, it is many times preferable to dilute the sample with a chemically inert support material such as alumina or silica, both of which, however, have strong absorptions in the infrared. Brimmer and Griffiths reported that the effect of an absorbing matrix is very complex. In general, the authors observed an apparent decrease in the absorptivity of analyte bands that occurred in regions of infrared absorption by the diluent. The authors postulated two possible causes. The increased average absorption of the sample can result in a decrease in the effective penetration depth of the radiation into the powder. Also, because the absorption index (k') for the diluent is nonzero in frequency regions where it absorbs, more specular reflectance from the front surface of the sample powder may be observed in those regions than in the case of a nonabsorbing diluent. The combined result is that the use of infrared-absorbing diluents may significantly distort analyte absorption bands, especially the intensity of one analyte band relative to another, compared to absorption bands observed in a transmission spectrum.

Recently, Frasier and Griffiths (14) investigated the effect of changes in the scattering coefficient s (eq 6) on the observed DRIFT spectra; specifically, the effect of the scattering coefficient on the infinite-depth criterion of the Kubelka–Munk expression. The scattering coefficient of KCl powder that had been pressed slightly, for samples of thicknesses greater than about 500 μm, was found equal to 25 cm^{-1}, and the scattering coefficient varied as a function of frequency (although by no more than about a factor of 2 over the mid-infrared range) and as a function of sample thickness. The calculations of Frasier and Griffiths indicate that the infinite depth approximation will not be valid for nonabsorbing samples until the sample thickness approached 1 cm.

However, if a sample does have infrared absorption bands, k' will decrease the effective depth of penetration and, over the frequency region of the absorption, the infinite depth approximation may be valid for samples much less than 1 cm thick.

Mandelis et al. (25) derived a treatment, based on Melamed's statistical approach (26), for calculating the diffuse reflectance and diffuse transmittance of powders with a particle size that is very large with respect to the wavelength of the incident radiation. Good agreement was obtained for samples made up of particles with $\langle d \rangle \sim 7\lambda$, where $\langle d \rangle$ is the average particle diameter and λ is the wavelength of the radiation; however, relatively poor agreement was found for smaller particles, where the KM expression becomes useful.

Polymer Studies Using DRIFT Spectroscopy

There are two different types of polymer investigations that have been carried out with DRIFT spectroscopy. The first type results from the study of a polymer powder. In this sort of investigation, the considerations regarding the sample itself are no different than those that would be involved for a typical organic analyte: particle size, sample homogeneity, and sample packing. Lee et al. (27) carried out just such a study on complexes of poly(vinyl phenol) (PVPh) and poly(2-vinyl pyridine) (P2VP). The complex precipitated from solution as a powder and gave very poor KBr pellet transmission spectra due to the large amount of scattering caused by the analyte powder. Using DRIFT spectroscopy, these investigators were able to study the interactions between the two species that comprise the complex; the spectra indicate the formation of a 1:1 complex. Figure 3 shows the 2200–3800-cm^{-1} region of the DRIFT spectrum of a complex of PVPh–P2VP, pure PVPh, and pure P2VP. The two broad, intense absorptions at 3525 and 3330 cm^{-1} for pure PVPh are associated with free and hydrogen-bonded O–H groups. These bands are not present in the complex; rather, a new broad feature centered at about 3000 cm^{-1} is apparent. This new absorption is assigned to a strongly perturbed O–H vibration; the perturbation is caused by interaction with the nitrogen atom of the P2VP.

Siesler (28) used silica powder to mimic glass fibers and treated the powder with γ-propyltrimethyloxysilane, a coupling agent used to increase reinforcement effectiveness in fiber reinforced plastics. This investigator was able to monitor the decrease in free OH (3740 cm^{-1}) on the glass surface and the growth of CH$_2$ (2900 cm^{-1}) and C=O (1700–1725 cm^{-1}) due to the coupling agent, as a function of the amount of coating.

Savolahti (29) used DRIFT to study the thermal degradation of barley protein. The sample was lyophilized or spray-dried. Samples were mixed with KBr and heated in a controlled environment cell (from Spectra Tech). Loss

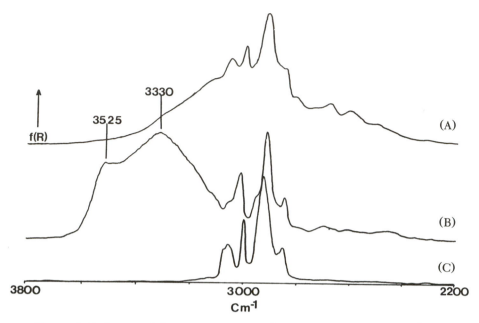

Figure 3. Scale-expanded DRIFT spectra in the range 3800 to 2200 cm⁻¹ of solid solutions in KCl (1:50 weight ratio): PVPh–P2VP polymer complex formed with 4:1 initial composition (A); pure PVPh (B); and pure P2VP (C). (Reproduced with permission from reference 27. Copyright 1984.)

of the olefinic C–H stretch absorption was assigned to oxidation of the double bond. The growth of a strong C=O absorption band was observed as was the loss of intensity associated with O–H, N–H, C–H, and C–O groups. The amide I and II bands decreased at 1650 and 1540 cm⁻¹, respectively.

Polymer Fibers. The second type of polymer investigation using DRIFT spectroscopy is carried out where grinding polymer fibers or films would change the chemistry and morphology of the system too drastically to allow spectroscopic correlation of the results with the unground sample.

In 1980, Maulhardt and Kunath (*30*) used DRIFT spectroscopy with their own optical device and a commercial FTIR to study a silane coupling agent on glass fibers. Contributions from the C–H, C=O, and C=C vibrations of the coupling agent were clearly seen, but useful information for frequencies less than about 1500 cm⁻¹ was not obtained because of the strong infrared absorptions of the silica substrate.

Investigation of coupling agents on glass fibers was of interest to Koenig and co-workers. Although much of their early work was conducted using transmission spectroscopy (*31*), Koenig and co-workers carried out a number of investigations using DRIFT spectroscopy to study polymer fibers and films.

Culler et al. (6) characterized several of the infrared methods that may be used to obtain infrared spectra of polymers, including DRIFT spectroscopy, and pointed out several of the advantages and disadvantages of DRIFT spectroscopy. These advantages include good sensitivity and little sample preparation; the disadvantages included the need for a consistent particle size for quantitative results, limited depth penetration, and that strongly absorbing modes in a material limit the sensitivity of the technique in spectral regions that those modes overlap. The authors reported initial spectra of a coupling agent on silica fibers that closely resembled those of Maulhardt and Kunath (30).

In 1984, the technique of using KBr overlayers to obtain DRIFT spectra of polymer fibers was introduced by McKenzie et al. (32), who investigated the coupling agent γ-aminopropyltriethoxysilane (γ-APS) on E-glass fibers. E-glass fiber mats were cut into circles to fit the DRIFT spectroscopy cell. Had the samples obeyed the Kubelka–Munk relation, it would have been possible to obtain the spectrum of the coupling agent on the fibers by subtracting a spectrum of the fibers alone from that of the fibers treated with γ-APS. Unfortunately, this method did not work well. Subtraction of the spectrum due to the E-glass substrate from the spectrum of E-glass treated with γ-APS yielded a spectrum in which some of the substrate bands had not been subtracted completely and some had been oversubtracted. This result was an obvious indication of non-Kubelka–Munk behavior, presumably due to anomalous dispersion. An additional problem that McKenzie et al. encountered was that rotation of the sample generated profound changes in the observed spectrum. As in their previous study (31) and the study by Maulhardt and Kunath (30), the useful range of the DRIFT spectra was 4000–1600 cm^{-1}. The lower limit is a significant limitation because it eliminates virtually all of what is typically referred to as the fingerprint region of the infrared spectrum from study. Use of a relatively thin (approximately 50-mg) layer of KBr on top of the sample scrambled the incident radiation and removed the observed rotational dependence of the spectra, as well as the apparent contribution of specular reflectance. Subtractions could be carried out over the range from 4000 to 900 cm^{-1} with good elimination of the E-glass bands and with clear spectral contributions from the coupling agent. The added KBr reduced the spectral intensities of all bands, but the reduction factor was greater for the substrate than for the coupling agent on the surface of the fiber. McKenzie et al. postulate that the KBr overlayer method makes the incident and reflected radiation isotropic, which removes the orientational dependence, and increases the average angle of incidence of the radiation to the fibers, which causes greater reflection at the fiber surface and less penetration into the fiber itself.

McKenzie and Koenig (33) compared the use of ground KBr overlayers on oriented fibers to the technique of averaging several different sample orientations (33, 34) for obtaining DRIFT spectra of glass fibers treated with a coupling agent. Each orientation yielded a different spectrum, but averaging several different sample orientations yielded linear KM plots as a function of surface treatment. Unfortunately, the standard deviation of the average computed in this way was as high as 35%, and the concentration breakpoint corresponding to association of the coupling agent could not be determined. McKenzie and Koenig then used coarse-ground (120–180-μm particle size) KBr overlayers to remove the orientation dependence of the observed spectra. The overlayers reduced the intensity of the observed bands by about a factor of 2, but eliminated the orientation dependence of the spectra and enabled the use of subtraction techniques to study the onset of association of the coupling agent.

Both Drift spectroscopy and photoacoustic spectroscopy (PAS) were used by Urban et al. (9) to investigate stretching-induced phase transitions of poly(butyl terephthalate) fibers and to characterize annealed and drawn fibers. The spectra obtained using the two methods were similar; the PAS studies had the added ability to characterize the orientation of surface species. Using DRIFT spectroscopy, the authors were able to characterize the alpha–beta phase transition observed upon drawing the fibers and observe the relaxation of the fibers upon annealing.

Chatzi et al. (35) used DRIFT spectroscopy with the KBr overlayer technique to study water absorbed in polyamide fibers (Kevlar 49). Use of gradually increasing amounts of KBr as the overlayer demonstrated the loss of absorbed water due to a nitrogen purge as a function of depth in the polymer fibers. The H_2O infrared absorptions were characterized with respect to the nature of the intermolecular interactions in the polyamide fibers.

Polymer Films.   Culler et al. (36) used the KBr overlayer technique to study polymer films. They investigated a model system consisting of a 1.5-μm poly(vinyl fluoride) (PVF) layer over a 29-μm-thick substrate of poly(ethylene terephthalate) (PET). Addition of a surface layer of KBr increased the relative contribution of the PVF layer in the spectra and decreased the contribution of the PET substrate. Fine KBr powder (average particle size less than 75 μm) had a more dramatic effect than course powder (average particle size 105–250 μm). The authors suggest that the presence of the KBr overlayer increases the average angle of incidence to the films and increases the amount of light reflected at the film interface, which makes the technique more sensitive to surface groups as KBr is added. The sensitivity of the method to a very thin film of a γ-APS coupling agent on PET was also demonstrated. The KBr overlayer technique brought out spectral contributions from the coupling agent, whereas the film on PET was transparent in transmission measurements. Culler et al. postulate that the infrared radiation may make several reflections inside the thin-surface film in this method that enhances the sensitivity of the technique.

DRIFT spectroscopy was compared to attenuated total reflectance and transmittance measurements by Cole et al. (37) for the study of a carbon–epoxy composite consisting of a woven carbon fiber material impregnated with an epoxy resin. The sample was prepared by cutting the prepreg (fiber-reinforcement impregnated with partially cured resin) into 1-cm × 1-cm squares that were placed in the sample cup. A sulfone index and an epoxide index were calculated from the observed spectra, parameters that are indicative of the hardener to epoxy ratio and the degree of polymerization, respectively. The method that gave the poorest results in terms of quantitative performance was DRIFT spectroscopy. Some of the spectra clearly showed the effects of a large contribution from Fresnel reflectance; that is, the derivative-shaped bands due to the anomalous dispersion. Several different methods to reduce the Fresnel reflectance with KBr overlayers were attempted. An opaque layer reduced the spectral contrast of the spectra to the extent that the method could not be used. Two other methods were used to generate very thin KBr overlayers, both of which increased the spectral contrast and reduced the apparent contribution of Fresnel reflectance, but neither of which improved the precision of the measurements to any great extent. The variation in the thickness of the surface resin layer, which contributes to variations in the Fresnel reflectance of the samples, was proposed to be the main reason for the lack of reproducibility. Cole et al. only used one orientation to acquire the spectra: and orientation with approximately 80% of the fibers aligned with the beam direction of the instrument.

Cole et al. also used DRIFT spectroscopy to measure the degree of crystallinity on the surface of a composite material (38). A prepreg (different from the one in the foregoing study) that consisted of a unidirectional arrangement of carbon fibers impregnated with high molecular weight polyphenylene sulfide (PPS) was investigated. No KBr overlayer and no rotation of the sample were used, and the orientation was with carbon fibers parallel to the beam direction of the instrument. Excellent spectra that resembled spectra obtained with transmission measurements were obtained. The spectrum of a sample with low crystallinity was subtracted from the spectrum of a sample with high crystallinity, and a pair of peaks in the resulting difference spectrum, whose ratio changed significantly as a function of annealing, was determined. Cole et al. obtained an excellent correlation between the intensity ratio and the enthalpy of crystallization for the prepreg that indicates that the intensity ratio is a good measure of the crystallinity of the composite.

Foams.   Chalmers and Mackenzie (39) carried out a variety of investigations aimed at illustrating the utility of DRIFT spectroscopy for industrial applications. One of the more interesting applications in their study was the use of DRIFT spectroscopy for the study of foams. Foams are sometimes difficult to study by ATR because the pressure placed on a foam sample is so critical in the ATR measurement. However, the DRIFT measurement was easily carried out by simply cutting the foam to the appropriate size, and the identification of a variety of foams was demonstrated.

Catalytic Studies Using DRIFT Spectroscopy

Hamadeh et al. (*40, 41*) were the first to develop a heatable–evacuable cell that could interface with DRIFT optics. This cell was used to study the infrared absorption spectroscopy of alumina-supported rhodium clusters under varying amounts of CO, which represented different coverages. The observed peaks in the DRIFT spectrum were assigned by comparison of the DRIFT spectra with previously assigned infrared spectra of rhodium carbonyl species. In a recent extension of these studies, Van Every and Griffiths (*42*) obtained detection limits approaching 10^{-6} monolayers of CO adsorbed on alumina-supported rhodium and observed that equilibration of the spectra with time is a complicating factor that depends on the pressure of CO to which the catalyst is exposed. Small amounts of CO equilibrate relatively rapidly on the surface of the catalyst whereas large amounts of CO equilibrate over long times, presumably because this equilibration involves diffusion into the pores of the support. Baseline changes in the spectra as a function of the pressure of admitted CO were also observed and assigned to pressure-induced changes in the sample scattering coefficient. The results of Van Every and Griffiths indicated that mercury cadmium telluride detectors are susceptible to conduction band saturation effects when hot samples are being observed, an effect that will lead to signal loss for high-temperature samples. They showed that deuterated triglycine sulfate or triglycine sulfate detectors do not exhibit this susceptibility.

With regard to using DRIFT spectroscopy as an in situ technique, Figure 4 is a schematic diagram of an environmental chamber that has been used in our laboratory to obtain in situ DRIFT spectra. The chamber is similar in design to commercially available cells, although most currently available cells are much more sophisticated in design than the cell shown. The sample sits on top of a post that can be heated via the cartridge heaters, and the temperature of the cell is monitored with the thermocouple. More sophisticated designs call for a thermocouple to be placed directly in contact with the sample, so that an accurate reading of the sample temperature can be obtained (*43, 44*). The cell allows for control of the internal atmosphere, either vacuum or a gas mixture, through connections to an external gas manifold and a vacuum pump. Commercially available cells allow for cooling the section of the cell containing the windows, because the window seals are usually the limiting consideration with regard to the ultimate temperature attainable. The flow path of any gases from an external gas manifold takes the

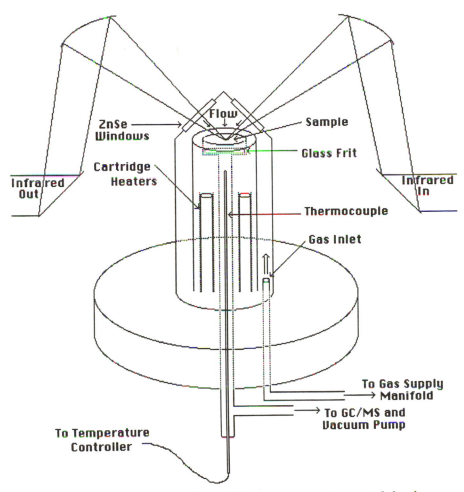

Figure 4. Schematic diagram of a DRIFT spectroscopy environmental chamber.

gases directly through the sample material. This flow-through of gases compresses the powder to a certain extent, which gives rise to enhanced band intensities compared with intensities observed without the flow. The samples should not, however, be mechanically pressed to any great extent, because compression prohibits gas flow through the sample (*40*). In addition, after flowing through the sample material, the gas stream can be sampled by gas chromatography (GC) or gas chromatography–mass spectroscopy (GC–MS), an application suggested by Hamadeh et al. (*40*) and implemented by Maroni et al. (*43*).

   Martin and Zabransky (*45*) studied various forms of the zeolite ZSM-5 (H–ZSM-5, Fe–ZSM-5, and Na–ZSM-5) and silicalite for the conversion of methanol to dimethyl ether. To study the adsorption process and products, neat samples of the zeolites in a commercial environmental chamber were used. Changes in the O–H stretching region ($3500–3750$ cm^{-1}) that occurred as methanol adsorbed on the zeolites were studied to determine which sites were being occupied by the methanol as it adsorbed as a function of temperature. At room temperature, adsorption of methanol onto the zeolites occurred at only one OH site; at 150° C, two sites were involved for the active zeolites H–ZSM-5 and Fe–ZSM-5, and only one site is involved for Na–ZSM-5.

Methanol fragmentation products were studied by observation of the infrared absorption pattern in the C–H stretching region ($2700–3100$ cm^{-1}). On H–ZSM-5, adsorption of methanol gave rise to infrared bands that could be assigned to the methoxide group at room temperature. At higher temperature (150 °C), the absorption pattern changed and the result could no longer be identified as a methoxide. Fe–ZSM-5 gives the same sequence of products. Methanol adsorption on Na–ZSM-5 and silicalite resulted in the formation of the surface methoxide only at room temperature and at 200 °C.

Thompson and Palmer (22) used in situ DRIFT spectroscopy to study the reaction of SO_2 with $CaCO_3$. $CaCO_3$ is used as an adsorbent for SO_2 in dry-injection flue gas cleanup systems. The evolution of adsorbed SO_2 as a function of time and temperature from sulfite to sulfate was investigated. Based on their assignment of the observed vibrational bands and on their previous infrared photoacoustic measurements on pure $CaCO_3$, Thompson and Palmer proposed a mechanism for the adsorption of SO_2 and the conversion of the physisorbed species to SO_3^{2-} or SO_4^{2-} depending on temperature.

One of the limitations of the DRIFT technique is the need to dilute strongly absorbing samples in a nonabsorbing matrix to avoid *reststrahlen* effects. In their DRIFT studies, Thompson and Palmer used NaCl as the diluent (1:10 $CaCO_3$:NaCl), and in their photoacoustic studies using pure $CaCO_3$, bands that were assigned to CaS_2O_3 were observed. These bands were not observed in the DRIFT studies. Because NaCl, as well as other alkali additives, is known to accelerate the conversion of $CaSO_3$ to $CaSO_4$, Thompson and Palmer point out that the need to use the diluent in the DRIFT studies is potentially a very serious drawback, especially at elevated temperatures.

Maroni and co-workers (43, 46, 47) studied a variety of molecular sieve catalysts by using DRIFT spectroscopy. To avoid problems associated with alkali halide diluents, pure zeolite materials were used and regions of the spectrum that can show contributions from *reststrahlen* effects (which includes much of the region below 2000 cm^{-1} for these types of samples) were not investigated. Investigations were limited to the O–H and C–H (or O–D and C–D) stretching regions (from approximately 2000 to 4000 cm^{-1}). Because much of the catalytic activity of these zeolite catalysts is determined by the nature of the acid sites, a significant amount of information could be obtained even with this limitation.

The differences in the number and type of O–H sites were studied by Maroni and collaborators on three different zeolites as a function of metal atom substitution (H–ZSM-5 compared to AFS–H–ZSM-5, a ferrisilicate zeolite) and as a function of aluminosilicate framework (H–ZSM-5 compared to H–offretite). DRIFT spectroscopy with subsequent gas chromatography–mass spectrometry (GC–MS) was then used for determination of products to compare the reactivity of the ferrisilicate zeolite AFS–H–ZSM-5 and its sodium exchanged form AFS–Na–ZSM-5 toward the conversion of methanol to higher hydrocarbons as a function of temperature and composition. The experimental setup included a provision for sampling the exhaust gases of the DRIFT environmental chamber with an on-line gas chromatograph and with a sample bulb for GC–MS analysis. The sodium-exchanged forms were found to be 2 orders of magnitude less reactive. The authors point out that probably the most effective use of the DRIFT technique in in situ measurements is not as an absolute technique, but rather as a monitor of changes that occur as a function of experimental variables. A DRIFT spectrum is obtained of some initial sample; then this spectrum is subtracted from subsequent spectra. The set of difference spectra shows changes in the infrared spectrum as a function of time, temperature, and reaction conditions in a very clear way. A stacked plot of these spectra is a dramatic method for showing spectroscopic changes as a function of the various experimental parameters.

Maroni and co-workers (46) used the same experimental arrangement to study the formation of ZSM-5 from the template tetrapropylammonium cation (TPA–ZSM-5) and to study the reaction of methanol, ethanol, and ethylene on H–ZSM-5 for the formation of higher hydrocarbons. Using the DRIFT technique in conjunction with on-line GC analysis of reaction products, the investigators were able to show that OH groups linking aluminum and silicon atoms (Si–OH–Al sites) are the sites at or near which the alcohol decomposition products adsorb, and that terminal OH groups bound to silicon atoms (SiOH sites) are labile but do not appear to be active in the catalysis. The shape of the C–H stretch absorption due to the organic fragment is indicative of the activity of the catalyst for a particular reactant. At temperatures $T > 673$ K, the reaction continues to generate products as shown by the GC results, but the surface-adsorbed species are not held tightly enough to remain on the surface after a nitrogen purge. The onset of a broad infrared absorption in the range 3000–3200 cm^{-1} is attributed to polycyclic aromatic coke precursors.

Iton et al. (47) used DRIFT spectroscopy to examination the stabilization and redox chemistry of Co(III) in aluminophosphate zeolites. They were able to study the oxidation of H_2 by the stabilized Co(III) at elevated temperatures through an increased OH absorption, and the 50 and 100 °C oxidation of NO to NO$^+$ by Co(III) through the infrared absorption of the nitrosyl cation at 2162 cm^{-1} and coordinated nitrosyl at 1674 and 1562 cm^{-1}.

The species adsorbed on a silver catalyst during the oxidation of acetaldehyde was analyzed by Kanno and Kobayashi (48) using DRIFT spectroscopy. The DRIFT spectra obtained from the reaction were compared with similar spectra for silver salts, such as silver acetate, silver lactate, silver ketenide, and silver carbonate. These comparisons, facilitated assignment of observed peaks in the infrared spectrum of the catalyst as being due to silver acetate intermediates, a strongly adsorbed acetaldehyde, and an adsorbed alcohol. The spectra obtained with the DRIFT technique were compared with spectra for similar samples obtained by transmission methods; the DRIFT method yielded much better spectra in the lower frequency region (1100 cm^{-1} > ν). Although Kanno and Kobayashi did dilute their catalyst samples in KBr during the actual oxidation reactions, which were carried out at elevated temperature (160 °C), they did not address the possibility of perturbation of the catalysis results due to the high concentration of alkali halide.

DRIFT spectroscopy was used by Fries and Mirabella (49) to study commercial TiCl$_3$–AlCl$_3$ catalysts used in the polymerization of propylene. They were able to monitor the decay of the 1640-cm^{-1} absorption due to the C=C group of gas-phase propylene and the simultaneous growth of an absorption at 1370 cm^{-1} due to a methyl deformation mode of polypropylene on the catalyst surface. The 1820-cm^{-1} band of propylene, which corresponds to the first overtone of the terminal CH$_2$ wag, provided a more reliable measure of the loss of propylene due to the lack of overlap by other spectral bands.

Smyrl and Fuller (50) used in situ diffuse reflectance spectroscopy to examine the reactions of neat coal samples. These authors point out that coal samples need no diluent to be studied by DRIFT spectroscopy, because coal effectively forms its own dispersion medium. Brimmer and Griffiths (24) suggested a reason for this effect, based on the idea that the coal particles absorb so strongly that the effective penetration depth is very small (on the order of 10 μm) so that a sample normally considered too strongly absorbing to study by DRIFT spectroscopy actually yields quite good spectra. Very careful DRIFT investigations by Smyrl and Fuller yielded well-resolved spectra for the entire mid-infrared region. Use of the acetylation reaction as a probe of the OH sites on the coal and use of the infrared results to determine exactly what types of esters were formed by the acetylation allowed Smyrl and Fuller to propose that the particular type of coal under study contained only phenolic-type OH groups and no alkyl-type OH groups. The method holds much promise for the study of coal OH-site distribution and coal chemistry in general.

The use of DRIFT spectroscopy for the study of carbon-supported metal catalysts, in particular metal carbonyls, was explored by Venter et al. (44,

51–53). Their initial investigations (*44, 51, 52*) focused on the adsorption of $Fe_3(CO)_{12}$, $Ru_3(CO)_{12}$, and $Os_3(CO)_{12}$ on amorphous carbon black. Their studies investigated, in a very careful manner, the effects of a variety of experimental parameters on the observed spectra, both theoretically and experimentally. The effects of metal loading, choice of background spectrum, various forms of baseline correction, and the diluent, both the dilution ratio and the type of diluent, were observed and the linearity of the Kubelka–Munk transform of the data for a range of metal loadings on these carbon-supported samples was established. This range depended on the dilution ratio, and the choice of dilution ratio involved a compromise between detection limitations and linearity of the transformed data. The final choice for dilution ratio was 1:200 by weight for carbon-supported catalyst to diluent. The choice for the diluent itself was a compromise between the effect of the diluent on the observed chemistry and the effect on the observed spectroscopy. Al_2O_3 and SiO_2 had the least effect on the chemistry but tended to adsorb water, which caused baseline drifts in the spectra and made the spectra hard to interpret. Ultimately, CaF_2 was chosen as the diluent.

The choice of material to generate the background spectrum also affected the final Kubelka–Munk spectra. Rather than use a nonabsorbing sample (i.e., pure diluent) as the background "sample", Venter et al. used a sample consisting of decarbonylated carbon black in the appropriate ratio, 1:200, in the diluent. Thus, the Kubelka–Munk spectra were calculated relative to a background that absorbed radiation at the same frequencies as the catalyst support. This background was not expected to have any effect on the linearity of the KM intensities for the bands due to the adsorbed CO-containing species, and the use of this material to generate the background spectrum yielded much better signal-to-noise ratios and better baseline correction.

In later investigations, Venter et al. (*53*) used the technique demonstrated in earlier studies to examine the thermal decomposition of five different metal carbonyls on carbon: $Fe_3(CO)_{12}$, $NEt_4[Fe_2Mn(Co)_{12}]$, $Mn_2(CO)_{10}$, $K[HFe_3(CO)_{11}]$, and $K[Fe_2Mn(CO)_{12}]$. DRIFT spectroscopy alone was used to measure the rate constant for decarbonylation as a function of temperature and to calculate the activation energy of decarbonylation for these five carbon-supported species under He and H_2. The decomposition kinetics of all of the species followed first-order kinetics, and the rates as a function of temperature were fitted well with the Arrhenius expression. The $Fe_3(CO)_{12}$ clusters were the least stable, with rate constants for decomposition of about 1 min^{-1} and E_a values of 75–88 kJ/mol. $Mn_2(CO)_{10}$ proved to be the most stable compound with decomposition rate constants of approximately 0.001 min^{-1} at 350 K and activation energy E_a values for the decomposition of 134–155 kJ/mol. The decomposition routes of the Mn-containing clusters were found to depend on the atmosphere under which the decomposition took place.

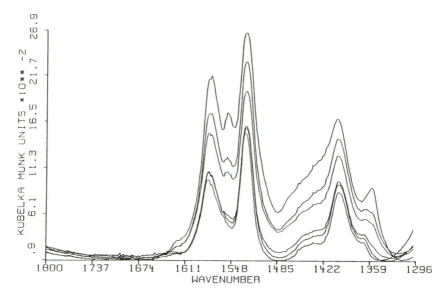

Figure 5. Evolution of the DRIFT spectrum of supported Cu(acac), as a function of metal loading. The lowest spectrum is that of a 0.92% wt/wt Cu sample on SiO₂ (Cab-O-Sil). The spectra of six different metal loadings are shown in sequence from bottom to top representing metal loadings of 0.92, 1.7, 2.5, 3.8, 4.5, and 5.6% Cu on SiO₂, respectively.

All of the foregoing data was generated by studying the infrared absorption of the CO species, nominally 1900–2100 cm^{-1}. The region of CH stretch absorption yielded some information concerning the types of products that might be expected. The K-containing catalysts showed significantly greater infrared absorption due to CH$_2$ (as opposed to CH$_3$) groups, presumably on the support surface. This information is consistent with earlier observations that the presence of potassium promotes chain growth in the products of the syngas reaction.

Kenvin and co-workers (54–56) showed that DRIFT spectroscopy is very useful for the study of supported catalyst precursors on oxide surfaces. Their study of Cu(acac)$_2$ on SiO$_2$ demonstrated that the DRIFT technique could be used to show the onset of multilayer formation of the supported complex and that subtraction of KM plots of the first-layer species from spectra of samples that had multiple layers of Cu(acac)$_2$ on the silica surface yielded a spectrum of the overlayers; that is, the Cu(acac)$_2$ species not in direct contact with the silica surface. Figure 5 shows a spectrum of the impregnated silica powder as a function of metal loading. The onset of the absorption at 1552 cm^{-1} is thought to indicate the presence of multiple layers of Cu(acac)$_2$ on the silica surface. Figure 6 is a spectrum that results from a subtraction of a less-than-a-monolayer spectrum from a spectrum of a sample containing

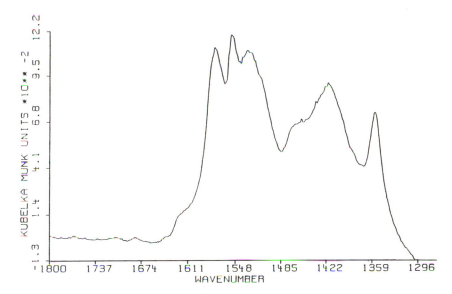

Figure 6. DRIFT spectrum that results from subtraction of 1.7% wt/wt Cu as Cu(acac)$_2$ on the silica spectrum from 5.6% wt/wt Cu on the silica spectrum.

more than one monolayer of complex. The result is in remarkably good agreement with that of pure Cu(acac)$_2$ as shown in Figure 7. Kenvin (55) applied the DRIFT technique to the study of a variety of similar complexes, including Pd(acac)$_2$, Pt(acac)$_2$, and Cr(acac)$_3$. The palladium and platinum complexes showed behavior similar to that shown by the copper system, although less adsorption for these complexes was observed. The chromium complex showed very little adsorption and no spectral changes for the material that did adsorb, which might be expected for the octahedral complex.

Summary

Diffuse reflectance spectroscopy in the mid-infrared range is a powerful technique for the study of certain types of samples, including polymers, heterogeneous catalysts, and novel materials. DRIFT has great potential for the study of a wide variety of processing problems and for in situ investigations, both as a stand-alone technique and in conjunction with GC and GC–MS. What must be kept in mind if useful results are expected are the approximations inherent in the technique such as optically dilute samples, the problems and the solutions associated with specular and diffuse Fresnel reflectance, and the need for reproducible sample preparation.

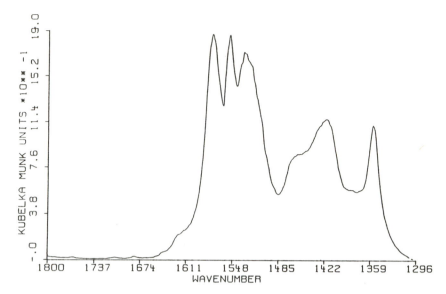

Figure 7. Drift spectrum of pure Cu(acac)₂ in KBr, 1:10 by weight.

References

1. Wendlandt, W. W.; Hecht, J. G. *Reflectance Spectroscopy*; Interscience: New York; 1966.
2. Kortum, G. *Reflectance Spectroscopy: Principles, Methods, Applications*; Springer: New York, 1969.
3. Willey, R. R. *Appl. Spectrosc.* **1976**, *30*, 593.
4. Fuller, M. P.; Griffiths, P. R. *Anal. Chem.* **1978**, *50*, 1906.
5. Fuller, M. P.; Griffiths, P. R. *Appl. Spectrosc.* **1980**, *34*, 533.
6. Culler, S. R.; Ishida, H.; Koenig, J. L. *Annu. Rev. Mater. Sci.* **1983**, *13*, 363.
7. Blitz, J. P.; Shreedhara Murthy, R. S.; Leyden, D. E. *Appl. Spectrosc.* **1986**, *40*, 829; Leyden, D. E.; Shreedhara Murthy, R. S. *Spectroscopy* **1977**, *2*, 28.
8. Childers, J. W.; Palmer, R. A. *Am. Lab.* **1986**, *18(3)*, 22.
9. Urban, M. W.; Chatzi, E. G.; Perry, B. C.; Koenig, J. L. *Appl. Spectrosc.* **1986**, *40*, 1103; Koenig, J. L. *Pure Appl. Chem.* **1985**, *57*, 971.
10. Story, W. C.; Masujima, T.; Liang, J.; Liu, G.; Eyring, E. M.; Harris, J. M.; Anderson, L. L. *Appl. Spectrosc.* **1987**, *41*, 1156.
11. Kubelka, P.; Munk, F. *Z. Tech. Phys.* **1931**, *12*, 593.
12. Kubelka, P.; Munk, F. *J. Opt. Soc. Am.* **1948**, *38*, 448.
13. Hecht, H. G. *Appl. Spectrosc.* **1980**, *34*, 161.
14. Fraser, D. J. J.; Griffiths, P. R. *Appl. Spectrosc.* **1990**, *44*, 193.
15. Brimmer, P. J.; Griffiths, P. R. *Appl. Spectrosc.* **1988**, *42*, 242.
16. Brimmer, P. J.; Griffiths, P. R.; Harrick, N. J. *Appl. Spectrosc.* **1986**, *40*, 258.
17. Yang, P. W.; Mantsch, H. J.; Baudais, F. *Appl. Spectrosc.* **1986**, *40*, 974.
18. Brimmer, P. J.; Griffiths, P. R. *Appl. Spectrosc.* **1987**, *41*, 791.
19. Hembree, D. M.; Smyrl, H. R. *Appl. Spectrosc.* **1989**, *43*, 267.
20. Martin, K. A.; Ferraro, J. R. *Appl. Spectrosc.* **1987**, *41*, 45.
21. Hawranek, J. P.; Jones, R. N. *Spectrochim. Acta, Part A* **1976**, *32*, 111.

22. Thompson, M. M.; Palmer, R. A. *Appl. Spectrosc.* **1988**, *42*, 945.
23. Yeboah, S. A.; Wang, S.-H.; Griffiths, P. R. *Appl. Spectrosc.* **1984**, *38*, 259.
24. Brimmer, P. J.; Griffiths, P. R. *Anal. Chem.* **1986**, *58*, 2179.
25. Mandelis, A.; Boroumand, F.; van den Bergh, H. *Spectrochim. Acta*, **1991**, *47A*, 943.
26. Melamed, N. T. *Appl. Phys.* **1963**, *34*, 560.
27. Lee, J. Y.; Moskala, E. J.; Painter, P. C.; Coleman, M. M. *Appl. Spectrosc.* **1986**, *40*, 991.
28. Siesler, H. W. *Mikrochim. Acta* **1988**, *I*, 319.
29. Savolahti, P. *Mikrochim. Acta* **1988**, *II*, 155.
30. Maulhardt, H.; Kunath, D. *Appl. Spectrosc.* **1980**, *34*, 383.
31. Ishida, H.; Naviroj, S.; Koenig, J. L. In *Physicochemical Aspects of Polymer Surfaces*; Mittal, K. L., Ed.; Plenum: New York, 1983; p 91.
32. McKenzie, M. T.; Culler, S. R.; Koenig, J. L. *Appl. Spectrosc.* **1984**, *38*, 786.
33. McKenzie, M. T.; Koenig, J. L. *Appl. Spectrosc.* **1985**, *39*, 408.
34. Xue, G.; Liu, S.; Jin, Y.; Jiang, S. *Appl. Spectrosc.* **1987**, *41*, 264.
35. Chatzi, E. G.; Ishida, H.; Koenig, J. L. *Appl. Spectrosc.* **1986**, *40*, 847.
36. Culler, S. R.; McKenzie, M. T.; Fina, L. J.; Ishida, H.; Koenig, J. L. *Appl. Spectrosc.* **1984**, *38*, 791.
37. Cole, K. C.; Pilon, A.; Nöel, D.; Hechler, J.-J.; Chouliotis, A.; Overbury, K. C. *Appl. Spectrosc.* **1988**, *42*, 761.
38. Cole, K. C.; Nöel, D.; Hechler, J.-J.; Wilson, D. *Mikrochim. Acta* **1988**, *I*, 291.
39. Chalmers, J. M.; Mackenzie, M. W. *Appl. Spectrosc.* **1985**, *39*, 634.
40. Hamadeh, I. M.; King, D.; Griffiths, P. R. *J. Catal.* **1984**, *88*, 264.
41. Hamadeh, I. M.; Griffiths, P. R. *Appl. Spectrosc.* **1987**, *41*, 682.
42. Van Every, K. W.; Griffiths, P. R. *Appl. Spectrosc.* **1991**, *45*, 347.
43. Maroni, V. A.; Martin, K. A.; Johnson, S. A. In *Perspectives in Molecular Sieve Science*; American Chemical Society: Washington, DC, 1988; p 85.
44. Venter, J. J.; Vannice, M. A. *Appl. Spectrosc.* **1988**, *42*, 1096.
45. Martin, K. A.; Zabransky, R. F. *Appl. Spectrosc.* **1991**, *45*, 68.
46. Johnson, S. A.; Rinkus, R.-M.; Diebold, R. C.; Maroni, V. A. *Appl. Spectrosc.* **1988**, *42*, 1369.
47. Iton, L. E.; Choi, I.; Desjardins, J. A.; Maroni, V. A. *Zeolites* **1989**, *9*, 535.
48. Kanno, T.; Kobayashi, M. *Mem. Kitami Inst. Tech.* **1985**, *17*, 75.
49. Fries, R. W.; Mirabella, F. M. *Transition Metal Catalyzed Polymerizations: Ziegler–Natta and Metathesis Polymerizations*; Quirk, R. P., Ed.; Cambridge University Press: Cambridge, England, 1988; p 314.
50. Smyrl, N. R.; Fuller, E. L., Jr. *Appl. Spectrosc.* **1987**, *41*, 1023.
51. Venter, J. J.; Vannice, M. A. *J. Am. Chem. Soc.* **1987**, *109*, 6204.
52. Venter, J. J.; Vannice, M. A. *Carbon* **1988**, *26*, 889.
53. Venter, J. J.; Chen, A.; Vannice, M. A. *J. Catal.* **1989**, *117*, 170.
54. Kenvin, J. C.; White, M. G.; Mitchell, M. B. *Langmuir* **1991**, *7*, 1198.
55. Kenvin, J. C. Ph.D. Thesis, School of Chemical Engineering, Georgia Institute of Technology, Atlanta, GA, 1990.
56. Mitchell, M. B.; Kenvin, J. C.; White, M. G. *Polym. Mater. Sci. Eng.* **1991**, *64*, 197.

RECEIVED for review July 15, 1991. ACCEPTED revised manuscript September 25, 1992.

Surface-Enhanced Raman Spectroscopy as a Method for Determining Surface Structures

One- and Two-Dimensional Polymeric Copper Azoles

Mary L. Lewis and Keith T. Carron*

Department of Chemistry, University of Wyoming, Laramie, WY 82071–3838

We report the results and interpretation of a surface-enhanced Raman spectroscopic (SERS) study of benzimidazole, benzotriazole, and benz-imidazole-2-thione on copper surfaces in aggressive media. We have found that SERS is an excellent tool for the study of corrosion inhibition on copper. SERS has enabled us to carry out in situ chemical analysis, isotopic substitution, and orientation determination for the inhibitors on copper. The results indicate an important role of the bridgehead heteroatom in imidazole compounds. We found that the progression of corrosion inhibition is benzimidazole < benzotriazole < benzimidazole-2-thione. Our orientation study of ben-zimidazole indicates that it is predominantly flat on copper surfaces.

$\mathbf{M}$ANY METHODS EXIST FOR DETERMINING THE BULK STRUCTURE OF poly-mers. One approach that has yielded very good descriptions of polymeric systems is vibrational spectroscopy (*1*). Often vibrational techniques are the only choice for structure determination because the amorphous nature of many polymers rules out the use of X-ray diffraction techniques. Both infrared and Raman spectroscopy provide a spectrum of the vibrational modes in molecular and polymeric systems. The compatibility of Raman

---

* Corresponding author.

0065–2393/93/0236–0377$06.00/0

spectroscopy with all phases of matter and complex environmental conditions makes it an excellent tool for in situ studies of polymers.

As the polymer films become thinner, problems begin to arise with sensitivity. The advent of Fourier transform infrared spectroscopy (FTIR) has overcome this problem for infrared spectroscopy (2). Recently developed optical multichannel systems based on charge-coupled devices (CCD) (3) and Fourier transform Raman spectroscopy (FT–Raman) (4) are alleviating the sensitivity problem for Raman spectroscopy. Currently, vibrational spectra can be obtained via either technique on monolayer films, but the instrumentation is very expensive and often the resulting spectra are of low quality due to the need for long integration times and sample degradation.

Surface-Enhanced Raman Spectroscopy

One technique that yields high-quality spectra of monolayer films is surface-enhanced Raman spectroscopy (SERS). The SERS effect arises from two sources: chemical enhancement (5) and electromagnetic enhancements.

Chemical Enhancement. Chemical enhancement occurs when the monolayer is composed of molecules that contain groups or atoms that can coordinate with the metal surface. The resulting surface complex can form charge transfer states with energy levels in the metal. This condition leads to an optical absorption and a surface-localized resonance Raman enhancement. Many of the molecules studied by SERS belong to this class of materials. For example, pyridine has been exhaustively studied as a probe for the SERS effect. Pyridine is a good lone-pair donating ligand for transition metals. Careful ultrahigh vacuum (UHV) studies have shown that it is possible to observe SERS from pyridine on substrates that, because of their morphology (smoothness), only show the chemical effect (6). On the other hand, some substrates show both chemical enhancement from chemisorbed pyridine and also multilayers of physisorbed pyridine. The long-range (multilayer) effect precludes the possibility of a chemical effect, and therefore, a through-space electromagnetic enhancement has also been proposed (7). Most substrates will show both forms of enhancement.

Electromagnetic Enhancement. In the experiments presented in this chapter we will be most concerned with the electromagnetic enhancement (8). Electromagnetic enhancement occurs when the surface contains roughness features smaller than the wavelength of light. The physics necessary to predict the enhancement was formulated at the end of the last century by Lorentz (9). Lorentz derived the solution for the response of a dielectric sphere in an electric field. If the radius of the sphere is much smaller than the wavelength of light, the solution to the Laplace equation leads to (*10*)

$$E_{in} = \frac{3}{\epsilon(\omega) + 2} E_0 \tag{1}$$

where E_{in} is the electric field inside the particle, E_0 is the electric field of the light incident on the sphere, and $\epsilon(\omega)$ is the dielectric function of the particle material. The purpose for Lorentz's derivation was the understanding of clouds and colloidal dispersions. In most of these systems $\epsilon(\omega)$ is real and positive. However, in free electron metals $\epsilon(\omega)$ can be negative and the imaginary part of $\epsilon(\omega)$ is small. Equation 1 demonstrates that a resonance will occur when $\epsilon(\omega)$ is -2. The electric field inside the particle does not reach infinity because of the imaginary part of $\epsilon(\omega)$. However, the simple equation does provide the rudiments for understanding the electromagnetic contribution to SERS.

Silver exhibits a resonance when $\epsilon(\omega) = -2$, around 380 nm, and has been shown to produce large enhancements in colloidal silver at this frequency (*11*). For the effect to be more generally useful the resonance must be shifted toward the lower frequencies where most lasers operate and other metals have resonances. For example, copper and gold both have large interband transitions in the blue–green region of the spectrum that increase the value of the imaginary part of the dielectric constant and damp out any possible resonances. As the particles are deformed from a perfect sphere into an ellipsoid the single sphere resonance is split into a resonance at high frequencies along the short axis and at low frequencies along the long axis. The plasmon resonance along the long axis is most often used in SERS and can be described by (*10*)

$$E_{in} = \frac{1}{1 + A(\epsilon(\omega) - 1)} E_0 \tag{2}$$

where A is a geometric factor accounting for the eccentricity of the particle (*10*):

$$A = \frac{1 - e^2}{e^2} \left[-1 + \frac{1}{2e} \ln \frac{1 + e}{1 - e} \right] \tag{3}$$

$$e^2 = 1 - \frac{b^2}{a^2} \tag{4}$$

where a is the major axis, b is the minor axis, and e is the eccentricity of the ellipsoid. The geometric factor A is $1/3$ for a sphere and becomes smaller as the particle becomes more eccentric. This decreasing factor has the effect of moving the resonance farther into the red spectrum.

Three additional factors have been added to equation 2 to correct for particle size effects. The first of these is electron scattering and it takes into account the depolarization of particles that can occur when the particle is smaller than the mean free path of an electron. This correction is added through ϵ in the form (12)

$$\epsilon = C \frac{\omega - i\gamma}{\omega(\omega^2 - \gamma^2)} \tag{5}$$

where ω is the frequency of the light, γ represents the damping constant, and c is the proportionality constant. The lifetime of the resonance is $1/\gamma$. For large particles, γ is determined by damping due to collisions of the free electrons with the metal lattice. When the particles become smaller than the mean free path of the electron, γ will increase (12):

$$\gamma = v_F \left[\frac{1}{l_\infty} + \frac{1}{l_{eff}} \right] \tag{6}$$

where v_F if the Fermi velocity of the electron, l_∞ is the mean free path length of an electron, and l_{eff} is the smallest radius of the particle.

The second correction takes into account that as the electrons are accelerated by the electric field of the light they also emit radiation at 90° to the incident radiation. This term is called radiation damping and places an upper limit on the particle size. This correction is added through the A term in equation 2 (13):

$$A'_{eff} = A - ik^3 \frac{2a^3}{9} \tag{7}$$

where k is the wave vector $2\pi/\lambda$ and a is the particle radius along the axis involved in the particle resonance. This contribution is imaginary because the emitted radiation is directed 90° to the incident light. The frequency position of the resonance is determined by the real part of ϵ; therefore, this term will not shift the position of the resonance.

The final term accounts for corrections to the assumption of a static electric field. Again this correction will occur at large particle sizes when the time-dependent electric field strength can vary across the particle. This term is denoted dynamic depolarization, and it too is added through the A term of equation 2 (14):

$$A''_{eff} = A'_{eff} - 1/3k^2a^2 \tag{8}$$

In this case, the contribution is real. Therefore, it causes a small frequency shift in the plasmon resonance as particles become large and also causes a slight damping of the resonance. The inclusion of these corrections yields a single equation that accurately matches the experimentally determined excitation profiles of isolated noble metal particles.

An important result from the derivation of equation 2 comes from the boundary conditions. The field perpendicular to the particle must obey (*15*)

$$D^{\perp}_{\text{in}_{r=a}} = D^{\perp}_{\text{out}_{r=a}} \tag{9}$$

whereas the field parallel to the surface obeys

$$E^{\|}_{\text{in}_{r=a}} = E^{\|}_{\text{out}_{r=a}} \tag{10}$$

because $D = \epsilon E$, a preferential enhancement of vibrations that contain a component perpendicular to the surface results. This preferential enhancement can be used as a basis for determining the orientation of molecules relative to the surface.

When the molecule contains a local C_{2v} symmetry, it is possible to express the vibrations clearly as modes that contain an xz or a yz component. In the C_{2v} point group the b_1 (α_{xz}) and b_2 (α_{yz}) modes can be used for orientation determinations. Because the modes perpendicular to the surface are along the z axis, a molecule that is perpendicular to the surface will show equal enhancement of the b_1 and b_2 modes. As the $\sigma_v(yz)$ plane of the molecule becomes parallel to the surface, the b_1 modes that contain an out-of-plane component will become preferentially enhanced. The azole compounds that we are studying all have a local C_{2v} symmetry.

Experimental Details

The azole compounds were purchased from Aldrich and purified by crystallization from ethanol. The copper substrates were 99.999% copper foil (Aldrich) with a thickness of 0.025 mm. The copper substrates were macroscopically roughened with a fine grade of sandpaper and then chemically etched in a 12% HNO_3 solution for 4 min with vigorous stirring. This procedure is similar to that developed by Miller et al. (*16*) and more recently by Carron et al. (*17*). The foil was then washed with distilled water and immediately immersed in a 2–3% ethanolic solution of the azole compound of interest. Upon removal from this solution the treated copper foil was rigorously washed with ethanol to remove any physisorbed species. The treated copper foil was then allowed to dry completely. We often observe oxides with our treated films, which indicates that the azole passivates the surface against further formation of copper–azole complexes. Therefore, our data indicate that the azoles are reacting with the first couple of monolayers of Cu_2O that form at a freshly etched copper surface (*18*). Preparations for the in situ experiment were the same as previously described. The treated copper foil was subsequently placed in a cuvette with water or a 3% solution of either sodium chloride, hydrochloric acid, or sodium hydroxide. Spectra were obtained immediately and again after 12 h without removal from solution.

The Raman spectra were obtained with a double monochromator (Jobin-Yvon Mole 1000), double monochromator, a photomultiplier tube (RCA 31034), and photon counting electronics (Ortec). Laser excitation was provided with a Kr^+ ion laser (Spectra Physics 2025). A backscattering geometry was used for all samples and a cylindrical lens was used to focus the laser to decrease the power density at the sample. All spectra were acquired with 647-nm radiation and laser powers of 100–150 mW. A 647-nm interference filter was used to remove most of the plasma lines. A plasma line at 235 cm^{-1} was used to maintain frequency calibration between the spectra. Data collection and storage were interfaced through a 386-20 microcomputer.

Results and Discussion

The multiple coordination site, variable heteroatom nature of the azole compounds that we have studied make their surface structure determination a formidable challenge to the surface spectroscopist. These compounds have been studied electrochemically and with electron spectroscopies (19–27). However, in both cases the results are incomplete and inconclusive due to the stringent ambient conditions that are associated with these techniques. The weak Raman scattering associated with aqueous solutions and air makes SERS an excellent technique for surface structure determination in realistic environments (28–31).

To determine the surface structures, we have employed synthesis of the bulk organometallic equivalent of the surface species, isotope substitution, in situ chemical modifications, and surface orientation studies with SERS. For clarity of presentation we will present our results for benzimidazole (BIMH), benzotriazole (BTAH), and benzimidazole-2-thione (SBIMH) separately. Chart I shows the structures of the azoles discussed in this chapter. Following the separate discussion of each compound, we will summarize our results with a comparison of each azole compound in terms of its ability to inhibit corrosion in aggressive media.

Benzimidazole. BIMH is the only compound for which a normal coordinate analysis has been performed. Furthermore, because Raman data were not available, it was necessary to use an incomplete set of frequencies for the normal coordinate analysis (32). This reference assigns the in-plane direction to the x axis, and therefore, a juxtaposition of the b_1 and b_2 modes must be made to correspond to the accepted convention. However, the incomplete analysis provides us with information that can aide in the analysis of some of the important features of the surface product of BIMH on copper. We have observed the following changes in BIMH upon deuteration. The 848- and 956-cm^{-1} bands associated with the bridgehead carbon of the imidazole ring either disappear or shift to lower frequencies. The 634-, 1136-, and 1496-cm^{-1} bands associated with the pyrrole nitrogen of BIMH either disappear or shift to lower frequencies. These same bands, which have been assigned to the pyrrole N−H vibrations, are not observed in the surface product. This observation indicates that the surface species is the anion of benzimidazole. The SERS spectrum of the surface product can be seen in Figure 1. Figure 1 also illustrates the strong oxidation of copper that occurs when the BIMH-treated samples are placed in water. Furthermore, these spectra show an important aspect of the oxidation of BIMH-treated copper.

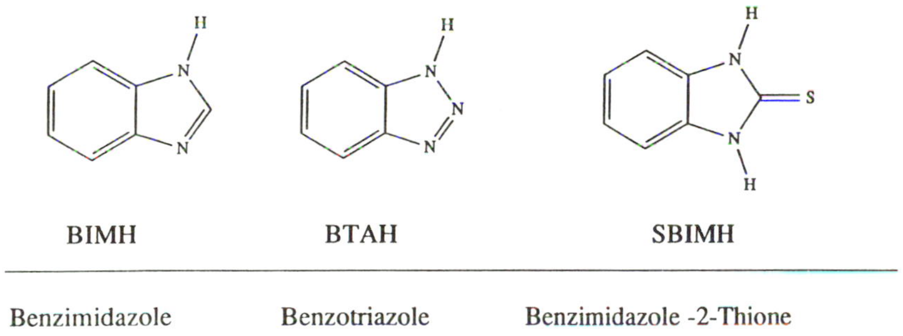

BIMH   BTAH   SBIMH

Benzimidazole   Benzotriazole   Benzimidazole -2-Thione

Chart I. Azole compounds investigated: benzimidazole (BIMH), benzotriazole (BTAH), and benzimidazole-2-thione (SBIMH).

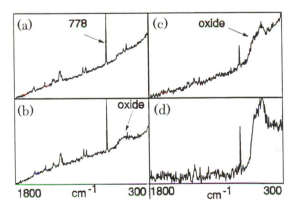

Figure 1. SERS spectra of benzimidazole–copper surface film in (a) air immediately, (b) air after 12 h, (c) water immediately, and (d) water after 12 h. Spectra were obtained with 150 mW of 647-nm laser irradiation, 5-cm^{-1} bandpass, 5-cm^{-1} step size, and 10-s integration time.

After 12 h the BIMH–Cu film is still present on the copper foil as can be seen from the presence of the 778-cm^{-1} band. We also observed that when untreated copper is exposed to water, the oxide bands are not present. Our interpretation of this is the removal of oxide due to its slight solubility in water. Apparently the treated sample retains its azole–Cu layer in water, but the oxidation occurs under the layer and is trapped. This situation indicates the occurrence of a directional transport of material. Oxygen is transported into the copper lattice, but $Cu^{+/++}$ is not solvated and conveyed in the opposite direction through the film.

The strong peak at 778 cm^{-1} is assigned as an imidazole ring-bending mode (33). In reference 32 this peak is assigned as an in-plane bend. However, the same reference assigns this peak to a normal coordinate with b_1 symmetry, which is indicative of an out-of-plane bend. Furthermore, a more recent normal coordinate analysis has assigned this 778-cm^{-1} peak to an out-of-plane mode (34). The surface selection rules for vibrations that contain components perpendicular to the surface can be used to find the orientation of the BIMH on the surface. Because the BIMH is expected to bind through both nitrogens of the imidazole ring, the molecule is expected to be lying flat on the surface. We have determined the orientation with the following formula (35):

$$R = \frac{I_{b_2}}{I_{b_1}} = \left[\frac{\sin(\theta)\cos(\phi) + \epsilon \sin(\theta)\sin(\phi) + \cos(\theta)}{\sin(\theta)\sin(\phi) + \epsilon \sin(\theta)\cos(\phi) + \cos(\theta)} \right]^2 \quad (11)$$

where R is the ratio of b_2 to b_1, θ is the axial or tilt angle with respect to the z axis, Φ is the rotational angle about the C_2 axis of the molecule, and ϵ is the dielectric constant of the metal. For copper $\epsilon = -12.2$ at 647 nm. We have used the 778-cm^{-1} band as the b_1 mode and the 1478-cm^{-1} b_2 vibration for the analysis. In bulk $Cu(BIM^-)_2$ spectra the 1478-cm^{-1} band is moderately strong. However, in the surface product this vibration is very weak. The ratio, R, of b_2/b_1 for SERS–bulk spectra for these peaks is 0.094. Two solutions to eq 11 were found. One solution gave $\Phi \approx 0°$ and a negative value for θ. Based on this we chose the more intuitive solution of $\theta = 5°$ and $\Phi = 14°$. These are average angles and are based on the weak 1478-cm^{-1} band in the surface spectrum. However, the analysis does verify the expected result that the BIMH is predominantly flat on the surface. We have reported in an earlier publication that it should be possible for an axial BIMH to be perpendicular to the surface (17). These results indicate that this site is most likely not occupied for BIMH. Tompkins et al. also found evidence from infrared (IR) reflectance studies that BIMH is mostly parallel to the surface (36, 37).

Figure 2 shows the effects of aggressive media on BIMH-treated copper. Most apparent in these spectra is the inability for BIMH to protect the

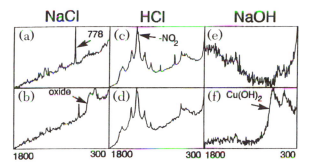

Figure 2. In situ SERS spectra of benzimidazole–copper surface film in (a) 3% NaCl immediately, (b) 3% NaCl after 12 h, (c) 3% HCl immediately, (d) 3% HCl after 12 h, (e) 3% NaOH immediately, and (f) 3% NaOH after 12 h. Spectra were obtained with 150 mW of 647-nm laser irradiation, 5-cm^{-1} bandpass, 5-cm^{-1} step size, and 10-s integration time.

copper against oxidation. The spectra were obtained in 3% solutions of the materials listed. The oxides of copper appear as two broad peaks around 585 and 520 cm^{-1} (38) that are the symmetric and asymmetric stretches of Cu_2O, respectively. In sodium chloride solution these oxide peaks can be seen after 12 h. As in water, the oxide appears to be trapped by the overlayer. In sodium hydroxide solution no bands from the BIMH–Cu film can be seen either initially or after 12 h. The surface film has been completely destroyed and after 12 h all that can be seen are two peaks associated with the copper hydroxide and copper oxide. The peak at 695 cm^{-1} is at a significantly higher wavenumber than that reported for the copper oxide; therefore, we have assigned it to a copper hydroxide vibration. The most significant changes in the surface product occur in hydrochloric acid solution, where the spectra become strongly enhanced and a new peak at 1380 cm^{-1} appears. This peak is characteristic of the symmetric nitro stretch (39) and appears to be due to the oxidation of at least one of the imidazole nitrogens in the presence of HCl. Further evidence for this is the complete loss of the 778-cm^{-1} imidazole peak after 12 h.

Benzotriazole. The surface products of BTAH in air and water are shown in Figure 3. In air, the 788-cm^{-1} peak is split in BTAH. By comparison to the BIMH molecule this peak has been assigned to the out-of-plane bending of the triazole ring. This splitting is not possible from symmetry lowering and has been interpreted as the presence of two BTAH–Cu species in the surface film (17). The possibility of the splitting being due to a mixture of Cu(I) and Cu(II) species on the surface has been ruled out because bulk samples of the surface product show only a Cu(I) species is present. Based on these data we believe the two peaks are due to BTAH anions on the surface that are bonded through either the two imidazole nitrogens or through all three nitrogens in the triazole ring. Some oxidation of the metal surface is observed in air and strong oxidation is observed for samples placed in water.

Our in situ study with BTAH is illustrated in Figure 4. It can be seen that BTAH is superior to BIMH as a corrosion inhibitor. In sodium chloride and hydrochloric acid solutions little or no oxidation is observed. No decomposition of the surface film is observed in the acidic solution as was the case for the BIMH surface film. In sodium hydroxide the film appears to be protective over short time periods. However, after 12 h the films appear to have decomposed. Oxide peaks are all that can be seen here. The copper hydroxide peaks are not observed as they were with the BIMH–Cu surface after 12 h.

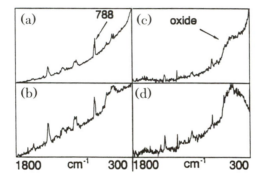

Figure 3. SERS spectra of benzotriazole–copper surface film in (a) air immediately, (b) air after 12 h, (c) water immediately, and (d) water after 12 h. Spectra were obtained with 150 mW of 647-nm laser irradiation, 5-cm^{-1} bandpass, 5-cm^{-1} step size, and 10-s integration time.

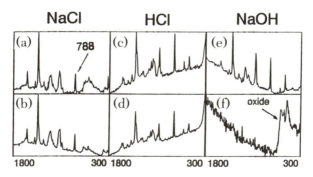

Figure 4. In situ SERS spectra of benzotriazole–copper surface film in (a) 3% NaCl immediately, (b) 3% NaCl after 12 h, (c) 3% HCl immediately, (d) 3% HCl after 12 h, (e) 3% NaOH immediately, and (f) 3% NaOH after 12 h. Spectra were obtained with 150 mW of 647-nm laser irradiation, 5-cm^{-1} bandpass, 5-cm^{-1} step size, and 10-s integration time.

Comparison of Figures 3 and 4 shows a large enhancement of the Raman signal in our aggressive media. Because we prepared the films prior to immersion in the media and no excess BTAH is present, the observed signal increase cannot be due to increased coverage of BTAH. This enhancement was observed before by Kester et al. (28). We interpret that the enhancement is due to the adsorption of anions at the interface and contributions from the hyperpolarizability contributing to the Raman signal. The observed Raman scattering in this case would be described by (40)

$$P = (\alpha + \beta E_{static}) E_{optical} \qquad (12)$$

where P is the induced dipole, α is the polarizability, β is the hyperpolarizability, E_{static} is the electric field due to the adsorption of anions, and $E_{optical}$ is the electric field of the incident radiation. The capacitive effect of the thin film of BTAH–Cu on the Cu surface should be significant enough to create large static electric fields. To test this hypothesis we repeated the experiment with noncoordinating anions, PF_6^- and ClO_4^-, and found that the signals were not enhanced as illustrated in Figure 5.

Kester et al. observed an intensity dependence due to the presence of Cl^- (28). They found that enhanced signals from BTAH on Cu could be obtained by anodizing the electrodes in the presence of laser illumination. Their interpretation is that special surface sites are formed when CuCl is present and the electrode is illuminated. This mechanism has been observed in other systems as well (41). In our case the surface was roughened and treated with BTAH prior to illumination and exposure to Cl^-. We also observed enhancements from OH^- in the case of SBIMH on Cu. Because it is unlikely that the illumination in our studies was roughening the electrode or changing the structure of the existing film, we interpret the Cl^- enhancement as being due to the formation of a large static field. Enhanced second-order optical processes by static fields localized at surfaces are well documented in the literature (42, 43).

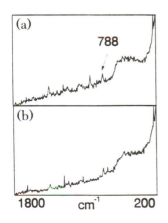

Figure 5. In situ SERS spectra of benzotriazole–copper surface film in noncoordinating ion solutions: (a) 3% KPF_6 and (b) 3% $NaClO_4$. Spectra were obtained with 150 mW of 647-nm laser irradiation, 5-cm^{-1} bandpass, 5-cm^{-1} step size, and 10-s integration time.

Benzimidazole-2-thione. SBIMH has the potential to bond to a copper surface through both of the imidazole nitrogens and the sulfur. The bulk and SERS spectra of SBIMH are illustrated in Figure 6. We have identified the C=S and an N–H vibration at 616 and 1196 cm^{-1}, respectively, through deuteration studies and comparisons with imidazole-2-thione (44). These bands are labeled in the bulk spectrum and are not present in the SERS spectrum, and we use this as evidence for a threefold bonding to the copper surface. Further evidence can be seen in Figure 7, which shows acidified copper substrate that was treated with SBIMH. We find that the N–H stretches appear when the substrate is exposed to the acidic solution. Final proof that the change is in the imidazole nitrogens comes from a similar study where we exposed the copper foil to DCl. In this case we see the appearance of new Raman bands due to the N–D stretch. This band is shifted to lower frequency by an amount similar to that for the deuterated

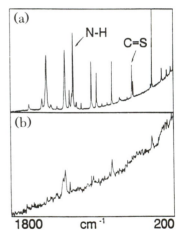

Figure 6. (a) The bulk spectrum of benz-imidazole-2-thione obtained with 150 mW of 647-nm laser irradiation, 2-cm^{-1} bandpass, 2-cm^{-1} step size, and 10-s integration time. (b) SERS spectrum of benzimidazole-2-thione–copper surface product obtained with 150 mW of 647-nm laser irradiation, 5-cm^{-1} bandpass, 5-cm^{-1} step size and 10-s integration time.

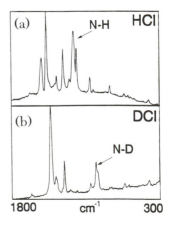

Figure 7. In situ SERS spectra of benzim-idazole-2-thione–copper surface film in (a) 3% HCl and (b) 3% DCl. Spectra were obtained with 150 mW of 647-nm laser irradiation, 5-cm^{-1} bandpass, 5-cm^{-1} step size, and 10-s integration time.

bulk sample. This in situ study distinctly illustrates the value of SERS in determining surface structures.

Figure 8 shows the in situ spectra of SBIMH on copper. We found that SBIMH was superior to both BTAH and BIMH with respect to corrosion prevention. Few oxides are observed, even in basic solutions. The improved corrosion inhibition appears to be related to the ability of SBIMH to remain surface-bound even in basic media whereas both BTAH and BIMH surface complexes decompose and the inhibitor is dissolved off the surface. We also observed unusual enhancements in sodium hydroxide and sodium chloride solution. The 1280-cm^{-1} peak, which is greatly enhanced, has been assigned to an amide III stretch. The origin of the enhancement under investigation at this time.

Summary

We have found the corrosion inhibiting strength of the three azole compounds studied to follow the order SBIMH > BTAH > BIMH. The ability to prevent corrosion appears to be related to the heteroatoms that are present. In particular, sulfur binds very strongly to copper surfaces and prevents desorption of the azole anion in basic solution.

Enhancements of the SERS signal can be seen in solutions containing coordinating anions. These enhancements can be explained by the presence of a large static electric field created by the adsorption of anions. Surface orientation for the BIMH surface product has been determined by using surface selection rules associated with SERS. Our initial study indicates that it is predominantly flat.

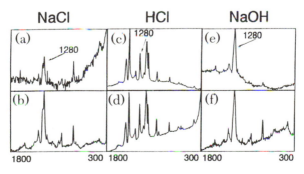

Figure 8. In situ SERS spectra of benzimidazole-2-thione–copper surface film in (a) 3% NaCl immediately, (b) 3% NaCl after 12 h, (c) 3% HCl immediately, (d) 3% HCl after 12 h, (e) 3% NaOH immediately, and (f) 3% NaOH after 12 h. Spectra were obtained with 150 mW of 647-nm laser irradiation, 5-cm^{-1} bandpass, 5-cm^{-1} step size, and 10-s integration time.

Future work will be directed at the determination of the mechanism of reaction of the surface products with aggressive media. As well, we will determine the surface orientation for the SBIMH and BTAH surface products.

Acknowledgment

The authors gratefully acknowledge the support of the University of Wyoming and the National Science Foundation EPSCoR program, Grant RII–8610680.

References

1. Painter, P. C.; Coleman, M. M.; Koenig, J. L. *The Theory of Vibrational Spectroscopy and Its Application to Polymeric Material*; Wiley: New York, 1982.
2. Bell, R. J. *Introductory Fourier Transform Spectroscopy*; Academic: Orlando, FL, 1972.
3. Pemberton, J.; Sobocinski, R.; Bryant, M.; Carter, D. *Spectroscopy* **1990**, *5*, 26.
4. Chase, D. B. *J. Am. Chem. Soc.* **1986**, *108(4)*, 7485.
5. *See*, for example, Billman, J.; Kovacs, G.; Otto, A. *Surf. Sci.* **1980**, *92*, 153.
6. Seki, H. *J. Chem. Phys.* **1982**, *96*, 9.
7. Rowe, J. E.; Shank, C. V.; Zwemer, D. H.; Murray, C. A. *Phys. Rev. Lett.* **1980**, *44*, 1770.
8. Wokaun, A. *Solid State Phys.* **1984**, *46*, 223.
9. Lorentz, H. A. *Wiedem. Ann.* **1880**, *9*, 641.
10. van de Hulst, H. C. *Light Scattering by Small Particles*; Dover: New York, 1981; p 71.
11. Kerker, M.; Wang, D.-S.; Chew, H.; Siiman, O.; Bumm, L. A. In *Surface Enhanced Raman Scattering*; Chang, R. K.; Furtak, T. E., Eds.; Plenum: New York, 1982; p 109.
12. Kraus, W. A.; Schatz, G. C. *J. Chem. Phys.* **1983**, *79(12)*, 6130.
13. Wokaun, A.; Gordon, J. P.; Liao, P. F. *Phys. Rev. Lett.* **1982**, *48(14)*, 957.
14. Meier, M.; Wokaun, A. *Opt. Lett.* **1983**, *8*, 581.
15. Jackson, J. D. *Classical Electrodynamics*; Wiley: New York, 1978; p 150.
16. Miller, S.; Baiker, A.; Wokaun, A. *J. Chem. Soc., Faraday Trans. 1* **1984**, *80*, 1305.
17. Carron, K.; Xue, G.; Lewis, M. *Langmuir* **1991**, *7*, 2.
18. Rhodin, T. N. *J. Am. Chem. Soc.* **1950**, *70*, 5102.
19. Cotton, J. B.; Scholes, I. R. *Br. Corros. J.* **1967**, *2*, 1.
20. Ogle, I. C. G.; Poling, G. W. *Can. Metall. Q.* **1975**, *14*, 37.
21. Notoya, T.; Poling, G. W. *Corrosion* **1976**, *32*, 223.
22. Chadwick, D.; Hashemi, T. *Corros. Sci.* **1978**, *18*, 39.
23. Fox. P. G.; Lewis, G.; Boden, P. J. *Corros. Sci.* **1979**, *19*, 457.
24. Siedle, A. R.; Velapoldi, R. A.; Erickson, N. *Appl. Surf. Sci.* **1979**, *3*, 229.
25. Yoshida, S.; Ishida, H. *J. Chem. Phys.* **1983**, *78*, 6960.
26. Yoshida, S.; Ishida, H. *J. Mater. Sci.* **1984**, *19*, 2323.
27. Xue, G.; Ding, J. *Appl. Surf. Sci.* **1990**, *40*, 327.
28. Kester, J.; Furtak, T.; Bevolo, A. *J. Electrochem. Soc.* **1982**, *129*, 1716.
29. Fleischmann, M.; Hill, I.; Mengoli, G.; Musiani, M.; Akhavan, J. *Electrochim. Acta* **1985**, *30*, 879.
30. Gardiner, D. J.; Gorvin, A. C.; Gutteridge, C. *Corros. Sci.* **1985**, *25(11)*, 1019.

31. Youda, R.; Nishihara, H.; Aramaki, K. *Electrochim. Acta* **1990**, *35*, 1011.
32. Cordes, M. M.; Walter, J. L. *J. Spectrochim. Acta* **1968**, *24A*, 1421.
33. Drolet, D. P.; Manuta, D. M.; Lees, A. J.; Katnani, A. D.; Coyle, G. J. *Inorg. Chim. Acta* **1988**, *146*, 173.
34. Mohan, S.; Sundaraganesan, N.; Mink, J. *Spectrochim. Acta* **1991**, *47A*, 1111.
35. Carron, K. T.; Hurley, L. G. *J. Phys. Chem.* **1991**, *95*, 9979.
36. Tompkins, H. G.; Sharma, S. P. *Surf. Interface Anal.* **1982**, *4*, 261.
37. Tompkins, H. G.; Allara, D. L.; Pasteur, G. A. *Surf. Interface Anal.* **1983**, *5*, 101.
38. Boerio, F. J.; Armogan, L. *Appl. Spectrosc.* **1978**, *32*(5), 509.
39. Dollish, F. R.; Fateley, W. G.; Bentley, F. F. *Characteristic Raman Frequencies of Organic Compounds*; Wiley: New York, 1974; p 42.
40. Garito, A. F.; Singer, K. D.; Teng, C. C. In *Nonlinear Optical Properties of Organic and Polymeric Materials*; Williams, D. J., Ed.; ACS Symposium Series 233; American Chemical Society: Washington, DC, 1983; pp 1–26.
41. Furtak, T. E. In *Advances in Laser Spectroscopy*; Garetz, B. A.; Lombardi, R., Eds.; Wiley: New York, 1983; Vol. 2, p 175.
42. Corn, R. M.; Romagnoli, M.; Levenson, M.; Philpott, M. *Chem. Phys. Lett.* **1984**, *106*, 30.
43. Richmond, G. L. *Chem. Phys. Lett.* **1984**, *106*, 26.
44. Shunmugam, R.; Sathyanarayana, D. N.; Anagnostopoulos, A. *Can. J. Spectrosc.* **1983**, *28*, 150.

RECEIVED for review July 15, 1991. ACCEPTED revised manuscript May 27, 1992.

Surface-Enhanced Raman Spectroscopy as a Method for Determining Surface Structures

Thiophenol at Group Ib Metal Surfaces

Keith T. Carron and Gayle Hurley

Department of Chemistry, University of Wyoming, Laramie, WY 82071–3838

Raman spectra of noble metal phenyl thiolates and the corresponding surface-enhanced Raman spectroscopy (SERS) spectra of the surface species on copper, silver, and gold are reported. The SERS spectra were used to obtain orientations of the thiophenol species at the noble metal surfaces. Orientations were determined through the electric field enhancement of vibrations normal to the metal surface. The model developed in this chapter allows both the azimuthal and axial angles of C_{2v} molecules at surfaces to be determined by performing the SERS measurement in media of differing indices of refraction. For silver and gold we found the axial angle $\theta = 85$ and $76°$, respectively. The azimuthal angle ϕ was found to vary from 32 to $0°$ from silver to gold. The SERS spectra of thiophenol adsorbed onto copper were too weak for accurate angle determinations with a surrounding media other than air. However, the SERS spectrum observed on copper in air does indicate a near perpendicular orientation.

PROPERTIES OF POLYMER ADHESION TO SURFACES are greatly affected by the structure of polymeric materials at interfaces. In this chapter we will present a method for the determination of orientation of a two-dimensional polymer composed of noble metal phenyl thiolates.

Organic thiol compounds form self-assembled monolayers on noble metal surfaces (1). Self-assembled monolayers can be formed from a large

0065–2393/93/0236–393$06.00/0

variety of organic thiols. An example of the importance of self-assembled monolayers is seen in the n-alkane thiols that form surfaces that have the lowest surface-free energies known (2). When the n-alkane is replaced by carboxylic or hydroxyl groups, it is possible to produce high free-energy wettable surfaces. The macroscopic properties of the films are determined in part by the nature of the organic functionality and in part by the self-assembled structures that form at the surface. The goal of this research is to develop a spectroscopic method by which the surface structure of organic thiols on noble metal surfaces can be determined. We chose thiophenol, phenyl sulfide, and phenyl disulfide as probes to develop and test our method of structure determination.

Self-assembly is believed to occur on noble metal surfaces due to Au–SR bond strengths that are weak enough to allow lateral surface diffusion, yet strong enough to prevent desorption under ambient conditions. Because much of the work has been on noble metal surfaces and the symmetric disulfide stretches in organic disulfides are not IR-active, many of the spectroscopic studies have used surface-enhanced Raman spectroscopy (SERS) as the spectroscopic tool (3–6). The first reported SERS study of thiols and disulfides on a surface was by Sandroff and Herschbach (3), who found that the surface products of thiophenol and phenyl disulfide on silver were identical. More recently, Sobocinski et al. (4) studied alkyl thiols at electrode surfaces using SERS and made comparisons between the surface interactions of alcohols and thiols. Another study of interest is that by Joo et al. (5) that indicates a photochemical cleavage of one of the C–S bonds in organosulfides on silver colloids. In our study we found that the formation of metal thiolates on roughened noble metal surfaces in air is not photochemical. However, our studies with colloids do confirm the results of photochemistry on colloids and roughened surfaces in aqueous salt solutions.

SERS spectra of thiophenol, phenyl sulfide, and phenyl disulfide were obtained on copper, silver, and gold surfaces. In all cases the surface products were similar regardless of the initial material. This result indicates that the metals are capable of breaking the thiol, sulfide, and disulfide bonds. The surface spectra also closely resemble the bulk Raman spectra of the metal phenyl thiolates.

In Chapter 14 we outlined the origin of the surface enhancement. We found that the general formula for the perpendicular and parallel electromagnetic contributions, $E_\perp$ and $E_\parallel$, to SERS is (7)

$$E_\perp = \frac{3\epsilon}{(1 + A(\epsilon - 1))} \tag{1}$$

$$E_\parallel = \frac{3}{(1 + A(\epsilon - 1))} \tag{2}$$

ϵ is the dielectric constant of the metal and A is a geometric factor equivalent to (8)

$$A = \frac{1 - e^2}{e^2}\left[-1 + \frac{1}{2e}\ln\frac{1 + e}{1 - e}\right]$$

$$e^2 = 1 - \left(\frac{b^2}{a^2}\right)$$

where a and b are the major and minor axes of the ellipsoid, respectively. The resonance condition shifts toward the red as ϵ becomes more negative, which can have drastic effects on the orientation analysis. For copper, silver, and gold at 647 nm, ϵ is about -12.2, -18.4, and -11.5, respectively (9), which means that the ratio of modes with a single perpendicular component will be enhanced 338:1 over purely parallel modes if the molecule is oriented perpendicular to a silver surface.

We have extended the orientation analysis to ascertain not only the tilt (or axial) angle of the z axis, but also to include the determination of the azimuthal rotation of the molecule about the z axis. Figure 1 is a depiction of our angle definitions. The analysis and experiments herein are for molecules with a local C_{2v} symmetry. In general, a_2 modes are very weak in aromatic ring systems because of the relatively small polarization changes that occur when the ring bends out of its plane. We are able to observe a moderately intense b_1 mode. In Chapter 14, the determination of orientation with b_1 and b_2 modes was demonstrated. The b_1 modes contain the α_{xz} polarizability vector and the b_2 modes are composed of the α_{yz} vector. Unfortunately, b_2 modes are very weak in the bulk and SERS spectra of phenyl thiolates (*see* Figure 3), which precludes accurate determinations of surface orientations. The a_1 modes are very strong in the Raman spectra of aromatic compounds, and can be difficult to use in orientation analyses because they

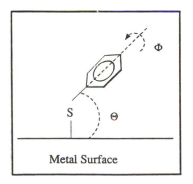

Figure 1. A definition of the axial and azimuthal angles. The axial angle θ is the angle between the C_2 axis and the surface. The azimuthal angle ϕ is the angle of rotation about the C_2 axis of the phenyl ring. In the text θ is equal to 0 when the molecule is oriented perpendicular to the surface; ϕ is 0 when the plane of the ring is parallel to the surface.

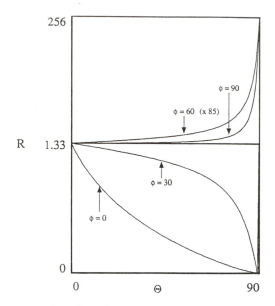

Figure 2. Changes in the value of R *as a function of* θ *for various values of* φ. *When* φ < 45°, R *goes to* 1/ε² *as* θ *goes to* 90°, *and when* φ > 45°, R *goes to* ε² *as* θ *goes to* 90°. *Our method of angle determinations fails when* φ = 45°. *This indeterminacy could be alleviated by using an* a_1 *mode, which contains an* $α_{xx}$ *component. For this figure we have assumed that the dielectric constant of the metal is* −16.

are composed of a linear combination of $α_{xx}$, $α_{yy}$, and $α_{zz}$. To solve orientations we use the totally symmetric a_1 stretch mode of the phenyl ring. Simple geometric arguments show that if this molecule is oriented with the ring perpendicular to the surface, then the intensity ratio of a_1/b_1 must be $(1.155)^2$, where, in this case, a_1 indicates only the ring breathing stretch. Equations 1 and 2 indicate that the intensity ratio should be $1/ε^2$ for parallel orientation. A more complete analysis shows that

$$R = \frac{I_{a_1}}{I_{b_1}} = \left[\frac{\sin(θ)\cos(φ) + ε\sin(θ)\sin(φ) + 1.155\cos(θ)}{\sin(θ)\sin(φ) + ε\sin(θ)\cos(φ) + \cos(θ)} \right]^2$$

where R is the ratio of relative intensities of the SERS bands in relation to the bulk or SERS bands with different surrounding dielectric constants. I_{a_1} and I_{b_1} correspond to the intensity of the a_1 and b_1 bands relative to the bulk metal thiolate or SERS spectrum with a different surrounding dielectric constant. Figure 2 shows a plot of R versus θ for different values of φ. The problem is that for a given value of R, several values of θ and φ can exist. A nontrivial solution can be found by either changing the excitation wavelength, thus ϵ, or by changing the dielectric constant of the surrounding media. We chose the second approach because resonance Raman effects are predicted due to photochemical degradation that was observed in some cases. The rigorous expression for the ϵ used in the previous equations is $\epsilon_{metal}/\epsilon_{surroundings}$. The new dielectric materials we have chosen are mixed hexanes and cyclohexane, which are not expected to adsorb to the surface and perturb the orientation of the phenyl thiolate. The mixed hexanes and cyclohexane used in our experiments have a dielectric constant of 1.90 and 2.03, respectively.

Experimental Details

The thiols, sulfides, and disulfides were purchased from Aldrich and used without further purification. The silver substrates were prepared by vapor deposition of silver onto calcium-fluoride-roughened slides. The silver substrate has produced enhancements of 1.7×10^5 for pyridine (10). Microscope slides were cleaned with a concentrated ammonia solution and cleaned with a 30-W radio-frequency plasma cleaner (Harrick) for 10 min. A vacuum coating unit (Edwards) operating at 10^{-6} mbar pressure was used for the vacuum depositions. The roughened substrate was prepared by first depositing a 600-nm-thick layer of calcium fluoride followed by 50 nm of silver. The depositions were carried out from a resistively heated molybdenum boat. The silver (Aldrich) was 99.9% pure and the calcium fluoride (Aesar) was optical grade. The film thickness and deposition rate were monitored with a quartz crystal microbalance. The silver was deposited at the rate of 0.2 nm/s. Gold (G. F. Goldsmith) substrates were prepared by using the preceding method from 99.999% pure gold.

The sample was dissolved in acetone and was spun-coated onto the substrates. The concentration of the thiol, sulfide, and disulfide was 0.01 M, and a 50-μL aliquot was used for the spin-coating.

Immersion was also tested as a viable application procedure. The spectral features were identical with spun-coat samples. However, the quality of the spectra was not as good as with spin-coating.

The copper substrates were prepared from 99.999% pure, 0.025-mm polycrystalline copper foil (Aldrich). These substrates were roughened by etching in 12% HNO_3 for 4 min under vigorous stirring. This procedure is similar to that developed by Miller et al. (11), and used recently by Carron et al. (12). The metallic copper samples were prepared by etching the copper with nitric acid, washing with distilled water, immediate immersion in a 0.01-M acetone solution of the molecule of interest, which had been warmed to 50 °C, and a final washing with acetone to remove any physisorbed reagent. This procedure is similar to that developed for coating copper with azole compounds (12).

The Raman spectra were obtained using a double monochromator (Jobin−Yvon Mole 1000), a photomultiplier (RCA 31034), and photon counting electronics (Ortec). Laser excitation was provided with an ion laser (Spectra Physics 2025 Kr +). Backscattering geometry was used for all samples, and a cylindrical lens was used to focus the laser to decrease the power density at the sample. The silver and gold SERS samples were spun at 1800 Hz to avoid any possible damage due to laser-induced heating. The copper foil samples were not spun. A 647-nm filter was used to remove the plasma lines.

Bulk Raman spectra were obtained on metal thiolates prepared in the laboratory. Silver phenyl thiolate (AgSPh) was prepared by mixing an excess of thiophenol in water with $AgNO_3$ (5). The resulting precipitate was washed with methanol. Gold phenyl thiolate was synthesized through the reduction of $HAu(III)Cl_4$ in the presence of thiophenol (13). The resulting precipitate was washed with methanol to remove the phenyl disulfide that results from oxidation of thiophenol by Au(III). Copper phenyl thiolate was synthesized by the addition of thiophenol to an aqueous solution of $CuCl_2$. Again, the product was washed with methanol to remove phenyl disulfide.

Results and Discussions

Silver. Figure 3 shows the Raman spectra of thiophenol and silver phenyl thiolate, and the SERS spectrum of the product of thiophenol reacted with a silver substrate. Comparison of thiophenol and the silver phenyl thiolate shows that there is a large intensity increase in the band at 1075 cm^{-1}. The low-wavenumber shoulder on the 1600-cm^{-1} peak has increased in intensity in the metal thiolates. There is also a large intensity increase in the 420-cm^{-1} vibration. Band assignments are tabulated in Table I and changes in intensity are tabulated in Table II.

The SERS spectrum shows the largest enhancement for the 420 mode, which is to be expected because mode 420 is largely composed of the α_{zz} tensor (14). Comparison of the SERS spectra of the surface product with the bulk metal thiolates shows several trends. The intensity ratio of the 1000-cm^{-1} peak relative to the 470-cm^{-1} peak is about unity. This is predicted for molecules that are perpendicular to the surface.

The orientation of thiophenol was determined by the procedure outlined earlier. We obtained SERS spectra in air, mixed hexanes, and cyclohexane. The results are tabulated in Table III. The results are in good agreement with X-ray studies of AgSPh that indicate that the phenyl ring is perpendicular to the layer of silver ions (15). One problem that can occur in SERS-based orientation studies is the assumption that the Raman spectrum of the chemisorbed species is equivalent to the bulk spectrum of the equivalent organometallic compound. The determination of the SERS spectrum in two

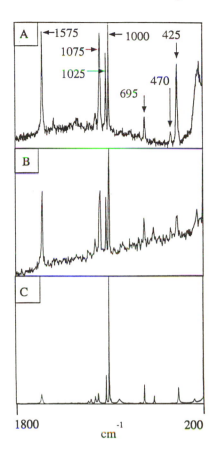

Figure 3. A Raman study of thiophenol on silver. (A) SERS spectrum of thiophenol on silver obtained with 30 mW of 647-nm laser irradiation, 3-cm^{-1} bandpass, 2-cm^{-1} step size, and 5-s integration. (B) A Raman spectrum of bulk AgSPh obtained with 5 mW of 647-nm laser irradiation, 3-cm^{-1} bandpass, 5-cm^{-1} step size, and 5-s integration time. (C) A Raman spectrum of thiophenol obtained with a depolarizer, 50 mW of 647-nm laser irradiation, 3-cm^{-1} bandpass, 2-cm^{-1} step size, and 2-s integration time.

solvents of differing dielectric constant allowed us to check this assumption and to make an orientation determination independent of the bulk material. The results from Table III indicate that, in this case, the assumption of equivalence of the bulk silver phenyl thiolate and the surface species is reasonably good. The changes in ϕ may be related to the relatively free rotation of the phenyl ring on the surface as opposed to the rigid structure found in the bulk.

Gold.   Our studies on gold are shown in Figure 4. A weak band corresponding to the Au–S stretch was found at 275 cm^{-1}, and the AuSPh was stable under prolonged irradiation at 647 nm. From Table II it can be seen that all modes are decreased in intensity relative to the 470-cm^{-1} mode. This relative decrease indicates that gold has an axial angle to the surface that is less than silver. By using equation 3 we found that the axial and azimuthal angles are 76.0 and 0 degrees, respectively, and excellent correlation to the predicted angles exists when either hexanes or cyclohexane was used.

Table I. Band Assignments

Thiophenol (Neat)		Thiophenol (SERS) $\Delta\nu$	Ag–Phenyl Thiolate $\Delta\nu$
$\Delta\nu$	Assignment[a]		
		230 (Ag–S)	250 (Ag–S)
278	(b_2)		335
414	(a_1, a_2)	420	420
464	(b_1)	470	480
616	(b_2)	620	620
698	(a_1)	695	700
734	(b_1)	745	740
834	(a_1)		835
914	(S–H band)		
988	(b_1)	985	
1000	(a_1)	1000	1000
1024	(a_1)	1025	1025
1069	(b_2)	1075	1075
1092	(a_1)		
1118	(a_1)	1110	1115
1156	(b_2)	1160	1160
1180	(a_1)	1180	1185
1270	(b_2)		1270
1328	(b_2)		
1380	(a_1)	1375	
1440	(b_2)	1440	
1478	(a_1)	1475	1475
1576	(a_1)	1575	1575
1584	(a_1, b_2)	1600	1600

[a] Band assignments are based on references 5 and 18.

Orientation determined independent of the bulk spectrum also matched the bulk values very well, which indicates that the surface species closely resembles the structure of Au(I)SPh synthesized from Au(III). The smaller ϕ with respect to silver can be attributed to increased steric hindrance with the surface as θ becomes smaller.

The intensities relative to the 470-cm^{-1} peak in the bulk are larger than those found with silver. We interpret this as an increase in bond polarizability in the phenyl ring due to the decreased charge withdrawal through the Au–S bond. For example, the relative intensity of the 1000-cm^{-1} band, and most other modes, approaches the value for free thiophenol as the metal is changed from copper to silver to gold (see Table II). This increase is clearly seen in the Raman spectra of the bulk metal phenyl thiolates. The similarity in the Raman spectrum of bulk AuSPh to thiophenol in comparison with its dissimilarities to silver and copper analogues can explain the differences in bonding that lead to self-assembly of thiols on gold. The phenyl thiolate species on gold most resembles thiophenol, and, therefore, can move laterally on surfaces most easily.

Table II. Changes in Intensity

Material	Frequency x (cm^{-1})	Bulk (I_x/I_{470})	SERS (I_x/I_{470})	I_{SERS}/I_{BULK}
Cu	420	0.54	3.67	6.83
	470	1.00	1.00	1.00
	695	3.54	1.00	0.28
	1000	10.0	9.83	0.98
	1025	2.77	3.33	1.22
	1075	2.54	2.00	0.78
	1575	2.38	1.50	0.63
Ag	420	2.20	6.38	2.90
	470	1.00	1.00	1.00
	695	2.70	1.85	0.68
	1000	10.6	8.54	0.81
	1025	5.60	6.08	1.08
	1075	6.40	7.69	1.20
	1575	7.65	8.08	1.05
Au	420	4.60	4.13	0.90
	470	1.00	1.00	1.00
	695	4.00	1.26	0.32
	1000	52.0	5.41	0.10
	1025	18.4	5.20	0.28
	1075	16.0	8.33	0.52
	1575	14.8	7.60	0.51
Thiophenol	414	25.5		
	464	1.00		
	698	31.8		
	1000	214.3		
	1024	47.8		
	1092	15.9		
	1584	15.9		

Copper.   The Raman spectra associated with our study of thiophenol on copper are shown in Figure 5. The Cu–S stretch was found at 270 cm^{-1}. The SERS spectra on copper were weak, and spectra of copper species in solvents were unobtainable. The background produced by the solvents prevented measurement of the 470-cm^{-1} peak. The ratio of I_{1000}/I_{470} is larger than gold and silver. Therefore, we would expect that thiophenol on copper is also oriented perpendicular to the surface. No photochemistry was observed with copper phenyl thiolates or the surface species.

Error Analysis.   The errors for angle determinations are shown in Table III. The error for silver is $\pm 4°$ for θ and $\pm 11°$ for ϕ. The error for gold is significantly smaller: θ is $\pm 1.5°$ and ϕ is $\pm 2°$. These errors are estimated from the noise level associated with the 470-cm^{-1} band. This peak is much smaller than the 1000-cm^{-1} peak, and, therefore, is the major source

Table III. Orientation of Thiophenol

Intensity Ratio	Ag, θ, ϕ	Au, θ, ϕ
I_{1000}/I_{470} (bulk)	10.6	52.0
$I_{1000}I_{470}$ (air)	8.09	6.82
I_{1000}/I_{470} (hexanes)	5.18, 85.4, 32	3.77, 76.0, 0
I_{1000}/I_{470} (cyclohexane)	4.91, 85.4, 32	2.71, 75.9, 0

NOTE: The orientation of thiophenol on Ag using the SERS spectrum in air as the reference shows $\theta = 81.3$, $\phi = 25$. The orientation of thiophenol on Au using the SERS spectrum in air as the reference shows $\theta = 77.0$, $\phi = 17.2$. The error in the angles for thiophenol on Ag are approximately $\pm 4°$ for θ and $\pm 11°$ for ϕ. The error in the angles for thiophenol on Au are approximately $\pm 1.5°$ for θ and $\pm 2°$ for ϕ.

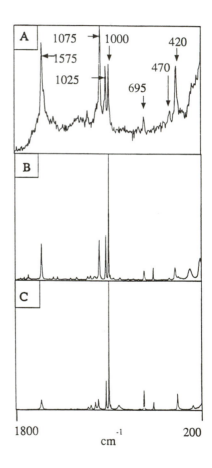

Figure 4. A Raman study of thiophenol on gold. (A) SERS spectrum of thiophenol on gold obtained with 30 mW of 647-nm laser irradiation, 3-cm^{-1} bandpass, 5-cm^{-1} step size, and 10-s integration. (B) A Raman spectrum of bulk AuSPh obtained with 50 mW of 647-nm laser irradiation, 3-cm^{-1} bandpass, 2-cm^{-1} step size, and 2-s integration time. (C) Same as Figure 3C.

of error. The smaller errors associated with gold are due to the much higher signal-to-noise ratio (S/N) in the AuSPh spectrum. The better S/N for this spectrum is due to the lack of photodegradation and the correspondingly higher laser powers that could be used. The observed errors are on the same order of those reported for IR reflection spectroscopy (16). The precision

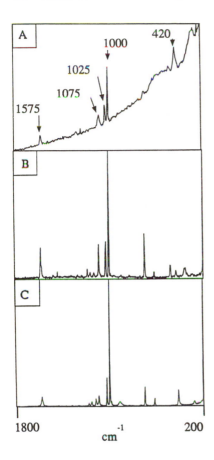

Figure 5. A Raman study of thiophenol on copper. (A) SERS spectrum of thiophenol on copper obtained with 10 mW of 647-nm laser irradiation, 5-cm⁻¹ bandpass, 5-cm⁻¹ step size, and 15-s integration. (B) A Raman spectrum of bulk CuSPh obtained with 20 mW of 647-nm laser irradiation, 3-cm⁻¹ bandpass, 3-cm⁻¹ step size, and 3-s integration time. (C) Same as Figure 3C.

found here is much better than the $\pm 32.5°$ found by Walls and Bohn (*17*), who used a depolarization method for angle determinations.

In general, Figure 2 shows that the S/N level of the spectra will be critical as ϕ approaches 45°. When ϕ is exactly equal to 45°, the described method fails due to the inability to locate an a_1 mode with a known amount of α_{xx} or an a_2 mode. As ϕ approaches 0, the R value is critical when θ is large. On the other hand, as ϕ approaches 90°, R is critical when θ is small. We anticipate that a complete analysis using potential-energy diagrams will result in more precise determination of angles because two or more very strong Raman bands can be used.

Summary

We have demonstrated a new procedure for the determination of molecular orientation at metal surfaces. This approach gives intuitively reasonable results for phenyl thiolates on noble metal surfaces. Our results indicate that thiophenol, phenyl sulfide, and phenyl disulfide all form metal phenyl thiolates with noble metal surfaces. In air the surface product was not observed to be photoactive under low laser powers and a time period of several hours. However, in colloids and continuous substrates in the presence of aqueous anions, photoactivity was observed (6).

SERS spectra obtained in mixed hexanes and cyclohexane indicate that the phenyl thiolate is oriented nearly perpendicular to the surface. There is a predisposition for thiophenol to tilt more for the different metals from copper to silver and to gold. The azimuthal angle, ϕ, around the C_2 axis varies between silver and gold. Observation that the angle ϕ approaches 0 as the tilt angle increases can be rationalized as increased steric hindrance around the C_2 axis as θ becomes smaller. The SERS spectra on copper were very weak, and accurate measurements in solvents were not obtained. However, relatively large a_1/b_1 ratios indicate that this species is oriented nearly perpendicular. The ability of this technique to obtain orientations without bulk spectra of the surface species will greatly aid in studies where the surface complex is weak and cannot be synthesized in large quantities.

Improvements in this approach are being explored. One very promising approach will be to use potential energy distributions to determine the exact amount of x, y, and z motion in a normal coordinate. This approach would allow other a_1 modes to be used in orientation analysis. In aromatic systems the a_1 modes are very strong, which would improve angle determination errors. The use of a_1 modes with an α_{xx} component would remove the redundancy associated with $\phi = 45°$ that is due to our current use of an a_1 mode that does not contain an x component.

References

1. *See*, for example, Bain, C.; Evall, J.; Whitesides, G. *J. Am. Chem. Soc.* **1989**, *111*, 7155. Whitesides, G.; Laibinis, P. *Langmuir* **1990**, *6*, 87. De Long, H. C.; Buttry, D. A. *Langmuir* **1990**, *6*, 1319.
2. Bain, C. D.; Troughton, E. B.; Tao, Y.; Evall, J.; Whitesides, G.; Nuzzo, R. G. *J. Am. Chem. Soc.* **1989**, *111*, 321.
3. Sandroff, C.; Herschbach, D. R. *J. Phys. Chem.* **1982**, *86*, 3277.
4. Sobocinski, R. L.; Bryant, M. A.; Pemberton, J. E. *J. Am. Chem. Soc.* **1990**, *112*, 6177.
5. Joo, H. J.; Kim, M. S.; Kim, K. *J. Raman Spectrosc.* **1987**, *18*, 57.
6. Lee, T. G.; Yeom, H. W.; Oh, S.; Kim, K.; Kim, M. S. *Chem. Phys. Lett.* **1989**, *163*, 98. Joo, T. H.; Yim, Y. H.; Kim, K.; Kim, M. S. *J. Phys. Chem.* **1989**, *93*, 1422.
7. Boettcher, C. J. F. *Theory of Electric Polarization*; Elsevier Scientific Publishing: London, England, 1973; Vol. 1, pp 74–82.
8. van de Hulst, H. C. *Light Scattering by Small Particles*; Dover Publications, Inc.: New York, 1981; p 71.
9. Johnson, P. B.; Christy, R. W. *Phys. Rev. B* **1972**, *6*, 4370.
10. Carron, K., Ph.D. Thesis, Northwestern University, Evanston, IL, 1985.
11. Miller, S.; Baiker, A.; Meier, M.; Wokaun, A. *J. Chem. Soc., Faraday Trans. 1* **1984**, *80*, 1305.

12. Carron, K. T.; Xue, G.; Lewis, M. L. *Langmuir* **1991**, *7*, 2.
13. Puddephatt, R. *The Chemistry of Gold*; Elsevier Scientific Publishing: Oxford, England, 1978; p 61.
14. Dollish, F. R.; Fateley, W. G.; Bentley, F. F. *Characteristic Raman Frequencies of Organic Compounds*; Wiley: New York, 1974; p 172.
15. Dance, I. G.; Fischer, K. J.; Herath Banda, R. M.; Scudder, M. L. *Inorg. Chem.* **1991**, *30*, 183. Dance, I. G. *Polyhedron* **1988**, *7*, 2205.
16. *See*, for example, Porter, M. D. *Anal. Chem.* **1988**, *60*, 1143A.
17. Walls, D.; Bohn, P. W. *J. Phys. Chem.* **1990**, *94*, 2039.
18. Scott, D. W.; McCullough, J. P.; Hubbard, W. N.; Messerly, J. F.; Hossenlopp, I. A.; Frow, F. R.; Waddington *J. Am. Chem. Soc.* **1956**, *78*, 5463.

RECEIVED for review July 15, 1991. ACCEPTED revised manuscript July 24, 1992.

Fourier Transform Infrared and Raman Studies of Coatings

Michael Claybourn[1] and Paul H. Turner[2]

[1]Research Department, ICI Paints, Wexham Road, Slough SL2 5DS, United Kingdom
[2]Bruker Spectrospin Ltd., Banner Lane, Coventry CV4 9GH, United Kingdom

Fourier transform (FT) IR reflectance measurements were used to monitor the top surface and back-interface curing processes of a polyurethane system. Problems associated with the reflectance technique are discussed. Additionally, the fundamental kinetic parameters for bulk film cure, obtained by nonisothermal FTIR measurements, were combined with a model for polymer network growth to predict the buildup in modulus of the curing film. The results were in excellent agreement with experimental data. Processing conditions for polyester coatings play a significant role in the behavior of the final film. FT Raman measurements were performed on coatings based on polyethylene terephthalate–isophthalate to demonstrate structural reordering of polymer chains during heat treatments.

SURFACE COATINGS UNDERGO A VARIETY of structural and chemical changes during initial processing and during their lifetimes. Curing, solvent behavior, heat treatments, deformation, and weathering can influence such factors as chemical and structural integrity, adhesion, hardness, and flexibility (*1–3*). The chemical and physical behavior of these systems must be understood so that predictions can be made concerning the subsequent properties of the protective coating. In addition, chemical, physical, and thermal stress will influence the integrity of the coating.

Vibrational spectroscopic techniques can give functional group information that can be linked to chemical and structural changes. Infrared measurements allow bulk and surface processes to be followed, although some

0065–2393/93/0236–0407$10.00/0

thought must be given to the physics of the measurement (discussed later). For the conventional Raman technique using visible laser excitation, many industrial materials are subject to fluorescence, which tends to obscure the Raman scattered radiation. This problem has largely been overcome by Fourier transform (FT) Raman spectroscopy using near-IR (NIR) excitation (4–6); this combined technique can be used for monitoring bulk processes in coatings.

Although the commercial implications are significant, comparatively little literature discusses the application of vibrational spectroscopic techniques to these problems. Clearly there are difficulties in sensitivity and discrimination for studying surface and interface effects, even with supposed surface techniques such as attenuated total reflectance (ATR) spectroscopy. However, characterization of processes at surfaces and interfaces of coatings is feasible. For example, surfactant enrichment at latex film interfaces has been investigated by ATR and photoacoustic FTIR spectroscopy (7); adhesion and interfacial failure between polymers have been characterized with rheophotoacoustic spectroscopy (8); and adhesion failure at the polymer–metal interface has been characterized by ATR spectroscopy (9). However, factors such as species enrichment or contamination, which govern surface and interface properties, are generally outside the sensitivity limits of these techniques.

Information about the bulk chemical and structural properties of coatings is easier to obtain by FTIR and Raman techniques because sensitivity is less of a problem. This information can help in understanding, for instance, how polymer cross-linking behavior or structural changes under processing conditions affect final coating properties.

For these types of measurements purely empirical kinetic data may be sufficient. However, the curing kinetics of a polymer or resin system can be strongly dependent upon such factors as the glass-transition temperature and the degree of cross-linking at any time during the cure. This dependence means that determination of the underlying kinetic parameters, such as rate constant, order of reaction, frequency factor, and activation energy, becomes difficult (10). These problems can be circumvented to some degree by using nonisothermal kinetic measurements so that the system is maintained above its glass-transition temperature (10).

Investigation of surface coatings by vibrational spectroscopic techniques clearly requires a broad range of approaches to give an understanding of the chemical and physical behavior in the bulk and at interfaces. For this investigation FTIR and FT Raman measurements were used for the analysis of curing and structural changes in polymer coatings, in particular, epoxy–polyester blends and polyurethanes. The optical properties of the coating in relation to FTIR reflectance measurements is discussed. In the analysis of these data, allowance for the optical interactions must be made, otherwise misleading conclusions can be drawn.

FTIR Reflectance Measurements on Polymer Coatings

FTIR reflectance techniques are widely used for characterizing coating systems. However, care must be taken with these types of measurements because they are susceptible to various optical effects that can make interpretation of the spectral data difficult. A basic understanding of the physics of these interactions is useful so that misleading conclusions are not drawn. Figure 1 shows the possible optical phenomena for light incident upon a protective coating on a reflecting substrate. The reflected light will consist of several components, namely, front-surface specular reflection (R_s), back-surface specular reflection (R_b), scattering from the top (and to a much lesser degree the bottom) surface (R_t), and diffuse scatter from scattering centers within the film (R_d). The measured reflectance, R_m, may be written as a sum of all of these:

$$R_m = R_s + R_d + R_t + R_b \tag{1}$$

Any combination of these reflectance components can occur depending upon the optical state of the sample. If all of these phenomena are superimposed, then the spectrum is highly distorted and difficult to interpret. As a consequence, in any reflectance measurement a sampling method that is optimized for one of these reflectance components must be employed. Figure 2 shows spectra for different types of coatings that give predominantly one type of reflected light. Clearly, the type of coating is significant for discrimination of an individual reflectance component.

The Diffuse Component.   The diffuse reflectance component arises when scattering centers are present in the matrix of the coating. This system

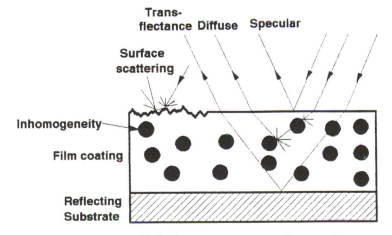

Figure 1. Reflectance of light from a coating on a reflecting substrate.

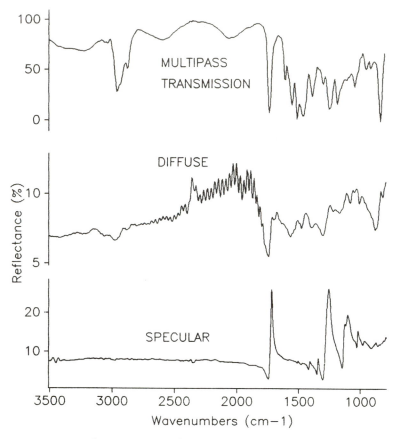

Figure 2. FTIR reflectance spectra from different types of coatings showing the possible spectral types.

will be inhomogeneous and have large discontinuities in refractive index at the interface of the host material and scattering centers. These discontinuities may be inorganic pigments, inclusions, extenders, crystallites, voids, or phase-separated components. As the scattered light passes through the sample, it is absorbed at characteristic frequencies for the scattering centers and for the host material. Eventually, the light escapes the front surface and is collected at the detector. The geometry of this system is described as a diffuse reflectance measurement and will be optimized with a diffuse reflectance accessory comprising a hemispherical collecting lens. If the measurement is performed with a specular reflectance accessory or an FTIR microscope operating in reflectance mode, then the specular component will be enhanced relative to the scattered component. In addition, the path length through the sample will be several micrometers. In effect this reflected light will comprise reflectance and transmittance at the polymer–inhomogeneity interface and multipass transmission through the matrix after scattering at this interface. In some cases the inhomogeneity has a similar refractive index to the matrix, so that little or no scattering occurs. The degree of refractive index mismatch as well as the concentration of scattering centers will strongly influence the intensity of the diffusely scattered component.

This approach is useful only for qualitative analysis of highly pigmented paint films. Figure 3 shows an example of a urethane-modified alkyd paint on a metallic substrate compared with the transmission spectrum of the same material. The reflectance spectrum is highly distorted and displays peak shifts of 20–30 cm^{-1} to higher wave numbers. In addition, the reflectance spectrum has interference fringes in spectral regions with high transmission. The distortion will be associated with superpositioning of the specular reflectance from the front surface. The distortion can cause ambiguities, and so the technique is used only for qualitative analysis at best.

Back-Surface Reflectance.   A back-surface reflection is essentially a double-pass transmission experiment, and subsequently a so-called transflectance spectrum is obtained. Interference between the front-and back-surface-reflected light gives rise to interference fringes; the fringe spacing is related to film thickness, d (*11*):

$$d = \frac{m}{2\sigma\sqrt{(n^2 - \sin^2 \phi)}} \tag{2}$$

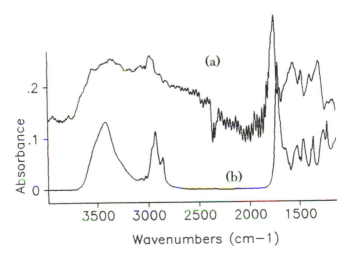

Figure 3. Reflectance (a) and transmission (b) spectra for a urethane-modified alkyd paint film.

where m is the number of fringes between the initial and final fringe, σ is the wave number difference between the initial and final fringe, ϕ is the angle of incidence, and n is the refractive index of the sample. Depending upon their magnitude, these fringes can obscure spectral detail, and it is best to avoid them. They can be removed, for example, by wedging the sample.

An additional problem with these types of measurements on metallic substrates is that dramatic changes in relative band heights can be observed (*12, 13*). These changes are due to an interference effect; the standing waves originating at the metal surface at all the different frequencies give rise to combinations of nodes and antinodes corresponding to minima and maxima in the reflectivity. For a given thickness of sample, nodes and antinodes (and hence, minima and maxima) will occur across the whole range of the reflectance spectrum. Consequently, for spectra from a material at different thicknesses, nodes and antinodes will occur at different wavelengths, and so spectral comparison can be misleading even for qualitative work because band ratios will change. Correction for this effect can be made as long as the optical properties of the material and substrate are known. Figure 4 shows the geometry of a measurement. The amplitude of the reflected light (r) is given by (*14*)

$$r = \frac{r_{12} + r_{12}e^{2i\beta}}{1 + r_{12}r_{23}e^{2i\beta}} \tag{3}$$

where i is r_{12} and r_{23} are the amplitudes of the reflected light from the air–film and film–metal interfaces, respectively, and for normal incidence are

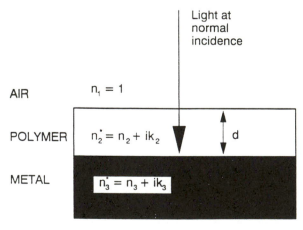

Figure 4. Optical geometry for a polymer coating of thickness d *on a metallic substrate.*

given by

$$r_{12} = \frac{n_2 - 1}{n_1 + 1} \qquad r_{23} = \frac{n_3 - n_2}{n_3 + n_2} \qquad (4)$$

where n_2 and n_3 are the complex refractive indices for the film and substrate, respectively (*see* Figure 4). At normal incidence, β is given by

$$\beta = \frac{2\pi \, dn_2}{\lambda} \qquad (5)$$

where d is the film thickness and λ is the wavelength of light. β describes the degree of attenuation of the light in the system. The measured reflectance R can be determined from

$$R = r \cdot r^* \qquad (6)$$

where r^* is the complex conjugate of r. Therefore, from a knowledge of the optical properties of the polymer and substrate the reflectance spectrum can be modeled to give quantitative data. This approach was employed by Packansky and co-workers (*15*) to model the composition and thickness of organic layered photoconductors. They also showed how this modeling approach can be used for a range of polymers on various metal substrates (*13*).

To show the result of this optical effect, Figure 5 displays spectra across a polyester film on a metal substrate. The film thickness was nominally 7 μm, but showed variations from 5 to 8 μm. The spectra are normalized to the intensity of the C=O stretching band at about 1730 cm^{-1}. Clear variations in band intensities are seen, and, without any knowledge of the physics of this optical interaction, misleading conclusions can be drawn.

Using this analysis, the variation in band ratios can be modeled. Figure 6 shows the variation in reflectance with film thickness for three bands at 1720, 1200, and 925 cm^{-1}; these bands are taken as arbitrary at different points in the spectrum to demonstrate the effect. What may be surprising is that with a comparatively small change in film thickness, the ratio can change significantly. Clearly, if this technique is used to obtain structural or chemical information about a polymer coating on a metallic substrate, then correction of the reflectance spectra must be performed to obtain meaningful data (*15*).

Top-Surface Specular Reflectance. The specular component (front-surface reflection) is often considered as a source of spectral distortion, particularly in diffuse reflectance measurements (*16, 17*). However, the technique can be made use of in cases where this reflected component is predominant. This situation arises when the light that penetrates the sample is either strongly scattered or strongly absorbed. Examples are very thick

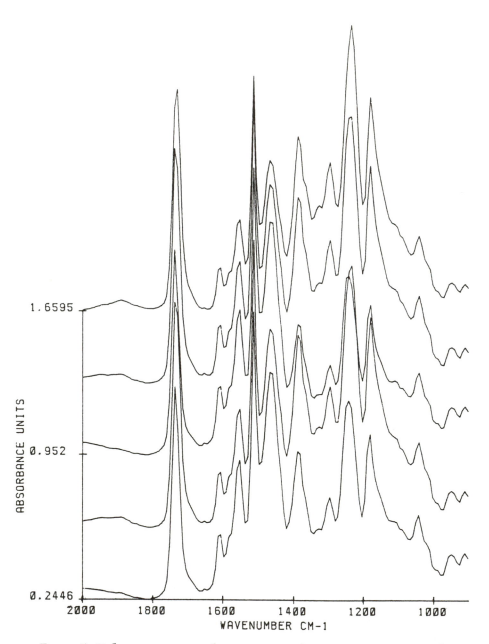

Figure 5. Reflectance spectra taken across a polyester coating on a metal substrate to show the variation in relative band intensities (spectra are normalized to C=O maximum).

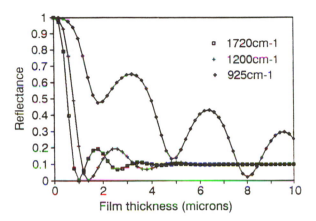

Figure 6. Variation in reflectance intensity with film thickness at 1720, 1200, and 925 cm^{-1}

samples so that back-surface-reflected light is absorbed, an optically rough back surface so that the light is scattered off the specular axis, and strongly absorbing components within the sample. The technique can easily provide analytical information, and additionally, if the measurement is performed with the IR microscope, then microscopic structural information can be obtained (*17*). The technique has been used to study carbon-filled polymers (*18, 19*) because any light penetrating the sample is absorbed and only the top-surface-reflected component is observed. However, one of the problems with this type of sampling is that the spectrum takes on a derivativelike shape. The shape arises from dispersion in the refractive index about resonant species in the material (*20*). However, the absorption index (related to absorption coefficient) can be recovered mathematically from the specular reflectance spectrum.

For normal incidence, the measured specular reflectance is given by

$$R = \frac{(n-1)^2}{(n+1)^2} \qquad (7)$$

where n is the complex refractive index of the material and takes into account both refraction (real) and absorption (imaginary). The imaginary part of the complex refractive index, k, is the absorption index and is related to the absorption coefficient α by

$$\alpha = 2\pi k \sigma \qquad (8)$$

where σ is the frequency in wave numbers. An additional effect to be considered is that light undergoes a phase change upon reflection at an interface; the reflection coefficient, r^*, is related to the reflection amplitude, r, and phase angle ϕ, by

$$r^* = |r|e^{-i\phi} \tag{9}$$

and the measured reflectance, R, is given by

$$R = |r|^2 \tag{10}$$

Separate expressions for n and k at wave number σ can be obtained from these equations:

$$n = \frac{1 - R}{1 - 2\sqrt{R}\,\cos\phi + R} \tag{11}$$

$$k = \frac{2\sqrt{R}\,\sin\phi}{1 + 2\sqrt{R}\,\cos\phi + R} \tag{12}$$

Therefore, if ϕ was determined at all frequencies, then n and k could be fully determined. This condition is possible because R and ϕ are related by the Kramers–Kronig (KK) integral, which enables ϕ to be determined from the measured reflectance (21):

$$\phi = \frac{P}{\pi}\int_0^\infty \frac{\ln(R)}{(\sigma - \sigma_0)}\,d\sigma \tag{13}$$

where P is the principal part of the integral and means that the singularity at $\sigma = \sigma_0$ is calculated as a Cauchy principal value. The integral is performed at each frequency σ_0 over all frequencies, σ. For any causal function, the KK integral describes the relationship between its real and imaginary parts (22), and from equation 9:

$$\ln(r^*) = 0.5\ln(R) - i\phi \tag{14}$$

Details of the origin of this integral have been discussed in the literature (20, 21). The Kramers–Kronig transform is now readily implemented by fast, commercially available software that uses established methods.

FTIR Reflectance Measurements. Although reflectance measurements have inherent problems, valuable data can be obtained. For these measurements FTIR specular reflectance was performed with a Bruker FTIR microscope attached to a Bruker IFS 48 spectrometer. This sampling geometry discriminates in favor of the specular reflected component. For the Kramers–Kronig transform, the data were transferred to a personal computer for processing using LabCalc (23). In addition ATR measurements for the polyurethane reaction at the substrate–coating interface were made with a Specac horizontal ATR accessory with a ZnSe ATR crystal.

FT Raman Measurements on Coatings

Clearly, the difficulties must be carefully considered in an FTIR reflectance measurement on a polymer coating. Raman spectroscopy offers an alternative approach and does not suffer from the problems associated with the IR reflectance method. However, for the conventional Raman technique using visible laser excitation, the types of materials that can be examined are limited because of sample fluorescence. This problem has been overcome by the use of NIR laser excitation, so that fluorescence has been, to a large extent, eradicated. The disadvantage in going to lower energy excitation giving weaker Raman intensity has been overcome by combining with FT instrumentation (5, 6), which gives high throughput. Additional problems are associated with Raman spectroscopy for studying processes in coatings, namely, film thickness and optical properties. For this work on unpigmented polymer films, the major factor affecting the signal-to-noise ratio is the film thickness because it significantly affects the Raman scattering volume.

The instrument used for these measurements was a Bruker IFS 66 FTIR with the FRA 106 Raman module attached. This instrument uses a Nd:YAG laser for excitation operating at 1064 nm. For the measurements samples were simply placed into the sample chamber of the instrument without any sample preparation (24).

Polyurethanes

Cross-Linking in Polyurethane Coatings.   The characterization of the cure processes of coating materials provides an essential base for developing models to describe film formation (25). This base gives a means for predicting the cure behavior as a function of certain variables such as formulation and temperature. Furthermore, the resulting properties of the cured film can be related to the cure behavior. Many coatings are based on a cross-linking reaction between isocyanate and hydroxyl functional resins (26, 27) to form a durable polyurethane film. In real coating systems, both reactants are usually multifunctional, and the alcohol may have both primary and secondary −OH groups. The resulting film is often highly cross-linked to give a high-performance protective coating.

The cure chemistry of isocyanate with hydroxy functional polymers to form polyurethane films is well characterized, and the kinetic parameters have been determined (28). From a technological point of view other parameters, such as polymer functionality, type of solvent, the rate of solvent loss, the curing profile through the film, and how these affect the properties of the subsequent coating, are of significance. To obtain realistic measurements that represent the behavior of the coating as it is used, film thicknesses prevent the use of IR transmission measurements. Reflectance methods are left. However, this situation does not limit us in sensitivity nor in being able to monitor the chemistry and kinetics of the reactions. However, the optical interactions described must be considered in any analysis.

Surface and Interface Curing Reactions. A method for obtaining surface curing information has been developed with a normal incidence specular reflectance measurement. The material for investigation is spread as a 200-μm film onto a black polymer substrate, in this case polypropylene; this step prevents any back-surface reflectance (17). The subsequent surface-reflectance measurement gives spectra with derivativelike line shapes that require Kramers–Kronig transformation to obtain absorption index spectra. Two isocyanate–hydroxy systems were investigated with differing cure rates (the functionalities differed by a factor of 3). Figure 7 shows the behavior of the isocyanate functionality with cure time at room temperature as determined from the N=C=O stretching band at 2275 cm^{-1}. Clearly the rate of reaction is not dependent upon the functionality of the polymers because the rates at the surface were almost identical. The probable reason for this property is that the surface is a dynamic system with very mobile species (e.g., solvent). In addition, the mobility of the polymers at the surface will be higher than in the bulk, so that during the period of the cure that was monitored, the cross-link density had little influence. An additional factor in this reaction will be atmospheric water, which will react with the surface

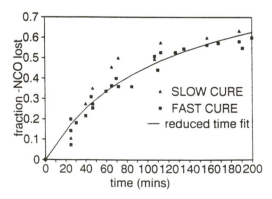

Figure 7. Surface cure of two isocyanate–hydroxy functional polymers with different functionalities.

isocyanate functionality. The solid line in Figure 7 is a reduced time-fit to the data giving the order of reaction (*29, 30*); this was found to be 2 as expected for this bimolecular reaction.

The surface cure plays a significant part in the properties of a coating once formed. If too much stress is built up because of extremes in cure rate at the surface, bulk, and back interface, then the adhesion properties of the coating will be poor. Therefore direct information on the cure profile is needed. To a large extent solvent loss plays an important role because it provides an environment that aids diffusion of polymer chains, in particular those with cross-linking functional groups. The process used to obtain coatings with good properties is generally empirical, without direct measurements on fundamental processes.

FTIR spectroscopy can give information about the reactions at the top surface and coating–substrate interface. Subsequently, the system can be modeled and the physical properties of the coating can be optimized. The reactions for a rapid curing system at the top surface were monitored as already described. The reaction mixture in butyl acetate solvent was spread onto a carbon-filled polypropylene panel at a thickness of 200 μm, and the processes were monitored with time by a normal incidence specular reflectance measurement in the FTIR microscope. For the reaction at the coating–substrate interface the sample was spread onto a horizontal ATR accessory at the same thickness as for the specular measurement. The reaction was then monitored with time under identical conditions to the top-surface measurement. Figure 8 shows the specular reflectance and calculated absorption index spectra for the top-surface cure measurement. The band at 1242 cm^{-1} due to the ester C–O of the solvent decays rapidly over 20 min. The appearance of the band at about 1520 cm^{-1} is due to the urethane linkage. For monitoring the solvent loss, the 1242-cm^{-1} band was used and not the C=O ester band because it was obscured by the C=O band of the polymer, and in addition, the urethane C=O increased during the process.

Figure 9 shows the spectra from the horizontal ATR measurement for the cure at the back surface. The spectra are normalized against the maximum band, which at $t = 0$ is the solvent C–O changing rapidly to a polymer band as solvent is lost. A problem with solvent loss for this measurement is that the average refractive index changes. This factor means that the intensity of the evanescent wave changes and then so does the absolute intensity of the ATR spectral bands. Consequently, the isocyanate band (2273 cm^{-1}) was normalized against the C–O–C band (1147 cm^{-1}) of the polymer. Figure 10 shows plots of the behavior of solvent and isocyanate during the curing reaction for the top surface and the coating–substrate interface. The solvent loss for the top and bottom surfaces was at a similar rate with almost complete loss after 20 min. The isocyanate decay over the period of solvent loss was rapid reducing by 70% at the top surface and by 40% at the bottom

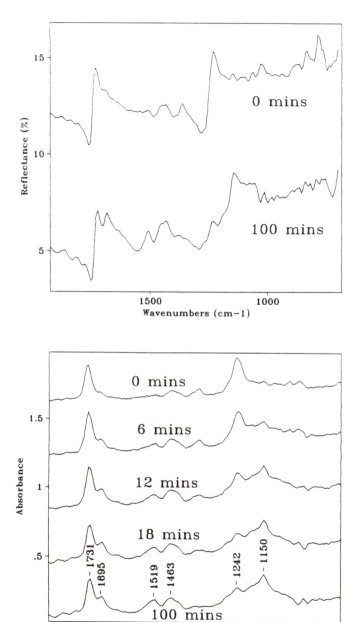

Figure 8. Specular reflectance spectra (top) and calculated absorption index spectra (bottom) for the polyurethane curing system.

surface. Subsequent isocyanate decay was slow as a result of buildup in cross-link density of the coating, which prevented interaction of reactive functional groups.

The difference in the rates of decay of the top and bottom surfaces are due to physical effects such as solvent loss, which create a dynamic system. At the top surface, where all the solvent in the film will eventually migrate and evaporate, the reaction will be enhanced compared to the bottom surface, where solvent will simply diffuse to the surface. This type of information concerning the cure profile helps in understanding the buildup of stress in the curing film. In this particular case the coating had very good adhesive and protective properties.

Isothermal versus Nonisothermal Kinetic Measurements. The curing mechanism is an important consideration in understanding and hence, predicting the behavior of coatings. The approach taken here was to obtain kinetic parameters with FTIR spectroscopy and to attempt to relate this information to changes in modulus by using a model for network growth. To obtain the underlying kinetics for the reaction to build this model, a comparison of isothermal and nonisothermal measurements was initially made to assess the applicability of both. A Bruker IFS 48 FTIR spectrometer was used to follow the curing reaction both isothermally and nonisothermally. The $-NCO$ and $-OH$ components were mixed in the appropriate ratio and coated onto a KBr disk. The disk was mounted onto a Specac variable-temperature cell that has a linear temperature programmer for controlled sample heating and thus allows both isothermal and controlled-temperature ramping experiments. The cure process was monitored by the decay of the isocyanate band at 2275 cm^{-1} both isothermally and at different heating rates.

For the isothermal measurements, whereby the extent of reaction, α, was monitored with time (t) at constant temperature, the kinetic behavior for an nth-order reaction can be described by the relation

$$\frac{d\alpha}{dt} = k(1 - \alpha)^{n} \tag{15}$$

where k is the rate constant and is thermally activated, following an Arrhenius expression:

$$k = Ae^{-(E/RT)} \tag{16}$$

where R is the gas constant; T is absolute temperature; and the kinetic

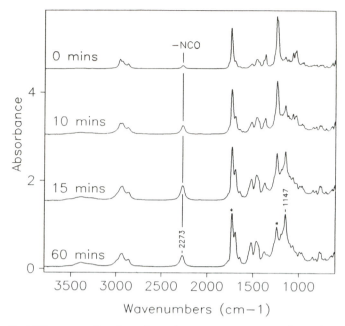

*Figure 9. ATR spectra for the polyurethane curing system. The * denotes the main solvent bands.*

parameters, namely, the activation energy (E), preexponential factor (A), and the order of reaction (n), can be determined by a series of isothermal measurements at different temperatures.

Isothermal measurements were performed at temperatures of 20, 80, 100, and 120 °C; the behavior of the band due to –NCO at each temperature is given in Figure 11. At 20 °C the reaction was very slow, and after 15 h the process had gone only to 15% completion. For each temperature, except 20 °C where the extent of reaction was too low, the apparent order of reaction of reaction was calculated by using the reduced-time method (*29, 30*). The orders of reaction for temperatures of 80, 100, and 120 °C were 5.7, 3.6, and 4, respectively. These values are anomalously high but are typical for diffusion-controlled reactions. In this case the controlling parameter for the cure process was the increasing glass-transition temperature (T_g). The results demonstrate the shortcoming of the isothermal method applied to polymer systems with final T_g values that are above ambient conditions. The different behavior at the different temperatures means that the activation energy for the reaction cannot be determined from these isothermal results.

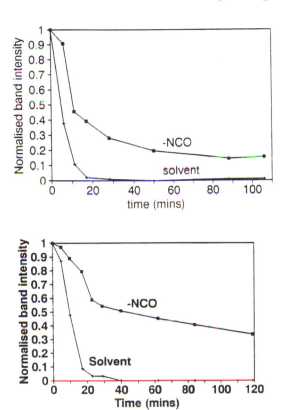

Figure 10. Behavior of the isocyanate and solvent for top surface (top) and coating–substrate interface (bottom).

Nonisothermal kinetic measurements on film curing processes can overcome these experimental problems. Techniques such as dynamic mechanical analysis (DMA), differential scanning calorimetry (DSC), and FTIR spectroscopy can provide a means for obtaining information for the kinetics of the cure using temperature scanning methods (25, 31–34). The effect of the T_g is avoided as long as the experimental temperature is always held above the T_g. To obtain the kinetic parameters from the rising-temperature FTIR results, we modified the approach originally devised by Ozawa (35). This analysis is based on a series of experiments performed at different heating rates. For this technique, the kinetic behavior is modeled by using an Arrhenius-type expression based on equations 15 and 16:

$$\frac{d\alpha}{dT} = \frac{(1-\alpha)^n}{b} A e^{(-E/RT)} \qquad (17)$$

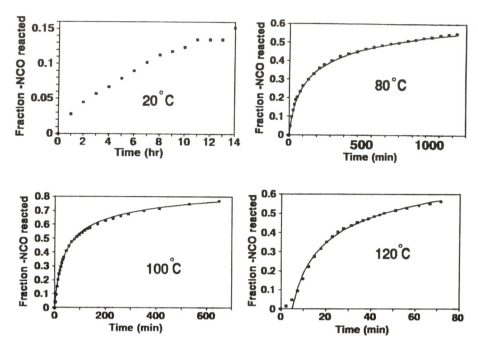

Figure 11. Fraction of isocyanate lost with cure time for isothermal measurements. (Reproduced with permission from reference 30. Copyright 1992 John Wiley.)

where $b = dT/dt$, that is, the heating rate. For a rising-temperature experiment (36)

$$g(\alpha) = \frac{AR}{bE} T^2 e^{-(E/RT)} I(E, T) \tag{18}$$

where

$$I(E, T) = 1 - \frac{2!}{(E/RT)} + \frac{3!}{(E/RT)^2} - \frac{4!}{(E/RT)^3} + \cdots \tag{19}$$

and

$$g(\alpha) = \int_0^\infty \frac{d\alpha}{(1 - \alpha)^n} \tag{20}$$

Taking $T = 20$ °C at time zero, thermal lag and resulting experimental errors can be avoided. Experiments were performed at heating rates of 0.5, 1, 2, 3,

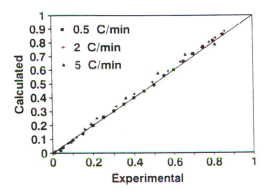

Figure 12. Predicted against experimental isocyanate fraction lost for the nonisothermal measurements. (Reproduced with permission from reference 30. Copyright 1992 John Wiley.)

and 5 °C/min. The experimental data were modeled by using equation 18. The fit between the experimental data and kinetic model is shown in Figure 12. There is excellent agreement, and the cure kinetics for this particular system can be described by the expression

$$\frac{d\alpha}{dt} = (1-\alpha)^{(2.9\pm0.1)}(6.6\pm0.4) \times 10^3 e^{(-(52\pm4)\times10^5)/RT} \qquad (21)$$

Although the reaction order is considerably lower than any of the isothermal experiments, it does not have the expected value of 2 that a bimolecular reaction should have. However, this lowering and the fact that n is independent of b provide additional evidence that the T_g has not influenced the kinetics of the reaction. Overall the results confirm that the nonisothermal approach is more reliable. If there were any influence of the T_g on the reaction rate, there would be significant deviations from the model, particularly at the slowest heating rates. The high value for the order of reaction probably arises because, once the gel point has been passed and a dense network established, then diffusion eventually has a predominating influence on the reaction kinetics.

We developed an approach that combines the kinetic parameters from nonisothermal measurements with a simple model for polymer network growth and simple rubber elasticity theory with the result that we can predict the rise in elastic modulus as measured by DMA (*30*). The correlation of the predicted against the measured rise in modulus for the polyurethane system described here is shown in Figure 13. This approach enables us to predict the behavior of coating materials from their formulation under different curing conditions. In general terms, this type of modeling gives a more predictive approach for optimizing commercial formulations.

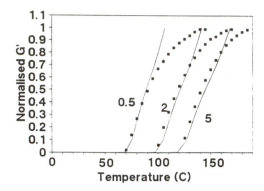

Figure 13. Theoretical fit (—) to the experimental (■) modulus obtained by DMA for the curing polyurethane. (Reproduced with permission from reference 30. Copyright 1992 John Wiley.)

Polyester Coatings

Structural Changes in Polyesters. The polyester poly(ethylene terephthalate) (PET) is extensively used commercially in the form of fibers and films because of its outstanding physical and chemical properties. It is important to have precise control over the processing of this material to give the optimum properties for specific applications. During the processing of PET such as drawing or annealing, the changes in the micro- and macroscopic structure of this polymer can strongly influence the final product properties. For example, the stress–strain behavior of heat-treated PET fibers (oriented and crystalline) can vary widely (*37*). In an effort to understand the behavior of PET, extensive studies have been performed (*38–53*) with a wide range of techniques including vibrational spectroscopy.

Changes in the level of crystallinity of PET have been associated with the relative amounts of the trans and gauche conformations of the glycol unit. In the amorphous state, PET has a predominantly gauch structure (*38–40, 45*), containing about 14% trans (*49*), whereas the crystalline phase has the trans conformation (*54*).

Changes in polymer structure can be characterized with Raman spectroscopy. The Raman band that relates to the trans structure type in PET occurs at 1096 cm^{-1} (*40, 55*); this band was shown to correlate linearly with polymer density and, hence, crystallinity. This band has been assigned to a mixture of stretching of ring and ethylene glycol C–C bonds and to stretching of the ester C–O bond (*43*). An alternative method for following structural changes relating to crystallinity is to monitor the C=O bandwidth, which varies by about a factor of 2.5, from 11 cm^{-1} (crystalline) to 26 cm^{-1} (amorphous); the dependence on density is also linear (*41, 56, 57*). In amorphous PET, there is a wide distribution of C=O orientation with respect to each other and to the phthalate unit. The functional group interactions under these varying environments give slightly different C=O stretching frequencies, which overlap to give a broad band. With increased ordering the C=O occupies a fixed orientation relative to the benzene ring with the result that the Raman band then narrows.

Other bands show increased intensity resulting from structural changes relating to ordering of the polymer chains, for example, at 857 and 278 cm^{-1} (*48, 55*). The 857-cm^{-1} normal mode contains substantial contributions from the breathing vibration of the benzene ring, the ring–ester stretching mode, the C–O stretch, rotation of CH$_2$ groups, and O=C–O angle deformation; however, this band appears to show a weaker dependence upon orientation for uniaxially oriented films (*58*). One problem is attempting to correlate Raman band intensities with density or crystallinity is that absolute Raman intensities cannot be used unless absolute Raman scattering cross sections are measured. To obtain the correlation, the relevant band is normalized against the band at about 633 cm^{-1} (*40*). This method is generally valid except for uniaxially oriented fibers, which show a clear nonlinear dependence (*55, 56, 58*).

Several infrared bands can be related to the trans and gauche conformers, and hence structural changes during the processing of PET can also be monitored by infrared techniques (*48–51*). The trans conformer gives bands at 1473, 1388, 988, 973, and 849 cm^{-1}, and bands for the gauche conformer are at 1454, 1372, and 898 cm^{-1}. Detailed assignments of these bands can be found in reference 48. The kinetics for the gauche–trans isomerization in PET were determined by FTIR spectroscopy (*51*).

Clearly, during the processing of PET, be it thermal or physical stress, changes in the structural properties of the polymer can be monitored by vibrational spectroscopic techniques. This work provides information about how the properties of the polyester component change when disorder is introduced into the polymer backbone by replacing some of the terephthalate component with isophthalate and additionally, how the rate of structural changes in the polyester may vary by blending with an epoxy.

Structural Properties of Poly(ethylene terephthalate) and Poly(ethylene terephthalate/isophthalate) (PETI) by FT Raman Spectroscopy.

Replacement of a proportion of the terephthalate units (1,4-substituted aromatic) in PET with isophthalate (1,3-substituted aromatic) introduces a more disordered structure to the polymer because the additional substituents are not linear. The polymer chains are bent at the positions of the isophthalate, and this configuration enhances chain entanglement and structural disorder. Above the T_g of the polymer, gauche–trans isomerism in the glycol units can occur. The degree of structural ordering and crystallinity in PET will depend upon this isomerism. However, with the introduction of isophthalate, the limit for the most ordered structure (i.e., with all *trans*-glycol units) will depend upon the amount of isophthalate component in the polymer. Ultimately, a minimum isophthalate composition will exist at which little or no ordering can be induced for a given processing condition.

The poly(ethylene terephthalate)–isophthalate ($PET_{1-x}I_x$) used for this study had varying molar ratios of the terephthalate and isophthalate components, with $x = 0$, 0.05, 0.1, 0.15, 0.18, 0.25, and 0.35. These amorphous, random copolymers (that is, random distribution of the isophthalate and terephthalate components) were obtained in pellet form. Figure 14 shows Raman spectra for this range of composition in the spectral region 1500–500 cm^{-1}. The additional features due to the isophthalate component occur at 1309, 1003, and 658 cm^{-1}. A plot of the integrated intensity of the 1003-cm^{-1} band normalized against the 633-cm^{-1} band as a function of percent isophthalate is shown in Figure 15; the dependence is clearly linear and shows direct correlation with this component.

The glass-transition temperatures for the full range of compositions from poly(ethtylene isophthalate) (PEI) to PET vary from 59 to 79 °C, although over the range of interest the variation is only about 7 °C (72 to 79 °C) (59). To induce the gauche–trans conformational change, it is necessary to anneal the samples above the T_g when the polymer chains become mobile. For the initial part of this work the polymer pellets for all the compositions of $PET_{1-x}I_x$ were annealed with time in an oven at 110 °C to obtain the structural transformation. FT Raman spectra were obtained from these samples for detailed analysis of the conformational changes in the polymer and the effect of varying the isophthalate–terephthalate ratio. Measurements were performed on the Bruker FT Raman system already described. Sampling was a simple case of containing the heat-treated pellets in glass vials and mounting these into the spectrometer. Glass is a very weak Raman scatterer and gives only a negligible contribution to the spectrum.

The spectral changes in the 1500–100-, 1800–1650-, and 3200–2850-cm^{-1} regions for PET and $PET_{0.82}I_{0.18}$ before and after heat treatment at 110 °C for 3 h are shown in Figures 16 and 17, respectively. The starting material for both of these compositions was predominantly amorphous, as indicated by the width of the C=O band at 1726 cm^{-1} (*41, 56, 57*), which was 26.5 cm^{-1} for PET and 26.2 cm^{-1} for $PET_{0.82}I_{0.18}$. This result was confirmed in both cases by the apparent absence of the band at about 1097 cm^{-1} due to the trans conformation associated with the ordered polymer

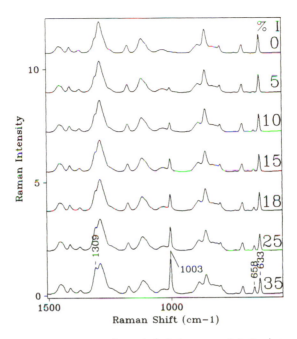

Figure 14. FT Raman spectra for poly(ethylene terephthalate)–isophthalate at different compositions.

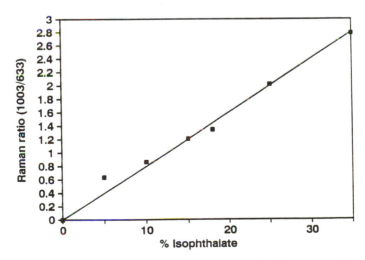

Figure 15. Raman intensity ratio of the isophthalate band (1003 cm^{-1}) against percent isophthalate in the polymer.

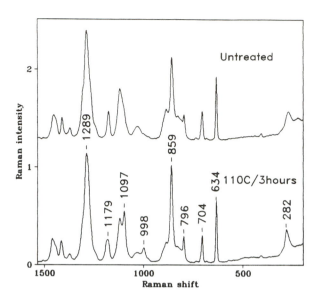

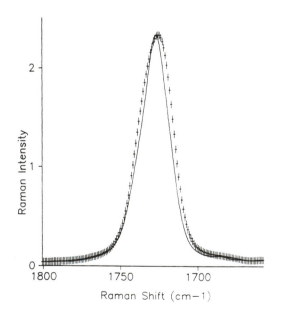

Figure 16. FT Raman spectra in different spectral regions for PET before and after a heat treatment at 110 °C. Key: +, untreated; and —, treated at 110 °C for 3 h.

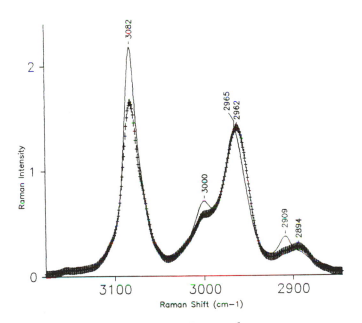

Figure 16. Continued.

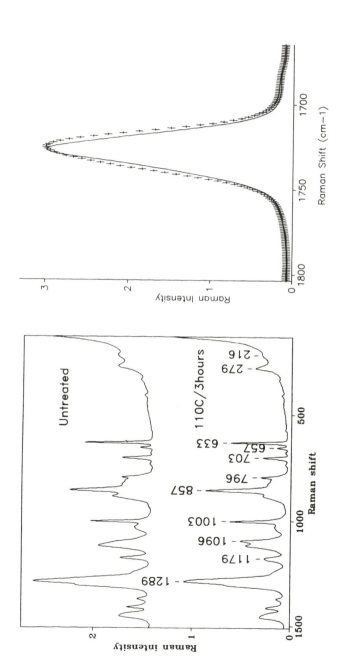

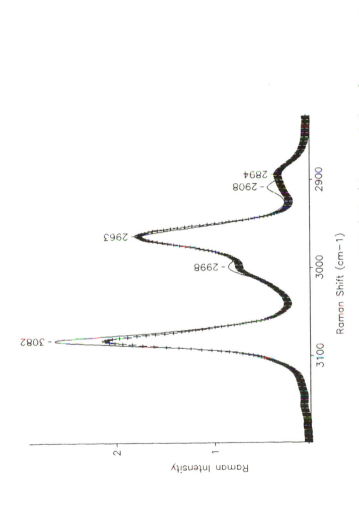

Figure 17. KFT Raman spectra in different spectral regions for PETI (18% I) before and after a heat treatment at 110 °C. Key: +, untreated; and —, treated at 110 °C for 3 h.

structure (*40, 55*). However, a clear shoulder occurs at about 1100 cm^{-1} at the low-wave-number side of the broad peak at 1120 cm^{-1}, so there appears to be a trans component in the starting material. This result will be expected because the amorphous structure will be a mixed composition of trans and gauche conformers to give the disordered structure; there is about 14% of the trans conformer in amorphous PET (*49*).

Additional changes in the spectra have not been studied in detail. These include an increase in the 859-cm^{-1} band whose normal mode has contributions from several different vibrations (*see* previous section). Also, extensive changes occur in the C–H stretching region, including sharpening of the bands at 2909, 3000, and 3082 cm^{-1}, which results from structural ordering for both the glycol and terephthalate units.

Figures 18–20 show the behavior of the C=O bandwidths and the normalized intensities of the bands at about 1096 and 857 cm^{-1} with annealing time for the different polyester compositions. At low isophthalate levels up to 10%, the rate of the gauche–trans conformational change does not vary. As the isophthalate component is increased, the facility to introduce order into the polymer is reduced because of the disordering effect of the isophthalate. For 25% and above there was no ordering over the time of the experiment. Figure 21 shows that the rates of change of the C=O bandwidth and intensities of the 1096- and 857-cm^{-1} bands are linearly related and gives confirmatory data that the 857-cm^{-1} band is suitable for monitoring

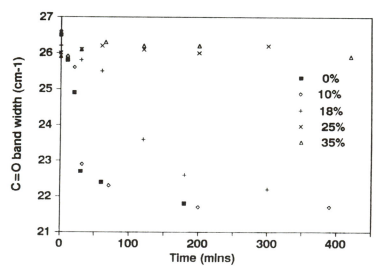

Figure 18. Behavior of the C=O bandwidth with annealing time at 110 °C for PETI at different percentages of isophthalate.

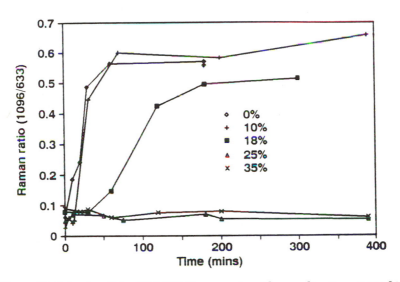

Figure 19. Behavior of the 1096/633 Raman ratio with annealing time at 110 °C for PETI at different percentages of isophthalate.

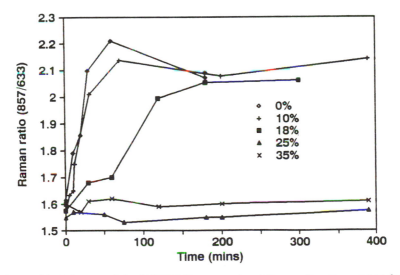

Figure 20. Behavior of the 857/633 Raman ratio with annealing time at 110 °C for PETI at different percentages of isophthalate.

structural changes in the polymer. Although the 857-cm^{-1}, 1096-cm^{-1}, and C=O bands are not related to the same functional groups, they do indicate an overall ordering in the polymer chain during the heating process. As mentioned, the 857-cm^{-1} band does not show good correlation for uniaxially oriented films. However, the material used in this case was amorphous, unoriented pellets.

Over the composition range of PET$_{1-x}$I$_x$ ($0 \geq x \geq 0.35$) studied, the gauche–trans conformational change in the glycol units occurred with the 110 °C heat treatment. The major contribution to the structural ordering in the polymer occurs across the ethylene terephthalate (ET) units; the gauche–trans conformational change over a section of polymer chain with a significant number of ET units will enhance the linearity over that region. In the ethylene isophthalate (EI) units, the gauche–trans isomerism will not have a significant effect on the linearity of the polymer chains because of the 1,3-substitution on the aromatic ring. Increasing the amount of EI in the polymer increases the disorder in the polymer. However, for the heat treatment used in this case, for low EI compositions (up to 10%), the degree and rate of crystallization remain constant. For the 18% EI composition, the rate drops significantly, although the final degree of crystallization (equilibrium after prolonged heating at 110 °C) is only slightly less than for PET. For higher EI concentrations, the disordering effect of the EI units plays a significant role, and in fact no detectable change in crystallinity was found.

Structural Properties of Polyester–Epoxy Films by FT Raman Spectroscopy.

The problem of reflectance measurements for films on reflective substrates was discussed earlier. To obtain structural information from FTIR measurements with the reflectance approach, a clear knowledge of the optical properties of the polymer must be known. It is a complex process to obtain accurate values for the optical properties over the spectral range of interest. Consequently, the preferred method for characterizing the films was FT Raman spectroscopy.

For this work we were interested in following the change in structural ordering of PETI (I = 18%) in a model film composed of a blend of the polyester and an epoxy resin. Changes in crystallinity can affect the protective properties of the film. Figure 22 shows spectra for the raw materials.

The films were applied to metal substrates at a thickness of about 10 μm and were heat-treated at 110 °C for periods up to 3 h. The spectra for the untreated and the 3-h treated samples are shown in Figure 23. The signal-to-noise ratio was not sufficient to allow accurate measurement of the C=O bandwidth. In addition the 1096-cm^{-1} band could not be monitored because it was obscured by a band from the epoxy resin; although a change was just detectable (*see* Figure 23 bottom). Therefore, the band at 855 cm^{-1} was

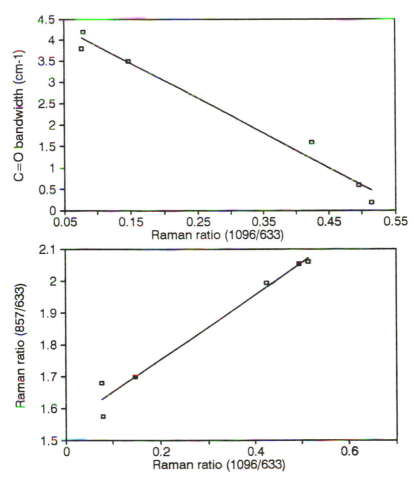

Figure 21. C=O bandwidth and 857/633 Raman ratio against 1096/633 Raman ratio.

monitored to determine ordering effects in the polyester. Figure 24 shows the behavior of this band with time of annealing. The absolute value for the band ratio (855/633) did not correlate with the previous work on the pellets because each overlapped with other bands from the epoxy resin. This condition meant that the integrated intensities had some contribution from neighboring bands. The rate of change appeared to occur at a similar rate, reaching a maximum after about 2 h for both the PETI pellets and the polymer blend films.

The potential for Raman spectroscopy to monitor structural changes, even in blended polymer films, is evident from these results. The physical and chemical properties of the film can then be related to the molecular

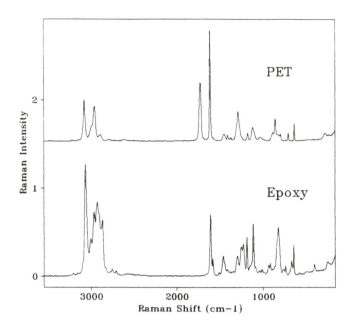

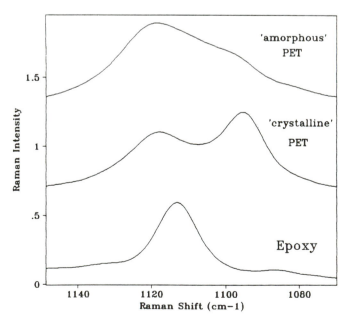

Figure 22. FT Raman spectra for PET and the epoxy.

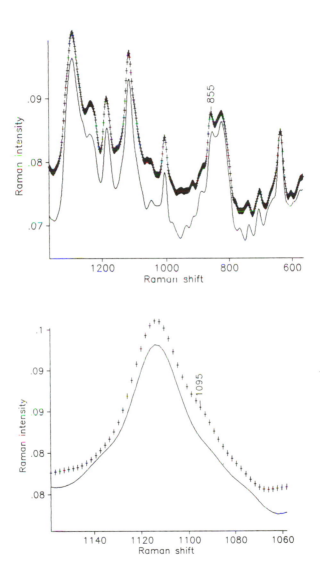

Figure 23. Top: PET–epoxy blend film (about 8 μm thick) showing the increase in the 857-cm^{-1} band. Key: —, untreated; and +, treated at 110 °C for 3 h. Bottom: expanded region around the 1096-cm^{-1} band.

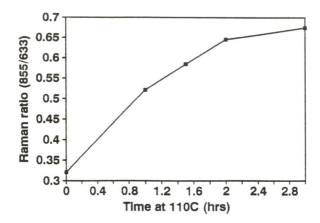

Figure 24. Variation of the 855/633 Raman ratio with annealing time at 110 °C.

information obtained by Raman spectroscopy. This ability is particularly important for optimizing the processing of the polymer material to give the desired properties.

Acknowledgments

We thank ICI Paints for their support and encouragement in this work, in particular D. M. Dick and M. Reading.

References

1. Ellis, W. H. *J. Coatings Technol.* **1983**, *55*, 63.
2. Provder, T. *J. Coatings Technol.* **1989**, *61*, 33.
3. Sullivan, W. F. *Prog. Org. Coatings* **1972**, *1*, 157.
4. Chantry, G. W.; Gebbie, H. A.; Hilsum, C. *Nature* **1964**, *203*, 1052.
5. Chase, B. *J. Am. Chem. Soc.* **1986**, *108*, 7485.
6. Hirschfeld, T.; Chase, B. *Appl. Spectrosc.* **1986**, *40*, 133.
7. Urban, M. W.; Evanson, K. W. *Polymer Commun.* **1990**, *31*, 279.
8. McDonald, W. F.; Urban, M. W. *J. Adhesion Sci. Technol.* **1990**, *4*, 751.
9. Spadafora, S. J.; Leidheiser, H. *J. Oil Colour Chem. Assoc.* **1988**, *71*, 276.
10. Claybourn, M.; Reading, M. *J. Appl. Polym. Sci.* **1992**, *44*, 565.
11. Harrick, N. J. *Appl. Optics* **1971**, *10*, 2344.
12. Briggs, L. M.; Bauer, D. R.; Carter, R. O. *Ind. Eng. Chem. Res.* **1987**, *26*, 667.
13. Packansky, J.; Waltman, R. J.; Grygier, R. *Appl. Spectrosc.* **1989**, *43*, 1233.
14. Born, M.; Wolf. E. *Principles of Optics*; Pergamon Press: New York, 1964.

15. Packansky, J.; England, C.; Waltman, R. J. *J. Polym. Sci. B: Polym. Phys.* **1987**, 25, 901.
16. Hembree, D. M.; Smyrl, H. R. *Appl. Spectrosc.* **1989**, 43, 267.
17. Claybourn, M.; Colombel, P. *Bruker Report*; Bruker Analytische Messtechnik GMBH: Karlsruhe, Germany, 1990; Vol. 1, pp 8–12.
18. Claybourn, M.; Colombel, P.; Chalmers, J. *Proc. Int. Workshop Fourier Transform Infrared Spectrosc.*; Vansant, E. F., Ed.; University of Antwerp Press: Antwerp, Belgium, 1990.
19. Claybourn, M.; Colombel, P.; Chalmers, J. *Appl. Spectrosc.* **1991**, 45, 279.
20. Lipson, S. G.; Lipson, H. *Optical Physics*; Cambridge University Press: Cambridge, England, 1981.
21. Harbecke, B. *Appl. Phys. A* **1986**, 40, 151.
22. Roth, J.; Rao, B.; Dignam, M. J. *Trans. Faraday Soc.* **1975**, 71(1), 86.
23. LabCalc software, Galactic Industries Inc., 395 Main Street, Salem, NH 03079.
24. Claybourn, M.; Agbenyega, J.; Hendra, P. J., Chapter 17 in this book.
25. Provder, T.; *J. Coatings Technol.* **1989**, 61, 32.
26. Saunders, J. H.; Frisch, K. C. *Polyurethanes: Chemistry and Technology*; Vol. I, Robert E. Krieger Publishing: Malabar, FL, 1962.
27. *Polyurethane Handbook: Chemistry, Raw Materials, Processing, Applications, and Properties*; Oertel, G., Ed.; Hanser Publishers: Munich, Germany, 1985.
28. Hernandez-Sanchez, F.; Vazquez-Torres, H. *J. Polym. Sci. A: Polym. Chem.* **1990**, 28, 1579.
29. Sharp, J. H.; Brindley, G. W.; Achar, B. N. N. *J. Am. Ceram. Soc.* **1966**, 49, 379.
30. Claybourn, M.; Reading, M. *J. Appl. Polym. Sci.* **1992**, 44, 565.
31. Snyder, R. W.; Wade Sheen, C. *Appl. Spectrosc.* **1988**, 42, 655.
32. Carlson, G. M.; Provder, T. *Proc. Water-Borne High Solids Coating Symp.* **1985**, 12, 44.
33. Carlson, G. M.; Neag, C. M.; Kuo, C.; Provder, T. *Polym. Prepr.* **1984**, 25, 171.
34. Provder, T.; Neag, C. M.; Carlson, G. M.; Kuo, C.; Holsworth, R. M. *Anal. Calorim.* **1984**, 5, 377.
35. Ozawa, T. *J. Ther. Anal.* **1970**, 2, 301.
36. Whitehead, R.; Dollimore, D.; Price, D.; Fatemi, N. S. *Proc. 2nd Europ. Symp. Therm. Anal.*; Dollimore, D., Ed.; Heyden: Aberdeen, Scotland, 1981; p 51.
37. Gupta, V. B.; Kumar, S. *J. Appl. Polym. Sci.* **1981**, 26, 1885.
38. Ward, I. M. *Chem. Ind. (London)* **1956**, 905.
39. Manley, T. R.; Williams, D. A. *Polymer* **1969**, 10, 339.
40. McGraw, G. E. *Polym. Prepr.* **1970**, 11, 1122.
41. Melveger, A. J. *J. Polym. Sci.: Polym. Phys. Ed.* **1972**, 10, 317.
42. Cunningham, A.; Ward, I. M.; Willis, H. A.; Zichy, V. *Polymer* **1974**, 15, 749.
43. Boerio, F. J.; Bahl, S. K.; McGraw, G. E. *J. Polym. Sci.: Polym. Phys. Ed.* **1976**, 14, 1029.
44. Purvis, J.; Bower, D. I. *J. Polym. Sci.: Polym. Phys. Ed.* **1976**, 14, 1461.
45. Ward, I. M.; Wilding, M. A. *Polymer* **1977**, 18, 327.
46. Jarvis, D. A.; Hutchinson, I. J.; Bower, D. I.; Ward, I. M. *Polymer* **1980**, 21, 41.
47. Hutchinson, I. J.; Ward, I. M.; Willis, H. A.; Zichy, V. *Polymer* **1980**, 21, 55.
48. Stokr, J.; Schneider, B.; Doskocilova, D.; Lovy, P.; Sedalek, P. *Polymer* **1982**, 23, 714.
49. Lin, S-B.; Koenig, J. L. *J. Polym. Sci.: Polym. Phys. Ed.* **1982**, 20, 2277.
50. Lin, S-B.; Koenig, J. L. *J. Polym. Sci.: Polym. Phys. Ed.* **1983**, 21, 1539.
51. Lin, S-B.; Koenig, J. L. *J. Polym. Sci.: Polym. Phys. Ed.* **1983**, 21, 2365.
52. Yazdanian, M.; Ward, I. M.; Brody, H. *Polymer* **1985**, 26, 1779.

53. Adar, F.; Noether, H. *Polymer* **1985**, *26*, 1935.
54. Daubeny, R.; Bunn, C. W.; Brown, C. J. *Proc. Roy. Soc.* **1954**, *226*, 531.
55. McGraw, G. E. In *Polymer Characterization: Interdisciplinary Approaches*;
    Craver, C. D., Ed.; Plenum Press: New York, 1972.
56. Melveger, A. J.; *Polym. Prepr.* **1972**, *13*(*1*), 180.
57. Bulkin, B. J.; Lewin, M.; DeBlase, F. J. *Polym. Mater. Sci. Eng.* **1986**, *54*, 397.
58. Purvis, J.; Bower, D. I. *J. Polym. Sci.: Polym. Phys. Ed.* **1976**, *14*, 1461.
59. Hsiue, G.; Yeh, T.; Chang, S. *J. Appl. Polym. Sci.* **1989**, *37*, 2803.

RECEIVED for review July 15, 1991. ACCEPTED revised manuscript June 25, 1992.

Fourier Transform Raman Spectroscopy in the Study of Paints

Michael Claybourn[1], Jonathan K. Agbenyega[2], Patrick J. Hendra[2], and Gary Ellis[3]

[1]Research Department, ICI Paints, Wexham Road, Slough SL2 5DS, United Kingdom
[2]Chemistry Department, University of Southampton, Highfield, Southampton S09 5NH, United Kingdom
[3]Consejo Superior de Investigaciones Científicas, Instituto de Ciencia y Technología de Polímeros, Calle Juan de la Cierva 3, 28006 Madrid, Spain

Fourier transform (FT) Raman spectroscopy with near-IR laser excitation was used successfully as an analytical tool for characterizing paint materials and processes. The limitations of the technique are discussed, in particular the problems of fluorescence, absorption, and heating; these effects are very much sample dependent. In addition we have shown the facility of the technique for studying waterborne systems with high sensitivity, in this case for latex systems; for this work an emulsion polymerization reaction of an acrylic system was monitored with reaction time to obtain the kinetic behavior for the process. Finally, autoxidation reactions during the cure of an alkyd system and its model compounds were monitored by both FT Raman and FTIR spectroscopy to gain a clearer understanding of the processes involved. FT Raman spectroscopy showed the configurational changes occurring for the C=C bond (cis–trans isomerism and conjugation) during autoxidation.

THE USE OF RAMAN SPECTROSCOPY for characterizing surface coatings has received little attention from the paints and coatings industry, unlike IR spectroscopy. The four basic reasons for this lack of interest are as follows:

1. The cost of instrumentation generally exceeds the budget of most industrial analytical laboratories.

2. If the sample under investigation absorbs strongly at the excitation and Raman wavelengths, then the Raman scatter will be too weak to be detectable.

3. Fluorescence, which may be inherent to a material or due to impurities, may obscure the weaker Raman scattered radiation.

4. Because only a relatively small number of industrial samples will give interpretable Raman spectra and do so only slowly and further, skilled operators are required, the cost per sample is very high (1).

Although conventional Raman spectroscopy presents problems, the technique can provide unique information and has potential advantages for investigating a broad range of analytical applications. For example, highly polarizable functional groups give a strong Raman scattering intensity, and groups with high polarity tend to give intense resonant absorption in the infrared region. Hence, structural and chemical properties of a material not easily characterized by IR may be characterized by Raman spectroscopy. For example, Raman spectroscopy has a higher sensitivity for the more symmetric bonds (e.g., C–C, C=C, N–N, S–S, and O–O).

Conventional Raman spectroscopy has been used in investigating paint systems is some cases. These include monitoring the cure of an alkyd resin (2) and the accelerated weathering of alkyd paints (3). However, the spectral data reported were poor because of the problems of sample fluorescence and high background response.

The problems with the conventional Raman technique using visible laser excitation have largely been overcome with the development of Fourier transform (FT) instrumentation using a near-IR (NIR) laser (4–9). This approach was originally proposed but not exploited by Chantry et al. (10). The use of lower energy excitation means that the Raman scattered radiation is much weaker in the NIR than in the visible range (the Raman intensity is proportional to the fourth power of the laser frequency). However, the étendue gain characteristic of the FT instrument somewhat outweighs this disadvantage because high instrument throughput is obtained (11). In addition, the lower excitation energy means that the electronic states attributable to the fluorescence effect are not readily accessible; however, fluorescence certainly is detected in NIR FT Raman instruments (see, for example, reference 12).

The earlier commercial FT Raman instruments were based on FTIR optical benches with the Raman facility provided by an add-on module. This type of system is a very flexible and powerful tool for performing molecular vibrational analyses because both FTIR and FT Raman spectra can be acquired on the same instrument. This addition to existing FT technology has meant that the cost of the total instrument is relatively low and certainly accessible to most analytical laboratories. The current technology includes dedicated FT Raman instruments with the optics optimized specifically for the measurement. The commercial instruments available are designed for safety, ease of use, little or no sample handling, no instrument alignment, and minor sample alignment. The result is that Raman spectra can now be routinely and rapidly measured by technical staff without extensive training. Rapid improvements in FT Raman technology have made the technique even more applicable for routine analysis as well as providing the facility for the detailed chemical and structural requirements of a research laboratory. Its broadening capabilities have brought it to the attention of potential industrial users in cases when product innovation depends upon a clear knowledge of the chemistry of a material or process. FT Raman spectroscopy can provide unique information and, as a complementary technique to FTIR spectroscopy, previously intractable problems can be approached. The instrument is sufficiently versatile to fulfill the needs of an analytical laboratory as well as one dedicated to research.

With the advent of FT Raman spectroscopy many analytical applications have been developed. Examples of these applications for the analysis of paint and polymer-related materials include chemical processes in paint systems (*13–15*), polymers (*16*), elastomers (*17*), polymer degradation (*18*), and organometallic N compounds (*19*). (Organocobalt compounds have been characterized (*20*); with the appropriate sensitivity, Raman spectroscopy may be useful for studying the behavior of cobalt catalysts.)

FT Raman spectroscopy has been clearly identified as an analytical tool with a potentially broad range of industrial application (*7, 12, 21, 22*). One of the key factors is the simplicity of sampling. Sampling accessories have been specifically designed for commercial FT Raman instruments. Furthermore, advances are being made in remote sampling using optical fibers (*23–26*); this approach is important in an industrial environment, for example, in process monitoring to obtain kinetic data or monitoring process end points. Other important features for industrial use are quantitative analysis (*27*) including the possibility of using chemometrics (*28, 29*) and microscopy (*30, 31*).

Advantages of FT Raman Spectroscopy for Studying Paint Materials

FT Raman spectroscopy presents several specific advantages for studying paint materials. The first is that little or no sample preparation is required; samples are examined in their raw state. This feature is also advantageous for rapid routine analysis and should prove powerful for troubleshooting.

Opaque samples such as pigments, extenders, and paint films are easily analyzed. Waterborne systems can be characterized because water is a weak

Raman scatterer. However, water does absorb in the NIR range so that the Raman radiation from the sample can be absorbed (this topic is discussed later). FT Raman spectroscopy gives the facility for following processes in latexes and direct analysis of waterborne paint systems without any sample preparation or problems of sensitivity.

The advantage of high sensitivity in environments where symmetric bonds (e.g., C–C, C=C, N–N, and S–S) are present has many important applications for following chemical processes such as emulsion polymerization, cross-linking reactions, and polymer network growth.

The various C=C species (trans, cis, vinyl, vinylidene, trisubstituted, and tetrasubstituted) can be distinguished. Of particular interest here is following in-chain and end-chain chemistry.

Films can be characterized easily. FTIR reflectance measurements on films tend to suffer from spurious optical effects that can make direct interpretation difficult (e.g., 32, 33); photoacoustic or attenuated total reflectance (ATR) spectroscopy are preferable FTIR techniques. However, FT Raman spectroscopy provides the capability of characterizing bulk effects in paints and coatings.

Small samples can be studied without the use of microscopy techniques because the laser beam can be focused down to about 100 μm. This property provides the facility for characterizing not only small samples but also defects in coatings arising from contaminants.

There is no interference from atmospheric CO_2. However, atmospheric H_2O will absorb the Raman radiation, but this effect is negligible particularly for purged optical benches as most FT instruments tend to be.

For specific analytical applications for coatings, technical limitations restrict the information that FTIR spectroscopy can provide, whereas FT Raman spectroscopy seems to offer a more feasible approach. However, FTIR and FT Raman techniques are clearly complementary because unique information can be obtained from each type of analysis. Therefore, a complete vibrational structure of a particular molecule can be built only by using both techniques.

Instrumentation

Two FT Raman instruments were used for these measurements: a prototype instrument based on a modified Perkin-Elmer 1710 FTIR spectrometer and a production Bruker spectrometer. The Perkin-Elmer instrument consisted of a Spectron Laser Systems model 301 Nd:YAG laser operating at 1.064 μm (9398 cm^{-1}) for excitation with a maximum output of 1 W and an InGaAs photodetector operating at room temperature. The optical design of this system gives a spectral range of 300–3500 cm^{-1} for detectable Raman scatter. This instrument was described in more detail elsewhere (11, 21). The

Raman spectra from the Perkin-Elmer spectrometer were corrected for instrument response. The optical layout of the Bruker FT Raman instrument is shown in Figure 1. It consists of the Bruker FRA 106 Raman module attached to the Bruker IFS 66 FTIR spectrometer. The laser was a diode-pumped Nd:Yag laser operating at the same wavelength as the Perkin-Elmer and with a maximum output of 300 mW. The detection system was a Ge photodiode cooled with liquid nitrogen. The spectral range for this instrument was $< 100{-}3500$ cm^{-1}. The Raman spectra from the Bruker system were not corrected for instrument response because it is almost flat over the spectral region 3300 cm^{-1} to the filter cutoff edge on the Rayleigh band (*34*).

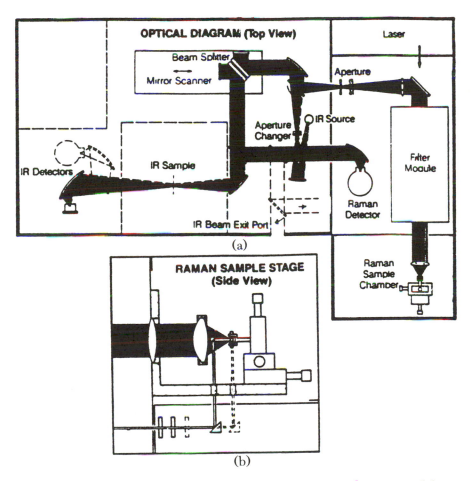

Figure 1. Part a: Optical layout of the Bruker spectrometer with Raman module attached. Part b: Cross-sectional view of the sample chamber.

The important feature for simple analytical measurements for both of these systems is the design of the sample chamber and sampling accessories. The exciting laser light is deflected onto the sample by a prism in each case, and the Raman scattered light is collected by the large collection lens in front of the sample; the inset in Figure 1 shows the Bruker sample chamber in cross-section. The Perkin-Elmer instrument used sampling accessories mounted on standard infrared 3- × 2-inch plates. The Bruker uses an alternative design, optimized for the Raman optics of this instrument with the sampling devices placed onto a spring-loaded optical mount. Specific designs of accessories for each instrument for liquids and solids sampling are available to maximize the signal-to-noise ratio (21). However, many samples require no specific design of accessory: For example, liquid samples could be held in glass vials because glass is both transparent and a weak Raman scatterer, and simply mounted in the instrument for the measurement. For liquids we found an improvement in the signal-to-noise ratio of about 3 by using a spherical liquid cell with an aluminized back-surface rather than a back-aluminized cuvette. For all of these sampling methods only minor realignment of the sample position was required in each case to optimize the Raman signal.

FT Raman Spectroscopy of Paint Materials: Advantages and Problems

Paint films are not ideal for study by FT Raman techniques because they are thin (typically 10−200 μm), and hence, the signal-to-noise ratio in the spectra tends to be low. To optimize the measurement, the signal averaging is prolonged and the power of the incident laser beam is increased so that spectra can easily be obtained. Figure 2a shows a FT Raman spectrum from a direct measurement on a 100-μm-thick polyester latex paint film on a metallic substrate. The major bands at 446 and 611 cm^{-1} are due to the TiO_2 pigmentation. The analogous FTIR transmission spectrum is shown in Figure 2b for comparison. These spectra clearly show one of the major problems with FTIR spectroscopy in that many inorganic materials give very broad IR absorption features and generally make any unequivocal identification of unknown inorganic materials difficult. The Raman bands from inorganic materials tend to be much sharper, and details from lower frequencies become accessible. Information such as pigment-to-binder ratios can be obtained from Raman measurements with appropriate calibration data. The signal-to-noise ratio of Raman spectra from paint films can easily be improved by scraping material from the film for examination in a Raman solids holder (21) (see Figure 3). A clear improvement in the signal-to-noise ratio by about a factor of 3 results. This improvement in the spectra is very much sample dependent because the penetration depth of the laser and the escape depth

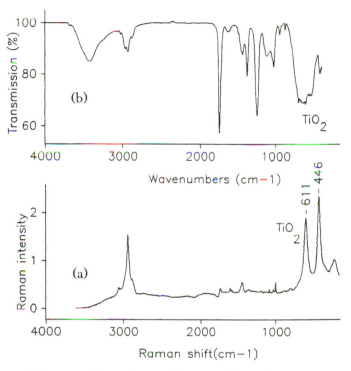

Figure 2. FT Raman (a) and FTIR (b) spectra from a latex paint containing TiO₂ pigmentation.

of the Raman scattered radiation will vary depending upon the components in the paint.

Metallic paints tend to give very poor Raman spectra for both the wet paint and the film. The poor spectral quality is due to the fact that the aluminum flakes, which have a "cornflake"-like shape, scatter and hence attenuate the Raman radiation. In addition, these types of paints often contain organic pigments that tend to fluoresce and also tinters containing carbon black that absorb the laser light and cause sample heating. Consequently, the Raman signal will be very weak with a prominent background response. Figure 4 shows Raman spectra taken from a metallic paint; the spectrum is very poor with few clearly defined bands. However, the fluorescence and heating effects can be reduced by pulsing the laser and taking the ratio of the signal against the intensity of the incident pulses (correction for pulse-to-pulse variation). The advantages of this approach for dealing with problems such as fluorescence and heating were described elsewhere (35). For this measurement the laser was pulsed at 1 kHz (matched to the instrument sampling frequency) operating at 70 mW. Figure 5 shows the

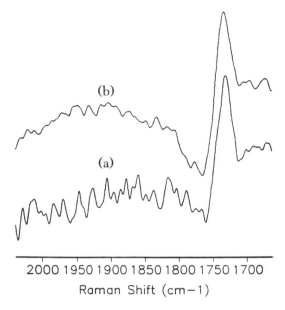

Figure 3. FT Raman spectra from a latex paint showing the relative signal-to-noise ratio for direct measurement on a paint film (a) and measurement of the sample in a Raman solids holder (b). The difference in signal-to noise ratio is about 3.

spectrum of a dried metallic paint taken in this way at a resolution of 8 cm^{-1} and with 100 scans using an extended range Ge detector operating at 77 K. The instrument used in this case was a prototype based on a modified Perkin-Elmer 1720 FTIR spectrometer (36). The resulting spectrum is improved with a much reduced background response.

Carbon black, at low levels, is a common pigment in paint films. Although spectra can be obtained from carbon-containing materials, the presence of carbon black in paints at levels as low as 1% (w/w) dry film destroys any spectral features. The reason for this problem is twofold. The carbon induces absorption, which severely attenuates the Raman spectrum produced. Unfortunately, absorption of the laser radiation also causes the sample to heat and in worst cases to burn. Burning can be reduced by spinning the sample. Figure 6 shows a Raman spectrum of a polyester coating containing about 1% carbon black. A direct measurement on the paint film gave a high background response with no spectral features evident. The spectrum shown in Figure 6 was obtained by loading the sample into a solids holder, the laser was set at low power and defocused, and the measurement was made with prolonged signal averaging (earlier measurements had shown the sample to burn even at low laser powers). However, there were no spectral features associated with the binder or pigments. The bands at about

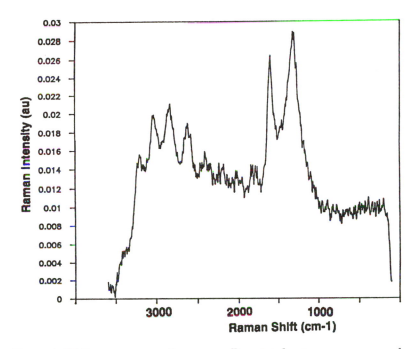

Figure 4. FT Raman spectrum from a metallic paint showing poor response due to fluorescence and heating.

1600 and 1309 cm^{-1} are thought to be due to Raman scatter from amorphous carbon (*37*). The apparent bands between 2000 and 3000 cm^{-1} are instrumental artefacts. Conventional methods for reducing the heating effect include dilution of the sample with KBr powder, which then acts as a heat sink; defocusing the laser beam so that localized heating is reduced; and reducing the laser power. All of these require increased signal averaging to improve the signal-to-noise ratio. FT Raman spectra of carbon black have been obtained by dilution in KBr (*38*). However, we were unable to obtain spectra for these particular types of paint systems because of the heating as well as inherent fluorescence from the pigmentation. Conventional Raman (*39*), photoacoustic FTIR (*40, 41*), and specular reflectance FTIR (*42, 43*) spectroscopies have been successful techniques for characterizing polymeric materials containing high levels of carbon (up to 40% w/w).

In addition to heating effects, fluorescence can be a serious problem for many paint samples, particularly those containing high levels of specific colored organic pigments. The result is a high background response, the noise of which obscures much of the Raman scattering. Figure 7 shows spectra from a range of different colored pigments and tinters. Different degrees of fluorescence are clearly observed depending upon the structure of the

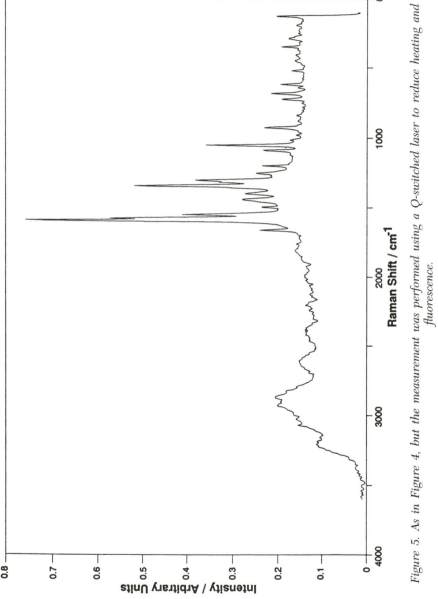

Figure 5. As in Figure 4, but the measurement was performed using a Q-switched laser to reduce heating and fluorescence.

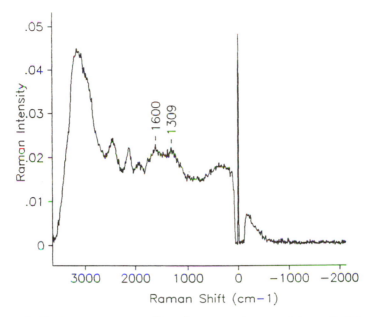

Figure 6. FT Raman spectrum of a polyester coating containing only 1% w/w dry film of carbon black.

electronic states in the material. Fluorescence can be reduced by burning out, by reducing the laser power and signal averaging for long periods, or by pulsing the laser at low power, as just described for the metallic paint.

The analytical capabilities of FT Raman spectroscopy for characterizing raw materials are clear, and in combination with FTIR techniques much better quality data can be obtained for such applications as quality control, troubleshooting production problems, and identifying contaminants. With its simple methods of sampling FT Raman spectroscopy will satisfy many routine spectroscopic problems. Obviously, is will not replace FTIR spectroscopy but rather act as a complementary approach or a means of tackling specific spectroscopic problems not easily solvable by FTIR spectroscopy.

Latex Systems

The Problems of Water.   The investigation of water-based systems is generally very difficult by FTIR spectroscopy because of the strong water absorption bands that obscure large portions of the spectral range. These problems can be overcome to some degree by using specialized sampling techniques such as ATR spectroscopy. This approach is useful for specific applications because spectral windows occur away from the water absorption, and kinetics of water-based processes can be monitored. However, when

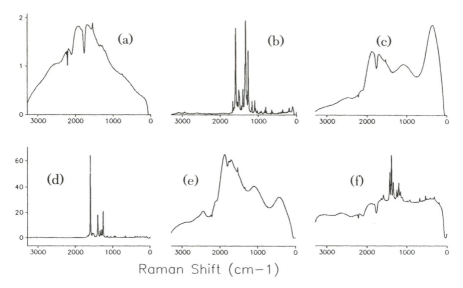

Raman Shift (cm−1)

Figure 7. Spectra taken from a range of organic pigments showing different degrees of heating and fluorescence: a, green, 100 scans, 30-mW laser power, 8-cm⁻¹ resolution; b, yellow, 20 scans, 300-mW laser power, 2-cm⁻¹ resolution; c, blue, 100 scans, 30-mW laser power, 8-cm⁻¹ resolution; d, dark yellow, 20 scans, 300-mW laser power, 2-cm⁻¹ resolution; e, dark blue, 100 scans, 30-mW laser power, 8-cm⁻¹ resolution; and e, purple, 100 scans, 230-mW laser power, 8-cm⁻¹ resolution.

components are phase-separated, care must be taken in the interpretation of data because in some instances, some components in a reaction can deposit preferentially on the ATR crystal and mask any processes being monitored (*44*).

One of the important features of FT Raman spectroscopy that has been little exploited is the facility to study water-based systems. Water is a very weak Raman scatterer, which means that such systems can easily be examined with only some degradation in spectral quality due to absorption of the Raman scattered radiation by water. Furthermore, Raman has greater sensitivity than IR spectroscopy for following processes involving C=C bonds; this sensitivity is particularly important for following an emulsion polymerization.

To show the effect that water might have on the FT Raman measurements, the spectral features in the NIR region were investigated. Figure 8a shows a NIR absorbance spectrum of water at a path length of 0.2 mm taken on a guided-wave-model 260 NIR spectrometer in the spectral range 1100–2400 nm. The range of the Raman measurement is indicated on this spectrum. Clearly the water absorption is significant over the spectral region in which the Raman measurements are performed. Figure 8b shows the absorption coefficient of water on the Raman shift scale. Absorption is very

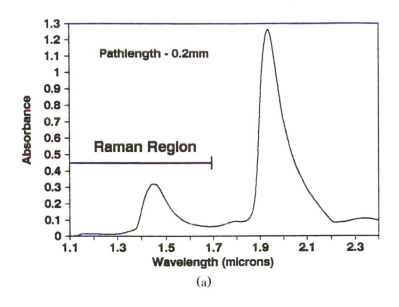

(a)

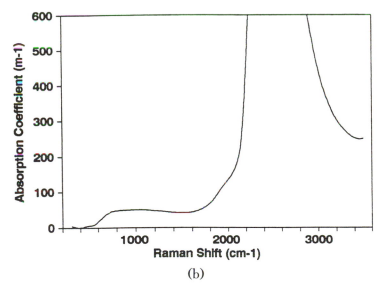

(b)

Figure 8. Part a: NIR absorbance spectrum of water (0.2-mm path length) showing the Raman spectral region. Part b: absorption coefficient plotted as a function of Raman shift.

strong for Raman shifts $> 2000\ \text{cm}^{-1}$, and this finding indicates that the Raman radiation in particular from C–H stretching modes will be attenuated. However, the peak absorption occurs at a shift of $2500\ \text{cm}^{-1}$, a region where few functional groups give a resonance. For shifts $< 1800\ \text{cm}^{-1}$, absorption is relatively weak, so that the Raman spectrum will not be strongly attenuated, though some deterioration in spectral quality will be observed. Figure 9 shows an FT Raman spectrum of water; the O–H stretching and bending modes are evident. The Raman scatter was weak, as expected for water. This work clearly indicates the problems that can arise from FT Raman spectroscopy of weak Raman scatterers at low concentrations.

Emulsion Polymerization.   The work described in this section was initially intended to assess the sensitivity of FT Raman techniques for latex analysis. The acrylic latex used for the measurement had a nominal particle size of 95 nm and a solids content of 35%. This content was sequentially diluted down to 2.2% solids for the measurements on the Bruker Raman system. The spectra shown in Figure 10 give a very good signal-to-noise ratio

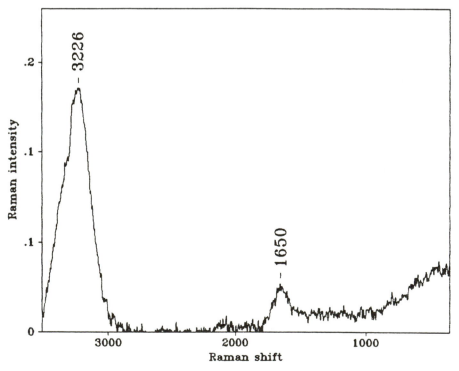

Figure 9. FT Raman spectrum of water, 2-cm^{-1} resolution.

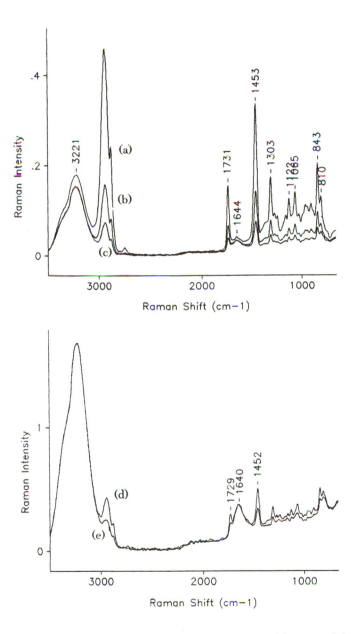

Figure 10. FT Raman spectra from an acrylic latex 35% (a), 17.5% (b), 8.8% (c), 4.4% (d), and 2.2% (e).

even at 2% solids. The broad feature at about 3200 cm^{-1} is due to water. These results suggest that the ultimate sensitivity for a detectable Raman signal is $\ll 1\%$ solids and clearly demonstrate the feasibility of following processes in latexes.

Polymer latexes are important technological systems in the coatings industry. Development of new applications of these waterborne systems has significant environmental and commercial implications. One method for the production of latexes is emulsion polymerization because it enables the synthesis of particles of controlled size and morphology. The presence of water in these systems has hampered the detailed quantitative analysis of the polymerization reaction. However, Raman spectroscopy can overcome this problem. In this case we monitored the emulsion polymerization reaction of an acrylic–methacrylic copolymer by following the behavior of the C=C bond with time of reaction. Raman spectroscopy was used previously as a means for following processes in bulk polymerization (45) and microemulsion polymerization (46) of styrene and methyl methacrylate (MMA).

Figure 11 shows spectra of typical acrylic monomers used in emulsion polymerization reactions. The spectral region between 1650 and 1630 cm^{-1} clearly indicates the Raman shifts of the different types of C=C bonds that can be monitored during a reaction. The band positions were 1639.3 cm^{-1} for the acrylic C=C in butyl acrylate (BA), 1641.0 cm^{-1} for the methacrylic C=C in MMA, and 1641.0 and 1649.1 cm^{-1} for the methacrylic and allylic C=C, respectively, in allyl methacrylate (AMA). These values are consistent with the values reported by Davison and Bates (47). This approach should enable different types of C=C bands to be monitored during any process involving these types of compounds.

To show the facility of FT Raman spectroscopy in being able to follow processes, an emulsion polymerization of a BA, MMA, and AMA copolymer latex reaction was followed. The acrylic and methacrylic components formed the polymer backbone, and the allylic monomers produced the cross-linking. FT Raman scatter is particularly attractive for this example because the bands due to C=C are strong. In the FTIR spectra these bands are not strong, and an additional problem of strong water absorption completely obscures the C=C stretching band.

Generally, the four components in an emulsion polymerization reaction are monomers (water immiscible), water, emulsifier, and initiator. The emulsifier enables the monomers to be dispersed in the water to form micelles. The polymerization process is propagated in the micelles, and monomers diffuse to the micelles through the aqueous phase.

To avoid any uncertainties in the particle initiation stage of the process, a seeded emulsion was initially produced; this step reduces the probability of reinitiation of new crops of particles during the process. The seeded emulsion was produced by adding BA, MMA, and AMA in the weight ratio 27:6:1 to

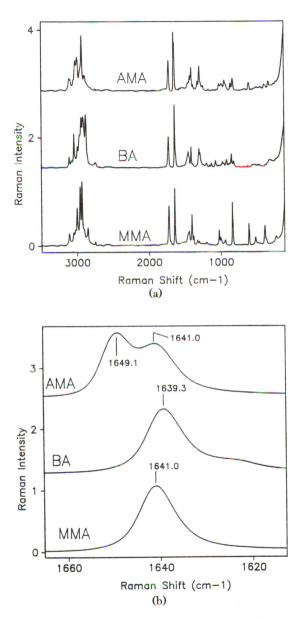

Figure 11. FT Raman spectra of MMA, BA, and AMA over the region 3500–100 cm^{-1} (a) and the C=C stretching region in detail (b).

demineralized water containing surfactant with the temperature maintained at about 75 °C. Ammonium persulfate initiator was added to this emulsion, and the reaction mixture was stirred. The result was a seeded emulsion with 5% solids content.

BA, MMA, and AMA in the ratio 27:6:1 in demineralized water containing surfactant were pre-emulsified. This mixture was fed into the reactor over a period of 130 min with the reaction mixture stirred continuously and kept at 75 °C. For monitoring this reaction, samples were drawn from the mixture into vials and shaken with inhibitor to stop the process. For the Raman measurement, 80-μL samples were transferred into the FT Raman liquid cell already described.

The disappearance of the C=C bond was monitored over the time of the addition and for a further 2 h of the subsequent hold period. The solids content after completion of the feed was 35%. Figure 12 shows the normalized C=C concentration plotted as a function of time. The C=C band was normalized against the C=O band at 1726 cm^{-1}. During the monomer feed, a rapid decrease in the C=C functionality occurred. After the monomer feed, the remaining 20% unreacted C=C reduced to 5% after the additional 2-h hold time. The Raman bands due to the different C=C bonds were not distinguishable because of the poor quality of the spectra, so that the relative rates of loss for each bond could not be determined. A plot of the solids content against the Raman band ratio 1450 cm^{-1} (CH$_2$)/3220 cm^{-1} (water) is shown in Figure 13. This technique clearly allows the determination of the solids content during a reaction directly from the Raman measurement.

The presence of water had a detrimental effect on the quality of Raman spectra, particularly in the early stages of the reaction when the solids content

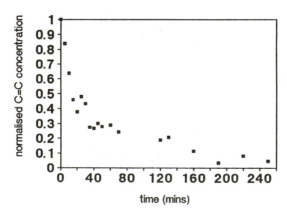

Figure 12. Plot of the C=C band intensity normalized against the C=O band (maximum normalized to 1) as a function of time of the reaction. (Reproduced with permission from reference 13. Copyright 1990 Pergamon Press.)

was low. Water absorbs radiation in the NIR region so that the detected Raman scattering is attenuated. At long path lengths in the sample this "self-absorption" leads to spectra of poor quality (*48*). At low solids content the path length is reduced, and the reduced path length gives spectral improvements, but the Raman scattering cross-section is also reduced, and this reduction gives poorer spectral quality. Consequently, aqueous solutions or suspensions of materials that are good Raman scatterers are easily accessible by Raman spectroscopy at relatively high concentrations. At low concentrations and in solutions containing poor scatterers, the technique is flawed unless shorter wavelength lasers are used. During the polymerization process the signal-to-noise ratio increases by about a factor of 10 because of the increased solids content of the reaction mixture.

All our Raman measurements for the polymerization process were made off-line. With the current state of fiber optic technology for remote process monitoring, these types of reactions could be performed on-line (*23–26*), and this development is sure to appear soon.

Autoxidation Studies of Alkyds

Conventional paint technology based on oil-modified alkyd resins still forms a significant proportion of currently marketed paint systems. The drying process of these materials to form robust, weather-resistant films relies on the ability of the unsaturation in the fatty acid component to undergo reaction with atmospheric oxygen to form hydroperoxides (*49, 50*). These compounds break down to form radicals, which are then responsible for the cross-linking reaction to form a hard, robust film (*51*):

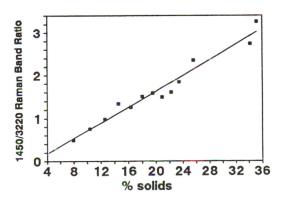

Figure 13. Water band (3220-cm^{-1}) normalized against the latex band (1450-cm^{-1}) as a function of solids content. (Reproduced with permission from reference 13. Copyright 1990 Pergamon Press).

$$R-CH=CH-CH_2-CH=CH-(CH_2)_nCOOR' \xrightarrow{O_2}$$

$$R-CH=CH-\underset{\underset{OOH}{|}}{CH}-CH=CH-(CH_2)_nCOOR' \longrightarrow$$

$$R-CH=CH-\underset{\underset{O^{\cdot}+\ ^{\cdot}OH}{|}}{CH}-CH=CH-(CH_2)_nCOOR' \longrightarrow$$
$$\qquad\qquad\qquad\qquad\qquad\qquad\text{cross-linked film}$$

The autoxidation process has been monitored by FTIR (13, 49) and photoacoustic FTIR (52, 53) spectroscopy. During the formation of hydroperoxide, both cis–trans isomerization and conjugation of the unsaturated linkages can occur (49, 54):

$$R-CH_2-CH=CH-CH_2-CH=CH-(CH_2)_nCOOR' \xrightarrow{O_2}$$
$$R-CH_2-\underset{\underset{OOH}{|}}{CH}-CH=CH-CH=CH-(CH_2)_nCOOR''$$
or
$$R-CH_2-CH=CH-CH=\underset{\underset{OOH}{|}}{CH}-CH-(CH_2)_nCOOR''$$

The Raman technique in principle is capable of detecting the configurational changes as they occur. After a short induction period, the C=C band decreases rapidly. The mechanism for this decrease is not clearly known; however, C–C cross-links, radical addition to C=C to form C–O–C cross-links, and the formation of cyclic peroxides have been postulated. Many additional chemical processes occur simultaneously as the fatty acid components of the alkyd undergo autoxidation. For example, fragmentation of the hydroperoxide can occur to form alkoxy radicals and carbonyl compounds (55, 56). Many byproducts are formed during autoxidation, and detailed work has been performed on alkyds and model systems (57, 58). The volatile aldehydes that are evolved during this process give the curing paint its characteristic pungent smell.

Autoxidation of an Alkyd Resin. Figure 14 shows FTIR and FT Raman spectra of the alkyd resin modified with soya bean oil used in this work; band positions have been identified (13). The C=C stretching band at about 1650 cm^{-1} is strong in the Raman spectrum but not easily detectable in the FTIR spectrum. The curing of alkyd resin based on soya bean oil with the appropriate cobalt driers was monitored by FT Raman methods over several weeks to follow the chemical processes. For this experiment, 200-μm films were spread onto glass plates with a block spreader. Solvents were

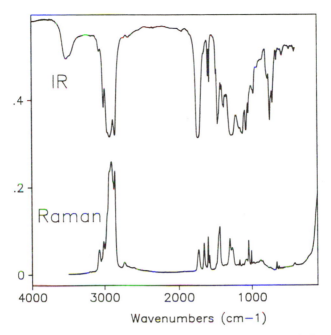

Figure 14. FTIR and FT Raman spectra of the alkyd resin modified with soya oil used in this work. (Reproduced with permission from reference 13. Copyright 1990 Pergamon Press.)

removed under nitrogen for 2 h to prevent onset of curing during this period. Samples were taken from the film for Raman analysis. Figure 15 shows FT Raman spectra taken over the period of the experiment. The band attributable to the C=C functional group at 1655 cm^{-1} in the Raman spectrum decayed over the period of the cure experiment. The intensity behavior of the C=C is shown in Figure 16; the initial apparent increase in unsaturation is thought to be due to the normalization of the band against the 1450-cm^{-1} C–H band and is not a real effect (discussed later). Some broadening of the C=C band is associated with the configurational changes expected during autoxidation. However, the different C=C components could not be identified from these measurements. The band at about 880 cm^{-1} that increases to a maximum after about 30 h is probably the O–H bending mode of the hydroperoxide. The nature of the cross-linking could not be elucidated from the Raman results. FTIR spectra taken during the autoxidation clearly show the C–O–C cross-links (1100-cm^{-1} region), –OOH formation (3420 cm^{-1}), and C=C loss (3010 cm^{-1}) (*see* Figure 17).

Curing of an Alkyd Resin Containing Pigment.   For comparison, the same resin with about 15% (w/w) TiO$_2$ was investigated. The

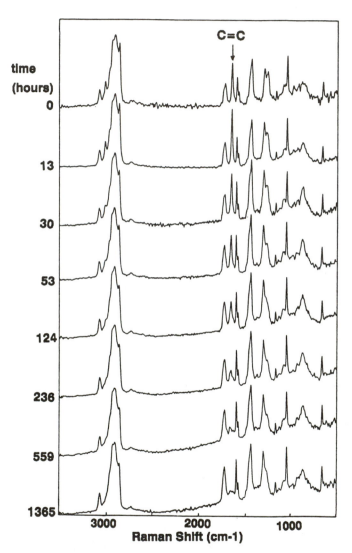

Figure 15. FT Raman spectra taken over the period of the cure reaction. (Reproduced with permission from reference 13. Copyright 1990 Pergamon Press.)

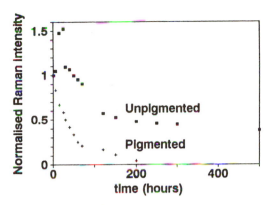

Figure 16. Plot of the C=C unsaturation as a function of cure time. (Reproduced with permission from reference 13. Copyright 1990 Pergamon Press.)

experiment was performed in the same way as for the unpigmented system. The signal-to-noise ratio was much improved compared to the unpigmented system as a result of Raman scatter enhancement by the TiO_2. (This improvement might well be unexpected, but it is characteristic of FT Raman measurements; turbid samples give better spectra than clear ones. The reasons for this effect are understood by all practitioners, but none of them agree!). The Raman spectrum is dominated by the bands at 608 and 442 cm^{-1} due to TiO_2 (*see* Figure 18). The change in the C=C band intensity during the cure is given in Figure 19 and plotted in Figure 16. Clearly no increase in unsaturation occurs in the early stages. The onset of cure occurs much earlier and reaches its limit much faster than the unpigmented resin. This result is expected because TiO_2 is thought to enhance the rate of autoxidation by providing active sites for the process, although the details are not clearly understood.

Autoxidation of Model Compounds. Curing of alkyd resins is usually too complex to permit drawing far-reaching generalizations concerning the mechanism involved in the autoxidation process. Therefore, it is necessary to study component systems such as methyl ester of oleic, linoleic, and other similar fatty acids that model the processes. In addition, these compounds can be obtained in a pure form. Our earlier work (*13*) dealing with the analysis of bands in the C=C stretching region was hampered by the contribution of bands due to other components. Problems also arose from the incomplete resolution of the C=C Raman bands. Therefore, we found it necessary to study a range of methyl esters of fatty acids that are important components of the alkyd resin.

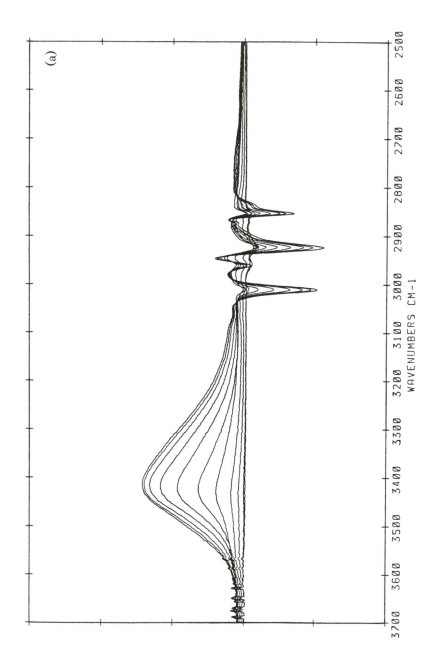

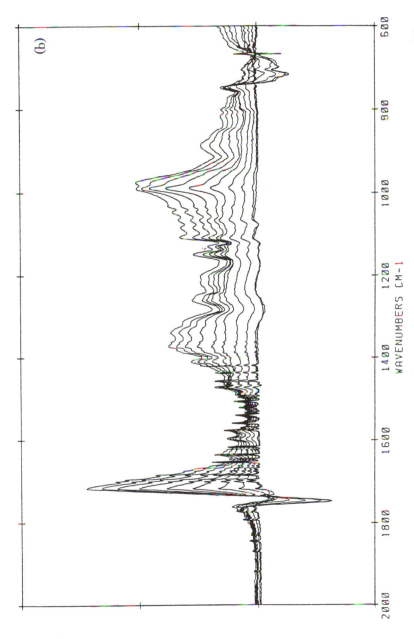

Figure 17. FTIR subtraction spectra showing (a) the formation of hydroperoxide and loss of unsaturation and (b) formation of ether cross-links during autoxidation. (Reproduced with permission from reference 13. Copyright 1990 Pergamon Press.)

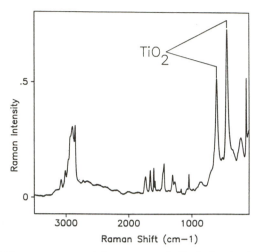

Figure 18. FT Raman spectra of the pigmented alkyd clearly showing the bands due to TiO$_2$ at about 440 and 610 cm^{-1}.

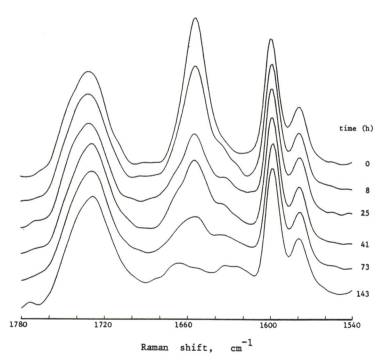

Figure 19. FT Raman spectra showing the behavior of the C=C band (1655 cm^{-1}) during cure. (Reproduced with permission from reference 13. Copyright 1990 Pergamon Press.)

Previous work on model compounds has helped in elucidating some of the complex processes such as identification of the volatile byproducts of the reactions (57). For this work we studied the curing processes of methyl ester of oleic, linoleic, and linolenic acids with both FT Raman and FTIR techniques. (These chemicals were supplied by Aldrich Chemical Company). These acids are the important fatty acid components in the alkyd described. A more in-depth discussion of the results is given elsewhere (58). These compounds have the following structures:

$$CH_3(CH_2)_7CH \overset{cis}{=} CH(CH_2)_7COOCH_3$$
methyl oleate

$$CH_3(CH_2)_4CH \overset{cis}{=} CHCH_2CH \overset{cis}{=} CH(CH_2)_7COOCH_3$$
methyl linoleate

$$CH_3(CH_2CH \overset{cis}{=} CH)_3(CH_2)_7COOCH_3$$
methyl linolenate

The FT Raman spectra for these materials are given in Figure 20. The effect of autoxidation on the different degrees of unsaturation in each species was investigated.

For the kinetic analysis, films of these samples were prepared by block spreading onto glass substrates, and samples were removed for examination by the FT Raman instrument.

Methyl oleate, having only one C=C bond, does not undergo conjugation accompanying the formation of hydroperoxide during autoxidation. However, some modification of the shape of the C=C band was noted, although even after 72 h no change in its overall intensity was seen. Even after 4 weeks in ambient conditions no increase in viscosity was detected, and in fact the film remained liquid at the end of this period. Figure 21 shows the FTIR spectra for the behavior during the cure process. The increase in noise is indicative of film thinning due to sample loss; scission reactions give rise to volatile saturated and unsaturated aldehydes (57), for example,

$$R-CH=CH-CH=CH-CH-(CH_2)_7-COOR' \xrightarrow{-HO^{\cdot}}$$
$$\underset{OOH}{|}$$

$$R-CH=CH-CH=CH-\}-CH-\}-(CH_2)_7-COOR'$$
$$\underset{O^{\cdot}}{|}$$

A   B

$\longrightarrow$ volatile aldehydes

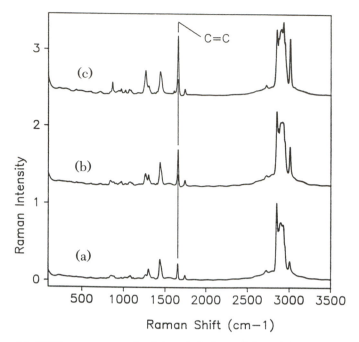

Figure 20. FT Raman spectra for (a) methyl oleate, (b) methyl linoleate, and (c) methyl linolenate.

Scission occurs at position A or B and gives rise to the formation of aldehydes, which give the curing paint its pungent smell.

Methyl linoleate has two nonconjugated C=C bonds, so that the facility to undergo cis–trans isomerism or conjugation does arise during autoxidation. After 2 weeks a hard film had resulted; this result suggests that a high degree of cross-linking over this period had occurred. Figure 22 shows FT Raman spectra taken at intervals over the period of the reaction. After 24 h new bands have appeared at 3050, 1599, 999, and 791 cm^{-1}. The very intense band at 999 cm^{-1} is typical of a cyclic, conjugated structure, although we have not yet identified this species. In addition to this change, a decrease in the level of unsaturation that occurred is associated with C–C cross-linking. Figure 23 shows the analogous FTIR spectra. Clear evidence is seen for hydroperoxide formation and ether cross-linking. Figure 24 shows plots of the behavior of the unsaturation at 3010 cm^{-1} and hydroperoxide at 3420 cm^{-1} taken from the FTIR data.

Methyl linolenate contains three nonconjugated C=C bonds. It underwent rapid cross-linking to form a hard film after 3 days of exposure to the atmosphere. Having three C=C bonds, methyl linolenate has two activated

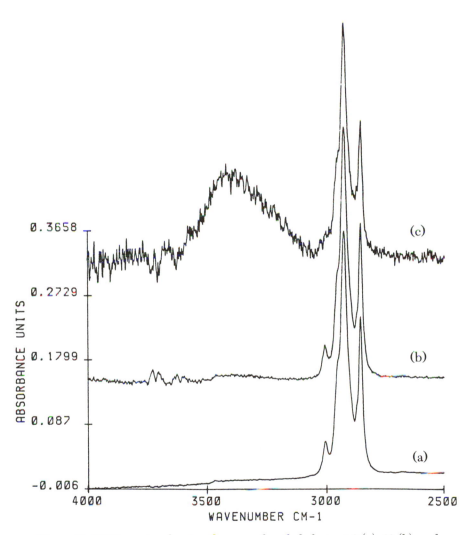

Figure 21. FTIR spectra showing the cure of methyl oleate at 0 (a), 10 (b), and 150 (c) h. (Reproduced with permission from reference 59. Copyright 1991 Pergamon Press.) Continued on next page.

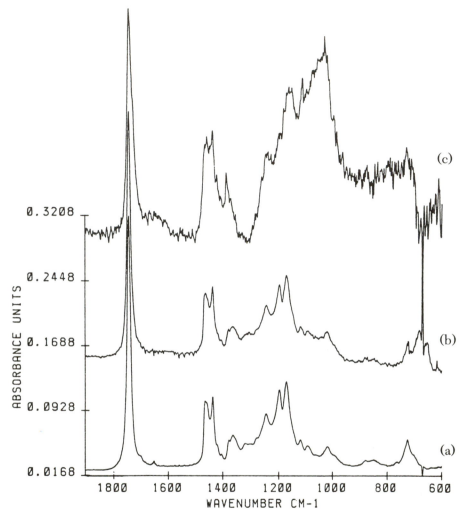

Figure 21. Continued

–CH$_2$– moieties, a feature that increases the rate of hydroperoxide formation and hence, the rate at which the ester cross-links. Figure 25 shows spectra in the C=C stretching region. Clear changes in conformation of this functional group are seen. As the cure proceeds, shoulders appear on either side of the original C=C band (1651 cm^{-1}) that are due to the cis conformation. The additional bands are thought to be due to trans (1670 cm^{-1}) and conjugated (1640 cm^{-1}) structures. The mechanism for this configurational change is well-known for the fatty acid materials (57). FTIR spectra taken over the cure are shown in Figure 26. In addition to hydroperoxide formation and

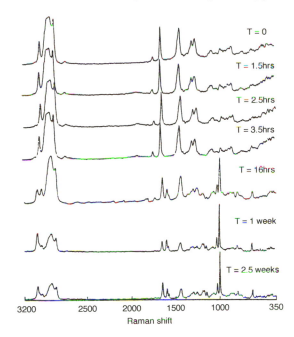

Figure 22. FT Raman spectra showing the cure of methyl linoleate. (Reproduced with permission from reference 59. Copyright 1991 Pergamon Press.)

cross-linking, an additional strong band at 877 cm^{-1} increases during the cure process; the nature of this species has yet to be elucidated. Figure 27 shows kinetic plots of these species. The increase in the 877-cm^{-1} band appears to occur at a similar rate to the hydroperoxide formation. This behavior suggests that the low-frequency band is due to the O$-$H bending mode of $-$OOH.

To summarize the results obtained from the fatty acid methyl ester investigation, the degree of unsaturation plays an influential role in the rate of autoxidation. Comparison of the rates of loss in unsaturation is shown in Figure 28. For methyl oleate, the unsaturation loss was very slow, and the rate of loss increased with increasing unsaturation. The bands around 1655 cm^{-1} have been assigned to trans (1670 cm^{-1}), cis (1655 cm^{-1}), and conjugated (1640 cm^{-1}) structures that were identified in the Raman spectra of methyl linolenate. These results are discussed in more detail elsewhere (*59*).

Clearly FT Raman spectroscopy is in its infancy. The sensitivity of the method, although adequate for most analyses, can be frustratingly low. However, instruments are rapidly improving in this respect as new detectors, lasers, and interferometers become available. Possibilities for the future were described recently (*60*), and some are already in evidence.

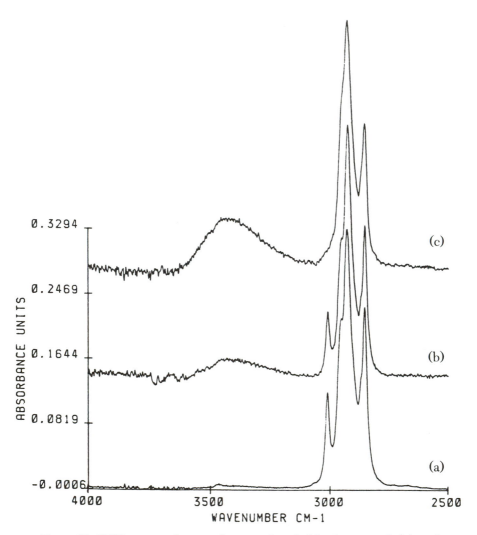

Figure 23. FTIR spectra showing the cure of methyl linoleate at 0 h (a), 15 h (b), and 100 h (c). (Reproduced with permission from reference 59. Copyright 1991 Pergamon Press.)

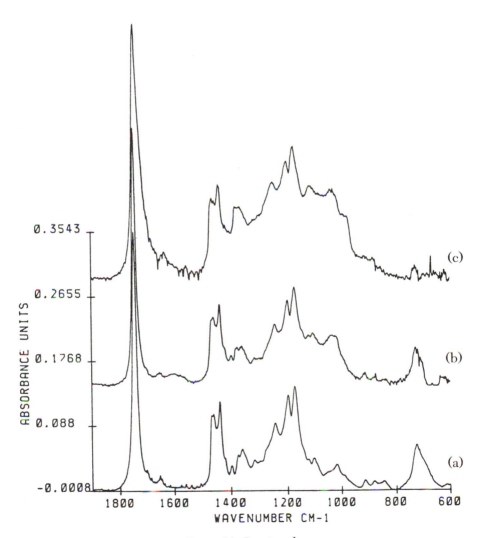

Figure 23. Continued

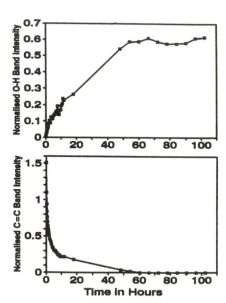

Figure 24. Behavior of the (top) hydroperoxide and unsaturation (bottom) for methyl linoleate from the FTIR results. (Reproduced with permission from reference 59. Copyright 1991 Pergamon Press.)

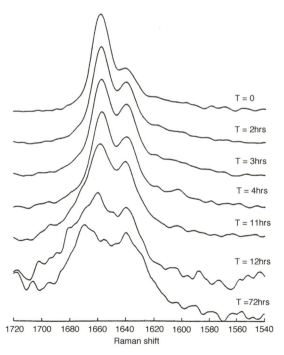

Figure 25. FT Raman spectra for the cure of methyl linolenate showing the behavior of the C=C stretching region. (Reproduced with permission from reference 59. Copyright 1991 Pergamon Press.)

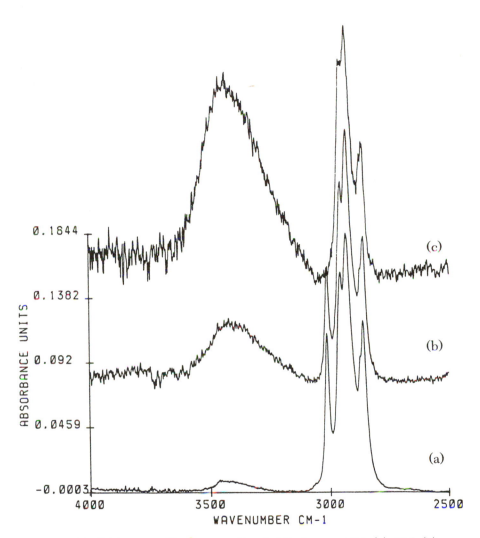

Figure 26. FTIR spectra for the cure of methyl linolenate at 0 h (a), 15 h (b), and 100 h (c). (Reproduced with permission from reference 59. Copyright 1991 Pergamon Press.) Continued on next page.

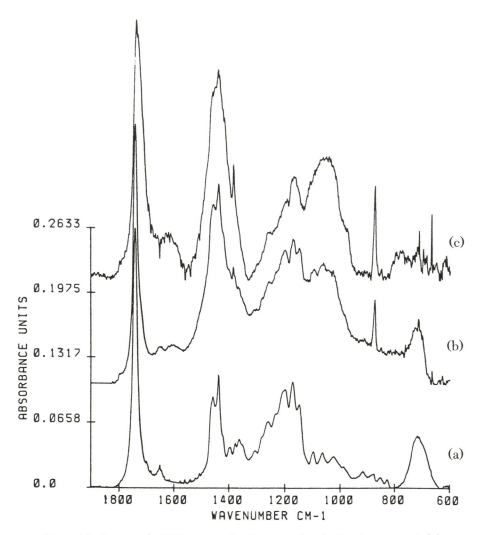

Figure 26. Continued. *FTIR spectra for the cure of methyl linolenate at 0 h (a), 15 h (b), and 100 h (c). (Reproduced with permission from reference 59. Copyright 1991 Pergamon Press.)*

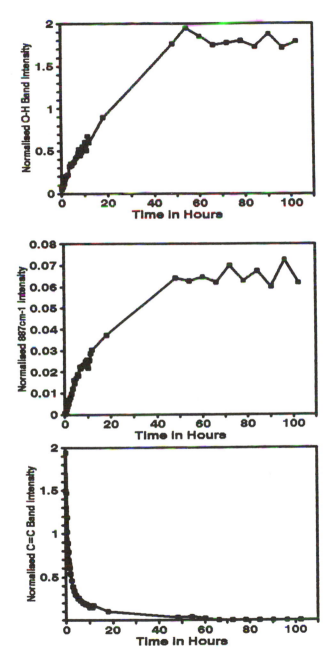

Figure 27. Behavior of the hydroperoxide (top), unsaturation (bottom), and 877-cm⁻¹ band (middle) for methyl linolenate from the FTIR results. (Reproduced with permission from reference 59. Copyright 1991 Pergamon Press.)

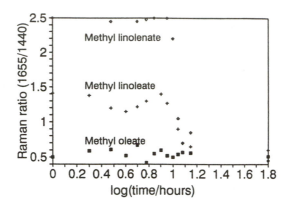

Figure 28. Behavior of the unsaturation in methyl oleate, methyl linoleate, and methyl linolenate during the cure. (Reproduced with permission from reference 59. Copyright 1991 Pergamon Press.)

Acknowledgments

J. K. Agbenyega thanks ICI Paints and the Science and Engineering Research Council for a CASE studentship. Thanks are also extended to P. Turner (Bruker Spectrospin Ltd., U.K.) for some of the Raman measurements reported here. Perkin-Elmer Ltd. and the Office of Naval Research (U.S. Navy) are also thanked for continuing generous support to the University of Southampton. We also express our gratitude to D. M. Dick of ICI Paints for encouragement in this work and to R. Ferguson for the NIR measurements on water.

References

1. Hendra, P.; Sweeney, M. *Spectrosc. World* **1991**, *3(2)*, 22.
2. O'Neill, L. A.; Falla, N. A. R. *Chem. Ind.* **1971**, 1349.
3. Jenden, C. M. *Polymer* **1986**, *27*, 217.
4. Chase, B. *J. Am. Chem. Soc.* **1986**, *108*, 7485.
5. Hirschfeld, T.; Chase, B. *Appl. Spectrosc.* **1986**, *40*, 133.
6. Zimba, C. G.; Hallmark, V. M.; Swallen, J. D.; Rabolt, J. F. *Appl. Spectrosc.* **1987**, *41*, 721.
7. Chase, B. *Anal. Chem.*, **1987**, *59*, 881A.
8. Parker, S. F.; Williams, K. P. J.; Hendra, P. J.; Turner, A. J. *Appl. Spectrosc.* **1988**, *42*, 796.
9. Hendra, P. J.; Mould, H. A. *Int. Lab.* **1988**, *18*, 34.
10. Chantry, G. W.; Gebbie, H. A.; Hilsum, C. *Nature London* **1964**, *203*, 1052.
11. Crookell, A.; Hendra, P. J.; Mould, H. M.; Turner, A. J. *J. Raman Spectrosc.* **1990**, *21*, 85.
12. Bergin, F. J.; Shurvell, H. F. *Anal. Proc.* **1989**, *26*, 263.
13. Ellis, G.; Claybourn, M.; Richards, S. E. *Spectrochim. Acta.* **1990**, *46A*, 227.

14. Agbenyega, J. K.; Claybourn, M.; Ellis, G. *Proc. Int. Conf. Raman Spectrosc.* (*ICORS*); Durig, J. R., Ed.; John Wiley: New York, 1990.
15. Hendra, P. J. *Lab. Practice* **1990**, *39*, 61.
16. Agbenyega, J.; Ellis, G.; Hendra, P.; Maddams, W.; Passingham, C.; Willis, H.; Chalmers, J. *Spectrochim. Acta* **1990**, *43A*, 197.
17. Jackson, K.; Loadman, M.; Jones, C.; Ellis, G. *Spectrochim. Acta* **1990**, *46A*, 217.
18. Williams, K.; Gerrard, D. *Eur. Polym. J.* **1990**, *26*, 1355.
19. Almond, M.; Yates, C.; Orrin, R.; Rice, D. *Spectrochim. Acta* **1990**, *46A*, 177.
20. Nie, S.; Marzilli, P.; Marzilli, L.; Yu, N. *J. Am. Chem. Soc.* **1990**, *112*, 6084.
21. Ellis, G.; Hendra, P. J.; Hodges, C. M.; Jawhari, T.; Jones, C. H.; le Barazer, P.; Passingham, C.; Royaud, I. A. M.; Sanchez-Blazquez, A.; Warnes, G. M. *Analyst* **1989**, *114*, 1061.
22. Church, S. P.; Stephenson, P. J.; Hendra, P. J. *Chem. Ind.* **1989**, 339.
23. Hendra, P. J.; Ellis, G.; Cutler, D. J. *J. Raman Spectrosc.* **1988**, *19*, 413.
24. Williams, K. P. J. *J. Raman Spectrosc.* **1990**, *21*, 147.
25. Williams, K. P. J.; Mason, S. M. *Spectrochim. Acta* **1990**, *46A*, 187.
26. Schrader, B. *Practical Fourier Transform Spectroscopy*; Ferraro, J. R.; Krishnan, K., Eds.; Academic Press: London, 1990; p 167.
27. Jawhari, T.; Hendra, P. J.; Willis, H. A.; Judkins, M. *Spectrochim. Acta* **1990**, *46A*, 161.
28. Smith, M. J.; Fuller, M. P. *Proc. Int. Conf. Raman Spectrosc.* (*ICORS*); Durig, J. R., Ed.; John Wiley: New York, 1990.
29. Smith, M.; Walder, F. *FTIR Spectral Lines* **1991**, *12(1)*, 16.
30. Sawatski, J.; Simon, A. *Proc. Int. Conf. Raman Spectrosc.* (*ICORS*); Durig, J. R., Ed.; John Wiley: New York, 1990.
31. Bergin, F. J. *Spectrochim. Acta* **1990**, *46A*, 153.
32. Packansky, J.; England, C.; Waltman, R. J. *J. Polym. Sci. B: Polym. Phys.* **1987**, *25*, 901.
33. Packansky, J.; Waltman, R. J.; Grygier, R. *Appl. Spectrosc.* **1989**, *43*, 1233.
34. Claybourn, M.; Turner, P. H., Chapter 16 in this book.
35. Cutler, D. J.; Mould, H. M.; Bennett, R.; Turner, A. J. *J. Raman Spectrosc.* **1991**, *22*, 367.
36. Cutler, D. J. *Spectrochim. Acta* **1990**, *46A*, 131.
37. Tuinstra, F.; Koenig, J. L. *J. Chem. Phys.* **1979**, *53*, 1126.
38. Turner, P. H., Bruker Spectrospin Ltd., U.K., private communication.
39. Katagiri, G.; Ishida, H.; Ishitani, A. *Carbon* **1988**, *26*, 565.
40. Carter, R. O. III; Paputa-Peck, M. C. *Appl. Spectrosc.* **1989**, *43*, 468.
41. Carter, R. O. III; Paputa-Peck, M. C.; Samus, M. A.; Kilgoar, P. C. *Appl. Spectrosc.* **1989**, *43*, 1350.
42. Claybourn, M.; Colombel, P.; Chalmers, J. *Appl. Spectrosc.* **1991**, *45*, 279.
43. Claybourn, M.; Colombel, P.; Chalmers, J. *Proc. Int. Workshop Fourier Transform Infrared Spectrosc.*; University of Antwerp Press: Antwerp, Belgium, 1990.
44. Claybourn, M., ICI Paints, unreported data.
45. Gulari, E.; McKeigue, K.; Ng, K. Y. S. *Macromolecules* **1984**, *17*, 1822.
46. Feng, L.; Ng, K. Y. S. *Macromolecules* **1990**, *23*, 1048.
47. Davison, W. H. T.; Bates, G. R. *J. Chem. Soc.* **1953**, 2607.
48. Jawhari, T., Ph.D. thesis, Southampton University, Southampton, England, 1989.
49. Hartshorn J. H. *J. Coatings Technol.* **1982**, *54*, 53.
50. Parker, N. A.; Weber, B. A.; Weenen, H.; Khan, J. A. *J. Am. Chem. Soc.* **1961**, *38*, 161.
51. Swern, D. *J. Am. Chem. Soc.* **1953**, *75*, 3135.
52. Salazar-Rojas, E. M.; Urban, M. W. *Prog. Org. Coatings* **1989**, *16*, 371.

53. Urban, M. W.; Salazar-Rojas, E. M. *J. Polym. Sci. A: Polym. Chem.* **1990**, *28*, 1593.
54. Leeves, N. J., Ph.D thesis, Royal Holloway College, London, 1985.
55. Frankel, E. N.; Nowakowska, J.; Evans, C. D. *J. Am. Chem. Soc.* **1961**, *102*, 5597.
56. Frankel, E. N.; Neff, W. E.; Seike, E. *Lipids* **1981**, *16*, 279.
57. Hancock, R. A.; Leeves, N. J.; Nicks, P. F. *Prog. Org. Coatings* **1989**, *17*, 321.
58. Hancock, R. A.; Leeves, N. J.; Nicks, P. F. *Prog. Org. Coatings* **1989**, *17*, 337.
59. Agbenyega, J. K.; Claybourn, M.; Ellis, G. *Spectrochim. Acta* **1991**, *47A*, 1375.
60. Petty, C. J.; Bennett, R. *Spectrochim. Acta* **1990**, *46A*, 331.

RECEIVED for review July 15, 1991. ACCEPTED revised manuscript May 28, 1992.

SPECTROSCOPIC APPROACHES TO POLYMERS IN SOLUTIONS AND POLYMER NETWORKS

Whereas metachromasy of aqueous dye–polyelectrolyte solutions can be used for quantitative assessments, intermolecular associations of polymers in water, photochemical and degradative reactions in polymers, and polymer–monomer interactions can be analyzed by various types of optical spectroscopy, including fluorescence, phosphorescence, and chemiluminescence. Although the concept of free network volume in polymers is essential and in spite of many theoretical treatments, relatively limited experimental data are available. Detailed applications of positron annihilation lifetime measurements to investigate in situ free volume changes in polymer networks are presented.

Fluorescence Studies of Polymer Association in Water

Mitchell A. Winnik[1] and Françoise M. Winnik[2]

[1]Department of Chemistry and Erindale College, University of Toronto, Toronto, Ontario, Canada M5B 1A1
[2]Xerox Research Centre of Canada, 2660 Speakman Drive, Mississauga, Ontario, Canada L5K 2L1

Water-soluble polymers containing hydrophobic substituents form micellelike clusters in water. For rigid cellulosic polymers, the association is interpolymeric with little evidence for intramolecular association of the substituents. For more flexible linear polymers, the nature of the interaction can depend upon chain microstructure and the location of the hydrophobic substituents. Those at the chain ends, as in C_n–PEO–C_n where PEO is poly(ethylene oxide), undergo strong interpolymeric associations to form a network linking micellelike clusters. Polystyrene–poly(ethylene oxide) (PS–PEO) diblock and PEO–PS–PEO triblock copolymers behave quite differently. They undergo a sharp association transition with increasing concentration to form spherical micelles with a very narrow size distribution. Their sizes and aggregation numbers seem to be well described by the star model of block copolymer micelles.

WATER-SOLUBLE NONIONIC POLYMERS bearing hydrophobic substituents undergo association in aqueous solution (*1*). This association process has a profound effect on the macroscopic properties of the solutions, such as viscosity and cloud point. One of the major technological applications of these types of materials is as viscosity modifiers for aqueous solutions in areas as diverse as coatings and enhanced oil recovery (*2*).

The nature of these interactions depends sensitively on the polymer microstructure as well as on the type and content of hydrophobic substituents. To develop an understanding of these structure–property relation-

0065-2393/93/0236-485$06.25/0
© 1993 American Chemical Society

ships, much more detailed information at the molecular level is needed. In the work described here, we employed fluorescence spectroscopy in conjunction with other methods to obtain this kind of information.

The polymers hydroxypropylcellulose (HPC) and poly(N-isopropylacrylamide) (PNIPAM) share the property that they precipitate from their aqueous solutions upon heating. This lower critical solution temperature (LCST) is 42 °C for HPC and 32 °C for PNIPAM. Both polymers are relatively nonpolar but are solubilized in water by hydrogen bonding. These H bonds are disrupted by heating, and this disruption leads to phase separation. One of the key differences between these polymers is that of chain stiffness, HPC being substantially more extended and less flexible than PNIPAM. The introduction of hydrophobic substituents on these polymers, either alkyl chains or aromatic chromophores, perturbs the hydrophobic–hydrophilic balance and has a number of interesting effects on polymer behavior in water. We examined a number of these features, relying heavily on fluorescence studies to reveal behavior at the molecular level.

Polystyrene–poly(ethylene oxide) (PS–PEO) diblock and PEO–PS–PEO triblock copolymers have a very different microstructure; the hydrophobic block of the chain is at one end or in the middle. Samples containing more than 50 wt% PEO dissolve in water and associate to form spherical micelles containing a dense PS core surrounded by a corona of solvent-swollen PEO chains (3, 4). A number of features of these micelles are of interest, such as the aggregation number, the effective size, and the concentration at which micelles first form (the critical micelle concentration or CMC). These same issues are also important for PEO containing two hydrophobic end groups. Here this apparently small change in microstructure leads to a complete change in the nature of the polymer association in water. In both systems, fluorescence techniques provide important information, although in the block copolymer micelles, light scattering plays a much more important role in determining micelle size.

Hydroxypropylcellulose (HPC)

Hydroxypropylcellulose (Figure 1) is soluble not only in organic solvents such as tetrahydrofuran and methanol, but also in cold water. Attaching a small number of hydrophobic substituents to this polymer has a significant effect on its properties in water. When HPC carries an average of one or fewer hydrophobic groups per chain, its solubility in water is hardly affected. However, a much larger change in the properties of HPC occurs when the level of labeling is increased to an average of two or more groups per chain. The effect is very noticeable in the pyrene-labeled polymer, HPC-Py/56 (1 Py per 56 glucose units, Figure 1). The solubility of this polymer in water is greatly reduced compared to that of the unlabeled polymer. This polymer

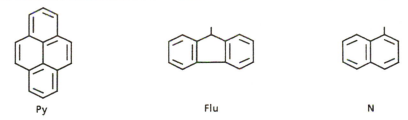

$(CH_3)_2CH - NH - CO - CH$
CH_2

$R - NH - CO - CH$
CH_2

$R' - NH - CO - CH$
CH_2

PNIPAM-Py:	$R = CH(CH_3)_2$	$R' = (CH_2)_4$-Py
PNIPAM-C_n:	$R = CH(CH_3)_2$	$R' = C_nH_{2n+1}$
PNIPAM-Py-N:	$R = CH(CH_3)$-N	$R' = (CH_2)_4$-Py

OH
OCH₂CHCH₃
CH₂

H   OH
OCH₂CHCH₃ H

H   OCH₂CHCH₃
OH

H   O CH₂CHCH₃
OH

CH₂
OCH₂CHCH₃
O – R

OH
OCH₂CHCH₃
CH₂

H   OH
OCH₂CHCH₃ H

H   OCH₂CHCH₃
OH

HPC-Py:	$R = (CH_2)_4Py$
HPC-Flu:	$R = CH_2$-Flu

Py   Flu   N

Figure 1. Structures of the hydroxypropylcellulose (HPC) and poly(N-isopropylacrylamide) (PNIPAM) derivatives.

undergoes extensive interchain aggregation in water, and various aspects of the heat-induced phase transition are perturbed. Through the use of several fluorescence techniques we demonstrated (5–7) that in such hydrophobically modified HPC: (1) interchain aggregation below the LCST occurs predominantly through association of the substituents, (2) the polymers form interchain aggregates even in extremely dilute solutions, and (3) the association between hydrophobic groups is destroyed when the solutions are heated through their LCST.

Aggregation of Labeled HPC-Py/56 in Cold Water.   Fluorescence spectra of HPC-Py/56 in methanol and in water are presented in Figure 2. The spectrum shows two bands, one due to locally excited pyrene (intensity I_M, "monomer emission") with the (0, 0) band located at 376 nm and a broad emission centered at 480 nm due to pyrene excimer emission (intensity I_E). For samples in methanol, identical excitation spectra are obtained for emissions monitored at 396 and 480 nm, and the maxima correspond to those in the UV absorption spectrum. Two differences are observed for aqueous solutions of HPC-Py/56. The emission spectrum exhibits a much stronger excimer band. In addition, the excitation spectra for the monomer and excimer are clearly different; the monomer's spectrum is shifted by about 3 nm. The excimer excitation spectrum corresponds to the UV spectrum of the sample. These features and other aspects of the spectroscopy of HPC-Py/56 lead to the conclusion that the excimer emission originates from aggregates of pyrenes that exist before excitation.

This type of phenomenon is unique to aqueous solutions of chromophore-labeled polymers. An important task is to distinguish whether the pyrene association occurs intramolecularly or between chromophores on different chains. To address this question we monitored the effect of polymer concentration on the ratio I_E/I_M. This ratio decreases somewhat with decreasing concentration, a feature indicating interchain contributions. However, the results of these experiments did not allow us to exclude the occurrence of intramolecular pyrene–pyrene association, nor to quantify the relative importance of each contribution. Another approach was devised to elucidate this point.

Aggregation of HPC-Py/438 and HPC-Flu/33 in Cold Water.   Energy-transfer experiments allow the detection of association between fluorescent labels attached to different polymeric chains. To proceed, we prepared solutions containing a mixture of two polymers, identical except for their fluorescent tags: One polymer carried a very small number of pyrene labels (HPC-Py/438); the second, fluorene labels (HPC-Flu/33, Figure 1). The two chromophores interact as donor (fluorene) and acceptor (pyrene) by nonradiactive energy transfer (NRET) (8). This Förster process originates from dipole–dipole interactions between the excited donor and the acceptor. The efficiency (E) of energy transfer depends sensitively on the separation distance R, where R_o is a characteristic distance, which, for rotationally averaged pairs, is the donor–acceptor separation distance for which the

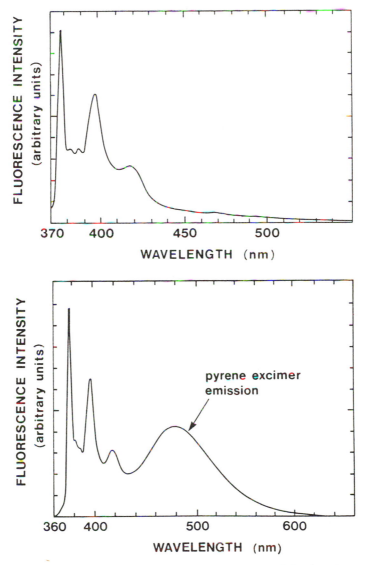

Figure 2. Fluorescence spectra of HPC-Py/56 in methanol (top) and in water (bottom).

energy-transfer efficiency is 50%. For the Flu–Py pair, $R_o = 32$ Å:

$$E = R_o^6 / \left(R^6 = R_o^6 \right) \tag{1}$$

When samples containing both polymers were excited at 290 nm, most of the light was absorbed by the Flu chromophore. The emission spectrum of this mixture exhibited not only features characteristic of Flu emission, but also a strongly enhanced Py monomer emission, superimposed upon the weak emission from directly excited pyrenes. These experiments (7) provide a proof of the existence of interpolymeric association via the hydrophobic substituents.

To examine this phenomenon in more detail, the influence of total polymer concentration on the efficiency of energy transfer was examined (7). First, for a given Py to Flu molar ratio, the total polymer concentration was decreased from 200 to 2 ppm (Figure 3). Significant energy transfer between HPC-Flu/33 and HPC-Py/438 takes place when the total polymer concentration is as low as 50 ppm. It becomes negligible only at concentrations lower than 5 ppm. Second, the effect of added unlabeled HPC on energy transfer was monitored. The amounts of HPC-Flu/33 and HPC-Py/438 were kept constant, while the total HPC concentration was increased from 200 to 5000 ppm. In all cases the results were time dependent. At high HPC concentration (> 2000 ppm) the efficiency of energy transfer decreased continuously with time and became negligible after 7 days. In the 200-to 500-ppm concentration range, the extent of NRET decreased, but the time dependence was more complex. These results indicate the influence of aging time on the aggregation properties of HPC in water. The association of hydrophobic chromophores is disrupted and, in the presence of HPC in high concentration, eventually destroyed. Evidence from other sources indicates that in concentrated solutions HPC itself forms an intricate network of polymolecular aggregates. In the labeled polymers the chromophores pre-

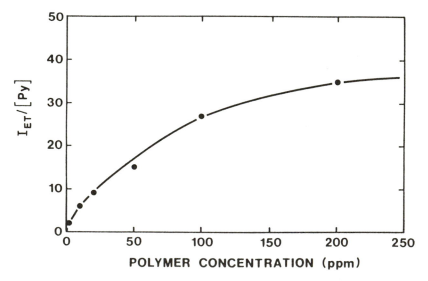

Figure 3. Reduced intensity of pyrene emission due to energy transfer ($I_{ET}/[Py]$) as a function of the total labeled HPC concentration for a mixture of HPC-Flu/33 and HPC-Py/438.

sumably can be accommodated in "hydrophobic pockets" within this network. This phenomenon bears some similarity with those observed when dilute solutions of labeled polymers are heated through their LCST.

Phenomena Associated with the Cloud Point of Aqueous Solutions of HPC-Py/56.

For HPC-Py/56 in water at 25 °C (Figure 2) we calculate from peak heights a value of I_E/I_M of 0.46. When this solution is heated, this intensity ratio undergoes several changes (7). First, it increases and reaches a maximum of 0.55 at 35 °C. Then, the ratio decreases sharply to its limiting value of 0.21 at 50 °C. The midpoint of the transition is ca. 42 °C (Figure 4). When the solution is cooled from 55 to 25 °C, a sharp increase in I_E/I_M occurs, and the midpoint of the transition occurs at ~ 41 °C. At all temperatures lower than 50 °C, the ratio I_E/I_M measured during cooling was smaller than that measured during heating. When the cooled sample was kept at 25 °C for several hours, its emission spectrum became identical to that of a solution that had not been subjected to the heating–cooling treatment. These results imply that the association between hydrophobic groups is dramatically

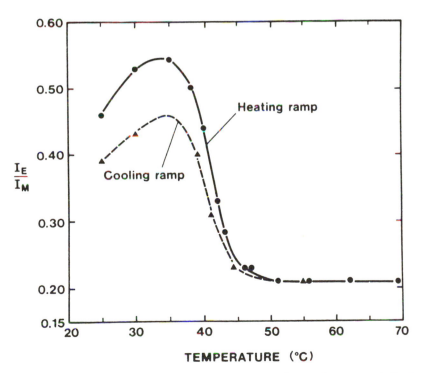

Figure 4. I_E/I_M *as a function of temperature for a solution of HPC-Py/56 (0.12 g/L) in water.*

disrupted during the phase transition that takes place at the LCST. Further-more, after cooling, the polymers undergo a very slow relaxation during which the Py groups reassociate.

These Py aggregates are probably stabilized by hydrophobic interactions. The nonpolar dimers are surrounded by a cage of highly organized water molecules tightly bound through hydrogen bonding. This formation of dimers or higher aggregates in water has a positive entropy and a positive enthalpy. The entropic term is dominant, rendering the free energy of dimer formation favorable (negative). As the solution is heated, a temperature is reached at which the free energy of Py dimer formation becomes positive. The Py groups separate and become accommodated within hydrophobic cavities in the polymer-rich phase. The disruption of Py dimers is initiated at tempera-tures slightly below the LCST of the solution. However, the temperature that triggers the disruption of the dimers is a characteristic of the polymer and not of the hydrophobic groups.

Poly(N-isopropylacrylamide) (PNIPAM)

Poly(*N*-isopropylacrylamide) and its fluorescently labeled analog, PNIPAM-Py/200 (Figure 1), exhibit solution properties quite similar to those of HPC and HPC-Py/56, respectively. They dissolve in organic solvents such as tetrahydrofuran and methanol and are soluble in cold water, but not in warm water (LCST = 32 °C). The fluorescence properties of PNIPAM-Py/200 parallel those of HPC-Py/56 (9). Below the LCST, fluorescence spectra of PNIPAM-Py/200 in water exhibit emissions due to locally excited Py and to Py excimers ($I_E/I_M = 0.36$ at 20 °C). As with HPC-Py/56, here also the Py excimer emission originates predominantly from preassociated pyrenes. How-ever, in PNIPAM-Py/200, evidence from concentration studies points toward a larger contribution of intramolecular aggregation compared to interchain association. This reluctance of PNIPAM to form interchain aggregates in water solutions below their LCST was reported also by Schild and Tirrell (*10*). When a solution of PNIPAM-Py/200 is heated through its LCST, the Py aggregates are destroyed. At 32 °C Py excimer emission is negligible ($I_E/I_M < 0.01$). The polymer-rich phase that separates above the LCST provides a hydrophobic environment in which pyrenes are solubilized and isolated from each other.

An intriguing issue concerns the changes in dimensions of a single polymer coil as it passes through its critical point at the LCST. Dynamic and static light-scattering measurements were reported by Fujishige (*11*), who found a sharp collapse transition. In a 13-ppm solution of PNIPAM ($M_W \sim 8 \times 10^6$) the hydrodynamic radius decreased from ~ 1000 Å at 30 °C to ~ 300 Å at 32 °C. Fluorescence studies can provide further insights into this heat-induced chain collapse. We carried out energy-transfer experiments (*12*) on a polymer (PNIPAM-Py/366-N/50) containing both donor (naphthalene, N) and acceptor (Py) groups. In these experiments we excited N groups selectively at 290 nm and followed changes in the fluorescence spectra as the solution temperature was raised from 20 to 40 °C. The naphthalene emission decreases slowly with increasing temperature, accompanied by a strong increase in the pyrene emission intensity. Heating above 32 °C has no further effect on the fluorescence intensity from either chromophore. The simultaneous decrease in total naphthalene emission and increase in total pyrene emission are indicative of an increase in the energy-transfer efficiency above the LCST, as a consequence of a contraction of the polymer coil allowing the chromophores to come into closer proximity.

An intriguing feature of these experiments (*12*) is that the increase in energy transfer occurs not at the LCST itself, but below this temperature and over a broad temperature range. This pattern is seen, for example, in a plot of the naphthalene quenching efficiency (E_N) by NRET as a function of temperature (Figure 5). Such a gradual increase is surprising, particularly because it monitors a property of a solution passing through a critical point. It implies that the average distance between the chromophores is gradually reduced as the temperature increases. This reduced distance may result either from a continuous decrease in the size of the polymer coil or from enhancement in density fluctuations, giving rise to a higher incidence of naphthalene–pyrene contacts. Although the spectroscopic changes observed

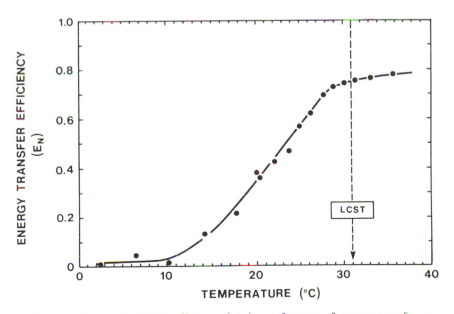

Figure 5. Plot of the NRET efficiency (E_n) as a function of temperature for an aqueous solution of PNIPAM-Py/366-N/50 (44 ppm).

cannot be described in conformational detail, they confirm that local contractions of individual polymer chains precede the global collapse in hydrodynamic radii (R_H) and the subsequent interpolymeric aggregation at the LCST of PNIPAM solutions.

Hydrophobically Modified Poly(N-isopropylacrylamides)

In the previous sections, we described polymers whose hydrophobic substituents were fluorescent chromophores that served as intrinsic monitors of hydrophobic association. When polymers contain alkyl substituents, introduced, for example, to modify surface adhesion or solution rheology, a different approach is needed to study the association of these substituents. For example, Ringsdorf et al. (13) prepared a series of PNIPAMs containing small amounts of pendant n-alkyl groups to enhance their binding to bilayer membranes. Their structures, PNIPAM-C_n/100 and PNIPAM-C_n/200, are shown in Figure 1. To study the properties of their solutions in water, we examined the fluorescence of trace amounts of probe molecules [10^{-7} to 10^{-6}/M pyrene or bis(1-pyrenylmethyl) ether, (dipyme)] added to these solutions (14).

The objective of these experiments was to confirm the presence of hydrophobic microdomains in aqueous solutions of the amphiphilic PNIPAM and to detect changes in the polarity and rigidity of these hydrophobic microdomains as a function of alkyl chain length and polymer concentration. In one set of experiments we monitored the ratio I_1/I_3 of the intensity of the (0,0) band (I_1) to that of the (0,2) band (I_3) of the pyrene emission under various conditions. This ratio is used routinely in the study of micelles (15). It takes a high value in polar media (16) (I_1/I_3 = 1.81 in water), and decreases with decreasing polarity (I_1/I_3 = 1.12 for Py solubilized in sodium dodecyl sulfate (SDS) micelles) (15). In aqueous solutions of the C_{14}- and C_{18}-PNIPAM amphiphiles the I_1/I_3 ratio decreases rapidly with increasing polymer concentration (Figure 6). In the PNIPAM-C_{10}/200 the ratio remains constant (1.80) over the entire polymer concentration range probed, a behavior identical to that for PNIPAM. This result implies that pyrene is not solubilized by PNIPAM or the C_{10}-PNIPAM copolymers.

The fluidity of the environment within a micelle or any such hydrophobic microdomain, often termed "microviscosity", was investigated through the use of a second fluorescence probe, dipyme (17, 18). Dipyme can fold to form an intramolecular excimer. The extent of excimer emission depends upon the rate of conformational change. This motion is resisted by the local friction imposed by the environment. As a consequence, the excimer-to-monomer intensity ratio, I_E/I_M, provides a measure of the microviscosity of the probe environment. In addition, the vibrational fine structure in the dipyme monomer emission is sensitive to the polarity of the probe microenvi-

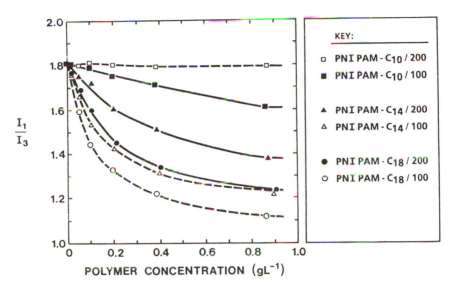

Figure 6. I_1/I_3 *vs. polymer concentration for aqueous solutions of PNIPAM-C_n in the presence of 6×10^{-7} M pyrene.*

ronment in much the same way as pyrene itself. The probe has an extremely low solubility in water, but it can be solubilized within hydrophobic microdomains. In the experiments described here it was added to solutions containing polymers in a concentration of 2g/L, which corresponds to alkyl chain concentrations of ca. 1.6×10^{-4} mol/L for PNIPAM-C_n/100 and 8.2×10^{-5} mol/L for PNIPAM-C_n/200, and care was taken to avoid the presence of dipyme microcrystals in the solution (*14*). Neither PNIPAM nor its C_{10} derivatives are effective at solubilizing significant amounts of dipyme at room temperature. This result indicates that neither polymer, below its LCST, has domains of significant hydrophobicity, consistent with the results of experiments with pyrene.

Fluorescence spectra of dipyme in a solution of PNIPAM-C_{18}/100 and in *n*-octylthioglucopyranoside (OTG) micelles are shown in Figure 7. Figure 7 shows that the extent of excimer emission is much smaller in the spectrum of dipyme in the polymeric solution than in the spectrum of dipyme in surfactant micelles. This result points to a rigid structure for the alkyl chain clusters formed in these solutions. Bilayer vesicles possess a less fluid microenvironment in their hydrophobic interior than the core of a simple surfactant micelle (*15*). We had anticipated that the alkyl chain clusters formed within the polymer would be similar to surfactant micelles in terms of their structure and their mobility. As yet little is known about the structure of these clusters, but an interesting and significant observation is that their

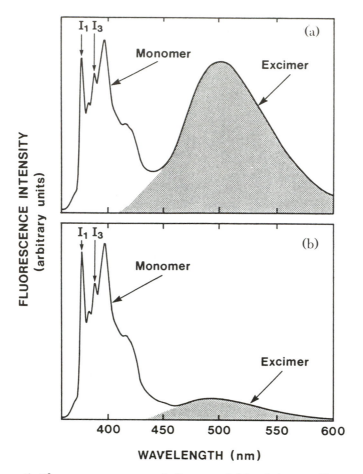

Figure 7. Fluorescence spectra of dipyme solubilized by micellar aqueous n-octylthioglucopyranoside (a) and by PNIPAM-C$_{18}$ (b).

microfluidity is more comparable to that of vesicles than micelles. A similar conclusion was drawn from fluorescence depolarization studies with diphenyl-hexatriene and perylene in these systems (*13*).

Dipyme emission spectra are essentially identical in both the C$_{14}$ and the C$_{18}$ polymers, a result indicating that the clusters formed are all very similar in their properties. The I_E/I_M values point to a very similar microfluidity, and the I_1/I_3 values indicate that in each type of cluster, the dipyme probe is located in a very similar and strongly hydrophobic environment. These ratios do not change when the polymer concentration is varied, a feature consistent with the idea that cluster formation is an intramolecular, single-polymer phenomenon. We cannot rule out contributions of interpolymer association.

These, however, would have to lead to structures with common properties across the entire range of polymer concentrations.

PS–PEO Block Copolymers

Polystyrene–poly(ethylene oxide) block copolymers sufficiently rich in PEO dissolve spontaneously in warm water (60 °C) to form micelles. Some samples less rich in PEO can be induced to form micelles by mixing a solution of the polymer in tetrahydrofuran (THF) with water followed by careful removal of the THF under vacuum. The remaining samples do not form stable micellar dispersions in water. For example, if a dilute THF solution of one of these polymers is mixed with an equal volume of water and slowly concentrated, the polymer precipitates (*19*). The phase diagram in Figure 8 summarizes our results. Low-molecular-weight polymers are more tolerant of a larger weight fraction of PS, and samples requiring THF to promote micelle formation represent borderline examples near the phase boundary.

The CMC of these micelles is too small to be determined by light-scattering methods. We developed a simple method for CMC determination based upon the use of pyrene as a fluorescent probe (*20, 21*). Py has a limiting solubility in water of ca. 6×10^{-7} M. In the presence of micelles, the probe partitions between the aqueous and micellar phases. This partitioning leads to a number of interesting changes in the fluorescence behavior, the most pronounced of which is the shift in the (0,0) band in the Py excitation spectrum from 332 nm (water) to 339 nm (micelle) (Figure 9). The intensity ratio $F = I_{339}/I_{332.5}$ is small and constant (F_{min}) at polymer concentrations (*c*) below the onset of micelle formation and levels off at a much higher value (F_{max}) at values of *c* at which nearly all the pyrene is located in the micelle phase.

A plot of $(F - F_{min})/(F_{max} - F)$ versus *c* shows curvature at low concentration, becoming straight at higher *c* (*21*). Extrapolation to *c* = 0 identifies the CMC of the system. This analysis is confirmed in a reciprocal plot of

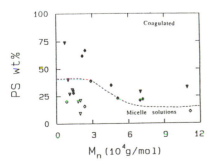

Figure 8. Phase diagram for PS–PEO diblock ($\triangledown$) and triblock ($\diamondsuit$) copolymers in water, plotted as weight percent PS composition vs. total molecular weight. Key: open points, samples that form micelles upon heating in water; half-filled points, micelles that form with the aid of tetrahydrofuran as a cosolvent; and filled points, no micelles form.

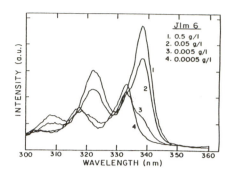

Figure 9. Excitation spectra of pyrene in water $(6 \times 10^{-7}\ M)$ as a function of block copolymer concentration.

the data (Figure 10) according to the expression

$$\frac{F_{max} - F}{F - F_{min}} = \frac{1000\ \rho_{PS}}{K_V\ \chi_{PS}\ (c - CMC)} \qquad (2)$$

where χ_{PS} is the weight fraction of PS in the copolymer and ρ_{PS} is the density of PS. Equation 2 is very sensitive, and incorrect values of CMC lead to curvature in the plot. From this analysis we find partition coefficients K_V of 3×10^5 (dimensionless) favoring partitioning of Py into the micellar phase (*21*).

Static light scattering normally provides straightforward information about M_W and the weight-averaged aggregation number, N_W, of block copolymer micelles, and dynamic light scattering (DLS) allows measurement of the hydrodynamic radius (*2–4*). Unfortunately, the tendency for PEO to undergo self-association in water promotes clustering of micelles. The fraction of micelles that form micelle clusters is very small, but their contributions to the light-scattering signals are substantial. In the DLS experiment, Laplace inversion methods can be used in the data analysis (*22*) to reveal the two populations of species present and to determine their respective hydrodynamic radii (R_H).

With the various di- and triblock copolymer samples in hand, these R_H values can be used to test one of the key predictions of the "star model" for block copolymer micelles, namely that

$$R_H \sim N_{PS}^{4/25}\ N_{PEO}^{3/5}\ Q \qquad (3)$$

where N_B is the degree of polymerization of a particular block and Q is a factor equal to 1.0 for diblocks and 0.29 for triblock copolymers. Results presented in Figure 11 indicate excellent agreement with the star model for the micelles, but not for the secondary aggregates. As a consequence we *assume* the validity of the star model in analyzing the DLS data to obtain M_W values and aggregation numbers for the micelles.

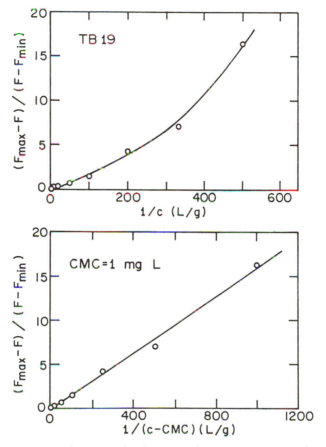

Figure 10. A plot of $(F_{max} - F) / (F - F_{min})$ *vs.* $1/c$ *and vs.* $1/(c - CMC)$, *showing the sensitivity of the data to micelle formation.*

Zimm plot analyses of SLS data yield the apparent weight-averaged molecular weight, M_W (app) of all the species present. Under our experimental conditions ($c \gg$ CMC) these comprise the micelles and the clusters

$$M_W(\text{app}) = W_M(\text{micelles}) + (1 - W_M) M_W (\text{clusters}) \qquad (4)$$

where W_M is the weight fraction of polymer present as micelles. To proceed with the analysis, we make three key assumptions: that the size distribution of micelles is narrow (i.e., $M_W \approx M_N$ for micelles), that the star model is valid, and that the density of the micelle clusters is proportional to that of the micelles themselves (*19*). The first two assumptions are strongly supported by the DLS experiment. The third assumption and the details of the data analysis are discussed in detail in reference 19. The essence of this analysis is

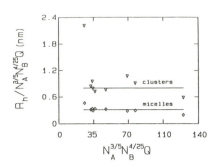

Figure 11. A plot of $R_H / QN_A^{4/25} N_B^{3/5}$ vs. $QN_A^{4/25} N_B^{3/5}$ for a series of PS–PEO diblock and triblock copolymers. The lower data fit the star model and correspond to the micelles. The upper data set is due to large secondary aggregates (clusters) of micelles.

Table I. Characteristics of Block Copolymer Samples and Their Micelles in Water

Sample Name	N_B (PS)	N_A (PEO)	M_n	N	R_H (nm)	R_{core} (nm)
Diblock						
DB23	110	400	28,700	290	23	12
DB40	36	236	14,100	120	14	5.6
DB41	36	450	23,600	120	19	6.0
jlm5	17	155	8,500	64	9	3.5
Triblock						
TB19	40	2 × 100	13,100	130	12	5.8
TB51	18	2 × 200	20,000	67	9	3.6
jlm4	41	2 × 180	20,000	130	13	6.0
jlm6	35	2 × 164	18,000	120	13	5.5
jlm11	47	2 × 240	26,000	150	18	6.8

that the proportionality assumption is not bad, but the proportionality constant for triblock copolymer micelles is different from that for diblocks. As a consequence, we can calculate W_M, M_n (micelle), and the mean aggregation number N_n for each micelle. Important characteristics of the polymers and the micelles are collected in Table I.

PEO with Hydrophobic End Groups

PEO polymers containing hydrophobic end groups are members of a class of polymers known as associative thickeners (ATs). These substances are added at about 0.5–2 wt% to latex paint formulation to modify the rheology and to reduce splatter. At higher concentrations the solutions will gel, presumably because the end groups associate to form micellelike aggregates, bridged by the network of PEO chains. A typical linear AT has the structure

$$RO-(DI-O-PEO-O-)_n DI-OR$$

1

where DI is a diurethane, formed typically from toluene diisocyanate or isophorone diisocyanate, and R is an alkyl or an alkylaryl (e.g., nonylphenyl) group. In many of these materials, the PEO has $M_n = 8000$ with a narrow molecular-weight distribution (MWD). Average n values range from 2 to 50. The final polymers, produced by condensation of the reactants, have a broad MWD. Rohm and Haas markets an associative thickener ($R = C_{12}H_{25}$) under the trademark RM-825.

Another class of associative thickeners is based upon cellulose derivatives. For example, Aqualon produces a hydrophobically modified hydroxyethylcellulose (HMHEC) containing a small amount ($< wt\%$) of alkyl chains (C_{12} to C_{24}, e.g., Natrasol 250 GR). Fluorescent probe experiments were employed (23) to study the networks formed by these polymers in water. The fluorescence data show that the hydrophobic groups of HMHEC associate to form clusters above a critical polymer concentration of 500 ppm.

Our experiments (24) employed a series of model materials ($R = C_{16}H_{33}$) prepared by the Bassett group at Union Carbide. We added these polymers to aqueous solutions containing a fixed concentration of pyrene (6×10^{-7} M) and examined the changes in the probe fluorescence as a function of c for two samples of **1** with $M_W = 40,000$ and $47,000$ (M_W/M_n = 1.3 and 1.5, respectively). To estimate the onset of association, we took advantage of the shift in the excitation spectrum of Py akin to that shown in Figure 2: By choosing $\lambda_{ex} = 338$ nm Py is selectively excited in a hydrophobic environment. A plot of fluorescence intensity I versus log c is sigmoidal (Figure 12). The rising portion of the curve indicates the onset of end-group association coupled with Py partitioning into the micellelike clusters (MLC's) that form.

A more elegant, but less general, approach to this problem was reported recently by Richey et al. (25), who prepared a derivative of **1** with 4-(1-pyrenebutyl) end groups (PyAT). Viscosity studies suggest that PyAT and **1** with $R = C_{12}H_{25}$ have end groups of comparable hydrophobicity. With Py groups serving as the hydrophobic substituents, fluorescence experiments report directly upon association without complications due to probe partitioning between phases. In very dilute solution, PyAT forms a small amount of excimer (concentration-independent) because of self-cyclization. At higher concentrations I_E/I_M increases with c due to intermolecular end-group association. The crossover for PyAT ($M_W = 60,000$; $M_W/M_n = 2$) occurs at $c = 5$ ppm. This value is in the same range as that inferred for our samples shown in Figure 12.

The most important information about these polymers is their aggregation number, the number of chain ends per MLC. We have as yet no information about whether a closed-association model, with a narrow distribution of ends groups per MLC, or an open-association model, involving aggregates whose size depends upon end-group concentration, provides a better description of association in this system.

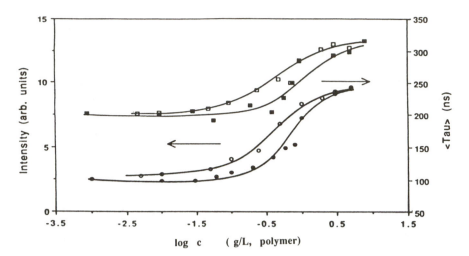

Figure 12. Plot of the fluorescence intensity (lower two curves) and mean fluorescence decay times (upper two curves) as a function of polymer concentrations for **1** *(R = $C_{16}H_{33}$). Key: open symbols, M_W = 40,000; and closed symbols, M_W = 47,000.*

To begin, mean aggregation numbers, N_n must be determined. Richey et al. (25) approached this problem by adding incremental amounts of PyAT to solutions of RM-825, keeping the total c fixed at 0.02 wt%. This approach enhances the probability of finding two or more Py per micelle, which should lead to excimer emission, as opposed to monomer emission from MLC containing only one Py. In this way the authors (25) calculated N_n = 6 end groups per micelle from analysis of their I_E/I_M data. This number depends upon the unfortunate assumption that monomer and excimer have identical quantum yields and that an excimer will be formed whenever two pyrenes occupy one MLC. We now know the second assumption is not correct (26).

We also used Py excimer emission to determine mean aggregation numbers. In the course of our experiments with pyrene as a probe (24), we noticed that over a limited range of polymer concentration, Py excimer emission could be observed. This observation is in many ways remarkable. With the bulk Py concentration held at 6×10^{-7} M, a strong partitioning of Py into a relatively small number of micellelike aggregates would be necessary to achieve local concentrations high enough for excimers to be formed. A plot of I_E/I_M versus log c is shown in Figure 13. The maximum in this plot can easily be understood: at low c most of the Py is in the aqueous phase. MLC's composed of end groups form as c increases, and Py partitions between the micellar and aqueous phases. At high c values most of the Py is located in the MCL's, but because the number of MCL's vastly exceeds the number of Py molecules, the probability of finding two Py in the same micelle is negligible, and I_E drops to zero.

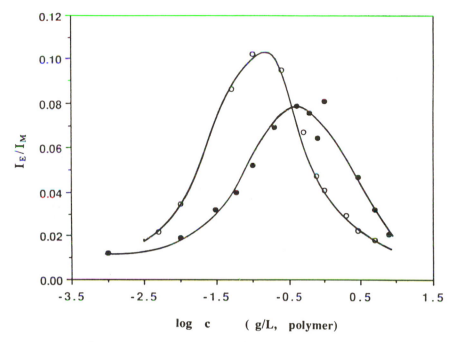

*Figure 13. The intensity ratio I_E/I_M as a function of polymer concentration for the two samples of **1** described in Figure 12.*

The maximum in the I_E/I_M versus c plot should occur at the point at which the number of MLC's is half the number of Py molecules. From this assumption and knowledge of the molecular weight of the polymers, mean aggregation numbers can be determined. These refer to the average number of chain ends per MLC. We find $N_n = 15$ for the sample with the lower molecular weight and $N_n = 25$ for the sample with the higher molecular weight. These numbers are small compared to those for normal nonionic surfactants, and somewhat larger than that calculated by Richey et al. for their polymer at much higher polymer concentration.

These types of experiments now need to be extended to a broader range of polymer concentrations, end groups, and chain lengths. In addition, more reliable M_n values are needed for the ATs to calculate the aggregation numbers. Values determined by size exclusion chromatography alone depend upon column calibration and other factors that make the three aggregation numbers reported here only estimates of their true values.

In summary, the essential features of associative thickener behavior in water are first, that the end groups associate to form micellelike aggregates; second, the onset of association occurs in the parts-per-million concentration range; and, finally, the mean aggregation numbers are much smaller than those of corresponding alkyl PEO nonionic surfactants.

Summary

Water-soluble polymer containing hydrophobic substituents form micellelike clusters in water. For rigid cellulosic polymers, the association is interpolymeric with little evidence for intramolecular association of the substituents. For more flexible linear polymers, the nature of the interaction can depend upon chain microstructure and the location of the hydrophobic substituents. Those at the chain ends, as in C_n–PEO–C_n, undergo strong interpolymeric associations to form micellelike clusters composed of a small number of end groups (6 to 25). Only at parts-per-million concentrations is there a dominance of intramolecular end-group interactions.

In the PNIPAM containing a random distribution of C_n substituents, hydrophobic association seems to be intramolecular in origin. No detectable association is seen when the chains are short (e.g., C_{10}), but pronounced association is seen for longer chains (e.g., C_{14} and C_{18}). These micellelike aggregates are more rigid, and presumably smaller in size, than traditional surfactant micelles. Upon heating solutions of hydrophobically modified PNIPAM and HPC above their respective LCSTs, the micelles are disrupted.

PEO–PS and PEO–PS–PEO block copolymers behave quite differently. If these polymers have a narrow MWD, they undergo a sharp association transition (a true CMC) with increasing concentration. They form spherical micelles with a very narrow distribution of sizes. Their sizes and N_n values seem to be well described by the star model of block copolymer micelles.

Acknowledgments

We thank the National Science and Engineering Research Council of Canada and the Province of Ontario for their financial contributions in support of this research.

We acknowledge the contributions of J. Venzmer and H. Ringsdorf to part of this work.

References

1. *Polymers in Aqueous Media: Performance through Association*; Glass, J. E., Ed.; Advances in Chemistry 223; American Chemical Society: Washington, DC, 1989.
2. (a) Emmons, W. D.; Stevens, T. E.; "Polyurethane Thickeners in Latex Compositions", U.S. Patent No. 4, 079, 028, 1978. (b) Hoy, K. L.; Hoy, R. C.; "Polymers with Hydrophobe Bunches", U.S. Patent No. 4, 426, 485, 1984.
3. (a) Tuzar, Z.; Kratochvil, P. *Adv. Colloid Interface Sci.* **1976**, 6, 201. (b) Tuzar, Z.; Kratochvil, P. *Colloids Surf.*, in press.
4. Riess, G.; Hurtrez, G.; Bahadur, P. *Encyclopedia of Polymer Science and Engineering*; 2nd ed.; Wiley: New York, 1985; Vol. 2, pp 324–434.
5. Winnik, F. M. *Macromolecules* **1987**, 20, 2745.
6. Winnik, F. M.; Winnik, M. A.; Tazuke, S.; Ober, C. K. *Macromolecules* **1987**, 20, 38.

7. Winnik, F. M. *Macromolecules* **1989**, *22*, 734; Winnik, F. M.; Tamai, N.; Yonezawa, J.; Nishimura, Y.; Yamazaki, J. *J. Phys. Chem.* **1992**, *96*, 1967.
8. Berlman, I. B. *Energy Transfer Parameters of Aromatic Molecules*; Academic Press: Orlando, FL, 1981.
9. Winnik, F. M. *Macromolecules* **1990**, *23*, 233.
10. Schild, H. G.; Tirrell, D. A. *Polym. Prepr. Am. Chem. Soc. Div. Polym. Chem.* **1989**, *30*(2), 350.
11. Fujishige, S. *Polym. J.* **1987**, *19*, 297.
12. Winnik, F. M. *Polymer*, **1990**, *31*, 2125.
13. Ringsdorf, H.; Venzmer, J.; Winnik, F. M. *Macromolecules* **1991**, *24*, 1678.
14. Winnik, F. M.; Winnik, M. A.; Ringsdorf, H.; Venzmer, J. *J. Phys. Chem.* **1991**, *95*, 2583.
15. Kalyanasundaram, K.; Thomas, J. K. *J. Am. Chem. Soc.* **1977**, *99*, 2039.
16. Dong, D. C.; Winnik, M. A. *Can. J. Chem.* **1985**, *62*, 2560.
17. Georgescauld, D.; Desmasez, J. P.; Lapouyade, R.; Babeau, A.; Richard, H.; Winnik, M. A. *Photochem. Photobiol.* **1980**, *31*, 539.
18. Zachariasse, K. A.; Vaz, W. L. C.; Sotomayor, C.; Kühnle, W. *Biochem. Biophys. Acta* **1982**, *688*, 323.
19. Xu, R.; Winnik, M. A.; Riess, G.; Croucher, M. D.; Chu, B. *Macromolecules* **1992**, *25*, 644.
20. Zhao, C. L.; Winnik, M. A.; Riess, G.; Croucher, M. D. *Langmuir* **1990**, *6*, 514.
21. Wilhelm, M.; Zhao, C. L.; Wang, Y.; Xu, H.; Winnik, M. A.; Riess, G.; Mura, J. L.; Croucher, M. D. *Macromolecules* **1991**, *24*, 1033.
22. Xu, R. L.; Winnik, M. A.; Hallet, F. R.; Riess, G.; Croucher, M. D. *Macromolecules* **1991**, *24*, 87.
23. Sivadasan, K.; Somasundaran, P. *Colloid Surf.* **1990**, *49*, 229.
24. Wang, Y.; Winnik, M. A. *Langmuir* **1990**, *6*, 1437.
25. Richey, B.; Kirk, A. B.; Eisenhart, E. K.; Fitzwater, S.; Hook, J. W. *J. Coatings Technol.* **1991**, *63*(798), 31.

RECEIVED for review May 22, 1991. ACCEPTED revised manuscript June 1, 1992.

Metachromasy: The Interactions between Dyes and Polyelectrolytes in Aqueous Solution

Roger W. Kugel

Department of Chemistry, Saint Mary's College of Minnesota, Winona, MN 55987

The literature on metachromasy is reviewed and two quantitative analytical methods for poly(acrylate-co-acrylamide) based on the metachromatic effect are proposed. The color-array method allows for the visual determination of polymer concentration by observing the color of toluidine blue O, cresyl violet acetate, or safranine O at various P/D (polymer acrylate residue to dye molecule) ratios. An abrupt color change is observed at P/D = 1. The complexation–extraction method is based on the removal of the dye Janus green B from solution by complexation with poly(acrylate-co-acrylamide) followed by extraction with 1,1,2-trichlorotrifluoroethane (Freon) solvent. Typical concentration ranges for both tests were 0–10 ppm polymer $(0–7.4 \times 10^{-5}$ M acrylate residue) and $1–6 \times 10^{-5}$ M dye.

$\mathbf{M}$ETACHROMASY IS THE COLOR CHANGE IN THE DYE ABSORPTION SPECTRUM that occurs when certain cationic dyes interact with anionic polyelectrolytes in aqueous solution. This metachromatic effect was first observed in the field of histology where it was observed that certain dye stains changed color when complexed by proteins or nucleic acids and could, therefore, be used to selectively stain various subcellular structures. The term "metachromasy" was coined by Ehrlich to describe the appearance of more than one color in tissue stained by a single dye (*1*). Since the discovery of this phenomenon, a large body of research has developed around the study of the nature of the interaction between dyes and polyelectrolytes, the effects of the interaction

0065–2393/93/0236–0507$07.75/0

on absorption or fluorescence properties of the dye, the types of dyes capable of interacting with polyelectrolytes and undergoing a metachromatic shift, the types of polyelectrolytes capable of serving as substrates for the metachromatic dyes, and finally the uses to which the phenomenon of metachromasy can be put. This chapter reviews the status of research in the field of metachromasy and proposes two new analytical methods for anionic polyelectrolytes based on this phenomenon.

Dyes and Polymers That Interact

The metachromatic effect has been observed between cationic dyes and anionic polyelectrolytes and between anionic dyes and cationic polyelectrolytes, although interactions of cationic dyes have been studied more commonly. Some of the cationic dyes that interact with anionic polyelectrolytes include acridine orange and its derivatives (2–7), methylene blue (8–11), toluidine blue (12, 13), brilliant cresyl blue (14), cresyl violet (15), ethidium bromide (16), a variety of cyanines (12, 17, 18), and phthalocyanine dyes (19) as well as a number of others. Anionic polyelectrolytes that have the capability to induce a metachromatic color change in cationic dyes are known as chromotropes. Polyelectrolytes that can serve as chromotropes for dyes include synthetic polyphosphates, polyacrylates, polysulfonates, polysulfates, carboxylated polysaccharides, polyphenols, and so forth, as well as a variety of naturally occurring proteins and nucleic acids. Actually, a relatively small fraction of the commercially available cationic dyes show a metachromatic shift in the presence of anionic polyelectrolytes. The dyes that do display metachromatic shift generally do so with a wide variety of polyanions, even exotic ones like 11-tungstocobaltosilicate (20).

Some restrictions on the polymer chain length of the chromotrope were discovered by Yamaoka et al. (21), who found a minimum critical range of polymer chain length of 7–20 monomer units in polyphosphate chromotropes for metachromasy to occur. Generally, however, the chemical nature of the anionic sites on the chromotrope has little effect on the metachromatic shift exhibited by a given dye (3, 22).

Only a few minor polyelectrolyte-dependent differences in metachromasy have been noted. Methylene blue, for example, showed an apparently greater tendency to aggregate when bound to poly(vinyl sulfate) than when bound to poly(styrene sulfonate) (11). This tendency resulted in a greater hypsochromic–metachromatic shift in the poly(vinyl sulfate) case than in the poly(styrene sulfonate). However, the same basic absorption peaks due to bound monomeric (662 nm), dimeric (610 nm), and higher aggregated forms (550 nm) of methylene blue were observed for both polyelectrolytes. Therefore, in this case, the different spectral changes observed for the two polyelectrolytes with methylene blue were the result of different amounts of the bound forms of dye rather than inherently different chemical interactions.

The Nature of the Metachromatic Interaction

Early studies on the nature of the metachromatic interaction suggested that the spectral shifts were due to aggregation of the dyes bound to the surfaces of the polyanions (7, 19, 23–29). Similar spectral changes were observed in solutions of the dyes themselves at high concentrations or low temperatures, and these changes were believed to be due to stacking of relatively planar dye molecules (30). Although the aggregation explanation is generally correct, the metachromatic interaction is understandably more complex than simple dye aggregation. Vitagliano (31) reviewed the research on metachromasy up to 1975.

Studies of the effect of ionic strength on metachromasy found that high ionic strength decreased the number of dye molecules attached to the polyelectrolyte, but had no effect on the spectrum of bound-dye species (15, 32, 33). These observations were interpreted to be the result of the displacement of individual dye cations from the polymer surface by electrolyte cations at high ionic strengths. However, because the bound-dye spectra in these studies were unchanged, the simple stacking theory of metachromasy is an imperfect model of polyelectrolyte–dye structure. Other studies examined how changing the dielectric constant of the solvent affected the extent of the metachromatic interaction (34, 35). Such studies found that lowering the dielectric constant of the solvent decreased metachromasy. This observation is understandable in terms of an aggregation model because if a dye is more soluble in a particular solvent, it will show less tendency to aggregate on a polymer surface.

In further studies, Pal and Ghosh (12) reported two types of metachromatic interactions of the dye pinacyanol chloride with synthetic polyanions. One type of interaction caused a sharp red-shifted band that was ascribed to regular stacking of dye molecules on the surface of the polyanion; the other type of interaction caused a broad multiple-banded spectrum that was ascribed to more random irregular interactions of the dye molecules. Regular stacking occurred at polymer residue-to-dye molar ratios (P/D) of 2, which indicated that when dye molecules were located on alternate anionic sites on the polymer, they would stack together in a regular arrangement, whereas at P/D ratios near 1 the stacking was more crowded and irregular.

Pal and Ghosh also studied the metachromasy of pseudoisocyanine chloride (PIC) in the presence of deoxyribonucleic acid (DNA) or various vinyl polyanion chromotropes (18). They found that many, but not all, polyanions caused a sharp red-shifted band (known as a *J* band) in the spectrum of PIC. This band was attributed to a particular staggered stacking arrangement of the PIC dye on the alternate anionic sites of the chromotrope that gave a staircaselike arrangement of the angular PIC dye molecules.

Shirai et al. (22) studied the metachromasy of methylene blue with poly(sodium acrylate) and found that the wavelength of the metachromatic band reflects the strength of the stacking of the bound dyes: the shorter the wavelength of the metachromatic band, the stronger the stacking of the bound dyes.

Scott (19) studied the metachromasy of cationic phthalocyanine dyes and found that these dyes were already aggregated in dilute aqueous solution and, therefore, already metachromatic even before any polyanion was added. Apparently, a metachromatic dye must be a dye that has a *tendency* to aggregate, but can be disaggregated if the solution is dilute enough.

In a summary of aggregation phenomena in xanthene dyes, Valdes-Aguilera and Neckers (36) proposed that parallel aggregation (stacking), which they called H-type aggregation, resulted in a metachromatic shift to the shorter wavelengths (hypsochromic shift) and linear aggregation (head-to-tail), which they called J-type aggregation, resulted in a metachromic shift to longer wavelengths (bathochromic shift). Conditions that favor one type of aggregation over another could provide structural information on the polymer and its complexation with the dye.

Evidence for the importance of the role of electrostatic interactions in the metachromatic effect was obtained by Carroll and Cheung (8), who studied the interaction of the cationic dye methylene blue with carboxylated starches. The thermodynamic behavior of the reaction indicated that the interaction was *primarily* electrostatic. The dye was adsorbed in the dimeric form and in a 1:1 ratio with carboxylate groups on the starch.

Gummow and Roberts (37) studied the metachromasy of anionic dyes on the polycationic chromotrope chitosan and developed a new theory for the origin of the metachromatic effect in general. According to this theory, metachromasy does not arise because of adsorption of ionic dye molecules to specific counterionic sites on the polymer backbone, but rather because of the increased concentration of dye molecules in the vicinity (electrostatic domain) of the polyelectrolyte molecules. The aggregation of dye molecules is then a natural consequence of this increased concentration and is similar to the aggregation observed at higher dye concentrations and driven by hydrophobic interactions between adjacent dye molecules in solution.

Although this electrostatic domain explanation is possible, it does not account for the sharp metachromatic color changes observed in quantitative studies at P/D ratios of 1. In other words, the stoichiometry of metachromasy suggests that the dye molecules coordinate to specific charged sites on the polyelectrolyte molecule.

Shirai et al. (9, 38) studied the metachromatic behavior of methylene blue with poly(vinyl sulfate) homologs. Evidence for saturation and reversal of the metachromatic effect at high P/D ratios was reported. Shirai et al. also observed that the metachromatic effect could be reversed by the addition of large excesses of KCl or urea, and they found evidence that both electrostatic and hydrophobic interactions contribute to the binding of cationic dyes to polyanions. A correlation between the binding strength of dye and the flexibility of the polyanion was observed. Shirai et al. concluded that the stacking of dye molecules is facilitated by conformational changes (coiling) of the polyanion as its charge is neutralized by complexation of the cationic dye molecules. The significance of this work for polymer studies is that metachromasy may be used as a probe of polymer conformation.

A number of other studies of the nature of the interaction between cationic dyes and anionic polyelectrolytes have been carried out (*11, 22, 39–47*). For example, Yamagishi and Watanabe (*40*) found little dependence on the alkyl chain length of *N*-alkylated acridine orange interacting with polystyrene sulfonic acid, Yun et al. (*41*) found evidence for both dye binding and aggregation in a study of the effects of salts and soluble organic compounds on metachromasy, and Shirai et al. (*22*) found that the magnitude of the metachromatic effect was directly related to the flexibility and charge density of the polymer whereas differences in the chemical nature of the anionic charged sites were less important.

The preceding metachromasy studies and, in particular, those of Shirai et al. (*22*) and Pal and Schubert (*29*), suggest that the interaction between ionic dyes and polyelectrolytes is the result of the interplay of three types of interactions:

1. The electrostatic interaction of ionic dye molecules with oppositely charged sites on the polyelectrolyte chain.

2. The hydrophobic interaction of dye molecules with each other and with nonpolar sites on the polyelectrolyte.

3. The interaction between pi-electrons on adjacent, adsorbed dye molecules.

Among these interactions, the third interaction probably has the greatest influence on the spectral properties of the adsorbed dye, because the pi-electrons of the dye are those directly involved in the light absorption and emission processes. Ionic dyes that show no observable metachromatic effect with oppositely charged polyelectrolytes can (and probably do) interact with the polymeric electrolytes, but they interact in such a way that the pi-electrons of the dye are relatively unaffected by the interaction. In other words, in nonmetachromatic dyes the bound-dye chromophore is indistinguishable from the free-dye chromophore. Some evidence for this hypothesis was found in the present study wherein certain cationic dyes that showed no metachromatic shift with poly(acrylate-*co*-acrylamide) in aqueous solution still formed complexes with and gradually coprecipitated with these polymers.

The Free-Energy Change of the Metachromatic Interaction

Metachromasy occurs when certain dyes and polyelectrolytes are present in the same solution because the interaction lowers the free energy of the system. Quantitative studies of metachromasy using the principal-component analysis technique were carried out (3, 4, 16, 39, 48, 49) and some equilibrium constants for the metachromatic interaction were measured. Some of these measurements indicated that the entropy of interaction played a significant role in the driving force for metachromasy (2, 35, 42). In other words, the entropy of the system was significantly increased by the interaction of the dye with the polyelectrolyte. For example, Nishida and Watanabe (35) measured the binding entropy of acridine orange with poly(sodium acrylate) in water and water–organic solvent mixtures. Values between 6 and 34 cal/mol-K for the entropy of binding, which generally increased with solvent dielectric constant, were obtained. Thus the entropy of interaction was the greatest in water. Furthermore, in aqueous solution the binding reaction was found to be endothermic, so the entire driving force for the metachromatic interaction came from the entropy term. This positive entropy change may be understood in terms of a greater flexibility of the polyelectrolyte chain when a number of ionic sites are neutralized by adsorbed dye molecules. It might also be expected that this partial charge neutralization will increase the solvation entropy.

Metachromasy and Energy Transfer

Shirai et al. carried out a number of studies of energy transfer in dyes bound to polyelectrolytes (11, 50–55). These studies focused on energy transfer in methylene blue adsorbed to various polyelectrolytes, such as poly(styrene sulfonate), poly(vinyl sulfate), polyacrylate, and polymethacrylate. The results suggest that polyanions can act as templates for energy transfer or chemical reaction to bring the molecular partners into close proximity for transfer or reaction to occur. Aggregation of methylene blue on the polyelectrolytes tended to spoil its energy-transfer potential, presumably because of the shorter excited-state lifetimes of dimers and higher aggregates (51, 53, 55). Poly(styrene sulfonate) and poly(vinyl sulfate) were found to be much more effective than polyacrylate and polymethacrylate at mediating excitation energy transfer between bound dyes (52). This mediation mechanism may be due to the presence of sulfur atoms in the poly(styrene sulfonate) and poly(vinyl sulfate) and the ability of the sulfur atoms to enhance the singlet–triplet intersystem crossing efficiency of methylene blue. In any event, energy-transfer rates were as much as 67 times higher in the presence of polyelectrolytes than in their absence.

Another interesting study of energy transfer and metachromasy was carried out by Baumgartner et al. (56). In this study the ability of methyl viologen to quench the fluorescence of 1-naphthylamine was inhibited by the presence of polyelectrolytes. In other words, the solution phase energy transfer that normally occurs between 1-naphthylamine and methyl viologen was reduced by the addition of poly(vinyl sulfonate) or poly(styrene sulfonate) electrolytes. The fluorescence of 1-naphthylamine, which is quenched by methyl viologen, is "unquenched" by further polyelectrolyte addition. In the absence of methyl viologen, however, the polyelectrolyte had no effect on 1-naphthylamine fluorescence. The results are interpreted in terms of a sequestering of the (charged) methyl viologen in the electrostatic domains of the polyelectrolyte. In this case, therefore, the interaction between the dye and the polyelectrolyte actually reduced the energy-transfer potential because the polyelectrolyte did not concentrate the energy donor molecule (1-naphthylamine) where the acceptor molecule (methyl viologen) resides.

In a classic experiment on energy transfer between DNA and adsorbed acridine dyes, Isenberg et al. (*57*) observed and studied delayed fluorescence of the DNA–bound dyes. Based on the assumption of an intercalation model of adsorption [the relatively flat acridine dyes slide between the base pairs of DNA (*58, 59*)], they concluded that when a DNA chromophore (base pair) absorbs light, some of the energy of the excited singlet state is converted into the lower energy triplet state, which is a long-lived state. The triplet energy can jump from base pair to base pair up and down the chain during its long lifetime until it decays spontaneously by DNA phosphorescence, degrades to heat, or, if dye is present, transfers to an intercalated dye molecule and triggers delayed fluorescence. This observation of triplet migration, sometimes called exciton transfer or triplet delocalization (*60*), also substantiated the intercalation model for acridine dye adsorption to DNA because triplet-to-singlet energy transfers from DNA to dye must take place over very short (contact) distances.

Metachromasy as a Structural Probe

A number of studies have examined the effects of structural properties of polyelectrolytes on metachromasy. Because in many instances the metachromatic effect is sensitive to structural changes, it may be used as a structural probe for polyelectrolytes. Typical polymer structural properties measured by metachromatic techniques include intrinsic viscosity, conformation, and chain length, but other properties may be probed as well. The studies covered here include both absorption and fluorescence techniques, and these topics will be used as a convenient, though somewhat arbitrary, division for the review.

Metachromatic Absorption Studies of Polymer Structure.
Absorption studies include those metachromatic experiments in which UV–visible absorption spectrophotometry is the primary tool for analysis. Other physical measurement techniques, such as dialysis or viscosimetry, are

often used in conjunction with the spectrophotometry but do not constitute the primary method of use. Structural parameters of the polyelectrolytes—conformation, chain length, charge density, hydrophobicity, tacticity, distance of charged sites from polymer backbone, etc.—may be varied systematically and the effects on the metachromasy of various dyes noted. Those parameters that have a consistent, predictable effect on the metachromasy of a particular dye may be probed by that dye in an unknown polymer system. Caution must be used in interpretation of the results of such experiments, however, because changes in metachromasy are fairly nonspecific. In other words, a variety of changes in structural parameters or even conditions of the experiment may cause similar metachromatic changes with a given dye. Care must be taken, therefore, to ensure that only one parameter has been varied before metachromatic changes are assigned to changes in that parameter.

Vitagliano (31) reported the effect of polymer conformation on the metachromasy of acridine orange. In this study, acridine orange was bound to polyacrylic acid and polymethacrylic acid at P/D ratios of 7600 and 8000 under acidic conditions at low degree of neutralization (a_N) values. Under these conditions the polyelectrolytes were assumed to be in a tightly coiled form because of their hydrophobicity, and the acridine orange showed the characteristic spectrum of monomeric binding to the polyelectrolytes. As base was added and a_N increased, the polymers underwent a conformational change (at $a_N = 0.1$–0.3) as they became more ionized and the acridine orange spectrum shifted to the characteristic spectrum of aggregate binding. In other words, in this system the adsorbed acridine molecules could migrate and aggregate more readily when the polyelectrolytes to which they were bound were in an open rather than coiled form. Other things being equal, under these conditions acridine orange could be used as a sensitive probe of polyacrylic acid or polymethacrylic acid conformation.

Shirai et al. also studied the effects of conformational changes of polyelectrolytes on metachromasy (9, 22, 38, 61). In these studies the interaction of methylene blue with various polyelectrolytes was examined spectrally. The results suggested that important factors for metachromasy are the flexibility and charge density of the polymer: the more flexible the polyanion or the higher its charge density, the greater was the tendency of adsorbed methylene blue to aggregate. Viscosity measurements (38) also indicated that conformational changes in the polymers could be induced by dye adsorption; that is, binding of cationic dyes to anionic polyelectrolytes neutralizes the charge of the polyanion and causes the polymer molecules to coil up. Methylene blue was found to aggregate more readily when adsorbed to polymers in the following order (22): poly(sodium maleate-co-vinyl alcohol) > poly(sodium acrylate) > poly(sodium methacrylate). The poly(sodium maleate-co-vinyl alcohol) might bind more effectively to the cationic dye because of its vicinal anionic sites, and the poly(sodium acrylate) might allow dye aggregation more effectively than the poly(sodium methacrylate) because of its higher flexibility. Thus, the ability of a polyanion to induce metachromasy in methylene blue might be used as a probe for flexibility or charge density of the polyanion.

In a further study of polymer conformation Guhaniyogi and Mandal (*10*) observed the metachromatic effect of the dye pinacyanol when it interacted with short-chain benzene-soluble polymers that had carboxylate end groups. In this case, the dye showed a metachromatic shift due to dimerization under conditions that allowed the intramolecular association of polymer end groups. Therefore, the observed spectral shift was used as a probe of polymer conformation in nonaqueous media.

In some studies the degree of polymerization of the polyanion also was found to affect the ability of a dye to engage in a metachromatic interaction with the polyanion (*9, 21, 38, 62*). For example, Yamaoka et al. (*21*) studied the metachromatic interaction of crystal violet with sodium polyphosphates of varying chain length. The metachromasy of this dye was found to increase sharply in a narrow range of chain lengths from 7 to 20 monomer phosphate units. An effect of degree of polymerization on metachromasy also was seen to a certain extent with carbon-backbone polyanions (*9*). However, above a critical minimum value, the degree of polymerization had an insignificant effect on the strength of the metachromatic interaction (*22*).

The effects of polymer chain tacticity were also studied using metachromasy (*22, 31, 63*). In a study of the interaction of acridine orange with various polyelectrolytes, for example, it was found that the binding strength was higher and the stacking coefficient was lower with isotactic polymethacrylic acid and poly(styrenesulfonic acid) than with the corresponding atactic polyacids (*63*). The results were interpreted in terms of a relatively stable helical conformation of the isotactic polyacids with the carboxylate groups facing outward. The dye can associate readily with these exposed anionic groups, but the helical conformation might be more difficult to disrupt to facilitate dye aggregation. It has also been suggested (*31*) that in the case of isotactic poly(styrene sulfonate) the arrangement of aromatic groups around the helical backbone chain may allow some kind of partial intercalation of dye molecules similar to that proposed for the interaction of acridine dyes with DNA (*64*).

Furano et al. (*65*) described the use of the metachromasy of acridine orange to probe the conformation of ribonucleic acid (RNA) in ribosomes. Comparison of the sensitivity of the extinction coefficient of acridine orange with the P/D ratio and the dye stacking coefficient K for a variety of ordered and disordered RNAs yielded a useful correlation. In particular, a highly ordered (double-helical) RNA gave a rapid loss in extinction of acridine dye with polymer addition and a low stacking coefficient, whereas a disordered RNA gave a more gradual loss in extinction and a high stacking coefficient. This test was applied to ribosomal RNA and led to the conclusion that RNA in the ribosomes had little double-helical structure. This conclusion, in turn, resulted in a general model of ribosomal structure. Similar results by Stone and Bradley established acridine orange as a structural probe for DNA conformation (66).

Recently Pal and Mandal (98) used the metachromatic interactions of 1,9-dimethyl methylene blue and pinacyanol with potassium alginate polymers to probe conformational changes of these polymers. Circular dichroism induced in the spectra of these dyes suggested that the dye molecules complexed with the polymers in a helical arrangement. These observations prompted Pal and Mandal to infer a left-handed helical conformation for potassium alginate in solution. General results of such experiments must be interpreted with caution, however, because high concentrations of dyes aggregated on a polymer conceivably could influence the conformation of the polymer.

Metachromatic Fluorescence Studies of Polymer Structure. Luminescence techniques in general have been used widely and successfully in the study of polymers and their properties. Many of these studies involve polymers that have covalently attached fluorescent groups or pendant side chains. Because metachromasy technically involves only the interaction between polymers and dyes that are *not* covalently attached, this chapter will not address structural studies of inherently fluorescent polymers or polymers covalently tagged with fluorescent probes. The reader is referred, instead, to excellent reviews on this subject (67–70). Fluorescence metachromasy as a polymer structural probe may be divided into two categories: interactions with naturally occurring polymers and interactions with synthetic polymers.

Naturally Occurring Polymers. Fluorescence metachromasia covered in an early review by Stockinger (71) also had its origins in the field of biological stain technology. Certain dyes, known as fluorochromes, bind to specific naturally occurring polymers or complexes and highlight subcellular structures where these polymer compounds are concentrated. This highlighting allows the subcellular structures to be viewed by fluorescence microscopy. Acridine orange presents a classic example of the usefulness of this method because, under the proper conditions, acridine orange complexed with DNA fluoresces yellow-green, whereas acridine orange complexed with RNA fluoresces red-orange. Therefore, the chromosomes in a cell nucleus (DNA) can be distinguished readily from the nucleolus (RNA) by using a single stain procedure. This dual fluorescence arises because of differences in the interactions between acridine and the two nucleic acids, and can be used as a probe for DNA or RNA. Other dyes serve as effective fluorochromes for proteins, carbohydrates, and lipids (72) in the fluorescent staining of biological structures. Some of the uses of these fluorescing molecules for the elucidation of naturally occurring polymer structures are summarized in the following paragraphs.

Pal (*24*) studied the fluorescence of acridine orange and rhodamine 6G in the presence of natural polyelectrolytes like DNA, chondroitin sulfate, and λ-carageenan. Absorption metachromasy and fluorescence quenching were observed to different extents for both dyes, and the effects were found to be reversable by the addition of salts, urea, or alcohols. These results were not surprising because both effects (metachromasy and quenching) were known to accompany dye aggregation at higher concentrations.

Nile red is a fluorescent dye that has an emission wavelength that is a sensitive function of solvent polarity. In hydrocarbons, Nile red fluoresces yellow-gold whereas in ethanol it fluoresces red (*73*). This fact was utilized in a novel way in a study in which Nile red was used as a polarity-sensitive fluorescent probe of hydrophobic protein surfaces (*74*). Interaction of Nile red with various types of proteins enhanced and blue-shifted the fluorescence of the dye to an extent related to interaction of the dye with the hydrophobic domains on the protein surfaces. The dye could be used to monitor partial denaturation of a protein and the concomitant exposure of hydrophobic groups by following the progressive shift in fluorescence wavelength and enhancement of fluorescence intensity. This use of Nile red as a denaturation probe was demonstrated with the protein ovalbumin.

A number of other studies have focused on the use of fluorescent probes to study proteins in solution (*75–77*). For example, 1-anilino-8-naphthalene sulfonate, dansylamide, and other fluorophores have been used to characterize and quantitate drug bonding sites on serum albumin by various competition studies (*78–80*). In addition, auramine O was used as a fluorescent probe for the quantitation and drug binding site analysis of α-acid glycoprotein (*81*). Such studies show the usefulness of certain fluorescent dyes in characterizing particular proteins including specific binding site analyses.

Synthetic Polymers. Levshin et al. (*82*) studied the interaction of rhodamine 6G (Rh 6G) with various synthetic sulfonate polymers. Observed interactions between the dye and polyelectrolytes were accompanied by typical metachromatic absorbance changes in the dye spectrum as well as fluorescence quenching, both of which reversed at high P/D ratios. The effects were more dramatic with poly(styrene sulfonate) than with poly(vinyl sulfonate), probably owing to the enhanced hydrophobic interaction between the dye and the aromatic side chain groups in the poly(styrene sulfonate). Additionally, a higher charge density on the polyelectrolyte magnified the fluorescence quenching effect that a polymer exerted on Rh 6G.

Fenyo et al. (*83*) as well as Braud (*84*) used the interaction of auramine O with various synthetic polycarboxylic acids to probe the structural parameters of the polymer. In their experiments, auramine O was found to be an effective and sensitive probe to establish the presence of hydrophobic regions in synthetic polymers or, at least, the existence of special compact conformations. Auramine O fluoresced intensely when it was placed in contact with such regions on a tightly coiled nonionized synthetic polyacid. The fluorescence disappeared, however, when the polymer unravelled or became ionized. Thus, in such polymer systems, the fluorescence of auramine O can be used as a direct measure of polymer conformation or degree of ionization.

Similar behavior was observed with crystal violet (47). This dye fluoresces strongly in interaction with undissociated polymethacrylic acid, but only weakly fluoresces when the polymer dissociates 40% or more. Thus crystal violet may also serve as a conformation–charge-density probe for polymethacrylic acid.

Muller and Fenyo (85) studied the fluorescence metachromasy of acridine orange in the presence of polyacrylic acid, polymethacrylic acid, and a polycondensate between 1,3-benzenedisulfonyl chloride and L-lysine. They found that acridine orange could be used as a sensitive probe of polymer conformation, especially in the case of the polycondensate. When the dye binds to a polymer that is in the coiled undissociated state, it binds in the monomeric form and its green fluorescence (535 nm) is enhanced. As the polymer unfolds with acid group dissociation, the green fluorescence disappears and is replaced by the characteristic red fluorescence (640 nm) of acrindine dimers and higher aggregates. The conformational state of the polymer can be assessed by the color of acridine orange fluorescence; hence, this dye may be used as a visual probe of polyion conformation.

In another study of polymer conformation by metachromasy, ethidium bromide and auramine O were found to be useful as fluorescent metachromatic probes of the conformational states of maleic acid–olefin copolymers (86). In this study, metachromatic results were correlated with viscosity and equilibrium constant measurements to verify the conformational states of the polymers, and the fluorescence of ethidium bromide was found to be sensitive to the structure of the coiled polymer. Possible similarities between the behavior of ethidium bromide in this system and its intercalation behavior with DNA were suggested.

Another study examined the interaction in aqueous solution of pyrene substituted with a positive side group and a polyanion as well as pyrene substituted with a negative side group and a polycation (44). In either case, the metachromatic results were the same: The pyrene absorption spectrum broadened and its principle fluorescence bands shifted from 377 and 400 nm to a broad band centered around 480 nm. The 480-nm fluorescence band was believed to be due to excimer fluorescence from pyrene dimers and higher aggregates. Thus, the fluorescence shift on contact with polyelectrolytes was taken as evidence for aggregation of polymer-bound pyrene moieties and could be used as yet another probe of polyelectrolyte conformation.

Metachromasy is useful as a structural probe for polyelectrolytes. The effect relies on the fact that spectral properties change when certain dyes aggregate or form clusters. The appearance of dimer absorption bands or excimer fluorescence bands signals the onset of dye molecule aggregation, which is facilitated by polyelectrolytes whose structural properties affect the strength of the polymer–dye interaction. The ability of a given dye to induce metachromasy is affected when a polymer changes structure. Thus metachromatic color changes are sensitive to polymer structural changes and may be used as effective structural probes.

Metachromasy as an Analytical Tool

It was observed very early in the history of metachromasy that the stoichiometry of the main interaction between a charged dye and an oppositely charged polyelectrolyte is 1:1 (8, 12, 13). That is, metachromasy, if it occurs, usually increases to a maximum as the P/D (polymer charged residue to dye molar concentration) ratio increases from 0 to 1. At a P/D ratio of 1 the metachromatic effect typically levels off, and at very high P/D values it reverses (7). If the metachromatic reaction is "clean" and stoichiometric (i.e., if the dye and polyelectrolyte react quickly, form a strong complex, and leave very little free dye in solution at the equivalence point), and if the complexed dye completely aggregates at P/D ratios less than 1, then the process may be used to quantify the concentrations of dye or polymer in the P/D range from 0 to 1. Some examples of such quantitative analyses are given in the following text.

Gormally et al. (90) developed a method that used the interaction between toluidine blue and carboxymethyl cellulose to measure the dye concentration over a wide range of polymer and dye values with good reliability. Some inherent advantages of this method over direct spectrophotometric or fluorometric techniques were discussed.

Metachromatic methods have been developed for the quantitative determination of many naturally occurring polyelectrolytes, such as arylsulfatase (91), carageenan (92), glycosaminoglycans (46), heparin (93), and others. Although these methods typically involve standard absorptive metachromatic shifts, Wu (94) described an interesting secondary method involving fluorescence metachromasy: The fluorescence intensity of certain polycyclic aromatic dyes significantly increased when the dyes interacted with various cationic or nonionic polymers. These complexes also had an affinity for certain naturally occurring biological polyanions, and the resulting ternary complexes could be used in various quantitative methods, such as fluorescence microscopy or flow cytometry. In this method the intermediate cationic or nonionic polymer served as a mordant to fix the fluorescent dye to the naturally occurring analyte.

Optical or spectrophotometric titration has been used to quantify the metachromatic interaction (2, 63). In this method the absorbance of the main visible band of the free dye in solution was monitored as a function of polymer (titrant) concentration or P/D ratio. An example of such a titration curve is shown in Figure 1. The titration curves for clean metachromatic

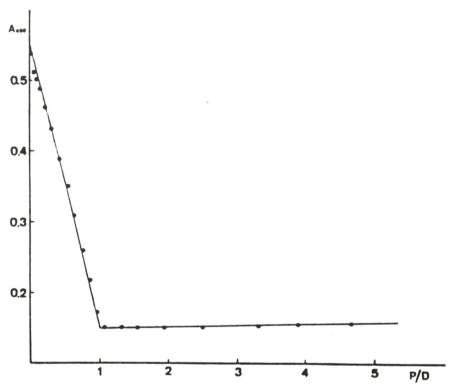

Figure 1. Spectrophotometric titration of acridine orange with isotactic poly(styrene sulfonate); dye concentration 10^{-5} M. The end point corresponds to the 1:1 ratio between dye concentration and polyelectrolyte equivalent (SO_{3^-} groups) concentration. (Reproduced with permission from reference 63. Copyright 1976.)

reactions are characterized by a linear decrease in absorbance in the P/D range 0–1 followed (ideally) by an abrupt change in slope with no further decrease in absorbance. The end point of such a titration is determined as the intersection of the two linear segments of the curve and typically occurs at P/D = 1.0. Note that the titration curve shown in Figure 1 was measured under ideal conditions: acridine orange interacted with isotactic poly(styrene sulfonate), a metachromatic reaction that is especially "clean". Titration curves measured under less optimum conditions (with weaker metachromatic dyes or less ordered or nonaromatic polyelectrolytes) will typically show nonlinearity in both segments of the curve or less abrupt end points. Furthermore, very often the metachromatic reaction is not instantaneous; spectral changes can take place over 30 min to 1 h (C. Pierce, Nalco Chemical Company, unpublished results), which makes exact quantitation very difficult. In addition, the metachromatic reaction is often accompanied by partial coagulation of the dye–polyelectrolyte complex that can raise the turbidity of the solution and further complicate spectral measurements.

Despite these complications, metachromasy has been proposed as a method of quantitation of polycarboxylate of polysulfonate concentrations down to the 1-ppm level (95) using pinacyanol or Nile blue A as the metachromatic dye. This method found reasonable linearity in the metachromatic calibration curves over narrow ranges of dye concentration. Furthermore, the method could analyze separately for polycarboxylates and polysulfonates in the same sample by adjusting the pH of the system.

Another method of quantitation for cationic polyelectrolytes was reported by Parazak et al. (96). This method utilized the interaction between anionic dyes and cationic polyelectrolytes. However, instead of requiring a metachromatic spectral change upon interaction, the method utilizes the fact that the neutral complex, which is somewhat water-insoluble, may be extracted from aqueous solution by a hydrophobic solvent. In this method a 1,1,2-trichlorotrifluoroethane (TCTFE; Freon) solvent was used, and the neutral complex was attracted to the TCTFE–water interface where it formed a weblike structure. The dye that remained in aqueous solution was then measured spectrophotometrically and the loss in absorbance was found to be directly proportional to the concentration of polyelectrolyte. This method was found to be sensitive down to 0.5 ppm of cationic polyelectrolyte.

Finally, Onabe (97) observed that the visible metachromatic color change of toluidine blue from light blue in the free cationic state to red-purple in reaction with anionic polyelectrolytes was abrupt and dramatic. This observation led to the suggestion that this color change could be used to visually detect the end point in a colloid titration; that is, toluidine blue could serve as a colloid titration indicator.

Color-Array and Complexation–Extraction Studies

Because the metachromatic spectral change occurs linearly over a narrow range of P/D values and because the metachromatic organic dyes generally have high molar absorptivities, the metachromatic effect has been proposed as a basis for the quantitative analysis of synthetic polyelectrolytes in the 1–10-ppm range in water (95, 96). Although the technique does work, its sensitivity to temperature, pH, ionic strength, and saturation make application difficult, especially in the field. A new set of standards must be created for every new water sample and sometimes reproducibility is difficult to attain on separate days even with the same water sample. Furthermore, a time dependence for aggregation has been observed in some instances wherein the metachromatic spectral change does not occur instantaneously, but develops gradually over several minutes or hours. In addition, over long time periods a gradual settling out of the "dyed" neutral polyelectrolyte

occasionally has been observed. The aforementioned effects certainly complicate the use of metachromasy for polymer analysis.

In our study, the use of the metachromatic effect for low-level polyelectrolyte analysis was reexplored with a new twist. Because the metachromatic
color change takes place over a narrow fixed P/D range, we varied the *dye*
concentration at a fixed polymer concentration and visually observed the dye
concentration that caused the spectral shift to take place. This process was
repeated at several standard polymer concentrations to generate a standard
color array (i.e., color of solution versus polymer and color of solution versus
dye concentration) against which a set of solutions with an unknown polymer
concentration and various known dye concentrations could be visually compared. Some cationic metachromatic dyes have spectral changes upon complexation with sodium poly(acrylate-*co*-acrylamide) that are significant enough
to be readily detectable visually. The behavior of these dyes was examined in
our study.

In addition, we found that the method of Parazak et al. (96) may be
applied to anionic polyelectrolytes such as sodium poly(acrylate-*co*-acryla-
mide). When certain cationic dyes reacted with this polymer, the resultant
complex could be extracted into a TCTFE–water interface. This extraction
process removed dye and polyelectrolyte from the aqueous phase in constant
proportion. Thus the loss in dye absorbance after complexation and extraction
was proportional to the initial concentration of polyelectrolyte. The polymer
concentration could be assessed either visually or spectrophotometrically by
this complexation–extraction procedure.

Experimental Details

Dyes and extraction solvents used in this study were obtained from Aldrich
Chemical Company and were used without further purification. Solutions
were buffered to pH 7.0 using a phosphate buffer. Sodium poly(acrylate-*co*-
acrylamide) samples were obtained from Nalco Chemical Company. UV–visible spectra were run on a spectrophotomer (IBM 9430). A copolymer of 70%
acrylic acid and 30% acrylamide with a molecular weight range of
25,000–40,000 g/mol was used.

Results

From 39 cationic dyes screened for their ability to manifest a visual
metachromatic effect, 3 dyes were chosen: toluidine blue O, cresyl violet
acetate, and safranine O. The structures of these dyes are shown in Chart I.
In this test, an array of solutions was made up with varying polymer and dye
concentrations. The polymer concentrations were 0, 2.5, 5.0, 7.5, and 10.0
ppm and the dye concentrations were 1, 2, 3, 4, 5, and 6×10^{-5} M. Thus,

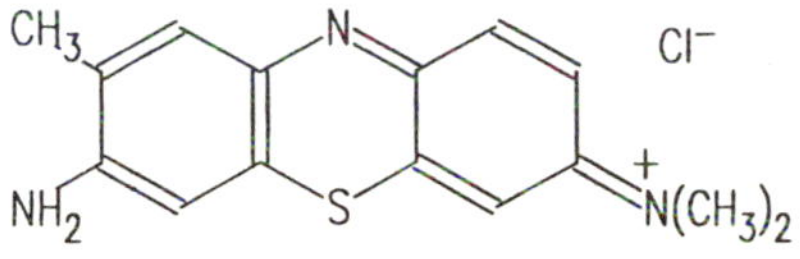

(a) toluidine blue O

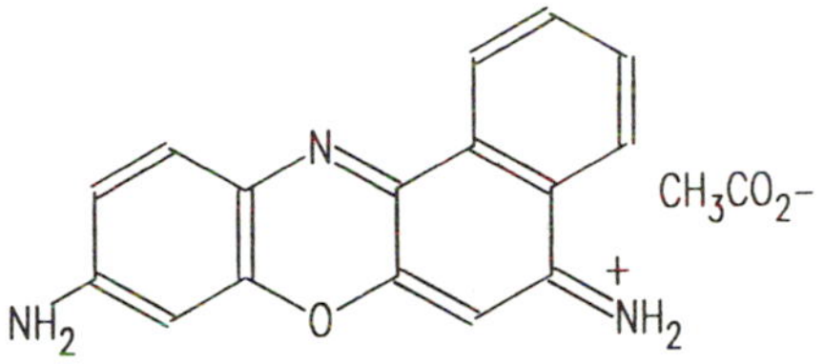

(b) cresyl violet acetate

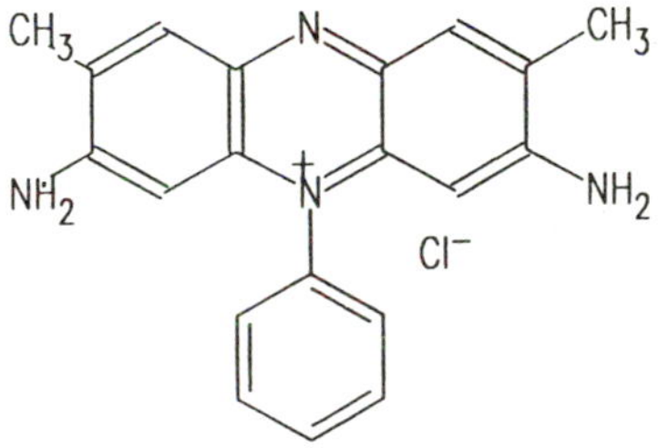

(c) safranine O

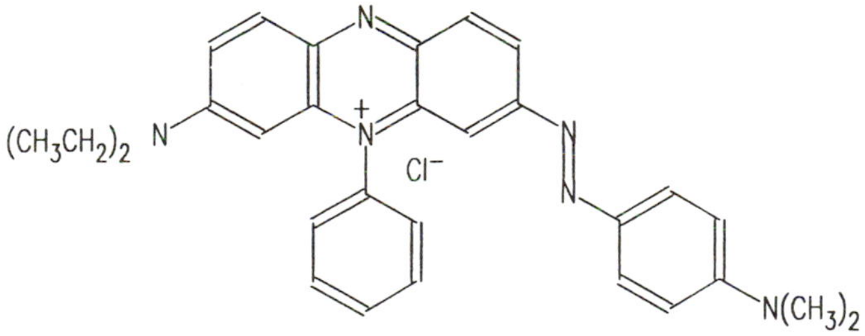

(d) janus green B

Chart I. Dyes used for color-array and complexation–extraction studies.

for each dye an array of thirty 100-mL solutions was prepared and photographed. Copies of these photographs are shown in Plates 1–3. In these color-array tests, the polymer concentration that caused a metachromatic shift in the dye increased with increasing dye concentration. When the molar concentrations of carboxylate residues in the polymer solutions were determined and divided by the molar dye concentrations, the resulting P/D ratios were calculated. A chart of P/D ratios for each solution in the array was made and is presented in Figure 2. A comparison of the P/D values on this chart and the colors in the two-dimensional arrays shows that the significant metachromatic color change occurs over a fairly narrow P/D ratio range near unity. This is consistent with previous work that established that the metachromatic effect was the strongest at P/D values of 1.0 (13).

The color-array results for toluidine blue O give the most dramatic visual metachromatic effect, the results for cresyl acetate also show a readily discernible color change, and the results for safranine O indicate very little effect on the photograph. (Note that color changes in the safranine O array were subtle but noticeable visually when the solutions were photographed. Unfortunately these color changes did not come through in the pictures.)

The peak absorbance values for solutions in each of the color-array tests are shown in Figures 3–5. These graphs indicate a general decrease in dye absorbance as polymer is added through P/D = 1.

Some of the metachromatic solutions were extracted with nonaqueous solvents, such as toluene, methylene chloride, or TCTFE. The nonaqueous solvent removed the dye–polymer complex from aqueous solution and the resulting aggregate was attracted to the interface between the two liquids, where it formed a weblike film. This colored film was readily seen when the two-phase mixture was shaken, but the film nearly disappeared in the interface when the mixture stood undisturbed. The phenomenon accentuated the differences between equivalent cationic dye solutions with and without 1.0 ppm of polyanion, and could be used to determine the concentration of polymer in solution mainly by depletion of the dye color in the aqueous phase. Such a method was proposed for anionic dyes complexing with polycations (96) and was found to be sensitive down to 0.5 ppm of polyelectrolyte.

The extraction results obtained in this study suggest the feasibility of application of this technique to the analysis of low levels of polyanions. Solutions that contain the dye Janus green B and sodium poly(acrylate-*co*-acrylamide) at P/D ratios near 1 were successfully extracted with 1,1,2-trichlorotrifluoroethane (Freon) and the aqueous layer was left nearly colorless. Plate 4 is a photograph of the solutions that resulted from an extraction experiment in which 1×10^{-5} M Janus green B was mixed with 0, 1, and 5 ppm of sodium poly(acrylate-*co*-acrylamide). As the figure clearly indicates, the complexation–extraction method is visually sensitive down to a polymer

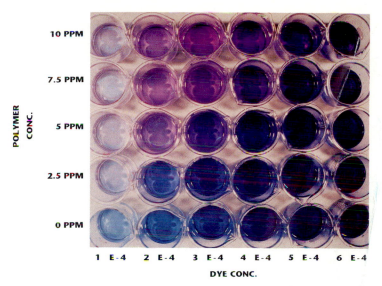

Plate 1. Color-array study of toluidine blue O with poly(acrylate-co-acrylamide); polymer = 0–10 ppm; dye = 1–6 × 10⁻⁵ M; pH = 7.

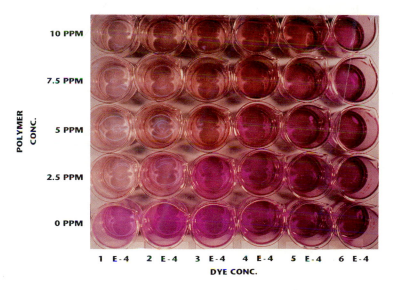

Plate 2. Color-array study of cresyl violet acetate with poly(acrylate-co-acrylamide); polymer = 0–10 ppm; dye = 1–6 × 10⁻⁵ M; pH = 7.

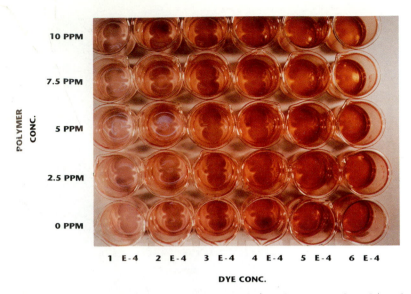

Plate 3. *Color-array study of safranine O with poly(acrylate-co-acrylamide); polymer = 1–10 ppm; dye = 1–6 × 10⁻⁵ M; pH = 7.*

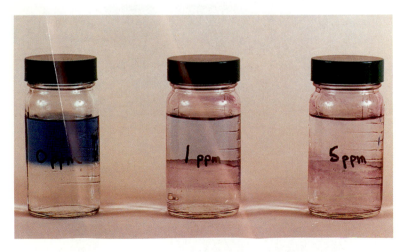

Plate 4. *Complexation–extraction results. Each bottle contains 25 mL of 1,1,2-trichlorotrifluoroethane (Freon; bottom layer) and 25 mL water with 1×10^{-5} M Janus green B and 0, 1, and 5 ppm of poly(acrylate-co-acrylamide) (top layer).*

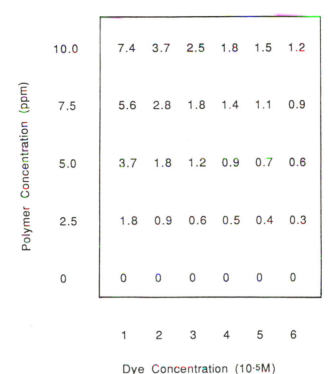

Figure 2. Chart of P/D ratio values for each of the solutions in the color-array studies shown in Plates 1–3.

concentration of 1 ppm or less. The structure of the dye Janus green B is shown in Chart I.

Discussion

The experimental results suggest that for dyes that exhibit a metachromatic shift with poly(acrylate-*co*-acrylamide), the color change occurs over a narrow P/D range near unity. The P/D ratio represents the number of anionic polymer residues per dye molecule in the system. In this copolymer, 30% of the residues are nonionic acrylamide, which is apparently metachromatically inactive. The observation of color change at P/D = 1 is consistent with many previous results that established P/D ratios near 1 for the maximum metachromatic effect. This stoichiometry also emphasizes the importance of the electrostatic interaction between the cationic dye and the anionic polyelectrolyte residue vis á vis the hydrophobic and pi-electron interactions.

Because metachromatic change occurs over a small range of P/D, it is difficult to use metachromasy for the linear analysis of polymer solutions. The results in the literature also suggest that as P/D increases beyond 1, the

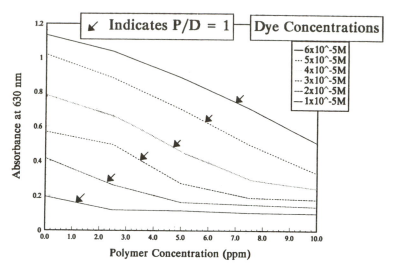

Figure 3. *Peak absorbance values for the toluidine blue O color-array studies; cf. Plate 1.*

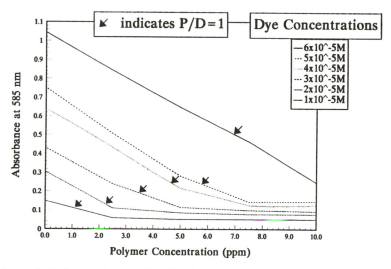

Figure 4. *Peak absorbance values for the cresyl violet acetate color-array study; cf. Plate 2.*

metachromasy saturates very quickly and begins to reverse at P/D greater than 10 (7). Dependent on the dye concentration, the linear useful range of spectral change versus polymer concentration is often too narrow to be used.

In our study, the narrowness of the metachromatic range was used to advantage. When polymer solutions were mixed with several concentrations

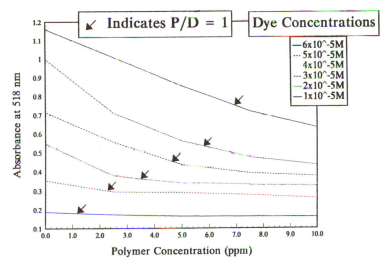

Figure 5. Peak absorbance values for the safranine O color-array study; cf. Plate 3.

of the same metachromatic dye, the solutions with polymer (anionic residue) concentration less than the dye concentration did not show a color change, whereas the solutions with polymer (anionic residue) concentration greater than the dye concentration did change color. In other words, color change occurred at P/D = 1 with the metachromatic color manifest at P/D > 1 and the free dye color seen at P/D < 1. Because the dye concentrations in these test solutions were known, the polymer (anionic residue) concentration could readily be determined. Furthermore, toluidine blue O and cresyl violet acetate gave pronounced visual color changes so that the analytical test could be carried out without a spectrophotometer or colorimeter, and a simple color wheel or chart could be used to determine polymer concentrations.

The preliminary extraction studies in which solutions that contained cationic dye and anionic polyelectrolyte were extracted with a nonaqueous solvent like TCTFE may also be applied to polymer analysis. Because the neutral dye–polymer complex migrated to the interface between the two immiscible liquids and was effectively removed from the aqueous layer, any remaining color in the water layer must be attributable to excess dye (presuming a strong interaction between dye and polymer so that complexation was complete). The polymer concentration in such systems can be determined spectrophotometrically (by analysis of the residual dye in the water layer using a standard curve) or visually (as before, by use of solutions of varying dye concentration to seek the maximum dye concentration for which no residual color is left in the aqueous phase after the extraction). Because the complexation exhibits 1:1 stoichiometry, this maximum dye concentration should be equal to the polymer (anionic residue) concentration in the solution. Again this method requires that a strong complex is formed between the dye and the polymer so that essentially all of the anionic sites are occupied by dye molecules at the equivalence point.

An additional advantage of the extraction method is that it only requires a strong complex form between the dye and the polymer, and it does not require a spectral shift or color change. Polyelectrolytes complex with some dyes that exhibit minimal spectral changes. If these dyes form strong enough complexes, they will also be viable candidates for an analytical method based on the extraction process.

Summary and Conclusions

Metachromasy, the color-changing ability of certain ionic dyes in the presence of oppositely charged polyelectrolytes, has been used to study the structure–property relationships of water-soluble polymers. Color changes in these dyes, detected both visually and spectrophotometrically, have been shown to be sensitive to their molecular environments and states of aggregation as they adhere to polymer surfaces. These metachromatic color changes have been correlated to polymer charge density, degree of ionization, hydrophobicity, conformation, molecular weight, tacticity, and chain flexibility. The effect is fairly nonspecific, however, and care must be taken when the results are interpreted.

Metachromasy also can be used as an analytical tool to measure the concentration of polyelectrolyte in aqueous solution. By quantitatively monitoring the spectral changes induced in a metachromatic dye with various concentrations of polyelectrolyte, a standard curve can be constructed. The method is fairly sensitive to pH, temperature, ionic strength, various impurities, and interaction time, however, so care must be taken to control all other experimental conditions so that a high degree of precision can be obtained.

The color-array and the complexation–extraction methods proposed herein are extensions of the metachromatic methods for quantitative analysis of polymers. These methods are based on the 1:1 stoichiometry of metachromasy rather than on the development of a standard curve and so are believed to be easier to use and inherently more accurate than traditional techniques. Furthermore, the color-array method is amenable to a visual determination of polymer concentration down to 1 ppm, which obviates the need for a spectrophotometer in field determinations. Toluidine blue O showed the most promise as a dye for use in the color-array method. The advantage of the complexation–extraction method is that it may be applied to determine the concentration of anionic polyelectrolyte using a nonmetachromatic dye. As long as a strong neutral complex forms between the cationic dye and the poly(acrylate-co-acrylamide) and this complex is extractable by a nonaqueous solvent such as TCTFE, the complexation–extraction method may be used for the quantitative analysis of the polymer. This method may also be applied visually, but more accurate results will be achieved using a spectrophotometer. The dye Janus green B showed the most promise for use in the complexation–extraction method.

Acknowledgments

The author is grateful to Thomas C. McGowen for carrying out much of the experimental work and to Nalco Chemical Company for providing poly(acrylate-*co*-acrylamide) samples.

References

1. Bradley, D. F.; Wolf, M. K. *Proc. Natl. Acad. Sci. U.S.A.* **1959**, 45, 944–952.
2. Vitagliano, V.; Costantino, L.; Zagari, A. *J. Phys. Chem.* **1973**, 77, 204–210.
3. Yamaoka, K.; Takatsuki, M.; Nakata, K. *Bull. Chem. Soc. Jpn.* **1980**, 53, 3165–3170.
4. Yamaoka, K.; Matsuda, T.; Murakami, T. *Bull. Chem. Soc. Jpn.* **1981**, 54, 3859–3860.
5. Ranadive, N. S.; Korgaonkar, K. S. *Biochim. Biophys. Acta* **1960**, 39, 547–550.
6. Ortona, O.; Costantino, L.; Vitagliano, V. *J. Mol. Liq.* **1989**, 40, 17–24.
7. Dey, A. N.; Palit, S. R. *Indian J. Chem.* **1968**, 6, 260–262.
8. Carroll, B.; Cheung, H. C. *J. Phys. Chem.* **1962**, 66, 2585.
9. Shirai, M.; Nagatsuka, T.; Tanaka, M. *J. Polym. Sci., Polym. Chem. Ed.* **1977**, 15, 2083–2095.
10. Guhaniyogi, S. C.; Mandal, B. M. *Makromol. Chem.* **1974**, 175, 823–831.
11. Shirai, M.; Tanaka, M. *J. Polym. Sci., Polym. Chem. Ed.* **1976**, 14, 343–351.
12. Pal, M. K.; Ghosh, B. K. *Makromol. Chem.* **1979**, 180, 959–967.
13. Singh, C. *Anal. Biochem.* **1970**, 38, 564–568.
14. Taylor, K. B. *Stain Technol.* **1961**, 36, 73–83; *Chem. Abstr.* **1962**, 56, 5060g.
15. Yamaoka, K.; Takatsuki, M. *Bull. Chem. Soc. Jpn.* **1981**, 54, 923–924.
16. Yamaoka, K.; Matsuda, T.; Shiba, D.; Takatsuki, M. *Bull. Chem. Soc. Jpn.* **1982**, 55, 1300–1305.
17. Bean, R. C.; Shepherd, W. C.; Kay, R. E.; Walwick, E. R. *J. Phys. Chem.* **1965**, 69, 4368–4379.
18. Pal, M. K.; Ghosh, B. K. *Makromol. Chem.* **1980**, 181, 1459–1467.
19. Scott, J. E. *Histochemie* **1970**, 21, 277–285.
20. Ram, S. N.; Yadav, K. D. S.; Sharma, B.; Singh, C. *Indian J. Chem.* **1989**, 28A, 1038–1041.
21. Yamaoka, K.; Takatsuki, M.; Yaguchi, K.; Miura, M. *Bull. Chem. Soc. Jpn.* **1974**, 47, 611–617.
22. Shirai, M.; Murakami, Y.; Tanaka, M. *Makromol. Chem.* **1977**, 178, 2141–2147.
23. Michaelis, L. *Cold Spring Harbor Symp. Quant. Biol.* **1947**, XII, 131.
24. Pal, M. K. *Histochemie* **1965**, 5, 24–31; *Chem. Abstr.* **1965**, 63, 10093h.
25. Blauer, G. *J. Phys. Chem.* **1961**, 65, 1457; *Chem. Abstr.* **1962**, 56, 5541g.
26. Bidegaray, J. P.; Vivoy, R. *J. Chim. Phys. Phys.-Chim Biol.* **1964**, 61, 1383–1390; *Chem. Abstr.* **1965**, 63, 7657a.
27. Chayen, R.; Roberts, E. R. *Sci. J. Royal Coll. Sci.* **1955**, 25, 50–56; *Chem. Abstr.* **1957**, 51, 5861e.
28. Levshin, L. V.; Slavnova, T. D.; Penova, I. V. *Izv. Akad. Nauk SSSR, Ser. Fiz.* **1970**, 34, 604–607; *Chem. Abstr.* **1970**, 73, 26582y.

29. Pal, M. K.; Schubert, M. *J. Phys. Chem.* **1963**, 67, 1821–1827.
30. Zanker, V. *Z. Phys. Chem.* **1952**, 199, 15.
31. Vitagliano, V. *NATO Adv. Study Inst. Ser., Ser. C* **1975**, 18, 437–466.
32. Shirai, M.; Nagaoka, Y.; Tanaka, M. *Makromol. Chem.* **1977**, 178, 1633–1639.
33. Shirai, M.; Nagaoka, Y.; Tanaka, M. *J. Polym. Sci., Polym. Chem. Ed.* **1977**, 15, 1021–1025.
34. Ortona, O.; Vitagliano, V.; Sartorio, R.; Costantino, L. *J. Phys. Chem.* **1984**, 88, 3244–3248.
35. Nishida, K.; Watanabe, H. *Colloid Polym. Sci.* **1974**, 252, 392–395.
36. Valdes-Aguilera, O.; Neckers, D. C. *Acc. Chem. Res.* **1989**, 22, 171–177.
37. Gummow, B. D.; Roberts, G. A. F. *Makromol. Chem., Rapid Commun.* **1985**, 6, 381–386.
38. Shirai, M.; Nagatsuka, T.; Tanaka, M. *Makromol. Chem.* **1977**, 178, 37–46.
39. Takatsuki, A.; *Bull. Chem. Soc. Jpn.* **1980**, 53, 1922–1930.
40. Yamagiski, A.; Watanabe, F. *J. Phys. Chem.* **1981**, 85, 2129–2134.
41. Yun, S. S.; Park, C. H.; Shin, D. H.; Lee, H. *Rep. Res. Inst. Phys. Chem., Chungnam Natl. Univ.* **1984**, 5, 5–11; *Chem. Abstr.* **1986**, 105, 210389w.
42. Shirai, M.; Nagatsuka, T.; Tanaka, M. *Makromol. Chem.* **1978**, 179, 173–179.
43. Shirai, M.; Yamashita, M.; Tanaka, M. *Makromol. Chem.* **1978**, 179, 747–753.
44. Herkstroeter, W. G.; Martic, P. A.; Hartman, S. E.; Williams, J. L. R.; Farid, S. *J. Polym. Sci., Polym. Chem. Ed.* **1983**, 21, 2473–2490.
45. Battacharyya, A. K.; Chakravorty, N. C. *Indian J. Chem.* **1990**, 29A, 32–37.
46. Pal, M. K.; Roy, A. *Indian J. Biochem. Biophys.* **1988**, 25, 368–372; *Chem. Abstr.* **1988**, 110, 3977z.
47. Stork, W. H. J.; deHasseth, P. L.; Schippers, W. B.; Kormeling, C. M.; Mandel, M. *J. Phys. Chem.* **1973**, 77, 1772–1777.
48. Yamaoka, K.; Takatsuki, M. *Bull. Chem. Soc. Jpn.* **1978**, 51, 3182–3192.
49. Takatonki, M.; Yamaoka, K. *Bull. Chem. Soc. Jpn.* **1979**, 52, 1003.
50. Shirai, M.; Nagatsuka, T.; Tanaka, M. *J. Polym. Sci., Polym. Chem. Ed.* **1978**, 16, 2411–2421.
51. Shirai, M.; Murakami, Y.; Tanaka, M. *J. Polym. Sci., Polym. Chem. Ed.* **1979**, 17, 2627–2631.
52. Shirai, M.; Ohyabu, M.; Tanaka, M. *J. Polym. Sci., Polym. Chem. Ed.* **1981**, 19, 1847–1854.
53. Shirai, M.; Ohyabu, M.; Ono, Y.; Tanaka, M. *J. Polym. Sci., Polym. Chem. Ed.* **1982**, 20, 555–563.
54. Shirai, M.; Ono, Y.; Tanaka, M. *Makromol. Chem.* **1983**, 184, 153–163.
55. Shirai, M.; Hanatani, Y.; Tanaka, M. *J. Macromol. Sci., Chem.* **1985**, A22, 279–292.
56. Baumgartner, E.; Bertolotti, S. G.; Cosa, J. J.; Gsponer, H. E.; Hamity, M.; Previtali, C. M. *J. Colloid Interface Sci.* **1987**, 115, 417–421.
57. Isenberg, I.; Leslie, R. B.; Baird, S. L.; Rosenbluth, R.; Bersohn, R. *Proc. Natl. Acad. Sci. U.S.A.* **1964**, 52, 379–387.
58. Lerman, L. S. *J. Mol. Biol.* **1961**, 3, 18.
59. Lerman, L. S. *Proc. Natl. Acad. Sci. U.S.A.* **1963**, 49, 94.
60. Bersohn, R.; Isenberg, I. *J. Chem. Phys.* **1964**, 40, 3175.
61. Shirai, M.; Nagatsuka, T.; Tanaka, M. *Chem. Lett.* **1976**, 291–294.
62. Yamaoka, K.; Takatsuki, M.; Miura, M. *Bull. Chem. Soc. Jpn.* **1975**, 48, 2739.
63. Vitagliano, V.; Costantino, L.; Sartorio, R. *J. Phys. Chem.* **1976**, 80, 959–964.
64. Schreiber, J. P.; Daune, P. M. *J. Mol. Biol.* **1974**, 83, 487.
65. Furano, A. V.; Bradley, D. F.; Childers, L. G. *Biochemistry* **1966**, 5, 3044.
66. Stone, A. L.; Bradley, D. F. *J. Am. Chem. Soc.* **1961**, 83, 3627–3634.
67. Winnik, M. A. In *Polymer Surfaces and Interfaces*; Feast, W. J.; Munro, H. S., Eds.; Wiley: New York, 1987; p 1.

68. Tazuke, S.; Winnik, M. A. In *Photophysical and Photochemical Tools in Polymer Science*; Winnik, M. A., Ed.; Reidel: New York, 1986; p 15.
69. Anufrieva, E. V.; Gotlib, Y. Y. In *Advances in Polymer Science*; Springer-Verlag: Berlin, Germany, 1981; Vol. 40, p 1.
70. Ghiggino, K. P.; Roberts, A. J.; Phillips, D. In *Advances in Polymer Science*; Springer-Verlag: Berlin, Germany, 1981; Vol. 40, p 69.
71. Stockinger, L. *Acta Histochem.* **1958**, *Suppl. No. 1*, 103–120; *Chem. Abstr.* **1962**, *56*, 5050e.
72. *H. J. Conn's Biological Stains*; 9th ed.; Lillie, R. D., Ed.; Williams and Wilkins: Baltimore, MD, 1977.
73. Greenspan, P.; Fowler, S. D. *J. Lipid Res.* **1985**, *26*, 781–789.
74. Sackett, D. L.; Wolff, J. *Anal. Biochem.* **1987**, *167*, 228–234.
75. Chignell, C. F. In *Methods in Pharmacology*; Chignell, C. F., Ed.; Appleton-Century-Crofts: New York, 1972; Vol. 2, p 33.
76. Wang, J. L.; Edelman, G. M. *J. Biol. Chem.* **1971**, *246*, 1185.
77. Sugiyama, Y.; Yamada, T.; Klapowitz, N. *Biochem. Biophys. Acta* **1982**, *709*, 342.
78. Birkett, D. J.; Myers, S. P.; Sudlow, G. *Mol. Pharmacol.* **1977**, *13*, 987.
79. Santous, E. C.; Spector, A. A. *Biochemistry* **1972**, *11*, 2299.
80. Hsu, P. L.; Ma, J. K. H.; Jun, H. W.; Luzzi, L. A. *J. Pharm. Sci.* **1974**, *63*, 27.
81. Suziyamma, Y.; Suzuki, Y.; Sawada, Y.; Kawasaki, S.; Beppu, T.; Iga, T.; Hanano, M. *Biochem. Pharmacol.* **1985**, *34*, 821–829.
82. Levshin, L. V.; Slavnova, T. D.; Penova, I. V.; Nazirov, B. *Zh. Prikl. Spektrosk.* **1973**, *18*, 416–421; *Chem. Abstr.* **1973**, *79*, 43645n.
83. Fenyo, J. C.; Braud, C.; Beaumais, J.; Muller, G. *J. Polym. Sci., Polym. Lett. Ed.* **1975**, *13*, 669–675.
84. Braud, C. *Eur. Polym. J.* **1977**, *13*, 897–901.
85. Muller, G.; Fenyo, J. C. *J. Polym. Sci., Polym. Chem. Ed.* **1978**, *16*, 77–87.
86. Fenyo, J. C.; Mognol, L.; Delben, F.; Paoletti, S.; Crescenzi, V. *J. Polym. Sci., Polym. Chem. Ed.* **1979**, *17*, 4069–4080.
87. Wilhelm, M.; Zhao, C.-L.; Wang, Y.; Xu, R.; Winnik, M. A.; Mura, J.-L.; Riess, G.; Croucher, M. D. *Macromolecules* **1991**, *24*, 1033–1040.
88. Winnik, F. M.; Winnik, M. A.; Ringsdorf, H.; Venzmer, J. *J. Phys. Chem.* **1991**, *95*, 2583–2587.
89. Valdes-Aguilera, O.; Pathak, C. P.; Neckers, D. C. *Macromolecules* **1990**, *23*, 689–692.
90. Gormally, J.; Panak, J.; Wyn-Jones, E.; Dawson, A.; Wedlock, D.; Phillips, G. O. *Anal. Chim. Acta* **1981**, *130*, 369–375.
91. Chen, H.; Wang, D.; Zeng, J.; Li, L. *Shanghai Dier Yike Daxue Xuebao* **1988**, *8*, 351–354; *Chem. Abstr.* **1988**, *110*, 150212z.
92. Guevener, B.; Gueven, K. C.; Peremeci, E. *Sci. Pharm.* **1988**, *56*, 283–286; *Chem. Abstr.* **1988**, *110*, 120991k.
93. Guler, E.; Guven, K. C. *Acta Pharm. Turc.* **1988**, *30*, 49–52; *Chem. Abstr.* **1988**, *109*, 204284z.
94. Wu, A. L. European Patent 0 231 127, 1987; *Chem. Abstr.* **1987**, *109*, 3431w.
95. Myers, R. R.; Fink, J. E. European Patent 0 144 130, 1984.
96. Parazak, D. P.; Burkhardt, C. W.; McCarthy, K. J. *Anal. Chem.* **1987**, *59*, 1444–1445.
97. Onabe, F. *Mokuzai Gakkaishi* **1982**, *28*, 437–444.
98. Pal, M. K.; Mandal, N. *Biopolymers* **1990**, *29*, 1541–1548.

RECEIVED for review July 15, 1991. ACCEPTED revised manuscript August 10, 1992.

Probe Spectroscopy, Free Volume Concepts, and Physical Aging of Polymer Glasses

J. E. Kluin[1], H. Moaddel[2], M. Y. Ruan[1], Z. Yu[2], A. M. Jamieson[2], R. Simha[2], and J. D. McGervey[1]

[1] Department of Physics and [2] Department of Macromolecular Science, Case Western Reserve University, Cleveland, OH 44106

Many properties of polymeric materials can be interpreted by using the free volume concept. Our particular efforts in this area have utilized the Simha–Somcynsky theory, which enables the computation of a free volume function from the experimental equation-of-state of the polymer. In recent years several groups have investigated the possibility of applying spectroscopic techniques as a direct probe of the structural disorder (free volume) in polymers. We review these efforts, which include fluorescence spectroscopy, electron spin resonance spectroscopy, and positron annihilation lifetime (PAL) measurements. We describe in detail the application of PAL analysis to investigate free volume changes in situ in bisphenol A polycarbonate subjected to mechanical deformation.

THE FREE VOLUME CONCEPT IS CENTRAL TO THE INTERPRETATION of many properties of polymeric materials. The term "free volume," applied to amorphous polymers, refers to the difference between the total volume of the material (V) and the volume occupied by the component molecules (V_{occ}): $V_f = V - V_{occ}$, where V_f denotes free volume. The idea that V_f plays a major role in determining the molecular mobility of the polymeric matrix has been widely applied to interpret various bulk phenomena. Examples include the temperature dependence of the viscoelastic relaxation times (1) of polymer liquids [viz. the Williams–Landel–Ferry (WLF) equation] and the physical aging phenomenon in the mechanical properties of polymer glasses (2).

0065–2393/93/0236–0535$06.25/0

Nevertheless, although the free volume concept can be defined quite precisely, its implementation as a quantitative measure of the structural disorder in polymers is a challenging problem. In this regard, we note that a statistical mechanical description of the bulk polymeric state by Simha and Somcynsky (3) incorporates a free volume function, $h(P, T)$, where h is the fractional free volume, P is the pressure, and T is the temperature. The free volume function can be computed from the experimental equation-of-state of the polymeric liquid, and, therefore, explicitly incorporates both entropic and enthalpic contributions to the structural disorder. In glass, the free volume depends on thermal history and the corresponding free volume function, $h(P, T, t_a)$, can be determined from volume relaxation data at aging time, t_a, on glasses of well-specified formation histories (4). Of note is the observation that quantitative connections can be made via the Simha–Somcynsky theory between pressure–volume–temperature (PVT) data for the glass and the time evolution of mechanical properties during isothermal physical aging (5).

Several investigations have explored the possibility that spectroscopic techniques can be used to probe directly the free volume in polymeric materials. Yu et al. (6) utilized the cis–trans conversion of substituted stilbenes, which were dispersed in the polymeric matrix and monitored by the change in UV absorption. The ease of interconversion decreases with increasing size of substituent groups (6). These workers were able to further demonstrate that the fraction of trans isomers formed decreases upon annealing in the glassy state (6). These observations are consistent with the idea that the stilbene derivative probes a subset of the distribution of free volume sites whose lower limit is determined by the molecular volume of the probe, and that the preponderance of these free volume sites decreases in the aging glass. By covalently attaching the stilbene moiety to the polymer at different locations, Yu et al. (6) were further able to show that the free volume at chain ends is larger than that in the interior of the chain. A particular feature of the stilbene probes is that a reasonably accurate estimation of the volume swept out during the cis–trans conversion can be made (6).

Fluorescence probes have been investigated by several groups. The fluorescence anisotropy that is a measure of the rotational diffusive motion of the probe is found to be strongly dependent on the probe size as well as on the molecular mobility of the polymer matrix in which it is located. If the probe is large, the fluorescence anisotropy is sensitive to the free volume of the matrix in which it is dispersed (7, 8). Thus, the temperature dependence of the fluorescence anisotropy is described by an equation of the WLF type (7, 8). This observation is consistent with an expression for the probe rotational correlation time, τ_c, of the form (7, 8)

$$\tau_c = \tau_c^0 \exp(BV_m/V_f) \tag{1}$$

where τ_c^0 is the correlation time of a freely rotating probe, V_m is the probe volume, and B is a system-dependent constant. If B and V_m are temperature independent, τ_c should exhibit WLF behavior characteristic of the polymer host. However, when the probe is small, its rotational motion is also influenced by sub-glass-transition temperature (sub-T_g) motion (7, 8).

In a preliminary way, Meyer has explored the potential of fluorescence anisotropy to follow free volume changes during isothermal physical aging of a polymer glass (9). The anisotropy of diphenylanthracene dispersed in poly(vinyl acetate) (PVAc) was isothermally monitored at $T_g - 2$, after a shallow quench from above T_g. The anisotropy increased with aging time consistent with an increase in τ_c due to the collapse of free volume. However, in our experience (9), the sensitivity of such measurements, which require taking ratios of the polarized and depolarized emission, is not sufficient to effectively monitor the comparatively small changes in free volume that accompany isothermal physical aging deep in the glassy state. Again, it is anticipated that the probe monitors a subset of the free volume distribution state that depends on probe size, V_m.

Likewise, the spectral linewidth of electron spin resonance (ESR) probes dispersed in polymers shows a temperature dependence determined by the mobility of the matrix. Specifically, the ESR spectrum of nitroxide probes shows a collapse associated with the point at which the mean rotational relaxation frequency becomes equal to the spin resonance frequency. A correlation has been noted between the temperature at which this occurs (T_{50G}) and the glass-transition temperature (10, 11). The subscript 50G indicates that the transition temperature is defined when the ESR line width equals 50 gauss.

Meyer (9) has investigated isothermal changes at $T_g - 7$ in the linewidth of the nitroxide probe, 4-(2-bromoacetamide)-2,2,6,6-tetramethyl-1-oxylpiperidine (BROMO), dispersed in PVAc following a quench from the equilibrium melt. At this temperature, BROMO is in the motionally restricted region where a "bimodal" spectrum is observed (11). As the sample annealed (physical aging), a small increase was observed (9) in the position of the high-field peak of the derivative spectrum, from which an increase with aging time of the average rotational correlation time τ_c was deduced. At aging times $t_a > 1$ h, we observed (9) $d \log \tau_c / d \log t_a$ = constant. Such behavior can be rationalized by free volume arguments used by Struik (2) and applied to an equation of the form of equation 1, provided the instantaneous free volume is far from both the initial quench and the final equilibrium values. However, as in fluorescence anisotropy, the linewidth variations that we observe during physical aging in the glassy state are exceedingly small (9), which makes ESR nitroxide radicals an ineffective probe for isothermal free volume changes in the glass. In addition, the probe again locates in a portion of the free volume distribution that depends on the probe size.

An alternative approach using fluorescence probe spectroscopy is to monitor the increase in emission intensity, F, that occurs when the mobility of the polymeric host matrix decreases. This increased emission intensity occurs because of a decrease in nonradiative deexcitation mechanisms. An equation has been proposed (12) that relates the emission intensity to the matrix free volume:

$$F/F_0 = \exp(bV_m/V_f) \qquad (2)$$

Here F_0 is the emission intensity of a freely rotating probe, V_m is the molar volume of the probe, and b is a system-dependent constant. We have studied (13) the emission intensity of auramine O dispersed in PVAc in the liquid and glass states. The fluorescence intensity shows a change in temperature coefficient at T_g that is consistent with that anticipated by free volume concepts applied to eq 2. Following a quench into the glass state from the melt, a time-dependent isothermal increase in fluorescence was observed (2, 13) that showed a logarithmic dependence on aging time: $d\log(F/F_0)/d\log t_a = $ constant. Again, using arguments given by Struik, this behavior appears to be consistent with equation 2, provided we are far from equilibrium (1, 13). Similar observations have been made for other probe–polymer combinations (14, 15). Most interestingly, a recent communication (16) describes the observation of nonlinear responses (asymmetry) in the isothermal emission of probes dispersed in polystyrene and poly(methyl methacrylate) following a two-stage thermal conditioning, which consisted of a quench into the glass at $T_g - T_1$, followed by a single temperature jump to $T_g - T_2 > T_g - T_1$. The shapes of the fluorescence relaxation curves are reported (16) to depend on the direction and magnitude of the jump in a fashion similar to the well-known behavior found in the bulk specific volume (17, 18).

The preponderance of the preceding evidence indicates that fluorescence and ESR probes are indeed sensitive to free volume effects in polymeric materials. However, the implementation of these techniques as a quantitative measure of the small changes in free volume associated with isothermal physical aging is difficult because of uncertainties with regard to probe size variations (6–10), poor sensitivity (9), and possible temperature dependence of the parameters B and b in eqs 1 and 2, respectively (13–16). The positron lifetime method offers the possibility of avoiding some, if not all, of these difficulties. This experiment involves bringing the polymer into contact with a radioactive positron source, typically ^{22}Na. Positrons emitted into the polymeric matrix become thermalized and suffer several possible fates (19–21): (1) annihilation as free positrons ($e^+ + e^- \rightarrow h\tau$); (2) bonding of an electron of opposite spin to form parapositronium (p-PS) and subsequent self-annihilation ($e^+ + e^- \rightarrow$ p-Ps $\rightarrow$ hν), and (3) bonding of an electron of parallel spin to form orthopositronium (o-Ps) and then annihilation by "pick-off" of a matrix electron ($e^+ + e^- \rightarrow$ o-Ps; o-Ps $+ e^- \rightarrow$ hν). The lifetimes associated with these processes can be measured by measuring the time intervals between the detection of the high-energy γ ray released when the positron is emitted and the low-energy gamma rays released on annihilation. The time intervals are quite distinct: $\tau_1 = 120$ ps, for $e^+ + e^- \to h\nu$; $\tau_2 \sim 400$ ps for $e^+ + e^- \to$ p-Ps $\to h\nu$; and $\tau_3 \sim 1000$–2500 ps for $e^+ + e^- \to$ o-Ps, o-PS $+ e^- \to h\nu$. Of particular interest is the observation (19–20) that the relatively long-lived species o-Ps can become trapped in regions of low electron density (i.e., a 'hole') and that the pick-off annihilation rate of o-Ps is very sensitive to the size of the hole.

A variety of studies of o-Ps annihilation characteristics in polymers confirm that they are sensitive to phenomena associated with changes in matrix free volume (20–24). These studies include observations of changes in the temperature coefficient of τ_3 at T_g (20, 21, 24) and reports (22–24) of changes in the relative fraction of o-Ps annihilation events, I_3, during isothermal physical aging in the glass following a quench from above T_g. Focusing on our own efforts in this area, we have carried out an extensive study of o-Ps annihilation in poly(vinyl acetate) (PVAc), a polymer whose PVT relationships have been extensively characterized both in the liquid and glassy states (25). Furthermore, from these data, the Simha–Somcynsky free volume function has been computed in the melt [$h(P, T)$] and for glasses of well-specified histories [$h(P, T, t_a)$] (25, 26). Consistent with studies on other polymers (20, 21), we observe (23) that τ_3 increases with temperature and exhibits a discrete change to a larger temperature coefficient at T_g. On the other hand, I_3 exhibits (24) a broad shallow maximum centered on T_g. Because I_3 is a measure of the probability of formation of o-Ps, it is reasonable to assume that I_3 is proportional to the number density of holes. Also, we can relate τ_3 to the volume of a hole. We utilize an equation proposed by Nakanishi et al. (27), following a model developed by Tao (19) for a spherical hole with radius R and a surface electron layer of thickness ΔR. The lifetime of a trapped o-Ps atom has been shown by quantum mechanical arguments to be (19, 27)

$$\tau_3 = 0.5\left[1 - R/R_0 + (1/2\pi)\sin(2\pi R/R_0)\right]^{-1} \qquad (3)$$

where $R_0 = R + \Delta R$. It was shown (27) that eq 3, with $\Delta R = 0.1656$ nm gives good agreement with the o-Ps lifetimes in molecular solids such as zeolites, where hole sizes have been independently estimated. In polymers containing a distribution of hole sizes, we expect to find a distribution of o-Ps lifetimes. However, typically it is possible only to estimate an average o-Ps lifetime (23), $\langle \tau_3 \rangle$, which therefore yields an average hole volume $\langle v_f \rangle = (4/3)\pi R^3$, where R is the average hole radius computed via eq 3. We have related the positron results to the free volume fraction via the equation

$$h = Cl_3 \langle v_f \rangle \tag{4}$$

where C is a coefficient that depends upon the rate of formation of o-Ps in holes. We determined C by comparing the o-Ps parameters versus $h(P, T)$ computed from PVT data at $T = T_g$ via the Simha–Somcynsky theory (25).

In previous work, we found that the temperature dependence of the product $I_3 \langle V_f \rangle$ for PVAc is in excellent agreement with that of the theoretical h for temperatures $T > T_g$, in support of our hypothesis. At high temperatures, $T \geq T_g + 60\ °C$ the product $I_3 \langle V_f \rangle$ exhibited a positive deviation from the theoretical h. We speculate that this may be due to the fact that the smaller free volume sites relaxed too fast to be sampled by the o-Ps (23). Below T_g, a quantitative comparison is rendered difficult because of the physical aging phenomenon. However, the free volume values computed from the τ_3 and I_3 appear to be of the correct order of magnitude (23). We note that to obtain a detailed interpretation of certain features of the o-Ps data, specifically the existence of isofree volume states in the glass produced by distinct thermal histories, which differ in I_3 and $\langle V_f \rangle$, it will clearly be necessary to incorporate a treatment of the free volume distribution into the theoretical interpretation of o-Ps annihilation. Recently it was demonstrated (S. Vleeshouwers, private communication, 1992) that a two-dimensional (2D) Monte Carlo simulation of a process of hole nucleation and hole cluster formation on a lattice, in which the total fractional hole free volume is set equal to the Simha–Somcynsky value at each temperature, reproduces the maximum observed in the temperature dependence of I_3 in PVAc. Essentially, the increase in I_3 with temperature below T_g is because of an increase in the number of holes (hole nucleation). Above T_g, the number of holes decreases with temperature because of hole clustering.

Our emphasis in this report is the application of positron lifetime spectroscopy to the investigation of changes in the free volume of polymeric solids produced by mechanical deformation. In our initial efforts we focused on the engineering thermoplastic bisphenol A polycarbonate. To enable these experiments, a miniature load cell was constructed to perform in situ positron lifetime measurements under well-specified strains (28).

Experimental Details

Positron Lifetime Spectrometer. The positron annihilation lifetime spectroscopy system consists of a vacuum chamber, BaF_2 and CsF γ-ray detectors, and a fast fast-coincidence detection system, based on EG & G Ortec nuclear instrumentation modules (NIM) modules. Data were collected on a personal computer analyse (PCA) multichannel analyzer (Tennelec). Uniform sample temperature is maintained by a temperature controller (model 805, Lake Shore Cryogenics) using two diode sensors that monitor the temperature at two different surface areas and control fluctuations in temperature during data acquisition to within ± 0.04 °C. To optimize the time resolution function and the detecting efficiency of the system, a cylindrical cesium fluoride crystal (1.5 × 1.5 in.) was used as a scintillator to detect the 1.27-MeV gamma ray that functions as the positron "birth" signal and a conical barium fluoride crystal (0.8 × 1 × 1 in.) to detect the 0.511-MeV gamma ray that serves as the "death" signal. The windows of the constant fraction differential discriminators, which select the energy ranges of the gamma rays, were set using a sodium-22 positron source with a polycarbonate sample. For a 15-μCi sample of ^{22}Na, the count rate was about 300 counts/s. The source, which contained up to 30 μCi of ^{22}NaCl, was deposited on a thin aluminum foil (1.7 mg/cm^2) within an area of diameter $\approx$ 2.5 mm. The foil was then folded into a 7-mm square shape and sandwiched between two rectangular pieces (1 × 1.3 × 0.4 cm) of the polycarbonate sample. The time resolution function in each case was determined in the process of computer fitting of the data and was compared with independent measurements using the prompt gamma rays of ^{60}Co. For ^{60}Co the full width at half maximum (FWHM) was consistently less than 230 ps. Time calibration was done by several methods. One method, which was based on the random coincidence rate, showed that the channel width of the multichannel analyzer was 10.3 ps. To test this result, we observed the gamma rays from the decay of ^{207}Bi and found a mean lifetime of 182 ± 5 ps for the 570-keV level of ^{207}Pb, which agreed with the published lifetime of 186 ps.

Data Analysis of Positron Annihilation Lifetime Spectra.
Each positron annihilation lifetime (PAL) spectrum was fitted to a sum of four exponentially decaying functions convoluted with the resolution function. The fitting was by means of the program Patfit-88. The resolution function is approximated as the sum of three Gaussian functions whose statistical weights and full widths at half maximum (FWHM) are determined by the fitting program to be those that give the best fit to the data. After the resolution function has been determined for one lifetime spectrum with a given source–sample assembly, the function is assumed to be the same for all subsequent lifetime spectra as long as the assembly remains in place. The resolution function consistently had a FWHM of about 230 ps, which is equal to that found from the ^{60}Co source. When a new sample is introduced, the resolution function is again found from the fitting procedure; minor changes in the tail of the function are sometimes found, probably because of a change in scattering from one detector to the other when the geometry changes slightly. After the resolution function is determined from analysis of the initial run on a sample, the parameters to be computed in each subsequent analysis are the position τ_0 for the starting point of each exponential decay, the four intensities (I_s, I_1, I_2, I_3), and the four decay constants (reciprocals of the

mean lifetimes $\tau_s, \tau_1, \tau_2, \tau_3$) for the exponential functions. The source component (I_s, τ_s) results from positrons that are annihilated before they reach the sample; almost all of these stop in the aluminum foil covering the source. From the known foil thickness we calculate that in our experiments about 7% of the positrons were annihilated in the foil. The positron lifetime in the aluminum foil is known to be $\tau_s = 180$ ps; hence the values of I_s and τ_s can be fixed in the curve fitting procedure.

The program Patfit-88 determines the remaining seven parameters, after subtracting the known source components. The starting time t_0, the three mean lifetimes, and two intensities are varied; the third intensity is then determined by requiring the sum of the intensities to be equal to 100%. Because of the multicomponent nature of the PAL spectrum, several distinct choices can be made in computing the fit parameters, which will result in systematic differences in the reported τ_3 and I_3 values. For example, during physical aging experiments, the two shortest lifetimes show variations that appear to be random; therefore, to minimize the effects of the random fluctuations on the values of τ_3 and I_3, these two lifetimes are held at their mean values in a subsequent analysis of all of the curves. Also, theoretically, the shortest lifetime component results from annihilation of p-Ps, whose mean lifetime is known to be 125 ps. In addition, the intensity of p-Ps must be proportional to that of o-Ps ($I_1 = 0.33 I_3$). Generally, it is not possible to obtain a good fit by keeping τ_1 fixed at 125 ps and using a reasonable intensity for this component. Therefore, I_1 or τ_1 can be allowed to vary, assuming that this component is the unresolved sum of p-Ps and another positron state in the polymer. Clearly, it follows that an unambiguous description of the o-Ps decay component requires specification of the parameters for the p-Ps and e^+ decays.

Sample Preparation. The bisphenol A polycarbonate specimens used in this investigation were typical commercial materials obtained from Bayer AG Company (Leverkusen, Germany). The glass-transition temperature was determined to be 150 °C by a differential scanning calorimeter (DSC) at a heating rate of 20 °C/min. Because the principal aim of this investigation is to characterize changes of free volume in the polymer glass, the thermal or mechanical history must be precisely controlled. Thus, for each aging experiment the annealing temperature was at $T_g + 5$ °C. One hour was required to increase the temperature of the sample from room temperature to the annealing temperature, where it was held for 30 min to erase the prior thermal history to equilibrium. After quenching to the measuring temperature (cooling rate 2 °C/min) and a further equilibration period of 20 min, the temperature of the entire sample was uniform and stable. For our study of the temperature dependence of the positron spectrum (29), we likewise took 1 h to increase the temperature to $T_g + 5$ °C, held for 30 min, then cooled to the measuring temperature at an average rate

2 °C/min, waited 10 min for thermal equilibration, and acquired spectral data for 1 h. Finally after a further annealing at $T_g + 5$ °C we increased the temperature to obtain positron spectra at temperatures to 200 °C.

Results and Discussion

Temperature Dependence of τ_3 and I_3 and Fractional Free Volume. The temperature dependence of I_3 and τ_3 for polycarbonate is shown in Figure 1a and b. In Figure 1c, we show the product $h/C = I_3\langle v_f \rangle$, where the average hole volume $\langle v_f \rangle = 4/3\pi R^3$ and R is obtained from eq 3. This comparison follows upon our previous study of poly(vinyl acetate) (23) in which we argued that I_3 is proportional to the number of holes per unit volume and hence $I_3\langle v_f \rangle$ is proportional to the fractional free volume. Generally, the experimental results show behavior similar to that observed earlier (23) for PVAc. A change in temperature coefficient of τ_3 is observed near $T_g = 150$ °C, which indicates an increase in temperature coefficient of the hole radius in the liquid state compared to the glass state. In addition, I_3 exhibits a weak maximum in the glass-transition region. For comparison, in

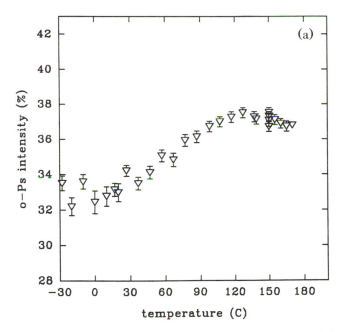

Figure 1. Temperature-dependence of orthopositronium annihilation in polycarbonate: (a) o-Ps intensity, I_3; (b) o-Ps lifetime, τ_3; (c) apparent free volume fraction, $h/C = I_3\langle V_f \rangle$. These values were extracted by fits in which τ_1 was constrained to $\tau_1 = 120$ psi. The solid lines are theoretical calculation of the Simha–Somcynsky free volume function. Continued on next page.

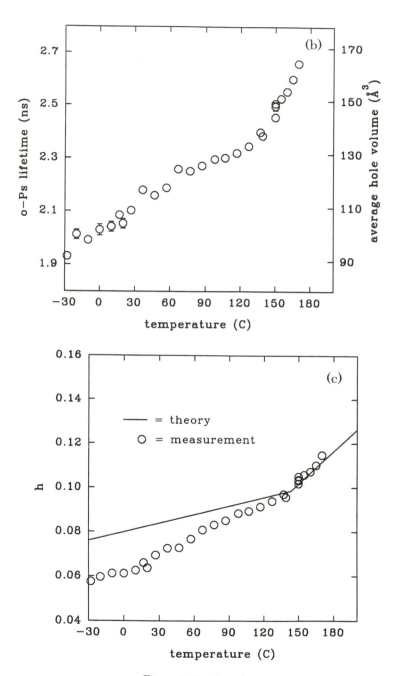

Figure 1. Continued

Figure 1c, the temperature variation of the Simha–Somcynsky free volume fraction for polycarbonate calculated from PVT data (J. E. Kluin, Z. Yu, J. D. McGervey, A. M. Jamieson, R. Simha, and K. Sommer, unpublished results) is shown. The theoretical function is arbitrarily matched to the positron values at the glass-transition temperature $T_g = 150$ °C. Again, as we reported previously for PVAc (23), good agreement with theory is observed for $T > T_g$; as the temperature is lowered further below T_g, increasingly poor agreement is observed.

Several points should be made here.

1. From analysis (J. E. Kluin, Z. Yu, J. D. McGervey, A. M. Jamieson, R. Simha, and K. Sommer, unpublished results) of simulated PAS spectra containing multiple o-Ps decay components derived from hole size distributions produced by the previously referenced Monte Carlo calculations (S. Vleeshouwers, private communication, 1992), we find that the average o-Ps lifetime $\langle \tau_3 \rangle$ derived from typical three-component fitting is the number average value $\langle \tau_3 \rangle = \Sigma n_i \tau_{3i}$.

2. We find that the corresponding o-Ps intensity $I_{3, \exp}$ is always numerically smaller than the number average; that is, $I_{3, \exp} < I_3 = \Sigma n_i I_{3i}$.

3. It is clear from these simulations (J. E. Kluin, Z. Yu, J. D. McGervey, A. M. Jamieson, R. Simha, and K. Sommer, unpublished results) that the anomalously large values of the p-Ps decay parameters, I_1 and τ_1, are due to overlapping contributions of comparatively short-lived o-Ps decay components.

4. The Monte Carlo simulations of hole size distributions based on the Simha–Somcynsky h value at all temperatures reproduce (30) the qualitative details of the temperature dependence of the o-Ps decay components in bisphenol A polycarbonate; that is, the increase in temperature coefficient of $\langle \tau_3 \rangle$ at T_g and the weak maximum in I_3. The deviation between the experimental and the predicted h values in Figure 1c is principally due to discrepancies in comparing $I_{3, \exp}$ and $I_{3, \text{calc}}$.

Isothermal Physical Aging in Strain-Free Glassy Polycarbonate. During isothermal physical aging in the glass at temperatures far below T_g following a quench from the equilibrium melt at 155 °C, we found that the o-Ps lifetime τ_3 remained almost constant with aging time, t_a, but the intensity, I_3, showed a significant decrease with t_a. These results are consistent with a previous positron study of polycarbonate (24) and, in fact, with observations of positron spectra in other polymers during physical aging (22, 23).

This behavior has been interpreted in terms of free volume changes that occur during physical aging; in particular, this behavior indicates a decrease in the number of holes (22–24). However, recently it was pointed out (31) that such decreases may also arise from prolonged exposure to e^+ radiation. Thus, in Figure 2 we compared the time-dependent variation of I_3 and τ_3 at 23 °C for an as-received polycarbonate specimen on constant exposure to e^+ radiation and again following a rejuvenation treatment in which it was heated to 155 °C and then quenched to 23 °C. In each case, τ_3 remained independent of time. For the as-received specimen, I_3 decreased very slightly, by about 0.5% over 60-h exposure, whereas the rejuvenated specimen shows an initially larger value that decreased and became indistinguishable from the as-received specimen after about 20 h. The difference may be ascribable to changes in free volume during isothermal physical aging following the quench. The small decrease in I_3 in the as-received polycarbonate must be a chemical effect due to e^+ irradiation.

Strain Dependence of Free Volume in Polycarbonate. Positron annihilation spectra were determined on a well-aged polycarbonate specimen at $T = 25$ °C subjected to increasing tensile strains up to 8%. Consistent with an independent experiment reported elsewhere (32), as shown in Figure 3a and b, application of strains up to a level of 3–4% produces an increase in τ_3, and hence in $\langle v_f \rangle$, but little change in I_3; that is in N. Above 4% strain, no further variation in $\langle v_f \rangle$ or in N is apparent until the macroscopic yield point is reached at 6%. The corresponding free volume fraction, h/C, computed from the positron data and shown in Figure 3c, increases up to 4% strain and then levels off. At 8% strain, two values of τ_3, I_3, and h/C are shown in Figure 3. These represent data taken, respectively, outside and inside the necking area of the specimen. There appears to be a distinct decrease in the o-Ps free volume in the necking region. However, it is interesting to remark that the free volume in the necking region remains larger than that of the initial undeformed well-aged polycarbonate. In Figure 4 is a stress–strain curve for our polycarbonate specimen that shows typical yield behavior at 6% strain. We also indicate on this curve the point in the pre-yielding region at which the leveling off occurs in the strain dependence of the free volume determined from the positron annihilation spectrum.

Several studies of the materials properties of polycarbonate show distinctive modifications in the pre-yielding regime. First, measurements of the bulk volume (33) are reported to show an increase with tensile strain, ϵ, up to 4%, at which point the density begins to increase with further strain. The initial volume increase is numerically consistent with that computed on the basis

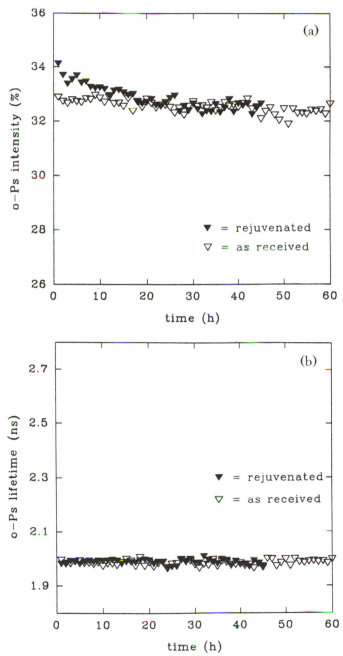

Figure 2. *Variation of orthopositronium annihilation in polycarbonate during isothermal physical aging at* T = 23 °C *in the glass following a quench from* T_g + 5 °C: (a) o-Ps intensity, I_3; (b) o-Ps lifetime, τ_3.

that the Poisson ratio, v, of polycarbonate, is smaller than 0.5:

$$\frac{\Delta V}{V_0} = (1 - 2v)\epsilon \tag{5}$$

For bisphenol A polycarbonate, $v = 0.385$. It is pertinent to note here that the relative increase in fractional free volume with strain, $\Delta h/h_0$, measured

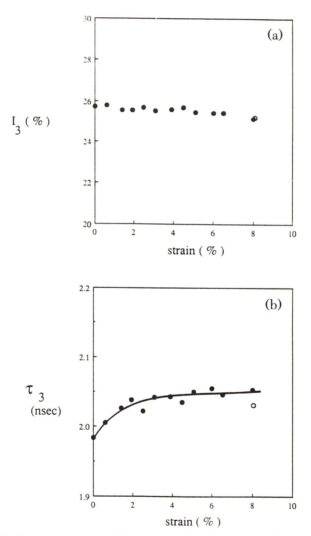

Figure 3. Orthopositronium annihilation in polycarbonate as a function of static tensile strain: (a) o-Ps intensity, I_3; (b) o-Ps lifetime, τ_3; (c) fractional free volume $h/C = I_3\langle V_f \rangle$. These values were generated (28) by unconstrained fits. The open circle at 8% strain was measured separately in the necking region.

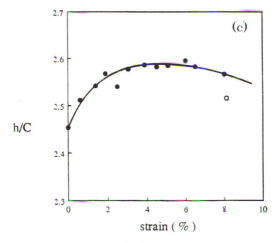

h/C

strain (%)

Figure 3. Continued

by the o-Ps annihilation spectrum, is substantially larger than the fractional increase in bulk volume, $\Delta V/V_0$, as shown in Figure 5. Thus the increase in free volume is larger than that produced by an affine expansion of the holes.

Second, a study (*34*) of the so-called strain-induced rejuvenation or reversal of the physical aging process in the mechanical properties of polycar-

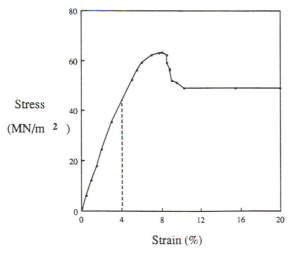

Stress

(MN/m ^{2})

Strain (%)

Figure 4. Tensile stress–strain curve for the polycarbonate sample used in the positron annihilation studies measured by tensile tester (Instron) at a crosshead speed of 2 mm / min according to ASTM test d 638-87b. The broken line indicates approximately where the apparent leveling-off occurs in the strain dependence of free volume measured by o-Ps annihilation (Figure 3).

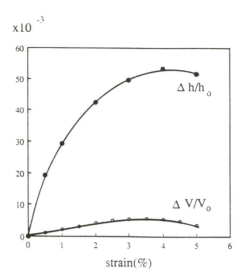

Figure 5. The percent increase in fractional free volume with strain, $\Delta h/h_0$, measured from o-Ps annihilation, compared with the percent increase in bulk volume, $\Delta V/V_0$, from reference 33.

bonate found characteristic changes when the applied strain reached 4%. Specifically, the storage and loss tensile moduli, E' and E'' were measured at 10 Hz during stress relaxation at static tensile strains from 1.2% to 6.5% at 50 °C. It was observed that E' and $1/E''$ decreased when the static strain was applied and then increased monotonically with time, which implies an initial decrease in rigidity, followed by a slow recovery in rigidity as the sample ages under strain. Both the magnitude of the initial change and the rate of the subsequent recovery increase uniformly with the applied strain up to 4%, and then level off. This behavior was attributed to a systematic initial increase in segmental mobility with applied strain up to 4%, followed by accelerated physical aging (34). Clearly, such an interpretation is quite consistent with our positron observations.

Physical Aging in Polycarbonate under Applied Strain. A positron annihilation lifetime study of free volume changes in a well-annealed polycarbonate specimen objected to a static tensile strain of 3.5% at 25 °C was carried out. The positron spectrum was recorded at 3-h intervals during stress relaxation and physical aging following the application of the strain for a total period of 200 h and again periodically every 3 h during further physical aging following release of the stress until a total of 340 h had elapsed. In Figure 6, we show variation of I_3, τ_3, and the fractional free volume h during this experiment. The horizontal line indicates the initial values of the un-strained specimen. Applications of strain causes an initial increase of τ_3, and hence $\langle v_f \rangle$, while I_3 (and hence N) remain constant within experimental error. During the physical aging process that accompanies stress relaxation, the fractional free volume h decreases, due to decreases in both $\langle v_f \rangle$ and N. Note, however, that a significant portion (about 50%) of the overall decrease in h is presumably due to e^+ radiation damage based on Figure 2. Upon

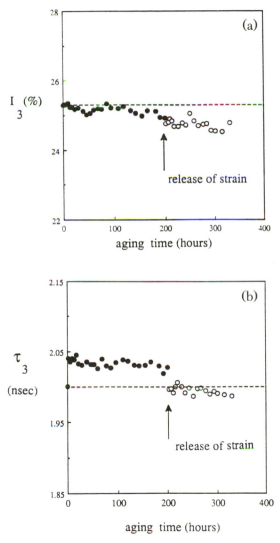

Figure 6. Variation of orthopositronium annihilation in polycarbonate during physical aging after the application of 3% static tensile strain. The solid circles denote data taken in the presence of constant 3% strain, during stress relaxation, and the open circles denote data taken following strain release. (a) o-Ps intensity, I_3; *(b) o-Ps lifetime,* τ_3; *(c) fractional free volume* $h/C = I_3 \langle V_f \rangle$.
Continued on next page.

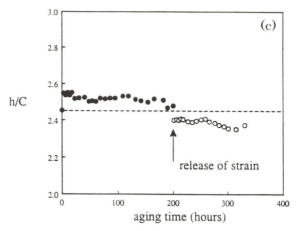

Figure 6. Continued

release of strain, further decrease of h occurs throughout the duration of the experiment until, after 340 h, h is smaller than the initial value. Apparently, the applications of strain to this well-aged polycarbonate specimen has reactivated the aging process, presumably because of the increase in size of free volume. However, it is interesting to note that Legrand et al. (*34*) reported a negative increment in the bulk volume of a polycarbonate specimen following long-term annealing at 3% strain and subsequent strain recovery. The implications is that the tensile strain permits a time-dependent reorientation or more efficient packing of chains even in the pre-yielding region.

Finally, we comment on our positron data in the necking region. Here we see a decrease in the measured free volume fraction, h/C, compared to the plateau value reached in the pre-yield region. This is consistent with the known development of extensive interchain orientation and macroscopic densification that accompanies the necking phenomenon. However, it is noteworthy that h/C remains above the value measured for well-aged amorphous polycarbonate. This observation supports the idea that cold drawing may increase the macroscopic densities because of enhanced interchain orientation and packing, and yet increase the free volume (*35, 36*). This concept has been advanced based on the observation that the rate of volume relaxation is enhanced by cold drawing (*35, 36*).

Summary

A variety of spectroscopic techniques have been investigated for their ability to detect changes in the free volume (molecular mobility) of bulk polymers. Many of these, for example, the fluorescence, ESR, and stilbene isomerization methods, are sensitive to the large changes in free volume that accompany changes in temperature, but are comparatively ineffective for following the small changes in free volume that occur, for example, during isothermal physical aging in the glassy state. All of the available techniques including PAL spectroscopy are limited in that under various circumstances, they are able to sample only a portion of the free distribution. With the PAL technique we were able to establish a quantitative relationship between the o-Ps lifetimes and intensities, and between the Simha–Somcynsky free volume function for PVAc and polycarbonate over a range of temperatures above T_g. However, more work needs to be done to obtain a complete understanding of the o-Ps annihilation characteristics. It is clear from our work and others (*37*) that the o-Ps decay component contains information on the free volume distribution.

Qualitatively, it has been demonstrated that the o-Ps spectrum can be used to investigate free volume changes in the glass produced by physical aging or mechanical strains. However, it is clear that the interpretation of such free volume changes is complicated by artifactual change in the o-Ps decay intensity produced by prolonged e^+ irradiation. Certain polymers appear to have heightened sensitivity to such effects (*30*). We find, that under isothermal conditions, the apparent free volume level of polycarbonate increases with application of static tensile strains up to 4%, after which it levels off. This behavior is consistent with observed enhancement of the aging rate of the mechanical properties in the presence of large strains. In the necking region, the free volume decreases but remains higher than the value for well-aged amorphous polycarbonate, despite the extensive chain orientation. Finally, the o-Ps spectrum of polycarbonate subjected to long-term aging at 3% strain, followed by further aging during strain recovery, indicates a negative change in free volume that is in agreement with published observations of a negative increment or bulk volume of specimens subjected to a similar mechanical history. This observation supports suggestions that the reorientation of chains is facilitated by the increased free volume associated with mechanical deformation and permits a more efficient interchain packing even in the pre-yielding region.

Acknowledgment

This research supported in part by the U.S. Army Research Office, grant number DAAL–03–90–6–0023, and by Dutch State Mines Research, B. V. Geleen, The Netherlands.

References

1. Ferry, J. D. *Viscoelastic Properties of Polymers*; Wiley: New York, 1980; Chapter 11, p 264.

2. Struik, L. C. E. *Physical Aging in Amorphous Polymers and Other Materials*; Elsevier: Amsterdam, The Netherlands, 1978.
3. Simha, R.; Somcynsky, T. *Macromolecules* **1969**, *2*, 342.
4. McKinney, J. E.; Simha, R. *Macromolecules* **1976**, *9*, 430.
5. See Simha, R. In *Frontiers of Macromolecular Science*; Taegusa, T.; Higashimoto, T.; Abe, A., Eds.; Blackwell Scientific Publishers: Oxford, England, 1989; p 307.
6. Yu, W. C.; Sung, C. S. P.; Robertson, R. E. *Macromolecules* **1988**, *21*, 355.
7. Jarry, J. P.; Monnerie, L. *Macromolecules* **1979**, *12*, 925.
8. Noel, C.; Laupretre, F.; Friedrich, C.; Leonard, C.; Halary, J. L.; Monnerie, L. *Macromolecules* **1986**, *19*, 201.
9. Meyer, E. F., III. Ph.D. Thesis, Case Western Reserve University, Cleveland, OH, August 1988.
10. Smith, P. M.; Boyer, R. F.; Kumler, P. L. *Macromolecules* **1979**, *12*, 61.
11. Cameron, G. G.; Miles, I. S.; Bullock, A. *Br. Polym. J.* **1987**, *19*, 129.
12. Loutfy, R. O. *Macromolecules* **1981**, *14*, 270.
13. Meyer, E. F.; Jamieson, A M.; Simha, R.; Palmen, J. H. M.; Booij, H. C.; Maurer, F. H. J. *Polymer* **1990**, *31*, 243.
14. Shmorhun, M.; Jamieson, A. M.; Simha, R. *Polymer* **1990**, *31*, 812.
15. Royal, J. S.; Torkelson, J. M. *Macromolecules* **1990**, *23*, 3536.
16. Royal, J. S.; Torkelson, J. M. *Macromolecules* **1992**, *25*, 1705.
17. Kovacs, A. J. *Fortschr. Hochpolym.-Forsch.* **1963**, *3*, 394.
18. Robertson, R. E.; Simha, R.; Curro, J. G. *Macromolecules* **1988**, *21*, 3216.
19. Tao, S. J. *J. Chem. Phys.* **1972**, *56*, 5499.
20. Stevens, J. R.; Mao, S. J. *J. Appl. Phys.* **1970**, *41*, 4273.
21. Malhotra, B. D.; Pethrick, R. A. *Macromolecules* **1983**, *16*, 1175.
22. McGervey, J. D.; Panigrahi, N.; Simha, R.; Jamieson, A. M. In *Proceedings of the Seventh International Conference on Positron Annihilation*; Jain, P. C.; Singru, R. M.; Gopinathan, K. P., Eds.; World Scientific Publishing: Singapore, 1985; p 690.
23. Kobayashi, Y.; Zheng, W.; Meyer, E. F.; McGervery, J. D.; Jamieson, A. M.; Simha, R. *Macromolecules* **1989**, *22*, 2302.
24. Hill, A. J.; Heater, K. J.; Agrawal, C. M. *J. Polym. Sci., Polym. Phys. Ed.* **1990**, *B28*, 37.
25. McKinney, J. E.; Simha, R. *Macromolecules* **1976**, *9*, 430.
26. Lagasse, R. R.; Curro, J. G. *Macromolecules* **1982**, *15*, 1559.
27. Nakanishi, H.; Wang, S. J.; Jean, Y. C. In *Proceedings of the International Conference on Positron Annihilation in Fluids*; Sharma, S. C., Ed.; World Scientific Publishing: Singapore, 1987; p 252.
28. Moaddel, H. M.S. Thesis, Case Western Reserve University, Cleveland, OH, May 1991.
29. Kluin, J. E.; Yu, Z.; Vleeshouwers, S.; McGervey, J. D.; Jamieson, A. M.; Simha, R. *Macromolecules* **1992**, *25*, 5089.
30. Vleeshouwers, S.; Kluin, J. E.; McGervey, J. D.; Jamieson, A. M.; Simha, R. *J. Polym. Sci., Polym. Phys. Ed.* **1992**, *30*, 1429.
31. Welander, M.; Maurer, F. H. J. In *Proceedings of the Ninth International Conference on Positron Annihilation*; Kajcsos, Zs.; Szeles, Cs., Eds.; Trans Tech Publishing: Switzerland, 1992; p 1811.
32. Ruan, M. Y.; Moaddel, H.; Jamieson, A. M.; Simha, R.; McGervey, J. D. *Macromolecules* **1992**, *25*, 2407.

33. Powers, J. M.; Caddell, R. M. *Polym. Eng. Sci.* **1972**, *12*, 432.
34. Legrand, D. G. *Thermochim. Acta* **1990**, *166*, 105.
35. Pixa, R.; Grisoni, B.; Gay, T.; Froelich, D. *Polym. Bull.* **1986**, *16*, 381.
36. Muller, J.; Wendorff, J. H. *J. Polym. Sci., Polym. Lett. Ed.* **1988**, *26*, 421.
37. Deng, Q.; Zandiehnadem, F.; Jean, Y. C. *Macromolecules* **1992**, *25*, 1090

RECEIVED for review July 15, 1991. ACCEPTED revised manuscript September 9, 1992.

Fluorescence Spectroscopy and Photochemistry of Poly(4-oxystyrenes) with Triphenylsulfonium Salts

Insight into the Photoinitiation of Chemically Amplified Resists

Nigel P. Hacker[1] and Kevin M. Welsh[2]

[1]IBM Research Division, Almaden Research Center, 650 Harry Road, San Jose, CA 95120-6099
[2]Advanced Technology Center, IBM General Technology Division, Hopewell Junction, NY 12533

The fluorescence spectroscopy of 4-oxystyrene polymers, the quenching of the fluorescence by triphenylsulfonium salts, and the photochemistry of the triphenylsulfonium salts in these polymers were studied. Three polymers were examined: poly(4-hydroxystyrene), poly (4-methoxystyrene), and poly[4-(tert-butoxycarbonyl)oxy]styrene]; the latter polymer is employed as a photosensitive polymer for resist applications in the presence of sulfonium salts. All three polymers fluoresce in the 300–340-nm region. Triphenylsulfonium salts quench the fluorescence from the polymers both in solution and the solid state by dynamic and static quenching mechanisms respectively. Photolysis of the sulfonium salts in the polymer films gave lower than expected cage to escape ratios (C/E) for a viscous medium, which is attributed to a sensitization process by the polymer. However, in-cage products are observed, which also implies a direct photolysis mechanism. Thus it is proposed that the process for photoacid production in 4-oxystyrene polymers occurs by a dual photoinitiation pathway that involves both the excited state of the sulfonium salt and the excited state of the polymer film. Performance of the sulfonium salt–oxystyrene polymer was assessed by a low irradiation dose–high dose thickness change comparison at various sulfonium salt concentrations. This thickness

0065–2393/93/0236–0557$06.00/0

test simulates photospeed but reduces errors introduced from the variation in dissolution characteristics caused by changing the sulfonium salt concentration. The thickness data follow the same trend as the percent excited state quenched by the static quenching mechanism with the exception that there is an additional direct photolysis component. The combined spectroscopic and photochemical studies also explain the dynamic range of photosensitivity of the resist that encompasses the UV absorption range of both the initiator and the polymer film.

THE USE OF ONIUM SALTS AS PHOTOINITIATORS for acid-catalyzed cross-linking reactions in polymer chemistry has found wide application (*1*). More recently, poly[4-[(*tert*-butoxycarbonyl)oxy]styrene] (PTBOC) has become vitally important in the electronics industry as an extremely sensitive photoactive polymer in formulation with an onium salt photoinitiator (*2*). Although the photochemistry of sulfonium and other onium salts has been extensively studied in solution (*1, 3*) and recent mechanistic investigations have shown the importance of solvent in-cage reactions and cage-escape reactions (*4–7*), there has been no substantial investigation of exactly how these polymers are rendered photosensitive by onium salts (*8*).

The initial reaction of onium salts in the PTBOC resist is photogeneration of Brönsted acid, which upon post-exposure bake catalyzes the removal (deprotection) of the TBOC group in the exposed areas and generates poly(4-hydroxystyrene) (PHOST). If a thin film of PTBOC is irradiated through a mask, then baked and developed with a solvent that selectively dissolves PHOST, a latent image can be obtained. Thus with this process, one incident photon generates acid that catalyzes multiple reactions and results in a "chemically amplified" photosensitive polymer. The acid-catalyzed process is very efficient, and chain lengths of 10^2–10^3 have been estimated (*9*). Although the acid-catalyzed process is efficient, it undergoes a termination process that prevents deprotection of the PTBOC in the unexposed areas, and this permits the development of images with features less than 1 μm.

There are some important aspects of the PTBOC–sulfonium salt resist that need to be characterized. For example, it is well known that the quantum yield for triphenylsulfonium salt photodecomposition decreases drastically in viscous solvents and polymer films (*10*), yet the photoresist exhibits remarkable sensitivity at low doses. Also, the polymer has a considerable absorbance in the deep UV that limits the amount of incident light absorbed by the onium salt photoinitiator. The ideal photoresist system is where the polymer has no absorbance at the wavelength emitted by the exposure tool and where the photoinitiator absorbs all of the incident light throughout the depth of the film (*11*). A maximum absorbance of about 0.4 at the required wavelength is thought to permit optimum light transmittance to the initiator at the bottom of the film (*12*). Despite the fact that the PTBOC–sulfonium salt resist is nonideal (i.e., the polymer absorbs at the wavelength of the incident light), it performs remarkably well as a deep UV photoresist with photospeeds of around 1 mJ/cm^2 at 248 nm, which is about 2 orders of magnitude faster than optimized nonchemically amplified resists.

We report here the fluorescence spectroscopy and photochemistry of poly(4-oxystyrene) derivatives with triphenylsulfonium salt in solution and as films. The goal of this study is to understand how the presence of these polymers affects the photoinitiation of acid and how the sulfonium salt functions in these polymers, and also to determine if there is a relationship between the spectroscopic properties, the photochemistry, and the performance of the PTBOC resist.

Experimental Details

Fluorescence Spectroscopy.   In a typical experiment, the samples were prepared by dissolving the appropriate amounts of polymer and quencher in propylene glycol methyl ether acetate (PGMEA) followed by filtration of 0.45 μm. The polymer films were formed by spin-casting on silicon wafers or quartz plates at 3500 rpm followed by a soft bake at 90 °C for 60 s. Thus, a 20-wt% solution of polymer–onium salt in PGMEA gave films with thicknesses of 0.8–1.0 μm.

Luminescence spectra were obtain on a spectrofluorometer (Shimadzu RF-540 or Perkin-Elmer LS5-B) using an interrogation wavelength of 290 nm, a source monochromator with a slit width of 2 mm, and a 290-nm narrow bandpass filter. The emission was monitored at the wavelength of interest through a 300-nm high-pass filter and a monochromator with a 2-mm slit width. The silicon wafers (or quartz plates) coated with the polymer thin film were placed in a specially modified solid sample holder at approximately a 45° angle to the source and a shutter between the source and the sample was employed to reduce the amount of exposure to actinic radiation. Luminescence intensity data were obtained from at least five different sites on the wafers. In data analysis, a density of 1.16 g/mL for the polymeric films was used. This density was assumed to remain constant with the addition of onium salt quencher.

Photochemistry.   Triphenylsulfonium salt–polymer films were prepared by dissolving the appropriate amounts of polymer and salt in a suitable solvent (PGMEA or dichloromethane), allowing the solvent to evaporate for 16 h, and finally drying overnight in a vacuum oven at 70 °C. Pieces of the films (0.1 g) were placed in quartz or Pyrex tubes and were irradiated for 30 min in a Rayonet reactor equipped with four bulbs, RPR2537A or RPR3000A, for λ = 254 or 300 nm, respectively. The reaction mixtures were mixed with acetonitrile (4 mL), sonicated, quenched with brine (10 mL), and extracted with hexanes (1 mL) containing *n*-tetradecane internal standard. The photo-products were identified by comparison of retention times with known concentrations of authentic samples in acetonitrile, which were subjected to the same workup procedure to compensate for extraction efficiencies and response ratios (*see* reference 6).

Resist Performance.   Thickness changes were recorded by exposing a wafer to a low radiation dose (1 mJ/cm^2 at λ = 248 nm), baking at 90 °C for 60 s, and comparing the thickness remaining to a similarly prepared film that was given a high radiation dose (100 mJ/cm^2) and processed identically. The thickness of the polymer films was measured before and after exposure by scratching the wafer with a razor blade and measuring the profile across the scratch at several different points on the wafer.

Results

Fluorescence Spectroscopy.   The three poly(4-oxystyrenes), PT-BOC, PHOST, and poly(4-methoxystyrene) (PMOST), all exhibited an emission in the 300–350-nm region with a broad tail toward the red spectrum upon excitation at λ = 290 nm (Figure 1). PTBOC and PHOST had maxima at λ = 304 and 307 nm, respectively, in acetonitrile, whereas the maximum for PMOST was red-shifted to 328 nm. To determine if the red tail from the

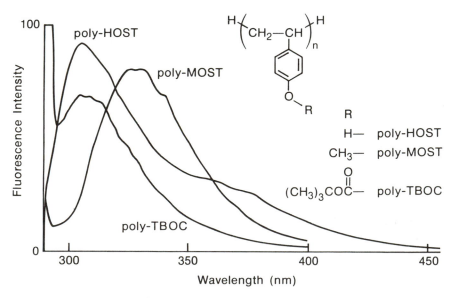

Figure 1. Comparison of PTBOC, PHOST, and PMOST fluorescence in acetonitrile solutions.

emission of each of these compounds was due to the polymer, the fluorescence spectrum of TBOC polymer was measured as a thin film (1 μm) and in acetonitrile solution, and was subsequently compared with the model compound for the repeating unit, *p*-cresol-BOC (Figure 2). The *p*-cresol-BOC gives a fluorescence emission at λ = 294 nm, whereas the TBOC polymer fluoresces at λ = 304 nm in solution and at λ = 308 nm as a film. In addition, the fluorescence spectra of the TBOC polymer, as a film and in solution, both exhibit the broad red-shifted tail, which is not detected from *p*-cresol-BOC. The broad red-shifted emission from each of the poly(4-oxystyrenes) is probably due to partial ordering of the polymer.

When triphenylsulfonium hexafluoroantimonate (TPSSb) was added to the acetonitrile solutions of each of the polymers, the wavelengths of the fluorescence maxima remained constant, but there was a marked decrease in luminescence intensity of the emissions. Figure 3 shows the effect of adding

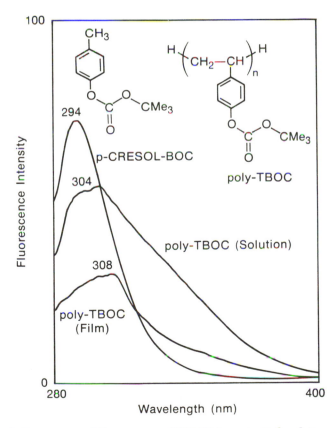

Figure 2. Comparison of fluorescence of PTBOC in acetonitrile solution and as a film with 4-cresol-BOC fluorescence in acetonitrile solution.

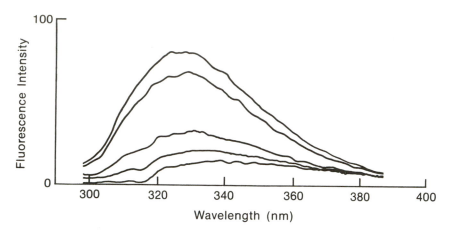

Figure 3. Fluorescence quenching of PMOST by triphenylsulfonium hexafluoro-antimonate (0.0020176, 0.0100878, 0.0201757, and 0.0403515 M) in acetonitrile. Spectra are uncorrected.

2.0×10^{-3}–4.0×19^{-2} M TPSSb to the fluorescence of PMOST in acetonitrile solution using an excitation wavelength of $\lambda = 290$ nm. The emission is almost completely quenched at the higher concentrations of TPSSb. For Stern–Volmer (dynamic) quenching of fluorescence,

$$\Phi_0/\Phi = 1 + k_q\tau[Q]$$

where Φ_0 and Φ are the quantum yields of fluorescence in the absence and presence of quencher, k_q is the bimolecular rate constant for quenching, τ is the lifetime of the emitting species, and $[Q]$ is the quencher concentration. The fluorescence intensities in the absence and presence of quencher, I_0 and I, are directly proportional to Φ_0 and Φ, respectively, and thus a plot of I_0/I versus TPSSb concentration should be linear with a gradient of $k_q\tau$ and an intercept of 1 (*13*, *14*). Figure 4 shows the plots of I_0/I for the three 4-oxystyrene polymers versus TPSSb concentration. All three plots are linear with intercepts at 1. Unfortunately, lifetimes for the 4-oxystyrene polymers are unknown; however, anisole ($\tau = 8.3$ ns) and phenol ($\tau = 2.1$–7.4 ns) can be considered as models for PMOST and PHOST, respectively (*15*). The bimolecular constant for diffusion in acetonitrile is 2×10^{10} M s^{-1}, and so for anisole $k_q\tau = 166$ M^{-1} and for phenol $k_q\tau = 42$–148 M^{-1}. The experimentally obtained values of $k_q\tau$ are 166 M^{-1} for PMOST, 90 M^{-1} for PTBOC, and 80 M^{-1} for PHOST. Thus the Stern–Volmer quenching constants for PMOST and PHOST with TPSSb are close to the diffusion-controlled rates. It is likely that PTBOC exhibits similar behavior; that is, TPSSb quenches the emission at diffusion-controlled rate.

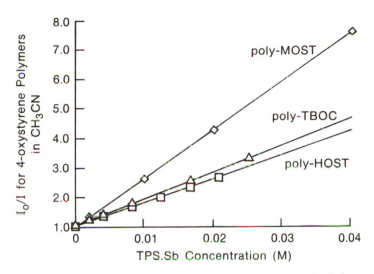

Figure 4. Plot of I_0/I *versus molar concentration for emission of poly (4-oxysty-rene) derivatives in acetonitrile solution in the presence of triphenylsulfonium hexafluoroantimonate.*

The quenching of the fluorescence of the 4-oxystyrene polymers by TPSSb as films was also measured. Using an excitation wavelength of 295 nm and monitoring the emission at 320 nm, the intensity is decreased by greater than a factor of 3 when the concentration of TPS is increased from 0 to 0.12 m, which corresponds to a loading of approximately 7% by weight of TPS in TBOC. At this loading level, the fraction of excited states quenched is approximately 68%. In each case the quenching of the emission of the polymer films did not result in linear Stern–Volmer plots. Perrin developed a model for solid-state (static) quenching:

$$\ln(I_0/I) = VN[Q]$$

where I_0 and I are the fluorescence intensities in the absence and presence of quencher, V is the volume of the active sphere, N is Avogadro's number, and $[Q]$ is the concentration of the quencher in the solid matrix (*14, 16*). The data presented in Figure 5 show that the observed luminescence quenching well approximates the Perrin model for static quenching. The radius, R, of the active quenching sphere is calculated from

$$R = \left[\frac{3V}{4\pi} \right]^{1/3}$$

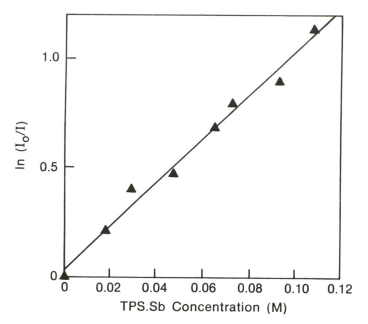

Figure 5. Plot of ln I_0/I *versus molar concentration for poly[4-[(tert-butoxy-carbonyl)oxy]styrene] emission at* $\lambda = 320$ nm *in the presence of triphenylsul-fonium hexafluoroantimonate (excitation at* $\lambda = 295$ nm).

The radius of the active sphere, calculated from the data in Figure 5, for PTBOC is found to be approximately 16 Å. Similar treatment of the emission quenching by TPS of the fluorescence from poly(4-hydroxystyrene) (pHOST) film results in a radius of 19 Å. These values seem to be higher than the expected 8–10-Å radius for pure static quenching (*14*) and may be the result of an additional, small dynamic component.

Photochemistry.   Irradiation ($\lambda = 254$ nm) of PTBOC films containing 0.1, 1.0, and 10.0% TPSSb gave 2-, 3-, and 4-phenylthiobiphenyls and diphenylsulfide (Scheme I). The ratios of the sum of the three phenylthiobiphenyl isomers to diphenylsulfide [the cage to escape ratio (C/E)] were 2.04:1 and 1.70:1 for 1 and 10% salt loadings, respectively (Table I, entries 1 and 2). These C/E values seem remarkably low for a viscous medium. For example, irradiation of poly(methylmethacrylate) (PMMA) films that contain similar loadings of TPSSb under identical conditions gives C/E values of 2.81–3.51:1 (Table I, entries 5 and 6). To determine if sensitization of TPSSb was the reason for the relative increase in diphenylsulfide formation, the films were irradiated at 300 nm where the polymer absorbs, but where the salt has only a very weak absorbance ($\epsilon < 1$). Under these conditions, at 1 and 10% loading, the cage to escape ratio decreases to 0.78:1 and 0.86:1, respectively

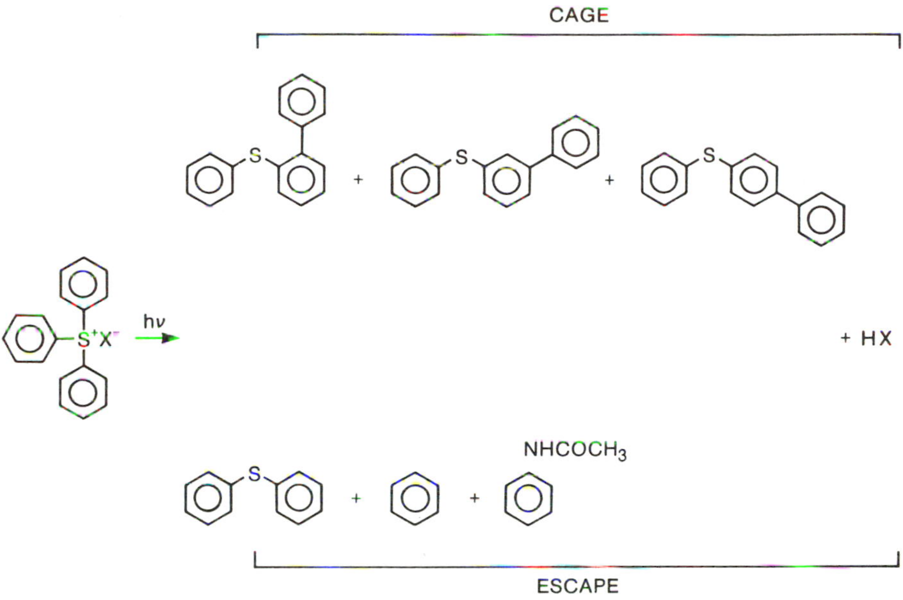

Scheme I. Photoproducts from irradiation of triphenylsulfonium salts.

Table I. Photoproduct Distribution from Irradiation of Triphenylsulfonium Salts (Concentration × 10^5 M)

Run No.	TPS (%)	WC Polymer	λ (nm)	Ph_2S	Ph–PhSPh	C / E
			Film			
1.	1.0	TBOC	254	0.95	1.95	2.04
2.	10.0	TBOC	254	3.96	6.74	1.70
3.	1.0	TBOC	300	0.70	0.55	0.78
4.	10.0	TBOC	300	5.95	5.07	0.86
5.	1.0	PMMA	254	1.92	6.75	3.51
6.	10.0	PMMA	254	6.57	18.46	2.81
			CH_3CN Solution			
7.	1.0	TBOC	254	13.24	10.42	0.79
8.	10.0	TBOC	254	110.9	82.8	0.75
9.	1.0	TBOC	300	1.28	trace	—
10.	10.0	TBOC	300	8.96	0.37	0.04
11.		0.01 M CH_3CN +0.1 M anisole	254	221.4	75.0	0.34
12.		0.01 M CH_3CN +0.1 M anisole	300	14.43	0.79	0.06

(Table I, entries 3 and 4). However, when the PTBOC–TPSSb films are dissolved in acetonitrile and irradiated under identical conditions the C/E values are 0.71–0.79:1 at 254 nm and 0.04 at 300 nm (Table I, entries 7–10). The normal C/E values for direct photolysis under these conditions in acetonitrile are 1.1:1 at λ = 254 nm and 0.35:1 at λ = 300 nm.

Lithographic Performance. A number of PTBOC–TPSSb films were evaluated for photosensitivity as a function of wavelength of the incident light and the concentration of the TPSSb photoinitiator. It was found that PTBOC–TPSSb is photosensitive over a larger dynamic range (200–300 nm) than the absorption range of the photoinitiator (210–280 nm). However, the photosensitivity range does seem to correlate with the combined absorptions of both the polymer and the photoinitiator. Thus the polymer participates in the photoacid generation process. The photosensitivity of the photoresist was evaluated at various concentrations of TPSSb by measuring the relative loss of film thickness after a low dose irradiation, versus a high dose on a similarly prepared film, followed by a postexpose bake. This technique eliminates the errors from the changes in dissolution characteristics of the resist that are introduced by increasing the photoinitiator concentration. The loss in film thickness is proportional to the relative amount of deprotection and has a good inverse correlation with lithographic photospeed.

Discussion

The combined spectroscopic, photoproduct, and lithographic performance studies indicate that the PTBOC backbone participates in the photodecomposition of the onium salt photoinitiator. The phenylthiobiphenyl photoproducts are formed by an in-cage fragmentation–recombination reaction that also results in production of Brönsted acid. However, the diphenylsulfide, benzene, and acetanilide are formed by reaction of the initially formed fragments with the solvent and may be considered as cage–escape products. The cage–escape reactions also generate Brönsted acid. The remarkable sensitivity of the PTBOC–TPSSb mixtures at λ = 300 nm, and the larger than expected amounts of cage–escape reaction products in both solution and as films at both photolysis wavelengths, λ = 254 and 300 nm, strongly indicates photosensitized decomposition of the TPSSb initiator. There are two types of sensitization that result in photodecomposition of triphenylsulfonium salts: triplet energy transfer and photoinitiated electron transfer.

Triplet sensitization occurs in the presence of ketone sensitizers with triplet energies > 74 kcal/mol to give 100% cage–escape product (17). For example, photolysis of triphenylsulfonium triflate in acetonitrile solutions containing acetone, indanone, acetophenone, or xanthone, gives diphenylsulfide, benzene, and acid. The absence of the in-cage phenylthiobiphenyls indicates that the initially formed intermediates from the excited state of the onium salt react with the solvent rather than recombine. The absence of the cage–escape product, acetanilide, suggests that phenyl cation is not an intermediate, and the detection of benzene suggests that phenyl radical is an intermediate. Thus the excited state of the sulfonium cleaves by homolysis to give a radical pair of intermediates that do not recombine. The triplet diphenylsulfinyl radical cation–phenyl radical pair fit the preceding criteria:

$$\text{sensitizer} \rightarrow [\text{sensitizer}]^* \rightarrow [\text{sensitizer}]^1 \rightarrow [\text{sensitizer}]^3$$

$$[\text{sensitizer}]^3 + Ph_3S^+X^- \rightarrow \text{sensitizer} + [Ph_3S^+X^-]^3$$

$$[Ph_3S^+X^-]^3 \rightarrow \overline{Ph_2S^+ \cdot Ph \cdot X}^{\,3} \rightarrow Ph_2S + PhH + H^+X^-$$

Although the 4-oxystyrene polymers are not ketones, they can all be considered as anisole derivatives. Anisole has a triplet energy of 80.8 kcal/mol and so triplet energy transfer is feasible (*15*).

Electron transfer occurs from the singlet excited state of the sensitizer, usually a polycyclic aromatic hydrocarbon (ArH), and proceeds via the triphenylsulfur radical–sensitizer radical cation in the solvent cage (*18*). Subsequent steps produce acid by escape reactions and by phenylation of the sensitizer radical cation:

$$ArH \rightarrow [ArH]^* \rightarrow [ArH]^1$$

$$[ArH]^1 + Ph_3S^+X^- \rightarrow \overline{ArH^{+\cdot} + Ph_3S \cdot X^-}^{\,1}$$

$$\overline{ArH^{+\cdot}Ph_3S \cdot X^-}^{\,1} \rightarrow \overline{ArH^{+\cdot} \cdot Ph \cdot Ph_2 \cdot SX^-}^{\,1}$$

$$\overline{ArH^{+\cdot} \cdot Ph \cdot Ph_2 \cdot SX^-}^{\,1} \rightarrow ArPh + Ph_2S + PhH + H^+X^-$$

If the photoinitiated electron transfer reaction occurs, the conditions of the Rehm–Weller equation must be satisfied (*19*, *20*). If anisole, which has a fluorescence quantum yield of 0.3, is used as a model for the 4-oxystyrene polymers, the requirement for electron transfer to occur is an exothermic reaction:

$$\Delta G = -[E^*_{ox} + E_{red}]$$

where E^*_{ox} is the excited state oxidation potential of the donor (PTBOC) and E_{red} is the reduction potential for the acceptor (TPSSb). The excited state oxidation potential is estimated from the energy of the excited state and ground-state oxidation potential of the donor; that is, $E^*_{ox} = S_1 - E_{ox}$ for the singlet excited state and $E^*_{ox} = T_1 - E_{ox}$ for the triplet excited state, where

S_1 and T_1 are the respective singlet and triplet excited state energies. For TPSSb, $E_{red} = -1.2$ V versus saturated calomel electrode (SCE) (21); for anisole, the model monomer for PTBOC, $E_{ox} = 1.35$ V vs. SCE (22) and $S_1 = 4.47$ eV and $T_1 = 3.50$ eV (15). Substituting these values in the preceding equation yields $\Delta G = -1.92$ eV (-44 kcal/mol) for S_1 and $\Delta G = -0.95$ eV (-22 kcal/mol) for T_1. Thus electron transfer is energetically favorable from both singlet and triplet excited states.

From the foregoing arguments it appears that sensitization can occur by three processes: triplet energy transfer, electron transfer from the singlet excited state, and electron transfer from the triplet excited state of the 4-oxystyrene polymers are all energetically favorable. The fluorescence spectroscopic studies should differentiate between the three possible modes of sensitization.

The detection of an emission of the 4-oxystyrene polymers from films and solutions, the observation of emission quenching, and its insensitivity to oxygen (a triplet excited state quencher), implicate the activity of the singlet excited state of the polymers. Although it could be argued that the emission from the polymer films should be insensitive to oxygen because of limited diffusion, the emission spectra of the films are measured close to the surface of the film where oxygen diffusion can occur. Also, for the triplet excited state to be responsible for the observed emission quenching, there would have to be an equilibrium between the singlet and triplet excited states of the 4-oxystyrene polymers. For the model compound, anisole, the difference between S_1 and T_1 (22 kcal/mol) is too large for this to occur and renders participation of the triplet excited state an unlikely mechanism for acid generation.

The mode of the emission quenching is also an interesting aspect of the sensitization process. Solutions of the polymers and TPSSb apparently behave as monomeric species and the fluorescence quenching gives normal Stern–Volmer plots, which implies a dynamic quenching mechanism that occurs at close to diffusion controlled rates. However, the emission quenching of the films of the 4-oxystyrene polymers appears to occur by a static quenching mechanism. Although dynamic quenching could occur by an exciton migration mechanism, it appears that quenching data better fit the Perrin formulation for static quenching. The radii of the active sphere for the films are 15–20 Å, which seems higher than the normal values of 8–10 Å that are expected for pure static quenching. Thus there may be an additional, small dynamic component, where the photon or exciton can migrate through 1–3 units of the polymer. This hypothesis is supported by the observation of the red-shifted emission tails in the polymers, which indicate a partial ordering of the monomer units in the polymers that may permit short-range energy migration.

The mechanism for the photolysis of PTBOC–TPSSb mixtures is shown in Scheme II. The incident light is absorbed by both the polymer and the

$$P + Ph_3S^+X^- \xrightarrow{h\nu} [P]^* + Ph_3S^+X^- + [Ph_3S^+X^-]^* + P$$

$$[Ph_3S^+X^-]^* \longrightarrow PhPhSPh + Ph_2S + HX$$

$$[P]^* + Ph_3S^+X^- \longrightarrow P^{+\cdot} + Ph_3S^\cdot + X^-$$

$$P^{+\cdot} + Ph_3S^\cdot + X^- \longrightarrow P^{+\cdot} + Ph^\cdot + Ph_2S + X^-$$

$$P^{+\cdot} + Ph^\cdot + X^- \longrightarrow P\text{-}Ph + HX$$

where P = poly[4-[(tert-butoxycarbonyl)oxy]stryrene]

Scheme II. Dual photoinitiation mechanism for photolysis of triphenylsulfonium salts in poly[4-[(tert-butoxycarbonyl)oxy]styrene].

initiator. The light absorbed directly by the initiator results in decomposition via the singlet excited state of TPSSb as previously described for direct photolysis of TPSSb in solution (6). The light absorbed by the polymer initiates and electron transfer to the initiator from the singlet excited state of the polymer, which yields diphenylsulfide and acid as previously described for sensitization with polycyclic aromatic hydrocarbons (18). This mechanism also accounts for the dynamic range of the lithographic performance of the PTBOC–sulfonium salt photoresist; that is, the absorption spectral range of both the polymer and the onium salt photoinitiator contribute to the observed photosensitivity. It must be stressed that the photoinitiation process must occur by both sensitized and direct photolysis pathways; that is, a dual photoinitiation process occurs. The polymer does not absorb 100% of the incident light and although elevated levels of the escape sulfide products are observed due to the sensitization mechanism, the in-cage phenylthiobiphenyls are significant products and indicate that direct photolysis also generates acid.

An additional factor in the lithographic performance of photosensitive polymers is the general observation that the required dose (photospeed) decreases as the concentration of photoinitiator is increased. Usually 1–10 wt% initiator is added to formulations and there is a general observation that the photosensitivity versus initiator loading is nonlinear. Typically the photospeed levels off at higher initiator concentration. Photospeed data are not presented because there are changes in dissolution properties associated with increasing the photoinitiator concentration in the polymer for unexposed films. To overcome this problem, thickness loss experiments were run. PTBOC is deprotected to PHOST in the presence of photogenerated acid, which represents a change from $C_{13}H_{17}O_3$ to C_8H_9O based on monomer units. Thus complete deprotection of a TBOC film results in approximately a 45% weight loss, which should give a corresponding film thickness loss. The film thickness loss for samples with varying photoinitiator concentration was

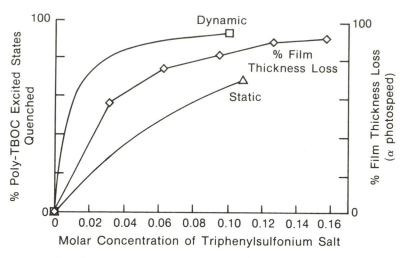

Figure 6. Plot of percent excited states quenched (static and dynamic quenching) and percent film thickness loss (low vs. high exposure dose) for PTBOC.

measured for samples at a low and high exposure doses, 1 and 100 mJ/cm^{-2}, respectively. The 100-mJ/cm dose gave the maximum film thickness loss for all the samples measured. The low dose gives the amount of thickness loss for typical resist processing conditions and the high dose gives the maximum thickness loss for a particular photoinitiator concentration. The thickness is proportional to the weight loss and thus proportional to the amount of deprotection. Thus the percent thickness loss for low dose versus high dose gives a measure of resist performance at various photoinitiator concentrations and eliminates artifacts caused by changes in dissolution characteristics of the resist. Photospeed, the measure of dose required to image a certain thickness resist film, decreases with increasing photoinitiator concentrations; that is, lower irradiation doses give the required amount of deprotection as the photoinitiator concentration increases. Conversely, film thickness loss increases with photoinitiator concentration and so there is an inverse relationship between photospeed and percent film thickness loss. The fact that both parameters correspond to the relative amount of deprotection is supported by the tendency of both photospeed and thickness loss to level off at high photoinitiator concentrations.

Figure 6 shows the relationship between resist performance based on film thickness loss versus concentration of initiator for a typical photoresist. Also shown in the same figure are the percent fraction of the quenched excited states of PTBOC versus TPSSb for static and dynamic quenching based on the data obtained experimentally. The dynamic quenching data show a sharp rise that rapidly reaches a plateau, whereas the rise is more gradual for static quenching. The photospeed data fall in between the dynamic and static quenching curves. There are a number of factors that could account for the differences between the observed thickness loss and percent excited states quenched. The photoproduct studies show that there is an additional direct photolysis mechanism that generates acid and contributes the photospeed data, but not the fluorescence quenching. This mechanism would give lower values for the percent quenching versus the percent thickness change. Also, the fluorescence spectroscopy quenching studies on the TBOC polymer films are static with a small dynamic component; that is, the Perrin formulation is an approximation for these polymers. It is also possible that there is some aggregation of the photoinitiator in the polymer film, a factor that has been used to explain differences yields for photoacid generation in poly(methacrylate) film (23). Although these factors lead to inaccuracies for modeling the photospeed data of the PTBOC photoresist, the observed trends are predicted by the Perrin model for static quenching.

Summary and Conclusions

The photolysis of TBOC–TPSSb formulations proceeds by a dual photoinitiation mechanism that involves both the singlet excited state of the polymer and the singlet excited state of the photoinitiator. Fluorescence spectroscopy has shown that the singlet excited state of the polymer is quenched by the initiator by a dynamic mechanism in solution and by a static mechanism in the film. The combined fluorescence quenching and photoproduct studies suggest that acid is generated by a photoinduced electron transfer reaction. In addition, there is evidence for direct photodecomposition of the initiator via its singlet excited state. The dynamic range of photosensitivity of the PTBOC photoresist formulation corresponds to the combined UV absorptions of both the polymer and initiator, and the photosensitivity follows the general trend of the amount of fluorescence quenching of the polymer by the initiator. The relative amount of reaction from sensitized versus direct photolysis depends not only on the relative absorbances of the initiator and the polymer, but also on reaction medium—film or solution. The combined spectroscopic, photochemical, and lithographic results indicate that not only is it important to increase the absorbance of the initiator relative to the polymer to optimize the sensitivity of a photoresist, but also that the polymer backbone can contribute to improved photosensitivity and that the photophysics and photochemical interaction between the polymer and initiator must be optimized.

References

1. Crivello, J. V. *Adv. Polym. Sci.* **1984**, 62, 1.
2. Willson, C. G.; Bowden, M. J. *CHEMTECH* **1989**, 19, 182.
3. Knapzyck, J. W.; McEwen, W. E. *J. Org. Chem.* **1970**, 35, 2539.

4. Dektar, J. L.; Hacker, N. P. *J. Chem. Soc., Chem. Commun.* **1987**, 1591.
5. Dektar, J. L.; Hacker, N. P. *J. Org. Chem.* **1990**, *55*, 639.
6. Dektar, J. L.; Hacker, N. P. *J. Am. Chem. Soc.* **1990**, *112*, 6004.
7. Dektar, J. L.; Hacker, N. P. *J. Org. Chem.* **1991**, *56*, 1834.
8. For a preliminary communication, *see* Hacker, N. P.; Welsh, K. M. *Macromolecules* **1991**, *24*, 2137.
9. McKean, D. R.; Schaedeli, U.; MacDonald, S. A. *J. Polym. Sci., Polym. Chem. Ed.* **1989**, *61*, 185.
10. Hacker, N. P.; Dektar, J. L. *Polym. Prepr.* (*Am. Chem. Soc., Div. Polym. Chem.*) **1988**, *29*, 524.
11. *Introduction to Microlithography*; Thompson, L. F.; Willson, C. G.; Bowden, M. J., Eds.; ACS Symposium Series 219; American Chemical Society: Washington, DC, 1983.
12. Gutierrez, A. R.; Cox, R. J. *Polym. Photochem.* **1986**, *7*, 517.
13. Stern, O.; Volmer, M. *Phys. Z.* **1919**, *20*, 183.
14. Turro, N. J. *Modern Molecular Photochemistry*; Benjamin-Cummings: Menlo Park, CA, 1978; p 318.
15. Murov, S. L. *Handbook of Photochemistry*; Marcel Dekker: New York, 1973.
16. Perrin, F. *Ann. Chem. Phys.* **1932**, *17*, 283.
17. Dektar, J. L.; Hacker, N. P. *J. Org. Chem.* **1988**, *53*, 1833.
18. Dektar, J. L.; Hacker, N. P. *J. Photochem. Photobiol A.* **1989**, *46*, 233.
19. Rehm, D.; Weller, A. *Ber. Bunsenges. Phys. Chem.* **1969**, *73*, 834.
20. Rehm, D.; Weller, A. *Isr. J. Chem.* **1970**, *8*, 259.
21. Wendt, H.; Hoffelner, H. *Electrochem. Acta* **1983**, *28*, 1453.
22. Lund, H. *Acta Chem. Scand.* **1957**, *11*, 1323.
23. Allen, R. D.; Schaedeli, U.; McKean, D. R.; MacDonald, S. A. *Polym. Mat. Sci. Eng.* **1989**, *61*, 185.

RECEIVED for review July 15, 1991. ACCEPTED revised manuscript June 26, 1992.

Polymer Characterization Using Singlet Oxygen Phosphorescence as a Spectroscopic Probe

Peter R. Ogilby[1], Maria P. Dillon[1], Yuanping Gao[1], Kai-Kong Iu[1], Marianne Kristiansen[1], Vicki L. Taylor[2], and Roger L. Clough[2]

[1]Department of Chemistry, University of New Mexico, Albuquerque, NM 87131
[2]Organic Materials Division (Org. 1811), Sandia National Laboratories, Albuquerque, NM 87185

Singlet molecular oxygen ($^1\Delta_g O_2$) can be produced in solid organic polymers by a variety of different methods. The phosphorescence of singlet oxygen can be monitored in both steady-state and time-resolved experiments, yielding valuable information on the structure and properties of (1) the polymer and (2) solutes dissolved in the polymer. With this spectroscopic probe, we are also able to comment on specific processes that can have important practical ramifications including, for example, the degradation of polymers. In addition, this probe provides a method by which oxygen diffusion coefficients can rapidly and accurately be determined for easily prepared polymer films.

SINGLET OXYGEN ($^1\Delta_g O_2$), THE LOWEST EXCITED ELECTRONIC STATE of molecular oxygen, can be produced in solid organic polymers by a variety of different methods. Once formed, singlet oxygen will follow one of three general deactivation channels:

$$^3\Sigma_g^- O_2 \xrightarrow{\text{formation}} {}^1\Delta_g O_2 \begin{array}{l} \xrightarrow[Q]{k_r} {}^3\Sigma_g^- O_2 + h\nu \qquad \text{radiative decay (phosphorescence)} \qquad (1) \\[2mm] \xrightarrow[Q']{k_q} {}^3\Sigma_g^- O_2 + \text{heat} \qquad \text{nonradiative physical quenching} \qquad (2) \\[2mm] \xrightarrow[A]{k_{rxn}} AO_2 \qquad \text{chemical reaction} \qquad (3) \end{array}$$

0065–2393/93/0236–0573$07.50/0

Singlet oxygen can return to the ground triplet state ($^3\Sigma_g^- O_2$) by either radiative (k_r, eq 1) or nonradiative (k_q, eq 2) decay. The rates for both processes depend on the surrounding medium including solutes (quenchers, Q) that may have been added to the polymer matrix. The singlet oxygen population can also be depleted by a chemical reaction with either a solute molecule or the polymer itself ((k_{rxn}), eq 3): Adducts (AO_2) such as an endoperoxide or hydroperoxide are formed rather than regenerating triplet oxygen. Although the quantum yield of phosphorescence is small ($\sim 10^{-5}$–10^{-4}, i.e., $k_r \ll k_q$), singlet oxygen can be detected by its emission at 1270 nm in both steady-state and time-resolved experiments. This phosphorescence can be used to characterize many properties of (1) solid organic polymers and (2) solutes dissolved in the polymer matrix. Furthermore, this spectroscopic probe can be used to examine a variety of specific events of practical significance. These events include the important processes of polymer degradation and oxygen diffusion in a polymer matrix.

Photosensitized Production of Singlet Oxygen

Singlet oxygen can be produced by energy transfer from a sensitizer (e.g., a dye molecule) added to the polymer matrix as a solute or from a chromophore that forms an integral part of the macromolecular structure (1–3). Although organic molecule singlet excited states ($^1M_1$) can sensitize the production of singlet oxygen (4, 5), they often fluoresce or rapidly undergo intersystem crossing (k_{isc}) to the triplet state ($^3M_1$) before a collision with oxygen can occur. This mechanism is certainly true in more rigid media where solute diffusion coefficients are small. (In designating organic molecule (M) electronic states, numerical superscripts identify the spin state, and subscripts identify either the ground (0) or first excited (1) state.)

$$^1M_0 \xrightarrow{h\nu} {}^1M_1 \xrightarrow{k_{isc}} {}^3M_1 \xrightarrow[{}^3\Sigma_g^- O_2]{k_o} {}^1\Delta_g O_2 + {}^1M_0 \qquad \text{formation} \quad (4)$$

$$^3M_1 \xrightarrow[{}^3\Sigma_g^- O_2]{k_o'} {}^3\Sigma_g^- O_2 + {}^1M_0$$

$$^1\Delta_g O_2 \longrightarrow \text{deactivation (Eqs 1–3)} \qquad\qquad (5)$$

$$k_{decay} = k_r[Q] + k_q[Q'] + k_{rxn}[A]$$

Triplet state lifetimes, however, are sufficiently long, and quenching by oxygen can be a very efficient process (3, 5, 6). The quantum yield of $^1\Delta_g O_2$, ϕ_Δ, depends on the sensitizer M and on the solvent, among other variables, and can range from 0 to 1.0 in a triplet state photosensitized reaction (7). Of course, if $^1M_1$ is quenched by oxygen to yield $^3M_1$, ϕ_Δ can be as large as 2.0 (4, 7).

In the triplet photosensitized process, we have shown that the rate of singlet oxygen formation is equivalent to the decay rate of its precursor, the sensitizer triplet state (5, 6); that is, dissociation of the oxygen–organic molecule complex that is formed when triplet oxygen interacts with the organic triplet state is not rate limiting. Therefore, in a time-resolved triplet state photosensitized experiment, the time-dependent behavior of singlet oxygen (as monitored by its phosphorescence) is properly expressed as a convolution of the intrinsic singlet oxygen decay function (k_{decay}) and the decay function of the triplet state sensitizer $\{k_{triplet} = (k_o + k'_o)[^3\Sigma^-_g O_2]\}$ (3). The triplet state synthesizer decay can conveniently be quantified in a flash absorption experiment. [The expression obtained upon solution of this convolution integral (eq 6) is identical to the expression obtained upon solution of the differential equation (eq 7) that characterizes the change, in time, of the $^1\Delta_g O_2$ concentration in a triplet photosensitized process (8).]

$$\left[^1\Delta_g O_2\right]_t = \frac{k_{triplet}\left[^3 M_1\right]_0}{k_{decay} - k_{triplet}}\left[\exp(-k_{triplet}t) - \exp(-k_{decay}t)\right] \qquad (6)$$

$$\frac{d\left[^1\Delta_g O_2\right]}{dt} = k_{triplet}\left[^3 M_1\right] - k_{decay}\left[^1\Delta_g O_2\right] \qquad (7)$$

In liquids, even at low oxygen concentrations, the rate of singlet oxygen formation is typically 2 orders of magnitude faster than the singlet oxygen decay rate. Thus, it is possible to quantify singlet oxygen formation and decay kinetics with simple nonlinear least-squares fits to the rising and falling portions, respectively, of the manifest (i.e., directly observed) time-resolved singlet oxygen phosphorescence signal (5, 6). In solid polymers, however, where solute diffusion coefficients can be very small, the bimolecular process that results in singlet oxygen formation (eq 4) can be slower than events that result in singlet oxygen decay (eq 5). To accurately quantify the singlet oxygen decay, it is therefore necessary to deconvolute the precursor decay kinetics from the manifest singlet oxygen phosphorescence signal. Under these conditions, even when the intrinsic lifetime of singlet oxygen ($1/k_{decay}$) remains constant, changes in the rate of singlet oxygen formation (eq 4) will appear as changes in both the rate of appearance and disappearance of the manifest singlet oxygen phosphorescence signal (3, 8).

In solid polymers, both the rate of singlet oxygen formation and the yield of singlet oxygen in a photosensitized reaction will depend on, among other factors, (1) the diffusion coefficient for oxygen as determined, for example, by the sample rigidity, the extent of polymerization, and the presence of plasticizers; (2) the concentration of triplet oxygen in the polymer sample; and (3) the sample temperature. The rate of singlet oxygen decay (eq 5) will be determined by, among other factors, (1) the atomic–molecular composition of the polymer matrix, (2) the sample temperature, and (3) the presence of quenchers either added as solutes to the host medium or that are an integral part of the macromolecule itself.

Effect of Polymer Rigidity. Gas diffusion in polymers is a thermally activated process that involves the movement of polymer segments. Macromolecular flexibility is affected, in part, by the strength of attractive interactions between adjacent polymer chains and the ease of rotation about bonds comprising the chain. For example, introduction of a low molecular weight solute in the polymer matrix can decrease attractive interactions between macromolecular chains, which decreases matrix rigidity and enhances gas diffusion. Additionally, alkyl group substitution along the macromolecular chain results in a sterically controlled increase in sample rigidity, with a corresponding decrease in gas diffusion. Thus, copolymerization of methyl methacrylate with increasing amounts of ethyl acrylate results in samples that become successively less rigid, as determined by Rockwell hardness tests and the glass-transition temperature (Figure 1). Similar data are observed when the rigidity of poly(methyl methacrylate) is successively decreased by the addition of a plasticizer such as dimethyl adipate.

The kinetics and yields of singlet oxygen, as monitored by singlet oxygen phosphorescence, reflect the trends obtained from the aforementioned more

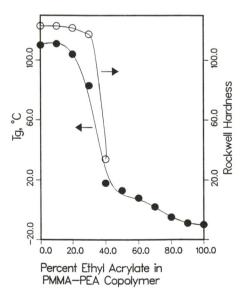

Figure 1. Glass-transition temperatures (T$_g$) and Rockwell hardness data for copolymers of methyl methacrylate and ethyl acrylate.

traditional macroscopic methods of polymer characterization. Specifically, as the polymer sample becomes less rigid and the oxygen diffusion coefficient increases, the frequency of encounters between triplet oxygen and triplet sensitizer increases, resulting in an increased rate of singlet oxygen formation. Within the context of the convolution integral previously discussed, and under conditions where k_{decay} is invariant (*vide infra*), these changes are reflected in both the rising and falling portions of the manifest time-resolved singlet oxygen signal (Figure 2). These manifest signals can be quantified by determining the half-life for phosphorescence signal decay. A plot (Figure 3) of this half-life as a function of the percent of added plasticizer or copolymer (i.e., sample rigidity) reflects the trend shown in Figure 1.

For samples that cover the plasticizer–copolymer range shown in Figure 2, deconvolution of the triplet decay function from the corresponding manifest singlet oxygen phosphorescence data yields intrinsic singlet oxygen lifetimes that are, to a first-order approximation, independent of sample rigidity. This result is also reflected in data recorded from silicone and hydrocarbon oils of differing viscosity (Tables I and II, respectively).

It is clear from the data in Tables I and II that the molecular events that contribute to the solvent-induced deactivation of singlet oxygen (*vide infra*, the section entitled Effect of Polymer Atomic–Molecular Composition), that is, the *intrinsic* lifetime, are largely independent of the macroscopic viscosity of the sample.

Effect of Oxygen Concentration.   The frequency of encounters between triplet oxygen and the triplet sensitizer can also be increased with an increase in the sample oxygen concentration. Changes in oxygen concentration are most easily accomplished and measured by equilibrating identical polymer samples with an ambient atmosphere that has a varied oxygen partial pressure. Within the context of the foregoing model, the rate of singlet oxygen formation should be greater at higher oxygen concentrations. This condition should be manifested as an increase in the rates of phosphorescence signal appearance and disappearance, which is indeed the case, as shown by the data in Figure 4. (For the time scale used in this particular plot, changes in the rate of signal appearance are not especially pronounced.)

Similarly, the absolute amount of singlet oxygen produced per incident photon absorbed by the sensitizer should increase with an increase in the ground state oxygen concentration. This situation is indeed reflected as an increase both in the peak intensity and the integral of the time-resolved phosphorescence signal (Figure 5). Of course, at the limit where all of the sensitizer triplet states produced are quenched by oxygen, the phosphorescence intensity will no longer increase with an increase in the oxygen concentration.

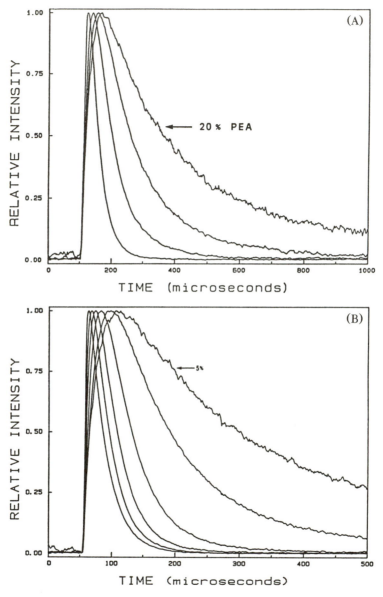

Figure 2. Time-resolved singlet oxygen phosphorescence signals, scaled to make all amplitudes equivalent. (A) The data were recorded as a function of percent ethyl acrylate content in poly(methyl methacrylate)–poly(ethyl acrylate) copolymers. Data for PEA–PMMA ratios of 20:80 (most rigid), 30:70, 40:60, and 70:30 (least rigid) are shown. (Reproduced from reference 3. Copyright 1989 American Chemical Society). (B) The data were recorded as a function of the percentage (by weight) of the plasticizer dimethyl adipate dissolved in poly(methyl methacrylate). Data for 5 (most rigid), 10, 15, 20, 25, and 30% (least rigid) plasticizer are shown.

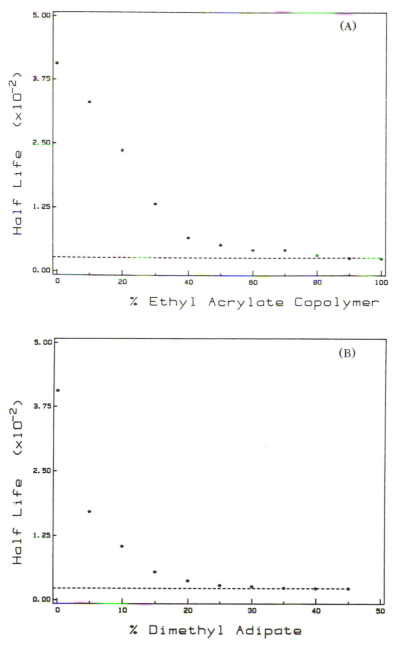

Figure 3. Half-lives for singlet oxygen phosphorescence decay (in microseconds) recorded as a function of percent ethyl acrylate in poly(methyl methacrylate)–poly(ethyl acrylate) copolymers (A) and percent dimethyl adipate dissolved in poly(methyl methacrylate) (B). In each case, the dashed line corresponds to data recorded from a liquid phase analog, methyl propionate. (Reproduced from reference 3. Copyright 1989 American Chemical Society.)

Table I. $^1\Delta_gO_2$ Lifetime (τ_Δ) in Silicone Oils of Differing Viscosity[a]

Viscosity (cS)	Phenazine[b] τ_Δ (μs)	2'-Acetonaphthone[c] τ_Δ (μs)
1	55.0	54.5
10	52.4	51.6
100	52.6	52.0
1,000	52.7	51.0
10,000	52.1	51.0
100,000	52.8	52.2

[a] Singlet oxygen was produced in two independent photosensitized experiments using phenazine and 2'-acetonaphthone as sensitizers, respectively. The laser energy was 1.2 mJ/pulse at 355 nm. Errors on the lifetimes are approximately ±0.5 μs.
[b] [Phenazine] = 1.1×10^{-4} M.
[c] [2'-Acetonaphthone] = 1.2×10^{-3} M.

Table II. $^1\Delta_gO_2$ Lifetime (τ_Δ) in Hydrocarbon Oils of Differing Viscosity[a]

Viscosity (cS)	Hydrocarbon Oil	τ_Δ (μs)
1	Cyclohexane	23.0
3.3	Hexadecane	22.9
26	Light mineral oil	21.8
75	Heavy mineral oil	21.9

[a] Phenazine (1×10^{-4} M) was used as a sensitizer. The laser energy was 1.6 mJ/pulse at 355 nm.

For polymer samples that have been deoxygenated and then subsequently exposed to an oxygen atmosphere, the aforementioned features of the time-resolved singlet oxygen phosphorescence signal (as well as the triplet–triplet flash absorption signal) can be monitored as a function of elapsed exposure time to the oxygen environment. This experiment yields information on oxygen permeation–diffusion coefficients. A more accurate way to ultimately determine a permeation coefficient, however, is to monitor changes in the singlet oxygen phosphorescence intensity in a steady-state experiment. The results of one such experiment are shown in Figure 6 in which the phosphorescence intensity is seen to increase subsequent to the exposure of a deoxygenated polymer film to an atmosphere of oxygen (9).

Similar data can also be recorded in the absence of a photosensitizer by irradiating into the polymer–oxygen charge-transfer absorption band (*vide infra*, the section entitled Production of Singlet Oxygen upon Charge-Transfer Band Photolysis). A desirable feature of the latter approach is that it is not necessary to incorporate sensitizer quenching kinetics into the analysis (9). Using well-established mathematical expressions (10), we have quanti-

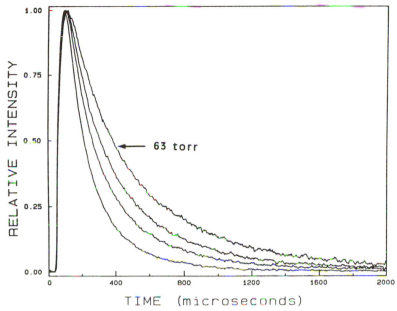

Figure 4. Time-resolved singlet oxygen phosphorescence signals in poly(methyl methacrylate) recorded as a function of the oxygen partial pressure with which the sample had been equilibrated. The data correspond successively to 63-, 200-, 315-, and 567-torr oxygen. The signals have been rescaled to make all amplitudes equivalent.

fied data, such as those in Figure 6, to yield oxygen diffusion coefficients (9). Thus, by using singlet oxygen phosphorescence as a spectroscopic probe, oxygen diffusion coefficients can be obtained rapidly (~ less than one minute) from easily prepared polymer films of small area (~5 mm × 5 mm). Furthermore, because this is a technique that depends on oxygen sorption into the polymer and not on detecting the amount of oxygen that permeates through the material, we are not susceptible to errors that can arise from "pin-holes" in the sample film.

It has been determined in liquid phase experiments that the decay rate of singlet oxygen (eq 5) is independent of oxygen concentration (3); this is also true in solid polymers. Deconvolution of the oxygen-concentration-dependent photosensitizer kinetics from manifest singlet oxygen phosphorescence signals yields intrinsic singlet oxygen lifetimes that are independent of the ambient oxygen partial pressure. This observation has also been confirmed in experiments where singlet oxygen was produced by an independent method (*vide infra*, Production of Singlet Oxygen upon Charge-Transfer Band Photolysis). These data are consistent with a rate constant for singlet oxygen quenching by triplet oxygen that is relatively small (~ 1×10^{3} sec^{-1} M^{-1} in liquids) (3).

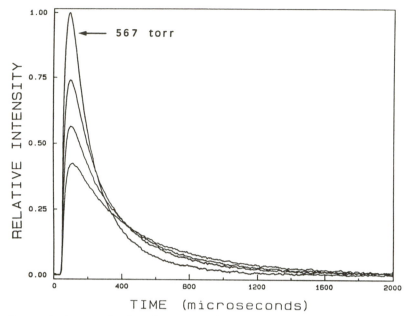

Figure 5. Time-resolved singlet oxygen phosphorescence signals in poly(methyl methacrylate) recorded as a function of the oxygen partial pressure with which the sample had been equilibrated. The data correspond successively to 567-, 315-, 200-, and 63-torr oxygen.

Effect of Sample Temperature.

We have already discussed several factors that can influence polymer rigidity. An increase in the polymer sample temperature also results in a decrease of matrix rigidity, which in turn can result in a substantial increase in gas diffusion rates. At high temperatures, inherent barriers to the various kinds of microscopic segmental movements characteristic of a specific polymer are more easily surmounted. Of course, at a sufficiently high temperature, accessible motions of the macromolecule involve large segments of the polymer backbone, for example, that can ultimately result in transforming an amorphous glass to a soft, rubbery solid.

Because gas diffusion in solid polymers is dependent on these microscopic, segmental movements, it follows that, over limited temperature ranges, the diffusion coefficient can be expressed in Arrhenius form (11, 12), and increases with an increase in temperature. With an increase in the oxygen diffusion coefficient, the sensitizer–oxygen encounter frequency likewise increases, resulting in an increased rate of singlet oxygen formation. Once again, within the context of the aforementioned convolution integral, the increased rate of singlet oxygen formation should appear as increases in the rate of both signal appearance and disappearance in the manifest time-

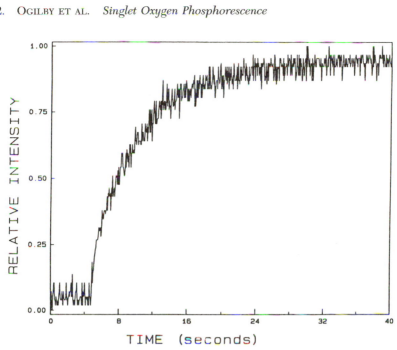

Figure 6. Steady-state singlet oxygen phosphorescence intensity as a function of elapsed exposure time to an oxygen atmosphere of 50 torr. The phosphorescence intensity increases immediately upon exposure of a degassed polystyrene film to oxygen (~ 44 μm thick cast on a glass slide). meso-Tetraphenylporphine was the singlet oxygen sensitizer. (Reproduced from reference 9. Copyright 1992 American Chemical Society.)

resolved singlet oxygen data. The data shown in Figure 7 confirm our expectation.

The intrinsic lifetime of singlet oxygen $(1/k_{\text{decay}})$ decreases only slightly with an increase in temperature in both liquids and polymers (3).

Effect of Polymer Atomic–Molecular Composition. In liquids, the lifetime of singlet oxygen is determined primarily by the ability of the surrounding medium (i.e., solvent) to accept the excitation energy of $^1\Delta_g O_2$ (22.5 kcal/mol, 7880 cm^{-1}) in a process of electronic-to-vibrational energy transfer (eq 2) (13–15). Solvent vibrational modes that are particularly efficient include the $C - H$ and $O - H$ stretch. A decrease in the concentration of $C - H$ and $O - H$ bonds and a corresponding increase in the concentration of bonds with a lower stretching frequency [e.g., $C - D$, $O - D$, or $C - X$ (X = halogen, oxygen)] results in an increase in the singlet oxygen lifetime. This latter parameter is therefore a sensitive measure of the atomic–molecular composition of the host medium.

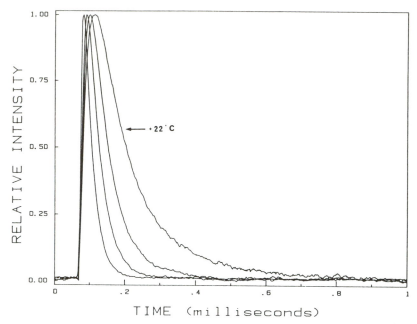

Figure 7. Time-resolved singlet oxygen phosphorescence recorded as a function of polymer sample temperature. Data are shown for poly (methyl methacrylate) samples at 22, 42, 60, and 90 °C. Signals have been rescaled to make all amplitudes equivalent. (Reproduced from reference 3. Copyright 1989 American Chemical Society.)

In polymers, we find that the singlet oxygen lifetime is similarly dependent on the composition of the host medium. This is dramatically shown in the data of Figure 8, where the time-resolved singlet oxygen phosphorescence was recorded from polystyrene and perdeuteriopolystyrene. In this case, the rate of singlet oxygen formation is unaffected by matrix deuteration; the triplet sensitizer decay kinetics, as measured in a flash absorption experiment, are identical for both samples. The large differences observed in the manifest phosphorescence signals, therefore, reflect the effect of solvent deuteration on the intrinsic singlet oxygen lifetime. Deconvolution of the precursor decay function yields singlet oxygen lifetimes that differ by a factor of ~ 12 (3). This number is consistent with solvent deuterium isotope effects on the singlet oxygen lifetime that have been recorded in liquid solutions (13) (also *see* data in Table III).

By deconvoluting the decay function of our singlet oxygen precursor from the manifest phosphorescence signal, we have determined intrinsic singlet oxygen lifetimes in a variety of solid polymers (Table III). Differences in the polymer rigidity–oxygen diffusion coefficient are reflected in the half-life values reported for both the manifest phosphorescence and triplet

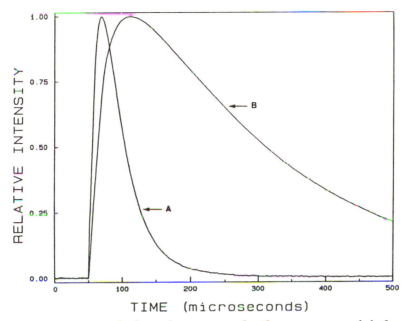

Figure 8. Time-resolved singlet oxygen phosphorescence recorded from polystyrene (A) and perdeuteriopolystyrene (B). (Reproduced from reference 3. Copyright 1989 American Chemical Society.)

sensitizer decay functions. For rigid polymers in which the oxygen diffusion coefficient is small, a combination of large manifest phosphorescence and triplet sensitizer half-lives results. [It should be noted that, in solid polymers, sensitizer triplet states rarely follow first-order decay kinetics (3). Although this does not pose a problem in our deconvolution routine, it is easier to report a "simplistic" half-life for the purpose of illustrating our points]. As previously discussed, however, it is clear that, to a first-order approximation, the intrinsic singlet oxygen lifetime is independent of polymer rigidity [e.g., poly(methyl methacrylate) and poly(ethyl acrylate) data].

Also reported in Table III are singlet oxygen lifetimes in liquid solvents chosen to represent discrete monomeric units of the polymer. The data indicate that the intrinsic singlet oxygen lifetimes in the polymer samples are slightly less than those in the liquid analogs. Although still under investigation, we believe these data may reflect two phenomena:

1. Impurities in the polymers may quench singlet oxygen to yield shorter lifetimes. Some of the polymers were used "as received" [e.g., Teflon AF, polycarbonate, poly(4-methyl-1-pentene), poly(α-methyl styrene)], and thus adventitious byproducts of the manufacturing process, for example, may quench singlet

Table III. Singlet Oxygen Lifetimes in Solid Polymers and Corresponding Liquid Analogs[a]

Polymer	Triplet Sensitizer $\tau_{1/2}$ $(\mu s)^{b}$	Manifest $\tau_{1/2}$ $(\mu s)^{b}$	Intrinsic τ $(\mu s)^{c}$	Liquid Analog	τ (μs)
Poly(methyl methacrylate)	600	360	22 ±3		
				Methyl propionate ($CH_3CH_2COCH_3$)	39 ±1
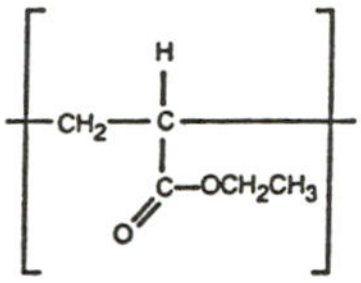 Poly(ethyl acrylate)	<1	21	31 ±1		
Polystyrene	27	56	19 ±2 / 22 ±2[e]		
				toluene	29 ±1
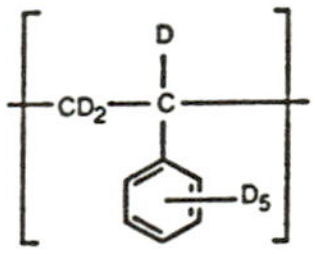 Perdeuteriopolystyrene	31	270	250 ±15	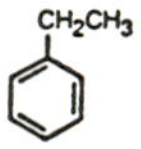 ethyl benzene	26 ±1
				Perdeuteriotoluene	285 ±15
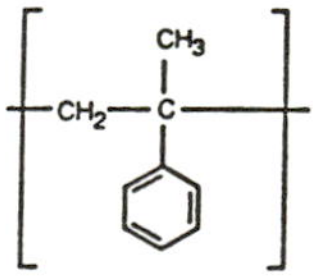 Poly(α-methylstyrene)	43	78	18 ±2	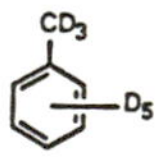	
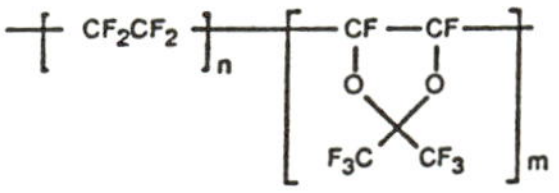 DuPont Teflon® AF Type 1600	<5	1180	1700 ±100	$BrCF_2CF_2Br$	49700 ±2000[d]

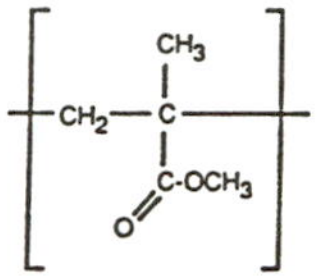

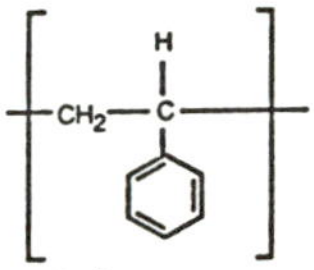

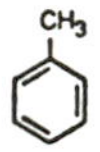

Table III. *Continued*

Polymer	Triplet Sensitizer $\tau_{1/2}$ $(\mu s)^b$	Manifest $\tau_{1/2}$ $(\mu s)^b$	Intrinsic τ $(\mu s)^c$	Liquid Analog	τ (μs)
[structure] Poly(4-methyl-1-pentene)	<1	12	18 ±2	$(CH_3)_2CHCH_2CH_2CH_3$ 2-methylpentane	34 ±1
[structure] Polycarbonate	43	91	29 ±2		
[structure] Poly(urethane) ND 3200(rigid)	180	175	14 ±2	$CH_3CH_2NHCOCH_2CH_3$ N-ethyl ethylcarbamate	23 ±1
Poly(urethane) ND 1100(semi-rigid)	156	132	16 ±2		
Poly(urethane) ND 2300(rubbery)	69	91	16 ±2		
[structure] Poly(dimethyl siloxane) "Silicone rubber"	<1	32	46±1^e	Low viscosity Silicone oil^g	54.7± 0.3

[a] Unless otherwise indicated, these data were obtained using a photosensitizer. With the exception of the dibromotetrafluoroethane data, all liquid τ values were determined in our laboratories.

[b] These numbers depend on the sample oxygen concentration, method of preparation, temperature, etc., and can vary over a wide range. Data reported are representative numbers for air saturated samples at ~20 °C. The numbers are presented principally to indicate whether the singlet oxygen precursor lifetime is much longer or shorter than that of singlet oxygen itself.

[c] Obtained by deconvoluting the sensitizer decay function from the manifest singlet oxygen phosphorescence signal [3]. For polymers in which the triplet sensitizer half-life is much shorter than the manifest singlet oxygen half-life, the intrinsic singlet oxygen lifetime (τ) can be obtained directly from the phosphorescence decay without deconvolution (i.e., as in liquids). In this case, the manifest half-life = $(\ln 2)\tau$.

[d] Reference 16.

[e] Obtained subsequent to photolysis into the polymer-oxygen charge-transfer absorption band. Under these conditions, it is not necessary to deconvolute the singlet oxygen precursor decay kinetics [21].

[f] Poly(urethanes) were prepared by using material provided by Cal Polymers, Inc. and the ND nomenclature system reflects their product literature. Although exact structures are proprietary, the hydrocarbon portions are entirely aliphatic.

[g] See Table I.

oxygen. For samples prepared by free-radical polymerizations [e.g., poly(methyl methacrylate)], residual monomer or local sites of unsaturation in the macromolecule chain could likewise decrease the singlet oxygen lifetime.

2. In the polymer, solvent cage effects may also increase the efficiency of events that result in singlet oxygen deactivation. Specifically, the more rigid solvent cage of the polymer may increase the number of "effective" deactivating collisions with either a quencher or the host medium.

For the optically transparent fluoropolymer (Teflon AF; Dupont), there is a much larger discrepancy between the liquid and polymer data. [Teflon AF is a copolymer of perfluoro-2,2-dimethyl-1,3-dioxole with, for example, tetrafluoroethylene (U.S. Patent 4,935,477). Homopolymers of tetrafluoroethylene, Teflon, cannot be readily studied because they contain microcrystalline regions that scatter light to yield a nontransparent material.] We believe this discrepancy is due in part to quenching phenomena that become important when singlet oxygen lifetimes exceed ~ 500 μs. Specifically, in this time domain, other transients generated in low concentration by the photolysis pulse (e.g., $O_2^- \cdot$), adventitious impurities, or the sensitizer itself become important singlet oxygen quenchers and can cause marked changes in observed lifetimes (13). The history of singlet oxygen lifetime measurements in solvents such as CS_2 and halogenated hydrocarbons reflects these effects, and values for the lifetime currently accepted as accurate are much longer than those determined earlier under conditions where unwanted quenchers were apparently more difficult to remove (13, 16). Indeed, the literature of organic photophysics is replete with analogous examples of the effects of quenchers on long-lived transients in liquids. Most notable are the problems that arise in determining accurate lifetimes for organic molecule triplet states. The singlet oxygen data for the bromofluorocarbon liquid shown in Table III were recorded under conditions that minimized these latter effects (e.g., low photolysis energy, ultrapure solvent) (16). Thus, although the singlet oxygen lifetime measured in the commercially available Teflon AF (used "as received") is shorter than in the liquid analog, it is likely that a longer lifetime could be obtained subsequent to extensive purification of the polymer and the use of lower sensitizer concentrations. Nevertheless, the value we report for this fluoropolymer sample in Table III is dramatically longer than lifetimes obtained from polymers with a high concentration of C – H bonds. Therefore, the data are consistent with a mechanism for singlet oxygen deactivation that is dependent on the atomic–molecular composition of the host medium (*vide supra*).

Although the polycarbonate data reported in Table III were obtained using an added solute (phenazine) as a singlet oxygen photosensitizer, we were also able to record and quantify a time-resolved singlet oxygen phosphorescence signal subsequent to photolysis of the neat polymer at 355 nm. The singlet oxygen lifetime obtained was identical to that recorded in the phenazine-sensitized reaction. Thus, these data clearly indicate that chromophores that form an integral part of the macromolecule can sensitize the production of singlet oxygen (also *see* discussion in the section Production of Singlet Oxygen upon Charge-Transfer Band Photolysis).

In many molecular systems, the decay rate of singlet oxygen is indeed governed by the process of electronic-to-vibrational energy transfer, as discussed before. In some cases, however, particularly for molecules that deactivate singlet oxygen very efficiently, other mechanisms for this induced deactivation process become important. For example, with amines, charge-transfer interactions play a key role in mixing the oxygen singlet and triplet states (*7, 17, 18*). Of course, singlet oxygen can also be deactivated by a chemical reaction (eq 3). The formation of (1) a hydroperoxide via the "ene" reaction with an olefin, (2) an endoperoxide in a $_\pi 2 +_\pi 4$ cycloaddition reaction with a polycyclic aromatic hydrocarbon, or (3) a dioxetane in a $_\pi 2 +_\pi 4$ cycloaddition reaction with an olefin are well-known examples (*1, 2*). By adding either physical or chemical singlet oxygen quenchers to a given polymer system, we are able to sufficiently perturb the kinetics of singlet oxygen in a controlled way and thus yield more accurate information about the role of the polymer in our photophysical system (*19, 20*). Furthermore, a complete understanding of these quenching processes can help us better determine the role played by singlet oxygen in the photooxidative degradation of the polymer (*19, 20*).

Production of Singlet Oxygen upon Charge-Transfer Band Photolysis

Singlet oxygen can be produced in polymers that are free of solutes or macromolecular chromophores that can act as sensitizers by photolysis into the oxygen–polymer charge-transfer (CT) absorption band (*21*). This feature in the absorption spectrum usually appears as a red shift in the absorption onset subsequent to dissolution of oxygen in the material (*21, 22*). As originally assigned by Tsubomura and Mulliken for organic liquids (*23*), this increase in optical density can be attributed to a transition from a ground-state oxygen–organic molecule (M) complex to a charge-transfer state $(M^+ \cdot O_2^- \cdot)$. Depending on the organic triplet state energy level, the CT state may either dissociate directly to form singlet oxygen or relax to a lower lying complex state prior to dissociation to form singlet oxygen (*7, 22*) (Figure 9).

The rate of singlet oxygen formation upon CT state dissociation is substantially faster than that observed in a photosensitized process where the sensitizer is dissolved at a low concentration in the solid matrix. In the

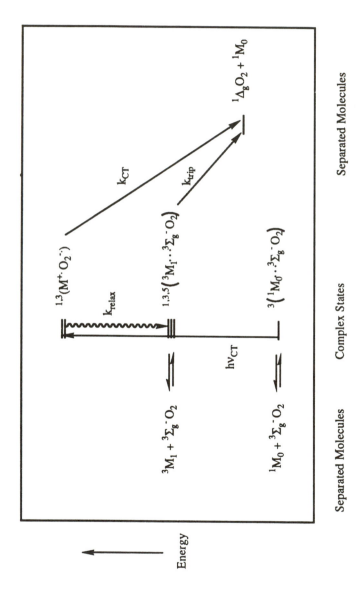

Figure 9. Schematic diagram showing relative energies and several transitions in an oxygen–organic molecule (M) system. (Reproduced from reference 7. Copyright 1991 American Chemical Society.)

photosensitized process, the encounter frequency of oxygen and the triplet sensitizer determines the overall singlet oxygen formation rate that, due to small diffusion coefficients, can be quite slow in rigid polymers (*vide supra*). Because the singlet oxygen formation rate is fast subsequent to CT band photolysis, it is not necessary to deconvolute the singlet oxygen precursor kinetics from the manifest phosphorescence signal to obtain an intrinsic singlet oxygen lifetime. Rather, a simple nonlinear least-squares first-order kinetic fit of the falling portion of the time-resolved signal is sufficient to quantify the data (*21*). Intrinsic lifetimes reported in Table III were obtained in this manner for two polymers [polystyrene and poly(dimethyl siloxane)]. In polystyrene, the data agree quite well with those obtained independently from a photosensitized process by using our deconvolution routine.

The pronounced difference in singlet oxygen formation rates for a $M - O_2$ CT versus sensitizer triplet state precursor can be a very sensitive probe of the purity of a polymer sample. For example, photolysis into the region where the polymer–oxygen CT band absorbs may also excite an impurity that can sensitize singlet oxygen formation. As a consequence of the convolution integral previously discussed, the presence of this impurity would be reflected as slower rates of phosphorescence signal appearance and disappearance. Similarly, the extent of polymer thermal–photochemical degradation to produce compounds that can sensitize singlet oxygen formation will be reflected in the time-resolved phosphorescence signal (Figure 10).

Other Applications of This Spectroscopic Technique: Future Work

Effects of Low Solute Concentration on Polymer Properties. We recently discovered that polystyrene samples doped with small amounts of either 9,10-diphenylanthracene or the saturated hydrocarbon cholestane yield singlet oxygen phosphorescence and triplet sensitizer flash absorption data that depend quite strongly on the amount of additive present (*20*). Specifically, upon exposure of a degassed sample to an atmosphere of air, the rate at which equilibrium with the ambient environment is achieved, as reflected in our spectroscopic data, is inversely proportional to the amount of additive present in the sample. Similar "antiplasticization" effects have been observed and discussed by others (*24*). Thus our spectroscopic probe is quite sensitive to what are apparently subtle solute-induced changes in polymer properties.

Distortion Polarizability of a Polymer. In the absence of an added quencher, singlet oxygen will return to the ground triplet state by one of two processes, both of which are dependent on the solvent (eqs 1–2). In

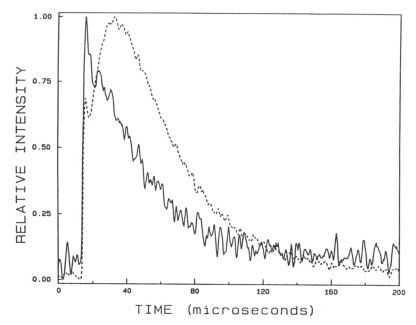

Figure 10. Time-resolved singlet oxygen phosphorescence recorded from polystyrene glasses subsequent to photolysis into the polymer–oxygen CT band. Key: sample prepared by free-radical polymerization, solid line; sample prepared by pressing polystyrene powder at 180 °C, where some degradation occurred to produce compounds that could sensitize the production of singlet oxygen, dashed line. (Reproduced with permission from reference 21. Copyright 1990.)

the first process, the surrounding medium will act as an energy sink in a nonradiative process of electronic-to-vibrational energy transfer (*vide supra*, the section entitled Effect of Polymer Atomic–Molecular Composition) (*13, 14*). This process is by far more efficient than the second process, the simple radiative decay of singlet oxygen (eq 1), and thus defines the overall lifetime of singlet oxygen and gives rise to a small quantum yield of phosphorescence (i.e., the singlet oxygen phosphorescence intensity is very small) (*15*). Solvent features that influence the nonradiative decay rate constant (e.g., $C-H$ bond concentration) do not affect the rate constant for radiative decay (k_r). We have shown, however, that k_r correlates well with the solvent distortion polarizability, which in turn is conveniently expressed as a function of the solvent refractive index (*15*). Thus, under conditions where either a relative or absolute quantum yield of singlet oxygen production can be determined, changes in the intensity of singlet oxygen phosphorescence as a function of the polymer can be ascribed to a change in the polarizability of the host medium. We thus have available a unique method of investigating electron distributions in a polymer.

Experimental Details

Instrumentation.   Detailed descriptions of our time-resolved singlet oxygen detection system and flash absorption spectrometer are published elsewhere (*3, 5, 6, 15, 22, 25*). Briefly, photolysis of either the sensitizer or the $M - O_2$ CT band is accomplished with a pulsed laser chosen for the desired excitation wavelength (Figures 11 and 12). Additional photolysis ("pump") wavelengths can be generated by stimulated Raman scattering

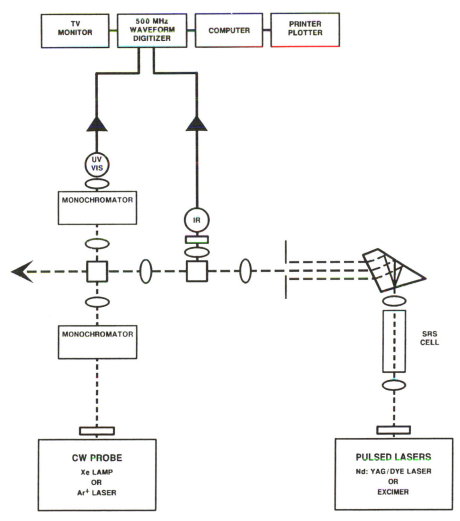

Figure 11. Block diagram indicating the key features of our time-resolved near-IR singlet oxygen detection system and UV–visible flash absorption spectrometer.

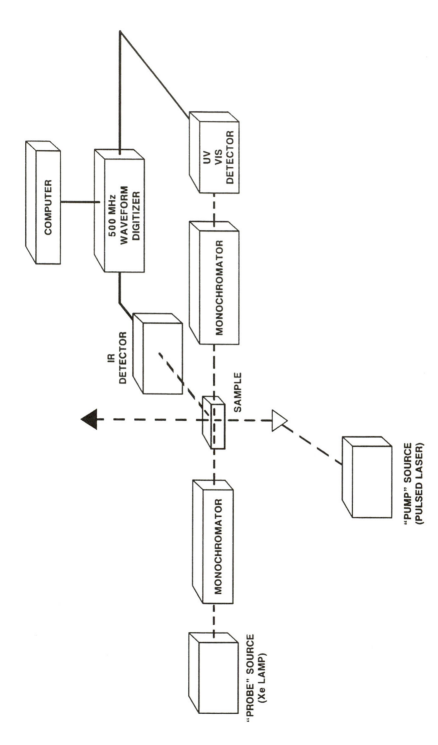

Figure 12. Block diagram that represents our time-resolved excitation–detection system in three dimensions.

(SRS) (*21*, *22*). The 1270-nm singlet oxygen phosphorescence is detected at right angles to the propagation direction of the pump laser. The optical signal is isolated by using bandpass and interference filters and is focused onto a germanium *p-n* junction detector (*25*). Amplified analog signals are digitized by using a waveform digitizer (Tektronix 7912 AD), and the data are stored for subsequent analysis on a personal computer (IBM). Flash absorption experiments of the sensitizer triplet state are accomplished using a 150 W Xe lamp as the probe source and a silicon photodiode as the optical detector (Figures 11 and 12).

A block diagram of our steady-state UV–visible–near-infrared (IR) luminescence spectrometer is presented in Figure 13. A 150 W Xe lamp (Photon Technologies International, Inc.) is used as the excitation source. The output intensity of this lamp is monitored with a silicon photodiode and used in a feedback loop to control the lamp current, thus insuring both long- and short-term stability. Monochromators (Oriel Model 77250) provide wavelength selection in the UV–visible spectral regions. A monochromator (Photon Technologies International, Inc. Model 01–001FAST; grating blazed at 1250 nm) is used to isolate the near-IR phosphorescence of singlet oxygen, which is monitored with a 77 K germanium detector (North Coast Model EO–817L). UV–visible fluorescence is detected with a photomultiplier tube (Hamamatsu Model R758). Signals from the optical detectors are processed by using a lock-in amplifier (Stanford Research Systems Model SR510) interfaced to a personal computer (IBM). The reference frequency for the lock-in amplifier is obtained by mechanically chopping the Xe excitation source. A second Xe lamp is available to independently photolyze the sample and thus initiate photodegradation, for example, while fluorescence–phosphorescence is being monitored.

Materials. Poly(methyl methacrylate), poly(ethyl acrylate), polystyrene, perdeuteriopolystyrene, and methyl methacrylate–ethyl acrylate copolymers were prepared by free-radical polymerizations. Radical inhibitors were removed from the monomers by either distillation or column chromatography. Mixtures of the monomer, radical initiator (typically 2,2′-azobisisobutyronitrile), and, if desired, any solutes (e.g., sensitizer, plasticizer, etc.) were degassed and sealed into a cylindrical ampule. The samples were heated under a variety of conditions (*3*). Resultant polymer rods were cut into disks with a diamond blade saw and polished with diamond paste polishing wheels.

Solid samples of Teflon AF (Du Pont), polycarbonate (polysciences), poly(α-methylstyrene) (polysciences), and poly(4-methyl-1-pentene) (polysciences) were prepared by pressing the commercially supplied polymer powder or beads in a heated mold under vacuum (∼ 3500 lbs in a Carver press at ∼ 180–200 °C). When desired, solutes were mixed with the powder or beads prior to pressing.

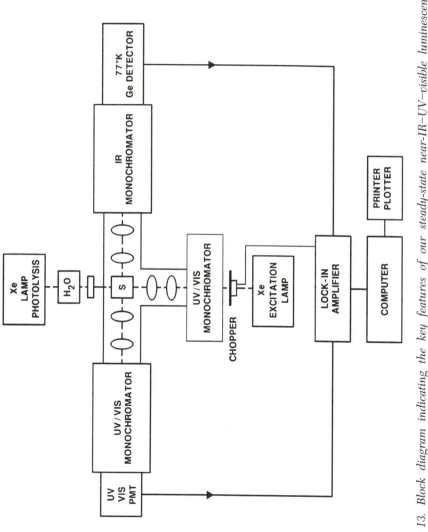

Figure 13. Block diagram indicating the key features of our steady-state near-IR–UV–visible luminescence spectrometer.

Urethanes (calthane ND; Cal Polymers, Inc.) were prepared according to the prescribed manufacturer's procedure. The poly(dimethylsiloxane) rubber was prepared by combining vinyl-terminated poly(dimethylsiloxane), an $(Si - H)$-bearing dimethylsiloxane cross-linker, and a chloroplatinum catalyst (21). Silicone oils (Petrarch, Inc.) were washed several times with methanol prior to use (21).

Polymer films were prepared by casting a solution of the polymer–sensitizer onto a quartz plate.

Conclusions

A variety of polymer properties can be investigated and characterized by using singlet oxygen phosphorescence as a spectroscopic probe.

Acknowledgments

We acknowledge the assistance of M. Malone (Sandia), who performed some of the polymer syntheses and measurements. This work was supported by grants from the National Science Foundation (CHE–8821324), U.S. Department of Energy (DE–AC04–76DP00789), and the donors of the Petroleum Research Fund administered by the American Chemical Society (20006–AC4).

References

1. Ranby, B.; Rabek, J. *Singlet Oxygen: Reactions with Organic Compounds and Polymers*; Wiley: New York, 1978.
2. *Singlet Oxygen*; Frimer, A. A., Ed.; CRC Press: Boca Raton, FL, 1985.
3. Clough, R. L.; Dillon, M. P.; Iu, K.-K.; Ogilby, P. R. *Macromolecules* **1989**, *22*, 3620–3628.
4. Dobrowolski, D. C.; Ogilby, P. R.; Foote, C. S. *J. Phys. Chem.* **1983**, *87*, 2261–2263.
5. Iu, K.-K.; Ogilby, P. R. *J. Phys. Chem.* **1987**, *91*, 1611–1617; erratum *J. Phys. Chem.* **1988**, *92*, 5854.
6. Iu, K.-K.; Ogilby, P. R. *J. Phys. Chem.* **1988**, *92*, 4662–4666.
7. Kristiansen, M.; Scurlock, R. D.; Iu, K.-K.; Ogilby, P. R. *J. Phys. Chem.* **1991**, *95*, 5190–5197.
8. Parker, J. G.; Stanbro, W. D. *Porphyrin Localization and Treatment of Tumors*; Doiron, D. R.; Gomer, C. J., Eds.; Alan R. Liss: New York, 1984; pp 259–284.
9. Gao, Y.; Ogilby, P. R. *Macromolecules*. **1992**, *25*, 4962–4966.
10. Crank, J. *The Mathematics of Diffusion*; Oxford University Press: Oxford, England, 1975.
11. Hopfenberg, H. B.; Stannett, J. In *The Physics of Glassy Polymers*; Haward, R. N., Ed.; Wiley: New York, 1973; pp 504–547.
12. Guillet, J. *Polymer Photophysics and Photochemistry*; Cambridge University Press: Cambridge, England, 1985; pp 55–56, 195–209.

13. Ogilby, P. R.; Foote, C. S. *J. Am. Chem. Soc.* **1983**, *105*, 3423–3430.
14. Hurst, J. R.; Schuster, G. B. *J. Am. Chem. Soc.* **1983**, *105*, 5756–5760.
15. Scurlock, R. D.; Ogilby, P. R. *J. Phys. Chem.* **1987**, *91*, 4599–4602.
16. Schmidt, R. *J. Am. Chem. Soc.* **1989**, *111*, 6983–6987.
17. Ogryzlo, E. A.; Tang, C. W. *J. Am. Chem. Soc.* **1970**, *92*, 5034–5036.
18. Furukawa, K.; Ogryzlo, E. A. *J. Photochem.* **1972**, *1*, 163–169.
19. Ogilby, P. R.; Dillon, M. P.; Kristiansen, M.; Clough, R. L. *Macromolecules*, **1992**, *25*, 3399–3405.
20. Clough, R. L.; Taylor, V. L.; Ogilby, P. R., manuscript in preparation.
21. Ogilby, P. R.; Kristiansen, M.; Clough, R. L. *Macromolecules* **1990**, *23*, 2698–2704.
22. Scurlock, R. D.; Ogilby, P. R. *J. Phys. Chem.* **1989**, *93*, 5493–5500.
23. Tsubomura, H.; Mulliken, R. S. *J. Am. Chem. Soc.* **1960**, *82*, 5966–5974.
24. Maeda, Y.; Paul, D. R. *J. Membrane Sci.* **1987**, *30*, 1–9.
25. Iu, K.-K.; Scurlock, R. D.; Ogilby, P. R. *J. Photochem.* **1987**, *37*, 19–32.

RECEIVED for review July 15, 1991. ACCEPTED revised manuscript May 19, 1992.

Triplet Excimer–Monomer Equilibria as Structural Probes in Pure and Molecularly Doped Polymers

Richard D. Burkhart and Nam-In Jhon

Department of Chemistry, University of Nevada, Reno, NV 89557

Solid films of vinyl aromatic polymers usually emit an excimeric type of phosphorescence at temperatures of 77 K or greater. At temperatures near 10 K these emission spectra are definitely nonexcimeric and at intermediate temperatures the two types of emission coexist. Using the assumption that triplet excimers are formed by trapping at an excimer-forming site, activation energies have been determined for both trapping and detrapping by measuring rates of decay of the excimeric and nonexcimeric emission. The activation energies for trapping of a given chromophore are independent of whether that chromophore is bonded to a chain backbone or present as a small molecule dopant in a polystyrene matrix. The activation energies for detrapping are potentially valuable as an indicator of microstructure for polymers containing a chromphoric group that is phosphorescent.

Investigations of triplet state phenomena are customarily carried out at 77 K because of the convenience of using liquid nitrogen as a coolant. When vinyl aromatic polymers, such as poly(N-vinyl carbazole) (PVCA) or poly(vinyl naphthalene), are studied in frozen glassy solvents at 77 K, the phosphorescence emission is usually very similar to that of the parent chromophore including the vibronic structure (1, 2). In the solid film state, on the other hand, phosphorescence spectra are broad, unstructured, and usually considerably red-shifted (3). It is generally accepted that these solid state spectra are due to the presence of triplet excimers (4).

Recent investigations show that even solid films of these polymers will emit a structured nonexcimeric type of spectrum if cooled to sufficiently low

0065–2393/93/0236–0599$06.00/0

temperatures (5). A simple three-step mechanism that can account for these observations includes a reversible triplet exciton-trapping step, relaxation of the trapped (excimeric) species, and relaxation of the nonexcimeric triplet. The relevant equations are

$$T_m + {}^1E_0 \rightleftarrows {}^3E^* + {}^1M_0 \tag{1}$$

$$T_m \rightarrow {}^1M_0 + h\nu_p \text{ and/or heat} \tag{2}$$

$$^3E^* \rightarrow {}^1E_0 + h\nu_e \text{ and/or heat} \tag{3}$$

where T_m is the nonexcimeric mobile triplet exciton, ${}^1M_0$ is an independent chromophore formed by energy transfer from T_m, ${}^1E_0$ is the excimer-forming site, ${}^3E^*$ is a triplet excimer, and the two photons emitted are either due to normal phosphorescence, $h\nu_p$, or to excimeric phosphorescence, $h\nu_e$. The reversible process indicated in eq 1 may compete with the step in eq 2 for the removal of mobile triplets from the system. Similarly, the reverse of eq 1 may compete with the step in eq 3 for the removal of ${}^3E^*$ from the system. Neither step 2 nor 3 is expected to have an appreciable activation energy, but both the forward and reverse of eq 1 may have.

The focus of attention in this work is on determing activation energies for the forward and reverse of step 1, and then relating these energies to the structural environment of the chromophores involved. Before we begin a description of the manner in which temperature effects on reaction 1 may be isolated from those on other photophysical processes, it is worth considering some of the motivations for conducting such an investigation.

A number of years ago, Cozzens and Fox (6) showed conclusively that triplet excitons migrate from chromophore to chromophore in polymeric systems and that certain aspects of the triplet photophysics in solid polymers depend on the rates of this exciton migration. Once this was pointed out, it became clear that opportunities existed for the use of exciton migration in energy or information transfer. However, the use of exciton migration in various technologies will be possible only if diverse aspects of exciton flow, such as rate, direction, and sense, can be controlled. The existence of excimer traps in polymer films is an obvious retarding influence for the rate of exciton flow. In fact, it has been shown that triplet exciton migration is, indeed, very rapid in an alternating copolymer of methyl methacrylate and 2-vinyl naphthalene (7), where the opportunity for excimer formation is minimized. It seems clear, therefore, that the rate of triplet exciton flow ought to be controllable by adjusting the effectiveness and density of excimer-forming sites. It is important, therefore, to inquire how these two aspects—effectiveness, on the one hand, and density, on the other—may be adjusted for vinyl aromatic polymers. It is especially important to determine how structural changes influence these properties.

In principle, the density of excimer-forming sites should be easy to adjust. In fact, good evidence for the control of this aspect of triplet excimer character has recently been presented in a study of copolymers of methyl methacrylate and carbazolyl ethyl methacrylate (8). The formation of triplet excimers in these copolymers was found clearly to be influenced by the density of carbazolyl groups in the system.

We have used the term "effectiveness" to describe the ability of an excimer-forming site to form and to retain an exciton with no detrapping until relaxation to the ground state occurs. The physical quantities of interest, therefore, are the activation energy for excimer formation (trapping) and the activation energy for excimer dissociation (detrapping). Presumably, the chemical makeup of a chromophoric group will influence these activation energies for the corresponding triplet excimer. Thus, synthetic work to modify chromophores and the accompanying physical measurements of the trapping and detrapping processes will, ideally, work hand-in-hand to achieve some measure of control. Before the synthetic work can be effective, however, reliable means to determine these activation energies must be at hand. Thus, the point of these investigations is to measure the activation energies for several different vinyl aromatic polymers and then to begin the construction of relationships between the thermal and the structural properties of the chromophore system.

Experimental Measurement of Triplet Excimer Trap Depth

A method for carrying out triplet excimer trap-depth measurements was suggested by noticing that the character of phosphorescence spectra from solid films of poly(N-vinyl carbazole) (PVCA) as well as other vinyl aromatic polymers is temperature-dependent. The basic observation was that at 77 K a definitely excimeric type of phosphorescence spectrum was found whereas at much lower temperatures near 15 K the spectrum is nonexcimeric in character. An example of this type of spectral transformation is shown in Figure 1. It was clear that triplet excimer formation required an activation energy that was evidently rather small judging by the low temperatures needed to achieve the excimer–nonexcimer transformation. A temperature dependence of this sort could be rationalized by at least two different mechanisms. If, for example, the migration of triplet excitons is a phonon-assisted process, then raising the temperature would be expected to increase the rate of triplet exciton migration and thereby increase the rate of trapping at excimer-forming sites. This mechanism assumes that the rate of the trapping process is controlled by the rate of exciton migration. In spite of the attractiveness of this mechanism, it does not satisfy all of the experimental observations as will be pointed out later. A more successful explanantion of this temperature dependence is based on the assumption that a finite activation energy exists for the formation of triplet excimers by trapping. Thus, below the transition temperature T_m disappears primarily by reaction 2. As the temperature is raised, eventually a point is reached such that reaction 1 can compete effectively with 2, and at that point the emission of excimeric phosphorescence begins.

This mechanistic sequence is supported by the behavior of Arrhenius graphs measured for the rate constant of reaction 2. These rate constants are relatively easy to measure because they are just the reciprocal of the phosphorescence lifetime. A typical example of the resulting Arrhenius graphs is presented in Figure 2, which traces the behavior of the rate

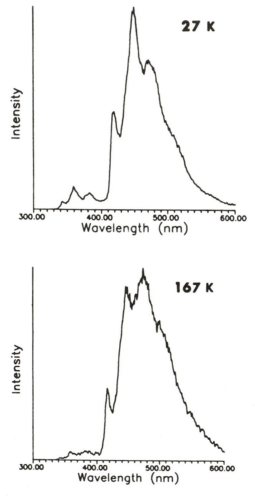

Figure 1. Phosphorescence spectra of N-(carboethoxy)carbazole *in a polystyrene matrix at various temperatures.*

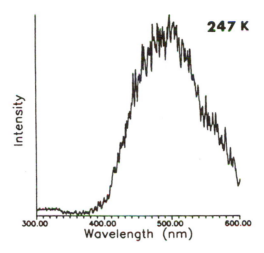

Figure 1. Continued

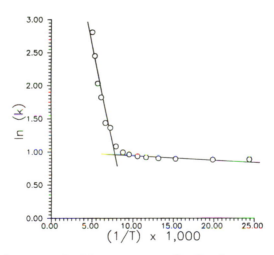

Figure 2. Arrhenius graph of the rate constant for phosphorescence decay of 5% triphenyl amine in a polystyrene matrix.

constant for phosphorescence decay of triphenyl amine (TPA) present as a dopant in a polystyrene matrix. In this particular experiment the phosphorescence intensity was monitored at the origin of the phosphorescence band and so it is definitely a nonexcimeric emission. The important point to note is that there is an extended temperature range over which the rate constant changes very little. Following this stable area, an abrupt increase in rate constant occurs over a small range of temperatures, and after this discontinuity the rate constants change much more rapidly with temperature. We have examined many different systems, either pure polymer films or molecularly doped polystyrene, and in each instance Arrhenius graphs of the type shown in Figure 2 are found. Also, for every system studied so far, the temperature range corresponding to the abrupt change in slope coincides with the temperature range where excimeric phosphorescence starts to replace nonexcimeric phosphorescence. In view of these facts, it seems clear that competitive processes are responsible for the disappearance of triplet states and that the process that has the higher activation energy is associated with the formation of excimers. For these reasons it is believed that the activation energy calculated from the steep slope of Arrhenius graphs such as Figure 2 will yield the activation energy for the forward process represented in reaction 1.

At temperatures large enough so that a distinct excimeric band is observed, one may also follow the rate of decay of excimeric phosphorescence with temperature. Arrhenius graphs for these experiments are quite similar to that displayed in Figure 2. That is, a discontinuity in slope occurs that separates a rather temperature-insensitive region from a region in which the rate of phosphorescence decay changes more rapidly with temperature. The interpretation here is exactly the same as before except now the competitive processes are eq 3 for the relaxation of excimeric triplets to the ground state versus the reverse of step 1, the detrapping of triplet excimers. Thus, the component of the Arrhenius graph that displays the larger slope will yield the activation energy for the reverse of reaction 1. Given in Table I is a collection of activation energies for trapping and detrapping for a number of different pure polymers and molecularly doped polystyrene (PS). Having obtained these data, the next major step is to provide a rational interpretation for them. Interpretation has to be done at several different levels of specificity, and we will start with a discussion of the trapping process.

Rate-Controlling Process in Triplet Trapping

Let us now consider in detail whether the rate of trapping is controlled by the rate of exciton diffusion or by processes associated with bond formation in the resulting excimer. In crystalline solids it is sometimes found that exciton migration is assisted by phonons associated with the vibrational motion of the lattice (9). In amorphous polymers there is no organized lattice structure, but low-frequency acoustic vibrations of the polymer chains will produce a quasiphonon behavior. The question is whether or not these quasiphonons can assist mobile triplet excitons in their migration from one chromophore to the next. An estimate of the energy of these quasiphonons was carried out by Frank and Harrah (10) for poly(2-vinyl naphthalene) (P2VN) and polystyrene. Unlike the present case, Frank and Harrah concluded that the rate-controlling step for singlet excimer formation was singlet exciton migra-

Table I. Activation Energies for Trapping and Detrapping of Triplet Excimers

Sample	E_{act}, Trapping (kJ/mol)	E_{act}, Detrapping (kJ/mol)	Reference
PVCA[a](shallow trap)	2.0	2.5	5
PVCA (deep trap)	2.0	12.1	5
PFCZ[b]	1.1	1.4	11
PACZ[c]	2.6	6.0	11
PDAPM[d]	5.1	4.6	12
NEC[e] in PS	1.7	7.4	11
MFCZ[f] in PS	1.7	7.4	11
DAPM[g] 12% in PS	5.0	3.6	13
DAPM 5% in PS	6.4	2.1	13
DAPM 1% in PS	6.9	1.3	13
TPA[h] 12% in PS	4.6	5.8	13
TPA 5% in PS	4.8	7.9	13
TPA 1% in PS	7.6	9.2	13

[a] Poly(N-vinylcarbazole).
[b] Poly[N-(vinyloxy)(carbonyl)carbazole].
[c] Poly[N-((allyloxy)carbonyl)carbazole].
[d] Poly[[4-N,N-diphenylamino)phenyl]methylmethacrylate].
[e] N-Ethylcarbazole.
[f] N-((vinyloxy)carbonyl)carbazole.
[g] [4-(N,N-Diphenylamino)phenyl]methylisobutyrate.
[h] Triphenylamine.

tion. In the case of singlet excitons, therefore, the activation energy for trapping will be the same as the activation energy for exciton migration. The rate of exciton migration is inversely proportional to the sixth power of the interchromophore separation distance, and it was suggested by Frank and Harrah (*10*) that low-frequency acoustic vibrations of the polymer chains would be effective in modulating these interchromophore distances. From measured values of the activation energies for trapping, the frequency of these effective acoustical modes could be determined. Values of 84 cm^{-1} (1.0 kJ/mol) for P2VN and 157 cm^{-1} (1.9 kJ/mol) for polystyrene were found. According to the analysis of Frank and Harrah (*10*) these vibration energies should equal one-half the activation energy for excimer formation. We notice in Table I that activation energies for small molecules doped into polystyrene vary considerably with the percentage of doping and also with the type of chromophore. For DAPM ([4-(*N*,*N*-diphenylamino)phenyl] methyl isobutyrate), which is essentially a triphenylamino group with a carboxylate residue attached to one of the phenyl rings, and for triphenyl amine itself, activation energies from 5.0 to 7.6 kJ/mol are found, which imply quasiphonon energies between 2.5 and 3.8 kJ/mol. On the other hand, for the carbazole chromophores, such as NEC (*N*-ethyl carbazole) and MFCZ [*N*-((vinyloxy)carbonyl)carbazole], much smaller activation energies on the order of 1.7 kJ/mol are found, which imply quasiphonon energies of 0.8 kJ/mol.

Thus, experiments with neither of these chromophores correctly reproduced the expected vibrational energies. Furthermore, if all of these activation energies were, in fact, associated with assistance by low-frequency acoustic modes of the polystyrene matrix, one would expect them to be similar. Instead one finds activation energies that vary considerably with the type of chromophore. One concludes, therefore, that it is the type of chromophore involved and not the matrix that determines activation energies for triplet trapping at excimer-forming sites. We hasten to note that this conclusion does not contradict that of Frank and Harrah (10) about the rate-controlling process for singlet trapping at excimer-forming sites.

It might be argued that this is an inappropriate comparison because the molecules involved in excimer formation are not covalently bonded to the polystyrene chain. It is found, however, that for polystyrene doped heavily enough so that the chromophore density is similar to the corresponding homopolymer, activation energies for trapping are essentially the same whether the chromophore is covalently bonded to a chain backbone or not. The implication again is that these activation energies are chromophore-dependent. That is, vibrations or rotations associated with the backbone structure or the pendant bonds are not relevant in determining the activation energies. As a result of the preceding considerations, we conclude that bond formation of the excimeric species is the rate-controlling process for reaction 1 and that the corresponding activation energy is associated with the rearrangement of electrons and atoms needed to accomplish this bond formation.

Many studies in the literature relate the intensity of phosphorescence from small molecule dopants present in polymer matrices to the temperature of the host polymer (11–14). In most of this work the augmentation of the diffusion rate of a quencher (usually oxygen) was related to the onset of a second-order transition of the polymer host. These studies led us to consider the possibility that the same mechanism might be responsible for transitions from nonexcimeric to excimeric phosphorescence. Our eventual conclusion was that the idea is untenable because the excimer–nonexcimer transition temperatures are a function of the dopant concentration and occur at similar temperatures in different polymer hosts. In addition, it was found that activation energies for triplet excimer formation are the same in polystyrene and in poly(methyl methacrylate).

Activation Energies for Detrapping and Their Relation to Polymer Structure

Let us first consider the data in Table I associated with the carbazole group: PVCA, PFCZ (poly[N-((vinyloxy)carbonyl)carbazole]), PACZ (poly[N-((allyloxy)carbonyl)carbazole]), MFCZ, and NEC. A striking feature of the data in Table I is the very close similarity of the activation energies for excimer formation varying between 1.1 and 2.6 kJ/mol. This is particularly interesting because three of these samples are homopolymers and two are molecularly doped polystyrene. Apparently the activation energy values, E_{act}, for trapping are independent of whether the chromophores are polymer-bound or independent molecules. On the other hand, studies of the activation energies for detrapping in these carbazole systems clearly indicate that some important distinctions are present. The deep trap of PVCA must be excluded from these comparisons because it is the only system for which evidence of dual excimer behavior is found. The detrapping activation energy of the shallow trap PVCA and of PFCZ are similar and not too different from the trapping activation energy. The activation energies for detrapping in PACZ and polystyrene films doped with NEC and MFCZ are much larger than the other samples.

The chromophore groups of the three homopolymers PVCA, PFCZ, and PACZ are attached differently to the backbone carbon atoms. In PVCA, of course, the carbazole nitrogen is only one bond removed from the backbone. The polymer that we call PFCZ is derived from poly(vinyl chloroformate) with the carbazole group replacing the chlorine atom. Thus, the carbazole nitrogen is removed from the backbone carbon by three bonds. In PACZ the allyl group replaces the vinyl group which results in four intervening bonds between the backbone carbon and the carbazole nitrogen. The nonexcimeric phosphorescence spectra of these three polymers are essentially the same, and we conclude that the different bonding modes do not appreciably affect the energetics of their respective triplet states. The differences in excimer detrapping behavior are therefore assignable to differences in the steric freedom of the carbazole groups. It appears that the commonality between PACZ and the two small molecule samples is steric freedom of the carbazole chromophore. Thus we conclude that for homopolymers and doped films containing the carbazole group, the activation energies for trapping are essentially the same, but the activation energies for detrapping depend on the presence or absence of steric crowding associated with the chromophore pair. In Table I there are additional results obtained with the triphenyl amino chromophore. In this case different levels of doping of the polystyrene matrix were used, which led to additional information about the interaction of excimer formation and structural properties.

Detrapping in Strained versus Nonstrained Excimers

The essential chromophore in DAPM is the triphenyl amino group, but it differs from triphenyl amine itself in that one of the phenyl rings is attached via a methylene group to a carboxylate center. Thus the nonexcimeric phosphorescence spectra of DAPM and TPA are essentially identical; the only difference between the two chromophores is the bulky group attached to one of the phenyl rings in DAPM. PDAPM is, of course, the homopolymer, corresponding to the monomeric model DAPM.

We notice that the activation energy for trapping in DAPM decreases with increasing dopant concentration. In the limit of 12% doping the activation energy for trapping is very similar to that of the homopolymer itself. We also notice, that the sums of the activation energies for trapping and detrapping in DAPM are nearly the same for each level of doping. The actual values are 8.6, 8.5, and 8.2 kJ/mol. The data suggest that whatever is causing the increase in activation energy as dopant level decreases may also be responsible for the corresponding decrease in activation energy of detrapping. As the dopant level increases, an increasing fraction of chromophore pairs, suitably oriented for excimer formation, is expected. Thus it is proposed that a relatively large fraction of excimer formed at high dopant concentrations are in a fully relaxed configuration. On the other hand, as the dopant level decreases, fewer chromophore pairs possess the optimum configuration for excimer formation and an increasing fraction of the resulting excimers are formed in an unrelaxed or strained configuration. The activation energy for detrapping these strained excimeric species will be less than that for the fully relaxed species and this corresponds to the decreasing activation energy for detrapping as dopant concentration goes down. An interesting corollary to this hypothesis is that the excimer lifetime must be less than the time required for a strained excimer to relax. Let us now apply this model to TPA to see if a similar hypothesis can rationalize those results.

The activation energies for trapping in PS films doped with TPA decrease with increasing dopant concentration, which is the same as the trend found for DAPM. However, the activation energies for detrapping also decrease with increasing dopant concentration; this is the reverse of the trend found for DAPM. As before, let us assume that excimers formed in more concentrated matrices tend to be less strained than those formed in dilute matrices. Because microscopic reversibility requires that the path followed for the trapping process be the same as that for detrapping, the data suggest that both strained and unstrained excimers have sufficient time to relax before a detrapping event can take place. Thus, the difference between TPA and DAPM is that in the TPA essentially all excimers detrap starting from the same or similar energy level, whereas in the DAPM strained excimers detrap starting from a higher energy level than unstrained excimers.

Summary

The early observations of phosphorescence spectra applied to polymeric systems have shown that nonexcimeric spectra from polymer-bound chromophores are very similar to spectra from chromophores that are unbound. Starting from this rather unremarkable observation, subsequent work demonstrated that the triplet states in polymeric systems are actually rather mobile and that excimeric triplets are commonly found in spectra of solid polymer films. These key observations have led to the unfolding of a rich tapestry of photophysical activity in the triplet manifold of polymeric systems. The work described here shows how these early seminal ideas can be extended further by investigating effects of changing temperature on rates of triplet state decay and on spectral band shapes. It is hoped that these results may eventually be useful in deducing structural aspects of polymers conducive to the formation of triplet excimers and, perhaps, lead to information about the structure of the excimers themselves. The results seem to suggest that there is a close relationship between the thermal properties of triplet excimers and the structure of the chromophores involved in excimer formation. It now seems likely that studies of the reversible excimer–nonexcimer equilibrium will provide valuable information about polymer microstructure. The use of temperature extremes to derive this information is highly recommended.

Acknowledgments

This work was sponsored by the U.S. Department of Energy under grant FG08–84ER45107. We are grateful to Dr. Sylvie Boileau for samples of MFCZ, PFCZ, and PACZ and to Professor Yasuhiko Shirota for samples of DAPM and PDAPM.

References

1. Burkhart, R. D.; Caldwell, N. J. *Photophysics of Polymers*; Hoyle, C. E.; Torkelson, J. M., Eds.; ACS Symposium Series 358; American Chemical Society: Washington, DC, 1987; p 242.
2. Vala, M. T.; Haebig, J.; Rice, S. A. *J. Chem. Phys.* **1965**, *43*, 886.
3. Klöpffer, W.; Fischer, D. *J. Polym. Sci., Polym Symp.* **1973**, *40*, 43.
4. Rippen, G.; Kaufmann, G.; Klöpffer, W. *Chem. Phys.* **1980**, *52*, 165.
5. Burkhart, R. D.; Chakraborty, D. K. *J. Phys. Chem.* **1990**, *94*, 4143.
6. Cozzens, R. F.; Fox, R. B. *J. Chem. Phys.* **1969**, *50*, 1532.
7. Burkhart, R. D.; Haggquist, G. W.; Webber, S. E. *Macromolecules* **1987**, *20*, 3012.
8. Ito, S.; Katayama, H.; Yamamoto, M. *Macromolecules* **1988**, *21*, 2456.
9. Morgan, J. R.; El-Sayed, M. A. *J. Phys. Chem.* **1983**, *87*, 2178 and earlier papers cited therein.
10. Frank, C. W.; Harrah, L. A. *J. Chem. Phys.* **1974**, *61*, 1526.
11. Burkhart, R. D.; Jhon, N. I.; Boileau, S. *Macromolecules* **1991**, *24*, 6310.
12. Burkhart, R. D.; Chakraborty, D. K.; Shirota, Y. *Macromolecules*, **1991**, *24*, 1511.
13. Burkhart, R. D.; Jhon, N. I. *J. Phys. Chem.* **1991**, *95*, 7189.
14. Guillet, J. *Polymer Photophysics and Photochemistry*; Cambridge University Press: Cambridge, England, 1985; pp 199–212.

RECEIVED for review July 15, 1991. ACCEPTED revised manuscript May 7, 1992.

Relations between Luminescence Emission and Physical Properties of γ-Irradiated Polypropylene

G. David Mendenhall[1], Ning Xu[1], and Stephen E. Amos[2]

[1]Department of Chemistry, Michigan Technological University, Houghton, MI 49931
[2]Himont Research and Development Center, 800 Greenbank Road, Wilmington, DE 19808

Eight formulated samples of isotactic poly(propylene) were γ-irradiated with ^{60}Co to 3 and 5 MR, and subsequent changes in the samples were followed during the next 14 days by spontaneous luminescence emission, color development, and then by tensile measurements. The spontaneous luminescence decayed rapidly at first, and was still measurable in most irradiated samples 5–10 days later. Subsequent heating of irradiated samples to 100 °C led to large increases in luminescence that slowly decreased at this temperature. The elongation at break of irradiated samples, with one exception, dropped after irradiation. The yellowness index of samples after irradiation remained about the same or increased, with one exception. General correlations between luminescence or color on one hand, and elongation at break on the other, could not be found. However, within a given formulation, the easily measured luminescence emission intensity at any time within 4 h after the end of the irradiation was monotonically related to the total radiation dose to the sample.

CHEMILUMINESCENCE FROM POLYMERS was first described by Ashby (*1*) in a published report on the diminished light emission from polypropylene (PP) at 150 °C when it was formulated with stabilizers. Ashby also observed synergistic effects of combinations of stabilizers with the same technique. Polypropylene has been a favored substrate for a number of subsequent investigations of chemiluminescence emission (2–6), in part because of the commercial importance of this polymer and in part because of its relatively strong chemiluminescence emission.

Recent interest in the subject has been connected with a concern for the stability of polypropylene after sterilization with ionizing radiation. In proprietary work, one of us (Mendenhall) correctly ranked eight formulated polypropylene samples toward their ultimate *thermal* stability at 150 °C by comparing their relative chemiluminescence emission at 150 °C for time durations of up to one week. This result inspired application of the technique to the measurement of damage caused by ionizing radiation. Ideally, the technique will provide an early indication of failure because it instantaneously measures the rate of a dynamic process associated with oxidative degradation.

The application of chemiluminescence is rather empirical, and there are a very large number of choices of pretreatment and examination conditions. Our approaches thus far are summarized in the following text.

When polypropylene is heated in the absence of oxygen, chemiluminescence is emitted from the thermal decomposition of accumulated peroxides and hydroperoxides in the samples (7). We heated 1-mg samples of formulated, γ-irradiated polypropylene in an evacuated capillary with an infrared laser (10.6 μm) and measured the visible light emitted from the samples with fiber-optic techniques (7). Linear correlations were found between total light emission measured in this way and the conventional peroxide value of the samples. The proportionality constants, however, were quite dependent on the formulation, a feature which limited the value of this approach.

Chemiluminescence emission from polypropylene is usually negligible from samples at ambient temperature, even from samples doped with free-radical initiators. After exposure to ionizing radiation, however, an easily measured luminescence is observed. Early studies by our group showed a qualitative correlation between the loss of impact strength and the very weak, residual light emission from formulated polypropylene samples a week after exposure to γ-irradiation (all at 25 °C) (8).

We also found that the rate of loss of impact strength of the samples at 60 °C could be correlated qualitatively with their chemiluminescence emission in air at 150 °C. The loss of impact strengths of samples stored at 25 °C was different than those stored at 60 °C, but the ranking according to their chemiluminescence emission at 25 °C was again in qualitative agreement.

These results were encouraging although the number of samples was limited, and the examination of the samples for light emission at 25 °C was only begun a week after the irradiation. Subsequently we studied the more intense light emitted within minutes after the exposure of polypropylene (and other hydrocarbons) to X-rays, and discovered that the color of the light was different than that of light from oxidation reactions carried out at elevated temperatures (9). This surprising fact, along with (1) the observation that the light emission did not depend on the presence of oxygen, and (2) its decay followed a power function ($I = At^{-n}$; I = intensity, t = time, and a, n = constants) made a convincing case that the luminescence was due to charge-recombination processes with the same characteristics observed earlier from a variety of polymeric substrates [10, 11]:

$$\text{polymer} \xrightarrow{h\nu} e^- + [\text{polymer}] + \xrightarrow{\text{slow}} [\text{polymer}]^* \xrightarrow{\text{fast}} h\nu'$$

where e^- = electron and $h\nu$ = photon.

In the present study we have measured the light emission from formulated polypropylene plaques that were irradiated in a conventional ^{60}Co source. Intermittent measurements of the luminescence (presumably from charge recombination) were made at ambient temperatures for several days after irradiation until it was too weak to measure. We then measured luminescence (presumably free-radical-induced oxyluminescence) from these same samples heated in air to 100 °C. Tests for impact strength and color development with irradiated sheets of PP were carried out for comparison.

Experimental Details

Isotactic polypropylene (Himont Pro-fax 6801) with a fractional (0.4) melt flow rate, containing approximately 0.01% of a phenolic processing stabilizer, was combined with calcium stearate and stabilizers by powder blending and extrusion. Extrusion compounding was carried out with a Brabender 3/4-in., single screw, 25:1 length:diameter, laboratory extruder with a 3:1 compression ratio mixing screw and a 3-in. final mixing zone comprised of pins. The extruder was run in air at a flat 230 °C temperature profile. Sample compositions are shown in Table I.

Compounded polymer was injection-molded into 3x3x0.041-in. plaques on a 1.5-oz Battenfeld injection molder in air. The injection molder temperature profile was 215–220–220–235 °C. The mold temperature was 110 °C.

Elongation at break was measured on die-cut tensile bars conforming to ASTM method D638 type IV. The tensile bars were pulled at a rate of 2 in./min.

The injection-molded plaques were irradiated in a 3-in. diameter pipe at the Phoenix Laboratory (University of Michigan, Ann Arbor) with an underwater ^{60}Co source with a dose rate of 1.108 MR/h. The plaques were separated from each other by strips of paper and stacked vertically in the pipe that was placed in the reactor for sufficient times to give doses of 3 and 5 MR.

The yellowness index was measured with a colorimeter (Colorquest) according to ASTM method D1925.

Table I. Compositions of Formulated Polypropylene[a]

Component	Sample No.							
	A	B	C	D	E	F	G	H
Pro-fax 6801 (PP)	100	100	100	100	100	100	100	100
Irganox 1010[b]		0.03	0.03					
Irganox 168[c]		0.10	0.10	0.10				
Tinuvin 770[d]			0.10	0.10	0.10			
Tinuvin 765[e]						0.10		
Tinuvin 770–bisNO[f]							0.095	
Calcium stearate		0.10	0.10	0.10	0.10	0.10	0.10	0.10

[a] Numbers in table refer to relative weight of ingredient in formulated mixture, and are approximately weight percent.
[b] Tetrakis-[methylene (3,5-di-*tert*-butyl)-4-hydroxyhydrocinnamate]methane.
[c] Tri-(2,4-di-*tert*-butyl) phenyl phosphite.
[d] Bis-(2,2,6,6-tetramethyl-4-piperidinyl) sebacate.
[e] Bis-(1,2,2,6,6-pentamethyl-4-piperidinyl) sebacate.
[f] Bis-(1-oxy-2,2,6,6-tetramethyl-4-piperidinyl) sebacate.

Measurement of Light Emission. Luminescence from disks of 1.0-in. diameter, cut from some of the plaques with a cork borer immediately after they were removed from the reactor, was measured within 5 min and again at increasing time intervals thereafter. It was of course not possible to examine all of the irradiated plaques immediately after irradiation. A few disks cut from different plaques with the same formulation, irradiated at the same time showed luminescence intensities that differed by amounts larger than the experimental error. This result suggested that samples may have been subjected to temperature inhomogeneities or difference in oxygen concentrations, although we did not have time to investigate this point further.

The light from the disks was measured after placing them in the sample well of a luminometer (Turner Designs, Inc., model TD-20e). The disks themselves were handled gently with cloth gloves and tweezers and placed in contact only with clean surfaces. During the fast, initial decay after irradiation, the luminescence was measured for 20 s after a 10-s delay. Subsequently the emission was measured with longer integration times, and the luminometer was thermostatted with circulating water at 25.0 °C.

The luminescence from the disks at 100 °C was measured by placing them into a small aluminum cylinder that was threaded to admit one end of a fiber-optic cable. The other end of the cable was inserted into the well of the luminometer. After measurement of (negligible) light from the samples at ambient temperature, the aluminum container, wrapped in a single sheet of polyethylene film, was placed in boiling water. The light emission was measured as consecutive 20-s readings, interrupted by 10-s delay, for the next 20 min.

The sensitivity of the luminometer was checked periodically with a $^{234}Th-^{234}Pa-UO_2^{+2}$ liquid scintillation standard (*12*).

Results

Values of elongation at break and yellowness index for irradiated and nonirradiated tensile bars are given in Table II.

The unstabilized samples (A and H) showed very large drops in strength with irradiation, whereas the other samples showed considerable retention of strength after 3 but not 5 MR. Interestingly the bis-nitroxide, which to our knowledge had not been tested previously, showed considerable protective ability at 3 MR. The yellowness index of the samples changed with irradiation, and showed the greatest increase with samples with phenolic stabilizer. The sample with the orange bis-nitroxide (G) showed much greater color initially, and it was the only sample whose index decreased with irradiation.

The initial luminescence readings from irradiated plaques are shown in Figure 1. The light emission decayed rapidly in the initial stages according to a power law, and was analyzed by least squares fit to a log–log plot. Because it is difficult to compare the samples in the initial stages when the emission is changing rapidly, we used the fitted parameters to calculate the emission intensity 1 h after the end of the irradiation of each sample for the purposes of comparisons.

Luminescence values at extended times appear in Table III. For the most part, the luminescence decays in this table are monotonic. Because the samples had been transported by plane and car, and were subsequently stored under ambient conditions, a more exact analysis of the decay curves is not warranted. Sample G (bis-nitroxide) shows irregular luminescence values,

Table II. Tensile and Color Measurements on Irradiated PP[a]

	Sample							
Dose (MR)	A	B	C	D	E	F	G	H
	Elongation at break (%)							
0	1359	1260	1596	1244	1372	1399	1228	1881
3	8	955	1250	1380	1184	1262	995	12
5	6	20	6	13	373	9	108	0
	Yellow Index[b]							
0	4.0	4.7	5.1	4.0	4.2	4.0	8.7	4.2
3	4.6	6.5	7.7	4.7	4.0	4.0	6.9	4.2
5	4.5	7.7	9.4	5.4	4.3	4.0	6.5	4.3

[a] Average of three bars, 13 days after irradiation. Samples were die-cut so that the long axis was in the flow direction of the polymer.
[b] ASTM D1925, 14 days after irradiation.

which were later traced to stimulation of delayed luminescence by fluorescent room lights.

For all samples, the recombination luminescence during the first 4 h after irradiation was higher for a 5- than for a 3-MR dose.

Except for samples B and H, the luminescence one day after irradiation was much higher for a dose of 5 than 3 MR. For later times the pattern was less distinct.

Ten days after the irradiation, the luminescence of nearly all samples had declined to very low levels, and additional information was then obtained by heating the samples to 100 °C. Plots of the resulting emission, ascribed to chemiluminescence, versus time were similar in shape for all irradiated samples. Representative plots are given in Figure 2. A plot of the maximum chemiluminescence versus dose is given in Figure 3 and shows a monotonic increase for all samples.

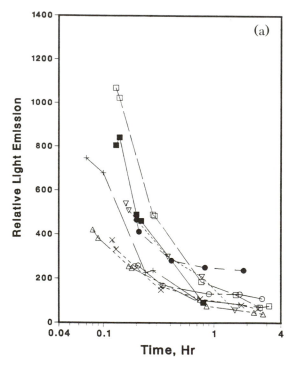

Figure 1. Initial luminescence decay curves from irradiated polypropylene samples. (a) 3 MR: open square, C; filled square, A; cross, F; inverted triangle, H; ×, E; triangle, B; open circle, D; closed circle, G. (b). 5 MR: filled square, A; open square, B; ×, E; open circle C; inverted triangle, G; filled circle, D; cross, F; triangle, H.

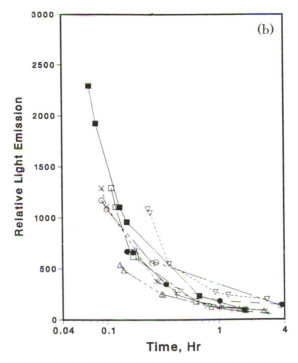

Figure 1. Continued

Empirical plots of calculated light emission 1 h after irradiation, the value of n in the power function, the maximum light emission at 100 °C, or yellowness versus percent of original elongation did not reveal any correlations, even when we restricted the data to samples with the lower radiation dose.

Discussion

The current study bears out some of the conclusions of our earlier studies with fewer samples. We expect, to a first approximation, that greater light emission from a sample will correspond to lesser stability. The charge-recombination luminescence that can be measured immediately after radiation treatment (Figure 1) or several day later (Table III), however, does not correlate with the relative loss of tensile properties. This result is in contrast to our earlier observations, which were made after irradiation of samples with different formulations (8). Fluorescence measurements might be a way to normalize the emission, but preliminary attempts to do this have not been promising.

Table III. Luminescence from Irradiated Polypropylene at Extended Times[a]

Time[b] (h)	Sample							
	A	B	C	D	E	F	G	H
			3 MR Irradiation					
21–26	20	31	53	47	33	34	72	34
47–49					24	20	79	21
55	16	14						
72–78	13	14	22	25	15	20	54	15
90–92			26	25	26	24	72	17
220–40	2	−1	−2	1	−2	−2	4	2
			5 MR Irradiation					
21–24	33	30	71	101	61	63	177	31
41–49			61	76	39	22	101	20
53–55	12	17						
72–78	22	35	43	61	17	16	55	6
90–97	36	44	18	16	49	2		
220–240	4	3	4	12	5	0	11	6

[a] Mean of three measurements with Turner luminometer. Values are last three digits on most sensitive setting. Average percent error is 3 units.
[b] Time or range of times of measurements from end of irradiation period.

Color development in irradiated samples is an undesirable feature from a practical standpoint. The irradiation of unstabilized sample A resulted in a small change of color index, which was completely suppressed by addition of calcium stearate (sample H). The larger changes of color index in some of the other samples after irradiation must involve reactions of additives. The large yellowness number of sample G and its large drop after irradiation is undoubtedly associated with disappearance of the orange bis-nitroxide. Although the luminescence from the irradiated sample with bis-nitroxide was higher because of the effect of room lights, the effect was too small to account for the mysteriously elevated levels of luminescence compared with the other samples.

Correlations aside, the bis-nitroxide was an effective stabilizer, second only to the parent amine (sample E) at the 5-MR dose (Table II). We should point out, however, that the formulated samples do not contain the same molal concentrations of stabilizers.

The only positive result of the study is the monotonic relationship between luminescence and radiation doses (Figures 1 and 3). Similar relationships were found with other samples of formulated polypropylene that were irradiated to very low levels and then made to emit light by laser heating (7). For most samples it appears that chemiluminescence under the conditions for the data in Figure 3 will distinguish between dose levels differing by more than about 0.4 MR. Moreover, the reproducability of the chemi-

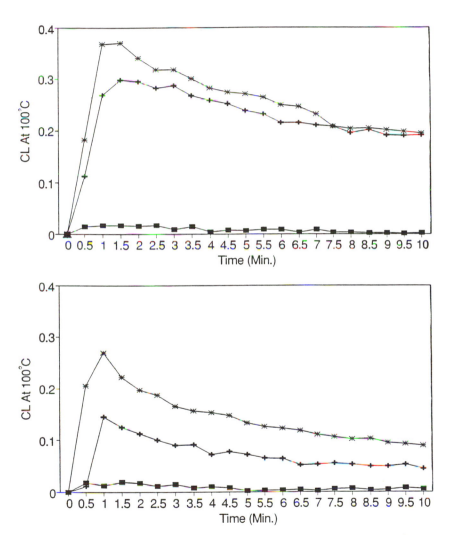

Figure 2. Curves of luminescence (CL) vs. time from polypropylene samples at 100 °C. Top, sample A; bottom, sample C. The intensities increase in each case in the order 0 (rectangle), 3 (cross), and 5 MR (asterisk).

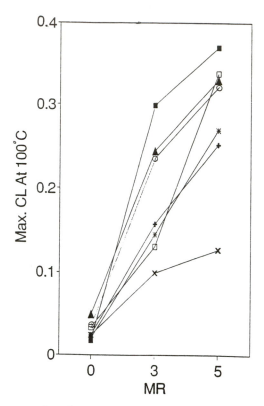

Figure 3. Maximum chemiluminescence at 100 °C vs. dose. Filled squares, A; open squares, D; filled triangles, G; open circles, H; asterisk, C; cross, B; ×, F.

luminescence measurements and their dependence on formulation are advantageous for quality control.

Acknowledgment

This work was supported by a grant from Himont USA, Inc. We thank Robert Blackburn (Phoenix Laboratory) and Carlos Saborio (Himont USA) for technical assistance, and Larry C. Thompson (University of Minnesota at Duluth) for some fluorescence spectra.

References

1. Ashby, G. E. *J. Polym. Sci.* **1961**, *50*, 99.
2. Schard, M. P. *Polym. Eng. Sci.* **1965**, *11*, 246 and previous papers in this series.
3. Barker, R. E., Jr.; Daane, J. H.; Rentzepis, P. M. *J. Polym. Sci.* **1965**, *A3*, 2033.
4. Rychly, J.; Matislova-Rychla, L.; Spilda, I. *Eur. Polym. J.* **1979**, *6*, 565.

5. Flaherty, K. R.; Lee, W. M.; Mendenhall, G. D. *J. Polym. Sci., Polym. Lett. Ed.* **1984**, *22*, 665–667.
6. Billingham, N. C.; George, G. A. *J. Polym. Sci., Polym. Phys. Ed.* **1990**, *28*(3), 257–265.
7. Mendenhall, G. D.; Cooke, J. M.; Byun, H. In *Advances in Polyolefins*; Seymour, R. B.; Cheng, T., Eds.; Plenum: New York, 1987; pp 405–435.
8. Mendenhall, G. D.; Agarwal, H. K.; Cooke, J. M.; Dziemianowicz, T. S. In *Polymer Stabilization and Degradation*; Klemchuk, P. P., Ed.; ACS Symposium Series 280; American Chemical Society: Washington, DC, 1985; pp 373–386.
9. Hu, X.; Mendenhall, G. D.; Becker, R. *Polym. Prepr. Am. Chem. Soc., Div. Polym. Chem.* **1990**, *31*, 323.
10. Mendenhall, G. D.; Agarwal, H. K. *J. Appl. Polym. Sci.* **1987**, *33*, 1259.
11. Guo, X.; Mendenhall, G. D. *Chem. Phys. Lett.* **1988**, *152*, 146.
12. Mendenhall, G. D.; Hu, X. *J. Photochem. Photobiol. A* **1990**, *52*, 285–302.

RECEIVED for review July 15, 1991. ACCEPTED revised manuscript August 3, 1992.

Investigation of Polycarbonate Degradation by Fluorescence Spectroscopy and Its Impact on Final Performance

Makarand H. Chipalkatti and Joseph J. Laski

GTE Laboratories Incorporated, 40 Sylvan Road, Waltham, MA 02254

A novel use of fluorescence spectroscopy as a sensitive, rapid, and nondestructive technique for the study of degradation in solid-state polycarbonate is described. The technique is especially suited for the study of degradation in aromatic polymers such as polycarbonate, but in principle it should have wider application. A comparison is made of emission and excitation spectra of bisphenol A polycarbonate with spectra of model compounds selected to represent previously established degradation products. The reported thermal and UV degradation products for polycarbonate in the presence and absence of air, respectively, are verified by a comparison of the emission and excitation peaks of the model compounds and the degraded polymer. The progress of thermal degradation as a function of processing temperature has been tracked by measuring the emission peak intensities of the degradation products and then correlating them with bulk performance measures of the polymer. The technique described is intended to serve as a rapid method of identifying the presence of known or proposed degradative pathways rather than as a substitute for more rigorous techniques of chemical and physical analysis. Potential applications of this technique as an aid in polymer processing as well as its extension to UV degradation studies and relationship with optical performance are also discussed.

THE INTERACTION OF THE STRUCTURE, MORPHOLOGY, AND PROCESSING conditions on the final properties of polymers is very critical in determining

0065-2393/93/0236-0623$06.00/0

the performance and reliability of plastic parts over extended periods of time. This interaction is especially important in products for high-precision applications where very narrow performance tolerances are specified. For instance, because of their unique status as transparent engineering plastics, polycarbonates, especially polycarbonate of bisphenol A, are in great demand as window or lens materials. In such applications, even small amounts of degradation induced during processing affect the mechanical performance of thin-walled molded parts as well as clarity and visual appeal.

In this chapter, the use of fluorescence spectroscopy to study the extent of UV and thermal degradation of polycarbonate is discussed. Like many other aromatic polymers, polycarbonate and its major degradation products have chromophores that absorb and emit electromagnetic radiation in a range conveniently addressed by commercial spectrophotometers. Our approach relies on the vast body of prior work that elucidates the degradation pathways and products of bisphenol A polycarbonate (1–3) by independent means. The emission and excitation spectra due to model compounds of known polycarbonate degradation products have been used as "signatures" of the emission spectra of key degradation products in the solid-state polymer. The intensity of emission is recorded as a function of exposure and is correlated with other macroscopic structure–property attributes, such as the dynamic modulus and viscosity and molecular weight distributions. Owing to its remarkable sensitivity to small traces of compounds and the influence of the molecular environment, UV fluorescence spectroscopy is ideal for monitoring subtle degradation reactions. Although aromatic polymers are most suitable for this analytical approach, other polymers with degradation products displaying significant fluorescence could also be investigated by similar means.

The value of the technique described here lies in its ability to monitor known degradation products. Like other methods of spectroscopy that deal with electronic transitions of a material, it is not as well suited to the elucidation of unknown chemical reactions as is vibrational spectroscopy. However, if the key degradation products are known and can be represented by model compounds, our technique can be a rapid and sensitive method of tracking the degradation. For example, in situations where a material has been exposed to undefined UV–thermal conditions, the technique can be used to determine if degradation is the result of UV or thermal degradation, or if both conditions are somewhat responsible. Similar resolution is not feasible in such a straightforward manner with traditional optical transmission measurements or mechanical analyses. Furthermore, the sensitivity of this technique and its ability to detect trace materials have a significant advantage over conventional methods. In a process control loop, the response time for corrective action to minimize degradation is dramatically reduced.

A brief review of the approach and experimental techniques of previous workers and representative examples are cited here to differentiate our approach. The study of degradation products and pathways for organic molecules in general and polymers in particular is often based on the use of infrared spectroscopy of samples exposed to UV radiation or thermal degradation. The classical approach relies on the complete alkaline hydrolysis of the degraded polymer and the chromatographic separation of the various products, followed by infrared spectroscopy (4, 5). Other techniques that involve luminescence phenomena in polymers for the purpose of analysis or monitoring chemical change (e.g., chemiluminescence due to photooxidation) have also been discussed in the literature (6). More recently, solution electronic spectra of polycarbonate before and after UV degradation were reported in an effort to elucidate the mechanisms and products of degradation (7). In this article Gupta et al. used gel permeation chromatography to follow chain-scission reactions, and exposure to UV was monitored via actinometric measurements of *o*-nitrobenzaldehyde rearrangement. Fluorescence data of various concentrations of polycarbonate solution as well as thin films were reported to demonstrate the effect of quenching by degradation products.

Model compounds have been used previously for comparison of solution UV absorption spectra with those of the degradation products (8) of flash photolysis of polycarbonate solution in 1,2-dichloroethane. Molecular weight distributions were monitored by gel permeation chromatography and intrinsic viscosity measurements. Thus, the ratio of rearrangement reactions to chain scission were compared. In yet another study (9), light-scattering of solutions of polycarbonate in chloroform was performed before and after thermal degradation to determine the weight average molecular weight. Photodecomposition was both induced and studied by irradiation of high- and low-density polyethylenes, polyvinyl chloride, polypropylene, and polycarbonate, with a 351.3-nm excimer laser (10). Laser-induced emission spectra as well as time-dependent quenching were used to assess the progress of degradation in the materials just named.

Although a variety of techniques, including UV absorption of polymers, have been applied to the study of degradation of polymers in general and polycarbonates in specific, our approach is relatively novel in that it is based on the measurement of UV fluorescence of the degradation products. Very little prior work presents (as we do) polycarbonate emission spectra and the evolution of its fluorescence spectra as a function of thermal and UV degradation. Our approach, therefore, provides a novel and sensitive method for the analysis of solid-state samples where a rapid nondestructive approach is desirable. This technique is particularly useful because it involves the analysis of degradation products that remain embedded in the matrix polymer without any need for the separation techniques discussed earlier. Furthermore, emission spectroscopy is conducted off the surface of the sample (hence minimal sample preparation is required) and is essentially nondestructive. Convenient correlation with the effect of processing history and final performance allows this approach to be used as a process control tool.

Processing of polymers often results in some amount of thermal- or shear-induced degradation (11). Such degradation may occur in the compounding stage or in injection molding, where the action of the extruder screw or the effect of pressure may result in mechanochemistry- or pressure-induced "depolymerization". The processing temperature significantly affects both the properties and structure of the polymers (12, 13) and results in thermal degradation via random chain scission, cross-linking, or oxidative degradation. Optical methods of processing, such as UV cure of adhesives and lithography, increasingly are being used. These treatments may also result in exposure of the substrate resin to UV and heat simultaneously or separately, and consequently, may affect both performance and reliability of polymer parts. The degree of degradation often may be too subtle to detect by conventional means such as infrared spectroscopy and yet be sufficient to affect long-term properties. In such cases fluorescence spectroscopy is ideally a rapid and extremely sensitive alternative once a thorough audit and labeling of spectral features of the polymer and its degradation products has been made.

Experimental Notes

All the instrumentation and compounds used are of the standard commercial variety, so only a few brief comments are made here about the experimental aspects of this work.

For the yellowness index measurements, the samples were injection-molded plaques of BPA–PC (Mobay Makrolon 2600–1000). With the exception of diphenyl carbonate of bisphenol A, which was synthesized and purified at GTE Laboratories as described by Pochan et al. (14), all the model compounds and the polymers were used as received from Aldrich, with no further purification.

All reported solution spectra were generated from as-prepared solutions without any attempt to exclude oxygen, and all solvents are of spectrophotometric grade used as received from Aldrich. Because our objective was not the identification of unknown compounds, but rather the tracking of emissions from known compounds, de-airing was not deemed essential.

All the reported spectra were corrected for instrument response (e.g., light-source intensity, grating characteristics, etc.) so that signal intensities at different wavelengths could be compared relative to each other. Although this renders the data suitable for quantitative measurements, no quantification was attempted here. Solid samples used for spectroscopic analysis were generally wiped with isopropyl alcohol once prior to analysis.

The emission and excitation spectroscopy were performed on a spectrofluorometer (SLM 8000C) with a 450-W xenon light source (wavelength range 200–900 nm) and a computer controller (IBM PC XT).

The gel permeation chromatography was carried out on a chromatography instrument (Millipore Maxima 820) with THF as the carrier solvent. The creep and complex viscosity measurements were made on a dynamic mechanical analyzer (DMA; Dupont 983).

Yellowness index calculations were based on visible transmission spectroscopy carried out on a spectrophotometer (Cary 17D). The transmission data were then converted to yellowness index values in accordance with ASTM D 1925–70.

Results and Discussion

Effect of Processing History.   Our study is based on comparisons of the emission and excitation spectra of pristine and degraded polymer with the spectra of suitable model compounds. A model compound is considered suitable if its emission and excitation spectra differ only slightly from what would be expected if it were incorporated into the polymer. This condition occurs if the only additional bonds required for incorporation into the polymer main chain are σ bonds, which only slightly perturb the π electron configuration in the relevant chromophoric species. In such a case, the spectra of the model compound are qualitatively identical to those of the degradation products embedded in the polymer, with peaks sometimes shifted by a few nanometers.

In the case of bisphenol A polycarbonate (BPA–PC), the model compound, diphenyl carbonate of BPA (DPC), was found to be a suitable model of BPA–PC electronic and molecular structure. The DPC as synthesized had to be substantially purified by several washings prior to preparation of a solution with no trace impurities that would be evident in the absorption or emission spectra. The validity of this model selection is clearly demonstrated in the solution emission and absorption spectra of Figure 1. The spectra for the polymer and model compound are very similar and they both share the same degree of symmetry between their respective emission and absorption spectra, which demonstrates a high degree of correspondence.

The identical principle also has been applied in the selection of model compounds for the various thermal and UV degradation by-products that are specifically discussed when appropriate during the course of this report.

In the transition from a solution sample to the solid state, a comparison is made between the model compound and the commercial BPA–PC sample. The two emission spectra (excitation at 280 nm) are compared in Figure 2. The similarities of the spectra of Figure 2 are quite remarkable; specifically, the match between the main "baseline" emission spectra of the polymer and the model compound. The additional peaks superimposed on the polymer emission at 330, 347, and 365 nm, respectively, are also notable.

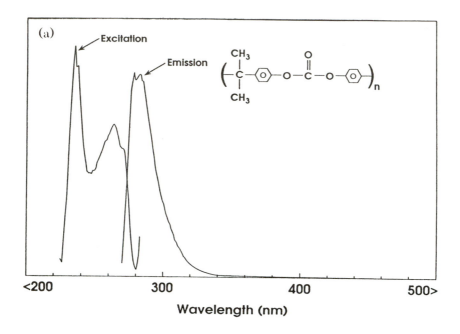

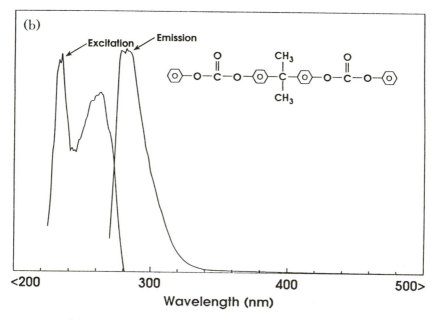

Figure 1. (a) The emission and excitation spectra of a dilute solution of BPA–PC in methylene chloride. (b) The emission and excitation spectra of the model compound diphenyl carbonate of BPA solution in methylene chloride. The molecular structure of the polymer repeat unit and the model compound are shown for convenient comparison of their active chromophoric groups.

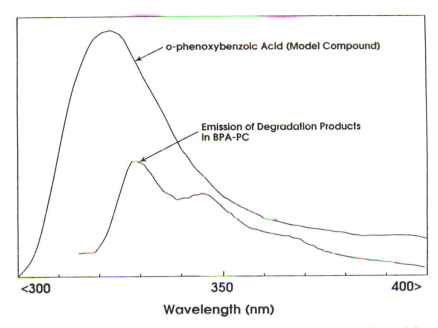

Figure 2. Comparison of the emission spectra of the degradation products of the as-molded BPA polycarbonate sample and the corresponding model compound, diphenyl carbonate of bisphenol A.

To identify the source of the "extra" emission peaks seen in the as-prepared sample of BPA–PCA, a careful review of the sample preparation procedure was made. Comparison of the emission spectra at each stage of the procedure and attempts to establish the point at which the fluorescent by-products were generated yielded the conclusion that the "extra" emission peaks were due to trace thermal degradation products present in the as-received samples.

With a view to further clarification of the nature of the emission peaks at 300, 347, and 365 nm, a series of samples was prepared by a process of vacuum-heating for 30 min followed by compression molding for 15 min at 200, 240, 260, 280, and 300 °C. The emission spectra of the five samples exposed to the thermal treatments and probed at an excitation wavelength of 314 nm are shown in Figure 3. All the samples were exposed to identical spectroscopic parameters and were superimposed on the same axes so that spectral intensities could be compared. A direct correlation is observed between the processing temperatures and the intensity of the emissions, which reinforces the view that the thermal degradation products are the source of the emissions and that fluorescence spectroscopy is a viable means of monitoring the extent of degradation once the "signature" emissions are clearly identified. It is interesting to note that commensurate increases in the

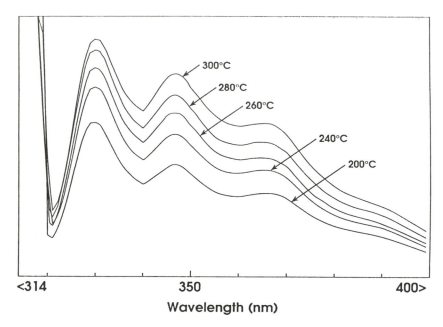

Figure 3. Emission spectra of as-molded BPA polycarbonate as a function of the processing temperature. Excitation wavelength used was 314 nm.

intensity of the peaks were observed whether the samples were heated in air or vacuum.

The relationship between the emission intensities and the progress of degradation is further underscored by examination of the molecular weight distribution of the molded samples listed earlier. The gel permeation chromatography data in Figure 4 demonstrate a progressive reduction in the average molecular weight of the polycarbonate samples as a function of processing temperature. In fact, the data lend credence to the mechanism of thermal degradation that proceeds through a process of random chain scission that leads to the formation of a phenyl *o*-phenoxybenzoate product. This mechanism is substantiated in the next section.

Thermal Degradation Schemes.   The generally accepted thermal degradation pathways for bisphenol A polycarbonate have been discussed in the literature in some detail (*15, 16*). A number of alternative mechanisms that may dominate or be absent, dependent on the specific thermal and environmental conditions (e.g., in air or nitrogen, presence of moisture), have been suggested. The thermal degradation of bisphenol A (BPA polycarbonate as proposed by Davis and Golden (*15*) occurs via a rearrangement of the BPA carbonate at the ester linkage to form an *o*-phenoxybenzoic acid. In a modification proposed by Bailly et al. (*16*), the possible involvement of a

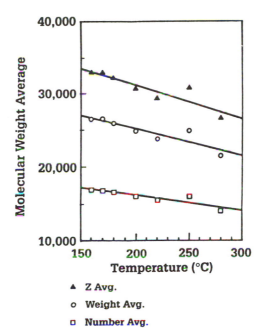

Figure 4. Molecular weight distribution of BPA polycarbonate samples as a function of temperature as seen by gel permeation chromatography. The z, weight, and number averages of molecular weight are shown as they influence different bulk properties of the polymer.

phenyl salicylate as an intermediate product was described. The final product, via a relatively rapid cross-linking mechanism, was shown to be phenyl *o*-phenoxybenzoate. This reaction occurs both in air and in vacuum. In addition, extensive chain scission was reported when volatile products with hydroxyl groups (such as low molecular weight phenols and water) are allowed to remain in the system (*17*). The generally accepted degradation sequences reported in the literature referred to earlier are summarized in Scheme I.

An inspection of the emission and excitation spectra of the degradation products listed in Scheme I suggests that the emission peaks of the degradation products are attributable mainly to phenyl *o*-phenoxybenzoate. In fact, a comparison of the excitation and emission spectra of the degradation products and *o*-phenoxybenzoic acid (model compound) shows extreme similarity (Figures 5 and 6). As stated earlier, small differences in the main spectral peak locations are believed to result from the attachment of the molecule to the polymer chain. Additional peaks at 347 and 365 nm are not identified yet (*18*). Based on the spectra of Figures 5 and 6, *o*-phenoxybenzoic acid is considered sufficiently similar to represent the benzoate product and is

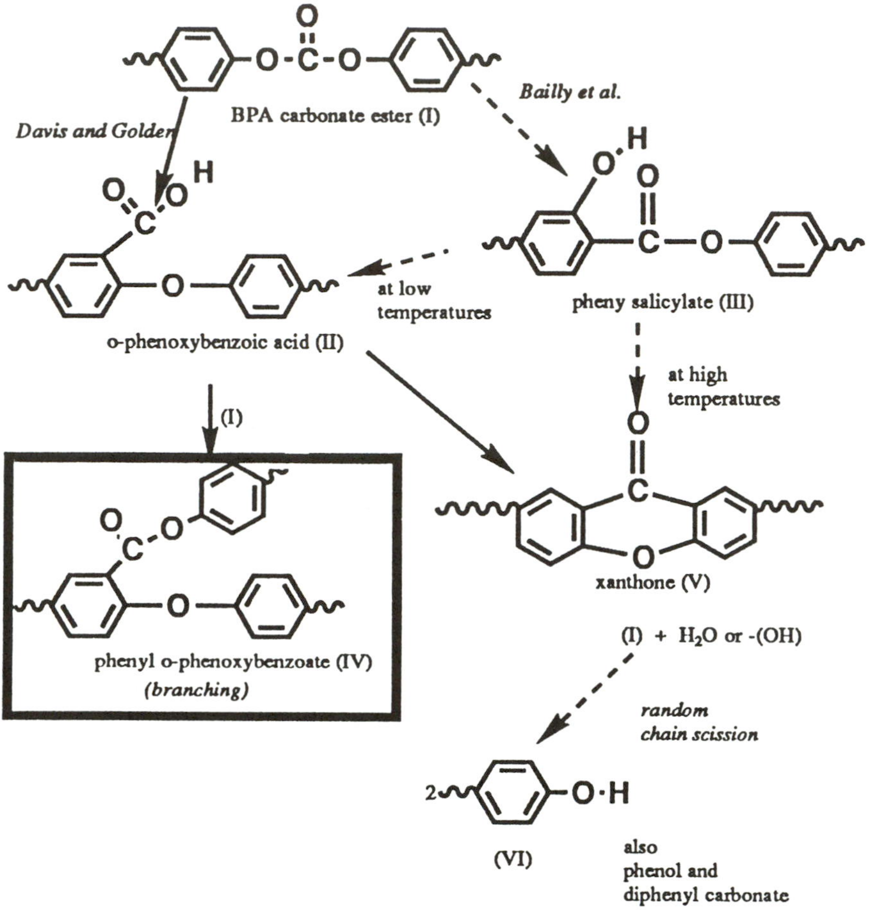

Scheme I. Review of reported BPA polycarbonate thermal degradation product sequences. Based on spectroscopic comparisons of reported and degradation products and thermally degraded polycarbonates, the phenyl o-phenoxybenzoate group was found to be reponsible for the "extra" emissions observed in thermally degraded BPA–PC.

treated as the model compound for the main thermal degradation product. Other potential degradation products, such as phenyl salicylate and xanthone, were also sought, but no spectral match was found in the degraded polymer.

Clearly, once the key degradation product of the polymer is identified and related to the corresponding model compound, the progress of degradation can be monitored quite conveniently. As shown in Figure 3, the emission intensity of the degradation product, in this case represented by the emission at 347 nm, is commensurate with the extent of degradation. Specifically, the concentration of phenyl o-phenoxybenzoate via the degradation pathway

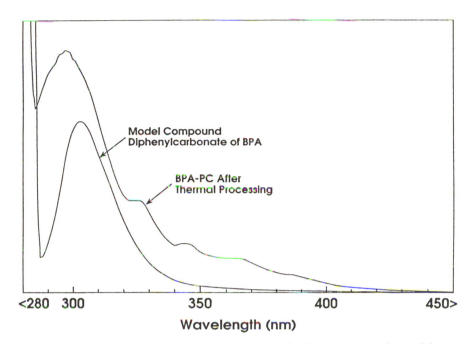

Figure 5. Comparison of the emission spectra of the degradation products of the as-molded BPA polycarbonate sample and the corresponding model compound, o-phenoxybenzoic acid.

(Davis and Golden mechanism or via phenyl salicylate route) increases with processing temperature. The intensity of emission is then correlated with significant bulk properties of interest as will be shown in a later section.

UV Degradation Schemes. The degradation mechanism of poly-carbonate due to UV irradiation at wavelengths greater than 250 nm also involves a rearrangement of the ester linkage, but results in a different product. Once again, our approach relies on the preceding literature for an enunciation of potential degradation products. The mechanism of UV degradation of polycarbonates in an inert atmosphere is quite well accepted in the literature. In the absence of oxygen, the reaction proceeds by the well-known Fries photorearrangement reaction (*1, 2*). The end product is 2,2′-dihydroxy-benzophenone via a phenyl salicylate intermediate (*4, 7, 8*). In fact, the dihydroxybenzophenone product produces the yellow coloration of UV-degraded polycarbonate. The UV degradation steps for polycarbonate are summarized in Scheme II. In the case of UV exposure in the presence of air, the reactions are a complex series of oxidative reactions, which are discussed in some detail in the prior literature (*19*). Reportedly, UV exposure in air involves a greater loss of carbonate groups than exposure in inert atmosphere

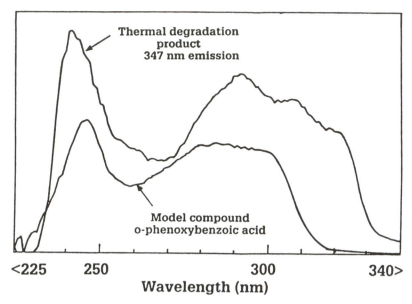

Figure 6. Comparison of the excitation spectra of the degradation products of the as-molded BPA polycarbonate sample and the corresponding model compound, o-phenoxybenzoic acid.

(and vacuum). Factor and Chu (*19*) also reported a surface layer that is a highly cross-linked dark layer with concentrations of hydroxyl, aromatic acid, and aromatic ketone groups higher than in the bulk.

Polycarbonates are extensively used as lens materials and as optical components because of their high transparency and good thermal properties. In such applications, the optical characteristics, rather than the mechanical properties, are measures of performance. Specifically, the visible transmission spectrum of lens or window materials and the yellowness due to degradation products is represented by a single weighted parameter (the yellowness index). This performance measure and some representative data are discussed briefly in the final section of this chapter.

For the study of UV degradation of polycarbonates, the 2,2'-dihydroxybenzophenone group was found to be representative of the main degradation product of the polymer, as is expected by examination of Scheme II. Once again, in the model compound selected (2,2'-dihydroxybenzophenone), the electronic configuration of the chromophoric species differs only slightly from that in the polymer. Spectra of other reported degradation products were also compared, but no spectral match was made. In Figure 7, the as-received BPA–PC emission spectra are compared with the spectra of the same sample after 16-h exposure to high-intensity UV radiation (λ = 254 nm) in nitrogen

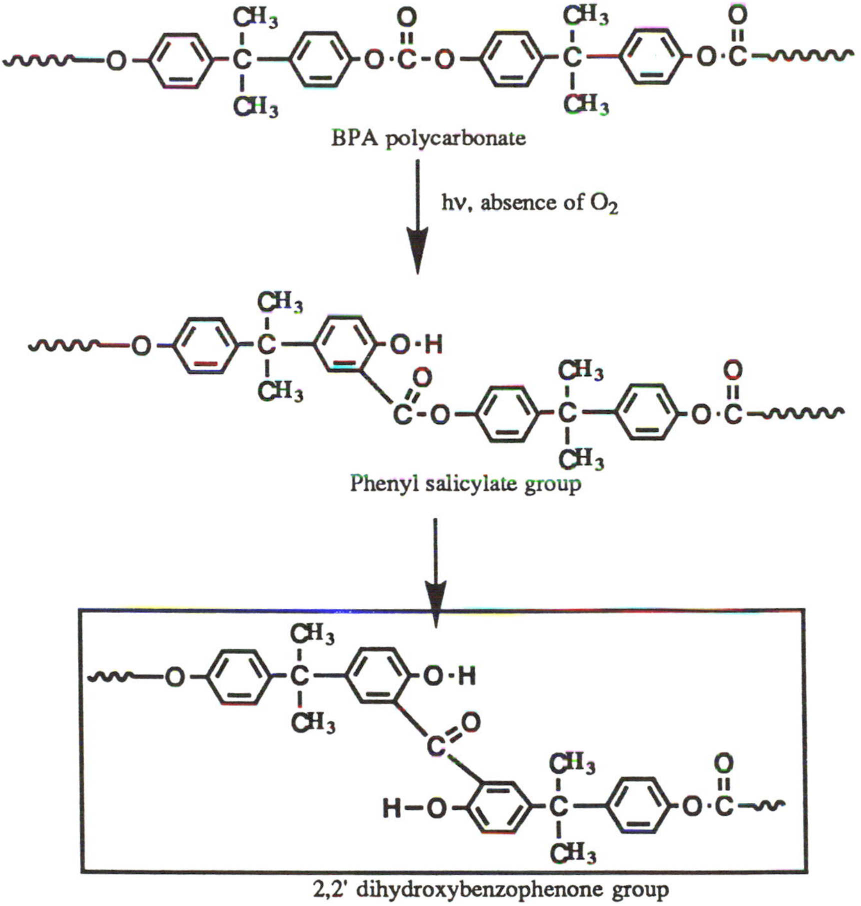

BPA polycarbonate

hv, absence of O$_2$

Phenyl salicylate group

2,2' dihydroxybenzophenone group

Scheme II. Schematic representation of UV degradation pathway for polycarbonate in the absence of O$_2$ or in an inert atmosphere.

at room temperature. A new emission is seen in the photolyzed sample at 390 nm that is directly attributable to the UV degradation product.

The emission and excitation spectra of dihydroxybenzophenone are compared with the corresponding spectra of the photolyzed samples and found to match quite precisely (390 nm emission) as shown in Figure 8. In the emission spectrum of the photolyzed sample, the emission for BPA–PC was not readily observed. The scattering signal of the excitation light, observed on the left side of the spectrum, is quite broad because of the very wide bandpass required to observe the weak emission of the degradation products. This scattering signal may have completely masked any emission in the left-hand region. In Figure 8 the complete absence of any inflection in the

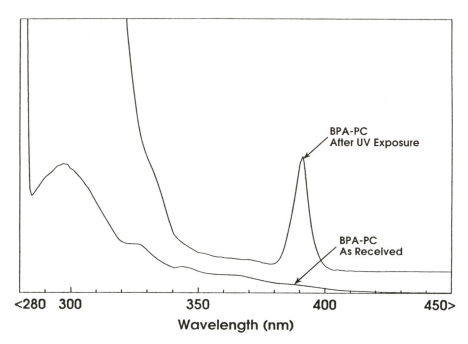

Figure 7. Comparison of the emission spectra of the as-received BPA–PC (Aldrich) with that of the same sample after 16 h of intense UV radiation (λ = 254 nm) in a nitrogen atmosphere.

scattering peak in the region where BPA–PC has its emission maximum is not shown. This absence of inflection may be evidence of an energy-transfer mechanism.

The emission spectra of solid 2,2'-dihydroxybenzophenone was obtained by using 290- or 297-nm wavelengths of excitation. The signal is relatively weak and requires precise control of the excitation wavelength. Also note that although the literature cited earlier (*19*) reports that the Fries photorearrangement reaction is quite limited in the presence of oxygen, we found traces of the 2,2'-dihydroxybenzophenone as evidenced by faint emissions that correspond to this product even for exposure in air (for 100-h exposure at 254 nm; data not shown). This observation has not been investigated in detail. In the study of the UV degradation of polycarbonate, 2,2'-dihydroxybenzophenone was the most appropriate model compound for the degradation product and its emission spectra were adopted as the spectral signature of the most significant fluorescent degradation product.

Applications to Processing and Final Performance.   Once an aromatic polymer such as polycarbonate has been characterized in detail and its degradation products identified, UV fluorescence spectroscopy can be

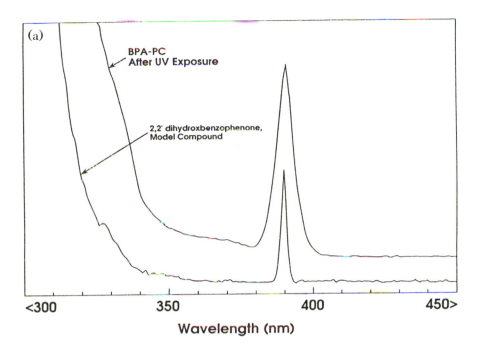

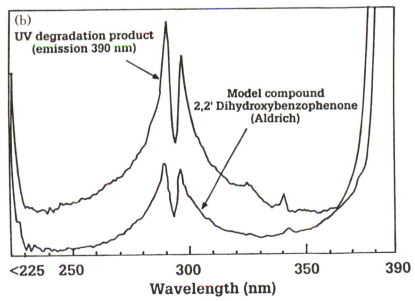

Figure 8. (a) Comparison of the emission spectra of solid-state samples of 2, 2′-dihydroxybenzophenone and BPA–PC that demonstrates congruence in emission peaks. (b) Comparison of the excitation spectra of the polymer and model compound.

used as a very rapid diagnostic technique for evaluating the extent of degradation. The intensity of the emission peaks of degradation products provides a qualitative means of establishing the level of change to the polymer. This method is particularly useful in cases where the actual exposure temperature cannot be measured directly, but can be inferred by the emission of degradation products. In such cases, the UV spectroscopy of solid polymer samples can be used as an aid in process control. This relationship is amply demonstrated by the emission spectra in Figure 3, where intensity is found to be roughly proportional to the processing temperature. The effect of UV and thermal aging or degradation and its impact on some final properties has been discussed previously (20).

The effect of thermal degradation during processing has been found to affect the rheological and bulk properties of polymers. Although for relatively small extents of thermal degradation, the effect on instantaneous or equilibrium properties (e.g., elastic modulus) may be small, thermal degradation can have quite a significant effect on nonequilibrium and molecular-weight-sensitive properties. The samples processed by vacuum-heating and compression molding (described previously) undergo the progressive degradation seen in Figure 3 and the reductions in molecular weight shown in Figure 4. Even in the case where the shear stresses were negligible (as in the preceding procedures) the changes in dynamic viscosity are very discernible as seen in Figure 9. The data show a progression of decreasing viscosity with increasing molding temperature. The variation of viscosity with testing temperature is the normally expected trend for glassy amorphous polymers.

Thermal degradation also influences the final performance of the material, particularly over longer times. Although the smaller extents of degradation induced during processing may not affect equilibrium properties such as the elastic modulus, long-term properties are indeed modified. For instance, dimensional stability under stress or creep for extended periods of time is an especially critical requirement for high precision parts. The impact of thermal degradation on creep compliance as a function of time is shown in Figure 10. The creep compliance of the polymer at a given time is seen to increase with higher molding temperatures and has the potential to result in a progressively less dimensionally stable material.

The role of UV and thermal degradation in the optical performance of polycarbonate was briefly alluded to earlier. Discussion of this topic in any detail is beyond the scope of this chapter, but it is referred to here for the sake of completeness. In applications such as lenses for automotive headlamps, one industrial measure of optical performance is based on a weighted parameter defined as the yellowness index and defined by a ASTM standard D 1925–70:

$$YI = \frac{100[1.28\,X - 1.06Z]}{Y} \qquad (1)$$

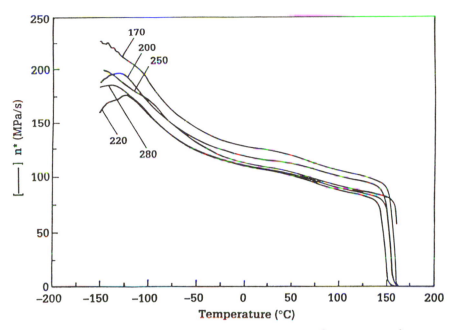

Figure 9. Effect of molding temperature on the complex viscosity of BPA polycarbonate as a result of thermal degradation. Complex viscosity data were generated on a dynamic mechanical analyzer (DuPont Instruments model 982).

where YI is the yellowness index, and X, Y, and Z are the CIE (Commission International de l'Eclairage) tristimulus values. Simply put, in clear polymers the yellowness index is a numerical representation of the visual perception of yellowness and is a function of the transmission intensities of red, blue, and green spectral bands. In general, the yellowness index increases with increasing UV degradation of polycarbonate. This effect is clearly shown for the case of commercial BPA–PC (Mobay Makrolon 2600–1000) in Figure 11. The polymer was exposed to the emission of a 450-W medium-pressure mercury lamp (Hanovia). The increasing yellowness has a detrimental effect on the optical performance of the polycarbonate both in terms of discoloration and loss of luminous transmittance of the polymer. Measurement of yellowness index of a material is perhaps the best way to gauge the overall discoloration of a material due to degradation, but it does not distinguish which degradative pathways are responsible. Clearly, our technique is more suited to making such distinctions.

Summary and Conclusions

The use of fluorescence spectroscopy as a rapid nondestructive technique for monitoring thermal degradation in polymers has been proposed and

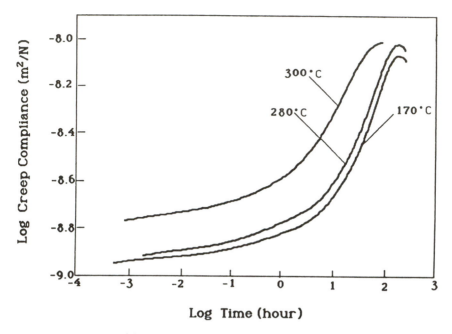

Figure 10. Impact of thermal degradation on the long term mechanical properties of BPA polycarbonate as seen on a plot of creep compliance versus time as a function of molding temperature. Creep compliance data were generated on a dynamic mechanical analyzer (DuPont Instruments model 983).

demonstrated. This approach avoids the repetitious use of wet chemistry as well as the separation techniques and analyses that have been used traditionally. However, it requires a knowledge of the main degradation products or pathways. An initial characterization of the emission and excitation spectra of the key degradation product is required. In this work, phenyl o-phenoxybenzoate was found to be the dominant fluorescing product of thermal degradation, whereas 2,2'-dihydroxybenzophenone was found to be the key by-product of UV degradation in the absence of oxygen. In the case of phenyl-o-phenoxybenzoate, the progress of thermal degradation was monitored by measuring intensities of the emission peaks of the as-molded polycarbonate. These intensities were found to increase with molding temperature and extent of thermal degradation. The thermal degradation also affected the bulk properties of the polymer, such as the complex viscosity and the creep compliance. In the case of UV degradation, the emission peak at 390 nm was found to be the signature of the main fluorescent photodegradation product. The detrimental effect of UV degradation on optical performance, as exemplified by an increasing yellowness index, was also demonstrated.

The approach presented here has been found to be a successful method applicable to process control and prediction of long-term and molecular-

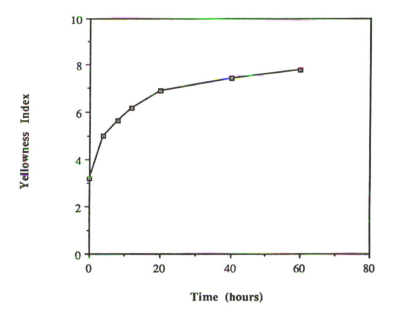

Figure 11. Yellowness index versus exposure time of commercial polycarbonate (Mobay Makrolon 2600–1000) to a 450-W medium-pressure mercury discharge lamp. Samples were kept at room temperature during exposure. Increasing yellowness is observable as a function of irradiation time.

weight-sensitive properties. However, because this technique relies on the clear identification of a key degradation product, the presence of impurities, additives, and fillers may complicate the issue by inducing unpredictable side reactions as well as possible energy transfers between excitable species.

The tracking of fluorescence spectra of degradation products as represented by model compounds and as demonstrated here is a comparatively novel approach to the study of polymer degradation. The reported spectra are also mostly solid-state spectra that have been successful in tracking degradation products, which are often present in such trace quantities as to be undetectable by traditional methods. Indeed, when UV exposures are performed in the solid state, the detection of degradation products by absorption spectroscopy and (in our experience) by attenuated total reflectance spectroscopy of solid samples is quite unsuccessful in all but the thinnest films. The success of solid-state UV emission spectroscopy is attributable to the localization of the degradation products in the surface layers of the polymer, which results in a higher concentration of chromophores at the solid surface and consequent higher intensity emissions. It is worth noting that this work also presents the fluorescence spectra of BPA–PC, a relatively unexplored aspect in the otherwise vast literature on the polymer. Finally, the technique demonstrated here for polycarbonate is broad in scope and can be applied to aromatic polymers, in general, because their emission and absorption characteristics are generally accessible by means of conventional spectrophotometers.

Acknowledgments

The authors extend their sincere thanks to their colleagues Elizabeth Yost, for the synthesis and purification of the model compound diphenyl carbonate of BPA, and Daniel J. Sandman, for his constructive comments and suggestions through all stages of this work.

References

1. Ranby, B. *Photodegradation, Photooxidation and Photostabilization of Polymers: Principles and Applications*; Wiley-Interscience: London, England, 1975.
2. Rabek, J. F. *Mechanisms of Photophysical Processes and Photochemical Reactions in Polymers: Theory and Applications*; John Wiley & Sons: Chichester, England, 1987.
3. *Chemical Kinetics*; Bamford, C. H.; Tipper, C. F. H., Eds.; Elsevier: Amsterdam, Netherlands, 1975; Vol. 14.
4. Bellus, D.; Hrdlovic, P.; Manasek, Z. *Polym. Lett.* **1966**, *4*, 1.
5. Pryde, C. A.; Hellman, M. Y. *J. Appl. Polym. Sci.* **1980**, *25*, 2573.
6. *Luminescence Techniques in Solid State Polymer Research*; Zlatkevich, L., Ed.; Marcel Dekker: New York, 1989.
7. Gupta, A.; Liang, R.; Moacanin, J.; Goldbeck, R.; Kliger, D. *Macromolecules* **1980**, *13*, 262.
8. Humphrey, J. S., Jr.; Shultz, A. R.; Jaquiss, D. B. G. *Macromolecules* **1973**, *6(3)*, 305.
9. Bartosiewicz, R. L.; Booth, C.; Marshall, A. *Eur. Polym. J.* **1974**, *10*, 783.
10. Ahmad, S. R. *J. Phys. D.* **1987**, *20(10)*, 1315.
11. Abbas, K. B. *Polym. Eng. Sci.* **1980**, *20(10)*, 703.
12. Bueche, F. *J. Appl. Polym. Sci.* **1960**, *4*, 101.
13. Glockner, G. *Plast. Kauc.* **1960**, *15*, 632.
14. Pochan, J. M.; Gibson, H. W.; Froix, M. F.; Hinman, D. F. *Macromolecules* **1978**, *11*, 165.
15. Davis, A.; Golden, J. H. *J. Am. Chem. Soc.* **1968**, *1*, 45.
16. Bailly, Ch.; Daumiere, M.; Legras, R.; Mercier, J. P. *J. Polym. Sci. Phys. Ed.* **1985**, *23*, 493.
17. Davis, A.; Golden, J. H. *J. Am. Chem. Soc.* **1968**, *1*, 40.
18. Berlman, I. *Handbook of Fluorescence Spectra of Aromatic Molecules*; Academic Press: Orlando, FL, 1979.
19. Factor, A.; Chu, M. L. *Polym. Degr. Stabiliz.* **1980**, *2*, 203.
20. Ram, A.; Zilber, O.; Kenig, S. *Polym. Eng. Sci.* **1985**, *25(9)*, 535.

RECEIVED for review July 15, 1991. ACCEPTED revised manuscript May 26, 1992.

SPECTROSCOPY AND THERMALLY INDUCED PROCESSES IN POLYMERS

The advantage of combining two or more techniques into one procedure comes from the fact that multilevel information can be obtained in one experiment. The combination of FTIR instrument with photoacoustic measurements, differential scanning calorimetry, and thermogravimetric analysis is an example of such multilevel investigations that perhaps will become common in the near future.

Photoacoustic Fourier Transform Infrared Response to Thermal and Interfacial Changes in Polymers

Marek W. Urban

Department of Polymers and Coatings, North Dakota State University, Fargo, ND 58105

Although photoacoustic Fourier transform infrared (PA FTIR) spectroscopy was initially developed as a surface technique, recent studies show that it can be used in the analyses of bulk polymer properties. This chapter highlights the recent advances in characterization of polymeric materials using PA FTIR spectroscopy. A particular emphasis is given to the molecular-level processes that occur during cross-linking of thermosetting polymers, thermal property changes during cross-linking, adhesion failure, and interfacial interactions. The results with the recently developed rheophotoacoustic FTIR (RPA FTIR) spectroscopy are presented and discussed in a view of polymer–polymer interactions.

THE PHOTOACOUSTIC PHENOMENON was observed for the first time in the 1880s, when Alexander Graham Bell noted that an audible sound was produced when sunlight modulated by a rotating wheel impinged on a sample placed in a closed brass cavity (*1, 2*). Bell was converting the electromagnetic energy of the sun to acoustic waves. This same conversion is the basis for photoacoustic spectroscopy; the basic principles are schematically depicted in Figure 1. The sample is placed in an acoustically sealed cavity and illuminated with IR modulated light. As a result of absorption of light followed by its reabsorption, a periodic heat is released to the sample surface, generating pressure changes and causing acoustic waves to occur in the surrounding gas. These periodic acoustic waves are detected by a sensitive microphone, converted to digital signals, and Fourier transformed.

0065–2393/93/0236–0645$06.00/0

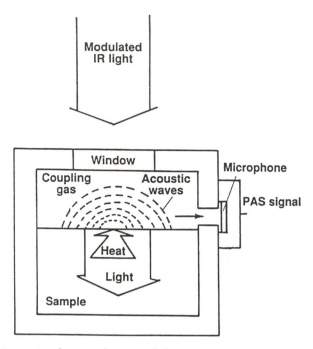

Figure 1. Schematic diagram of photoacoustic signal production.

This phenomenon was neglected for almost 90 years after Bell's discovery until the early 1970s, when Rosencwaig (3) along with others realized that it could be a useful source of molecular-level information, as the vibrating atoms forming a molecule or crystal absorb IR light that corresponds to the bonding energies between the atoms. Several experiments demonstrated the potential of the technique, but poor sensitivity, particularly when trying to measure the response of a condensed-phase surface, was a major drawback.

Today the vibrations can be heard with a combination of better equipment and Fourier transform techniques. The sensitive microphones needed for photoacoustic spectroscopy usually respond to pressure fluctuations that range in length from 5 to 5×10^{-4}. The signal from the microphone is converted into an electrical signal, and with Fourier transform techniques, a photoacoustic IR spectrum is produced.

The currently accepted photoacoustic theory (3) relates the amount of electromagnetic radiation absorbed by a sample to the amount of heat produced at the sample surface in the following manner. The intensity (I) of the incident beam is defined as

$$I = 1/2/I_o(1 + \cos \omega t)$$

where I_o is the initial intensity, ω is the interferometer modulation frequency, and t is time. When incident modulated light impinges upon the sample surface, the light intensity will be attenuated at the depth x. $I(x)$, the intensity at depth x, will reach a value

$$I(x) = I_o(1 - n)\exp(-\beta x)$$

where n is the refractive index, β is the absorption coefficient, and the amount of light absorbed within the thickness x, $E(x)$, will be

$$E(x) = \beta I(x) = \beta I_o(1 - n)\exp(-\beta x)$$

The depth of optical penetration, μ_{opt}, will be given by

$$\mu_{opt} = 1/\beta(\nu)$$

where $\beta(\nu)$ is the absorption coefficient, whereas the efficiency of heat transfer will be determined by the thermal diffusion coefficient of the sample and the modulation frequency of the incident radiation related to each other by the following relationship:

$$a_s = [\omega/2\alpha]^{1/2}$$

where a_s is the thermal diffusion coefficient of the sample and α is the diffusivity defined as k (thermal conductivity)/[ρ (density) $\times$ C (specific heat)]. The thermal diffusion length (μ_{th}), that is, the distance from the surface from which heat can reach the surface, is related to α through

$$\mu_{th} = 1/a_s = [2\alpha/\omega]^{1/2}$$

Thus, the amount of periodic heat transferred to the surface, $H(x)$, is given by

$$H(x) = E(x)\exp(-a_s x) = \beta I_o(1 - n)/\exp[(\beta + a_s)x]$$

Because the intensity of the photoacoustic signal is a function of optical and thermal properties, inherent difficulties in sorting out the two processes exist. The quantities involved in the generation of photoacoustic Fourier transform IR (PA FTIR) signals are presented in Figure 2 and include two primary components: optical absorption length (μ_{opt}) and thermal diffusion length (μ_{th}).

In a broader sense, photoacoustics is the generation of acoustic waves or other thermoelastic effects by almost any type of energetic radiation, from radiofrequency to X-rays, electrons, protons, ions, and other particles. As a

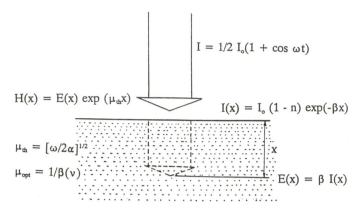

Figure 2. Generation of photoacoustic signal. Key: I, *intensity of the incident beam;* H(x), *heat intensity at depth x;* E(x), *light intensity at depth x;* μ_{opt}, *optical absorption length;* μ_{th}, *thermal diffusion length;* n, *refractive index;* β(ν), *absorption coefficient;* α, *thermal diffusivity; and* ω, *interferometer modulation frequency [for Michelson interferometer* $\omega = 2\Pi V\nu$ *where* V *is mirror velocity* (cm/s) *and* ν *is wavenumber* (cm^{-1})].

consequence, quite a substantial amount of the reported experimental work shows the use of PA in diverse applications, not only in chemistry and physics, but also in biology, medicine, and other disciplines. Many natural and synthetic materials, such as gels, oils and suspensions, fibers, textiles, and coatings, cannot be studied by conventional spectroscopic techniques in a nondestructive manner. Photoacoustic spectroscopy can be used to study molecular-level processes, including polymer network formation, adhesion, degradation, and interdiffusion. The molecular-level processes that govern the formation of materials are responsible for durability and performance. The advances in the analysis of polymeric surfaces and theoretical foundations in the field of PA FTIR were reviewed (4) elsewhere. This chapter focuses only on new concepts that PA FTIR spectroscopy can offer in polymer analysis, including polymer network formation, polymer interfacial interactions, and adhesion.

Polymer Network Formation

Substantial mass flow occurs in a developing network system, especially during the early stages of reactions responsible for network formation. If dispersed inorganic particles are present in a network that is still not cross-linked, the distribution of these particles may change as a result of the mass flow. In studies of alkyds containing inorganic particles (5, 6), we showed that the diminishing intensities of the bands attributable to the inorganic phase occur as a result of the particle stratification process at the

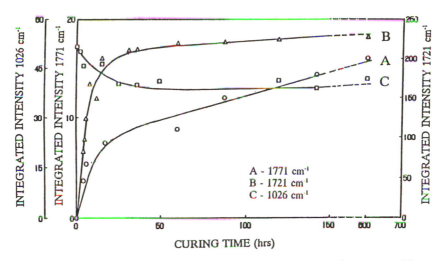

Figure 3. In situ studies of pigmented alkyd curing: integrated intensity of free C=O (A), hydrogen-bonded (B), and Si=O (C) bands plotted as a function of time. (Reproduced with permission from reference 5. Copyright 1990 Wiley and Sons.)

early stages of network formation, and the extent of stratification depends on the degree of hydrogen bonding between the organic binder and the inorganic particles. Integrated intensities of the hydrogen-bonded carbonyl bands at 1721 cm^{-1} increase as a function of cure time, whereas the band at 1026 cm^{-1} due to the Si–O vibrations of the inorganic phase (in this case, talc) decreases at the same rate. This effect is illustrated in Figure 3, which depicts the intensity changes of the Si–O, hydrogen-bonded, and free C=O bands as a function of time. The Si–O and the hydrogen-bonded C=O bands level off after about 25 h. This time coincides with the time required for a complete cure of the system.

This apparent relationship between the intensities changes with time for the hydrogen-bonded carbonyl and Si–O stretching vibrations. The inorganic particles are covered with hydroxyl groups, which possess an affinity for hydrogen bonding with carbonyl groups. As a result, a fraction of the hydrogen-bonded inorganic particles are held by the binder, while the unbonded ones can slowly migrate within the forming polymer network until a high enough density exists to prevent further motion.

In an effort to determine the driving force responsible for this behavior, a coating was deposited on a substrate and allowed to cross-link in an upside-down position. Although the intensity of the hydrogen-bonded carbonyl band in our PA FTIR spectrum followed the same trend as before, the band due to Si–O stretching vibrations increased, a result indicating that the inorganic-phase particles were now moving toward the surface of the coating.

This behavior is believed to be due to the stratification of inorganic particles in the organic phase and may be primarily governed by gravitational forces that compete with the hydrogen bonding. This stratification is schematically depicted in Figure 4.

During the cross-linking reactions that occur in amorphous network formation, usually low-molecular-weight cross-linker molecules react with oligomers. Although transmission FTIR spectroscopy can be used to monitor bond formation, low concentrations of the bonds that are actually reacting make such measurements difficult and, in many cases, impossible. In addition, the simultaneous making and breaking of energetically similar bonds may result in heavy spectral overlap of the bands in transmission spectra. Yet, these often difficult-to-detect chemical changes result in surprisingly drastic physical property changes. After all, a liquid mixture becomes a solid polymer.

Earlier, we demonstrated the sensitivity of PA FTIR spectroscopy to monitor cross-linking reactions (7, 8). Figure 5 illustrates the spectroscopic results of the study dealing with the cross-linking of OH-terminated poly(dimethylsiloxane) (PDMS) and triethoxysilane (TES) and monitoring the cross-linking reactions using photoacoustic and transmission measurements.

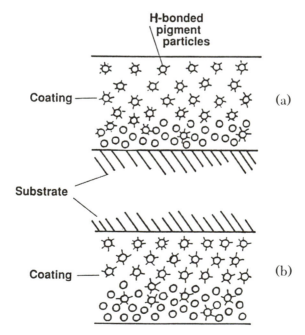

Figure 4. Schematic representation of pigment stratification in the alkyd coating: in right-side-up position (a) and in the upside-down position (b). (Reproduced with permission from reference 5. Copyright 1990 Wiley and Sons.)

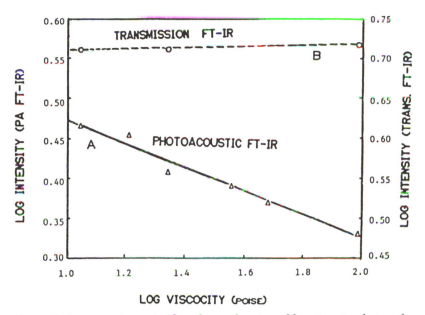

Figure 5. Log intensity ratio plotted as a function of log viscosity during the cross-linking process of the PDMS–TES system: A, PA FTIR detection; and B, transmission FTIR detection. (Reproduced from reference 7. Copyright 1989 American Chemical Society.)

Although the intensities of IR bands recorded as a function of viscosity (or time) in the transmission measurements do not change, the photoacoustically detected bands diminish as the reaction progresses.

The observed sensitivity enhancement of PA FTIR spectroscopy can be attributed to two factors: First, ethanol is produced as a result of the OH-PDMS and TES reaction, which significantly contributes to the gas-phase photoacoustic effect, and second, the photoacoustic effect is functioning as a calorimetric method that measures the amount of electromagnetic radiation converted to heat and therefore will be sensitive to the thermal property changes.

During the network formation, as mobile cross-linker molecules are being "frozen" by their reaction with oligomer molecules, the entropy of the cross-linking system decreases. This decreased entropy results in the heat capacity and thermal conductivity changes. One can draw an analogy to freely moving gas molecules that have translational, rotational, and vibrational degrees of freedom, but when frozen, only vibrational degrees of freedom are left. If this was an ideal gas, the reduction of heat capacity would go from $\frac{3}{2}RT$ ($\frac{1}{2}RT$ for each degree; where R is the gas constant and T is temperature) to $\frac{1}{2}RT$; $\frac{1}{2}RT$ accounting for vibrational degrees of freedom that will contribute to the overall gas heat capacity. Similarly, cross-linker and oligomer molecules become immobilized during cross-linking and lose partially their translational and rotational degrees of freedom. These losses will affect the thermal diffusivity of the newly formed network. In addition, substantial density changes will further affect photoacoustic intensity because the molecules go from nonbonding to bonding distances. Novel theoretical and experimental approaches that deal with these issues are being investigated (9, 10).

Although elevated temperature PA FTIR measurements were developed in the mid-1980s, limited sensitivity of the existing instrumentation did not allow simple temperature measurements unless a specially designed cell was built for that purpose. If the thermal property changes indeed affect photoacoustic intensity, then PA FTIR spectroscopy should be sensitive to heat capacity changes as a polymer is heated above its glass-transition temperature. Above that temperature, heat flow diminishes as a result of the increased free volume that causes lower heat conductivity. In an effort to demonstrate that PA FTIR spectroscopy is also sensitive to thermal property changes, Figure 6 illustrates PA FTIR intensity changes due to carbonyl stretching vibrations in polybutyl methacrylate plotted as a function of temperature. The parallel differential scanning calorimetry (DCS) curve shows a striking similarity. In DSC measurements a drop of the heat flow is

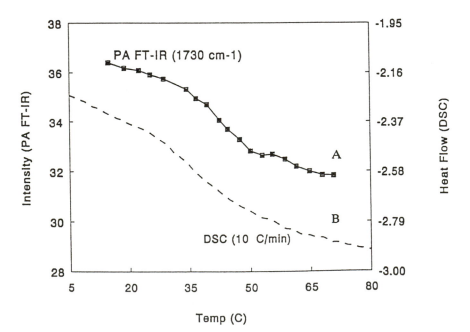

Figure 6. A comparison of PA FTIR intensity changes (A) and differential scanning calorimetry (B) for polybutyl methacrylate (M.W. 100,000) heated above the glass-transition temperature (T_g = 47 °C).

attributed to the fact that the sample was heated above its glass-transition temperature. The PA FTIR curve shows the same drop as that detected in the DCS experiment. The apparent similarity of the curves illustrates the fact that the heat capacity changes with temperature can also be monitored in temperature PA FTIR experiments, (Huang, J.-B.; Urban, M. W., unpublished results).

In some PA FTIR studies (*11*) the analysis of adsorbed gases on solid surfaces was the primary objective. In the majority of those experiments, apparently no distinction was made between photoacoustic spectra of physi–chemisorbed species and the gas-phase molecules. The main problem encountered with those measurements was that the IR light "saw" both species and, unless there was a frequency difference between gas and absorbed molecules, could not distinguish among them (*11*). To overcome this problem, a "photoacoustic umbrella" was developed to cover a solid sample in such a way that IR light could "see" the gas phase only. This approach was used in the recent quantitative studies of diffusion of small molecules from such polymers as poly(vinylidene fluoride) (PVDF) (*12, 13*).

Interfacial Stresses and Adhesion

Rheophotoacoustic (RPA) FTIR spectroscopy is another technique that has been explored (*14, 15*). The cell depicted in Figure 7 is a miniature stress–strain device built into the photoacoustic FTIR cell. Such cell configuration allows monitoring molecular-level events in materials as stress is applied. Although rheooptical FTIR spectroscopy has been around for awhile (*16*), rheophotoacoustics is distinctly different and more sensitive to processes other than bond cleavage or bond formation. If stress is induced in a material, in addition to the "normal" photoacoustic spectrum, additional acoustic signals will be produced owing to the energy release due to physical changes like cracking, phase changes, or deformations. An analogous process is the cracking of ice on a pond, also generating audio signals.

One of the interesting applications of RPA FTIR spectroscopy is in the field of adhesion. A common problem in the thin-film technology and coatings industry is adhesion to polymeric substrates. To illustrate the feasibility of using RPA FTIR spectroscopy to monitor or measure adhesion on a molecular level, polyethylene films were coated with a siloxane coating, and a series of RPA FTIR spectra were recorded at various degrees of elongation caused by applying forces to the substrate only. This process is illustrated by a sample configuration in Figure 8. The spectral changes are illustrated in Figure 9. As the substrate is stretched 8.4%, an increase in the intensity of the band at 2965 cm^{-1} is observed. This behavior is attributed to the thinning of the siloxane coating as it conforms to the substrate. As the substrate is stretched further to 16.8% elongation (Figure 9C), a strong

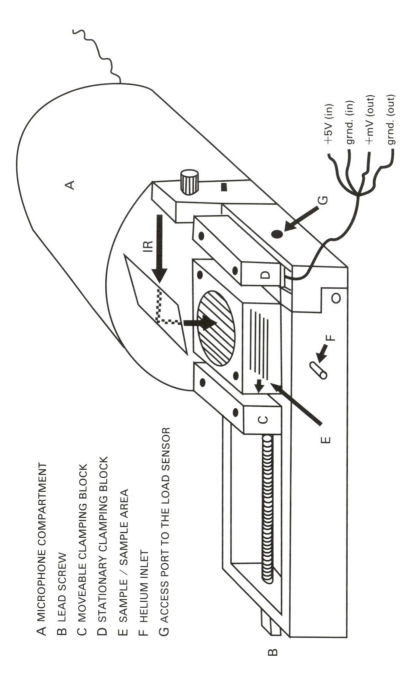

A MICROPHONE COMPARTMENT

B LEAD SCREW

C MOVEABLE CLAMPING BLOCK

D STATIONARY CLAMPING BLOCK

E SAMPLE / SAMPLE AREA

F HELIUM INLET

G ACCESS PORT TO THE LOAD SENSOR

Figure 7. Schematic diagram of PA FTIR cell adapted for rheophotoacoustic FTIR measurements. (Reproduced with permission from reference 14. Copyright 1989 Society for Applied Spectroscopy.)

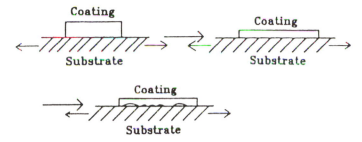

Figure 8. Schematic representation of the elongational process of the polymeric substrate–coating double layer.

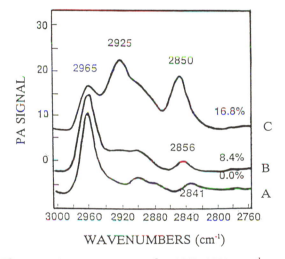

Figure 9. Rheo-PA FTIR spectra in the 3050–2800-cm^{-1} region of the siloxane-coated PE at various stages of elongation: A, 0.0%; B, 8.4%; and C, 16.8%.

increase in the polyethylene bands at 2935 and 2850 cm^{-1} is detected. Because conforming of the siloxane coating cannot be responsible for such drastic intensity changes, interfacial shear forces cause a separation of the two phases, resulting in an interfacial failure. The heat generated at the coating–substrate interface can escape directly through the interfacial voids, generating acoustic waves at the interface without passing through the coating. Parallel scanning electron microscopy experiments confirmed the formation of interfacial microvoids (*15*), whereas X-ray analysis indicated that no crystallinity changes are responsible for the RPA FTIR spectral changes.

Polymers exposed to elongational or other external stresses may experience crystallization or crystallinity changes (*17, 18*). The extent of these changes depends upon initial crystallinity, polymer glass-transition temperature, melting point, and the extent of elongation resulting from the applied external stresses. In an effort to establish the effect of pending CH_3 groups on PDMS adhesion, the same experimental procedures were employed on polypropylene (PP) (19). This substrate exhibits relatively high melting point and usually stress-induced crystallization occurs at elongations exceeding 40–50% (20). Using the rheophotoacoustic FTIR approach, we monitored the interfacial changes on PDMS–PP. The presence of polypropylene, however, may affect not only adhesion and physical interactions, but also molecular-level compatibility with PDMS. Stretching PP up to 20% alone in the rheophotoacoustic cell does not affect intensities of the RPA FTIR spectra, but the results of the same experiments on PP coated with PDMS are quite different and are shown in Figure 10. The band intensities throughout the entire elongational experiment increase until about 10% elongation is reached. Above this elongation, the bands level off, indicating again the void formation at the interface. Although the crystallinity changes in PP itself during 0–20% elongation are possible, X-ray diffractograms shown in Figure 11 indicate no

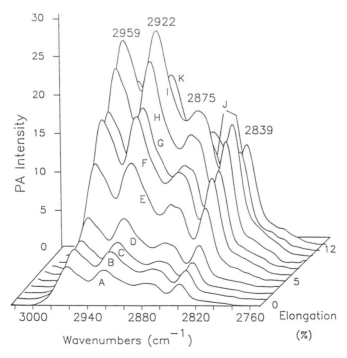

Figure 10. Rheo-PA FTIR spectra in the 3000–2760-cm^{-1} region of the siloxane-coated PP at various stages of elongation: A, 0.0%; B, 0.83%; C, 1.67%; D, 2.5%; E, 4.17%; F, 5.0%; G, 5.83%; H, 6.66%; I, 8.33%; J, 9.16%; and K, 11.67%.

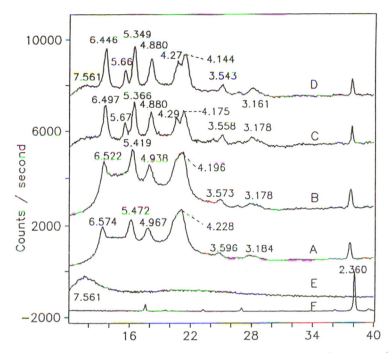

Figure 11. X-ray diffractograms of polypropylene, 0% elongation (A); polypropylene, 20% elongation (B); polypropylene–PDMS double layer, 0% elongation (C); polypropylene–PDMS double layer, 20% elongation (D); PDMS only, 0% elongation (E); and Al backing (F).

detectable changes. However, the situation changes drastically when the PDMS–PP double layer is exposed to the same experimental conditions. The analysis of X-ray diffractograms shown in Figure 11 indicates that in addition to the peaks characteristic of isotactic PP, two additional X-ray peaks corresponding to 5.6 and 4.2 Å d-spacings are due to crystalline syndiotactic PP. Based on RPA FTIR and X-ray measurements, the following scenario may occur at the PDMS/PP interface: When initially liquid PDMS is deposited on PP surface, non-cross-linked at this stage PDMS components (oligomer, MW = 3400; and tetrafunctional cross-linker) penetrate the PP surface. As a result, the surface may swell and become partially solubilized. However, if the hypothesis that the syndiotactic stereoisomer recrystalizes is correct, then the amorphous stereoisomer would be preferentially solvated by the oligomer and cross-linker. Analysis of the literature data indicates that, indeed the total solubility parameters for PP and PDMS are δ = 9.3 and 9.5 $(cal/cm^3)^{1/2}$ (*21*). The solubility parameters are similar, and hence liquid PDMS is capable of solvating PP surface, giving PP molecules segmental mobility. This segmental mobility allows the solubilized phase to mix with the liquid PDMS and diffuse to the interfacial regions. Upon cross-linking, PDMS molecules immobilize the entire interfacial region and cause amorphous syndiotactic PP to recrystallize. The argument of preferential solubilization of the syndiotactic regions is also supported by the enhanced solubilization of syndiotactic isomer; namely, the syndiotactic form is soluble below 40 °C, whereas other forms require higher temperatures (22).

Potential of Photoacoustics

The progress made in listening to atomic vibrations over the past 15 years has been impressive. Even Bell himself should be pleased with the current developments. Today the spectra of any material in its native state can be obtained, regardless of shape, color, or morphology. Although photoacoustic spectroscopy can be carried across the entire electromagnetic spectrum, IR appears to be quite a useful region. PA FTIR spectra can be used to study the surface protection of polymeric materials, wood, paper, fibers, composites, and other materials. Degradation processes that occur in polymeric materials as a result of exposure to environmental conditions like ultraviolet radiation or acid rain may be followed. Surface-depth profiling applications (23, 24) along with studies of the surface orientation effect (25–27) and enhancement of the photoacoustic signal due to interstitial gas expansion in zeolite channels (28) were reported.

The versatility of photoacoustics in answering a broad spectrum of fundamental questions such as thin-film composition at various depths or surface functionality of substrates has shown encouraging data and future potential. This technique, originally developed for the surface analysis, now emerges into a new era of research, revealing the coating–substrate bonding mechanisms and factors affecting adhesion or cross-linking. Because photoacoustics in such mode is a nonequilibrium process, it will allow future experimentalists to attack thermodynamic processes that are not in equilibrium. This work is in progress. New cell designs (M-Tech, Digilab, and others) with improved microphones, high-power sources, and more sensitive electronics will make photoacoustics faster and give an enhanced signal-to-noise ratio. This enhancement in sensitivity, along with the developments of new experimental approaches, should open new avenues to further understand molecular-level processes responsible for physical changes in materials.

Acknowledgments

The author is thankful to the 3M Company (St. Paul, MN) for a partial support of the studies on photoacoustic FTIR spectroscopy and all graduate students and postdoctoral fellows who were involved in the various stages of this work.

References

1. Bell, A. G. *Am. J. Sci.* **1880**, *20*, 305.
2. Bell, A. G. *Philos. Mag., B* **1881**, *5(11)*, 510.
3. Rosencwaig, A. *Photoacoustics and Photoacoustic Spectoscopy*; Wiley and Sons: New York, 1980, and references therein.
4. Urban, M. W.; Gaboury, S. R.; McDonald, W. F.; Tiefenthaler, A. M. *Polymer Characterization: Physical Property, Spectroscopic, and Chromalographic Methods*; Craver, C. D.; Provder, T. E., Eds.; Advances in Chemistry 227; American Chemical Society: Washington, DC, 1990, and references therein.
5. Urban, M. W.; Salazar-Rojas, E. M. *J. Polym. Sci., Part A: Polym. Chem.* **1990**, *28*, 1593.
6. Salazar-Rojas, E. M.; Urban, M. W. *Prog. Org. Coat.* **1989**, *16*, 371.
7. Urban, M. W.; Gaboury, S. R. *Macromolecules* **1989**, *22*, 1486.
8. Gaboury, S. R.; Urban, M. W. *Proc. Polym. Mater. Sci. Eng.* **1989**, *60*, 875.
9. Huang, J.-B.; Urban, M. W., in preparation.
10. Huang, J.-B.; Urban, M. W., in preparation.
11. Eyring, E. M.; et al., *Appl. Spectrosc.* **1983**, *37*, 131.
12. Ludwig, B.; Urban, M. W. *Polymer*, in press.
13. Ludwig, B. W.; Urban, M. W. *Polymer*, in press.
14. McDonald, W. F.; Goettler, H.; Urban, M. W. *Appl. Spectrosc.* **1989**, *43(8)*, 1387.
15. McDonald, W. T.; Urban, M. W. *J. Adhes. Sci. Technol.* **1990**, *4(9)*, 751.
16. Siesler, H., Chapter 2 of this volume.
17. Mark, J. E.; Eisenberg, A.; Graessley, W. W.; Koenig, J. L. *Physical Properties of Polymers*; American Chemical Society: Washington, DC, 1984.
18. Mark, J. E.; Erman, B. *Rubberlike Elasticity: A Molecular Primer*; Wiley Interscience: New York, 1988.
19. Mirabella, F. M. *J. Polym. Sci., Polym. Phys. Ed.* **1984**, *22*, 1293.
20. MacDonald, W. M.; Urban, M. W., in preparation.
21. *Polymer Handbook*; 3rd ed.; Brandrup, J.; Immergut, E. H., Eds.; John Wiley and Sons: New York, 1989, VII/519.
22. Youngman, E. A.; Boor, J. *Macromol. Rev.*, **1967**, *2*, 1169.
23. Urban, M. W.; Koenig, J. L. *Appl. Spectrosc.* **1987**, *40(7)*, 994.
24. Yang, C. Q.; Ellis, T. J.; Breese, R. R.; Fateley, W. G. *Polym. Mater. Sci. Eng.* **1985**, *53*, 169.
25. Urban, M. W.; Koenig, J. L. *Appl. Spectrosc.* **1986**, *39*, 1051.
26. Urban, M. W.; Koenig, J. L. *Appl. Spectrosc.* **1986**, *40(6)*, 851.
27. Urban, M. W.; Koenig, J. L. *Anal. Chem.* **1988**, *60(21)*, 2408.
28. Wang, P. L.; Eyring, E. M.; Huai, H. *Appl. Spectrosc.*, **1991**, *45*, 883.

RECEIVED for review July 15, 1991. ACCEPTED revised manuscript May 20, 1992.

A Combined Differential Scanning Calorimetry–Fourier Transform Infrared Approach for the Study of Polymeric Materials

David A. C. Compton, David J. Johnson, and Jay R. Powell

Digilab Division, Bio-Rad, 237 Putnam Avenue, Cambridge, MA 02139

A combination of the standard laboratory techniques differential scanning calorimetry (DSC) and reflectance Fourier transform infrared (FTIR) microspectroscopy is described. Both the FTIR and DSC analyses were under the direct control of one computer, and the overall operation of the instrument is described. This simultaneous DSC–FTIR technique gives spectroscopic and thermodynamic information about a solid or liquid sample undergoing thermal modification. DSC measures the exothermic and endothermic responses of the samples, whereas the FTIR analysis observes changes in chemical and physical composition. The curing reaction of epoxy samples and the phase transitions of poly(ethylene terephthalate) (PET) are used to illustrate the potential of the combined DSC–FTIR method for the determination of polymer structure and properties. In the PET example, changes in the infrared spectrum were used to deduce that recrystallization in one sample of PET occurred even though the DSC curve showed no strong inflection at that temperature.

THE STUDY OF POLYMERIC MATERIALS BY THERMAL ANALYSIS has become increasingly popular in modern analytical laboratories. Differential scanning calorimetry (DSC) directly measures endothermic or exothermic behavior of a material as a function of temperature and provides valuable information about the thermal properties and composition of a sample under investiga-

0065–2393/93/0236–0661$06.00/0

tion. Properties such as heat of cure, glass-transition temperature, percent crystallinity, melting point, and degree of cure may all be calculated with the DSC technique (1).

Concurrently, many laboratories study the same types of materials by infrared spectroscopy (2) that they examined by thermal analysis. In particular, Fourier transform infrared (FTIR) spectroscopy is a powerful tool for analysis of polymeric materials. The currently available FTIR spectrometers can provide a wealth of information about the composition, structure, crystallinity, and other properties of samples that have sizes ranging from a few centimeters down to a few micrometers (in conjunction with an infrared-transmitting microscope) (3).

Both DSC and FTIR analyses are well suited to the study of the thermal behavior of materials. The DSC instrument monitors changes in heat flow as a function of temperature, but is unable to identify the chemical nature of observed transitions and occasionally even may fail to observe a transition. The FTIR spectrometer is well suited to obtain information about the chemical properties of a sample. The availability of special accessories for FTIR allows the study of samples at nonambient temperatures (2), but these accessories generally do not allow for any thermal information (such as heat flow) to be recorded about the sample, except for its temperature. Thus, during a variable-temperature infrared spectroscopy study, any physical changes that take place must be inferred from changes observed in the infrared spectrum. Such spectral changes may be very subtle, in which case the physical interpretation of the changes can be ambiguous.

Performing simultaneous FTIR and DSC analysis offers great promise to overcome the limitations of the individual techniques for the study of the thermal properties of materials. The combination of DSC and FTIR allows the analyst to monitor structural changes in the material as the sample passes through various thermal transition states and, potentially, to obtain sufficient information by spectroscopic means to assign the different states to physical or chemical phenomena.

In 1986, two groups of workers reported studies that combined FTIR and DSC (4–6), but such work was limited by the equipment available at that time. To obtain the infrared spectral data, both groups used transmission spectroscopy, which placed severe restrictions on the types of samples that could be studied. Improvements to FTIR instruments, in both software and hardware terms, enable us to report experiments that show great promise for routine operation. Reflectance infrared spectroscopy allows examination of a broader variety of samples and makes more experimental information obtainable.

In all of the DSC–FTIR experiments reported, including those in this report, the DSC was a hot-stage microscopy cell (Mettler FP84). This particular DSC was used because it is very small, is designed to fit under a microscope objective, and has a sample cup accessible to incoming infrared radiation. The various reported DSC–FTIR studies differed significantly in the manner in which the microscopy cell was mounted in the FTIR spectrometer.

Koberstein and co-workers (6) reported an experiment where the DSC was simply placed directly in the main infrared beam of the spectrometer sample compartment. To accomplish this direct sample placement, the DSC sample cup was removed and a thin sample of polyurethane film (mounted on a KBr window) was placed over the hole. The simultaneous DSC data were recorded on a thicker sample of polyurethane placed in the DSC reference cup. This approach, unfortunately, means that the DSC and the FTIR experiments examined specimens in different environments, and, as is well known, the thermal properties of a sample are often dependent on the sample morphology.

To obtain a better signal-to-noise ratio on the FTIR data, Mirabella used an infrared transmitting microscope accessory (4) to condense the beam into the small opening of the DSC sample cup, which yielded much higher optical throughput. The infrared beam was transmitted through the sample and the DSC cup, facilitated by special cups fashioned from sodium chloride or potassium chloride crystals. Again, it was necessary to examine a thin film of each sample to avoid excessive infrared absorbances. This approach made it possible to monitor the structural changes in polypropylene (4) and polyethylene (5) during melt and recrystallization, and to study the degradation of poly(ethylene vinyl alcohol) (5).

During measurement of a transmission infrared spectrum, the absorbance of the strongest bands of interest should not exceed about 0.8 absorbance units. To maintain acceptable absorbance levels, a typical polymer sample must be ~ 10 μm or less in thickness. Samples this thin generally have insufficient mass to generate an adequate DSC curve. Consequently, both of the groups that previously reported DSC–FTIR work (4–6) placed a much thicker sample of the same material in the cup that is normally the reference cup. This procedure resulted in a DSC trace with the y axis reversed from the normal convention.

Our present research was performed in a significantly different manner. The same DSC (Mettler FP84) was employed, but we obtained the spectral data using an infrared microscope operating in the *reflectance* mode. The infrared beam was directed down from the microscope objective onto the sample in the DSC cup. Then the reflected energy was collected by the same microscope objective and focused onto the infrared detector. Infrared spectra were recorded continuously during a standard DSC experiment. In addition, software was written to operate both the DSC and FTIR simultaneously from one data station and to combine the experimental data in one set of data files.

There are several advantages to this experimental arrangement. It was not necessary to fashion DSC sample cups from a material (like potassium bromide) that transmits infrared radiation; instead, standard aluminum cups were employed. In many cases, it was possible to study samples that had the normal thickness for the DSC experiment, and the design of modern infrared-transmitting microscope accessories allowed spectra to be collected in the reflectance mode with excellent signal-to-noise ratio. Finally, the use of a single data station for the data collection from both DSC and FTIR significantly improves the ability to correlate the results from both instruments. Because only one sample is studied by both techniques simultaneously, the DSC curve can be compared directly to various spectroscopic parameters, such as peak frequency, peak ratios, band width, etc., all as a function of temperature.

Experimental

All spectroscopic data were collected using an FTIR spectrometer (Bio-Rad FTS 40) equipped with a KBr beam splitter, high-temperature ceramic source, and infrared transmitting microscope accessory (UMA 300A). To give sufficient working depth with the DSC cell in place on the microscope stage, a 15 × Cassegrain objective was utilized for both infrared analysis and viewing with visible light. Spectra were collected continuously at 8-cm^{-1} resolution during the DSC experiment.

A microscopy cell (Mettler FP84 TA) was used to heat the samples and to obtain the DSC data. The temperature was programmed to ramp from 25 to 280 °C at 10 °C per minute. Operation of the microscopy cell was controlled by the data station (Bio-Rad 3200) which passed instructions to a central processor (Mettler FP80). The central processor performed direct control of the DSC cell and fed data from the DSC experiment back to the data station. Thus, the data station was collecting data from both the FTIR optical bench and the DSC microscopy cell simultaneously, which allowed for the direct comparison of the optical and thermal results on the same time or temperature basis.

To allow for the infrared beam to reach the sample in the DSC cup, the glass cover slip that is normally positioned above the cell was removed. If necessary, this cover could be replaced with a thin window fashioned from an infrared transmitting material, but that was not done for this work.

Results and Discussion

During collection of a reflected infrared beam, a variety of phenomena that depend on the surface, geometry, and phase of the material under investigation may occur. Two interactions between the infrared beam and sample were observed in this research. In some cases, the sample is relatively transmissive to infrared radiation, and the beam transmits down through the sample, reflects off the aluminum cup, and passes back through the material.

This type of analysis is called reflection–absorption spectroscopy, and it can only be applied to a thin film of material in the sample pan. As previous workers (4–6) have done, the bulk of the sample must be placed in the reference side of the sample cup. Reflection–absorption behavior was exhibited by the samples of epoxy resin that we examined. Two advantages are gained by use of the reflectance mode of collection instead of the transmission mode (4–6) for this class of sample: (1) a special sample cup (which may require disposal after a single curing experiment) need not be fashioned and (2) formulations with varying amounts of filler all can be studied in reflectance.

A different behavior—specular reflectance of the light—was noted with a sample of poly(ethylene terephthalate) (PET) that was optically opaque. In that case, because the bulk of the energy was reflected from the top surface of the sample, the whole sample was placed in the sample cup. The advantage of this procedure is that both the infrared spectrum and DSC trace can be obtained from the same sample. In addition, experiments can be performed on samples that cannot be physically examined as a thin film, such as samples that are heavily filled. However, the infrared spectrum that is obtained by specular reflectance exhibits distorted or derivatized band shapes and requires a software correction to produce a normal absorbance spectrum. This correction, the Kramers–Kronig transformation (3), will be described in more detail later during the discussion of the results obtained for PET.

Epoxy Study. Thin films of uncured amine-activated epoxy formulations were placed in the sample pan of the microscopy cell and heated from 25 to 280 °C at 10 °C per minute. Changes in the structure of the epoxy as a function of temperature were simultaneously recorded by infrared spectroscopy. The reaction mechanism of cure initially involves the reaction of a cycloaliphatic primary amine activator (4-amino-4-methyl-cyclohexene-methaneamine) with the epoxide group of the resin [2-di-[4-(2,3-epoxy-1-propoxy)-1-phenyl] propane] to produce a secondary amine. The secondary amine further reacts with an additional epoxide group to form a tertiary amine. Further reaction, catalyzed by water, hydroxyl, and tertiary amine concentration, continues the cross-linking activity. This same system has been studied extensively by combined thermogravimetric analysis (TGA) and FTIR (7), as well as by DSC–FTIR (8). The research using TGA–FTIR demonstrated utility to quantitate the amine–resin ratio and to determine qualitatively the thermal history of the polymer after cure.

The curing of an epoxy is an exothermic reaction. Research (8) shows that the shape of the DSC exothermic peak changes considerably when the amine–resin ratio is varied. The DSC trace observed when a mixture of 35 parts per hundred parts resin (phr) amine (41% over the stoichiometric primary and secondary amine concentration) is heated shows a single-peak exotherm at about 140 °C. When the same epoxy is prepared with 17 phr amine (31% under the stoichiometric level of primary and secondary amines) and is heated, a double-peak exotherm that has a second, broader peak centered at about 200 °C is generated. The extra peak in the DSC trace for the second sample indicates a significant change in the reaction mechanism. By coupling the FTIR to the DSC, these changes were studied and differentiated.

During each run, 128 scans were coadded per spectrum and 30 spectra were collected. Four spectra generated at various temperatures during a typical experiment are shown in Figure 1. Comparison of the spectra shows a number of regions where differences are observed as a function of temperature. These differences can be studied easily as either peak intensities or frequencies.

As examples of typical results, the relative intensities of a pair of bands at 3030 and 3048 cm^{-1} change significantly during the reaction and, hence, can be used to monitor the degree of cross-linking. It is not an easy task to assign these bands to particular components of the reaction mixture, but they are almost certainly due to epoxy (3050–3030 cm^{-1}) and aromatic C–H stretching (3080–3010 cm^{-1}) (9). The 3048-cm^{-1} band decreases in intensity as the reaction proceeds, so it probably arises from the epoxy that is being consumed. A plot of this peak ratio as a function of temperature for both mixtures shows that the rate of reaction is slightly faster with the over-stoichiometric mix. Band shifts in the infrared spectra may also be used to monitor the progress of the reaction. For example, Figure 2 shows a plot of the frequency of the absorbance band near 1295 cm^{-1}. This band shifts steadily to lower frequencies as a function of cure. However, it is interesting that both sets of infrared data (peak ratios and peak frequencies) show a steady change throughout the reaction, whereas the DSC curve indicates a strong reaction at certain temperatures.

Not only does the rate of reaction vary, but the reaction mechanism itself changes as a function of activator–resin ratio. Figure 3 shows a series of three spectra generated when the over-stoichiometric mix system was heated. Even though this band sits on the normal broad, featureless band due to hydroxyl, we have assigned the band at 3350 cm^{-1} in the spectra to the N–H stretch of a secondary amine (3500–3300 cm^{-1}; reference 9). The presence of this band at 139 and 210 °C indicates that even at elevated temperatures, the dominant reaction is primary amine to epoxy-producing secondary amine. Figure 4 shows that when less activator is used in the mix, as with the 17 phr activator system, the 3350-cm^{-1} band is noticeably absent at elevated temperatures. We interpret this to indicate that the reaction between secondary amine and remaining epoxide groups becomes significant due to the lack of primary amine.

The preceding interpretation explains the differences between the DSC data obtained for the two samples. The single-peak character of the sample that contains the over-stoichiometric mix is almost completely due to an

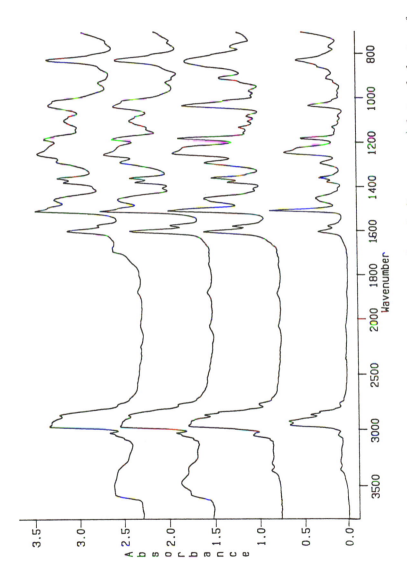

Figure 1. Transmission IR spectra generated at 25, 82, 170, and 249 °C (bottom to top) during the heating of an epoxy sample containing 35 phr amine curing agent.

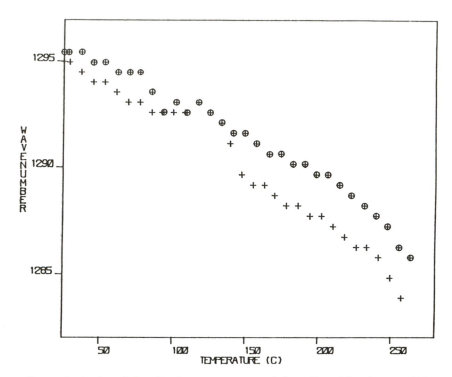

Figure 2. A plot of the absorbance maximum of the infrared band near 1290 cm⁻¹ as a function of temperature for epoxy samples containing 35 phr amine curing agent (+) and 17 phr activator (⊕).

absence of any other reaction except that of the primary amine, which adds to epoxide and generates a secondary amine group. The DSC trace for the other sample, however, has a double-peak nature where the first peak is due to consumption of primary amine. Because excess epoxide groups remain after the majority of the primary amine has been consumed, the less reactive secondary amine is able to react and give rise to the second DSC peak. Similar preference for the primary amine reaction with the use of aromatic curing agents is well documented (10). It has been shown that in aromatic amine curing agents, the secondary amine reaction with epoxide has approximately one-tenth the reaction rate constant of that for a primary amine reacting with epoxide. In our work, it is apparent that the cycloaliphatic amine shows the same preference for primary amine reaction.

Results from this study indicate that the simultaneous DSC–FTIR technique may be used to successfully monitor changes in reaction mechanisms during the cure of epoxy systems as a function of activator–resin ratio. Relative changes in the rates of reaction may also be monitored.

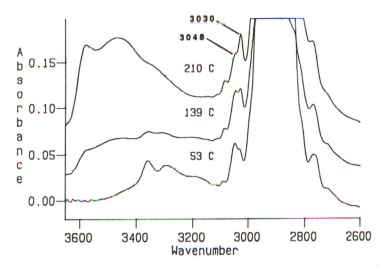

Figure 3. Expanded portion of the IR spectra generated during the cure of 35 phr amine curing agent mixture at 53, 139, and 210 °C.

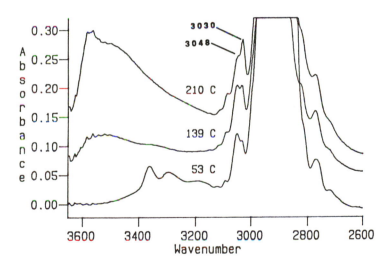

Figure 4. Expanded portions of the IR spectra generated during the cure of the 17 phr amine curing agent mixture at 53, 139, and 210 °C.

PET Research.   Two samples of PET were analyzed by DSC–FTIR to monitor the structural changes of the material as it was heated and cooled through glass-transition temperature and multiple melting endotherm. These samples contained different types of nucleating agent, which was believed to lead to different thermal properties. The samples were heated in the microscopy cell from about 40 to 280 °C at 10 °C per minute. Infrared spectra (256 scans coadded per spectrum) were simultaneously collected by the specular reflectance mode of analysis. As mentioned earlier, due to the reflective surface of the PET, the generated IR spectra exhibited distorted or derivitized band shapes. This phenomenon occurs when a material undergoes a significant change in refractive index in frequency regions of strong IR absorbances. The lower portion of Figure 5 shows an observed spectrum (in units of percent reflectance). The upper portion of Figure 5 shows the observed spectrum after a Kramers–Kronig correction was applied. The Kramers–Kronig correction separated the index-of-refraction component from the extinction coefficient component to produce a "K" spectrum, which has the appearance of a normal absorbance spectrum.

DSC curves for PET normally show a glass-transition temperature near 80 °C (*11*) and a multiple melting endotherm near 255 °C. One sample showed these two transitions as well as a sharp exotherm at 115 °C, due to recrystallization. The DSC curve for this sample, labeled sample X, is shown as the top plot in Figure 6. The DSC curve of sample Y showed the normal melt, but no other inflections due to either the glass transition or recrystallization (*see* top of Figure 7).

Significant changes in the spectra were observed when the two samples were heated. As an example, the temperature behavior for the 1100-cm^{-1} region of the spectrum of sample X is shown in Figure 8, where the bands at 1118 and 1098 cm^{-1} are plotted in ascending order with temperature increase. The 1118-cm^{-1} shoulder increases rapidly in intensity during

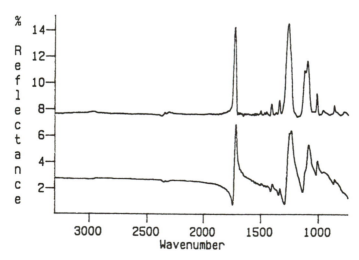

Figure 5. Observed reflectance IR spectrum of PET at 234 °C prior to Kramers–Kronig correction (bottom) and absorbance spectrum obtained after the correction (top).

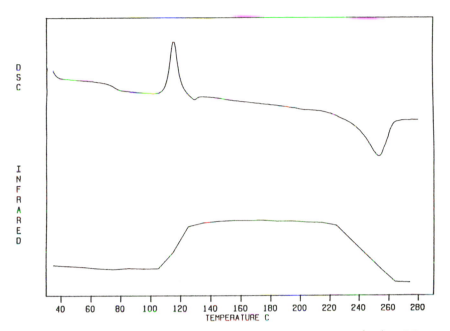

Figure 6. DSC curve for PET sample X heated at 10 °C per minute (top) and the ratio of the infrared peaks at 1118 and 1098 cm⁻¹ as a function of temperature (bottom).

recrystallization at about 120 °C. Figure 9 shows the effect of further heating, but now the peaks are plotted in descending order with increase in temperature for convenience, because the shoulder at 1118 cm^{-1} was observed to decrease during the melt. To show the temperature behavior of this spectral region more clearly, the 1118–1098 intensity ratios for both samples are plotted as a function of temperature in the lower curves in Figures 6 and 7 (below the corresponding DSC curve).

The 1118–1098 intensity ratio is very revealing. It shows that both samples do indeed go through a recrystallization near 120 °C. In the case of sample X, this transition is fast and is accompanied by an exotherm. In the case of sample Y, however, the transition is slower and less well defined, and the DSC curve shows no evidence of an exotherm.

Originally these changes in relative intensity of the band pair at 1100 cm^{-1} were believed to be related to rotational isomerism of the ethylene glycol segments in PET (*12*). The two bands at 1118 and 1100 cm^{-1} were assigned to the trans and gauche conformers of the glycol segment, respectively. Although the position is not conclusive, normal coordinate calculations indicate that the observed changes in relative intensity are due to symmetry and resonance characteristics of the aromatic ring framework instead (*12*).

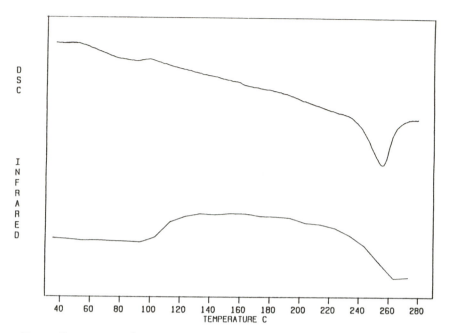

Figure 7. DSC curve for PET sample Y heated at 10 °C per minute (top) and the ratio of the infrared peaks at 1118 and 1098 cm^{-1} as a function of temperature (bottom).

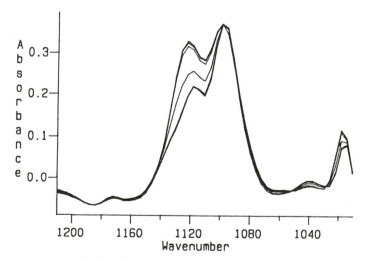

Figure 8. Corrected infrared spectra for PET sample X at 85, 95, 105, 115, 125, 135, and 145 °C (bottom to top) as the sample was heated.

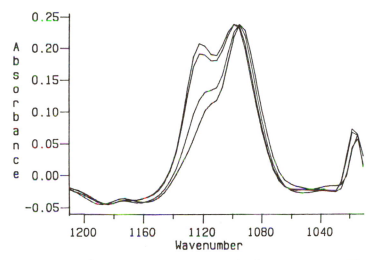

Figure 9. Corrected infrared spectra for PET sample X at 214, 234, 254, and 274 °C (top to bottom) as the sample was heated.

The calculations indicate that the PET framework is relatively rigid, even in the melt. Our observation that the 1100-cm^{-1} band increases in intensity during the recrystallization and then decreases to its original level on melting upholds the argument that these bands are not due to different conformations, but that their intensities arise from a more complex molecular mechanism. The classical way that conformational energy differences are calculated shows that the change in relative intensity of a conformer pair should be a smooth function of T^{-1} (temperature). In any case, whatever causes these changes, examination of spectra during the recrystallization from the melt show that these changes are reversible.

Summary and Conclusions

It is apparent from this research and earlier work performed by Koberstein and Mirabella (4–6) that the DSC–FTIR technique provides a powerful tool for the characterization of polymeric materials. By coupling an FTIR spectrometer to the DSC, structural information about the sample as it passes through thermal transitions may be gained in the same way that TGA–FTIR gives information about evolved gases. These types of information should provide additional insight into polymer behavior in various thermal environments. In this work all of the spectra were collected by use of the reflectance mode of infrared microspectroscopy instead of the transmission mode used previously (4–6). There are several advantages to the reflectance experimental arrangement. It was not necessary to fashion DSC sample cups from a material (like potassium bromide) that transmits infrared radiation; instead, standard aluminum cups were employed. In many cases it is possible to study samples that have the normal thickness for the DSC experiment, even when they are opaque to the infrared beam. Finally, the design of modern infrared transmitting microscope accessories allows for spectra to be collected in the reflectance mode with excellent signal-to-noise ratio. A further advance was gained by performing both DSC and FTIR experiments under the control of a single computer, which enabled easy direct comparison of the types of experimental data. As demonstrated here, spectroscopic data, such as peak ratios and peak frequencies, can be directly compared to the DSC curve, which avoids any ambiguity about the correlation of the various plot abscissae.

By coupling the FTIR to a DSC we were able to gain some insight into the final molecular structure and curing steps of cycloaliphatic polyamine cross-linked DGEBA (diglycidyl ether of bisphenol A) polymers. Our results suggest that the system under study reacts in steps, much the same way as for most epoxy–aromatic amine polymers. In such systems, it is thought that steric hindrance effects, due to an aromatic substituent adjacent to the amine nitrogen, makes the secondary and primary amines react at different rates. Most systems using *aliphatic* primary and secondary amines seem to have reaction rates that are indistinguishable (9). In no way do we dispute that most aromatic and aliphatic amine-cured epoxies show these differences. However, by theorizing a distinct two-step reaction (8) in our study, we suggest that steric hindrance factors influence epoxy–cycloaliphatic polyamines in much the same manner as epoxy–aromatic polyamine reactions.

References

1. *Thermal Characterization of Polymeric Materials*; Turi, E., Ed.; Academic: Orlando, FL, 1981.
2. *Laboratory Methods in Vibrational Spectroscopy*, 3rd ed.; Willis, H. A.; van der Maas, J. H.; Miller, R. G. J., Eds.; Wiley: Chichester, England, 1987.
3. Krishnan, K.; Hill, S. L. In *Practical Fourier Transform Infrared Spectroscopy: Industrial and Laboratory Chemical Analysis*; Ferraro, J. R.; Krishnan, K., Eds.; Academic: San Diego, CA, 1990; Chapter 3, pp 103–165.
4. Mirabella, F. M. *Appl. Spectrosc.* **1986**, *40*, 417–420.
5. Mirabella, F. M. In *Infrared Microspectroscopy*; Messerschmidt, R. G.; Harthcock, M. A., Eds.; Marcel Dekker: New York, 1988; Chapter 6, pp 85–92.
6. Koberstein, J. T.; Gancarz, I.; Clarke, T. C. *J. Polym. Sci., Polym. Phys. Ed.* **1986**, *24*, 2487–2498.
7. Johnson, D. J.; Compton, D. A. C.; Cass, R. S.; Canale, P. L. In *Proceedings of the 17th North American Thermal Analysis Society Conference*; North American Thermal Analysis Society: Colonia, NJ, 1988; Vol. II, pp 574–579.

8. Johnson, D. J.; Compton, D. A. C.; Canale, P. L. *Thermochim. Acta* **1992**, *195*, 5–20.
9. Socrates, G. *Infrared Characteristic Group Frequencies*; Wiley: Chichester, England, 1980.
10. *Epoxy Resins, Chemistry and Technology*; Mai, C. A., Ed.; Marcel Dekker: New York, 1988; pp 301–304.
11. *Polymer Handbook*, 3rd ed.; Brandrup, A., Ed.; Wiley: New York, 1989.
12. Painter, P. C.; Coleman, M. M.; Koenig, J. L. *The Theory of Vibrational Spectroscopy and Its Application to Polymeric Materials*; Wiley-Interscience: New York, 1982; pp 505–510.

RECEIVED for review July 15, 1991. ACCEPTED revised manuscript May 28, 1992.

Thermogravimetric Analysis–Infrared Spectroscopy

A Technique To Probe the Thermal Degradation of Polymers

Charles A. Wilkie[1] and Martin L. Mittleman[2]

[1]Department of Chemistry, Marquette University, Milwaukee, WI 53233
[2]BP American Research, 4440 Warrensville Center Road, Cleveland, OH 44128

The coupling of an infrared spectrometer to equipment for thermogravimetric analysis can help to develop a mechanistic understanding of the degradation mechanism of a polymer in the presence or absence of an additive. Because the additive can have an important effect upon the applications of the polymer, this technique permits correlation of its end uses with structural changes that may occur as a result of polymer–additive interaction. In this paper, we examine the effect of two additives, perfluorinated ionomer (Nafion-H) and manganese(II) chloride, on the thermal degradation of poly(methyl methacrylate) and propose mechanisms to account for the volatile products that evolve from these systems during thermal degradation.

ADDITIVES CAN HAVE AN IMPORTANT EFFECT ON THE PROPERTIES of polymers; they may be used as plasticizers, antioxidants, light stabilizers, and flame retardants. Each of these additives has some important effect on the potential end uses of the polymer. Because these additives can markedly affect the properties of the polymer, they must also affect its structure in some way. The particular additives of interest here are flame retardants. Flame retardant additives may function in the vapor phase or in the condensed phase. Vapor phase retardants generally are believed to generate radicals that combine with the radicals of the flame, thus removing them from the combustion zone; any chemical reactions occur only after thermal degradation and hence do not effect the structure of the polymer. In the condensed phase, some chemical reactions do occur between the additive and the polymeric substrate. These reactions can alter the structure, and hence the properties, of the substrate (1).

The thermal degradation of poly(methyl methacrylate) (PMMA) has been studied by a great many workers. Two recent series of papers have focused new attention on this process. Kashiwagi and co-workers (2–5) implicated weak links in the polymer as the principal sites from which degradation may occur. This research group showed that PMMA prepared by a radical process degrades in three distinct steps, whereas anionically polymerized material degrades in only a single step. This single step for the anionically prepared polymer is an end-chain scission process, which is typical of many polymers. The highest temperature degradation step for the radically prepared polymer occurs at the same temperature as that observed for the anionically prepared polymer and is ascribed to the same process. The other two steps are believed to be the result of the cleavage of weak links in the polymer chain; these are specifically described as head-to-head linkages and unsaturated end groups (5).

Manring (6–9) has postulated that the weak links are less important than previously thought and that degradation is begun by the cleavage of the carbomethoxy group from the main chain of the PMMA. Monomer is the principal product that is observed when PMMA is thermally degraded. Trace quantities of other products, notably CO_2, CO, CH_4, and CH_3OH, are also observed, but these products are truly present in very small amounts and are not easily seen. The initial step in the Manring degradation pathway is always the cleavage of the carbomethoxy group from the polymer chain with the formation of a carbomethoxy radical and a radical along the main chain. The carbomethoxy radical may degrade to produce CO_2 and a methyl radical or CO and a methoxy radical. The methyl and methoxy radicals can abstract hydrogens from the polymer and yield methane and methanol, respectively. The main chain radical will degrade to give monomer.

The thermal degradation of PMMA in the presence of additives has been extensively studied recently by the McNeill group in Scotland and our group at Marquette University. The McNeill group has examined the reaction of PMMA with silver acetate, ammonium polyphosphate, and zinc and cobalt bromides. The group at Marquette University has examined reactions of PMMA with red phosphorus and Wilkinson's catalyst.

McNeill and co-workers have examined the effect of silver acetate (10), ammonium polyphosphate (11), and zinc bromide (12, 13) on PMMA degradation by the use of thermal volatilization analysis (TVA). When silver acetate is used as an additive with PMMA, significant destabilization of the polymer is seen, presumably caused by the diffusion of acetoxy radicals into the PMMA chain, which initiates chain scission. A similar process is observed in a study of the degradation of PMMA in the presence of PVC. It is postulated that the chlorine atoms produced by PVC degradation initiate rapid degradation of the PMMA chain (*14*). It may be concluded that the presence of radicals is deleterious to polymer stability because they can initiate chain scission.

In a blend of PMMA and ammonium polyphosphate (*11*), the major product of thermal degradation is monomer, but significant amounts of other products, notably methanol, CO, CO_2, dimethyl ether, and char are also produced. It was suggested that the degradation of ammonium polyphosphate produces the strong acid, polyphosphoric acid, and that this catalyzes the hydrolysis of ester groups on the PMMA to give some charring.

The combination of zinc bromide and PMMA (*12*, *13*) seems to significantly retard depolymerization of the polymer. The initial step appears to be the coordination of zinc to the carbonyl carbon of the PMMA. This zinc complex can lose methyl halide and ultimately form a zinc salt.

In this laboratory we have been concerned with developing a mechanistic understanding of the reactions of PMMA with various additives. The belief is that if one can predict, in detail, how a polymer and a variety of additives will chemically interact, then an additive to perform a specific function can be designed. Initially we investigated the reaction of PMMA and red phosphorus. This investigation was motivated by reports that indicated some efficacy for red phosphorus as a flame retardant for oxygenated polymers (*15*). However, there were no reports that delineated the course of the reaction. Our investigation showed that red phosphorus attacks the carbonyl moiety of the PMMA with the formation of methyl methoxy phosphonium ions and an intramolecular anhydride (*16*, *17*). Because attack occurs at a carbonyl site, an additive known to interact with a carbonyl may prove to be an effective retardant. As such, we chose to use Wilkinson's catalyst, $ClRh(PPh_3)_3$. Reaction proceeds between PMMA and Wilkinson's catalyst by an oxidative insertion of the rhodium species into a carbon−oxygen bond of the polymer; both intra- and intermolecular anhydrides are formed and extensive char formation occurs. It is significant that the limiting oxygen index (LOI) increases by 6 points when the rhodium compound is physically combined with PMMA (*18*, *19*).

There is a significant difference between degradation carried out in a static system, such as a sealed tube, and a dynamic system, such as thermogravimetric analysis. In a sealed tube reaction, the degradation products are contained and may undergo further reaction, whereas in a dynamic system the products are swept out of the system as quickly as they are formed so that further reaction is prevented. The degradation of PMMA in a sealed tube leads to significant quantities of volatile products. When monomer alone is heated under identical conditions, smaller quantities of these volatile products are obtained, which indicates that these products arise not only from degradation of polymer, but also from some degradation of monomer.

In this chapter we will summarize our recent studies on the interaction of Nafions and manganese(II) chloride with PMMA and offer a new interpretation of the manganese(II) chloride reaction. Although these studies have utilized both sealed tube reactions and thermogravimetric analysis (TGA) coupled to infrared (IR) spectroscopy, the focus of this paper is on the development of a mechanistic understanding of the reaction by TGA–IR. Because the sealed tube reaction is a static system and the TGA–IR experiment is a dynamic system, it is not too surprising that the results are somewhat different in the details; however, they still provide the same overall conclusion. TGA–IR is a valuable technique for investigating the mechanistic aspects of the reaction between a polymer and its additives because it provides both temporal and temperature resolution of the thermal degradation processes.

Experimental Details

TGA–IR was performed using a thermogravimetric analyzer (supplied by Omnitherm Corporation) coupled to a Fourier transform infrared spectrometer (Digilab FTS-60). Specific details of this integrated system have been provided elsewhere (20, 21). Of significance is the fact that the TGA interface and slave processor are both controled by a single data station, which also simultaneously controls infrared data collection. The complete system runs under the direction of a single macro.

TGA sample sizes for this study ranged from 20 to 25 mg and the samples were heated at 20 °C/min under an inert gas purge of 50 cm³/min. Gases evolved from the heated sample were transferred to an IR gas cell via a glass transfer line heated to 210 °C. This line had an overall length of 47 cm and an inner diameter of 0.2 cm. The stainless steel gas cell had a 10-cm path length and 0.6-cm inner diameter; its temperature was maintained at 235 °C.

Spectroscopic data were collected using the Fourier transform infrared spectrometer equipped with a KBr beam splitter and a wide-band liquid nitrogen-cooled mercury–cadmium telluride (MCT) detector. Spectra were collected at 8-cm^{-1} resolution, coadding 16 scans per spectrum. This resulted in a temporal resolution of 4 s, more than sufficient for the gradual gas evolution rates characteristic of most TGA profiles. The problems and possibilities for quantification of evolved gases using the spectral information from TGA–IR have been reported (22). The present work is concerned only with qualitative identification.

Results and Discussion

Degradation of Perfluorinated Ionomer (Nafion-H). The structure of Nafion-H consists of a poly(tetrafluoroethylene) (PTFE) backbone with pendant sulfonic acid groups. Before the thermal degradation of Nafions can be understood, it is important to have some insight into the thermal degradation process for PTFE. PTFE is one of the most thermally stable linear polymers. Its thermal stability is attributed to the high $C-F$ bond strength and the shielding effect of the very electronegative fluorines. The thermal degradation of PTFE commences at about 450 °C and is believed to proceed by random chain scission with the formation of difluorocarbene. This reactive carbene leads to the observed monomeric tetrafluoroethylene (TFE) and oligomeric products (23–25).

There is little degradation of Nafion-H in a TGA experiment (26) below 280 °C in an inert atmosphere (Figure 1a). A small 5% weight loss occurs below this temperature and the only gases that are detected are H_2O, SO_2, and CO_2. In an analogous experiment, Ehlers et al. (27) subjected poly(arylene sulfonate) to thermal degradation in vacuum and observed that SO_2 was evolved by cleavage of the $C-S$ bond; the maximum evolution occurred between 250 and 350 °C.

A TGA weight loss of 7% occurs between 280 and 335 °C. The evolution of SO_2 and CO_2 increases throughout this region whereas that of water decreases. SiF_4 (1026 cm^{-1}), CO, HF, substituted carbonyl fluorides (1957 and 1928 cm^{-1}), and absorbances in the $C-F$ stretching region also appear over this temperature range. An IR spectrum of the gases evolved at 367 °C is shown in Figure 2. Of particular note are the bands just below 2000 cm^{-1} attributable to carbonyl fluorides. SiF_4 is not a primary product of the reaction, but rather it arises from the attack of evolved HF on glass. Thus when SiF_4 is observed, the formation of HF is indicated and the actual product will be identified as HF. At the highest temperatures, 355–560 °C, 88% of the sample volatilizes. The amounts of SO_2 and CO decrease dramatically at 365 °C and are no longer of consequence. The major absorbances in this temperature region are due to HF, carbonyl fluorides, and $C-F$-containing species. A mechanism that accounts for these observation is presented in Scheme I.

The $C-S$ bond is initially broken, which produces a CF_2 radical, SO_2 and a hydroxyl radical (eq 1). An alternative explanation is that this occurs in two steps: the initial formation of an SO_3H radical, followed by cleavage to form SO_2 and the OH radical. In either case, the fluorocarbon radical can then lose two difluorocarbenes (eqs 2 and 3), which produces an oxygen-based radical. This radical can subsequently lose a substituted carbonyl fluoride (eq 4) and carbonyl fluoride (eq 5). The remaining PTFE-like backbone will then degrade to tetrafluoroethylene monomer and oligomers.

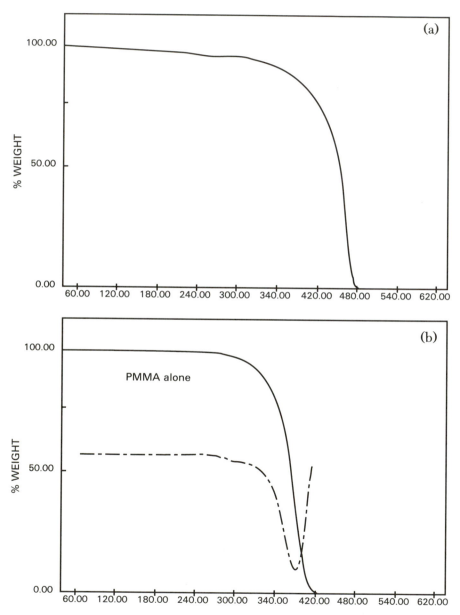

Figure 1. TGA curve for Nafion-H (a), PMMA (b), and a blend of the two (c). These curves were obtained under an inert atmosphere at a scan rate of 20 °C.

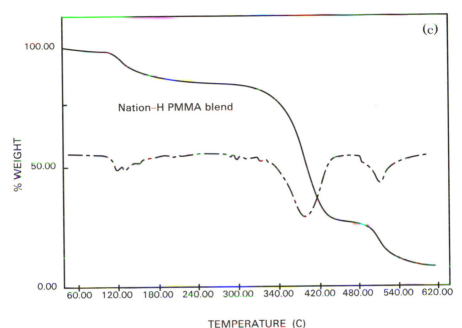

Figure 1. Continued

Verification of this mechanism comes from an examination of the degradation of the potassium salt of the sulfonic acid, Nafion-K, prepared by soaking a sample of Nafion-H in aqueous KOH. Because the thermal stability of amine arenesulfonates is greater than that of the corresponding sulfonic acids (28), cleaving the C–S bond should be more difficult for Nafion-K than it is for Nafion-H. This hypothesis is supported by the fact that when Nafion-K is subjected to TGA–IR investigation, there is no weight loss until 390 °C, some 100 °C higher than that observed for Nafion-H. SO_2 is not observed; the only products are HF and fluorocarbon oligomers that would be expected from PTFE.

Interaction of Nafion-H and PMMA. TGA–IR studies of a blend prepared by casting a PMMA film onto a Nafion-H film reveal a significant difference between the individual components and the blend (26). The TGA of Nafion-H is shown as Figure 1a, PMMA is shown in Figure 1b, and the blend is shown in Figure 1c. Both PMMA and Nafion-H have completely volatilized at 500 °C, whereas the blend has about 10% residue remaining at 600 °C. The TGA curve of the blend indicates that degradation occurs in three stages. In the first stage, from 120 to 265 °C, 11% of the sample is volatilized. The gases evolved during this stage are water, which is retained by the Nafion, and chloroform, the solvent from which the PMMA is cast onto

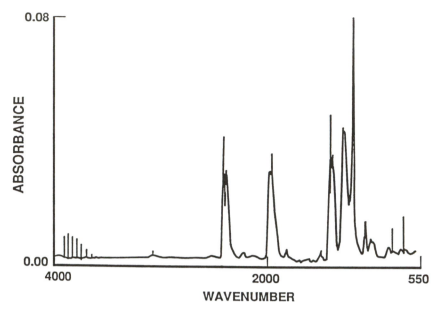

Figure 2. Infrared spectrum of the gases evolved from Nafion-H at 367 °C.

the Nafion. The evolution of the gases is merely the result of a physical thermal desorption.

The actual onset of chemical degradation of this blend begins in the second stage of degradation, between 265 and 430 °C during which 52% of the sample is volatilized. This onset is some 80 °C higher than is seen for PMMA alone, which indicates a significant interaction between PMMA and Nafion-H. The first gas observed at 265 °C is SO_2; its evolution begins to decrease near the end of the region. Monomeric methyl methacrylate is first observed at 270 °C, grows in intensity until 415 °C, and then significantly decreases. Monomer evolution is monitored by the carbonyl absorption at 1750 cm^{-1}. Methyl methacrylate appears to be the major product evolved in this region. Evolutions of HF, CO, and CO_2 are also apparent throughout this temperature range. The spectrum at 367 °C is shown in Figure 3. The lack of absorbances in the 1900–2000 cm^{-1} region indicates that carbonyl fluorides are not produced from the blend although they are formed when Nafion-H is pyrolyzed alone.

The gases evolved over the third region, 430–575 °C, are the same as those observed in the degradation of Nafion-H alone. However, the carbonyl fluorides that are observed in Nafion-H by itself at 280 °C are not seen in the blend until 450 °C and then only in very small amounts. When the residue produced at 400 °C is examined by IR, anhydride absorbances are observed.

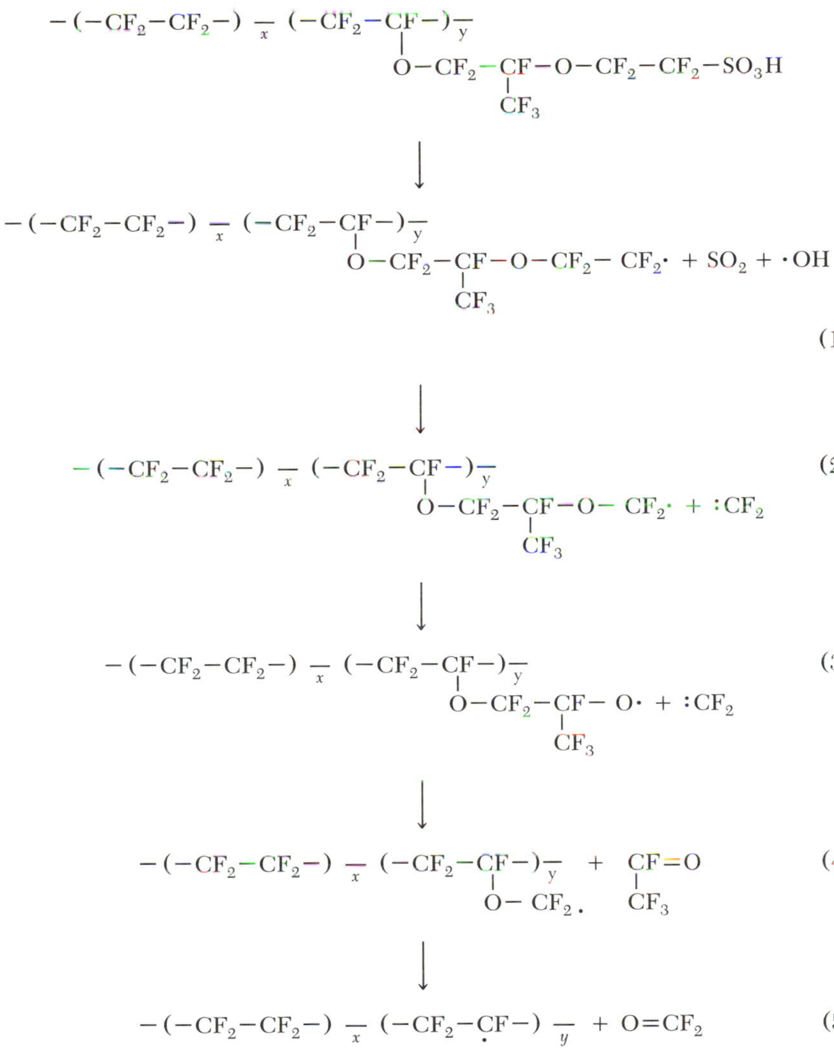

$$-(-CF_2-CF_2-) \overline{}_x \; (-CF_2-CF-)\overline{}_y$$
$$O-CF_2-CF-O-CF_2-CF_2-SO_3H$$
$$CF_3$$

$$\downarrow$$

$$-(-CF_2-CF_2-) \overline{}_x \; (-CF_2-CF-)\overline{}_y$$
$$O-CF_2-CF-O-CF_2-\,CF_2\cdot \; + SO_2 + \cdot OH$$
$$CF_3$$

$$\tag{1}$$

$$\downarrow$$

$$-(-CF_2-CF_2-) \overline{}_x \; (-CF_2-CF-)\overline{}_y \tag{2}$$
$$O-CF_2-CF-O-\,CF_2\cdot \; + \; :CF_2$$
$$CF_3$$

$$\downarrow$$

$$-(-CF_2-CF_2-) \overline{}_x \; (-CF_2-CF-)\overline{}_y \tag{3}$$
$$O-CF_2-CF-\,O\cdot \; + \; :CF_2$$
$$CF_3$$

$$\downarrow$$

$$-(-CF_2-CF_2-) \overline{}_x \; (-CF_2-CF-)\overline{}_y \; + \; CF=O \tag{4}$$
$$O-\,CF_2\cdot \qquad CF_3$$

$$\downarrow$$

$$-(-CF_2-CF_2-) \overline{}_x \; (-CF_2-CF-) \overline{}_y \; + \; O=CF_2 \tag{5}$$

Scheme I

A possible mechanism to account for these results is presented in Scheme II.

Because SO_2 is the first product observed, the initial step in the degradation of the blend must be the same as that for Nafion-H: cleavage of the $C-S$ bond to generate a fluorocarbon radical, SO_2, and hydroxyl radical. If radical propagation continued along the chain as it does for Nafion-H alone, the formation of TFE and carbonyl fluorides would be expected.

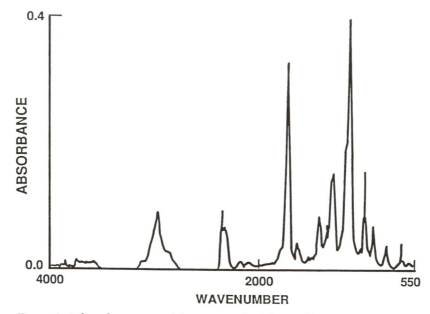

Figure 3. Infrared spectrum of the gases evolved from a blend of Nafion-H and PMMA at 367 °C.

Although some TFE is evolved, it does so at higher temperatures and in smaller amounts than seen for the degradation of the Nafion alone. This decrease in the amount of TFE and the absence of carbonyl fluorides suggests that radical propagation is halted and that instead these radicals react rather than undergoing further degradation. It is likely that both the CF_2 radical and the oxygen-based radical react with PMMA, although Scheme II shows only reaction with fluorocarbon radical. The fact that a large amount of CO_2 is produced indicates that some radical sites are generated on the polymer backbone by loss of carbomethoxy groups. These radicals would be expected to degrade in a manner similar to PMMA. The formation of anhydrides must be attributed to reactions of the hydroxyl radical. These may interact with the ester functionality and form methanol and anhydride.

Verification again comes from the examination of the blend of PMMA with Nafion-K. The TGA–IR investigation of this blend shows that the only gases produced are those expected for the individual components. No gases evolved that are formed by any chemical reactions between the components.

Degradation of a PMMA–MnCl$_2$ Blend. The work of McNeill et al. on $ZnBr_2$ and $CoBr_2$ (*12, 13*) suggested that simple species such as these may coordinate to PMMA and rearrange to give a salt with the evolution of methyl halide. The formation of methyl halide is a disadvantage for a flame retardant because methyl halide will act as fuel, but salt formation

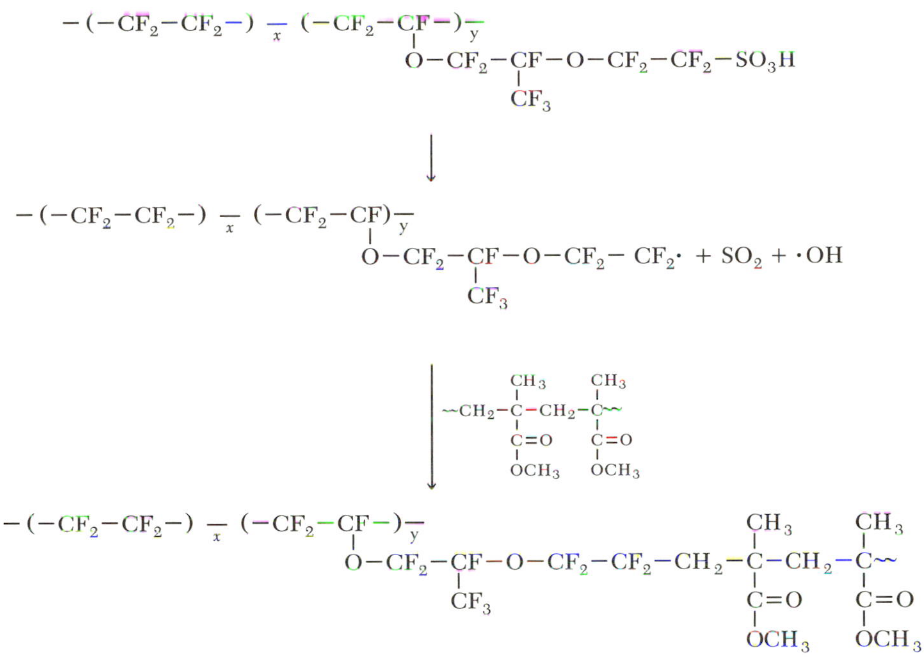

$$-(-CF_2-CF_2-) \frac{}{x} (-CF_2-CF-)\frac{}{y}$$
$$O-CF_2-CF-O-CF_2-CF_2-SO_3H$$
$$CF_3$$

$$\downarrow$$

$$-(-CF_2-CF_2-) \frac{}{x} (-CF_2-CF)\frac{}{y}$$
$$O-CF_2-CF-O-CF_2- CF_2\cdot + SO_2 + \cdot OH$$
$$CF_3$$

$$\begin{array}{cc} CH_3 & CH_3 \\ | & | \\ \sim CH_2-C-CH_2-C\sim \\ | & | \\ C=O & C=O \\ | & | \\ OCH_3 & OCH_3 \end{array}$$

$$-(-CF_2-CF_2-) \frac{}{x} (-CF_2-CF-)\frac{}{y}$$
$$O-CF_2-CF-O-CF_2-CF_2-CH_2-C-CH_2-C\sim$$
$$CF_3 \qquad\qquad C=O \qquad C=O$$
$$OCH_3 \qquad OCH_3$$

Scheme II

seems to be a desirable goal. Learning how to form a salt without the concomitant evolution of volatiles may prove advantageous as a flame retardant. Accordingly we commenced a systematic investigation of the reaction of a variety of simple transition metal salts with PMMA, beginning with the reaction of PMMA and $MnCl_2$ (29).

There are five distinct regions of thermal degradation in the TGA curve of a blend of $MnCl_2$–PMMA (Figure 4). Below 145 °C the only product observed is water — the result of physical desorption from manganese chloride. Even though anhydrous material was used to prepare this blend, $MnCl_2$ is sufficiently hygroscopic that water is retained. Initial thermal degradation of the blend occurs between 145 and 215 °C with the evolution of 9.5% of the sample as monomer. Following the work of Kashiwagi (2–5), the degradation in this region might be ascribed to the presence of weak links in the polymer, whereas the work of Manring (6–9) ascribes this degradation to the presence of radicals that may initiate degradation. Because the PMMA itself does not degrade at this temperature, the observed degradation must be ascribed to the presence of $MnCl_2$.

The second region of degradation, 215–345 °C, is characterized by the evolution of methyl chloride and methanol as well as the continued formation of monomer. 10.5% of the sample is lost in this region. The spectrum of the

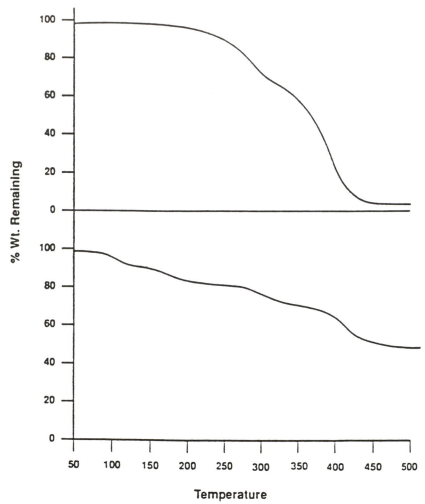

Figure 4. TGA of PMMA (top) and a blend of MnCl$_2$ and PMMA (bottom).

gases evolved at 303 °C is shown in Figure 5. This spectrum shows the presence of monomer, methanol, and methyl chloride absorbing near 1750, 1034, and 733 cm^{-1}, respectively. Methyl chloride must arise by the interaction of the ester methyl with the chloride from MnCl$_2$. An additional 19% of the sample volatilizes between 345 and 455 °C with the evolved gases identified as CO, CO$_2$, CH$_4$, HCl, and an aliphatic acid. The final region of weight loss, 455–630 °C, accounts for 5% of the sample. The only gas evolved in this temperature regime is HCl.

A possible pathway for the degradation is presented in Scheme III. The initial step is the coordination of the manganese ion to the carboxyl carbon.

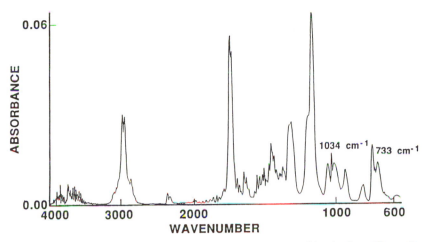

Figure 5. Infrared spectrum of the gases evolved from a blend of $MnCl_2$ and PMMA at 303 °C.

At low temperatures this complex loses a chlorine atom to initiate PMMA degradation. At a somewhat higher temperature it rearranges with the evolution of methyl chloride, by a concerted process, to produce a manganese salt. Recall that in the degradation of a PVC–PMMA sample, the presence of chlorine atoms was postulated to initiate degradation. These results suggest that the presence of radicals is not itself a problem if these radicals can be consumed in some concerted pathway. The manganese salt may coordinate with another carboxyl fragment to form an ionomer. Monomer, CO_2, CO, CH_4, and CH_3OH all arise by the normal degradation pathway proposed by Manring for PMMA degradation (9). The degradation pathway that we have previously proposed for this reaction is much more complicated; the pathway reported herein was suggested by the very recent work of Manring (9).

To investigate the nonvolatile products of the reaction, the reaction mixture can be heated to the desired temperature under an inert gas flow and the residue can be isolated. When the PMMA–$MnCl_2$ mixture is heated to 500 °C under these condition, MnO_2 is isolated from the reaction system as identified by X-ray powder pattern analysis. MnO_2 is also produced by the thermal degradation of manganese glutarate, so it is likely that the manganese salt or ionomer degrades to this product.

Summary and Conclusions

Condensed-phase flame retardants, which can have a significant effect on the properties of polymers, must occur by some chemical interaction between the additive and the substrate. The nature of this chemical interaction may be conveniently probed by TGA–IR to delineate a complete mechanistic path-

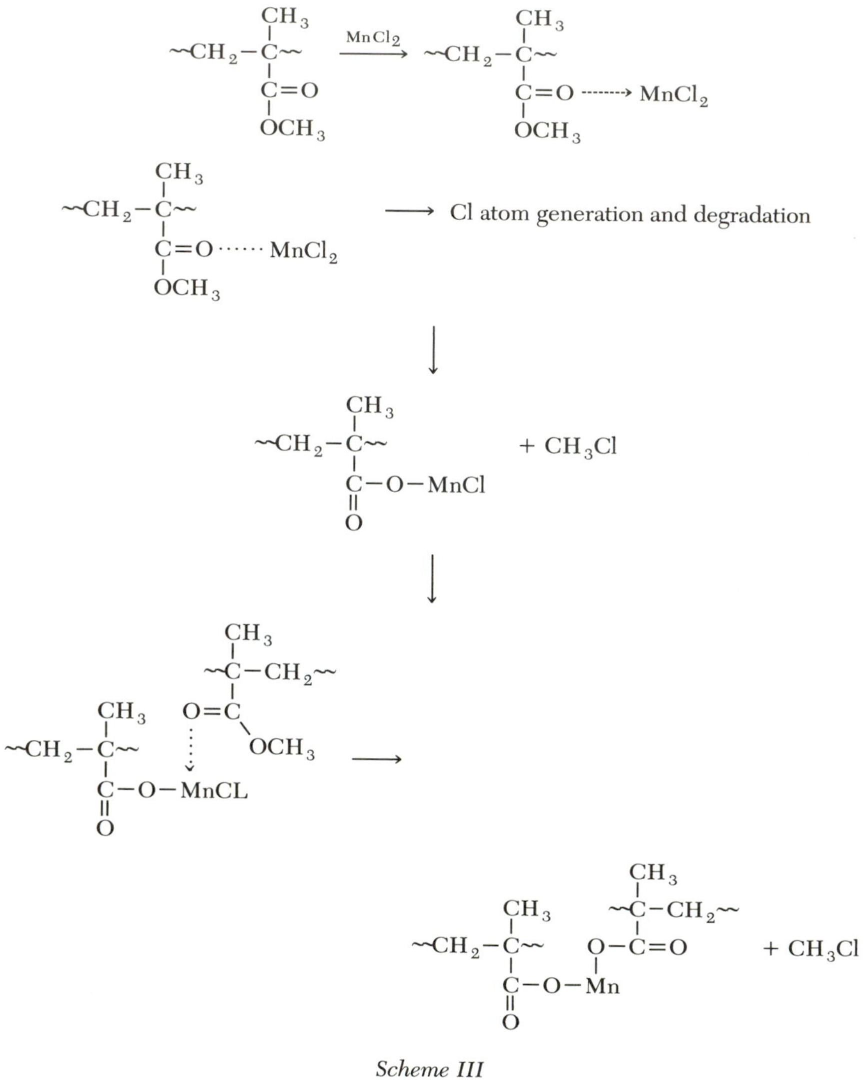

$$\begin{array}{c}CH_3\\|\\\sim CH_2-C\sim\\|\\C=O\\|\\OCH_3\end{array}\xrightarrow{MnCl_2}\begin{array}{c}CH_3\\|\\\sim CH_2-C\sim\\|\\C=O\cdots\cdots\rightarrow MnCl_2\\|\\OCH_3\end{array}$$

$$\begin{array}{c}CH_3\\|\\\sim CH_2-C\sim\\|\\C=O\cdots\cdots MnCl_2\\|\\OCH_3\end{array}\longrightarrow \text{Cl atom generation and degradation}$$

$$\downarrow$$

$$\begin{array}{c}CH_3\\|\\\sim CH_2-C\sim\\|\\C-O-MnCl\\\|\\O\end{array}\quad + CH_3Cl$$

$$\downarrow$$

$$\begin{array}{cc}&\begin{array}{c}CH_3\\|\\\sim C-CH_2\sim\\|\\O=C\\\quad\setminus\\\quad OCH_3\end{array}\\\begin{array}{c}CH_3\\|\\\sim CH_2-C\sim\\|\\C-O-MnCL\\\|\\O\end{array}&\end{array}\longrightarrow$$

$$\begin{array}{cc}&\begin{array}{c}CH_3\\|\\\sim C-CH_2\sim\\|\end{array}\\\begin{array}{c}CH_3\\|\\\sim CH_2-C\sim\\|\\C-O-Mn\\\|\\O\end{array}&\quad O-C=O\end{array}\quad + CH_3Cl$$

Scheme III

way for the interaction of the polymer and the additive. This important technique offers the opportunity to obtain both temporal and temperature resolution of the degradation process. A mechanistic understanding of the reaction pathway can eventually permit the design of an additive that undergoes a specific reaction with a polymer, which enables it to perform as an effective flame retardant.

References

1. Schnabel, W. *Polymer Degradation*; Hanser International: Munich, Germany, 1981.
2. Kashiwagi, T.; Harita, T.; Brown, J. E. *Macromolecules* **1985**, *18*, 131.
3. Harita, T.; Kashiwagi, T.; Brown, J. E. *Macromolecules* **1985**, *18*, 1410.
4. Kashiwagi, T.; Inaba, A.; Hamins, A. *Polym. Degrad. Stab.* **1989**, *26*, 161.
5. Kashiwagi, T.; Inaba, A.; Brown, J. E.; Hatada, K.; Kitayama, T.; Masuda, E. *Macromolecules* **1986**, *19*, 2160.
6. Manring, L. *Macromolecules* **1988**, *21*, 528.
7. Manring, L. *Macromolecules* **1989**, *22*, 2673.
8. Manring, L. *Macromolecules* **1989**, *22*, 4652.
9. Manring, L. *Macromolecules* **1991**, *24*, 3304.
10. Jamieson. A.; McNeill, I. C. *J. Polym. Sci., Polym. Chem. Ed.* **1978**, *16*, 2225.
11. Camino, G.; Grassie, N.; McNeill, I. C. *J. Polym. Sci., Polym. Chem. Ed.* **1978**, *16*, 95.
12. McNeill, I. C.; McGuiness, R. C. *Polym. Degrad. Stab.* **1984**, *9*, 167.
13. McNeill, I. C.; McGuiness, R. C. *Polym. Degrad. Stab.* **1984**, *9*, 209.
14. McNeill, I. C. *Dev. Polym. Degrad.* **1977**, *1*, 171.
15. Peters. E. N. In *Flame Retardancy of Polymeric Materials*; Kuryla, W. C.; Papa, A. J., Eds.; Marcel Dekker: New York, 1979; Vol. 5, p 113.
16. Wilkie, C. A.; Pettegrew, J. W.; Brown, C. E. *J. Polym. Sci., Polym. Lett. Ed.* **1981**, *19*, 409.
17. Brown, C. E.; Wilkie, C. A.; Smukalla, J.; Cody, R. B., Jr.; Kinsinger, J. A. *J. Polym. Sci., Polym. Chem. Ed.* **1986**, *24*, 1297.
18. Sirdesai, S. J.; Wilkie, C. A. *J. Appl. Polym. Sci.* **1989**, *37*, 863.
19. Sirdesai, S. J.; Wilkie, C. A. *J. Appl. Polym. Sci.* **1989**, *37*, 1595.
20. Compton, D. A. C.; Johnson, D. J.; Mittleman, M. L. *Res. Dev.* February **1989**, 142.
21. Compton, D. A. C.; Johnson, D. J.; Mittleman, M. L. *Res. Dev.* April **1989**, 68.
22. Mittleman, M. L. *Thermochim. Acta* **1990**, *166*, 301.
23. Wunderlich, B. In *Thermal Characterization of Polymeric Materials*; Turi, E. A., Ed.; Academic: Orlando, FL, 1981; p 221.
24. Errede, L. A. *J. Org. Chem.* **1962**, *27*, 3425.
25. Tutiya, M. *J. Appl. Phys.* **1969**, *8*, 1356.
26. Wilkie, C. A.; Thomsen, J. R.; Mittleman, M. L. *J. Appl. Polym. Sci.* **1991**, *42*, 901.
27. Ehlers, G. F. L.; Fisch, K. R.; Powell, W. R. *J. Polym. Sci., Polym., Part A–1* **1969**, *7*, 2969.
28. Kaczmarek, T. D.; Phillips, D. C.; Smith, J. D. B. *Microchem. J.* **1977**, *22*, 15.
29. Wilkie, C. A.; Leone, J. T.; Mittleman, M. L. *J. Appl. Polym. Sci.* **1991**, *42*, 1133.

Received for review July 15, 1991. Accepted revised manuscript September 16, 1992.

Thermal Degradation of Cotton Cellulose Studied by Fourier Transform Infrared–Photoacoustic Spectroscopy

Charles Q. Yang[1] and James M. Freeman[2]

[1]**Department of Textiles, Merchandising and Interiors, The University of Georgia, Athens, GA 30602**
[2]**Department of Chemistry, Marshall University, Huntington, WV 25755**

Cotton textiles are exposed to many types of heat damage during manufacture and consumer use. Therefore, it is important to understand the nature of thermal oxidation and degradation in cotton textiles. In this research, cotton fabrics heated at three different temperatures (180, 210, and 240 °C) were studied using Fourier transform infrared–photoacoustic spectroscopy. Ketone, aldehyde, carboxylic acid, ester, anhydride, and unsaturated hydrocarbon structures were identified at different stages of thermal oxidation. Anhydride was first formed in the near surface of the cotton fabric during this process. At 180 °C, the oxidation products (ketone, aldehyde, carboxylic acid, and ester) were homogeneously distributed between the near surface of the fabric and the bulk. At 240 °C, however, more cotton cellulose was oxidized in the near surface than in the bulk. The acceleration of thermal oxidation of cotton cellulose by increasing temperature was also observed.

T HE ANALYSIS OF THERMAL OXIDATION OF COTTON CELLULOSE under different conditions has long been of interest to chemists (*1–4*). Cotton textiles are exposed to many types of heat damage during manufacture and consumer use, such as singeing, excess heat setting in finishing, overdrying, and ironing (*5*). Therefore, it is important to understand the nature of thermal oxidation

0065–2393/93/0236–0693$06.00/0

and degradation in cotton fabrics. Knowledge about thermal degradation of cotton cellulose is also useful for the development of fire-retardant cotton fabrics (6) and the preservation of historical artifacts (7).

In the past, infrared spectroscopy has been used to study the degradation of cotton cellulose at elevated temperature (7–12). In this research, Fourier transform infrared–photoacoustic spectroscopy (FTIR–PAS) was used to study the thermal oxidation and degradation of cotton fabrics. Since the early 1980s when photoacoustic detection was extended to the mid-infrared region, FTIR–PAS has attracted considerable interest for material characterization (13). In the past, FTIR–PAS has been successfully used in our research as a near-surface analytical technique for a variety of chemically modified textile fabrics, yarns, and fibers (14–9).

A textile sample is first ground into a powder using a Wiley mill. Both the textile sample and the powder sample are analyzed by FTIR–PAS. Comparison of the photoacoustic infrared spectrum of the textile sample and the spectrum of the corresponding powder sample enables evaluation of the distribution of chemical species between the near surface and the bulk of the textile sample. In this research, the distribution of oxidation products between the near surface of a cotton fabric and its bulk was determined by this method.

Experimental Details

Instrumentation. A Fourier transform infrared spectrometer (Nicolet 20 DXB) with a photoacoustic cell (MTEC Model 100) was used to collect all the spectra. Resolution for all the spectra was 8 cm^{-1}, and the average number of scans was 500. Carbon black was used as a reference material and helium was used as the purging and conducting gas in the photoacoustic cell. The mirror velocity was 0.139 cm/s; no base-line correction or smoothing function was used.

Materials. The cotton fabric was a desized, bleached cotton print cloth (Testfabrics style 400). Cotton fabrics were heated in a radiant gravity oven. The temperature variation was within ±3 °C. Powdered samples were obtained by grinding the fabric samples in a Wiley mill to pass a 40 mesh screen.

Results and Discussion

The photoacoustic infrared spectra of the cotton fabric heated at 180 °C for different times are shown in Figures 1 and 2. The development of a carbonyl band at 1730 cm^{-1} is observed in the initial stages of thermal oxidation (Figure 1). The intensity of the carbonyl band at 1730 cm^{-1} increased as the

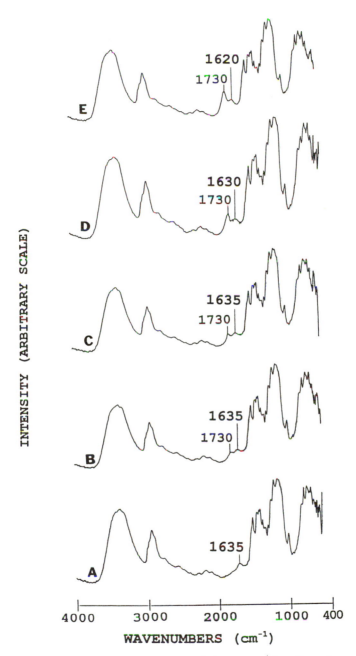

Figure 1. Photoacoustic infrared spectra of the cotton fabric heated at 180 °C for different times (hours): A, 0; B, 6; C, 15; D, 48; E, 72.

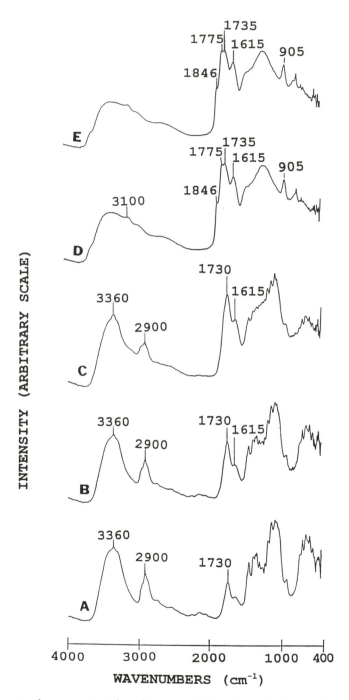

Figure 2. Photoacoustic infrared spectra of the cotton fabric heated at 180 °C for different times (hours): A, 123; B, 225; C, 295; D, 415; E, 515.

exposure time was increased. When the cotton fabric was heated from 123 to 295 h, the 1730-cm^{-1} band intensity continued to increase while the intensities of the bands around 2900 cm^{-1} (aliphatic C—H stretching) and the bands in the 1430–1310-cm^{-1} region (aliphatic C—H bending) were reduced (Figures 2A–C). The bands in the 1430–1100-cm^{-1} region (aliphatic C—H bending and C—O stretching) broadened and started to overlap in Figure 2C. The infrared spectroscopic data indicated that both the primary and secondary alcohols were oxidized to form carbonyls at this stage. A band at 1615 cm^{-1} also developed together with the increase in intensity of the 1730-cm^{-1} band during this period (Figure 2). The band at 1635 cm^{-1} associated with bending of cellulosic hydroxyls was overlapped by the newly developed band at 1615 cm^{-1} (Figures 2A and B). When the exposure time reached 415 h, the bands around 2900 cm^{-1} were no longer discernible, whereas a weak band at 3100 cm^{-1} (unsaturated C—H stretching) appeared (Figure 2D). Also in Figure 2D the bands in the 1430–1100-cm^{-1} region are observed to overlap to form one broad band. Also, a decrease in the intensity of the broad band at 3360 cm^{-1} (O—H stretching) in Figure 2D indicates that dehydration occurred within the cellulose molecules. Three new bands at 1846, 1774, and 905 cm^{-1} were also observed in Figures 2D and E.

To determine the nature of the chemical changes caused by continued heating, the cotton fabric heated at 180 °C for 48 h was treated with a 0.1-M aqueous solution of NaOH for 5 min. The photoacoustic infrared spectrum of the cotton fabric thus treated showed a decrease in the 1730-cm^{-1} band intensity and the formation of the 1615-cm^{-1} band, which is due to carboxylate carbonyl (Figure 3B). Treatment of the fabric with NaOH converted carboxylic acid to carboxylate. A decrease in the 1730-cm^{-1} band intensity in Figure 3B indicated that the 1730-cm^{-1} band in Figure 3A was partially contributed by carboxylic acid carbonyl. The same phenomenon was also observed for the cotton fabric heated at 180 °C for 72 h (Figures 3D and E). The fabric heated at 180 °C for 48 h was also treated with sodium borohydride in ethanol at reflux for 1 h. The infrared spectrum of the fabric thus treated (Figure 3C) shows a further slight decrease in the 1735-cm^{-1} band intensity due to the reduction of carbonyls of ketone and aldehyde. A weak band at 1730 cm^{-1} in Figure 3C, which remained after the treatment of the fabric by sodium borohydride, was likely due to ester carbonyl.

Three carbonyl bands at 1846, 1775, and 1735 cm^{-1} were shown in the infrared spectrum of the cotton fabric heated at 240 °C for 20 h (Figure 4A). When the fabric was treated with distilled water at 25 °C for 15 min, the two carbonyl bands at 1846 and 1775 cm^{-1} disappeared completely (Figure 4B). An increase in the 1735-cm^{-1} band intensity and broadening in the hydroxyl stretching band at 3360-cm^{-1} were also observed in Figure 4B. It can be concluded that the two bands at 1846 and 1775 cm^{-1} are due to an anhydride. Treatment of the fabric with water converted the anhydride to the corresponding carboxylic acid. As a result, the two anhydride carbonyls at

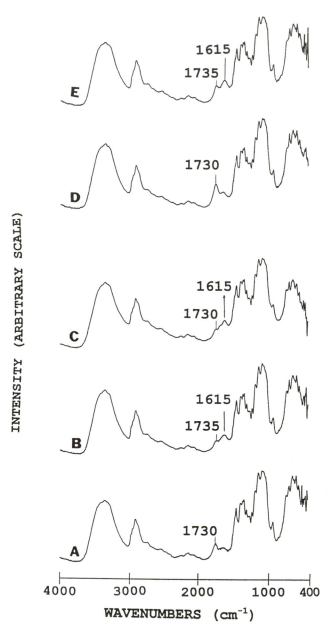

Figure 3. Photoacoustic infrared spectra of the cotton fabric heated at 180 °C for different times: A, 48 h, untreated; B, 48 h, treated with 0.1-M NaOH; C, 48 h, treated with NaBH₄; D, 72 h, untreated; E, 72 h, treated with 0.1-M NaOH.

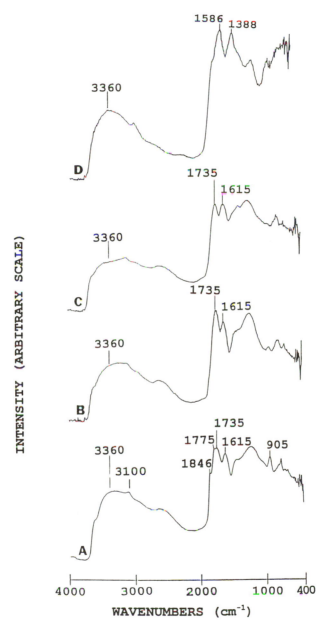

Figure 4. Photoacoustic infrared spectra of the cotton fabric heated at 240 °C for 20 h: A, untreated; B, treated with water; C, treated with 0.1-M NaOH; D, treated with 0.1-M NaOH in methanol at reflux.

1846 and 1775 cm^{-1} disappeared, while the carboxyl carbonyl band at 1735 cm^{-1} increased its intensity. The broadening of the 3360-cm^{-1} band is also due to the formation of carboxyl groups. The band at 905 cm^{-1} in Figure 4A is probably associated with the C–O–C bending of the anhydride. This band disappeared when the fabric was treated with water (Figure 4B).

The frequencies of the two carbonyl bands of some unstrained anhydrides and cyclic anhydrides reported in the literature are listed as follows (20):

Acetic anhydride	1824, 1748
Caprioc anhydride	1825, 1760
Succinic anhydride	1865, 1780
Maleic anhydride	1848, 1790
Phthalic anhydride	1845, 1775
Naphthalene-1,2-dicarboxylic anhydride	1848–1845, 1783–1779

The influence of ring strain induces a shift of the carbonyl bands to higher frequencies, whereas α, β conjugation results in a lowering of the two carbonyl frequencies (20). The anhydride formed in the cotton, which has carbonyl stretching frequencies at 1846 and 1775 cm^{-1}, appears to be an unsaturated (possibly aromatic) five-member cyclic anhydride. The band at 905 cm^{-1} due to the C–O–C bending in Figure 4A also indicates that the anhydride formed in the cotton was probably a cyclic anhydride (21). When the cotton heated at 240 °C for 20 h was treated with a 0.1-M solution of NaOH in ethanol for 1 min at room temperature, the intensity of the 1735-cm^{-1} carbonyl band was reduced whereas the 1615-cm^{-1} carbonyl band intensity was increased in the infrared spectrum (Figure 4C). The anhydride carbonyl bands at 1846 and 1775 cm^{-1} also disappeared in Figure 4C. Treatment of the fabric with NaOH at room temperature converted both the anhydride and the carboxyl to carboxylates.

The fabric heated at 240 °C for 20 h was treated in a 0.1-M solution of NaOH in methanol at reflux for 30 min. The carbonyl bands 1846, 1775, and 1735 cm^{-1} disappeared in the spectrum of the fabric thus treated (Figure 4D). Two strong bands at 1586 and 1385 cm^{-1} associated with the asymmetric and symmetric stretching of carboxylate carbonyl were shown in the same spectrum (Figure 4D). It can be concluded that the band at 1735 cm^{-1} in Figure 4C, which was not changed by the treatment of the fabric with NaOH at room temperature, was due to ester carbonyl. When the fabric was treated in a NaOH solution at reflux, all three types of carbonyls (e.g., anhydride, carboxyl, and ester) were converted to carboxylate. It appears that at this later stage of thermal oxidation, the aldehyde and ketone identified at an earlier stage of degradation were further oxidized to carboxylic acid, ester, and anhydride.

The cotton fabric heated at 240 °C for 20 h was also treated in a 0.1-M solution of bromine in carbon tetrachloride for 30 min at room temperature.

The intensity of the 1615-cm^{-1} band was reduced and the 3100-cm^{-1} band disappeared in the spectrum of the fabric thus treated (Figure 5). This observation demonstrated that the 1615-cm^{-1} band has contributions from the aliphatic C=C structures. Addition of bromine to the C=C structures resulted in the reduction in the 1615-cm^{-1} band intensity and disappearance of the 3100-cm^{-1} band (aliphatic unsaturated C−H stretching). The band at 1615 cm^{-1} and the shoulder at 1578 cm^{-1} in the spectrum of the bromine-treated cotton fabric (Figure 5) were possibly due to aromatic ring structures (22). The unsaturated aliphatic structures were probably formed by the dehydration in the cellulosic molecules during the thermal degradation process, because a reduction in the O−H stretching band at 3360 cm^{-1} was observed in the spectra of the cotton fabric heated at 180 °C (Figures 2D and E). The oxidation products of cotton cellulose and the corresponding stretching band frequencies are summarized as follows:

Aldehyde and ketone carbonyl	1730–1735
Carboxylic acid carbonyl	1730–1735
Ester carbonyl	1735
Cyclic anhydride carbonyl	1846, 1775
Aliphatic and aromatic C=C	1615
Unsaturated C−H (aliphatic)	∼ 3100

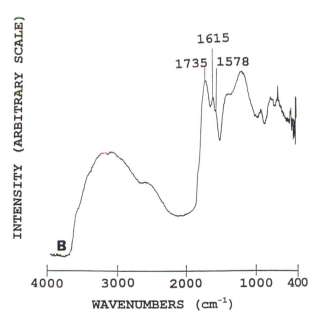

Figure 5. Photoacoustic infrared spectra of cotton fabric heated at 240 °C for 20 h and treated with bromine in CCl$_4$.

The acceleration of the thermal degradation of cotton cellulose by increasing temperature was demonstrated in Figure 6. Three cotton fabrics were heated for 6 h at different temperatures. When the fabric was heated at 180 °C, little oxidation resulted as shown by a very weak carbonyl band at 1730 cm^{-1} in Figure 6A. The carbonyl band at 1730 cm^{-1} was still weak in

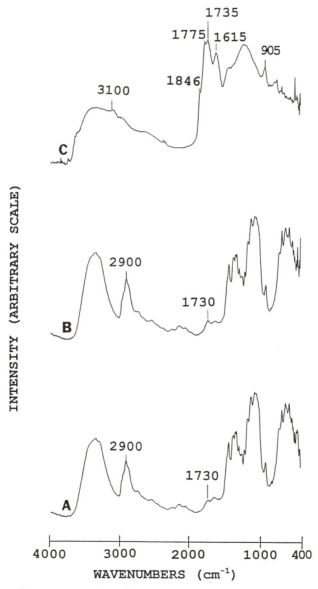

Figure 6. Photoacoustic infrared spectra of cotton fabric heated for 6 h at different temperatures (°C): A, 180; B, 210; C, 240.

the spectrum of the fabric heated at 210 °C (Figure 6B). When the fabric was heated at 240 °C, however, carbonyl bands of anhydride, ester, carboxylic acid, and bands due to C=C stretching and unsaturated C−H stretching were presented in the infrared spectrum (Figure 6C).

The distribution of the oxidation products between the near surface of the cotton fabric and its interior was also investigated using FTIR–PAS. The cotton fabric heated at 180 °C for different times was ground into powders. The carbonyl band intensities in the infrared photoacoustic spectra of the powders (Figure 7) appeared to be the same as those of the fabric samples (Figure 1). Little difference was seen between the spectra of the fabric samples and the spectra of the corresponding powder samples for the cotton fabric heated at 180 °C up to 225 h (Figure 8).

In a PAS experiment, modulated IR radiation absorbed by a sample is first converted to heat. When the heat propagates to the sample surface, and subsequently into the gas within a photoacoustic cell, it causes pressure variation and generates an acoustic signal. When a sample's thickness is larger than the thermal diffusion length, only the heat generated within the first thermal diffusion length from the sample's surface can propagate to the surface and generate photoacoustic signals (23). Because the thermal diffusion length of cotton in the mid-infrared region is in the range of a few micrometers for the optical velocity used in this research (0.278 cm/s) (14), the photoacoustic infrared spectrum of a fabric sample represents a few micrometers of the near surface of the fabric. When the fabric was ground into a powder, the near surface and bulk were mixed. Because the diameter of the cotton yarn in the fabric is approximately 300 μm, the photoacoustic infrared spectrum of the powder sample represents mainly the bulk. The similarity between the spectra of the fabric samples and the spectra of the powder samples indicated that the oxidation of cotton cellulose was homogeneous between the near surface and the bulk.

Differences are seen between the spectrum of the cotton fabric heated at 180 °C for 415 h and the spectrum of the corresponding powder (Figures 9A and B). The two anhydride carbonyl bands at 1846 and 1775 cm^{-1} shown in the spectrum of the fabric sample (Figure 9A) were too weak to be recognizable in the spectrum of the corresponding powder sample (Figure 9B). The band at 905 cm^{-1} associated with the C−O−C bending of the anhydrides appeared to be much weaker in the spectrum of the powder (Figure 9B) than in the spectrum of the fabric (Figure 9A). It is obvious that the anhydride was first formed in the near surface. The same phenomenon was also seen in the fabrics heated at 210 °C and 240 °C, respectively.

The photoacoustic infrared spectra of the cotton fabric heated at 240 °C for 2.5 h and the corresponding powder are presented in Figure 10. The carbonyl band at 1735 cm^{-1} in the spectrum of the fabric (Figure 10A) appears to be more intense than the same band in the spectrum of the powder (Figure 10B). Evidently, the cotton cellulose in the near surface of the fabric had a higher degree of oxidation than the cotton cellulose in the

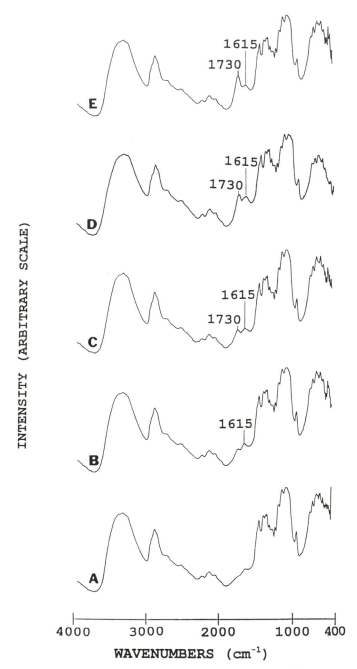

Figure 7. Photoacoustic infrared spectra of the powders of the cotton fabric heated at 180 °C for different times (hours): A, 0; B, 6; C, 15; D, 48; E, 72.

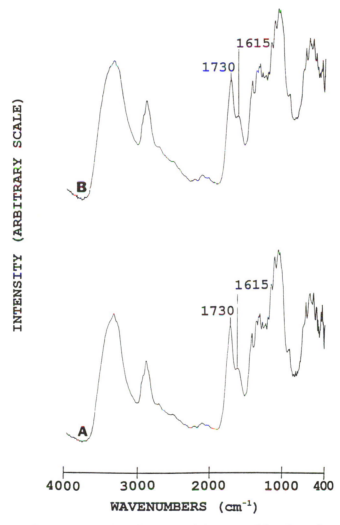

Figure 8. Photoacoustic infrared spectra of the cotton fabric heated at 180 °C for 225 h: A, fabric; B, powder.

bulk of the fabric. It was noted that when the cotton fabric was heated at 180 °C for 225 h, the spectrum of the fabric sample (Figure 8A) appeared to be similar to the spectrum of the powder sample (Figure 8B). The infrared spectroscopic data of the cotton fabric heated at 240 °C (Figure 10) indicated that more oxidation products (ketone, aldehyde, carboxylic acid, and ester) were formed in the near surface of the fabric than in the bulk. The inhomogeneous oxidation of cotton cellulose between the near surface and the bulk at a higher temperature is probably due to the acceleration of

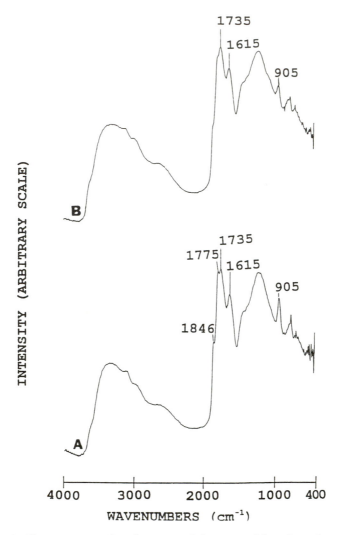

Figure 9. Photoacoustic infrared spectra of the cotton fabric heated at 180 °C for 415 h: A, fabric; B, powder.

oxidation at the higher temperature. When the fabric was oxidized at 180 °C, the rate of oxidation gas slow, which permitted enough oxygen to diffuse from the surface of the fabric into its bulk. As a result, the degree of oxidation in the near surface and in the bulk was similar, as demonstrated in Figures 7 and 8. When the fabric was oxidized at 240 °C, however, the oxidation of cotton cellulose was drastically accelerated. The amount of oxygen diffusing from the surface of the fabric into the bulk was not enough to supply the rapid oxidation of the cotton cellulose in the bulk. Only the cotton cellulose in the near surface was oxidized with abundant oxygen. Consequently, the

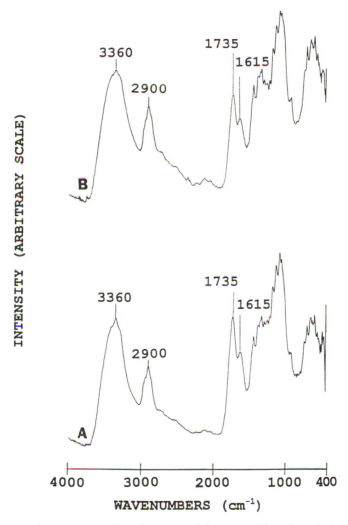

Figure 10. Photoacoustic infrared spectra of the cotton fabric heated at 240 °C for 2.5 h: A, fabric; B, powder.

degree of oxidation of cotton cellulose in the bulk of the fabric was lower than that in the near surface.

Summary

1. Ketone, aldehyde, carboxylic acid, ester, anhydride, and unsaturated carbon–carbon structures were identified as the oxidation products during different stages of the thermal degradation of cotton cellulose at various temperatures.

2. The formation of anhydride occurred first in the near surfaces of the cotton fabrics during the thermal degradation process at all three temperatures studied (180, 210, and 240 °C).

3. The oxidation products (ketone, aldehyde, carboxylic acid, and ester) were homogeneously distributed between the near surface and the bulk of the fabric when the fabric was oxidized at 180 °C. When the fabric was oxidized at 240 °C, however, more oxidation occurred in the near surface of the fabric than in the bulk throughout the degradation process.

References

1. Doree, C. *The Methods of Cellulose Chemistry*, 2nd ed.; Chapman and Hall: London, England, **1950**; p 216.
2. Waller, R. C.; Bass, K. C.; Roseveare, W. E. *Ind. Eng. Chem.* **1948**, *40*, 138.
3. Garten, V. A.; Weiss, D. E. *Rev. Pure Appl. Chem.* **1957**, *7*, 69.
4. Hebeish, A.; El-aref, A. T.; El-alfi, E. A.; El-rafie, M. H. *J. Appl. Polm. Sci.* **1979**, *23*, 453–462.
5. Merkel, R. S. *Analytical Methods for a Textile Laboratory*, 3rd ed.; Weaver, J. W., Ed.; American Association of Textile Chemists and Colorists: Research Triangle Park, NC, 1985; p 41.
6. Shafizadeh, F. *Adv. Carbohydr. Chem.* **1968**, *23*, 419–444.
7. Cardamore, J. M.; Gould, J. M.; Gordon, S. H. *Textile Res. J.* **1987**, *57*, 235–239.
8. Higgins, H. G. *J. Polym. Sci.* **1958**, *28*, 645–648.
9. Hofman, W.; Ostrowski, T.; Urbanski, T.; Witanowski, M. *Chem. Ind.* (*London*) **1960**, *23*, 95–97.
10. Holmes, F. H.; Shaw, C. J. G. *J. Appl. Chem.* **1961**, *11*, 210–216.
11. O'Connor, R. T. In *Analytical Methods for a Textile Laboratory*, 2nd ed.; Weaver, J. W., Ed.; American Association of Textile Chemists and Colorists: Research Triangle Park, NC, 1968; p 295.
12. Miller, B.; Gorrie, T. M. *J. Polym. Sci.* **1971**, *36*, 3–19.
13. Coufal, H.; McClelland, J. F. *J. Mol. Struct.* **1988**, *173*, 129–140.
14. Yang, C. Q.; Bresee, R. R.; Fateley, W. G. *Appl. Spectrosc.* **1987**, *41*, 889–896.
15. Yang, C. Q.; Bresee, R. R. *J. Coated Fabrics* **1987**, *17*, 110–128.
16. Yang, C. Q.; Perenich, T. A.; Fateley, W. G. *Textile Res. J.* **1989**, *59*, 562–568.
17. Yang, C. Q.; Bresee, R. R.; Fateley, W. G. *Appl. Spectrosc.* **1990**, *44*, 1035–1039.
18. Yang, C. Q. *Appl. Spectrosc.* **1991**, *45*, 102–108.
19. Yang, C. Q. *Ind. Eng. Chem. Res.* **1992**, *31*, 617–621.
20. Bellamy, L. J. *The Infrared Spectra of Complex Molecules*; Chapman and Hall: London, England, 1975; p 144.
21. Silverstein, R. M.; Bassler, G. C.; Morrill, T. C. *Spectrometric Identification of Organic Compounds*, 5th ed.; Wiley: New York, 1991; p 121.
22. Painter, R.; Snyder, M.; Starsinic, M.; Coleman, M.; Kuehn, D.; Davis, A. *Appl. Spectrosc.* **1981**, *35*, 475.
23. Rosencweig, A. *Photoacoustics and Photoacoustic Spectboscopy*; Wiley: New York, 1980; pp 93–98.

RECEIVED for review May 14, 1991. ACCEPTED revised manuscript May 30, 1992.

Polymer Analysis and Surface Modifications

Ionization of polymer surfaces under ultrahigh-vacuum conditions formulates the principles of SALI (surface analysis by laser ionization). K^+ ionization of desorbed species (K^+ IDS) can be useful in molecular weight determination and may serve to establish functionality and distribution of oligomers. Selective and highly localized energy sources are often employed in polymer synthesis and polymer surface modifications. These techniques include radio frequency (RF) and microwave plasmas and ultrasonic energy. Several applications of these approaches to inorganic polymer synthesis and polymer surface reactions are presented.

Characterization of Polymer Building Blocks by K⁺ Ionization of Desorbed Species

William J. Simonsick, Jr.

Marshall R & D Laboratory, E. I. du Pont de Nemours and Company, Philadelphia, PA 19146

K^+ ionization of desorbed species (K^+IDS) with mass spectrometric detection is a valuable tool for solving a wide variety of industrial problems that are not rationalized by traditional analytical techniques. K^+IDS provides molecular weight data on organic compounds in the form of $[M]K^+$. Based on molecular weight information and a knowledge of starting materials, the components in complex mixtures can be elucidated. K^+IDS will be used to identify and quantify ingredients in raw materials, oligomer functionality and distribution, and the reaction products of an organic synthesis. With the information gained by K^+IDS, one can optimize product yields of reactions and processes, and gain insight and understanding into how specific compounds affect end-use properties. The method is simple, usually requires no sample preparation, takes little operator time, and can be performed with popular quadrupole mass spectrometers.

THE FINAL PROPERTIES of architecturally designed polymers and cross-linked networks are dependent on the starting materials. The identity and quantity of all constituents that are used as building blocks is important in structure–property relationship (SPR) studies. Moreover, a thorough characterization of these building blocks prior to a product scaleup can prevent the financial consequences caused by polymer gelation in a reactor or large-scale synthesis of nonfunctional material. Furthermore, many of the compounds used to protect these polymers and networks against environmental damage can also affect the final properties. We will show the utility of potassium ionization of desorbed species with mass spectrometric detection (K^+IDS) (1–3) for the characterization of these building blocks. The K^+IDS technique requires only 10 min to perform, which makes it useful preventive medicine.

Many of the starting materials used in the preparation of current high-performance automotive coatings are not chromophoric and, therefore, are not amenable to easy analysis by high-pressure liquid chromatography (HPLC). Unfortunately, many of these building blocks exceed the volatility range of gas chromatography. The traditional spectroscopic methods, such as infrared and nuclear magnetic resonance (NMR), provide complementary information to the molecular weight data afforded by K^+IDS.

Although K^+IDS cannot be used to measure the resultant physical properties of polymers or networks, it is extremely useful for probing the contents of mixtures up to about 2000 Daltons (Da). In this chapter, we will show the applicability of K^+IDS for the characterization of an acrylic cross-linker, a functionalized oligomer synthesized by group-transfer polymerization (4), an epoxy resin, a low-molecular-weight polyester, isocyanate cross-linkers, and a polymerizable ultraviolet (UV) stabilizer.

Over the past 10 years the molecular-weight range amenable to commercial mass spectrometers has increased dramatically. Advances in instrument design and alternative ionization schemes have allowed the collection of intact molecular-weight data on polymers in excess of 10,000 Da. Soft ionization strategies have made intact molecular-weight data on compounds of low volatility almost commonplace. K^+IDS is an ionization method that produces pseudomolecular ions in the form of $[M]K^+$ with little or no fragmentation. Intact organic molecules are desorbed by rapid heating (5). In the gas phase the organic molecules are ionized by potassium attachment. Potassium ions are produced by thermionic emission of K^+ from an aluminosilicate matrix that contains K_2O. K^+IDS is easily performed on commercial quadrupole mass spectrometers. The equipment necessary for K^+IDS is less expensive than the more common soft-ionization techniques, such as field desorption (6, 7), laser desorption (8–10), secondary-ion mass spectrometry (11, 12), and plasma desorption (13). We have found only slight differences when the molecular-weight data afforded by K^+IDS are compared to any of the aforementioned ionization strategies up to 2000 Da.

Experimental Details

All K^+IDS experiments were performed on a quadrupole mass spectrometer (Finnigan model 4615B GC/MS). An electron impact source configuration operating at 200 °C and a source pressure of $< 1 \times 10^{-6}$ torr was used in all experiments.

The detailed experimental procedure for K^+IDS has been described elsewhere (*3*); however, a brief explanation will be given here. Commercially available (Finnigan, San Jose, CA) direct-exposure probe filaments are coated with a 10% w/w slurry of Al_2O_3, $2KNO_3$, and $2SiO_2$ in acetone. The coated filaments are dried and conditioned by heating the filament to approximately 500 °C under vacuum. Neat samples (0.1 mg) or 10% w/w solutions (0.2 μL) are deposited onto a stainless steel ribbon adjacent to the conditioned filament. The filament and sample are inserted into the ion source of the mass spectrometer and a 1.3-A current is applied to the filament. The application of a high current to the filament causes resistive heating of the filament with subsequent K^+ emission from the aluminosilicate matrix. The sample, which is in close proximity to the filament, is radiatively heated and intact large organic molecules are desorbed. The organic molecules collide with K^+ and are cationized. For these studies, the mass spectrometer was scanned from 100–1000 Da/s unless otherwise noted. Under these conditions, we obtained between three and five scans per sample that were averaged to yield the reported spectra. Ions are seen as $[M]K^+$, the mass of the analyte, plus 39 Da, the mass of potassium.

In the full scan mode the K^+IDS technique is about an order of magnitude less sensitive than traditional electron impact mass spectrometry. A 10-μg sample provides enough signal to clearly see a polymer distribution. However, using selected ion monitoring, we can obtain polymer distributions on as little as 100 ng of material. In the following investigations, we did not attempt to optimize the sensitivity of K^+IDS.

Results and Discussion

Acrylic Cross-Linker.   We used K^+IDS and supercritical fluid chromatography (*14*) to characterize tripropylene glycol diacrylate (TPGDA). The structure of the material is given in **I**. TPGDA is synthesized with acrylic acid (AA) and tripropylene glycol. TPGDA is difunctional and is used to cross-link acrylic formulations. The degree of cross-linking is key to the final properties; therefore, the contents of the cross-linker are important. Furthermore, to make a quality product it is important that the constituents of the raw materials do not vary. Therefore, for the purpose of quality assurance–quality

MW = 300 g/mol
$[M]K^+$ = 339 Da

I

control (QA–QC) it is important to have a method to quickly analyze commercial TPGDA.

TPGDA has a molecular weight of 300 g/mol and is amenable to gas chromatography; however, any higher molecular-weight compounds would exceed the volatility range of gas chromatography. Furthermore, the lack of a strong chromophore makes HPLC characterization difficult. Figure 1 (top) is the K[+]IDS mass spectrum of neat commercial TPGDA. There are several peaks present. The base peak seen at 339 Da is due to potassiated TPGDA. The peak seen at 285 Da is 54 Da less than the TPGDA, which we attribute to monofunctional material (**II**). The loss of one AA moiety (72 Da) coupled to a gain of water (18 Da) results in a net loss of 54 Da. The peak at 411 Da is 72 Da higher in molecular weight than the TPGDA. Because AA weighs 72 Da, we attribute the 411 peak to an additional AA attached through the vinyl group (**III**). However, other isomers can yield a [M]K[+]ion at 411 Da. Unfortunately, K[+]IDS does not distinguish between isomers. In such cases, ^{13}C NMR in conjunction with the K[+]IDS spectrum has provided isomer elucidation (*14*). The peak at 585 Da is 246 Da higher in molecular weight than the TPGDA. Because the molecular weight of the acrylic acid tripropylene glycol adduct is 246 Da, we attribute the peak at 585 Da to **IV**. Species

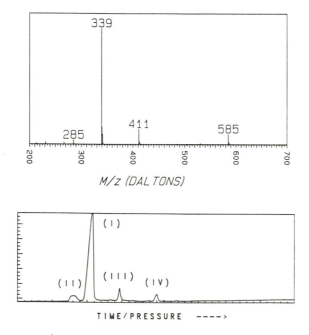

*Figure 1. Top: K[+]IDS mass spectrum of TPGDA cross-linker (**I**). Bottom: Supercritical fluid chromatogram of TPGDA. Mobile phase conditions and program are 100 °C carbon dioxide at 200 atm for 10 min ramped to 415 atm at 10 atm/min. Column-10M biphenyl-30. Detector-flame ionization at 350 °C.*

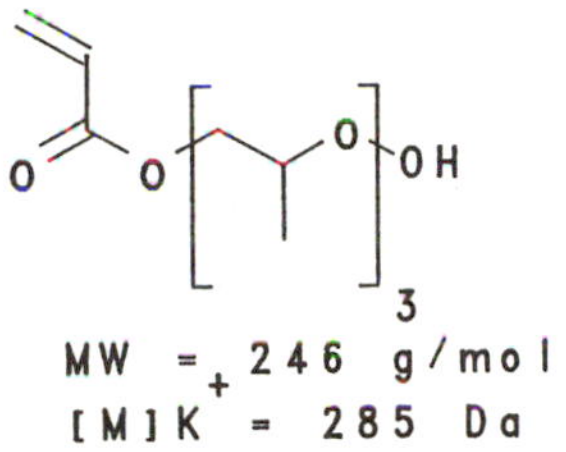
O
O
O
OH
3
MW = 246 g/mol
[M]K+ = 285 Da

II

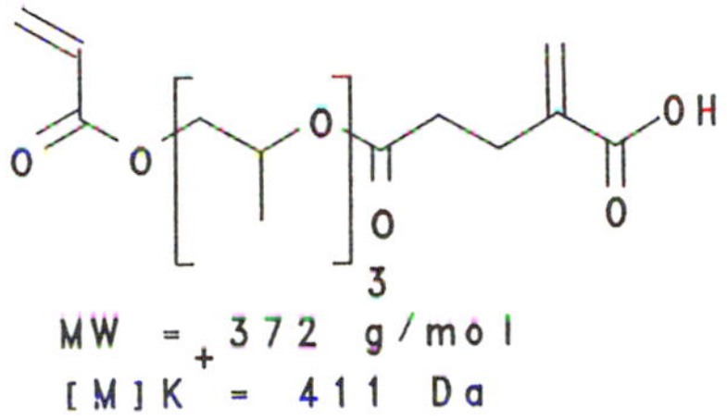
O
O
O
OH
O
O
3
MW = 372 g/mol
[M]K+ = 411 Da

III

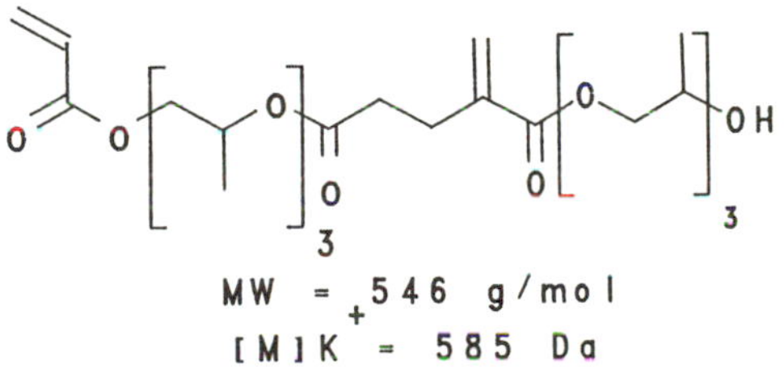
O
O
O
O
O
O
OH
3
3
MW = 546 g/mol
[M]K+ = 585 Da

IV

IV still affords cross-linking via the acrylate moiety. It is important to identify and quantify the ingredients in cross-linkers for both product reproducibility (QA–QC) and for SPR studies. K^+IDS provides the tool to characterize the cross-linker, particularly for the higher molecular weight adducts.

Under K^+IDS conditions, the sample under study may be exposed to excessive temperatures that yield artifacts. To insure that the sample integrity was preserved, we chose a chromatographic method to support our K^+IDS results. To show that the larger ions seen in the mass spectrum were not due to the reassociation of thermal fragments, we used supercritical fluid chromatography (SCF) with a CO_2 mobile phase and flame ionization detection (*15*). The sample integrity is preserved by use of this chromatographic technique. Figure 1 (bottom) is the SFC chromatogram of the TPGDA cross-linker. The presence of four major peaks corroborates the K^+IDS data. We did not identify each peak; however, SFC separates compounds based on their solubility in supercritical CO_2, which generally decreases with increasing molecular weight, hence our peak assignments. Both K^+IDS and SFC show the TPGDA comprise about 80% of the mixture.

Oligomer Functionality. We have used K^+IDS to identify the products of a group-transfer polymerization (4). Oligomeric poly(methyl methacrylate) (PMMA) was synthesized to a degree of polymerization of 4 using an epoxy initiator and tetrabutylammonium metachlorobenzoate (TBACB) as the catalyst. The epoxy initiator provides a site for subsequent reaction; therefore, it is critical that the epoxy groups be preserved. Run 1 was quenched with methanol and subjected to K^+IDS analysis; Figure 2 (top) shows the resulting mass spectrum. Run 2 was conducted in the same manner, except four times as much TBACB catalyst was added prior to the methanol quench. Figure 2 (bottom) shows the K^+IDS mass spectrum from this second run. Note that the high- and low-intensity components in these two spectra are reversed.

The primary envelope in Figure 2 (top), comprised of 483, 583, . . . , is attributed to the desired product (**V**). The 100-Da spacing between oligomer peaks is due to the methyl methacrylate (MMA) repeat unit. The primary envelope in Figure 2 (bottom), of ions with mass 341, 441, 541, . . . , corresponds to components with the structure containing only PMMA (**VI**). These components are formed by transesterification of the epoxide group initiator, resulting in the unwanted loss of this functionality. The ion observed in both spectra at 309 Da arises from the substituted cyclohexanaone moiety (**VII**), which in turn is generated by a backbiting chain termination mechanism that occurs to a greater extent at higher catalyst concentrations.

Because K^+IDS is a relatively new technique, we sought some corroborative evidence for our interpretation of these analyses. Wet chemical analysis did in fact show that the product of run 1 had an epoxy content approximately 2.5 times that from run 2. Moreover, glycidol, the byproduct of the transesterification reaction, was detected in run 2 at 2.5 times the level as in run 1. These results support our conclusions based on the K^+IDS analyses that about 20% of the epoxy groups are lost via transesterification when a normal catalyst concentration is used; substantially more epoxy groups are lost with the formation of unwanted materials at higher catalyst levels.

Epoxy Resins. The molecular weight information obtained by K^+IDS was used to characterize oligomeric (< 1000 Da) epoxy resins. The rich structural information provided by K^+IDS assisted in the identification

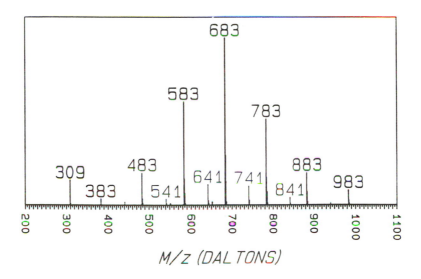

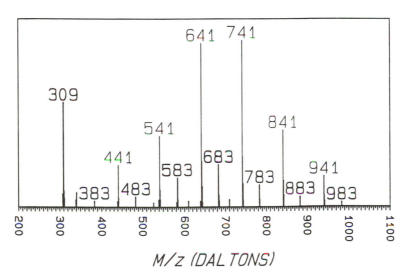

Figure 2. Top: K^+IDS mass spectrum of oligomeric PMMA synthesized with a normal catalyst dose. Bottom: K^+IDS mass spectrum of oligomeric PMMA with a 4 × normal catalyst load.

of many of the compounds contained in a synthetic reaction product. Several monofunctional, difunctional, and chlorine-containing oligomers were identified using K^+IDS. The structure of a bisphenol A ethoxylated epoxy resin is given in **VIII**. The desired difunctional material is critical to extending a polymer chain with other multifunctional materials. The K^+IDS mass spec-

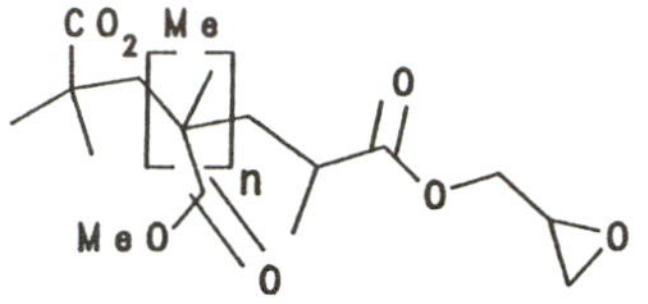

$[M]K^+$ - 3 8 3 ,   4 8 3 ,   5 8 3 . . .

V

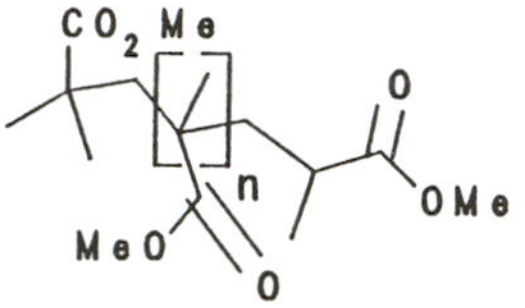

$[M]K^+$ - 3 4 1 ,   4 4 1 ,   5 4 1 . . .

VI

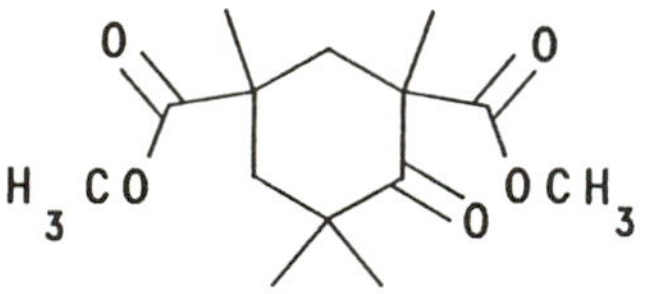

$[M]K^+$ = 3 0 9   Da

VII

trum of the reaction mixture targeted at the synthesis of **VIII** is seen in Figure 3.

The primary envelope consisting of m/z = 555, 599, 643 Da, . . . , results from potassium cationization of **VIII**. The envelope, which is composed of ions m/z = 445, 499, 543 Da, . . . , results from potassium cationization oligomers that contain only one epoxide ring termed monofunctional (*see* **IX**). Monofunctional material is undesirable because once the epoxide is reacted, no further chain extension can proceed. The structures that give rise to the envelope of 529, 587, 645 g/mol, etc., display a significant M + 41 ion. The intensities (> 30%) are much higher than expected from ^{41}K and

[13]C isotope contributions. This pattern is consistent with the isotope pattern expected from a monochlorine-containing species. Epichlorohydrin, used in the preparation of the epoxy resin, is the chlorine source. We attribute this envelope to **X**. The presence of chlorine can adversely affect the activity of certain catalysts. There are also other much less abundant envelopes whose

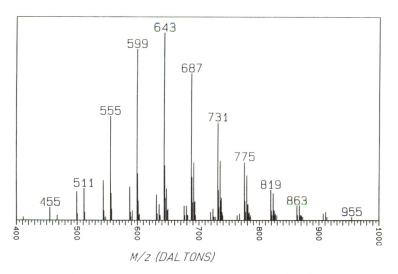

$m/z = 379$ Da $+ 44(n+m)$

VIII

$m/z = 323$ Da $+ 44(n+m)$

IX

$m/z = 471$ Da $+ 44(n+m)$

X

Figure 3. K[+]IDS mass spectrum of an ethoxylated bisphenol A epoxy resin.

molecular weight correspond to combinations of the aforementioned species such as dihydroxy-terminated species. In one 10-min experiment, the diepoxy, monoepoxy, and chlorine-containing oligomers were identified and quantified.

Polyurethanes. Polyurethane coatings offer a host of attractive properties, such as excellent chemical and abrasion resistance and superior weathering ability as measured by gloss retention. Polyurethane coatings can be produced by the reaction of a polyol with a multifunctional isocyanate. The polyol, such as a polyester, can be low molecular weight ($< 10{,}000$ g/mol). The isocyanate cross-linker is usually highly functional and lower molecular weight (< 1000 g/mol). The use of low-molecular-weight highly functionalized materials for polyurethane coatings is a popular method to raise the solids content of automotive coatings, thereby reducing the volatile organic content (*16*). K^+IDS is a useful technique to characterize all of the chemical moieties contained in both the polyester polyol and the multifunctional cross-linker. The specific ingredients of each component affect the final properties of the urethane coating.

Figure 4 is the K^+IDS mass spectrum of a neopentyl glycol (NPG) isophthalic acid (IA) polyester polymerization taken in the early stages. Based upon the molecular weight data afforded by K^+IDS and a knowledge of starting ingredients, structure assignments are possible. IA weighs 166 Da and NPG weighs 104 Da. Each ester linkage is accompanied by a loss of water (18 Da). Therefore, we attribute the base peak at 845 Da ion to the [NPG–IA–NPG–IA–NPG–IA–NPG]K^+ moiety, [104 Da + 166 Da + 104

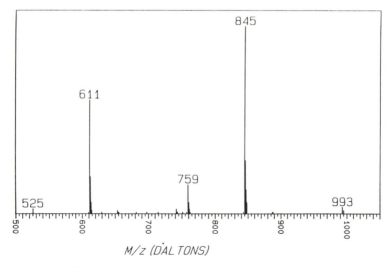

Figure 4. K^+IDS mass spectrum of a low-molecular-weight polyester resin taken in the early stages of polymerization.

Da + 166 Da + 104 Da + 166 Da + 104 Da] − [6 × 18 Da] + 39 Da = 845 Da. In a similar fashion, the other significant ions can be assigned structures. Table I summarizes our structure assignments.

We load an excess of polyol to insure that the majority of the polyester chains are hydroxyl terminated. The hydroxyl-terminated moieties are amenable to K^+IDS; however, the acid-terminated oligomers sometimes undergo dehydration reactions, which complicates data interpretation. For example, the ion seen at 741 Da is 18 Da less than the hexamer. We attribute the dehydration to an artifact of K^+IDS. To minimize competing reactions we can convert acid end groups to their corresponding methyl- or trimethyl-silyl esters. In general, we have found that esters desorb more efficiently than the corresponding acids.

Hexamethylene diisocyanate (HDI) is a popular aliphatic cross-linker used in the preparation of polyurethanes. Depending on the conditions under which the HDI was synthesized, a different distribution of products can result. Furthermore, depending on the specific isocyanate mixture used in a polyurethane film preparation, different final properties will result. Most of the constituents in commercial grade isocyanates differ in molecular weight and can therefore, be distinguished by K^+IDS.

Figure 5 shows the K^+IDS mass spectra of two HDI mixtures that were synthesized under different conditions. HDI(I) seen in the top spectrum of Figure 5 had been stored in a refrigerator for over 10 years, whereas HDI(II), seen in the bottom spectrum of Figure 5, was a freshly synthesized batch targeted at producing primarily the isocyanurate trimer (**XI**). The HDI batches contain several common ions, but their relative amount differ significantly. HDI, ($OCN − (CH_2)_6 − NCO$), weighs 168 Da. Therefore, the ion seen at 207 corresponds to the monomer. The 375-Da ion is due to the dimer, the 543-Da ion to isocyanurate trimer (**XI**), the tetramer appears at 711 Da, and the pentamer at 879 Da. The base peak in HDI(I) appears at 349 Da, which is 26 Da less than the dimer. We attribute a loss of 26 Da to a replacement of the isocyanate group ($NCO = 42$ Da) by a primary amine ($NH_2 = 16$ Da). Therefore, we attribute the peak at 349 Da to the $OCN − (CH_2)_6 − NH − C(O) − NH − (CH_2)_6 − NCO$ urea. Likewise, the peak at 517 Da is due to the biuret of HDI (**XII**). The ions seen at 225 and 407 Da in

Table I. Proposed Structures for Ions Seen in K^+IDS Mass Spectrum of an IA–NPG Polymerization

Ion (Da)	Proposed Structure
525	$[NPG–IA–NPG-IA]K^+$
611	$[NPG–IA–NPG–IA–NPG]K^+$
759	$[NPG–IA–NPG–IA–NPG–IA]K^+$
845	$[NPG–IA–NPG–IA–NPG–IA–NPG]K^+$
993	$[NPG–IA–NPG–IA–NPG–IA–NPG–IA]K^+$

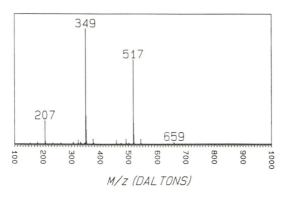

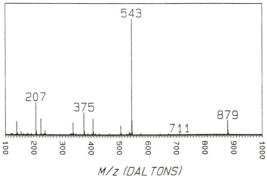

*Figure 5. Top: K⁺IDS mass spectrum of aged hexamethylene diisocyanate cross-linker. Bottom: K⁺IDS mass spectrum of HDI that was formulated to maximize the isocyanurate moiety (**XI**).*

Figure 5 (bottom) are attributed to monomeric HDI that has been reacted with one molecule of water to form carbamic acid and dimeric HDI that has reacted with two molecules of water.

Polymerizable Stabilizers.   The usable lifetimes of polymeric materials are lengthened by the addition of stabilizers. Specifically, degradation caused by UV radiation can be slowed by the addition of benzotriazole-type UV stabilizers. The commercially available stabilizers are relatively non-volatile so they do not vaporize during a thermal cure or exude from the polymer over time.

Another approach to minimize stabilizer volatilization or exudation is the use of polymerizable stabilizers (*17*). Polymerizable stabilizers prepared from functionalized monomers have advantages over conventional stabilizers. Thin-film coatings benefit from polymer bound stabilizers for protection against the harmful effect of increased temperature during curing. Furthermore, once these polymerizable stabilizers are incorporated into the polymer backbone, long-lasting UV stability, oxidative stability, and gloss retention are the recognized payoffs (*17*). However, the successful synthesis of such materials can be difficult and can produce a host of undesirable byproducts. We have found K^+IDS to be extremely valuable for the analysis of polymerizable stabilizer syntheses.

Figure 6 is the K^+IDS mass spectrum of the reaction product of 2(2-hydroxy-5-methylphenyl) 2H-benzotriozole (261 g/mol) and *N*-(methylol) methacrylamide (115 g/mol). The peak seen at 383 Da contains an even number of nitrogen molecules and the isotope pattern is indicative of a one

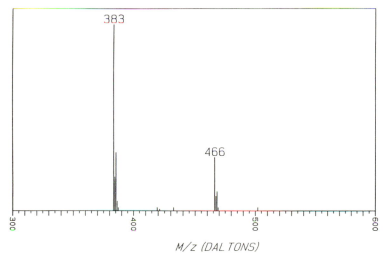

*Figure 6. K^+IDS mass spectrum of the polymerizable UV–stabilizer (**XIII**) synthesis.*

chlorine molecule pattern ($[M]K^+/[M + 2]K^+ = 3/1$). Thus, we propose that the 383-Da ion results from addition of a molecule of N-(methylol) methacrylamide to the benzotriazole coupled with the evolution of methanol: [261 Da + 115 Da − 32 Da] + 39 Da = 383 Da. Using K^+IDS we cannot distinguish the specific substitution site or isomer; however, both UV–visible spectroscopy and ^{13}C NMR spectroscopy can distinguish these isomers once K^+IDS narrows the possibilities. We attribute the 383-Da ion to the desired product (**XIII**).

The ion seen at 466 Da has an odd number of nitrogen molecules and possesses a one chlorine isotope pattern. Addition of N-(methylol) methacrylamide to the unreacted hydroxyl moiety coupled with methanol evolution would yield a $[M]K^+$ of 466 Da. Therefore, we attribute this ion to **XIV**. The use of the difunctional material (**XIV**) in a polymer synthesis would give an unacceptable product. Structure **XIV** is a difunctional methacrylate that would cross-link a resin, rapidly build molecular weight, and possibly gel the reaction. K^+IDS provides a rapid molecular weight check of a targeted synthesis. Typically, we use ^{13}C NMR in conjunction with K^+IDS to elucidate reaction pathways and products (*14*). Once we narrow the possibilities of products produced in an organic synthesis using K^+IDS, ^{13}C NMR assignments and other methods that can distinguish isomers are much easier.

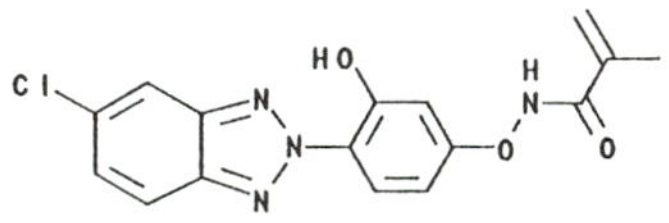

MW = 334 g/mol

$[M]K^+$ = 383 Da

XIII

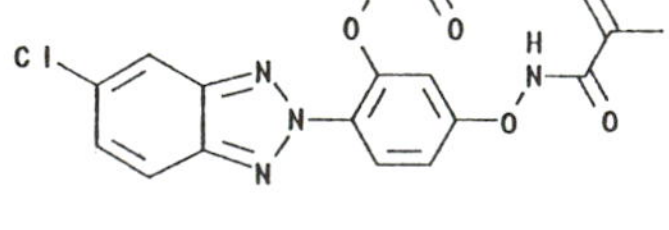

MW = 427 g/mol

$[M]K^+$ = 466 Da

XIV

Summary

We have demonstrated the utility of K^+IDS for the characterization of compounds with relatively low volatility in the molecular weight range of about 200–1000 Da. This molecular weight regime encompasses oligomeric building blocks and cross-linkers whose structures can dramatically affect the final properties of polymers and films. K^+IDS is widely applicable and inexpensive to perform on popular commercial equipment. The experiment requires a mere 10 min to perform, which includes sample preparation. The molecular weight information afforded by K^+IDS is useful for structure elucidation. Therefore, we use K^+IDS for QA–QC applications, troubleshooting, and as a quick screen for organic syntheses. Once the total ingredient package is known, structure–properties studies can be better conducted.

Acknowledgments

Maryann Silva performed most of the K^+IDS experiments discussed in this article. Lance L. Litty is acknowledged for providing the chromatographic data. O. Vogl and A. Sustic synthesized the polymerizable stabilizer. W. Anton synthesized the functionalized oligomers. M. C. Grady furnished the polyester. The author thanks Donnetta Moss and Susan Marie Schaner for their clerical assistance in the timely preparation of this manuscript.

References

1. Bombick, D.; Pinkston, J. D.; Allison, J. *Anal. Chem.* **1984**, *56*, 396.
2. Bombick, D.; Allison, J. *Anal. Chem.* **1987**, *59*, 458.
3. Simonsick, W. J., Jr. *J. Appl. Polym. Sci.: Appl. Polym. Symp.* **1989**, *43*, 257.
4. Webster, O. W.; Hertler, W. R.; Sogah, D. Y.; Farnham, W. B.; Rajan Babu, T. V. *J. Am. Chem. Soc.* **1983**, *105*, 5706.
5. Beuhler, R. J.; Flanigan, E.; Green, F.; Freidman, L. *J. Am. Chem. Soc.* **1974**, *96*, 3390.
6. Prokai, L. *Field Desorption Mass Spectrometry*; Marcel Dekker: New York, 1989.
7. Lattimer, R. P.; Schulten, H. R. *Anal. Chem.* **1989**, *61*, 1201A.
8. Cotter, R. J.; Honovich, J. P.; Olthoff, J. K.; Lattimer, R. P. *Macromolecules* **1986**, *19*, 2996.
9. Ijames, C. F.; Wilkins, C. L. *J. Am. Chem. Soc.* **1988**, *110*, 2687.
10. Nuwaysin, L. M.; Wilkins, C. L.; Simonsick, W. J., Jr. *J. Am. Soc. Mass Spectrom.* **1990**, *1*, 66.
11. Benninghoven, A.; Rudenauer, F. G.; Werner, H. W. *Seconday Ion Mass Spectrometry: Basic Concepts, Instrumental Aspects, Applications and Trends*; Wiley: New York, 1987.
12. Bletsos, I. V.; Hercules, D. M.; Van Leyden, D.; Benninghoven, A. *Macromolecules* **1987**, *20*, 407.
13. Macfarlane, R. D.; Torgerson, D. F. *Science* **1976**, *191*, 920.

14. Simonsick, W. J., Jr; Adamsons, K. *J. Appl. Polym. Sci.: Appl. Polym Symp.* **1991**, *48*, 389.
15. Lee, M. L.; Markides, K. E. *Analytical Supercritical Fluid Chromatography and Extraction*; Chromatography Conferences, Inc.: Provo, UT, 1990.
16. Armour, A. G.; Wu, D. T.; Antonelli, J. A.; Lowell, J. H. Presented at the 186th National Meeting of the American Chemical Society, Symposium on the History of Coatings, Washington, DC, September 14, 1989; Paper 38.
17. Dickstein, W.; Vogl, O. *J. Macromol. Sci. Chem.* **1985**, *22*, 387.

RECEIVED for review July 15, 1991. ACCEPTED revised manuscript September 8, 1992.

Surface Analysis by Laser Ionization Applied to Polymeric Material

S. M. Daiser[1] and S. G. MacKay[2],[*]

Physical Electronics Division, Perkin-Elmer Corporation, 6509 Flying Cloud Drive, Eden Prairie, MN 55344

Single-photon ionization surface analysis by laser ionization (SPI–SALI) coupled with a static primary ion beam or a soft laser beam for desorption is a promising new method for the characterization of polymer surfaces. SPI–SALI combines nonresonant photoionization of desorbed or sputtered neutral atoms and molecules with analysis by time-of-flight mass spectrometry in ultrahigh vacuum. Stimulated desorption is accomplished by an ion source (Cs^+, Ar^+, or Ga^+), electron gun, or laser-induced desorption (CO_2 or Nd:YAG). Single-photon ionization using 118-nm light is useful for identifying polymer material, such as polyethylene glycol, polystyrene, polydimethyl siloxane, and many others, by the ease of the monomer identification.

ANALYSIS OF SOLID SURFACES BY POSTIONIZATION TECHNIQUES has received much attention over the past several years (*1*). These techniques include surface analysis by laser ionization (SALI) (*2*), sputter-initiated resonance ionization spectroscopy (SIRIS) (*3*), and surface analysis by laser ionization of sputtered atoms (SARISA) (*4*). The use of postionization techniques for surface analysis has received widespread interest because of the increased sensitivity provided relative to the more traditional surface analytical techniques such as X-ray photoelectron spectroscopy (XPS). Postionization tech-

---

[1] Present address: Berger and Partner, Arabella Str. 33, 8000 Munich, Germany.
[2] Present address: 3M Company, 3M Center, Building 201–2S–16, St. Paul, MN 55144.
[*] Corresponding author.

niques also provide more reliable quantification compared to secondary ion mass spectrometry (SIMS).

Postionization methods have been used to study a variety of interesting material systems. When a surface is bombarded by a primary ion beam (Figure 1), the yield for secondary neutral emission is usually orders of magnitude higher than the yield for secondary ion emission (5). Laser ionization of desorbed neutrals in the gas phase above the sample surface is the basis for the SALI postionization technique. Decoupling the sputtering process from the ionization process and ionizing the dominant neutral species above the sample surface avoids the variations in ionization probabilities ("matrix effects") that often plague SIMS measurements.

Laser-based photoionization techniques fall into two categories: nonresonant-enhanced ionization and resonant-enhanced ionization. The use of nonresonant-enhanced ionization (SALI) is the most promising method when quantification and general applicability to a variety of material systems are considered.

Surface Analysis by Laser Ionization

SALI is a three-step process that involves a probe beam, photoionization, and mass analysis. Figure 2 depicts the SALI process. Under vacuum, the surface of interest is irradiated by a pulsed probe beam of ions, electrons, or photons.

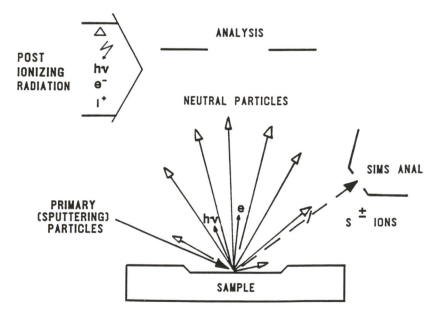

Figure 1. Sputter-induced formation of ions and neutral particles.

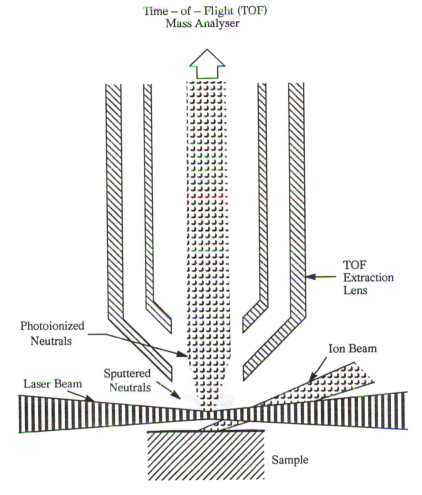

Figure 2. Technique description of the SALI process.

This irradiation causes the desorption of a small amount of surface material. The neutral species released in this process are nonresonantly ionized by a high-intensity pulsed ultraviolet laser passing in close proximity to the sample surface. The resulting photoions are then accelerated, focused, and allowed to drift in a field-free region for time-of-flight (TOF) mass analysis.

Desorption Beam.   Surface removal for sampling involves the movement of atoms and molecules from the top surface layer into the vapor phase. The fact that the ionization step is decoupled from the surface-removal step implies a great deal of flexibility and control in the type and conditions of the energetic beam chosen to stimulate desorption. For elemental analysis of inorganic materials, typically a 50–100-μm Ar^+ or Cs^+, or a submicrometer diameter Ga^+ beam at several kiloelectronvolts is used to enable submonolayer or "static" analysis by pulsing the beam and keeping the total dose extremely low ($< 1 \times 10^{13}$ ions/cm^2). The small-spot Ga^+ beam is well suited to quantitative chemical mapping with submicrometer spatial resolution. For bulk polymers or thin polymer films, energetic electrons or another laser beam sometimes result in superior mass spectra relative to pulsed ion-beam sputtering. Even thermal desorption can be used to investigate the temperature dependence of thermally sensitive samples.

Nonresonant Photoionization.   The two forms of nonresonant (and therefore nonselective) photoionization that are used for SALI are illustrated in Figure 3.

For elemental analysis, a powerful pulsed laser that delivers focused power sensitivities greater than 10^{10} W/cm^2 is used for multiphoton ionization (MPI). Typically all the species within the laser focus volume are ionized without the need for wavelength tuning and regardless of chemical type.

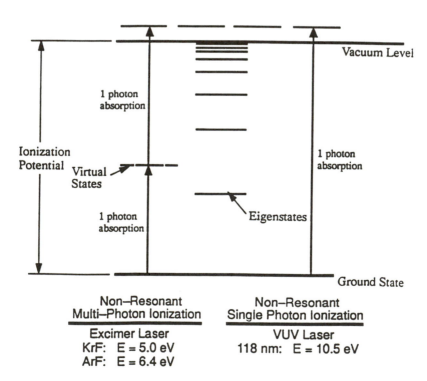

Figure 3. SALI ionization process.

Nonresonant photoionization yields the desired uniformity of detection probability essential for quantification. For molecular analysis, a "soft" (i.e., nonfragmenting) photoionization is needed and is supplied by vacuum ultraviolet (VUV) light with wavelengths in the range of 115–120 nm (10.5 eV). Photoionization at this wavelength is achieved by single-photon ionization (SPI). Because relative photoionization cross-sections for molecules do not vary greatly in this wavelength region, semiquantitative raw data result. Improvement in quantification for both photoionization methods is achieved with calibration. Sampling the majority neutral channel means much less stringent requirements for calibrants than for direct ion production from surfaces by energetic particles. Relaxed calibrant requirements are especially important for the analysis of unknown bulk polymer materials.

A detailed description of the SPI–SALI process is given by Pallix et al. (6).

Time-of-Flight Mass Spectrometric Analysis. Two features of the TOF mass spectrometer make it an ideal choice for a pulsed ionization measurement. First, there is an inherent multiplex advantage in that all masses arrive at the detector for each laser pulse. This advantage is very valuable from a sensitivity point of view when unknowns may be present. The second main feature is that TOF mass spectrometers have very high efficiencies; transmission frequently is in the range of 0.3 to 0.7.

The TOF mass spectrometer used in this study incorporates a special electrostatic mirror called a reflectron that acts as an energy-focusing device to achieve high mass resolution (7). The reflectron works because the higher kinetic energy ions penetrate deeper into the reflecting electrostatic field and thus take longer to turn around. The shorter drift time of the faster particles is closely matched by tuning the instrument to the longer turnaround time required. The instrument is set so that fast and slow ions of the same mass arrive almost simultaneously at the detector. Another feature of the reflectron-based analyzer is that the reflectron electrostatically discriminate against secondary ions that are formed at the sample surface and thereby eliminate a potential source of background. Secondary ions are formed in a region of higher potential and have higher kinetic energies than the photoions that are formed in the laser focal volume above the surface. The reflecting grid is adjusted to an intermediate potential that reflects the desired photoions and does not reflect the secondary ions. Another advantage of the reflectron analyzer is that with the laser off, TOF–SIMS measurements, which allow for mass resolutions $> 10,000$, can be taken.

Modes of Analysis

SALI applies to two methods of postionization: multiphoton and single-photon ionization (MPI and SPI). Each method can be utilized in one of the three modes of analysis: survey analysis, depth profiling, and chemical mapping. Both the survey and mapping modes are used in polymer material characterization.

Survey spectra using the SPI mode are used primarily for monomer identification in and modifications to polymer films as well as bulk polymer materials. Furthermore, SPI survey spectra can be used for the quantification of surface components in inorganic and organic materials that have a typical parts per million detection limit (8). The extremely high sensitivity of the SPI mode in application to organic materials is one of the most promising features of SALI. In certain cases, such as thin films of polycyclic 2,3,7,8-tetrachlorodibenzo-p-dioxin and 7-methylguanin, attomole sensitivities ($< 10^{-15}$ mol or 10^{-6} monolayer over 1 mm^2) have been found (9).

SALI mapping is a sensitive and quantitative method for characterization of the spatial distribution of elements in both organic and inorganic materials. SALI mapping is currently limited to elemental (rather than molecular) analysis because it is not practical to use SPI in this mode. This limitation is due to the low laser repetition rate (10–50 Hz) of the Nd:YAG lasers used for SPI analysis and the fact that useful yields for SPI–SALI are typically 10^{-5}. At this low duty cycle, only neutral species with high useful yields ($> 10^{-2}$) can be detected with reasonable sensitivity in a practical time frame for a 128 × 128 map.

Polymer Applications

Polymer Films.   The first example is a thin film (a few monolayers) of polyethylene glycol (PEG) that was dissolved in methanol and cast onto a silicon wafer. The SPI–SALI analysis used ion sputtering to desorb the polymer material. The primary beam conditions were as follows: 7-kV Ar$^+$, 2-μA dc current, 10-μs pulse width, and an ion dose of 5 × 10^{12} ions/cm^2. Photoionization by SPI was achieved using 118-nm light. The laser pulse width produced was ~ 8 ns, and the spectra were recorded using 10-ns time-resolution steps. The laser repetition rate was 10 Hz, which yielded an average analysis time of 1 min. No charge compensation was needed for this analysis.

The dominant feature in the SPI–SALI spectrum of the PEG film (Figure 4) is the monomer peak at $m/z = 44$ (M $= - CH_2 = CH_2 - O -$). In addition to the strong monomer peak, there are peaks at $m/z = 88$ and 132 that correspond to the dimer and trimer, respectively.

In comparison, Figure 5 shows the TOF–SIMS spectrum of an identical sample over the same mass range. Unlike the SALI spectrum, the TOF–SIMS spectrum shows only a small monomer peak with an intense M + H$^+$ peak at a mass-to-charge ratio $m/z = 45$. This result seems to indicate that the more characteristic monomer peak preferably sputters as a neutral species. The soft

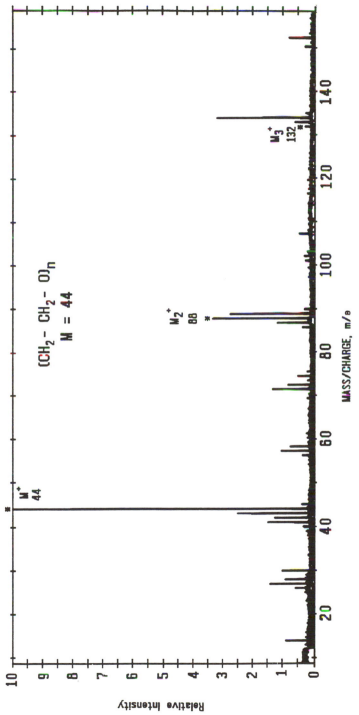

Figure 4. SPI–SALI spectrum of a thin polyethylene glycol (PEG) film on Si acquired with 500 pulses of 118-nm VUV postionization radiation, 5 E12 ions/cm² Ar⁺ probe beam.

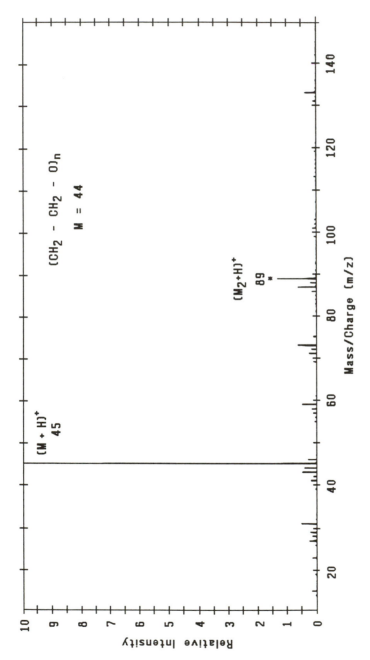

Figure 5. TOF–SIMS spectrum of a thin polyethylene glycol (PEG) film on Si acquired with 200,000 pulses of Cs⁺ ion source, 1.6 E10 ions/cm².

ionization of the SPI process is evident in the high intensities of the dimer and trimer ions relative to the monomer ion in the SALI spectrum. To complete the comparison, Figure 6 displays the static SIMS spectrum of the same polymer film taken on a quadrupole SIMS instrument. The static SIMS spectrum shows similarities with the TOF–SIMS spectrum in that the intensity of the monomer peak is also very weak and the fragmentation pattern is similar. The experimental conditions in TOF–SIMS and static SIMS were similar to SPI–SALI except that there was no laser for postionization (real secondary ions were analyzed). In the TOF–SIMS mode, the ion gun was pulsed with a very short pulse width of 1.0 ns, whereas in the static SIMS, the ion gun was operated in the dc mode with an ion dose of 5×10^{12} ions/cm^2. The fully interpreted static SIMS spectra, including individual peak assignments, can be found in *Static SIMS Handbook of Polymer Analysis* (*10*).

The ease of monomer identification and the relatively clean spectrum in SALI are the result of the soft ionization produced by using SPI. The result of this type of ionization process is that there is fragmentation from both the desorption step and the ionization step. The contribution of the ionization step in many cases is minimal.

The superiority of SPI–SALI over conventional techniques like TOF–SIMS or static SIMS in terms of fragmentation and ease of monomer identification has already been measured and confirmed on a variety of different polymers, such as polystyrene (*6*), poly(tetrafluoroethylene) (Teflon) (*11*), and poly(methyl methacrylate) (PMMA) (*6*) as well as polymer blends (*12*).

Bulk Polymers. The first example is a 1-mm-thick film of poly(dimethylsiloxane) (PDMS; silicon rubber sheet) with an average molecular weight between cross-links of 5726. The SALI experimental conditions were 7-kV Ar$^+$, 2-μA dc current, 5-μs pulse width, and 5×10^{12} ions/cm^2.

Charge compensation for the PDMS sample was performed using a pulsed neutralizer–extractor assembly in which 12-eV electrons were used to flood the sample surface.

Figure 7 is the SPI–SALI spectrum for a thick film of PDMS. The dominant peak in the low-mass range is again the monomer at $m/z = 74$ [M = Si(CH$_3$)$_2$O]. A second scan, taken to extend the mass range, is shown in Figure 8. The observed distribution is a series of peaks at mass intervals of 74 (SiCH$_3$O) with an average weight of 355 mass units. The presence of a lower molecular weight species on the surface, which could be either a thermal degradation product or a non-cross-linked low molecular weight siloxane species, is indicated. Figure 9 is an expanded region of the higher mass range spectrum that includes the (M)$_4$ + SiCH$_3$O$^+$ peak at $m/z = 355$. The mass resolution, defined as $M/\Delta M$, where ΔM is the full width at half maximum, is approximately 1500 at mass 355.

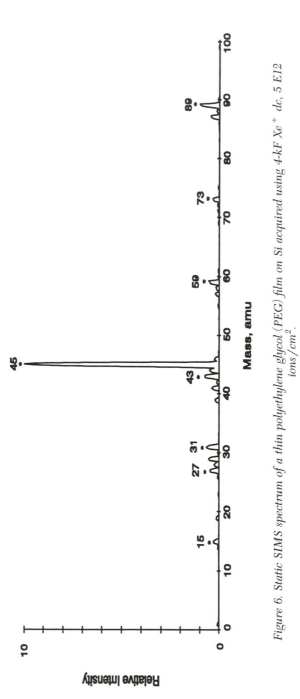

Figure 6. Static SIMS spectrum of a thin polyethylene glycol (PEG) film on Si acquired using 4-kF Xe$^+$ dc, 5 E12 ions/cm^2.

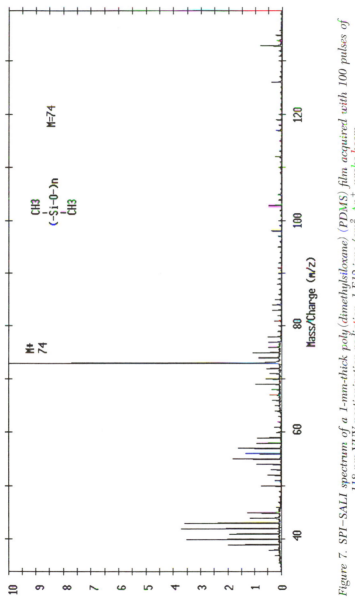

Figure 7. SPI–SALI spectrum of a 1-mm-thick poly(dimethylsiloxane) (PDMS) film acquired with 100 pulses of 118-nm VUV postionization radiation, 1 E12 ions/cm² Ar⁺ probe beam.

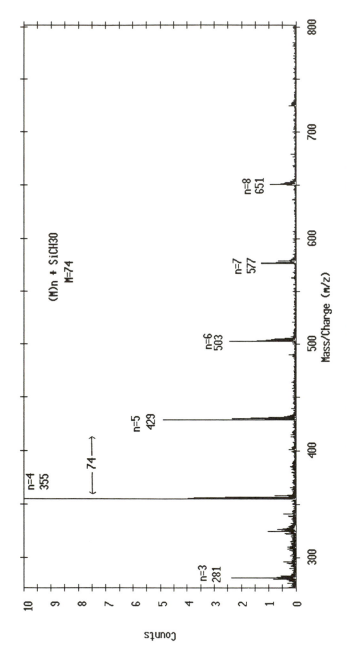

Figure 8. Extended mass range of the SPI–SALI spectrum of PDMS.

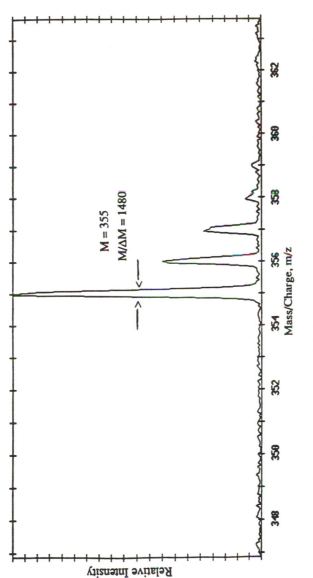

Figure 9. Expanded ration of the SPI–SALI PDMS spectrum. Mass resolution was determined by dividing the centroid mass value by the full width at half maximum.

Another example of the SPI–SALI analysis for bulk polymer materials is a thick piece of polystyrene. Figure 10 shows the ion-induced SPI–SALI spectrum of the polystyrene sample. Ion bombardment was performed using 5-keV Ar$^+$ at a total ion dosage of 7×10^{11} ions/cm^2. The SALI spectrum again has an intense monomer peak at $m/z = 104$ (M = C$_8$H$_8$), unlike TOF–SIMS or quadrupole static SIMS spectra of the same material (13). Other characteristic polystyrene peaks are observed in the SALI spectrum at $m/z = 115$ (C$_9$H$_7$), 91 (C$_7$H$_7$), and 78 (C$_6$H$_6$). The dominance of the monomer peak is even more pronounced when the ion beam (used for desorption) is replaced by a laser beam. Figure 11 shows the laser-induced SPI–SALI spectrum of the same sample. This spectrum shows minimal damage to the polymer. The prevailing peak is the monomer peak at $m/z = 104$ (C$_8$H$_8$). For laser-induced desorption, the third harmonic of a standard Nd:YAG laser at 355 nm was focused onto the surface. The SALI postionization mode using the 118-nm line was the same as in the ion-induced SALI experiments. There is clear evidence that photon-induced desorption, coupled with subsequent photon postionization, further reduces the polymer sample damage caused by ion-induced desorption SALI. Damage is reduced because, in contrast to ion sputtering, there is no nuclear momentum transfer involved. However, when a laser is used as the probe beam, SALI is no longer a surface-specific technique.

Future Directions

Recent progress in our laboratory has improved the resolution to the TOF analyzer and should increase the mass resolution achievable by SPI–SALI. Further improvements could result from using the ultrashort pulse mode of the Nd:YAG laser, which can produce laser pulse widths of 2.5 ns. The increase in mass resolution along with the ease of analysis will establish SPI–SALI as a powerful tool for surface analysis of insulating materials. The analytical versatility of this technique will be exploited in studies using the various desorption methods (ion sputtering versus electron-stimulated desorption versus laser desorption). The ability to analyze the surface of thick insulating materials with minimal sample preparation coupled with the ease of monomer identification makes this an extremely attractive technique for polymer surface characterization. Future study will involve the quantification aspects of SPI–SALI applied to modified polymer surfaces and copolymer systems.

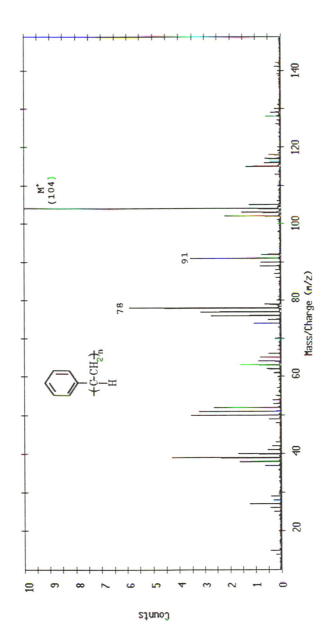

Figure 10. Ion-induced SPI–SALI spectrum of bulk polystyrene acquired using 100 pulses of 118-nm VUV postionization radiation, 1 E12 ions/cm² Ar⁺ probe beam.

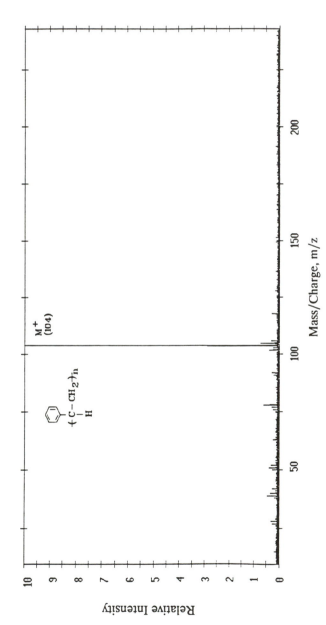

Figure 11. Laser-induced SPI–SALI spectrum of bulk polystyrene acquired using 100 pulses of 118-nm VUV postionization radiation and 100 pulses of 355-nm VUV probe beam.

Acknowledgments

The authors thank Dr. David G. Welkie of the Physical Electronics Division for his assistance in the interpretation of the data. The authors also acknowledge the cooperation of Dr. Christopher H. Becker of SRI International, who initiated the SALI technology.

References

1. Gruen, D. M.; Pellen, M. T.; Calaway, W. F.; Young, C. E. *Laser Post-Ionization Sputtered Neutral Mass Spectroscopy*; Benninghoven, A.; Huber, A. M.; Werner, H. W., Eds.; John Wiley and Sons: Chichester, England, 1988; p 789.
2. Becker, C. H.; Gillen, K. T. *Anal. Chem.* **1984**, *56*, 1671.
3. Parks, J. E.; Schmitt, H. W.; Hurst, G. S.; Fairbanks, W. M., Jr. *Thin Films* **1983**, *8*, 69.
4. Pellin, J. J.; Young, C. E.; Calaway, W. F.; Gruen, D. M. *Surf. Sci.* **1984**, *144*, 619.
5. Storms, H. A.; Brown, K. F.; Stein, J. D. *Anal. Chem.* **1977**, *49*, 2023.
6. Pallix, J. B.; Schüle, W.; Becker, C. H.; Huestis, D. L. *Anal. Chem.* **1989**, *61*, 807.
7. Mamyrin, B. A.; Karataev, V. I.; Shmikk, D. V.; Zagulin, V. A. *Sov. Phys. JETP (Engl. Transl.)* **1973**, *37*, 45.
8. Becker, C. H.; Gillen, K. T. *J. Vac. Sci. Technol., A* **1985**, *3*(3), 1347.
9. Schüle, U.; Pallix, J. B.; Becker, C. H. *J. Am. Chem. Soc.* **1988**, *110*, 2323.
10. Carlson, B. A.; Katz, W.; Newman, J. G.; Van Ooji, W. J.; Moulder, J. F. *Static SIMS Handbook of Polymer Analysis*; Perkin-Elmer, Physical Electronics: Minneapolis, MN, 1991; p 38.
11. Schüle, W.; Pallix, J. B.; Becker, C. H. *J. Vac. Sci. Technol., A* **1988**, *6*(3), 936.
12. Pallix, J. B.; Schüle, U.; Becker, C. H. *Surface Analysis of Bulk Organic Polymers by Single-Photon Ionization*; Benninghoven, A.; Evans, C., Eds.; John Wiley and Sons: New York, 1990; p 207.
13. Carlson, B. A.; Katz, W.; Newman, J. G.; Van Ooji, W. J.; Moulder, J. F. *Static SIMS Handbook of Polymer Analysis*; Perkin-Elmer, Physical Electronics: Minneapolis, MN, 1991; p 26.

RECEIVED for review July 15, 1991. ACCEPTED revised manuscript September 1, 1992.

Spectroscopic Characterization of Films Obtained in Pulsed Radio-Frequency Plasma Discharges of Fluorocarbon Monomers

Charles R. Savage[1], Richard B. Timmons*[1], and Jacob W. Lin[2]

[1]Department of Chemistry, Box 19065, The University of Texas at Arlington, Arlington, TX 76019–0065
[2]Polytronix Inc., 805 Alpha Drive, Richardson, TX 75081

The structure of plasma-polymerized films obtained using pulsed radio-frequency (rf) duty cycles is shown to vary significantly with the type of rf duty cycle employed. The controllability and tailoring of surface compositions as functions of the rf duty cycle are illustrated for processes carried out with two fluorocarbon monomers, namely, hexafluoropropylene oxide (C_3F_6O) and hexafluoropropene (C_3F_6). The variations in surface composition with rf duty cycles are documented via X-ray photoelectron spectroscopy and Fourier transform infrared spectroscopy analyses of the films obtained. Progressive and substantial changes in the molecular composition of the plasma-deposited films as functions of the rf duty cycles employed were achieved with both monomers. The experimental results reveal a relatively high level of compositional control of polymers in the plasma polymerization of these monomers as a function of the rf duty cycle employed. In general, there is a progressive increase in the extent of polymer cross-linking with increasing rf duty cycle as evidenced by spectroscopic characterization of the films obtained in this study.

T HE USE OF PLASMA TECHNIQUES TO INITIATE POLYMERIZATION PROCESSES continues to exhibit rapid growth. The magnitude of the current activity in

---

* Corresponding author

plasma polymerization was illustrated in terms of the scope and diversity of topics presented at the recent Symposium on Plasma Polymerization and Plasma Interactions with Polymeric Materials (1). As demonstrated at that symposium, the focus of the majority of recent studies lies in the application of plasma polymerization methods to achieve surface modifications. These modifications are obtained by thin-film deposition or reactive interactions of plasma-generated intermediates with surface atoms and molecules.

The use of plasma polymerization as a surface-modification technique offers a number of advantages over various conventional coating procedures. Among these advantages are the fact that an unusually wide range of monomers are available (e.g., even CH_4 can function effectively as a "monomer" in plasma polymerization), relatively pinhole-free films can be deposited, film thickness is usually controllable over the important range from tens to hundreds of angstroms, and surface modifications are achieved in a simple and relatively inexpensive one-step deposition process. Yasuda (2) has provided a detailed description and analysis of the important facets of plasma polymerization reactions.

To date, the overwhelming majority of plasma polymerization investigations have involved continuous-wave (CW) plasma operation, invariably at the radio frequency (rf) of 13.56 MHz. In general, experimentation with important plasma variables (e.g., absorbed rf power, monomeb flow velocity and pressure, etc.) permits identification of reaction conditions in which uniform films can be obtained from a given monomer. Again, Yasuda (2) has provided an excellent analysis and discussion of the influence of various plasma variables on the dynamics of the polymerization process and the quality of the resultant films.

An important aspect of CW plasma polymerizations is the extensive fragmentation of the monomer that occurs under CW conditions. Indeed, it is this extensive fragmentation that permits saturated molecules such as CH_4 and C_2H_6 to be employed as monomers in these systems. At the same time, it is necessary to recognize that extensive monomer fragmentation translates to a relative lack of controllability in the molecular composition of the films that are obtained. For example, the relative indiscriminate reorganization of atoms under CW plasma polymerization conditions was aptly demonstrated in a recent CW study of the plasma polymerization of several carbonyl-containing monomers. Plasma polymerization of acetaldehyde, acetone, and 2-butanone yielded remarkably similar surface films despite variations in the molecular structures and carbon–hydrogen–oxygen content of these molecules (3). In a similar vein, it is often noted that CW plasma polymerization of certain monomers results in molecular compositions that frequently bear relatively little resemblance to those obtained in the conventional polymerization of a reactive monomer [e.g., acrylonitrile (4) and tetrafluoroethylene (5)]. To a certain extent, reduction of the absorbed rf power can, with some monomers, reduce the extent of gas-phase fragmentation and thus generate a closer similarity between the plasma and the conventionally polymerized monomer as demonstrated, for example, with styrene (6). However, control of plasma-deposited film composition via absorbed rf power has several inherent limitations. These limitations include the fact that a minimum absorbed rf power is required to ignite and maintain the plasma and uniform films (i.e., those devoid of powder or oillike deposits) are typically obtained only over relatively restricted power regimes for a given monomer under specified flow conditions. Additional approaches to controlling film compositions in CW plasma polymerizations include depositions with reduced substrate temperatures (7) and variations in the positions of the substrates located downstream of the plasma (8).

The purpose of the present study was to introduce an added dimension to the *controllability* of plasma-achieved surface modifications with respect to the molecular composition of the deposited films. We hoped this added controllability could be achieved via pulsed rf plasma discharges as opposed to the standard CW operation. The rationale behind this approach centered on a recognition that reactive intermediates (i.e., radicals and ions) produced during the plasma-"on" periods may undergo decay mechanisms during the plasma-"off" period that are significantly different from decay mechanisms observed under typical CW conditions. For example, relatively high reactive intermediate concentrations during the plasma-on periods tend to enhance the importance of recombination or disproportionation reactions that are second order with respect to reactive intermediates. On the other hand, at some point during the off period, first-order reactions in reactive intermediates will be dominant as the concentration of these reactive species drops. Thus, competitive processes of the type

$$R_1 + R_2 \rightarrow \text{products} \tag{1}$$

$$R_1 \, (\text{or } R_2) + M \rightarrow \text{products} \tag{2}$$

where R_1 and R_2 represent reactive intermediates and M is a monomer, would be expected to favor a relatively higher contribution of reaction 1 compared to reaction 2 under plasma-on compared to plasma-off conditions. Furthermore, if such effects are indeed noted, the actual molecular compositions of the polymers that are obtained might be tunable to a certain extent with respect to the type of rf duty cycle employed (i.e., the ratio of on time to on + off times). In connection with this idea, it is significant to note that the limited literature data that compared fixed duty cycle pulsed and CW plasma polymerizations support the fact that marked differences in film growth rates and residual radical concentrations are observed for selected monomers (2, 9). Additionally, Nakajima et al. have experimented with the use of pulsed-rf plasmas (10).

This chapter reports on a detailed study of variable duty cycle pulsed rf plasma polymerizations as a route to achieving a higher level of molecular compositional control of the plasma-initiated surface modifications obtained. This topic is explored herein with respect to systematic variations in rf duty cycles employed in the plasma polymerization of several monomers, while all other plasma variables are kept constant. The molecular compositions of the polymer films obtained as functions of rf duty cycle were monitored spectroscopically using X-ray photoelectron spectroscopy (XPS) and Fourier transform infrared (FTIR) spectroscopy. The combination of these spectroscopic approaches provides clear and abundant evidence for significant variations in the molecular composition of the plasma-deposited films as functions of the rf duty cycles employed. These initial results are encouraging with respect to achieving controllability of the molecular composition of the surface modifications obtained in plasma polymerizations. Some of the implications and potential applications of such surface molecular compositional control are considered briefly at the end of this chapter.

Experimental Details

All plasma polymerizations were carried out in a cylindrical Pyrex glass reactor that was 10 cm in diameter and 30.5 cm in length. Radio-frequency power was provided to this reactor by two concentric metal rings located at either end of the reactor. The reaction system and associated electronics are shown in Figure 1. The rf circuit included a function generator (Wavetech Model 166), a pulse generator (Tetronik Model 2101), a rf amplifier (ENT Model A300), a frequency counter (HP Model 5381A), a wattmeter (Bird; to measure absorbed and reflected power), and a capacitor–inductor matching network used to tune the circuit to minimize reflected power. An oscilloscope, previously calibrated against the wattmeter under CW conditions, was employed in tuning the circuit to minimize reflected power under rf pulsed operation. All data reported in this study represent runs carried out at a rf frequency of 13.56 MHz. An MKS butterfly valve controller (Baratron Model 252A) was used to both monitor and control pressure in the reactor.

A standardized procedure was adopted for all film depositions. Substrates that consisted of polished Si and KCl disks were placed on top of an inverted 20-mL beaker located in the center of the reaction chamber. The Si samples were cleaned via sonification and rinsing prior to placement in the reactor. The system was then evacuated to a pressure ~ 1 μm, after which the sample was subjected to a 10-min CW argon plasma at an Ar pressure of 0.7 torr and rf absorbed power of 200 W. This Ar pretreatment was conducted to ensure surface cleanliness of the substrates. Subsequently, the Ar flow was terminated and the reactor system was evacuated to background pressure before admission of the reactant monomers. Between runs the reaction

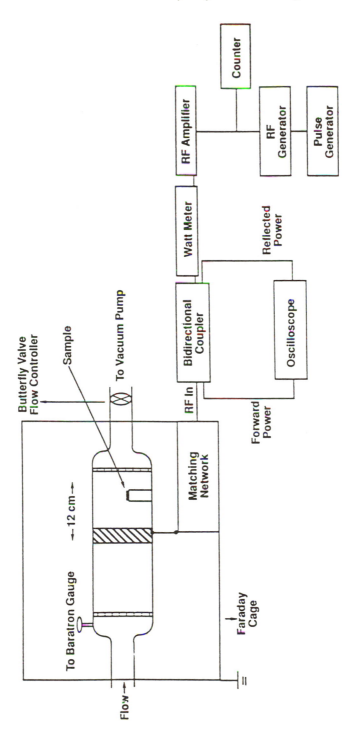

Figure 1. Schematic diagram of the pulsed rf plasma components and the flow-tube reactor.

chamber was cleaned using an O_2 plasma to remove surface depositions on the chamber walls.

Because the focus of this study was the investigation of the effect of rf duty cycle on film compositions obtained in pulsed plasma polymerizations, most depositions of a given monomer were carried out using constant rf power and constant monomer pressure and flow velocity. In this way, we were able to isolate rf duty cycle effects on the chemical composition of the films obtained. The actual reaction variables employed with each of the monomers studied are identified in the following text.

Hexafluoropropylene (C_3F_6) and hexafluoropropylene oxide (C_3F_6O) were employed as monomers in this study. The gaseous C_3F_6 and C_3F_6O reactants were obtained from PCR Inc. and they had stated purity in excess of 98%, which was confirmed by gas chromatography–mass spectroscopy analysis.

The XPS spectroscopic characterizations were carried out using a spectrometer (Perkin-Elmer PHI 5000) equipped with an X-ray source monochromator. The X-ray source was Al $K_\alpha = 1486.6$ eV. A pass energy of 17.90 eV giving a resolution of 0.60 eV with Ag ($3d_{5/2}$) was employed. The XPS results of fluorine-containing monomers (C_3F_6O and C_3F_6) were shifted (i.e., standardized) by centering the F (1s) peak at 689 eV. An electron flood gun was employed to neutralize charge buildup on the insulator-type films involved in this work. The electron flood gun was operated under conditions that provided optimum resolution of the C (1s) peaks. Typical conditions for the neutralizer were 21.0-ms emission current and 3.0-eV electron energy.

Film deposition rates were determined by measurement of the film thicknesses using a profilometer (Tencor Alpha Step 200) after each run. A diamond-tipped pen was employed to scribe a thin scratch on the plasma-deposited films, after which the film thicknesses were determined. Silicon substrates were employed in the film-thickness measurements.

Results

Substantial changes in the molecular composition of plasma polymerized films were noted as a function of employed rf duty cycles for both of the monomers. These results are presented with respect to each of the reactants in the following sections.

Hexafluoropropylene Oxide.   Plasma polymerizations of hexafluoropropylene oxide (C_3F_6O) were carried out at a monomer pressure of 0.43 torr, a flow velocity of 9.6 cm^3/min (standard temperature and pressure), and an absorbed rf power of 300 W. Polymerizations were carried out over a wide range of rf duty cycles. In one set of experiments, a constant rf on time (e.g., 10 ms) was employed and the off time was varied systematically from 20 to as long as 1000 ms. In a second set of experiments, a constant off time (i.e., 200 ms) was employed while the on time was varied systematically over the range from 10 to 70 ms. Additionally, several other on–off ratios were studied as identified in the following paragraph.

In both experimental sequences (i.e., constant on–variable off time and constant off–variable on time), marked changes in the molecular composition of the plasma-deposited films were noted with variation of the rf duty cycle. The compositional changes are documented by both XPS and FTIR analyses of these films. The XPS results are summarized in Figures 2–4. Each of these spectra shows high-resolution scans of the C ($1s$) binding energy. Figure 2 depicts the variation of the C ($1s$) region obtained for films deposited at a constant plasma-on time of 10 ms as a function of a progressively longer off time that corresponded to values of 20, 60, 100, and 1000 ms (curves a–d, respectively). From many previous XPS studies of fluorocarbons, the separate peaks in these spectra can be assigned to the following functional groups (*11, 12*): CF_3, 294 eV; CF_2, 292 eV; $CFCF_n$, 290 eV; CF, 289 eV; CCF_n, 287.5 eV. Plasma polymerized fluorocarbon films generally exhibit four prominent peaks in that only a single broad peak centered in the 290–289-eV binding region is observed. This observed peak corresponds to unresolved CF-substituted and nonfluorine-substituted groups. Because we are interested in the correlation of trends in group functionalities with varying rf duty cycles, the XPS spectra are interpreted in terms of CF_3, CF_2, CF, and C groups. The C group refers to C bound to CF_3 and CF_2. Additionally a C ($1s$) peak at ~ 285 eV (observed only in the 10 ms on–20 ms off and CW runs) is typical of a graphitic or SiC carbide structure. Analysis of Figure 2 reveals a clear shift in film composition with increasing off time toward a progressively more dominant CF_2-type structure.

The second set of experiments, carried out with a constant off time of 200 ms and systematic variation of the on time over the interval of 10, 20, 30, 40, and 70 ms, also reveals progressive changes in molecular film composition as functions of rf duty cycle. These results are presented in Figure 3 in terms of two stacked plots in which each spectrum was normalized with respect to the peak heights of the CF_3 carbons. Analysis of Figure 3 reveals a clear, progressive diminution of the intensity of the CF_2 carbons relative to the CF_3, CF, and C functionalities with increasing plasma-on times.

Finally, a series of runs carried out at a constant plasma on–off ratio of 1/20 but with varying on–off pulse times produced the C ($1s$) XPS results shown in Figure 4. Again, dramatic changes in the relative proportion of CF carbon atom functionalities are noted over the on–off sequence of 1/20, 10/200, 20/400, and 100/2000 (in milliseconds). The 1/20 sample had no observable coating as shown by XPS analysis. The XPS spectra for the 10/200, 20/400, and 100/2000 runs are shown in Figure 4.

The molecular compositional changes achieved in film structure as functions of the rf duty cycle employed in these plasma polymerizations are further confirmed by FTIR analysis. An example of these changes is shown in Figure 5 for a series of plasma polymerizations carried out at rf duty cycles of on–off ratios (in milliseconds) of 10/20, 10/200, and 10/400 (curves a, b, and c, respectively). The strong absorption noted at ~ 1200 cm^{-1} is characteristic of CF stretching vibrations. As shown in this figure, the broad, relatively featureless absorption band at 1200 cm^{-1} obtained at the 10 ms on–20 ms off duty cycle is replaced with a clear doublet structure as the off portion of the duty cycle is increased. This well-resolved doublet, indicative of a highly ordered solid-state structure, is remarkable when compared with

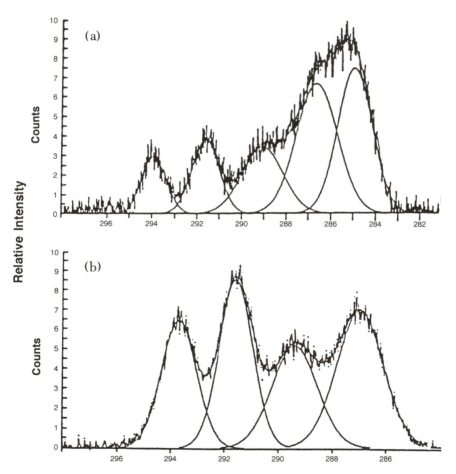

Figure 2. High-resolution C (1s) XPS results obtained with C_3F_6O showing the change in molecular structure of the plasma-deposited surface films as a function of the pulsed rf duty cycle employed. Curves a–d represent rf on–off duty cycles of 10/20, 10/60, 10/100, and 10/1000 (in ms), respectively.

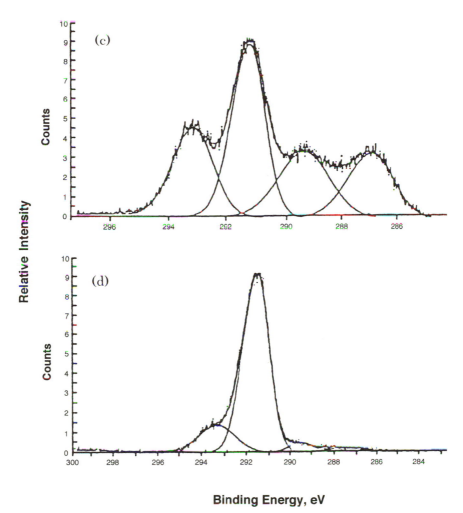

Figure 2. Continued

typical infrared (IR) absorption spectra of fluorocarbons obtained under CW plasma polymerization conditions (2). The frequencies of these two doublets are at 1207 and 1153 cm^{-1}. In light of the aforementioned XPS C (1s) results for 10 ms on–variable off time (Figure 2), it seems reasonable to assign the 1153-cm^{-1} band to a CF stretch associated with CF_2 groups, because it is clearly the CF_2 functionality that grows with increasing off times. Furthermore, it is significant to note that we observe a constancy in the ratio of the 1153–1207-cm^{-1} peaks at off times greater than 400 ms. This constancy in film molecular composition at very long rf off times is confirmed by the XPS

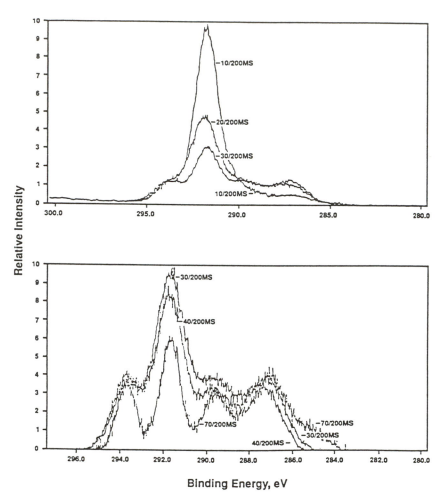

Figure 3. High-resolution C (1s) XPS results obtained with C_3F_6O showing the progressive variation in plasma-deposited surface films for runs at constant off time of 200 ms and on times of 10, 20, 30, 40, and 70 ms are shown. Plots were normalized with respect to the high binding energy 294-eV peak (CF_3).

analysis, which reveals virtually no change in composition over the on–off sequences (in milliseconds) of 10/400, 10/600, 10/800, 10/1000, and 10/1200.

The XPS and FTIR spectroscopic characterizations reveal that variation of the rf duty cycle in the plasma polymerization of C_3F_6O results in clear changes in the surface film compositions obtained. The characteristic feature of these changes is that a decrease in the ratio of on–off times results in less cross-linking of the films obtained (i.e., a decrease in C and CF groups with

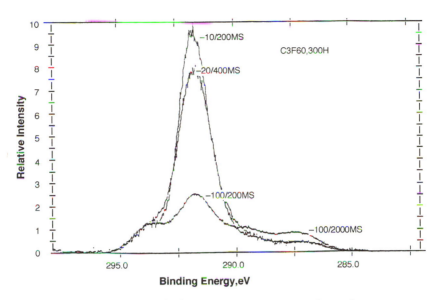

Figure 4. High-resolution C (1s) XPS results with C_3F_6O obtained at a constant on–off duty-cycle ratio of 1/20 but with variable on–off pulse widths of 10/200, 20/400, and 100/2000 (in milliseconds) as shown. Plots were normalized as in Figure 2.

progressive growth of the CF_2 functionality). In fact, the films obtained at duty cycles involving relatively small on–off plasma duty-cycle ratios and sufficiently long off times (i.e., longer than 100 ms) are consistent with the formation of a highly linear CF_2 polymer, which is relatively similar to that obtained in the conventional polymerization of C_2F_4 to form poly(tetrafluoroethylene) (Teflon). Also note that the variation from a cross-linked to a linear polymer with decreasing rf duty cycle is confirmed by physical properties of the films obtained. For example, there is a noticeable decrease in film hardness with decreasing rf duty cycle.

Overall, as documented in Figures 2 through 5, the variation of rf duty cycle in the pulsed plasma polymerization of this monomer provides a means to control tunability of the molecular composition of the plasma-produced polymers. In the case of C_3F_6O reactant, this surface-composition tunability encompasses very wide changes in the relative proportions of the carbon–fluorine group functionalities produced.

An important feature of the dynamics of C_3F_6O pulsed plasma polymerization is noted with respect to film growth rate as a function of rf duty cycle. This film growth rate is useful to the understanding of the variation in plasma-produced film compositions as functions of rf duty cycle employed. An interesting example of the film growth rate–rf duty cycle relationship is shown in Figure 6. In this diagram, the film deposition rate (expressed in

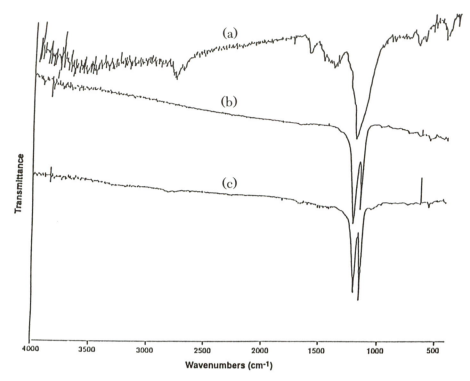

Figure 5. FTIR absorption spectra obtained from the plasma deposition of C_3F_6O as a function of the rf duty cycle employed. The curves represent data obtained for rf duty cycles of 10/20, 10/200, and 10/40 (in milliseconds) as shown in curves a, b, and c, respectively.

milliangstroms per Joule absorbed rf energy) is plotted as a function of the rf off time for a series of runs carried out with a constant on time of 10 ms. The film-deposition rate is striking in that a rapid *increase* in film growth rate is noted with *increasing* off time. As this result clearly demonstrates, significant polymerization that leads to film deposition occurs during the off portion of the rf duty cycle. In fact, note that under the reaction conditions employed in this study, no film deposition was obtained when the plasma was operated under a CW condition. The lack of CW film deposition is implied by the data shown in Figure 6.

As shown in Figure 7, a plot of film deposition rate as a function of plasma-on time at a constant off time of 200 ms is also revealing. In this series, the deposition rate goes through a sharp maximum at 10 ms on–200 ms off as the rf duty cycle is increased at a constant off period of 200 ms.

The results shown in Figures 6 and 7 are clearly supportive of major differences in film deposition rates during on and off periods of the pulsed plasma polymerization. We believe that the rapid increase in film deposition

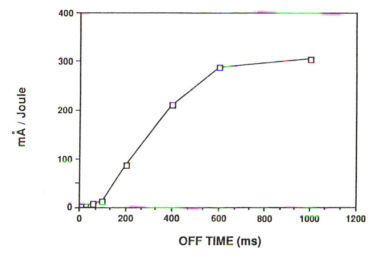

Figure 6. Film deposition rate obtained during the pulsed rf plasma polymerization of C_3F_6O as a function of the plasma-off duration with a constant on time of 10 ms. The deposition rate is expressed in terms of milliangstroms per Joule of absorbed rf energy.

rate during the plasma-off periods (Figure 6), coupled with the simultaneous growth in CF_2 functionality in the polymer films with increasing off periods (cf. Figure 2) is indicative of a free-radical-chain process that is operative in this system. We speculate that reactive intermediates R (i.e., ions or radicals) that are produced during the plasma-on periods can initiate a reaction of the type

$$R + CF_3-\underset{\underset{O}{\diagdown\diagup}}{\overset{\overset{F}{|}}{C}}\underset{}{\overset{\overset{F}{|}}{C}}-F \longrightarrow RCF_2CF_2\cdot + CF_2O \qquad (3)$$

by reaction with undissociated C_3F_6O monomer. Subsequently, the $RCF_2CF_2\cdot$ radical produced in reaction 3 can continue the polymerization process by the reaction

$$RCF_2CF_2\cdot + CF_3-\underset{\underset{O}{\diagdown\diagup}}{\overset{\overset{F}{|}}{C}}\underset{}{\overset{\overset{F}{|}}{C}}-F \longrightarrow R(CF_2)_3CF_2\cdot + CF_2O \quad (4)$$

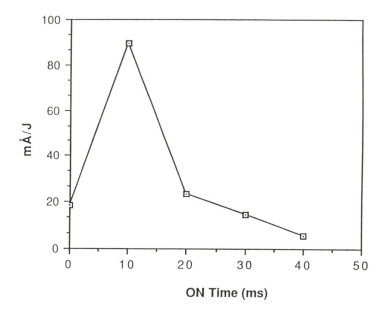

Figure 7. Film deposition rate obtained during the pulsed rf plasma polymerization of C_3F_6O for runs with a constant off time of 200 ms expressed as a function of the on times employed.

The continued reaction of $R(CF_2)_3CF_2 \cdot$ with the monomer represents a chain reaction that leads to the formation of polymeric $R(CF_2)_nCF_2 \cdot$ units and a resultant CF_2-type linear polymer. It is important to note that CF_2O is an exceptionally stable species $[\Delta G_f^0(CF_2O) = -624 \text{ kJ/mol}$ where G_f^0 is the fracture energy] at 298 K (*13*), which makes reactions of types 3 and 4 favorable processes in terms of free-energy considerations. This type of chain growth is strongly supported by the complete absence of oxygen (as shown by both XPS and FTIR analysis) in all films obtained from C_3F_6O polymerization.

Although the foregoing radical-chain process could also be operative during plasma-on periods, the higher free-radical concentrations present during on periods will favor radical–radical-type processes that lead to a relative loss in product specificity and thus more highly cross-linked structures as observed at higher on–off rf duty cycle ratios (Figure 3).

In addition to the aforementioned film-forming processes, the experimental evidence is also strongly suggestive of the presence of ablation induced and plasma-on induced surface rearrangement processes that are competitive with the film deposition step. For example, as noted previously, virtually no film is obtained when the deposition process is carried out under

CW conditions. Furthermore, polymerizations carried out at a fixed plasma-on–plasma-off ratio of 1/20 and varying pulse lengths resulted in a significant variation in the proportion of CF groups as shown in Figure 4. In this figure, a sharp diminution of CF_2 groups is noted with increasing on times and fixed on–off ratios. This systematic change in film composition can be rationalized in terms of plasma-induced molecular rearrangements that result from impact of high-energy species with the initially deposited surface films. Alternatively, the results in Figure 4 might be indicative of preferential ablation of CF_2 groups during the plasma-on period. Because these changes occur at a fixed on–off ratio of 1/20, it seems clear that some type of inductive (i.e., time-delay) process is involved in the creation of the reactive species responsible for the plasma-induced molecular rearrangement or ablation processes. The occurrence of a time-delayed ablation process during the plasma-on period is also strongly confirmed by the deposition rate data shown in Figure 7 in which a sharp decrease in the rate of film formation is noted at increasing on times longer than 10 ms.

Perfluoropropylene.   Plasma polymerizations were carried out using perfluoropropylene (C_3F_6) at a monomer pressure of 0.43 torr and a flow velocity of 9.6 cm^3/min (standard temperature and pressure). In contrast to the 300-W rf power used in the C_3F_6O studies, the rf power employed in the majority of runs was 200 W. This lower power was employed with C_3F_6 because it produced better quality films at higher deposition rates.

Systematic studies of the effect of variations in rf duty cycle on the plasma polymerization of C_3F_6 were carried out with all other reaction variables kept constant. As in the case of C_3F_6O, two time sequences were used to obtain film depositions: a sequence of a constant on time and variable off time as well as a series of experiments with a constant off time and variable on time. In addition, experiments at a fixed on–off ratio of 1/2 but with variable on–off pulse widths were also carried out.

XPS analyses of the C (1s) region of the plasma polymerized films that were obtained with this monomer are shown in Figures 8–10. Figure 8 summarizes results for a constant on time of 10 ms and off periods of 20, 100, and 200 ms. The results are shown as a stacked plot, normalized relative to the CF_3 group peak heights. As revealed in this figure, there is a marked increase in the relative proportion of CF_2 groups (vis á vis other CF and C groups) with increasing off times. This behavior parallels that observed with C_3F_6O monomer (cf. Figure 2). The C (1s) XPS spectra for films obtained at constant off and variable on times are shown in Figure 9. In this series, results obtained at on times of 10, 20, and 100 ms and constant off time of 200 ms are shown as curves a, b, and c, respectively. Again, the results parallel those from C_3F_6O (cf. Figure 3) in which the CF_2 molecular fraction decreases substantially and systematically as the on time increases. Finally, the C (1s) XPS spectra obtained from films deposited at a constant on–off

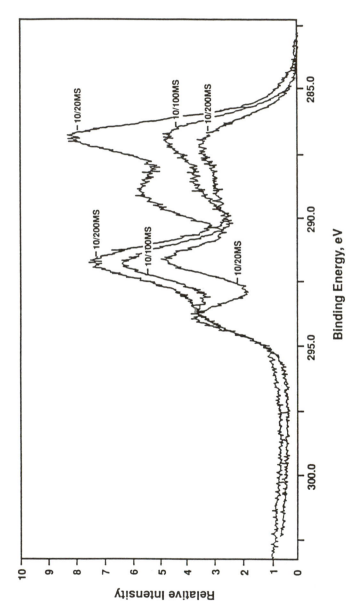

Figure 8. High-resolution C (1s) XPS results obtained with C_3F_6 showing the molecular structure variation of the plasma-deposited surface films as a function of the rf duty cycle employed. Curves shown represent data obtained for rf on–off duty cycles on 10/20, 10/100, and 10/200 (in milliseconds), as indicated. Plots were normalized with respect to the high binding energy 294-eV peak (CF_3).

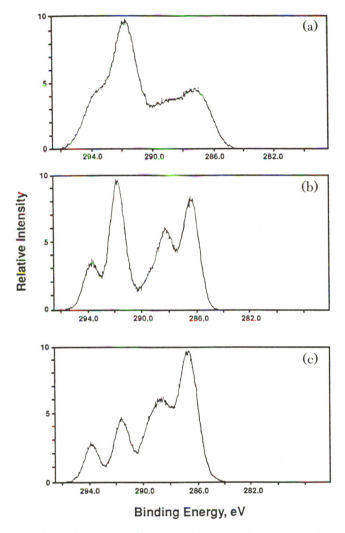

Figure 9. High-resolution C (1s) XPS results obtained with C_3F_6 showing the variation in molecular structure of films deposited during pulsed rf plasma polymerizations with a constant off time of 200 ms and variable on times of 10, 20, and 100 ms. Curves a, b, and c represent results obtained with rf on–off duty cycles of 10/200, 20/200, and 100/200, respectively.

duty cycle ratio (in this case 1/2) and with variations in pulse widths are shown in Figure 10. Again, a substantial change in film composition was noted with respect to carbon functionality with a progressive increase in C (1s) at 287 eV relative to the other C (1s) peaks noted at on–off values of 0.1/0.2, 10/20, and 1000/2000 (expressed in milliseconds). The result is

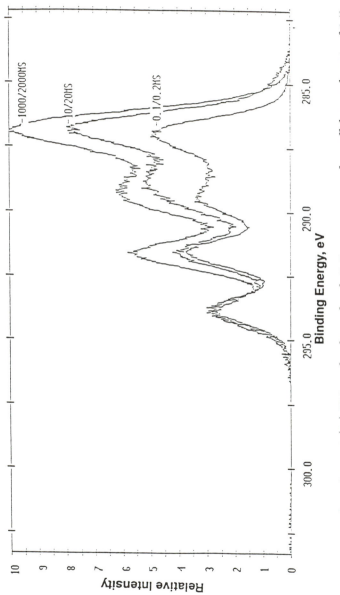

Figure 10. High-resolution C (1s) XPS results obtained with C_3F_6 at a constant rf on–off duty cycle ratio of 1/2 with variable on–off pulse widths of 0.1/0.2, 10/20, and 1000/2000 (in milliseconds) as shown. The stacked plots were normalized as in Figure 8.

again similar to that observed with C_3F_6O as shown in Figure 4, where the on–off ratio was 1/20.

The FTIR spectra obtained in C_3F_6 plasma polymerization are also clearly supportive of marked changes in the molecular structure of the deposited films as functions of the pulsed rf duty cycle employed. This variation in film composition is illustrated in Figure 11 for films obtained from C_3F_6 at rf on–off duty cycles of 10/20 and 10/100 (in milliseconds) as curves a and b, respectively. Again the single broad band ~ 1200 cm^{-1} obtained under 10/20 (in milliseconds) conditions, indicative of the presence of a wide range of CF stretching frequencies (i.e., a randomized structure), begins to reveal a more highly structured composition under pulsing conditions in which the ratio of on–off time is reduced. In view of the XPS data in Figure 8, it is clear that the growth in the peak at ~ 1060 cm^{-1} can be attributed to the increasing prominence of CF_2 groups formed with decreasing values of the on–off rf duty cycle.

The film-deposition rates obtained as a function of the pulsed rf duty cycle variations are shown in Figures 12–14. Figure 12 shows the film-deposition rates obtained at a constant on time of 10 ms and variable off time up to 200 ms. In the case of C_3F_6, the film deposition rate exhibits a sharp maximum at 20 ms off as the off time is increased. This behavior is quite different from that noted for C_3F_6O under comparable pulsing conditions (Figure 6). A plot of deposition rate for a constant off time of 200 ms and variable on time is shown in Figure 13. This graph also exhibits a maximum, although the maximum observed is significantly less pronounced than that observed in the comparable plot for C_3F_6O (Figure 7). Figure 14 shows the deposition rate as a logarithmic function of on time for a series of runs carried out at a rf duty cycle on–off ratio of 1/2. As shown by these data, there is a steady increase in film deposition rate with increasing on and off pulse times.

The variations in deposition rates per absorbed joule with rf duty-cycle changes shown in Figures 12–14 are not easily rationalized in terms of a specific reaction mechanism. The initial increasing film deposition rate with increasing off time in Figure 12 is indicative of the occurrence of continued polymerization during the off period. However, unlike the results obtained with C_3F_6O, the deposition rate decreases with increasing off times at off times longer than 20 ms. We believe that a possible explanation of this phenomenon is that a chain radical polymerization process of the type

$$R + C_3F_6 \rightarrow R(CF_2)_2CF_2 \cdot \qquad (5)$$

that is analogous to reaction 4 with C_3F_6O occurs during the off period. However, the activation energy for reaction 5 may be higher that that of reaction 4. Pulsed plasma operations under conditions in which a long off time is operable will produce a lower average gas temperature than that

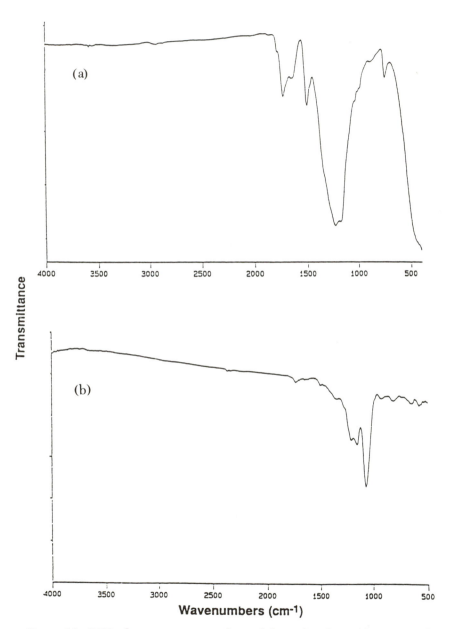

Figure 11. FTIR absorption spectra obtained from the plasma deposition of C_3F_6 as a function of the rf duty cycle employed. Curves a and b represent data from runs carried out at rf duty cycles of 10/20 and 10/100 (in milliseconds), respectively.

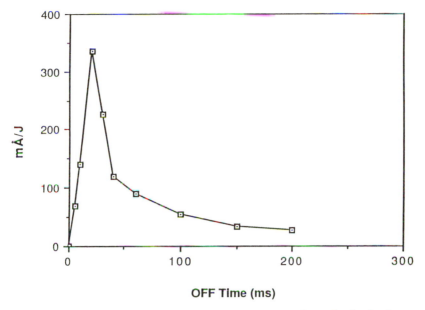

Figure 12. Film deposition rate obtained during the pulsed rf plasma polymerization of C_3F_6 for runs at a constant on time of 10 ms expressed as a function of the off time. Deposition rate is expressed in terms of milliangstroms per joule of absorbed energy.

achieved at short on times. The lower gas temperature, in turn, decreases the extent of film formation achieved during the off period. The initial portion of the curve in Figure 12 represents contributions to the film growth rate during both the on and off periods. It appears that 20 ms off represents the optimum time for polymerization during the off portion of the duty cycle.

However, as in the case of C_3F_6O polymerization, it is clear that both ablation- and reactive-induced surface rearrangements occur in the C_3F_6 pulsed plasma polymerization. The occurrence of ablation processes is clearly evident in the film growth rate data shown in Figure 13 in which a decrease in deposition rate is noted at long on times. The film growth rate must continue to decrease at on times longer than those shown in this figure because virtually no film is obtained under CW conditions. The occurrence of surface-induced molecular rearrangements during the on periods is evidenced by the XPS spectral changes observed at constant on–off ratios and varying pulse lengths (Figure 10). It is clear that there is a progressive shift toward a more highly cross-linked structure [i.e., an increase in the relative importance of C ($1s$) with binding energy of 287 eV] with increasing lengths of the on time pulse widths. This surface-induced rearrangement, which promotes increased polymer cross-linking during long on periods with C_3F_6

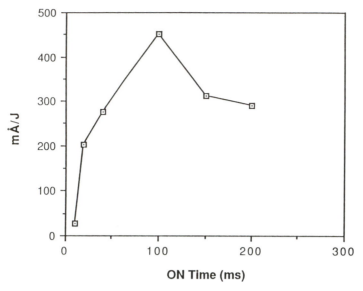

Figure 13. Film deposition rate obtained during the pulsed rf plasma polymerization of C_3F_6 for runs with a constant off time of 200 ms expressed as a function of the on times employed.

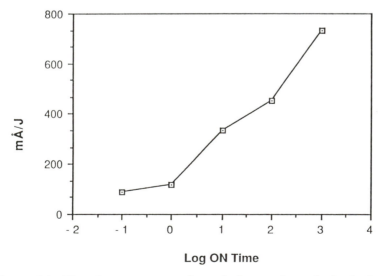

Figure 14. Film deposition rate obtained during the pulsed rf plasma polymerization of C_3F_6 for runs with a constant on–off ratio of 1/2 but with variable on–off pulse widths. Deposition rate (in milliangstroms per joule absorbed rf energy) are expressed as a function of the log on time.

monomer, is similar to, but significantly more pronounced than the rearrangement observed with C_3F_6O.

Discussion

Ideally, we would like to be able to predict, a priori, the molecular composition that will be obtained in the plasma polymerization of a given monomer under specified reaction conditions. Despite an impressive number of studies of the plasma polymerization of numerous molecules, this goal of molecular structure prediction of plasma-derived polymers is simply not available at present. This nonpredictability reflects the complex variety of processes initiated by the plasma. Overall, as documented in many literature reports on CW plasma polymerizations, there is clearly an indiscriminate nature to the plasma polymerization processes.

The major objective of this work was to explore the use of pulsed rf plasma discharges as a means by which a higher degree of selectivity and control could be introduced with respect to the molecular composition of polymeric materials obtained during rf plasma operation. From the results presented in this chapter, it is clear that a new level of molecular compositional control of resulting polymers is indeed attainable as a function of variations of the pulsed rf duty cycles employed. We intend to investigate the generality of these findings by extending this study to a wide range of additional monomers.

Based on the results of this study, several important conclusions are justifiable at this time. The first conclusion is that variation in the pulsed rf plasma duty cycle provides an avenue for variation of surface compositions over significant ranges in molecular structure. As demonstrated in this work, this variability is achieved using a single monomer. Subsequent studies will be conducted with mixed monomers, in which the tunability of molecular surface compositions could well be extended over significantly wider ranges.

A second conclusion is that the film-forming dynamics in these pulsed plasma depositions are obviously complex and represent a variety of competing processes. The complexity of the deposition processes is illustrated in terms of the deposition rates per joule as a function of the rf duty cycles as reported, coupled with the spectroscopic results of changes in molecular composition. During plasma-on periods, deposition−ablation−radical-induced surface rearrangements clearly occur. During plasma-off periods, additional film deposition is obtained. This plasma-off film formation can be very significant (e.g., as in C_3F_6O) and it represents polymerization in which a higher degree of specificity is shown with respect to the surface compositions achieved. We believe it is this difference in polymer compositions obtained during on and off plasma periods that provides the basis for tailoring surface structures as demonstrated in this work. Ultimately, as more results are made available from pulsed rf plasma studies, we hope to be able to anticipate and predict the nature of the surface compositional changes that might be achieved as functions of the plasma duty cycles employed for a particular reactant system.

Finally, we wish to acknowledge explicitly that controllability and tunability of the molecular composition of surfaces represent an important goal that has obvious applications to a wide range of situations. Practical examples of the need for surface compositional control include such areas as improved biocompatibility of materials, improved permselectivity of membranes, better passivation and antireflective coatings for optical materials, and improved hydrophobic or hydrophilic surfaces for various applications. Based on the spectroscopic characterization of the plasma-induced surface modifications described in this chapter, it is felt that the pulsed rf plasma polymerization technique demonstrates great potential with respect to the ultimate goal of tailoring surface molecular composition for specific applications.

Acknowledgment

This material is based, in part, on work supported by the Governor's Energy Management Center, State of Texas Energy Research in Applications Programs under Contract Number 003656–005.

References

1. Symposium on Plasma Polymerization and Plasma Interactions with Polymeric Materials, Division of Polymeric Materials: Science and Engineering, ACS National Meeting, Spring 1990, Boston, Massachusetts.
2. Yasuda, H. *Plasma Polymerization*; Academic: Orlando, FL, 1985.
3. Chilkoti, A.; Ratner, B. D.; Briggs, D. *Chem. Mater.* **1991**, *3*, 51.
4. Bhuiyan, A. H.; Bhoraskar, S. V. *J. Mater. Sci.* **1989**, *24*, 3091.
5. Wydeven, T.; Golub, M. A.; Lerner, N. R. *J. Appl. Polym. Sci.* **1989**, *37*, 3343.
6. Evans, J. F.; Prohaska, G. W. *Thin Solid Films* **1984**, *118*, 171.
7. Lopez, G.; Ratner, B. *ACS Polym. Mater. Sci. Eng.* **1990**, *62*, 14.
8. O'Kane, D. F.; Rice, D. W. *J. Macromol. Sci., Chem.* **1976**, *A10(3)*, 567.
9. Yasuda, H.; Hsu, T. *J. Polym. Sci., Chem. Ed.* **1977**, *15*, 81.
10. Nakajima, K.; Bell, A. T.; Shen, M. *J. Appl. Polym. Sci.* **1979**, *23*, 2627.
11. Clark, D. T.; Shuttleworth, D. *J. Polym. Sci., Polym. Chem. Ed.* **1980**, *18*, 27.
12. Wang, D.; Chen, J. *J. Appl. Polym. Sci.* **1991**, *42*, 233.
13. *JANAF Thermochemical Tables*; 2nd ed.; U.S. Department of Commerce, National Bureau of Standards: Washington, DC, 1971.

RECEIVED for review July 15, 1991. ACCEPTED revised manuscript September 9, 1992.

Comparative Raman and Infrared Vibrational Study of the Polymer Derived from Titanocene Dichloride and Squaric Acid

Melanie Williams[1], Charles E. Carraher, Jr.*[2], Fernando Medina[2], and Mary Jo Aloi[2]

[1]**Motorola, Inc., Plantation, Florida 33317**
[2]**Departments of Chemistry and Physics, Florida Atlantic University, Boca Raton, Florida 33431**

The vibrational bands associated with the cyclopentadiene (Cp) are similar in location and intensity to the bands found for the polymer formed from the condensation of titanocene dichloride and squaric acid. This similarity is consistent with the Cp ring remaining in the original angular sandwich structure. Bands associated with the Cp ring can be treated through the use of normal coordinate analysis on the basis of a "local" C_{5v} (fivefold axis of symmetry) symmetry. Selected bands associated with the squaric acid portion are displaced from bands generally due to the presence of ring strain and delocalization of electrons. The combined use of Raman and infrared spectroscopy allows the correct assignment of these bands.

$\mathbf{C}$HARACTERIZATION OF MANY METAL-CONTAINING POLYMERS is complicated by the additional avenues of reaction that can be taken by the metal-containing moiety, lack of specific knowledge concerning many metal-containing candidate monomers, and the generally poor solubility and processability of such products. Determination of the dependence of and relationship between monomer and polymer properties is helpful to ascertain the extent to which

* Corresponding author

0065–2393/93/0236–0769$06.00/0

known monomer properties can be used to describe the polymeric properties
of such monomer-containing materials.

The purpose of the present study is twofold: (1) To determine the extent
to which vibrational band assignments of the monomers titanocene dichloride, **1**, and squaric acid, **2**, can be applied to the polymeric product formed
from condensation of these two reactants and (2) to provide additional
structural data to allow differentiation between two basic structural possibilities. Briefly, three major structures are possible. Structure **3** is eliminated
based on data presented herein; **4** is eliminated based on molecular weight
data. For the product studied here, a weight average molecular weight that
corresponds to an average chain length of 140 was determined utilizing light
scattering photometry. Finally, physical characterization data are consistent

with **5**. Data that eliminate **3** and support **5** are presented. An additional structural question involves the type of bonding of the Cp groups into the titanium. The Cp ring can be bonded through a pi bond as pictured in **6** or through a sigma bond as pictured in **7**. Sigma bonding of the Cp ring requires a 1, 2 migration of the metal atom. Such low-energy migrations require only about 10 kcal/mol. Again, spectral data that are consistent with the connection of the Cp rings to the titanium atom through sigma bonds are given (*1–3*). Additional structural characterization is reported in reference 1.

Experimental Details

Infrared Spectra. Infrared spectra were recorded using a Fourier transform infrared (FTIR) spectrometer (Mattson Alpha-Centauri) equipped with KBr and mylar optics and a deuterated triglycine sulfate (DTGS) detector. Atmospheric water vapor was removed from the spectrometer by purging with dry nitrogen. The spectrometer was calibrated with polystyrene film (mid-IR) and water vapor (far-IR). All spectra in the 500–200-cm^{-1} range were recorded as mineral oil (Nujol) mulls between polyethylene plates. All spectra were recorded at an instrumental resolution of 4 cm^{-1} using 32 scans.

Raman Spectra. Raman spectra were obtained with the 514.5-nm line of an argon-ion laser, a double monochromator (Spex 1403), and photon counting techniques. The instrumental resolution for a typical spectrum was 6.7 cm^{-1}. All spectra were recorded at room temperature (23 °C). The region below 200 cm^{-1} could not be examined due to the high intensity of the Rayleigh line. Polarization studies were attempted, but were unsuccessful due to the nonordered nature of the polymer.

Results and Discussion

The assignment of the bands that arise from the cyclopentadienyl (Cp) ring vibrations can be readily accomplished by comparison with the monomer titanocene dichloride (Cp$_2$TiCl$_2$) and other titanocene derivatives because

6   **7**

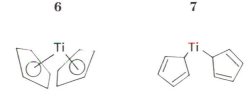

the frequency of the ring vibrations remains relatively constant upon substitution of the halide ligands (4–7). The assignment of the bands that arise from the cyclopentadienyl ring vibrations was based on previous studies of Cp_2TiCl_2 derivatives, as well as studies on ferrocene, ruthenocene, and dicyclopentadienyltin derivatives (4–9).

Previous studies on Cp_2TiCl_2 showed that each ring can be treated on the basis of a "local" C_{5v} (fivefold axis of symmetry) symmetry. Under C_{5v} symmetry, each ring gives rise to 24 normal mode vibrations (the number of degrees of freedom is given by the equation $3n-6$, where n is the number of atoms) distributed as $3A_1 + A_2 + 4E_1 + 6E_2$, where the E modes are doubly degenerate. Under symmetry considerations the A, E, and E_2 modes are Raman active whereas the A_1 and E_1 modes are IR active. Therefore, 7 normal vibrations are expected in the IR, whereas 13 normal vibrations are expected in the Raman spectra (4–9). The observed bands and their assignments for the ring Cp vibrations of Cp_2TiCl_2 and the squarate polymer are given in Table I.

Most of the assignments for the A_1 and E_1 modes are consistent with previous assignments for biscyclopentadienyltitanium (IV) titanocene derivatives (2, 10) in similar matrices, but differ somewhat from the assignments of

Table I. Observed Infrared and Raman Frequencies for $(Cp_2Ti(C_4O_4))_n$ and Cp_2TiCl_2 Obtained in This Study: Bands Arise from Cyclopentadienyl Ring Vibrations

Descripion of Mode[a]	Symmetry Species	Cp_2TiCl_2 Infrared[b] (cm^{-1})	Squarate Polymer	
			Infrared[b] (cm^{-1})	Raman[b] (cm^{-1})
CH stretch	A_1	3103m	3095	c
CH o.p. bend	A_1	821vs	822s	825w
Ring breathing	A_1	1130w	1149vw	1147ms
CH i.p. bend	A_2	1271vw	1267vw	—
CH stretch	E_1	—	—	—
CH i.p. bend	E_1	1015s	1018m	1020m
CH o.p. bend	E_1	871m	880m	880w
CC stretch	E_1	1440s	1455s	1460m
CH stretch	E_2	—	—	—
CH i.p. bend	E_2	1197vw	—	1190w
CH o.p. bend	E_2	1074w	1084w	1050m
CC stretch	E_2	1364w	1360w	1380w
CC i.p. bend	E_2	927w	—	—
CC o.p. bend	E_2	597w	595vw	605w
359 + 597 = 956		957w		

[a] o.p., out of plane; i.p., in plane.
[b] v, very; s, strong; m, medium; w, weak.
[c] Region above 2000 not recorded.
SOURCE: Portions of the data presented in this table are reproduced with permission from reference 28. Copyright 1990 Plenum Press.

Balducci et al. (*6*) for the matrix-isolated spectra of biscyclopentadienyl dihalides. Tentative assignments for the E_2 mode, therefore, have been made and are given in Table I.

The E_2 modes should be forbidden in the IR, but selection rules are not always strictly obeyed. Thus several weak infrared bands with frequencies near corresponding Raman frequencies are assigned to E_2 modes. Additionally, weak bands at 1271 (Cp$_2$TiCl$_2$) and 1267 cm^{-1} (squarate polymer) are assigned to an A_2 mode. This band has been observed consistently in cyclopentadienyl derivatives even though it is forbidden in both the infrared and Raman spectra (*6, 8, 9*).

There is little difference between the position and intensity of the Cp ring vibrations in Cp$_2$TiCl$_2$ and in the squarate polymer. This observation suggests that the Cp$_2$Ti moiety retains its original angular sandwich structure, with the $_n^5$Cp rings tetrahedrally coordinated about the titanium atom. A change in structure would require a change in the orientation of the rings relative to one another. This orientation is the same as that found in the biscyclopentadienyltitanium (IV) dihalides, and all of the biscyclopentadienyl derivatives of titanium (*11*). Thus the IR data are consistent with Cp bonding do the titanium atom through a pi-type bond as pictured in **7**.

The remaining bands in this region of the spectrum can be assigned to the squarate portion of the molecule. A number of these assignments are facilitated by the results of previous studies on related molecules (*7, 12–16*). There are two bands in the infrared spectrum above 1600 cm^{-1}, which suggests the presence of uncoordinated oxygens (or free carbonyls) (*17*). The band at 1792 cm^{-1} in the Raman spectrum is very strong, which is consistent with a symmetric stretching mode because a totally symmetric fundamental generally results in a strong Raman line (*14, 18*). The IR band that corresponds to this fundamental appears at 1793 cm^{-1} as a moderately intense band. In contrast, the band at 1690 cm^{-1} is very weak in the Raman spectra and appears as a strong band in the infrared spectrum at 1693 cm^{-1}; this indicates an asymmetric stretching mode. These bands are assigned to $C=O$ stretching modes.

The symmetric stretch occurs at a higher frequency than the asymmetric stretch because, in a symmetric stretch, the ring bond is forced to compress. This compression results in a larger force constant with an accompanying vibration at a higher frequency. On the other hand, the asymmetric stretch causes less strain, because the ring carbons can move to adjust to the motion, and the carbonyl carbons can keep their distance constant as shown in **8** and **9**.

The band at 1555 cm^{-1} in the infrared spectrum is the second most intense band in the spectrum. The position and intensity of this band suggest a symmetric ($C=C$) stretch coupled with a $C-O$ stretch. This band appears in the Raman spectrum at 1557 cm^{-1} with a weak shoulder at 1565 cm^{-1}. A similar band has been observed in other metal-containing derivatives (*12,*

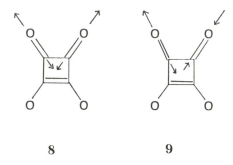

8   9

19–22) and organic derivatives of squaric acid (*23*), as well as many metal carboxylates (*14, 24*). The frequency of the C=C stretch is lower than the frequency of other cycloalkenes as a result of a longer C=C bond distance in the four-membered rings. This lower frequency results in a decreased force constant. For instance, an isolated double bond usually vibrates in the $1680–1620 \text{cm}^{-1}$ range (as do other cycloalkenes), whereas in cyclobutene this stretch is observed at 1566 cm^{-1} (*25*). The frequency of the C=C stretch in the polymer is greater than the corresponding frequency in squaric acid (1513 cm^{-1}). This difference may be a result of *p-pi–d-pi* bonding, which would increase the electronegativity of the oxygen atom.

The most intense band in the infrared spectrum appears at 1405 cm^{-1}, but is absent in the Raman spectrum. This band is assigned to an asymmetric C–O stretch coupled to a C=C stretch, which is in agreement with the position and intensity of the C–O stretch found in organic derivatives of squaric acid (*23*). A band at this position is also characteristic of metal carboxylates and usually is assigned to a symmetric O–C–O stretch (*14, 24*). The assignment of the symmetric stretch at 1555 cm^{-1} to a higher frequency than the asymmetric stretch at 1405 cm^{-1} is again due to the ring strain effects mentioned earlier. The weak bands at 1067 (IR) and 1075 cm^{-1} (Raman) are assigned to C–C stretching modes, in accordance with their intensity and position.

A characteristic vibration in all ring systems is the ring breathing mode. The band that corresponds to this vibration is intense in the Raman spectrum and is infrared active only in molecules that do not contain a center of symmetry (*25*). Because nearly all of the bands in the IR and Raman spectra of the polymer are coincident, no center of symmetry is present. A ring breathing vibration should, therefore, be observed in the IR spectrum. The ring breathing mode usually gives rise to a vibration in the $950–1000 \text{ cm}^{-1}$ range in cyclobutene derivatives, but is found at considerably lower frequencies (550–750 cm^{-1}) in the oxocarbons (26, 27). The only band in the Raman spectrum that can be assigned to a symmetric ring breathing mode is the strong band at 755 cm^{-1}. The corresponding infrared band is observed as a medium-intensity band at 743 cm^{-1}. This assignment is consistent with other metal-containing derivatives of squaric acid (*see* references 12, 21, and 27).

At least two C=O deformations are expected in the 550–750 cm^{-1} region. However, coincidental degeneracies may occur. Additionally, because there are several bands below 700 cm^{-1}, some of these bands may be hidden. The one band that could be attributed to a C=O deformation occurs at 681 cm^{-1} (IR) and as a weak band in the Raman spectrum at 685 cm^{-1}. This band is assigned to an asymmetric C=O bending mode.

The vibrational frequencies of a number of metal-containing–squarate-containing complexes have been studied (12, 21, 27). In these complexes, there are no bands in the infrared spectrum above 1600 cm^{-1}. In addition, there is a very strong, broad band between 1400 and 1700 cm^{-1} that corresponds to a combination of C$\cdots$C + C$\cdots$O and C=O stretching modes. This band is absent in the diketosquarate derivatives and is replaced by bands that correspond to C=C, C–O, and C=O stretching modes. The positions of the bands for the complexes listed in references 12, 21, and 27 remain fairly constant for a given vibration, which suggests similarities in the structures of these complexes.

Reported infrared frequencies for diketosquarate–metal complexes that contain uncoordinated carbonyl groups that are analogous to the structure proposed here for the squarate polymer are given in references 19, 20, 23, and 26. In these complexes, there are several bands above 1600 cm^{-1} that are attributed to C=O stretching modes. A strong band appears between 1500–1600 cm^{-1} in the spectra of all of these derivatives. This band corresponds to a C=C or C=C + C–O stretching mode.

The presence of bands consistent with the presence of the alkene moiety and IR information that indicates a lack of symmetry are evidence for polymer structure as pictured in **5** and are not consistent with a structure as described in **3**. Further, the absence of unassigned bands is consistent with a product of form **5** with little contributions from other structures.

In summary, vibrational frequencies found for the titanocene–squarate polymer are analogous to frequencies found in the respective monomers and are consistent with the lack of influence of the polymeric nature on these vibrational modes. This finding indicates that, at least for some polymers, vibrational assignments derived for monomers can be used to identify vibrational bands found in the corresponding polymer. Also, vibrational band assignments are consistent with the polymer being of a general structure as pictured in **5** with the Cp rings bonded through pi bonds to the titanium metal atom.

References

1. Williams, M.; Carraher, C.; Medina, F.; Aloi, M. *Poly. Mater.* **1989**, *61*, 227.
2. Semmingsen, D. *Acta. Chim. Scand., Ser. A* **1973**, *27*, 3961.
3. Ballhausen, C. J.; Dahl, J. P. *Acta. Chim. Scan., Ser. A* **1961**, *15*, 1333.
4. Fritz, H. P., *Adv. Organomet. Chem.* **1964**, *1*, 239.
5. Maslowsky, E. *Vibrational Spectra of Organometallic Chemistry*; Wiley: New York, 1973; Chapter 3.
6. Balducci, G.; Bencivenni, L.; DeRosa, G.; Gigli, R.; Martini, B.; Cesaro, S. N. *J. Mol. Struct.* **1980**, *64*, 163.
7. Druce, P. M.; Kingston, B. M.; Lappert, M. F.; Spalding, T. R.; Srivastava, R. C. *J. Am. Chem. Soc.* **1969**, *91A*, 2106.
8. Lippincott, E. R.; Nelson, R. D. *Spectrochim. Acta* **1958**, *10*, 307.
9. Harrison, P. G.; Healy, M. A. *J. Organomet. Chem.* **1973**, *51*, 153.
10. Seymour, R. B.; Carraher, C. E. *Polymer Chemistry, An Introduction*; Marcel Dekker: New York, 1968.
11. Wailes, P. C.; Coutts, R. S. P.; Weigold, H. *Organometallic Chemistry of Titanium, Zirconium, and Hafnium*; Academic: Orlando, FL, 1974.
12. West, R.; Niu, H. Y. *J. Am. Chem. Soc.* **1963**, *85*, 2589.
13. Habenschuss, M.; Gerstein, B. C. *J. Chem. Phys.* **1974**, *61*(3), 1.
14. Bellamy, L. J. *The Infrared Spectra of Complex Molecules*; 2nd ed.; Methuen: London, 1968.
15. *Oxocarbons*; West, R., Ed.; Academic: Orlando, FL, 1980.
16. Miller, F. A.; Kiviat, F. E.; Matsuhara, I. *Spectrochim. Acta, Part A* **1968**, *24*, 1523.
17. Doyle, G.; Tobias, R. S. *Inorg. Chem.* **1968**, *7*(12), 2484.
18. Tobin, M. C. *Laser Raman Spectroscopy*; Chemical Analysis Series 35; Wiley: New York, 1971.
19. Long, G. *Inorg. Chem.* **1978**, *17*(10), 2702.
20. Condren, S. M.; McDonald, H. O. *Inorg. Chem.* **1973**, *12*, 57.
21. Schwering, H. U.; Olapinski, H.; Jungk, E.; Weidlein, J. *J. Organomet. Chem.* **1974**, *76*, 315.
22. Wrobleski, J. T.; Brown, D. B. *Inorg. Chem.* **1979**, *18*(10), 2738.
23. Cohen, S.; Cohen, S. G. *J. Am. Chem. Soc.* **1966** *88*, 1533.
24. Rao, C. N. R. *Chemical Applications of Infrared Spectroscopy*; Academic: Orlando, FL, 1963; Chapter 7.
25. Avram, M.; Mateesca, M. *Infrared Spectroscopy*; Wiley-Interscience: New York, 1972.
26. Baglin, F. G.; Rose, C. B. *Spectrochim. Acta, Part A* **1970**, *26*, 2293.
27. Ito, M.; West, R. *J. Am. Chem. Soc.* **1975**, *97*, 2580.
28. Sheats, J.; Carraher, C.; Pittman, C.; Zeldin, M.; Currell, B. *Inorganic and Metal-Containing Polymeric Materials*; Plenum: New York, 1990.

RECEIVED for review July 15, 1991. ACCEPTED revised manuscript September 13, 1992.

Analysis of Gas–Plasma-Modified Poly(dimethylsiloxane) Elastomer Surfaces

Attenuated-Total-Reflectance–Fourier Transform Infrared Spectroscopy

Scott R. Gaboury and Marek W. Urban*

Department of Polymers and Coatings, North Dakota State University, Fargo, ND 58105

Attenuated-total-reflectance–Fourier transform infrared (ATR–FTIR) spectroscopy is a surface-sensitive technique that is particularly suitable for the analysis of plasma-treated silicone elastomers. This chapter discusses the recent advances in gas–plasma treatments of silicone elastomer surfaces and surface analysis using ATR–FTIR spectroscopy. A particular focus is given to silicone elastomer surfaces that are plasma treated with non-film-forming gases such as argon, nitrogen, carbon dioxide, oxygen, and ammonia. The non-film-forming nature of the aforementioned gases leads to surface functionalization by incorporation of such gases into the film surface, as observed for ammonia, oxygen, and carbon dioxide gases. Furthermore, silicone elastomer surfaces can be functionalized through reactions of the surface atoms, as demonstrated for argon and nitrogen.

GAS–PLASMA SURFACE MODIFICATIONS OF POLYMER SYSTEMS have been widely studied and used to enhance adhesion or biocompatibility as well as other properties. The scope of adhesion studies of plasma-modified polymers typically deal with the enhancement of surface adhesion (1–3), improvement

---

* Corresponding author.

of bonding between polymers and fillers (4–6), or increased composite strength (7). One of the areas that recently received increased attention is the enhancement of polymer biocompatibility. This compatibility is usually accomplished by gas–plasma surface treatments (8–11) of polymers that have good mechanical properties, but poor surface biocompatibility. One of the most widely used biopolymers is cross-linked siloxane rubber, which, despite its inertness, needs to be surface-modified to function in biologically active environments. Although several studies (12–15) have utilized gas–plasma surface modifications, the issue of surface structure–property relationships remains questionable. The advantage of gas–plasma modification lies in its ability to favorably alter surface properties without adversely affecting the bulk mechanical properties of the polymer. Energy in the radio and microwave frequency ranges is often used as a means for either surface plasma polymerization or surface plasma modification. For surface plasma modifications of polymer surfaces, cold plasmas are in a thermal nonequilibrium reaching electron temperatures of 20,000–30,000 °C, whereas the bulk gas temperatures remain below 200–300 °C (16). Consequently, the electrons as well as other gas particles have sufficient energy to cleave common organic bonds, but thermal degradation of bulk polymer is marginal because of the low bulk gas temperature. Although the types of plasmas generated by radio and microwave frequencies are in general similar, there are differences in equipment. Due to the nature of radiation, the use of radio frequency plasma chambers allows a greater degree of flexibility in terms of geometry and the size of plasma chambers and electrodes. Microwave plasmas, however, generate a more uniform distribution of energy and, therefore, more homogeneous surface treatments (17).

To understand the chemical changes that occur as a result of plasma modifications, it is essential to characterize the molecular level changes and, on that basis, to further advance the understanding of surface chemistry and structure–property relationships on surfaces. Because vibrational spectroscopy has proven to be a sensitive technique, the primary focus of this study will be on spectroscopic monitoring of the structures that develop on silicone elastomer surfaces as a result of gas–plasma modifications.

Characterization of Plasma-Treated Polymers

The changes that occur on polymeric surfaces as a result of gas–plasma exposure have a fundamental significance on surface properties. Therefore, it is important to use surface characterization methods that are selective and sensitive to the structural changes resulting from the treatments. A list of several other surface techniques utilized in the study of plasma-modified surfaces is given in Table I. Although X-ray photoelectron spectroscopy (XPS), secondary-ion mass spectrometry (SIMS), and ion-scattering spectroscopy (ISS) are surface-sensitive techniques, which yield molecular level

Table I. Techniques for Analysis of Plasma-Modified Polymers

Technique	Depth (nm)	References
ATR–FTIR spectroscopy	100–10,000	24, 26, 29, 30, 35
X-ray photoelectron spectroscopy (XPS)	< 10	38–40
Secondary-ion mass spectroscopy (SIMS)	< 1	30, 41
Ion-scattering spectroscopy (ISS)	< 0.5	42, 43
Transmission electron microscopy (TEM)	150 max. thickness	44, 45
Surface potential difference	—	46
Contact angle (surface energy)	—	9, 44, 47
Adhesion enhancement	—	1, 48

information, the use of high-vacuum conditions and careful surface preparation is required. Transmission electron microscopy (TEM), surface potential difference, contact angle, and adhesion measurements yield important information about material properties that are altered by plasma modification, but do not provide molecular level information about the species formed on the surface. Because of instrumental advancements in Fourier transform infrared (FTIR) spectroscopy, ATR–FTIR spectroscopy has become a highly sensitive technique. Due to good crystal–sample contact in polymers with low glass-transition temperatures, such as silicone elastomers, ATR–FTIR is particularly suited. Because this technique requires no high-vacuum environment, which can often disrupt weakly bonded surface species, and no special sample preparation, it is very attractive and useful in the studies of polymer surfaces.

In the ATR–FTIR experiment, the sample is placed in intimate contact with an infrared transparent crystal. When infrared light is passed through the crystal at an angle above the critical angle, the light is internally reflected except for a small portion of the light that is absorbed by the sample surface. The light reflected from the crystal–sample interface carries information about the vibrational energy of chemical bonds present on the sample surface. Typically, ATR–FTIR spectroscopy allows the detection of vibrational spectra for surface layers in the range of 0.1 to 10 μm in thickness. For quantitative analysis of surfaces, it is necessary to consider the depth of penetration that depends upon the refractive index of the crystal and incident angle of the light. Furthermore, to account for optical effects during the measurements, Kramers–Kronig transformations must be utilized (*18–20*). The effect of band separation and intensity changes on the ATR analysis and spectroscopic interpretation has been described recently (*21*). Because this chapter focuses on the actual use of ATR–FTIR spectroscopy for surface analysis, readers interested in theoretical principles should refer to the original literature (*22*).

Inert Gas–Plasma Modifications

The use of inert gases in the presence of plasma environments can substantially affect surface properties of polymers. Unlike other gas treatments, inert gases are not incorporated into the polymer surface, but surface reactions result from rearrangements caused by interactions with the highly energetic gas particles (23).

Argon Treatment. ATR–FTIR spectroscopy and dynamic mechanical thermal analysis (DMTA) were used to study the effects of argon–, carbon dioxide–, and ammonia–plasma modifications on poly(dimethylsiloxane) (PDMS) (24). A radio frequency (rf) plasma chamber with a steady gas flow configuration was used for the plasma treatments. The extent of the surface changes was characterized by the relative increase of the carbonyl band at 1725 cm^{-1} ratioed to the PDMS bulk band at 1412 cm^{-1} as shown in Figure 1. In the power range from 10 to 200 W, the surface carbonyl concentration increases steadily until a power of 50 W is reached. Beyond 50 W, the carbonyl band decreases, which is attributed to two competing processes: surface modification and surface ablation. Below 50 W there is a faster increase in the rate of surface modification than in surface ablation, whereas above 50 W the surface ablation becomes more rapid than the surface modification. Even though argon gas was used, such inert gas–plasma surface treatments lead to the formation of two carbonyl bands at 1725 and 1720 cm^{-1}. Although the origin of the oxygen that yields carbonyl formation is unknown, there are various possible sources: (1) When the chamber is vented to the air, atmospheric oxygen may react with radicals formed on the PDMS surface during the plasma treatment or (2) oxygen may come from residual air in the plasma chamber or air trapped in the PDMS network. In addition, the network contains oxygen in the polymer backbone and in the filler that may dissociate and contribute to surface oxidation. DMTA analysis shows that argon–plasma treatment of the PDMS reduces the storage modulus. Because the storage modulus is directly related to cross-link density (25), these studies also show that the decrease of storage modulus as a result of argon–plasma modification leads to the decrease of PDMS cross-link density.

In the most recent studies (26), microwave argon– and nitrogen–plasma treatments of PDMS have formed Si–H functionality on the surface of PDMS. ATR–FTIR surface spectra in the 2300–1500-cm^{-1} region of the untreated and nitrogen– and argon–plasma-treated PDMS samples that contain no SiO$_2$ filler are shown in Figure 2, traces A, B, and C, respectively. The striking appearance of bands at 2158 cm^{-1} upon plasma treatment indicates the formation of new surface species. The band at 1726 cm^{-1} (Figure 2, traces B and C) is due to carbonyl species and results from surface oxidation, whereas the band at 2158 cm^{-1} is attributed to the Si–H stretching mode (27). As illustrated in Figure 3, the Si–H stretching mode at 2158 cm^{-1} is accompanied by the Si–H bending mode observed at 912 cm^{-1}. In an effort to further confirm the Si–H formation, a series of PDMS samples with various nitrogen–plasma exposure times were analyzed. With increasing nitrogen–plasma exposure time, the 2158- and 912-cm^{-1} bands

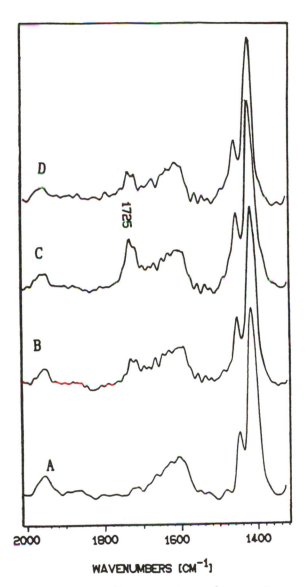

Figure 1. ATR–FTIR spectra in the 2000–1350-cm^{-1} region of argon–plasma-treated PDMS: A, untreated; B, 10 W; C, 50 W; D, 200 W.

increase at approximately the same rate, which provides further evidence that both bands are due to the formation of the same species, namely, Si–H. At this point, however, the mechanism for the Si–H formation is not fully understood. The bond energies of the methyl silicones shown in Table II (*28*) suggest that due to the lower carbon–silicon and carbon–hydrogen bond

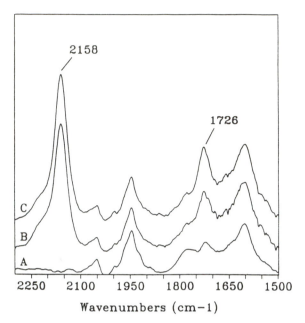

Figure 2. ATR–FTIR spectra of PDMS in the 2300–1500-cm^{-1} region: A, untreated; B, nitrogen–plasma treated; C, argon–plasma treated.

energies, these bonds are preferentially cleaved by the excited gas particles. This process yields reactive silicon sights on the polymer backbone as well as reactive carbon and hydrogen fragments in the gas phase. If the bonds are cleaved, the reactive silicon sites may recombine with the reactive hydrogen-containing species. The primary reason for the foregoing assessment is that the silicon–hydrogen bond energy is higher than the silicon–carbon bond energy. Therefore, more stable species are formed when silicon bonds to hydrogen. Furthermore, because the fragments cleaved from the polymer are primarily CH_3 groups that can further dissociate in the gas phase, there is an optimum hydrogen:carbon ratio of 3:1 in the gas phases. This excess of hydrogen species compared to carbon species leads to a statistically higher chance of hydrogen reacting with the silicon. Finally, the hydrogen:carbon ratio in the gas phase may be further increased by the cleavage of hydrogen species from carbon atoms that remain attached to the polymer chain. A schematic representation of the process is shown in Chart I.

Chemically Active Gas–Plasma Modification

Carbon Dioxide.   Carbon dioxide–plasma treatments of PDMS surfaces lead to the formation of multiple carbonyl as well as alkene groups (*24*). This change is demonstrated by the appearance of carbonyl bands at 1725,

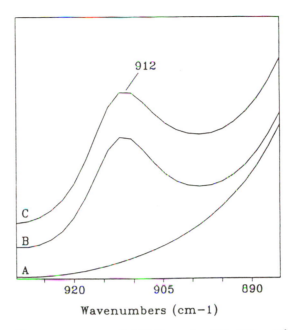

Figure 3. ATR–FTIR spectra of PDMS in the 930–885-cm^{-1} region: A, untreated; B, nitrogen–plasma treated; C, argon–plasma treated.

Table II. Energies of Methyl Silicone Bonds

Species	Energy (eV)
Si – C	3.3
Si – O	4.7
Si – H	4.3
C – H	3.3

1700, and 1675 cm^{-1} and an alkene band at 1596 cm^{-1} in the ATR–FTIR spectra. The 1700-cm^{-1} band is assigned to hydrogen-bonded carbonyl groups, most likely to the Si—OH functional groups that are formed as a result of the gas–plasma treatment. The Si—OH formation is demonstrated by a broad band centered at 3400 cm^{-1}. The band at 1596 cm^{-1} is attributed to the formation of vinyl groups.

Oxygen. Oxygen–plasma modification of PDMS generates a substantial amount of surface hydroxyl functionalities (*29, 30*). However, the nature of the species to which the hydroxyl groups are bonded is not agreed upon. For example, it was observed that the formation of two broad hydroxyl bands

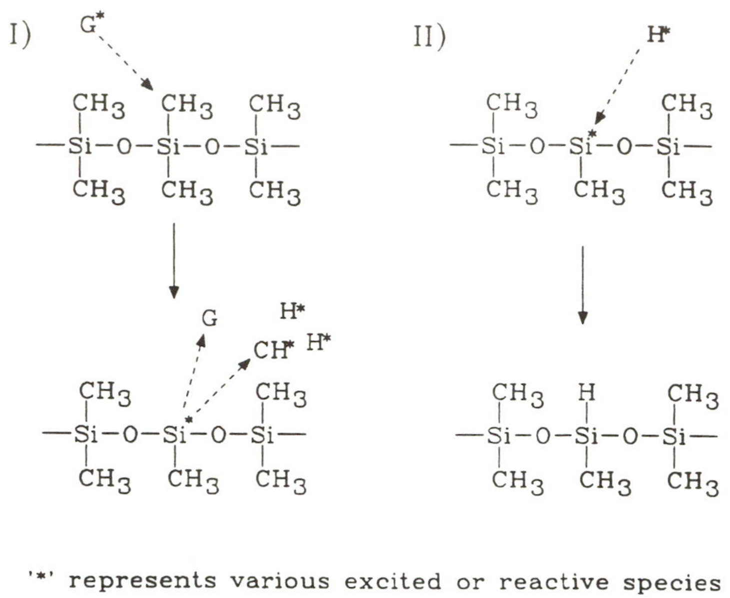

'*' represents various excited or reactive species

Chart I. A tentative mechanism of Si − H formation on PDMS surfaces.

centered at 3425 and 3225 cm^{-1}, both of which were assigned to hydrogen-bonded hydroxyl groups (29). The 3425-cm^{-1} band was attributed to intramolecular bonding, whereas the 3225-cm^{-1} band was due to intermolecular bonding. Several structural possibilities were illustrated for each bonding situation. Because C−H stretching bands showed only minor intensity decreases during plasma treatment, it was proposed that the majority of hydroxyl groups are bonded onto the pendant carbons. However, based on such small C−H frequency changes, it was proposed that a fraction of the hydroxyl groups reacted onto the silicone atoms after pendant methyl groups were cleaved off. Exposure of the oxygen–plasma-treated PDMS to deuterium oxide caused a rapid exchange of hydrogen for deuterium that was manifested by a shift of both hydroxyl bands by approximately 900 cm^{-1}.

In the recent studies on oxygen–plasma-treated PDMS, the preceding findings appear to be controversial (30). Although ATR–FTIR spectra again showed substantial hydroxyl group formation, as evident by the presence of the 3400-cm^{-1} band, all C−H stretching, bending, and deformation modes decreased in intensity as a result of oxygen–plasma treatment. Complementary XPS and SIMS measurements led to the deduction that the loss of CH$_3$ groups is the result of the formation of hydroxyl groups in the plasma environment that are bonded directly to silicon atoms in the form Si−CH$_2$−OH (30), rather than to carbon atoms as was proposed in the previous studies (29).

Although the conclusions drawn from the recent study (*30*) about the nature of the hydroxyl bonding on the PDMS surface appear to be accurate, the assessment that this study disproves previous work (*29*) seems unfounded. It is well known that the changes of plasma parameters, such as chamber geometry, vacuum conditions, substrate preparation, and others, can drastically affect the surface treatment and, therefore, the structures that develop as a result of such surface modifications (*31*). In view of the experimental procedures employed and a comparison of the experimental parameters used in each study, it is quite apparent that the exposure time, plasma power, and pressure substantially vary between the two studies. Therefore, although a comparison between the two studies is of little use, each study contains useful information.

Ammonia. Numerous studies have been conducted on the ammonia–plasma modification of polymer surfaces. The primary driving force was the desire to form reactive $N-H$ species for future surface-grafting reactions. Indeed, studies have shown that the use of ammonia–plasma treatment leads to amine- or amide-functionalized surfaces (*32–34*).

Recently, we studied the effects of ammonia–plasma on PDMS and PDMS-containing chlorofunctional impurities (*35*). During the ammonia–plasma treatment of PDMS that contained nonbonded chlorofunctional impurities, ammonia was incorporated into the polymer surface in the form of amide functionality. The incorporation was demonstrated by the appearance of an amide carbonyl band at 1653 cm^{-1}. As is well known, during gas–plasma surface modification, a thin surface layer of the substrate is often etched or ablated away. After plasma treatment, a fraction of the ablated material often remains on the substrate in the form of nonbonded surface layers. For that reason, ammonia–plasma-pretreated PDMS films were washed in distilled–deionized (DDI) water to remove any residues of nonbonded material. ATR–FTIR spectra, shown in Figure 4 were collected before (trace B) and after washing (trace C). With the exception of a slight intensity decrease in the $1800-1600$-cm^{-1} region, the spectrum of the ammonia–plasma-treated PDMS appears almost identical to the spectrum of the same sample before washing. This observation was attributed to the removal of a fraction of ablated material from the surface. The presence of the 1653-cm^{-1} band after washing indicates that the amide functionality formed as a result of ammonia–plasma exposure was chemically bonded to the PDMS surface.

When a chlorofunctional impurity such as the chlorofunctional radical initiator 2,4-dichlorobenzoyl peroxide is used, the effect of ammonia–plasma surface modifications is significantly different (*35*). Similarly to the previous experiments, PDMS cross-linked with this initiator was ammonia–plasma-treated and ATR–FTIR spectra were collected before and after water washing. The resulting spectra are shown in Figure 5. The spectrum of the unwashed PDMS shows the formation of the amide carbonyl groups with a

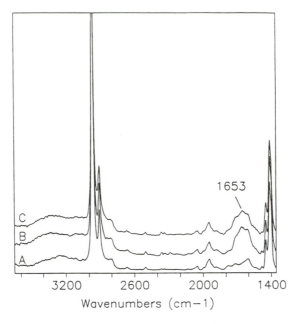

Figure 4. ATR–FTIR spectra in the 3600–1400-cm⁻¹ region of PDMS with no chlorine-containing molecules: A, untreated; B, ammonia–plasma treated; C, ammonia–plasma treated and washed.

band at 1653 cm^{-1} as well as two bands of significant intensity at 3150 and 3053 cm^{-1} (Figure 5, trace B). However, the spectra of the washed samples indicate a complete elimination of the 3150- and 3053-cm^{-1} bands along with an approximate 50% intensity reduction of the band due to amide carbonyl groups at 1653 cm^{-1} (Figure 5, trace C). These changes result from ammonium chloride formation on the PDMS surface during ammonia–plasma modification. The frequency shift of the bands for ammonium chloride on the PDMS surface as compared to the bands of solid ammonium chloride is due to the adsorption of the ammonium chloride on the PDMS surface (36). These experiments indicated that PDMS substrates free of chlorofunctional molecules can be amide-functionalized utilizing ammonia–plasma treatment and that the amide groups are chemically bonded to the PDMS surface. However, the presence of the chlorine-containing initiator causes the formation of surface ammonium chloride, which partially inhibits bonding of the amide groups to the PDMS surface. Although additional experiments followed by spectroscopic analysis substantiated these findings, the exact mechanism of ammonium chloride formation on the PDMS surface has not been addressed. Furthermore, the distribution of NH$_4$Cl within the PDMS surface layers may be at the angstrom level or well below the surface (micrometer level). This distribution may also influence surface treatments.

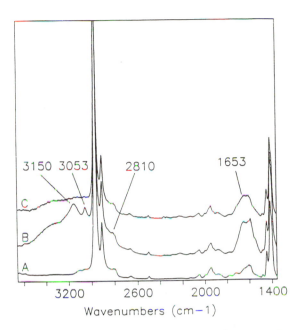

Figure 5. ATR–FTIR spectra in the 3600–1400-cm^{-1} region of PDMS with a chlorine-containing initiator: A, untreated; B, ammonia–plasma treated; C, ammonia–plasma treated and washed.

As the previous section illustrated, even small quantities of chlorine-containing species may generate undesirable surface properties. Therefore, the presence of nonbonded molecules, such as processing agents or small molecule residues, may further enhance problems associated with gas–plasma surface modifications. Freon is a well-known cleaning agent that is used to remove processing agents because it exhibits good solvating power and high volatility (37). In an effort to establish the effect of such cleaning practices, residual 1,1,2-trichloro-1,2,2-trifluoroethane (Freon TS, DuPont) in the PDMS network was examined as another chlorine-containing impurity (35). ATR–FTIR spectra of the gas–plasma-treated PDMS before surface washing again showed intense ammonium chloride bands, which indicates that freon trapped within the polymer network contributes chlorine to the formation of ammonium chloride (Figure 6, trace C). Furthermore, the amide carbonyl band at 1653 cm^{-1} is absent and a new carbonyl band at 1765 cm^{-1} is detected. After the surface was washed with water, the spectra indicate that all surface species that were formed concurrently with the ammonium chloride formation were water-soluble and removable. ATR–FTIR spectra of the plasma-treated and washed PDMS surfaces appear to be virtually identical to the spectra of the substrate before plasma treatment (Figure 6, traces A and B). Apparently, ammonium chloride completely inhibits the generation of bonded

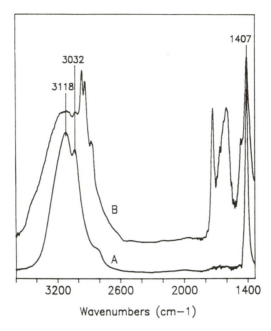

Figure 6. Transmission FTIR spectra in the 3600–1400-cm⁻¹ region of ammonium chloride (A) and material washed from the ammonia–plasma-treated PDMS (B).

surface amide functionality as well as the formation of other surface-bonded species.

The conclusion based on these studies is that PDMS substrates free of chlorine-containing molecules can be amide-functionalized by ammonia–plasma modification. However, the presence of chlorine-containing molecules, such as a radical initiator or residual chlorinated cleaning solvents, leads to the formation of surface ammonium chloride. The ammonium chloride layers inhibit the development of surface-bonded amide groups. Our preliminary experiments indicate that the degree of amide formation is inversely proportional to the amount of ammonium chloride formed on the surface (S. R. Gaboury and M. W. Urban, unpublished work in progress).

Summary

Although, in general, only limited surface spectroscopic dada are available for the gas–plasma-modified PDMS elastomers, ATR–FTIR spectroscopy apparently provides a means for analysis of the structures that develop as a result of energetic plasmons. In this chapter, the most recent advances, which use rf and microwave (mw) gas–plasma treatments, were reviewed. Ar–rf plasma leads to the formation of oxidized layers, whereas Ar–mw plasma generates useful Si–H surface groups. Although NH_3–rf plasma can be used to introduce surface amide functionality, it should be kept in mind that even residues of nonbonded chlorine-containing species may generate water-soluble layers of ammonium chloride. Apparently, these layers inhibit the amide group formation of PDMS surfaces.

References

1. Kaplan, S. L.; Rose, P. W. *Plast. Eng.* **1988**, *44*, 77.
2. DeLollis, N. J.; Montoya, O. *J. Adhes.* **1970**, *3*, 57.
3. Liston, E. M. *J. Adhes.* **1989**, *30*, 199.
4. Shah, K.; Beatty, C. L. *Org. Coat. Appl. Polym. Sci. Proc.* **1982**, *47*, 232.
5. Schreiber, H. P.; Wertheimer, M. R.; Lambla, M. *J. Appl. Polym. Sci.*, **1982**, *27*, 2269.
6. Schreiber, H. P.; St. Germain, F. *J. Adhes. Sci. Technol.* **1990**, *4*, 319.
7. Nguyen, H. X.; Weedon, G. C.; Chang, H. W. *Int. SAMPE Sypm. Exhib.* **1989**, *34*, 1603.
8. Cohn, D. *Polymers in Medicine III*; Migliaresi, C., Ed.; Elsevier Science: Amsterdam, Netherlands, 1988; p 43.
9. Triolo, P. M.; Andrade, J. D. *J. Biomed. Mater. Res.* **1983**, *17*, 129.
10. Hoffman, A. S. *J. Appl. Polym. Sci., Appl. Polym. Symp.* **1988**, *42*, 251.
11. Hoffman, A. S. *Adv. Polym. Sci.* **1984**, *57*, 141.
12. Yang, C.; Wnek, G. E. *Polym. Mater. Sci. Eng.* **1990**, *62*, 601.
13. LeGrand, D. G. *J. Polym. Sci.* **1969**, *B7*, 579.
14. Hoffman, A. S.; Horbett, T. A.; Ratner, B. D.; Hanson, S. R.; Harker, L. A.; Reynolds, L. O. *Biomaterials: Interfacial Phenomena and Applications*; Cooper, S. L.; Peppas, N. A., Eds.; Advances in Chemistry 199; American Chemical Society: Washington, DC, 1982; Chapter 6.
15. Ziegel, K.; Erich, F. *J. Polym. Sci., Polym. Chem. Ed.* **1970**, *8*, 2015.
16. Boenig, H. V. *Encyclopedia of Polymer Science and Engineering*; 2nd ed.; Wiley Interscience: New York, 1988; Vol. 11, p 248.
17. Bell, A. T. *Techniques and Applications of Plasma Chemistry*; Hollahan, J. R.; Bell, A. T., Eds.; Wiley Interscience: New York, 1974; Chapter 10.
18. Urban, M. W.; Huang, J. B. *Appl. Spectrosc.*, **November 1992**.
19. Bertie, J. E.; Eysel, M. M. *Appl. Spectrosc.* **1985**, *39*, 392.
20. Dignam, M. J.; Mamiche-Afara, S. *Spectrochim. Acta* **1988**, *44A*, 1435.
21. Huang, J. B.; Urban, M. W. *Appl. Specrtrosc.* **1992**, *46(6)*, 1014.
22. Harrick, N. J. *Internal Reflection Spectroscopy*; Wiley Interscience: New York, 1967.
23. Coopes, L. H.; Gifkins, K. J. *J. Macromol. Sci., Chem.* **1982**, *17*, 217.
24. Urban, M. W.; Stewart, M. T. *J. Appl. Polym. Sci.* **1990**, *39*, 265.
25. Ikeda, S. *Prog. Org. Coat.* **1973**, *1*, 205.
26. Gaboury, S. R.; Urban, M. W. *Polym. Commun.* **1991**, *32(13)*, 390.
27. Socrates, G. *Infrared Characteristic Group Frequencies*; Wiley: New York, 1980.
28. Eaborn, C. *Organosilicon Compounds*; Butterworth: London, 1960.
29. Hollahan, J. R.; Carlson, G. L. *J. Appl. Polym. Sci.* **1970**, *14*, 2499.
30. Morra, M.; Occhiello, E.; Marola, R.; Garbassi, F.; Humphrey, P.; Johnson, D. *J. Colloid Interface Sci.* **1990**, *137*, 11.
31. Yasuda, H. *Plasma Polymerization*; Academic: Orlando, FL, 1985.

32. Hollahan, J. R.; Stafford, B. B. *J. Appl. Polym. Sci.* **1969**, *13*, 807.
33. Lub, J.; van Vroonhoven, F.; Bruninx, E.; Benninghoven, A. *Polymer*, **1989**, *30*, 40.
34. Nguyen, L. T.; Sung, N-H.; Suh, N. P. *J. Polym. Sci., Polym. Lett. Ed.* **1980**, *18*, 541.
35. Gaboury, S. R.; Urban, M. W. *J. Appl. Polym. Sci.* **1992**, *44*, 401.
36. Folman, M. *Trans. Faraday, Soc.* **1961**, *57*, 2000.
37. Scheffer, H. R. *Metallography* **1973**, *27*, 315.
38. Dwight, D. W.; McGrath, J. E.; Wightman, J. P. *J. Appl. Polym. Sci., Appl. Polym. Symp.* **1978**, *34*, 35.
39. Ratner, B. D. In *Biomaterials: Interfacial Phenomena and Applications*; Cooper, S. L.; Peppas, N. A., Eds.; Advances in Chemistry 199; American Chemical Society: Washington, DC, 1982; p 9.
40. Turner, N. H. *Anal. Chem.* **1988**, *60*, 377R.
41. van Ooij, W. J.; Michael, R. S. *Polym. Mater. Sci. Eng.* **1988**, *59*, 734.
42. Arefi, F.; Montazer-Rahmati, P.; Andre, V.; Amouroux, J. *J. Appl. Polym. Sci., Appl. Polym. Symp.* **1990**, *46*, 33.
43. Vargo, T. G.; Gardella, J. A.; Salvati, L., Jr. *J. Polym. Sci., Polym. Chem. Ed.* **1989**, *27*, 1267.
44. Nowlin, T. E.; Smith, D. F. *J. Appl. Polym. Sci.* **1980**, *25*, 1619.
45. Nadiger, G. S.; Bhat, N. V. *J. Appl. Polym. Sci.* **1985**, *30*, 4127.
46. Momose, Y.; Ohaku, T.; Chuma, H.; Okazaki, S.; Saruta, T.; Masui, M.; Takeuchi, M. *J. Appl. Polym. Sci., Appl. Polym Symp.* **1990**, *46*, 153.
47. Inagaki, N.; Tasaka, S.; Kawai, H. *J. Appl. Polym. Sci., Appl. Polym. Symp.* **1990**, *46*, 399.
48. DeLollis, N. J.; Montoya, O. *J. Adhes.* **1971**, *3*, 57.

RECEIVED for review July 15, 1991. ACCEPTED revised manuscript September 8, 1992.

Ultrasonic Synthesis and Spectroscopic Characterization of Poly(phthalocyanato) Siloxane

Bert J. Exsted and Marek W. Urban*

Department of Polymers and Coatings, North Dakota State University, Fargo, ND 58105

A new synthetic method for the preparation of poly(phthalocyanato) siloxane is described. The method utilizes ultrasonic energy in the conversion of dichlorosilicon phthalocyanine monomer, a sodium chalcogenide, and molecular moisture to form the cofacially stacked poly(phthalocyanato) siloxane polymer $[Si(Pc)O]_n$. Of four sodium chalcogenides investigated, sodium telluride was found to be the most effective polymerizing agent with an n value (degree of polymerization) greater than 45 repeating units. Sonication reactions were carried out at room temperature conditions extending from 1 min to 8 h in length. For comparative purposes, poly(phthalocyanato) siloxane was also prepared thermally via the traditional synthetic route. Electronic and vibrational band assignments of the prepared phthalocyanine monomers and their respective polymers are presented and assigned to the inherent structure of the macrocycle. The role of water in the sonication process is also discussed.

COFACIAL ASSEMBLY OF METALLOMACROCYCLES was pioneered in the late 1950s by Elvidge and Lever (*1*) and Joyner and Kenney (*2*) for phthalo-cyaninogermanium and phthalocyanomanganese(IV) complexes. Soon thereafter, it was discovered that dihydroxysilicone phthalocyanine (Si(Pc)(OH)$_2$) could also form a stacked, planar phthalocyanine moiety upon thermal dehydration (*3*). The resulting poly(phthalocyanato) silicon oxide, $[Si(Pc)O]_n$,

* Corresponding author.

0065-2393/93/0236-0791$06.00/0

upon iodine doping, has since gained wide attention as the cornerstone of a new class of electrically conductive macromolecules (4–7).

In traditional two-step thermal polymerization of poly(phthalocyanato) siloxane extreme parameters are required: for example, high vacuum (10^{-3} torr) and temperature (440 °C) conditions over extensive periods of time (12 h) (8). As a consequence, much work on the development of $[Si(Pc)O]_n$ has been impeded. The apparent need for a more convenient synthetic technique should be addressed. In this chapter, we report a novel one-step, room temperature method for the synthetic preparation of $[Si(Pc)O]_n$ that uses ultrasonic energy.

Utilization of ultrasonic waves in chemical applications dates back to the mid-1920s when Loomis and Richards carried out the first studies of high-frequency sound wave ($\geq$ 20 kHz) effects on organic and aqueous solutions (9). In the following decades, however, very little research utilizing ultrasonic energy in the procurement of new chemical compounds was investigated. Recently, however, sonochemistry has gained interest in the area of heterogeneous synthetic reactions, the majority of which involve the chemical interaction between metallic powders and functionally active carbon or silicon compounds (10, 11). In the area of polymer synthesis and modification, however, only limited applications of ultrasonic energy have been explored (12). In an effort to address this issue, we report the use of ultrasonic irradiation in the synthesis of poly(phthalocyanato) siloxane polymer, $[Si(Pc)O]_n$.

Experimental Details

Synthetic Procedures. *Synthesis of Si(Pc)Cl₂.* Silicon phthalocyanine dichloride was prepared by a modification of the procedure of Lowery et al. (13). Under nitrogen, 16.45 mL (0.143 mol) of freshly distilled silicon tetrachloride (Petrarch Systems) and 165 mL of dry quinoline solvent (Aldrich) was syringed into a 500-mL three-neck round bottom flask equipped with a nitrogen inlet, thermometer, heating mantel, mechanical stirrer, and dry-ice condenser (3-heptanone–dry ice; − 38 °C) fitted upon a West-type condenser. The solution was brought to reflux. When the solution reached 200 °C, 15.000 g (0.103 mol) of 1,3-diiminoisoindoline (Aldrich) was added. The resulting solution was refluxed at 220 °C for an additional 45 min and then slowly cooled to room temperature. Upon cooling, 40 mL of chloroform was added to the resulting dark violet–blue reaction mixture in an effort to facilitate the workup procedure. The crude product was centrifuged in portions and Büchner filtered with quinoline, chloroform, and acetone, respectively. The resulting purple, crystalline powder was stored in a vacuum oven at 120 °C under 10^{-3} torr. *Analytically calculated (%)* for $C_{32}H_{16}N_8SiCl_2$: C, 62.85; H, 2.64; N, 18.32; Cl, 11.59. *Found*: C, 52.86; H, 2.49; N, 16.73; Cl, 10.55.

Ultrasonic Synthesis of [Si(Pc)O]ₙ.   Inside an argon glovebox, sodium telluride (Cerac, Inc.; 0.142 g; 8.180×10^{-4} mol) and (phthalocyanato)silicon dichloride (0.500 g; 8.180×10^{-4} mol) were added to an oven-dried 50-mL three-neck round bottom flask. The flask was transferred to a hood, purged with nitrogen (N_2), and submerged into a common ultrasonic laboratory cleaner (Bransonic model 2200) filled with deionized water. Approximately 10 mL of freshly distilled tetrahydrofuran (THF) dried over benzophenone and sodium was syringed into the flask. The resulting mixture was sonicated at room temperature over periods ranging from 1 min to 8 h in length, and cavitation was observed in the reaction flask. The resulting poly(phthalocyanato) silicon oxide product was Büchner filtered with tetrahydrofuran, water, and acetone, respectively, and dried in a vacuum oven (120 °C; 10^{-3} torr). Similar products, with lower degree of polymerization values were obtained by substituting 8.180×10^{-4} mol of sodium sulfide (Pfalz and Bauer; 0.064 g) and sodium selenide (Cerac, Inc.; 0.103 g), respectively, in place of the sodium telluride. Sodium oxide (Atomergic Chemetals Corp.), however, appeared to be ineffective. *Analytically calculated* (%) for $C_{32}H_{16}N_8SiO$ upon 8 h of $Si(Pc)Cl_2$ ultrasonication with sodium telluride: C, 69.05; H, 2.90; N, 20.13; Cl, 0.00. *Found*: C, 64.24; H, 2.14; N, 11.96; Cl, 0.07.

Synthesis of Si(Pc)(OH)₂.   (Phthalocyanato) silicon dihydroxide was prepared in a slightly modified procedure from that reported by Davison and Wynne (*14*). Into a 250-mL three-neck round bottom flask equipped with a thermometer, condenser, heating mantel, stir bar, and plate was added 1.500 g (2.45×10^{-3} mol) of (phthalocyanato) silicon dichloride, 75 mL of 2-M aqueous sodium hydroxide, and 15 mL of pyridine cosolvent. After 12 h the hydrolyzed (phthalocyanato) silicon product was recovered by Büchner filtration with distilled deionized water and acetone, respectively. *Yield*: 1.292 g (91.7%) *Analytically calculated* (%) for $C_{32}H_{18}N_8SiO_2$: C, 66.89; H, 3.16; N, 19.50; Cl, 0.00. *Found*: C, 60.68; H, 3.24; N, 17.85; Cl, 0.00.

Thermal Synthesis of [Si(Pc)O]ₙ.   Poly(phthalocyanato) siloxane, was prepared by the traditional thermal synthetic procedure outlined by Joyner and Kenney (*3*). (Phthalocyanato) silicon dihydroxide was placed into a pyrolysis quartz tube and the axially functional monomer was heated to 440 °C in a Lindberg heavy-duty tube furnace under a continuous vacuum (10^{-3} torr) for a period of 8 h. The resulting poly(phthalocyanato) siloxane was likewise recovered as a dark purple powder. *Analytically calculated* (%) for $C_{32}H_{16}N_8SiO$: C, 69.05; H, 2.90; N, 20.13; Cl, 0.00. *Found*: C, 64.64; H, 2.79; N, 18.92; Cl, 0.00.

H₂(Pc). Nonmetallated phthalocyanine monomer (98% purity; β form) was purchased from the Aldrich Chemical Company, Inc. and spectroscopically analyzed as received without further modification.

Analytical Methods. ***Elemental Analysis.*** Elemental analyses (C, H, N, and Cl) were performed by Desert Analytics Organic Microanalysis of Tucson, AZ.

Infrared Spectroscopy. Photoacoustic Fourier transform infrared (PA–FTIR) spectra were collected on a spectrometer (Digilab FTS-10M) continuously purged with purified air (free of hydrocarbons, carbon dioxide, and water; Balston Filter Products). The single-beam spectra of phthalocyanine samples enclosed in a helium-purged photoacoustic cell were recorded at a resolution of 4 cm^{-1} and ratioed against a carbon black reference. All spectra were transferred to an AT compatible computer and analyzed with the aid of Spectra Calc software (Galactic Industries).

Optical Spectroscopy. Solution spectra of Si(Pc)Cl$_2$, Si(Pc)(OH)$_2$, and both sonically and thermally prepared [Si(Pc)O]$_n$ were recorded on a multiple-cell diode array UV–visible spectrophotometer (Hewlett-Packard 8451A) equipped with a deuterium lamp. Pure tetrahydrofuran (THF) solvent was utilized as a reference.

Nuclear Magnetic Resonance Spectroscopy. In an effort to access the role of the solvent in the ultrasonication procedure, ^{1}H and ^{13}C NMR spectra of untreated tetrahydrofuran solvent and tetrahydrofuran solvent (10 mL) sonicated for 8 h in the presence of sodium telluride (0.142 g; 8.180 × 10^{-4} mol) were recorded on a 400-MHz Fourier transform NMR spectrometer (JEOL GSX-400). Approximately 0.05 mL of each sample was dissolved in 1 mL of deuterated chloroform (CDCl$_3$). Chemical shifts are reported in measurements relative to tetramethylsilane (TMS).

Results and Discussion

To circumvent the traditional two-step synthetic procedure of poly(phthalocyanato) siloxane, [Si(Pc)O]$_n$, we developed a new ultrasonic synthetic procedure that requires only one step. Both reaction routes are schematically depicted in Chart I. Note that the formation and isolation of the Si(Pc)(OH)$_2$ intermediate, previously required in the traditional two-step polymerization procedure of poly(phthalocyanato) siloxane, has been eliminated (3). Hence, dichlorosilicon phthalocyanine (Si(Pc)Cl$_2$), the precursor of Si(Pc)(OH)$_2$, can be utilized directly via the one-step ultransonication reaction pathway.

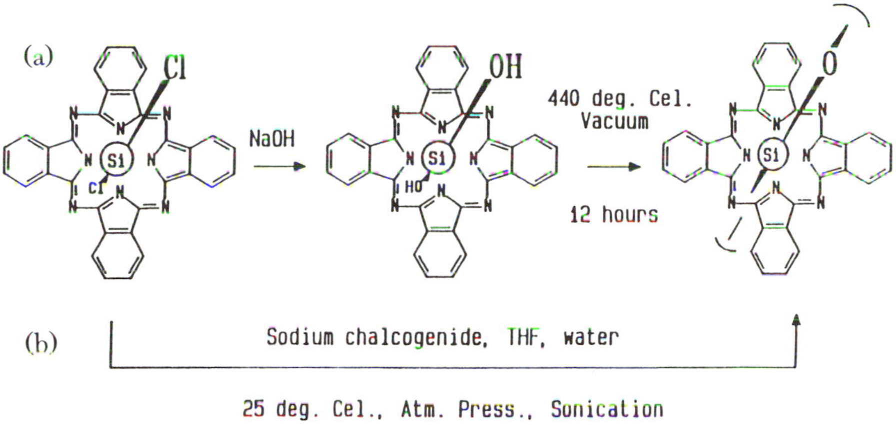

(a)

NaOH

440 deg. Cel.
Vacuum

12 hours

(b)

| Sodium chalcogenide, THF, water |

25 deg. Cel., Atm. Press., Sonication

Chart I. Synthetic schemes for the preparation of poly(phthalocyanato) si-loxane via the traditional thermal condensation route (a) and a new ultrasonic procedure (b).

Before we analyze our infrared absorption spectral results, let us first set the stage by defining the bands most relevant to both the monomeric and polymeric phthalocyanine macromolecular structures.

Because of the vast number of infrared absorptions displayed by the phthalocyanine moiety, the first spectroscopic reports of vibrational modes inherent to the phthalocyanine macromolecule were complicated by a numeric scheme that was commonly substituted for standard vibrational assignment tables (15). In addition, vibrational analyses were often disrupted by bands attributed to the dispersive media, for example, mineral oil (16, 17). Later, sublimation techniques for phthalocyanine deposition on the surface of potassium bromide crystals were developed to simplify the data interpretation (18). Each of these IR techniques, however, was rather time-consuming and, in many cases, altered the structure of the phthalocyanine moiety (15). Fortunately, a more convenient method of vibrational spectroscopy now exists that allows infrared spectra of opaque insoluble powders to be obtained via the use of a Fourier transform infrared spectrometer equipped with a photoacoustic cell (PA FTIR) (19–21). This vibrational spectroscopic technique is also nondestructive to the sample; thus the sample can be retrieved in its original form. Subsequently, all vibrational spectra were recorded by this PA technique.

PA–FTIR vibrational data of $Si(Pc)Cl_2$ and $Si(Pc)(OH)_2$ monomers, along with their respective ultrasonic and thermally synthesized $[Si(Pc)O]_n$ polymers are reported in Table I. In addition to the reported silicon phthalocyanine vibrational bands, characteristic infrared frequencies of a nonmetallated phthalocyanine, $H_2(Pc)$, are included to make a more complete listing of bands that are attributed solely to the phthalocyanine moiety and bands

Table I. Infrared Bands of $H_2(Pc)$, $Si(Pc)Cl_2$, $Si(Pc)(OH)_2$, and Both Thermally and Sonically Polymerized $[Si(Pc)O]_n$[a]

Vibrational Mode		Band No.[b]	β H_2–Pc	$Si(Pc)Cl_2$	$Si(Pc)(OH)_2$	Thermal $[Si(Pc)O]_n$	Sonic[c] $[Si(Pc)O]_n$
C–C	Def.$_{ro}$	1			450w		
Si–Cl	Str.			464[d] vs			
C–C	Def.$_{ro}$	2	487, 493m	507w			
				532[d] s	528[d] m	528[d] m	530[d] m
C–C	Def.$_{ro}$	3	555m	574m	574m	574m	574m
				609[d] w	615[d] w		
C–C	Def.$_{ro}$	4	613s	648m	644m	644w	646w
			684[e] s		673[d] w		
C–H	Def.$_o$	5	719vs	694[d] m	702[d] w	700w	692w
C–H	Def.$_o$	6	729, 734vs	734s	732vs	727vs	734vs
N–H	Def.$_o$	7	748vs	—	—	—	—
C–H α	Def.$_o$	8	767s	761s	759vs	758s	759s
C–H β	Def.$_o$	9	779s	775w	779m	773w	773w
				785m			
Cl–H	Def.$_o$	10	—	808[d] m	806sh	804[d] w	804[d] w
Si–O	Str.				831s	831w	831w
C–H α	Def.$_o$	11	—	866[d] m	871w	866[d] w	869[d] w
C–H β	Def.$_o$	12	880sh	883m	—	—	—
C–H	Def.$_o$	13	873vs	914s	912m	910s	910s
		14	947vs	947w	950w	947sh	947sh
β		15	956m				
				960[d] m	974[d] w		
				987[d] w	993[d] w		
Si–O—Si	Str.					999[d] bd	999[d] bd
				1004[d] w	1004[d] w		
N–H	Def.$_i$	16	1006vs	—	—	—	—
C–H	Def.$_i$				1022[d] w		
C–H	Def.$_i$				1051[d] sh		
C–H	Def.$_i$	17	—	1062vs	1068vs		
C–N	Str.	18	1093vs	1082vs	1078, 1089vs	1082s	1082s
C–H	Def.$_i$	19	—	—	—		
C–H	Def.$_i$	20	1118vs	1124vs	1122, 1134[d] s	1122s	1122s
C–H	Def.$_i$	21	1156m	1165s	1166m	1165m	1165m
C–H β	Def.$_i$	22	1163m	—	1175w		
				1240[d] w	1222[d] w		
C=N	Def.		1249m		1249[e] w		
C–C	Str.	23	1276s	1292s	1292s	1288s	1288s
N–H	Def.	24	1303s	—	—	—	—
N–H	Def.	25	1323s	—	—	—	—
		26	1334vs	1338vs	1336vs	1336vs	1336vs
			1342[e] sh	1342[d] sh			
					1352[d] sh	1352[d] sh	1352[d] sh
C–C	Str.	27	1436s	1431s	1431s	1427vs	1427vs
		28	1458w	1473	1471m	1471m	1471m
		29	1477w	1486w	1487w	—	1485w

Table I. Continued

Vibrational Mode		Band No.[b]	β H_2–Pc	$Si(Pc)Cl_2$	$Si(Pc)(OH)_2$	Thermal $[Si(Pc)O]_n$	Sonic[c] $[Si(Pc)O]_n$
C=N	Str.	30	1502m	1533s	1519s	1518s	1517s
C=C	Str.	31	1597sh	1596sh	1595sh	1595w	1595w
C=C	Str.	32	1606, 1614m	1610s	1608m	1614m	1614m
N–H	Str.	33	3277s	—	—	—	—
O–H	Str.		—	—	3500bd	3500bd	3500bd

[a] Symbols and abbreviations: Str. = stretching; Def. = Deformation; Def.$_{ro}$ = out-of-plane ring deformation; Def.$_o$ = out-of-plane deformation (e.g., wagging and twisting). Def.$_i$ = in-plane deformation (e.g., scissoring and rocking); s = strong; m = medium; w = weak; bd = broad; sh = shoulder; v = very;
[b] Band numbering scheme (1–33) is in accordance with reference 15.
[c] $Si(Pc)Cl_2$–Na_2Te sonicated in THF for 8 h.
[d] = additional peaks found and reported by Marks (reference 8).
[e] = additional peaks found and reported by Kobayashi (reference 16).

that are related to the chelated metal atom and its axial substituents. Although the infrared absorption modes of detection differ, the obtained photoacoustic IR group frequencies correlate closely with absorption bands, previously cited in the literature, that utilized more laborious and time-consuming transmission infrared techniques (8, 16, 18, 22). The traditional numbering scheme initially used by Sidorov and Kotlyar (15), but seldom assigned, is included in Table I to provide a comprehensive listing of the vibrational bands of the phthalocyanine macrocycle. Not all vibrational bands are listed, however, because more than 50 normal vibrational modes are observed in the infrared spectrum of the highly symmetrical phthalocyanine macrocycle. Hence, Table I is limited primarily to band frequencies between the region of 1620–450 cm^{-1} that is most commonly cited in the literature for the analysis of characteristic phthalocyanine vibrational bands and is also sensitive to the formation of cofacially stacked polymers.

Analysis of the photoacoustic vibrational data listed in Table I reveals that the nonmetallated phthalocyanine, $H_2(Pc)$, with a point group symmetry of D_{2h} displays several band splittings (bands 2, 6, and 32) that distinguish it from its highly symmetrical metallated D_{4h} counterparts. The displacement of $H_2(Pc)$ from a horizontal plane symmetry is shown in Chart II. In addition to band numbers assigned as a result of the $H_2(Pc)$ molecular symmetry, N–H vibrational modes (bands 7, 16, 24, 25, and 33) are also assigned.

The silicon phthalocyanine dichloride monomer, $Si(Pc)Cl_2$, is observed to display vibrational bands characteristic of metallated phthalocyanines with a D_{4h} point group symmetry, except an additional vibrational mode at 464 cm^{-1} is readily observed. This infrared absorption has been assigned directly to the Si–Cl stretching frequency that is positioned axially off the phthalocyanine ring (8). Although the hydroxy-functional silicon phthalocyanine monomer, $[Si(Pc)(OH)_2]$, displays virtually the same vibrational bands as its

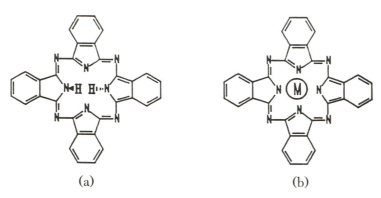

(a)   (b)

Chart II. Structures of nonmetallated [D_{2h}] (a) and metallated [D_{4h}] (b) phthalocyanines.

precursor, $Si(Pc)Cl_2$, several apparent differences between their vibrational spectra are noted. These spectral changes include the following:

1. The formation of a broad $-OH$ stretching band at approximately 3500 cm^{-1}.

2. A downfield shift of the $C=N$ intraligand vibrational stretching band (band 30) from 1531 to 1519 cm^{-1} upon replacement of the chloride to a hydroxide axial substituent.

3. Splittings of bands 18 and 20 due to symmetry distortions by $Si(Pc)(OH)_2$ from the D_{4h} point group.

4. The formation of a strong antisymmetric $Si-O$ stretching mode at approximately 831 cm^{-1}, characteristic of the dihydroxy-functional silicon phthalocyanine monomer.

Now that several characteristic spectral features of the axial-functional silicon phthalocyanine monomers have been established, let us return to the main thrust of this chapter and spectroscopically follow the ultrasonic and thermal polymerization of poly(phthalocyanato) siloxane from its respective $Si(Pc)Cl_2$ and $Si(Pc)(OH)_2$ monomeric precursors. As illustrated in Figure 1, thermal dehydration of the $Si(Pc)(OH)_2$ monomer for 8 h at 440 °C (trace d) diminishes the characteristic antisymmetric $Si-O$ stretch at 831 cm^{-1} of the hydroxy-functional monomer (trace b). Coincident with the diminishing vibrational band is the formation of a broad, new band centered at approximately 1000 cm^{-1}. Formation of this broad vibrational band at 1000 cm^{-1} was previously attributed to the antisymmetric $Si-O-Si$ stretch of a poly(phthalocyanato) siloxane, $[Si(Pc)O]_n$ (23). Recent investigations by Dirk et al. (8) verified the assignment of the polymeric $Si-O-Si$ stretching

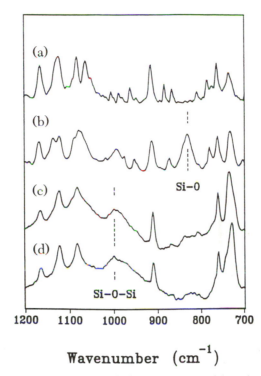

Figure 1. *PA—FTIR spectra of Si(Pc)Cl₂ monomer (a), Si(Pc)(OH)₂ mono-mer (b), sonically polymerized [Si(Pc)O]ₙ, (c) and thermally polymerized [Si(Pc)O]ₙ (d).*

vibration by noting its downfield shift of approximately 50 cm^{-1} upon replacement of the bridging oxygen atoms with ^{18}O, which subsequently forms the isotopic [Si(Pc)^{18}O]$_n$ polymer. Broadening of this vibrational transition has been correlated with increases in the molecular weight of the polymer (*24*). Similar in respect to the PA—FTIR spectrum of the thermally prepared macrocyclic siloxane polymer, the infrared spectrum of ultrasonically prepared poly(phthalocyanato) siloxane (Figure 1, trace c) displays identical features, even though it is prepared directly from the silicon phthalocyanine dichloride precursor, Si(Pc)Cl$_2$ (Figure 1, trace a; Si(Pc)Cl$_2$/Na$_2$Te sonicated in THF for 8 h).

An additional band effected by the ultrasonic polymerization pathway of the Si(Pc)Cl$_2$ monomer is that of the intraligand C=N stretching vibration at 1535 cm^{-1} (band 30). This vibrational mode was previously noted to shift downfield from 1533 to 1519 cm^{-1} upon replacement of the chloride substituents of Si(Pc)Cl$_2$ with hydroxyl substituents in the case of the monomeric hydrolysis reaction; *see* Table I. The same trend appears to be

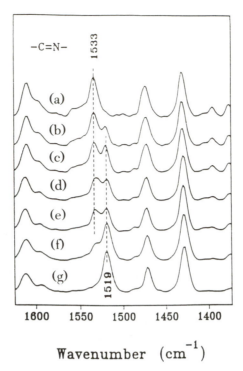

-C=N-

1533

(a)

(b)

(c)

(d)

(e)

(f)

(g)

1519

1600   1550   1500   1450   1400

Wavenumber (cm^{-1})

Figure 2. PA–FTIR spectra of Si(Pc)Cl$_2$ sonicated with sodium sulfide over periods of 0 min (a), 1 min (b), 5 min (c), 30 min (d), 1 h (e), 2 h (f), and 8 h (g).

evident in the ultrasonic polymerization process. Figure 2 monitors this axial ligand replacement effect as a function of sonication time. As illustrated in Figure 2, trace g, a minimum of 8 h of sonication with sodium chalcogenide is required for complete ligand substitution to occur. Vibrational overlap of the intraligand C=N group frequency (band 30) of all the ultrasonically treated samples (Figure 3, traces b, c, and d), in conjunction with the C=N vibrational modes of the thermally prepared Si(Pc)(OH)$_2$ monomer and [Si(Pc)O]$_n$ polymer (Figure 3, traces e and f, respectively), tend to indicate that the axial chloride substituent subsequently has been replaced. This observation is supported by the substantial absence of chlorine in the elemental analysis data (presented earlier in the experimental section) for both the sonically and thermally prepared [Si(Pc)O]$_n$ polymers.

Complementary to Figure 3, the PA infrared spectra of several ultrasonically polymerized species that display a broad vibrational mode at 1000 cm^{-1} are shown in Figure 4. This band corresponds to the broad Si−O−Si antisymmetric stretch of the thermally prepared poly(phthalocyanato) siloxane. Chalcogenide bridging of the silicon phthalocyanine moieties (e.g.,

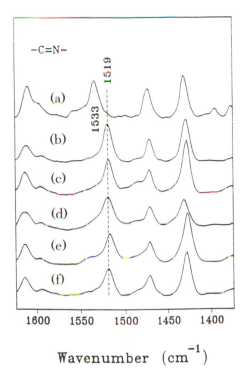

$$\text{Wavenumber} \ (\text{cm}^{-1})$$

Figure 3. PA–FTIR spectra of Si(Pc)Cl₂ (a), Na₂S-treated (b), Na₂Se-treated (c), and Na₂Te-treated (d) Si(Pc)Cl₂ ultrasonically treated for 8 h. Spectra of Si(Pc)(OH)₂ (e) and its thermally polymerized product, [Si(Pc)O]ₙ (f) are also shown.

Si−S−Si, Si−Se−Si, or Si−Te−Si) would be expected to display a Hookean behavior and shift the stretching vibration downfield to frequencies lower than 1000 cm^{-1}. This shift, however, is not observed as illustrated in Figure 4, traces b, c, and d. Subsequently, these spectral data indicate that a Si−O−Si bridged phthalocyanine oligomer or polymer has indeed been formed.

In an effort to account for the presence of oxygen in the ultrasonic formation of the poly(phthalocyanato) siloxane, several spectroscopic studies were undertaken. One possible source of oxygen that is responsible for the Si−O−Si cofacial stacking of the phthalocyanine rings was the sonication medium, tetrahydrofuran (THF; C_4H_8O). As illustrated in Figure 5, axial polymerization of Si(PC)Cl₂ into a siloxane chain resulted only when THF was utilized as the sonication medium (Figure 5, trace d), whereas substitution of the THF sonication medium with a non-oxygen-containing solvent, such as hexane (Figure 5, trace b), resulted in virtually no vibrational changes with respect to the original starting reagent, Si(Pc)Cl₂ (Figure 5, trace a).

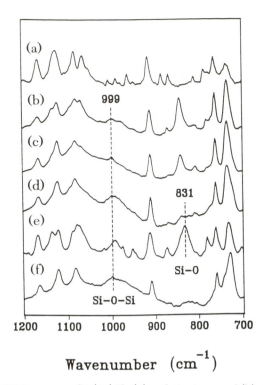

Figure 4. PA–FTIR spectra of Si(Pc)Cl₂ (a) and Na₂S-treated (b), Na₂Se-treated (c), and Na₂Te-treated (d) Si(Pc)Cl₂ sonicated for 8 h. Spectra of Si(Pc)(OH)₂ (e) and its thermally polymerized product, [Si(Pc)O]ₙ (f) are also shown.

^{13}C NMR studies of THF sonicated for a period of 8 h in the presence of sodium telluride, however, revealed that no new chemical shifts were detected, as shown in Table II, thereby indicating that the complete chemical structure of the heterocyclic solvent, THF, remains intact with no loss of oxygen. Similar ^{1}H NMR studies of the chalcogenide-treated THF confirmed the inactive role of THF as a direct oxygen source; *see* Table II. ^{1}H NMR, however did detect a new chemical shift at approximately 2.5 ppm in both the nontreated and chalcogenide-treated THF. A small addition of water to each sample dramatically enhanced the intensity of the peak at 2.5 ppm, which indicates that molecular water was the agent responsible for this chemical shift.

Although THF did not directly account for the source of oxygen in [Si(Pc)O]ₙ, the presence of water if THF could subsequently serve as the primary source of oxygen. Hence, the role of water in the ultrasonication reaction was investigated by initially charging a minute concentration of water (2% by volume) into the THF sonication medium. The formation of the

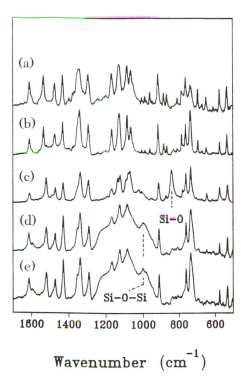

Wavenumber (cm^{-1})

Figure 5. PA–FTIR spectra of Si(Pc)Cl$_2$ monomer (a), Si(Pc)Cl$_2$/Na$_2$Te soni-cated in hexane (b), Si(Pc)Cl$_2$/Na$_2$Te sonicated in THF/H$_2$O (c), Si(Pc)Cl$_2$/Na$_2$Te sonicated in THF without any post treatment, and Si(Pc)Cl$_2$/Na$_2$Te sonicated in THF and worked-up via Büchner filtration (e).

monomeric Si—O vibration stretch (Figure 5, trace c), characteristic of the Si(Pc)(OH)$_2$ monomer, was readily observed, thereby indicating that water plays a direct role as an oxygen source. However, the concentration of water present in the THF is believed to be a determining factor in whether the resulting ultrasonication product will be monomeric or polymeric in its structural nature.

In view of the foregoing considerations, the proposed mechanism for the ultrasonic formation of [Si(Pc)O]$_n$ is schematically presented in Chart III. Recent investigations by Arya et al. (25) on a series of silatrane molecules, which mimic the chemical structure of the silicon phthalocyanine complex, tend to support the existence of a silicon–chalcogenide intermediate. How-ever, the presence of molecular water (moisture) lead to the immediate rearrangement of the transient silicon–chalcogenide intermediate (e.g., a silanone: Si=S or Si=Se) to form a more stable, cyclic, six-membered siloxane ring structure. Due to geometrical restraints imposed upon the planar phthalocyanine moiety by its bulky aromatic rings, it is postulated in

Table II. ^{13}C and ^{1}H NMR Data of the Sonication Media^a

Sample	δ (ppm)
^{13}C NMR of THF	67.4 (singlet, 1C; α-methylene)
	25.1 (singlet, 1C; β-methylene)
^{13}C NMR of Na$_2$Te–THF^b	67.4 (singlet, 1C; α-methylene)
	25.1 (singlet, 1C; β-methylene)
^{1}H NMR of THF	3.60 (multiplet, 2H; α-methylene)
	2.58 (singlet)
	1.72 (multiplet, 2H; β-methylene)
^{1}H NMR of Na$_2$Te–THF^b	3.60 (multiplet, 2H; α-methylene)
	2.58 (singlet)
	1.72 (multiplet, 2H; β-methylene)
^{1}H NMR of THF–H$_2$O	3.60 (multiplet, 2H; α-methylene)
	2.58 (singlet, 1H; H$_2$O)
	1.72 (multiplet, 2H; β-methylene)

^{a} Dissolved in CDCl$_3$, plus 1% TMS.
^{b} Sonicated 8 h.

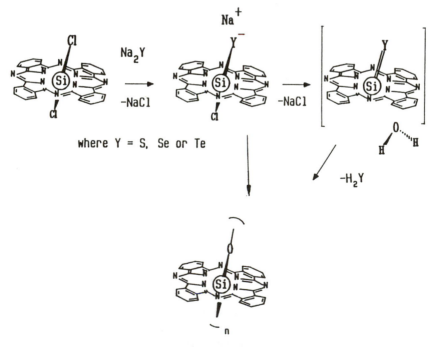

Chart III. A proposed mechanism for the ultrasonic synthesis of poly(phthalocyanato) siloxane.

our case that trimerization of the short-lived silanone phthalocyanine inter-mediates becomes inhibited. As a consequence, the proposed silanone inter-mediates (e.g., Si=S, Si=Se, Si=Te), upon nucleophilic displacement with water, may readily rearrange to form a linear, noncyclic poly(phthalocyanato) silicon oxide oligomer–polymer. Attempts to isolate this short-lived silanone phthalocyanine intermediate (26), however, have been unsuccessful to date as evidenced by the PA–FTIR spectrum in Figure 5, trace e, of the ultrasonically prepared $[Si(Pc)O]_n$ polymer, which was synthesized and ana-lyzed under dry conditions (inert nitrogen gas) with no post purification procedures. Although it is also possible that water may react with the intermediates to form siloxane-like compounds, more experiments are needed to confirm a precise mechanism of polymerization.

To provide further evidence for the ultrasonic formation of a linear poly(phthalocyanato) siloxane, UV–visible spectra of several phthalocyanine monomers and polymers were likewise recorded; their corresponding elec-tronic excitation wavelengths (λ_{max}) are compiled in Table III. Typical solution optical spectra of monomeric phthalocyanines are reported to exhibit two characteristic $\pi \rightarrow \pi^*$ intraligand transitions, which are shown in Figure 6. The electronic band transition near 350 nm is commonly referred to as the Soret (or B) band, whereas the transition at ca. 680 nm is cited as the Q band. The Q band is responsible for the inherent greenish blue color of phthalogen dyestuff (27). As illustrated in Figure 6, trace a, splitting of the lower-energy Q-band in the red region occurs in the case of the nonmetal-lated phthalocyanine, $H_2(Pc)$, due to $D_{4h} \rightarrow D_{2h}$ symmetry changes. This phenomenon results from the two hydrogen atoms being less symmetrical with respect to the four central ring nitrogen atoms of a metallated phthalo-cyanine structure. Subsequently, the degenerate lowest unoccupied molecu-lar orbital (LUMO), π^*, energy level of the nonmetallated phthalocyanine macrocycle is split into two different molecular orbital energy states (28).

In contrast to the nonmetallated $H_2(Pc)$, metallated $Si(Pc)Cl_2$ and $Si(Pc)(OH)_2$ phthalocyanine monomers dispersed in THF, along with their respective ultrasonically and thermally prepared polymers, $[Si(Pc)O]_n$, do not

Table III. Number Average Molecular Weight Data for Both Sonically and Thermally Synthesized $[Si(Pc)O]_n$

Compound	Absorption Maxima (nm)
$H_2(Pc)$	348, 594 (weak), 652, 688
$Si(Pc)Cl_2$	335, 364, 622 (weak), 692
$Si(Pc)(OH)_2$	334 (weak), 366, 626, 692
$[Si(Pc)O]_2$ sonic[a]	342, 442 (weak), 630, 662
$[Si(Pc)O]_2$ thermal[b]	345, 445, 636, 666

[a] $Si(Pc)Cl_2$–Na_2Te sonicated in THF for 8 h.
[b] Thermal condensation (440 °C) of $Si(Pc)(OH)_2$ for 8 h.

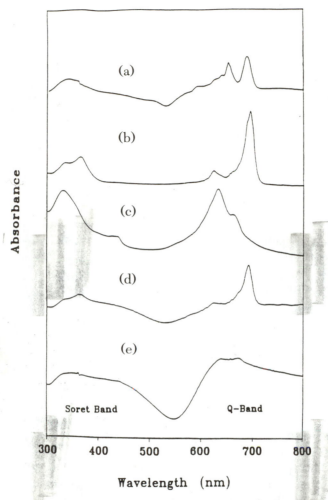

Figure 6. *Electronic absorption spectra of $H_2(Pc)$ (a), $Si(Pc)Cl_2$ monomer (b), sonically polymerized $[Si(Pc)O]_n$ (c), $Si(Pc)(OH)_2$ monomer (d), and thermally polymerized $[Si(Pc)O]_n$ dispersed in THF (e).*

display Q-band splittings as depicted in Figure 6, traces c and e, respectively. This observation suggest that the monomeric silicon phthalocyanine macromolecule with its highly symmetrical D_{4h} point group symmetry subsequently is conserved upon polymerization, forming a linear $[Si(Pc)O]_n$. Hence, the axial siloxane bridges of the ultrasonically prepared $[Si(Pc)O]_n$ are believed to be similar with respect to the structural conformation of its linear thermally polymerized counterpart. It should also be noted that a distinct blue (hypsochromic) shift of approximately 40 nm is observed upon polymerization of both the monomeric phthalocyanines reagents, regardless of the mode of preparation. This shift has previously been explained in terms of decreased electron density within the phthalocyanine ring upon π-orbital stacking, which subsequently lowers the highest occupied molecular orbital (HOMO) energy level of the phthalocyanine and raises the energy required to undergo an electronic transition (29).

Determination of the various degrees of polymerization for both sonically and thermally polymerized poly(phthalocyanato) siloxane have been performed quantitatively by infrared end-group analysis (8) and are reported in Table IV. In this method, the area of the Si$-$O vibrational mode for both the monomeric, Si(Pc)(OH)$_2$, and polymeric, [Si(Pc)O]$_n$, species are ratioed against the area of an invariant C$-$H out-of-plane deformation of the phthalocyanine moiety (band 6). Extents of polymerization (or n values) calculated in this manner previously have been shown to be accurate within 10% of the molecular weight values determined by tritium labeling (8). As noted in Table IV, ultrasonically treated species display cofacial stacking to various degrees. Increased degree of polymerization values appears to follow the chalcogenide series of $Te^{-2} > Se^{-2} > S^{-2} > O^{-2}$. This trend corresponds inversely to the bond strength of the silicon chalcogenide series $O^{-2} > S^{-2} > Se^{-2} > Te^{-2}$ as reported in the literature (30). Because sodium telluride is the most effective polymerizing agent of the four sodium chalcogenides investigated with a degree of polymerization value of 48, the observed polymerization trend may be attributed to the instability of the silicon chalcogenide intermediate. Being the least stable of the four prospective silicon chalcogenide intermediates, the silicon$-$telluride intermediate would be expected to be more rapidly displaced by molecular water in the formation of a linear, cofacially stacked poly(phthalocyanato) siloxane product with a resulting degree of polymerization value that is very comparable in respect to its thermally prepared predecessor ($n = 40$).

PA$-$FTIR and UV$-$visible spectra of both thermally and ultrasonically prepared poly(phthalocyanato) siloxanes display coincident spectral features.

Table IV. Optical Absorption Spectra of H$_2$(Pc), Si(Pc)Cl$_2$, Si(Pc)(OH)$_2$, and Both Thermally and Sonically Polymerized [Si(Pc)O]$_n$

[Si(Pc)O]$_n$ Synthetic Technique	Temp., (°C)	Time, (hours)	n Value by IR Analysis
Sonic & Na$_2$O[a]	25	8	—
Sonic & Na$_2$S	25	8	2
Sonic & Na$_2$Se	25	8	4
Sonic & Na$_2$Te	25	8	48
Thermal & vacuum[b]	440	8	40

[a] Pa$-$FTIR spectrum was identical to Si(Pc)Cl$_2$.
[b] Prepared at 10^{-3} torr.

Their identical vibrational and electronic spectra provide substantial evidence that a new, one-step synthetic procedure now exists for the preparation of poly(phthalocyanato) siloxane that utilizes ultrasonic energy.

Summary

To circumvent the traditional thermal two-step synthetic procedure for the preparation of poly(phthalocyanato) siloxane, an alternative one-step ultrasonic synthetic procedure has been developed. Ultrasonication of $Si(Pc)Cl_2$ in the presence of a sodium chalcogenide with molecular moisture results in the formation of poly(phthalocyanato) siloxane polymer. Of the four different sodium chalcogenides investigated (Na_2O, Na_2S, Na_2Se, and Na_2Te), sodium telluride was the most effective polymerizing agent. FTIR and UV–visible data indicate that a linear, cofacially stacked poly(phthalocyanato) siloxane with various degrees of polymerization can be prepared via this new synthetic route. In addition, this new ultrasonic synthetic procedure can successfully be carried out at atmospheric and room temperature conditions, thereby providing a viable alternative to the traditional two-step polymerization technique, which requires a high vacuum and extreme temperatures.

Acknowledgment

The authors are grateful to the National Science Foundation (EPSCoR program) for supporting this work.

References

1. Elvidge, J. A.; Lever, A. B. P. *Proc. Chem. Soc.* **1959**, *195*.
2. Joyner, R. D.; Kenney, M. E. *J. Am. Chem. Soc.* **1960**, *82*, 5790.
3. Joyner, R. D.; Kenney, M. E. *Inorg. Chem.* **1962**, *1*, 717.
4. Diel, B. N.; Inabe, T. I.; Kannewurf, C. R.; Lyding, J. W.; Marks, T. J.; Schoch, K. F., Jr. *J. Am. Chem. Soc.* **1983**, *105*, 1551.
5. Anderson, A. B.; Gordon, T. L.; Kenney, M. E. *J. Am. Chem. Soc.* **1985**, *107*, 192.
6. Kundalkar, B. R.; Marks, T. J.; Schoch, K. F., Jr. *J. Am. Chem. Soc.* **1979**, *101*, 7071.
7. Dirk, C. W.; Inabe, T. I.; Lyding, J. W.; Kannewurf, C. R.; Marks, J. J. *J. Polym. Symp.* **1983**, *70*, 3.
8. Dirk, C. W.; Inabe, T. I.; Schoch, K. F., Jr.; Marks, T. J. *J. Am. Chem. Soc.* **1983**, *105*, 1539.
9. Loomis, A. L.; Richards, W. T. *J. Am. Chem. Soc.* **1927**, *49*, 3086.
10. Boudjouk, P. In *High-Energy Processes in Organometallic Chemistry*; Suslick, K. S., Ed.; ACS Symposium Series 333; American Chemical Society: Washington, DC, 1987; Chapter 13.
11. Cracknell, A. P. *Ultrasonics*; Wykeham: London, 1980.
12. Matyjaszewski, K.; Yenca, F.; Chen, Y. L. *Polym. Prepr.* (*Am. Chem. Soc., Div. Polym. Chem.*) **1987**, *28*(2), 222.

13. Lowery, M. K.; Starshak, A. J.; Esposito, J. N.; Krueger, P. C.; Kenney, M. E. *Inorg. Chem.* **1965**, *4*, 128.
14. Davison, J. B.; Wynne, K. J. *Macromolecules* **1978**, *11*, 186.
15. Sidorov, A. N.; Kotlyar, I. P. *Opt. Spectrosc.* **1961**, *11*, 92.
16. Kobayashi, T.; Kurokawa, F.; Uyeda, N.; Suito, E. *Spectrochim. Acta, Part A* **1970**, *26*, 1305.
17. Ebert, A. A., Jr.; Gottlieb, H. B. *J. Am. Chem. Soc.* **1952**, *74*, 2806.
18. Steinbach, F.; Joswig, H.-J. *J. Chem. Soc., Faraday Trans. I* **1979**, *75*, 2594.
19. Urban, M. W. *Prog. Org. Coat.* **1989**, *16*, 321.
20. Urban, M. W. *J. Coat. Technol.* **1987**, *59*(745), 29.
21. Exsted, B. J.; Urban, M. W. *Polym. Prepr.* (*Am. Chem. Soc., Div. Polym. Chem.*) **1990**, *31*(2), 663.
22. Pinzuti, L.; Schurvell, H. F. *Can. J. Chem.* **1966**, *44*, 125.
23. Esposito, J. N.; Sutton, L. E.; Kenney, M. E. *Inorg. Chem.* **1967**, *6*, 1116.
24. Smith, A. L. *Analysis of Silicones*; Wiley: New York, 1974; p 275.
25. Arya, P.; Boyer, J.; Carré, F.; Corriu, R.; Lanneau, G.; Lapasset, J.; Perrot, M.; Priou, C. *Angew. Chem. Int. Ed. Eng.* **1989**, *28*, 1016.
26. Barton, T. J.; Paul, G. C. *J. Am. Chem. Soc.* **1987**, *109*, 5292.
27. Moser, F. L.; Thomas, A. L. *The Phthalocyanines*; CRC Press: Boca Raton, FL, 1983; Vol. 1.
28. Lever, A. B. P. In *Advances in Inorganic Chemistry and Radiochemistry*; Emeléus, H. J.; Sharpe, A. G, Eds.; Academic: Orlando, FL, 1961; Vol. 7, p 28.
29. Cuellar, E. A.; Marks, T. J. *Inorg. Chem.* **1981**, *20*, 3766.
30. Armitage, D. A. In *The Chemistry of Organic Silicon Compounds*; Patai, S.; Rappoport, Z., Eds.; Wiley: New York, 1989; Chapter 23.

RECEIVED for review July 15, 1991. ACCEPTED revised manuscript October 6, 1992.

INDEXES

Author Index

Affiliation Index

Subject Index

A

Copy editing: Cheryl L. Kranz
Indexing: Deborah H. Steiner
Production: Margaret J. Brown
Acquisition: A. Maureen Rouhi
Cover design: Ronna Hammer

Typeset by Technical Typesetting Inc., Baltimore, MD
Printed and bound by Maple Press, York, PA

Highlights from ACS Books

Good Laboratory Practice Standards: Applications for Field and Laboratory Studies
Edited by Willa Y. Garner, Maureen S. Barge, and James P. Ussary
ACS Professional Reference Book; 572 pp; clothbound ISBN 0–8412–2192–8

Silent Spring Revisited
Edited by Gino J. Marco, Robert M. Hollingworth, and William Durham
214 pp; clothbound ISBN 0–8412–0980–4; paperback ISBN 0–8412–0981–2

The Microkinetics of Heterogeneous Catalysis
By James A. Dumesic, Dale F. Rudd, Luis M. Aparicio, James E. Rekoske,
and Andrés A. Treviño
ACS Professional Reference Book; 316 pp; clothbound ISBN 0–8412–2214–2

Helping Your Child Learn Science
By Nancy Paulu with Margery Martin; Illustrated by Margaret Scott
58 pp; paperback ISBN 0–8412–2626–1

Handbook of Chemical Property Estimation Methods
By Warren J. Lyman, William F. Reehl, and David H. Rosenblatt
960 pp; clothbound ISBN 0–8412–1761–0

Understanding Chemical Patents: A Guide for the Inventor
By John T. Maynard and Howard M. Peters
184 pp; clothbound ISBN 0–8412–1997–4; paperback ISBN 0–8412–1998–2

Spectroscopy of Polymers
By Jack L. Koenig
ACS Professional Reference Book; 328 pp;
clothbound ISBN 0–8412–1904–4; paperback ISBN 0–8412–1924–9

Harnessing Biotechnology for the 21st Century
Edited by Michael R. Ladisch and Arindam Bose
Conference Proceedings Series; 612 pp;
clothbound ISBN 0–8412–2477–3

From Caveman to Chemist: Circumstances and Achievements
By Hugh W. Salzberg
300 pp; clothbound ISBN 0–8412–1786–6; paperback ISBN 0–8412–1787–4

The Green Flame: Surviving Government Secrecy
By Andrew Dequasie
300 pp; clothbound ISBN 0–8412–1857–9

For further information and a free catalog of ACS books, contact:
American Chemical Society
Distribution Office, Department 225
1155 16th Street, NW, Washington, DC 20036
Telephone 800–227–5558

Bestsellers from ACS Books

The ACS Style Guide: A Manual for Authors and Editors
Edited by Janet S. Dodd
264 pp; clothbound ISBN 0—8412—0917—0; paperback ISBN 0—8412—0943—X

The Basics of Technical Communicating
By B. Edward Cain
ACS Professional Reference Book; 198 pp;
clothbound ISBN 0—8412—1451—4; paperback ISBN 0—8412—1452—2

Chemical Activities (student and teacher editions)
By Christie L. Borgford and Lee R. Summerlin
330 pp; spiralbound ISBN 0—8412—1417—4; teacher ed. ISBN 0—8412—1416—6

Chemical Demonstrations: A Sourcebook for Teachers,
Volumes 1 and 2, Second Edition
Volume 1 by Lee R. Summerlin and James L. Ealy, Jr.;
Vol. 1, 198 pp; spiralbound ISBN 0—8412—1481—6;
Volume 2 by Lee R. Summerlin, Christie L. Borgford, and Julie B. Ealy
Vol. 2, 234 pp; spiralbound ISBN 0—8412—1535—9

Chemistry and Crime: From Sherlock Holmes to Today's Courtroom
Edited by Samuel M. Gerber
135 pp; clothbound ISBN 0—8412—0784—4; paperback ISBN 0—8412—0785—2

Writing the Laboratory Notebook
By Howard M. Kanare
145 pp; clothbound ISBN 0—8412—0906—5; paperback ISBN 0—8412—0933—2

Developing a Chemical Hygiene Plan
By Jay A. Young, Warren K. Kingsley, and George H. Wahl, Jr.
paperback ISBN 0—8412—1876—5

Introduction to Microwave Sample Preparation: Theory and Practice
Edited by H. M. Kingston and Lois B. Jassie
263 pp; clothbound ISBN 0—8412—1450—6

Principles of Environmental Sampling
Edited by Lawrence H. Keith
ACS Professional Reference Book; 458 pp;
clothbound ISBN 0—8412—1173—6; paperback ISBN 0—8412—1437—9

Biotechnology and Materials Science: Chemistry for the Future
Edited by Mary L. Good (Jacqueline K. Barton, Associate Editor)
135 pp; clothbound ISBN 0—8412—1472—7; paperback ISBN 0—8412—1473—5

For further information and a free catalog of ACS books, contact:
American Chemical Society
Distribution Office, Department 225
1155 16th Street, NW, Washington, DC 20036
Telephone 800—227—5558